The Arsenal of Eighteenth-Century Chemistry

Nuncius Series

Studies and Sources in the Material and Visual History of Science

Series Editors

VOLUME 10

The titles published in this series are listed at *brill.com/nuns*

The Arsenal of Eighteenth-Century Chemistry

The Laboratories of Antoine Laurent Lavoisier (1743–1794)

By

Marco Beretta
Paolo Brenni

TUTA SUB AEGIDE PALLAS · 1683 ·

BRILL

LEIDEN | BOSTON

Cover illustrations: (*front*) Séguin being weighed on a large set of scales (ca. 1790; Fig. 23, p. 44), and Séguin, seated, with his head under a glass canopy that is illuminated by a candle (ca. 1790; Fig. 22, p. 44). Courtesy of the Wellcome Library. See also p. 43 in this volume; (*back*) Two gasometers (1787; Figs. 65) and two pneumatic throughs (ca. 1780; Figs. 66). © MAM/Photos Franck Botté. See also pp. 340–347 in this volume.

The Library of Congress Cataloging-in-Publication Data is available online at https://catalog.loc.gov
LC record available at https://lccn.loc.gov/2022023147

Typeface for the Latin, Greek, and Cyrillic scripts: "Brill". See and download: brill.com/brill-typeface.

ISSN 2405-5077
ISBN 978-90-04-40869-2 (hardback)
ISBN 978-90-04-51121-7 (e-book)

This book is printed on acid-free paper and produced in a sustainable manner.

To the memory of Maurice Daumas (1910–1984) and Douglas McKie (1896–1967)

Contents

Foreword

The substantial collection of Antoine Laurent Lavoisier's apparatus is not the only surviving collection of eighteenth-century chemical apparatus and instrumentation, but it is without question the most important. This is for several reasons. First is the towering importance of Lavoisier, who is unquestionably the central figure of the age which changed the course of the development of chemistry, and to whom the much-used term 'chemical revolution' can be most closely applied. Secondly is the size and range of the collection, larger on both counts than any other group from the age of Enlightenment. Thirdly is the manner in which the material can be associated with particular experiments and demonstrations described by Lavoisier in his publications and manuscripts. Fourthly is the way in which the apparatus and instruments can be studied from the point of view of their design and manufacture and which provides evidence about this important relationship between scientist and instrument maker.

In an historical assessment of the work of any scientist, the evidence provided by printed works, manuscripts and material culture needs to be considered, which together can provide the most comprehensive, critical and balanced picture of the process of conducting science. For historians involved in research on Lavoisier, most of his books and many of his papers have been reprinted with scholarly commentary. The Comité Lavoisier has nearly completed a complete edition of the correspondence, the first volume of which appeared in 1955. Secondary literature considering these primary resources is vast. What has been missing up to now has been a means of assessing the material evidence, preserved for many years in the Musée des Arts et Métiers in Paris. The joint team of Marco Beretta and Paolo Brenni could not be surpassed, with their experience and judgement of Lavoisier's thoughts and practical approach to experimentation. That will be obvious to all who study the fruits of their labours, which have been conducted over the past five years, in this volume.

The Science History institute, established in Philadelphia, has been delighted to provide a modest grant to assist publication. The match is perfect: the Institute concentrates on the history of chemistry, engineering and life sciences, and possesses a huge library of primary and secondary sources in these fields, together with a considerable collection of archival material, and a museum possessing significant numbers of instruments and other apparatus. Each year the Institute invites applications for its visiting fellowship program. This work will act as a stimulus to those who wish to see material culture occupy its rightful place as providing evidence for history of science studies.

Robert G.W. Anderson
President and CEO, Science History Institute
May, 2020

Preface

The collection of Lavoisier's instruments, minerals, manuscripts, and books was inventoried in 1794, during the French Revolution, following the Convention nationale's order that the properties and estate of the members of the Ferme Générale be confiscated. In 1795 all of Lavoisier's belongings, particularly his books, instruments, tools, chemical substances, and minerals, were deposited in various republican institutions such as the École des mines, the Muséum d'histoire naturelle and the Conservatoire des arts et métiers.[1]

It is only thanks to this confiscation that a precise picture of Lavoisier's laboratory and its elements can be reconstructed. From the valuable related documents we know that Lavoisier owned some 10,000 pieces of chemical apparatus, 4,000 minerals and fossils, 3,000 books, and 50 boxes of manuscripts. In 1796, when most of this material was returned to his widow, Marie Anne Lavoisier immediately felt it was her duty to preserve the collection as a memorial to her husband, in commemoration of his achievements.[2] Madame Lavoisier held a salon at her beautiful residence in the Rue d'Anjou (today Faubourg St. Honoré), which was frequented by scientists such as Pierre Simon Laplace, Jean Baptiste Biot, and Joseph Louis Gay-Lussac, and by philosophers such as Pierre Jean Georges Cabanis, Benjamin Constant, and François Guizot. As reported by one of the privileged members of this group, Adrien Delahante, Marie Anne Lavoisier's house became a sort of reliquary where the memory of her husband was honoured with the greatest devotion. In 1805 Marie Anne Lavoisier married Benjamin Thompson, Count of Rumford, and after their amicable divorce in 1809 some of Lavoisier's instrument ended up in Rumford's collection while some of Rumford's were retained by Marie Anne Lavoisier, and these are now part the collection preserved at the Musée des arts et métiers.[3]

After Madame Lavoisier's death in 1836, and for the following two decades, her heirs donated parts of the collection of manuscripts, books, instruments, and minerals to the following French libraries and institutions: the Académie des sciences, the Bibliothèque de l'Institut, the Conservatoire des arts et métiers, and the Muséum d'histoire naturelle in Paris; the Muséum Lecoq and the Bibliothèque municipale in Clermont Ferrand; the Archives départementales du Loir et Cher in Blois; and the Bibliothèque Muncipale d'Orléans in Orléans.

Lavoisier's heirs still retained a considerable number of manuscripts, instruments, books, and other objects which were not sold until the 1950s. One part of these were acquired by French libraries and museums, another by the English collector Denis I. Duveen, who then sold his extensive Lavoisier collection to Cornell University Library in 1963.

The fate of Lavoisier's instruments was particularly problematic. In 1837 Léon de Chazelles, Marie Anne Lavoisier's heir, donated some 30 instruments to the Académie des sciences, which were never exhibited before they were moved to the Conservatoire des arts et métiers in 1864, and only then began to attract some public attention. It was thanks to their first public display in 1879 that the chemist Pierre Truchot drafted a report of his visit to the château de la Canière, where Lavoisier's heirs kept a large collection of instruments. It was only after the end of World War II, however, that Lavoisier's instruments attracted the attention of scholars and donors more widely. In the late 1940s Pierre Dupont, CEO of the American chemical company Dupont, asked the British historian of science Douglas McKie to locate the collection and to negotiate its acquisition with the heirs as well as the possibility of the creation of a Museum Lavoisier in France. In 1952, after complex negotiations, McKie managed to persuade the heirs to sell the instruments collection. It was bequeathed to the Conservatoire des arts et métiers as Pierre Dupont's generous donation. The collection consisted of more than 400 objects, most of them instruments, but the Museum was too small to be able to display them to the public. During the process of acquisition McKie photographed most of the instruments, drew up a preliminary inventory, and hoped to prepare a scientific catalogue together with Maurice Daumas, who was

1 The historical vicissitudes of the collection are illustrated in chapter 5 of the present volume.

2 No standardised way to refer to Madame Lavoisier has been settled upon in the past two-and-a-half centuries. During her life time she was known as 'Madame Lavoisier', as the 'Comtesse de Rumford', and as the 'Comtesse de Lavoisier Rumford'. However, in 1789 she signed her engravings for the *Traité élémentaire de chimie* as 'Paulze Lavoisier'. Since the nineteenth century, the historiography has preferred to call her 'Madame Lavoisier'. The reasons for this choice are not entirely clear, and since recent historiography has developed a more perceptive view on the complex relations between gender and science in eighteenth-century France, we consider it more appropriate to choose a different moniker. Unfortunately, library records are not particularly helpful: the Library of Congress uses the very long form 'Marie Anne-Pierrette Paulze-Lavoisier', while the Bibliothèque nationale de France prefers that of 'Marie Anne de Lavoisier'. We have decided to opt for Marie Anne Lavoisier.

3 Hereafter abbreviated MAM.

then curator of the MAM. However, this project was never carried out, probably because of a lack of resources. In the late 1950s the most important instruments from the Lavoisier collection were exhibited in the MAM's 'salle Lavoisier', but the majority of objects were kept in the Musée's warehouses, first in the Rue Saint-Martin, and after the creation of the Reserve in 1997, in Saint-Denis. For almost half a century this splendid and unique collection remained hidden and attracted little attention.

The present study, which is the result of more than 20 years of intermittent work, brings the project initiated by McKie and Daumas in 1952 to conclusion. It is dedicated to their memory. Our work aims to both offer a historical reconstruction of Lavoisier's laboratory and to provide the first scientific catalogue of all his surviving instruments. Accordingly, the volume is divided into two complementary parts: the first part, authored by Marco Beretta, is a historical study of Lavoisier's laboratory and its contexts, while the second part, co-authored by Marco Beretta and Paolo Brenni, provides a detailed description of Lavoisier's collection of instruments. The inventories of Lavoisier's laboratories and instruments prepared in 1794, 1836, and 1864 are reproduced at the end of the first part to provide additional documentary information. The first part also includes a biographical dictionary of Lavoisier's instruments makers and suppliers of chemicals.

A project of this size and complexity could not have been realised without the support and the generous help of numerous institutions, scholars, and friends. A first outline of the project was presented in 1998 when, thanks to the support of Dominique Ferriot, Elise Picard, and Bruno Jacomy, the authors began to work with both the collection and with McKie's inventory. The project would have never been started without the interest expressed by and the early support of Dominique Ferriot. A change of director at the MAM brought the project to a standstill although in the subsequent two decades we gained useful information and expertise in the course of our cataloguing projects devoted to Volta's and Nollet's collections of instruments (PB) and the exhibit of the chemical room of the Museo Galileo (MB).

In 2014 and 2015 different propitious circumstances provided the possibility to resume the original project. In 2015 one of us (MB) spent his research leave in Paris to work on a book devoted to Lavoisier's laboratories. In the same year Robert G.W. Anderson, then President of the Chemical Heritage Foundation in Philadelphia, highly encouraged the expansion of the scope of this work, and helped find the necessary support at the MAM to resume the project. Catherine Cuenca of the MAM immediately showed a great interest in our proposal for a catalogue of the Lavoisier collection, and has been a key figure in making it possible since 2015. Yves Winkin, then director of the MAM, seconded our project and provided the necessary funding to get it started. Thanks to these fortuitous circumstances the project took off at the end of 2016, and its funding supported the preliminary preparation of the catalogue, the restoration of the damaged instruments,[4] and the photographic documentation of Lavoisier's collection.[5] We coordinated the scientific catalogue with the restoration, photographing, and internal inventorying of the instruments. We would like to thank the new director of the MAM, Marie-Laure Estignard, Florence Calame-Levert, *Cheffe du département des patrimoines du MAM* (up to January 2021) and Pascale Heurtel, *Adjointe à l'administrateur général pour le patrimoine, l'information et la culture scientifique du MAM* for their support in the final stages of this project, and for providing us with the necessary licences for using the photographs. During our stay in Paris we benefitted from the generous and untiring assistance of the MAM staff. In particular we would like to thank Denis Pruvrel for his invaluable and efficient assistance during our numerous visits to the Reserve and to the Musée. We also would like to thank Franck Botté, Florence Calame-Levert, Anne Laure Carré, Remi Catillon, Frederique Desvergnes, Cyrille Foasso, Thierry Lalande, Elisabeth Lefevre, Lucie Lohier-Fanchini, Sylvie Maillard, Agnès Meynard, Nathalie Naudi, Diego Nunez and Perrine Stark of the MAM for their generous help in answering and solving the numerous questions and problems that we came across.

Throughout these years we had several opportunities to discuss this project with Patrice Bret of the Comité Lavoisier, whom we thank for his constant interest in the progress of our research. We also thank Patrice for his thorough reading of our manuscript and his most helpful suggestions.

We are grateful to Francesca Antonelli, Christine Lehman and the anonymous reviewers who read the manuscript and saved us from making many mistakes.

We would like to thank the copy-editors of this volume, Jonathan Mandelbaum and Anke Timmermann, for their invaluable effort to make our texts clearer. Part I was copy edited by Anke Timmermann, supported by Brill; and part II by Jonathan Mandelbaum, supported by the Science History Institute.

This work would never been published in the present form without the generous assistance of Stefan Einarson and Brill's editorial staff.

4 Coordinated by Remi Catillon.

5 Coordinated by Franck Botté and the photothèque of the MAM.

In the last phases of our research we have benefited of the support of Italian Ministry of University and Research (MUR), PRIN 2017 ("Material and Visual Culture of Science: A longue durée Perspective" Prot. 201727TRJX_001).

For their help, assistance, and support we also would like to thank Marianne Ageberg (Rudbecksskolan – Örebro), Maria Asp (Center for the History of Science at the Royal Swedish Academy of Sciences – Stockholm), Bruno Belhoste (Comité Lavoisier), Bernadette Bensaude-Vincent (Université de la Sorbonne), Ilva Beretta (Serravalle Pistoiese), Andrée Bergeron (Labex – Paris), Ulrika Bergman (Tekniska museet – Stockholm), Charlotte Bigg (CNRS – Paris), Robert Bud (Science Museum), Anne Camuset (Musée Mairie de Dijon), Elena Canadelli (Università di Padova), Stefano Casati (Museo Galileo), Pierre-Jacques Chiappero (Muséum national d'histoire naturelle), John Christie (Oxford), Bernard Clavreuil (Paris), Peter Corina (Cornell University Library – Division of Rare and Manuscript Collections), Pietro Corsi (University of Oxford), Elena Danieli (Università di Bologna), Elizabeth Denton (Université Paul Valéry – Montpellier), Matthew Eddy (Durham University), Marie Ekman (Chalmers tekniska högskolas bibliotek), Danielle Fauque (Université de Paris-Sud), Roberto Ferrari (Buenos Aires), Laurent Ferri (Cornell University Library – Division of Rare and Manuscript Collections), Paolo Galluzzi (Museo Galileo), Camille Gaumant (Musée Bargoin of Clermont-Ferrand), Florence Greffe (Archives de l'Académie des sciences), Liliane Hilaire-Pérez (EHESS – Paris), Roald Hoffmann (Cornell University), Ernst Homburg (Maastricht University), Bruno Jacomy (Lyon), Fredéric Jarrousse (Comité Lavoisier), Ursula Klein (Max Planck Institute for the History of Science – Berlin), Anders Larsson (Göteborgs universitetsbibliotek), Alessandra Lenzi (Museo Galileo), Laura Linke (Cornell University Library – Division of Rare and Manuscript Collections), Christoph Meinel (Universität Regensburg), Eisha Neely (Cornell University Library – Division of Rare and Manuscript Collections), Stéphane Pelucchi (Muséum Lecoq), John Perkins (Oxford), the late Jean Pierre Poirier, David Pullins (The Metropolitan Museum), Carsten Reinhardt (Universität Bielefeld), Dinni Rolfo (Museo Galileo), Sara Schechner (Harvard University), Andrea Scotti (Camogli), Lynsey Sczechowicz (Hagley Museum and Library), Anthony J. Turner (Paris), Véronique Van de Ponseele (Muséum national d'histoire naturelle), and Brigitte Van Tiggelen (Science History Institute).

Marco Beretta, Paolo Brenni

Post scriptum

It is with the greatest sadness that shortly after the manuscript of this book was submitted to the publisher, my co-author Paolo Brenni died suddenly of natural causes. This book acts as a memorial to a great and influential historian of science. *Marco Beretta*

Abbreviations

LC	Antoine-Laurent Lavoisier, *Correspondance* (Paris: Albin Michel, Belin, Académie des sciences, Hermann, 1955–2012), 7 vols.
LO	Antoine-Laurent Lavoisier, *Œuvres* (Paris: Imprimerie impériale, Imprimerie nationale, 1862–1893), 6 vols.
MAM	Museé des Arts et Métiers (Paris)

PART 1

Lavoisier's Laboratory Networks

(Marco Beretta)

Introduction

In the study of the history of scientific instruments, the greatest challenge is to both put them into their historical context and to understand the ways they were used. These efforts are made even more arduous by the fact that scientific instruments were part of collections that are now, for the most part, dispersed. The present study is devoted to the richest surviving collection of eighteenth-century chemical instruments but their number (ca. 600 items) account for less than 10% of the original composition of Lavoisier's laboratory. Many important instruments have been either lost or, more probably, destroyed. In contrast to this, Lavoisier's collection of manuscripts (around 4000 items), which is preserved in French and US's libraries and archives, has suffered a far less significant loss and, with the exceptions of few laboratory notebooks, the most important records of his life and works have survived. The contrasting situation between the relative richness of literary records on the one hand and the scarcity of material sources on the other is a common feature of the history of science but, at least until recently, this contradiction has been found unproblematic.

During the eighteenth century, chemistry was an eminently experimental science, and the lack of sources testifying to the concrete and material way by which experiments were set up and performed is quite frustrating. Also, the surviving instruments are not easy to read and the assessment of their reliability has often been based more on conjectures than on compelling evidence. Yet, Lavoisier's collection of instruments is remarkable not only for the quality of many of them but, above all, for the number of items that have survived. Given such a wealth and variety of instruments, it was clear to us that their scientific catalogue had to be complemented by a reassessment of Lavoisier's laboratory and of his networks of collaborators.

With this aim in view, the first part of this work provides the reader with a detailed historical survey of the development of Parisian chemical laboratories during the eighteenth century (Chapter 1), followed by Lavoisier's approach to instrument making (Chapter 2), his privileged sites of experiments (Chapter 3), a comparative view of the costs of chemical laboratories during the eighteenth century (Chapter 4) and the vicissitudes of Lavoisier's collection of instruments after his death and up to the present time (Chapter 5).

By putting the instruments in context and by following as close as possible their genealogy and fate, the ramified extension of Lavoisier's network and strategies are seen in a new light.

Lavoisier made experiments in several sites and laboratories and the first parts attempt to explore this variety in some detail. Lavoisier's famous laboratory at the Arsenal was occasionally used by the Académie des sciences as an official venue for experiments. Reciprocally, Lavoisier and his assistants brought instruments and apparatus to the Académie in order to make public experiments during the Academy's official meetings. The mobility of large apparatus and experiments implied the assistance of a wide range of makers and laboratory assistants, whose number rapidly grew and enlarged Lavoisier's network well beyond the academic circles. The historical part of this study aims at offering a map of this complex network of relations, sites, and experimental practice orbiting around Lavoisier's multifold research programs.

The examination of the surviving instruments combined with the analysis of a vast array of manuscript sources has made a rich population of artisans, many of whom assisted Lavoisier on a daily basis, emerge. Such an extended web of social and professional relations invited us to go beyond the biographical perspective and to approach Lavoisier's network within a prosopographical framework.[1] Lavoisier and the community attending his laboratory have become the object of a collective biography from which social relations and mutual interests give us a picture illustrating the transformation of the experimental practice witnessed at the Arsenal during the last decades of the eighteenth century. Within this approach, the traditional controversies on Lavoisier's role in the Chemical Revolution and on the supposedly rhetoric background of his quantitative method have been substituted by questions that only laterally address this historiography. By exploring, instead, the reasons why, when and by whom a certain instrument was made, and when and how it was used, we decided to deal with a context in which Lavoisier is just a part of a much larger group of actors, the contributions of whom shed a light not only on the Arsenal but, more generally, on the Parisian community of chemists and artisans.

This historical contextualization of the collection is followed by several documentary appendixes containing the transcriptions of inventories, the most important of which dating 1795–1796, illustrating the original composition of

1 Stone (1971).

 | DOI:10.1163/9789004511217_002

Lavoisier's laboratory and the disposition of instruments in various rooms of his residences and, after 1804, of the residence of his widow Marie Anne Lavoisier. An inventory of the first instruments acquired by the Conservatoire des arts et métiers in 1866 is extremely important in order to understand how such a selection ultimately conditioned the subsequent historiographic views on Lavoisier's apparatus.

The first part ends with a biographical dictionary of Lavoisier's instrument makers and suppliers of chemicals. This list of short biographical notices, drawn from both the signed instruments and manuscript sources, offers an overall view of the prosopographical network illustrated in both parts of the present study.

The second part of this work presents a descriptive catalogue of Lavoisier's collection of instruments and chemicals. When possible we have gone beyond the simple description of the instruments and we have examined the historical context of their production and the records reporting to their use. Privileging a more analytical approach, this part is complementary to the historical narrative.

As a whole, this book can therefore be used both as a contribution to the history of Lavoisier's laboratories and as a reference work devoted to his collection of instruments.

CHAPTER 1

The Chemical Laboratories in Paris (1743–1776)

1.1 Chemistry Comes to Paris

Between 1743 and 1794, the years across which the life of Antoine Laurent Lavoisier unfolds, Paris experienced social, cultural, and economic changes of such importance that they spread throughout and exerted an influence on Europe with surprising rapidity.[1] In these exceptional circumstances, the chemical arts played a remarkable role: chemical discoveries rung in important changes with repercussions that reached even the lowest strata of society.[2] It is, thus, not surprising that, from the 1780s onwards, references to the 'chemical revolution' became ever more frequent, both with reference to theory and to innovative practices and inventions in the arts, crafts, and manufacturing. However, especially during the first half of the century, the major chemical discoveries and theoretical achievements had been attained outside the French capital. It was in the early 1750s, thanks to the interest shown in the subject by Denis Diderot and other eminent contributors to the *Encyclopédie*, that a systematic campaign of translation and dissemination of foreign chemical literature was undertaken, and attracted an unprecedented attention both to the theory and practice of chemistry.[3]

The eighteenth century was an exciting period in the history of chemistry. A greater number of metals, earths, alkalis, and gases was discovered than in the entire preceding history of chemical science, and this contributed to the collapse of older philosophies of matter. The main discoveries were: cobalt (1735), platinum (1748), nickel (1751), bismuth (1753), magnesium (1754), hydrogen (1766), nitrogen (1772), barium (1772), oxygen (1773–1774), chlorine and manganese (1774), methane (1777), molybdenum (1778), tungsten (1781), tellurium (1782), strontium (1787), uranium (1789), titanium (1791), yttrium (1794), chromium (1797) and beryllium (1798).[4]

Important as they were, these discoveries were not the result of new theoretical ideas but rather reflected an increased accuracy in analytical methods and apparatus. Technical progress was made both in and outside the laboratory. The new gases, for instance, were isolated in laboratories as well as (in the case of methane) discovered in marshes. The different circumstances of the discoveries of the new elements indicate that a wide range of types of apparatus was available, from the portable pocket laboratory to large-scale chemical equipment.[5]

The history of the development of chemical laboratories and technologies in the eighteenth century is rather complex due to the variety of actors and institutions involved. Sites, techniques, and materials varied, sometimes substantially, according to the needs of the different chemical specialties and crafts. Although fire was a dominant element in chemical practice, the specific furnaces, tools, combustibles, reagents, and spaces differed from trade to trade – glassmaking, for instance, required particular furnaces, tools, combustibles, and reagents quite different from those used in pharmacy, dyeing, tanning, assaying, mining, smelting, or distilling.

Amidst the complexity of this overall picture, the chemical trades made much progress, and many able artisans distinguished themselves with inventions and ideas that gave chemistry an unprecedented status.[6] In a draft memorandum devoted to the usefulness of the applied sciences presented to the Académie royale des sciences de Paris between 1716 and 1727, one of the most authoritative French scientists, René-Antoine Ferchault de Réaumur, pointed out the central role of chemistry in the growth of several trades.

> Chemistry, the investigations of which seem rather frivolous to those who do not know its true purpose, could become one of the most useful parts of the Académie. [...] The conversion of iron into steel, the method of plating or whitening iron to make tinplate, the conversion of copper into brass, three great industries which the State lacks, are in the province of chemistry. It is the business of chemistry, also, to investigate the mineral substances used in dyeing and ores and minerals. Glassworks, pottery works, faïencers, porcelain works – industries which all require improvement – also concern it.[7]

1 On the social changes see Daumard, Furet (1961) and Roche (1981). On the interactions between science and the city during the eighteenth century see Van Damme (2005) and Belhoste (2011).

2 Perkins (2010).

3 *Encyclopédie* (1751–1772).

4 Weeks (1933).

5 On portable pocket laboratories, see Smeaton (1966).

6 Klein (2012).

7 "La chimie, dont les recherches paroissent assez vaines à ceux qui ne connoissent pas son véritable objet, pourroit devenir une des

 | DOI:10.1163/9789004511217_003

Réaumur, who gave new impetus to Jean-Baptiste Colbert's monumental project of a comprehensive *Description des arts et métiers*, published in 1722 *L'art de convertir le fer forgé en acier*, a treatise of chemical technology which gained him a prize of 12,000 *livres* and paved the way for subsequent publications in the series. Moreover, in the next few decades Parisian chemical trades expanded their activities considerably and contributed greatly to the remarkable economic expansion of the French capital.[8] In 1765 Jèze's famous guide of the city listed 110 trades together with the statutes of the guilds and, for some, their financial details.[9] Daniel Roche has estimated the number of members of the Parisian guilds to have been around 40,000 in the middle of the century. This number would rise to 120,000 – more than one fifth of the entire population – if all those engaged with activities connected to the trades are taken into account.[10] Towards the end of the century Lavoisier claimed that the Parisians' annual expenditure reached the remarkable sum of 260 million *livres*, by far the highest in France and one of the highest in Europe.[11] In spite of the strict regulations of the guilds, the economic expansion of the trades was accelerated by important innovations and, in a few cases, important technological inventions.[12] It is not surprising, therefore, that the chemical trades were dominating the scene in the second half of the eighteenth century: grocers, *vinaigriers*, pharmacists, distillers, (Fig. 1) metal workers (*forgerons*), glassmakers, *fayencers*, enamellers, mirror makers, *vitriers*, dyers, tanners, goldsmiths, tinsmiths, smelters, *salpêtrièrs*, and all the trades connected with the production of food and wine greatly benefitted from the application of chemical knowledge to their art, and filled the Parisian marketplace with a remarkably rich array of new commodities.[13] Trades were not the only factor contributing to the growing prosperity of the French capital. From an enquiry

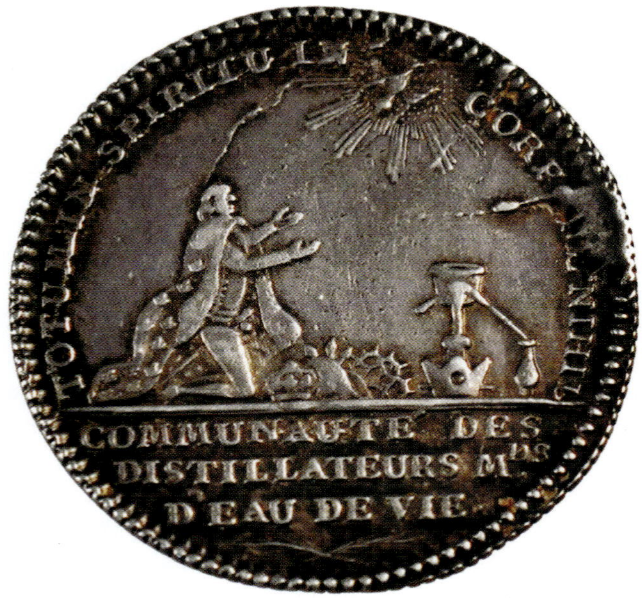

FIGURE 1 Token of the Parisian distillers (1731) picturing a master with alembic, and accompanied by the motto '*Totum in spiritu in corpore nihil*'
PRIVATE COLLECTION

commissioned after the Revolution we know that in 1790 there were 937 manufactories in Paris, one third of which related to chemistry.[14] The number of workers employed by these proto-industrial enterprises varied from 150 to 1,000, and most received investments from enthusiastic aristocrats. In the middle of the century Gobelins, Saint-Gobain, and Sèvres (all long established Parisian firms) began to employ experts who could help improve the quality of their products (tapestries, glass and soft porcelain).[15] Upon the suggestion of the Bureau de Commerce or of the Académie royale des sciences, renowned academic scientists and chemists such as Jean Hellot, Jean Nollet, Pierre-Joseph Macquer and Jean Darcet were recruited as scientific directors or experts, and made responsible for introducing important technical innovations.[16] For example, a new type of furnace for the production of porcelain was conceived by Macquer while at Sèvres, and this was adopted in Parisian research laboratories as a standard equipment the early 1760s.[17] In addition to these more traditional manufactories, there were more recent ones which helped transform the landscape of Parisian chemistry. Javel, founded in 1777 with an investment of the Count

plus utiles parties de l'Académie […]. La conversion du fer en acier, les manières d'étamer ou blanchir le fer pour faire le fer blanc, la conversion du cuivre en léton, trois grandes manufactures qui manquent au Royaume, sont du ressort de la chimie. Il luy appartient aussy de faire les recherches sur les matières minérales emploïées pour les teintures, sur les mines, sur les minéraux; les verreries, les potteries, les fayenceries, les ouvrages de porcelaine, manufactures qui demandent touttes à être perfectionnées entrent aussy dans son objet". Réaumur (1716–1727), pp. 104–105.

8 On the Parisian trades and their effort to expand see Bertucci (2017). I have examined the case of glassmaking in Beretta (2012a) and Beretta (2014a).

9 Jèze (1765), pp. 210–237.

10 Roche (1981), p. 73.

11 Lavoisier (1791), p. 146.

12 Franklin (1905); Ballot (1923); Sée (1925); Coornaert (1968).

13 Spary (2014).

14 Gille (1963), pp. 58–62.

15 While not situated in the city itself, Saint-Gobain and Sèvres were just on the outskirts; see Gillispie (1980).

16 On Macquer, see Lehman (2012) and Lehman (2019).

17 The "Macquer furnace", named after his inventor, is also mentioned in Lavoisier's collected works on several occasions. For textual search in LO see Beretta (1999–2010).

d'Artois, was a large-scale production of sulphuric and nitric acids, hydrogen, and – following Berthollet's discovery – a potassium hypochlorite-based bleach known as the *eau de Javel*, which would revolutionise the textile industry. Equally important were the Périer brothers, the manufacturers of laminated iron and copper pipes[18] who also introduced Watt's steam engine for the distribution of Seine's water to a significant number of domestic dwellings on the Parisian faubourgs.[19] Turgot's reform of the gunpowder production and the 1775 foundation of the *Régie des poudres* at the Arsenal prompted remarkable innovations and, not least thanks to Lavoisier's contributions, a successful combination of large-scale industrial production and chemical research. Other manufactories for paint, glass bottles, and sodium carbonate made the chemical arts more prominent in the capital. This increased presence and the corresponding pollution provoked the population to protest, and thus the first attempt to control the side effects of industrialisation.[20]

The primary innovations in chemical laboratories of the eighteenth century occurred in specific areas of applied research. The facilities in which contemporary chemists operated differed significantly from each other, and ranged from public sites used for demonstrations and teaching purposes to private laboratories where the most capable students were trained in sophisticated experiments and even, at times, research.

Although the apparatus of several chemical specialties saw some innovation, it was pharmacy which had the greatest influence on the features of the eighteenth-century chemical laboratory. There were several reasons for this.[21] First of all, pharmacists were by far the largest professional group practicing chemistry in the French capital. In 1756 there were 130 apothecary shops in Paris alone.[22] Moreover, laboratory practice was part of the apothecaries' daily routine. Their work included the manipulation of chemicals and the accurate preparation of remedies, as well as accurate measurement, and thereby contributed to the establishment of standards for apparatus, weights, and measures. In addition to running their private laboratories and shops, pharmacists taught chemical courses in medical faculties, botanical gardens, and other academic establishments. In these institutions apothecaries introduced large audiences to relatively simple chemical operations, and thus transformed chemistry into an extremely popular science. Finally, pharmacy was of social relevance, as evidenced by the remarkable fact that out of the 120 apothecary shops active in Paris in 1776, 22 were run by the widows of deceased pharmacists.[23] These female practitioners continued the role played by women in the history of ancient and Renaissance alchemy, as well as their long-established skills in the preparation of cosmetics and perfumes.[24]

In the second half of the eighteenth century, also thanks to the prominence of apothecaries, chemistry became so fashionable that the annual attendance of courses in Paris reached 3,130 students in 1781.[25] Although significant progress was made in other European cities, especially in Germany, Parisian chemists appropriated the innovations very rapidly and quickly adapted their courses and the design of their laboratories to accommodate the latest discoveries.[26] The French context offers some of the most representative examples of chemical laboratories before the discovery of gases.

One of most famous successful chemistry teachers of the eighteenth century was the Parisian apothecary Guillaume-François Rouelle (Fig. 2).[27] He delivered courses at the amphitheatre of the Jardin du roi where he was employed from 1742 as a demonstrator of chemistry. Since his laboratory was at some distance from the amphitheatre, his experimental demonstrations on stage were necessarily limited. Nevertheless, his courses became so popular that were attended every year by hundreds of students – including foreigners and women – and he had an unprecedented reputation. Rouelle also delivered private advanced courses for selected students, in which more complex experiments were carried out, in his shops in the Place Maubert and, after 1746, in the Rue Jacob. Hence Rouelle divided his professional life between two establishments: the Jardin du roi and his private laboratory.

At the Jardin du roi he lectured before large audiences (sometimes several hundred students at a time), introducing them to the notions of vegetable, animal, and mineral chemistry, conducting chemical experiments, and even demonstrating a few scientific tricks for their diversion. In an outline of Rouelle's private course published in 1759 we find the following programme:

> [t]he plants, animals, and minerals are the subjects of these experiments; they are divided into three

18 Lavoisier purchased some of these for his famous experiments on the decomposition of water in 1785.
19 Payen (1969).
20 Guillerme (2007); Le Roux (2011).
21 Simon (2005).
22 Bouvet (1937), p. 264.
23 *État* (1776), pp. 138–46.
24 See for instance Ray (2015).
25 Lehman (2008); Perkins (2010).
26 On Germany see Hufbauer (1982); Klein (2007).
27 Rappaport (1960); Rappaport (1961); Partington (1961–70), vol. 3, pp. 73–76.

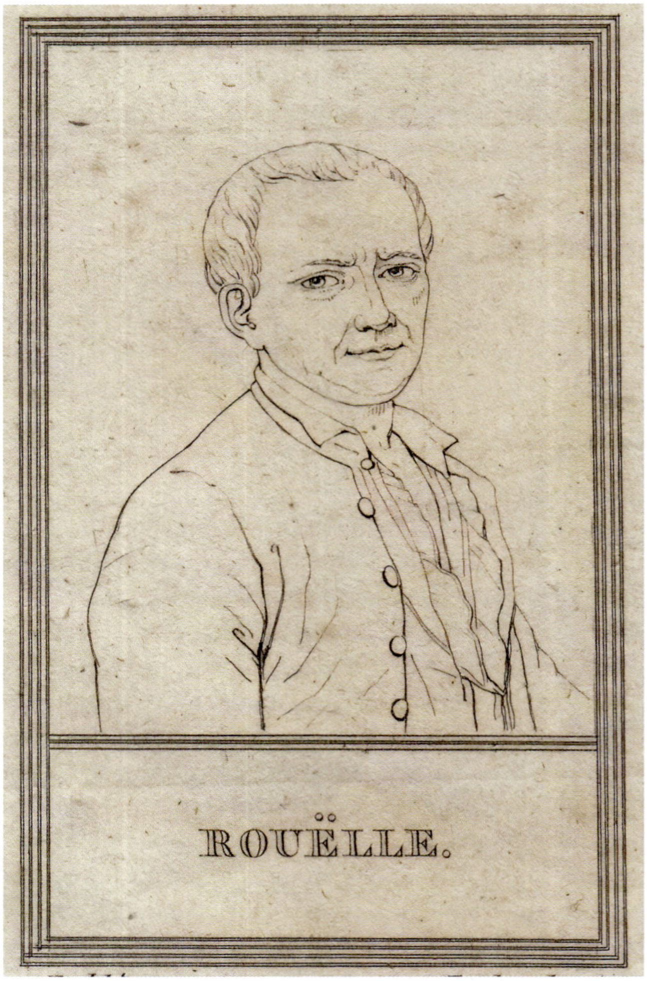

FIGURE 2 Engraved portrait of Gabriel-François Rouelle from Michaud (1811–1837)
PRIVATE COLLECTION

> parts. *Part one*: examples will be given of the various ways of analysing plants, of extracting from them oils, essential salts, and fixed salts, and of making extracts from them; we will then proceed to fermentation and spiritous liquors [...]. The various combinations of vegetable substances will be shown, such as the oils with alkalis and tartaric salts rendered soluble by other salts or by earths.
>
> *Part two*: animals are the subject of the second part. They will be analysed and several substances derived from them [...]. English phosphorus, which up to the present has been known to only very few persons, and which is the product of the final violence of fire, will be the subject of most unusual experiments at the end of this part.
>
> *Part three*: in this part, there will be treated the substances derived from the bowels of the earth, which are called minerals. This part is one of the most prolific in experiments. The analyses of bitumens, nitre, [etc.] [...] will be conducted with exactitude, to them will be joined the recomposition of these salts [...]; fermentations hot and cold, the change in colour, detonations and the production of flame are all effects of these mixtures that will be discussed. [...] We shall conclude with the vitrification of metals, the coloured glasses, the volatilisation of metals and several unusual procedures which throw light on the nature of substances.[28]

In his private laboratory in the Rue Jacob Rouelle met a handful of students who wished to pursue chemistry at a more advanced level. This private and highly selective course focused on systematic experimental research, and did not involve the spectacles with which he enlivened his lectures at the Jardin du roi. On Sundays Rouelle delivered lectures in his laboratory to an illustrious group of friends and colleagues, among them Réaumur, Bernard de Jussieu, Henri-Louis Duhamel de Monceau, and François Venel as regulars.

In the course of his career Rouelle published very little, but he was a brilliant teacher, as testified by his students. In August 1770, shortly after Rouelle's death, Denis Diderot's biographical note described him as the founder of chemistry in France and acknowledged that his grasp of theoretical chemistry, combined with his eloquence and teaching style, led to an unprecedented interest in the science, despite his occasionally maladroit laboratory

28 "Les Plantes, les Animaux & les Minéraux, font l'objet de ces Expériences & en font les trois Parties. Première Partie. On donnera des Exemples des différentes manières de faire l'analyse des Plantes d'en tirer les huiles, les sels essentiels, les sels fixes & d'en faire les extraits de-là on passera à la fermentation & aux Liqueurs spiritueuses [...]. On fera voir différentes combinaisons de substances végétables telles que les huiles avec les sels alkalis & les sels tartareux rendus solubles par d'autres sels ou par des terres. Seconde Partie. Les Animaux font l'objet de cette Partie. On fera leurs analyses & celles de plusieurs substances qui en viennent [...]. Le Phosphore d'Angleterre qui jusqu'ici n'a été connu que d'un très-petit nombre de personnes & qui est le produit de la dernière violence du feu terminera cette Partie avec quelques expériences des plus Singulières fur cette matière. Troisième Partie. Dans cette Partie on traitera des substances qu'on retire des entrailles de la Terre, qu'on appelle Minéraux. Cette Partie est des plus fécondes en expériences singulières. Les analyses des Bitumes, du Nitre, [etc.] [...], seront suivies avec exactitude on y joindra la recomposition de ces Sels [...]; les fermentations chaudes & froides; le changement des couleurs; les détonations; la production de la flâme, sont tous des effets décès mélanges dans le détail desquels on entrera. [...] On finira par la vitrification des Métaux par les verres colorés, par la volatilisation des Métaux & par plusieurs procédés singuliers, qui donnent des lumieres sur la nature du corps". Rouelle (1759).

demonstrations and his enduring belief in alchemy.[29] When Rouelle appeared on the scene there were only three chemistry laboratories in Paris, but when he died there were nearly two hundred.[30] Diderot's admiration stemmed from his immediate experience of Rouelle's work, for he had followed his course for three years and later used his chemical knowledge in developing arguments in his philosophical works.[31] Moreover he prepared – probably with an eye to their eventual publication – a comprehensive recension of Rouelle's lectures.[32] Quite separately, since varying manuscript copies of lecture notes taken in Rouelle's course were in wide circulation, his family planned the publication of a definitive edition to stop the proliferation of unauthorized and inaccurate versions:

> [t]he reputation which the late M. Rouelle built up through his courses has led those who attended them to assemble all that they could note down of his lectures: several of his disciples have even taken the trouble to collect the greater part of these scattered fragments and to put them together in bound form, which were passed from hand to hand, and which have more than once been the object of attempts to some day produce printed copies. His family justly fears that some time someone will succeed in publishing these lectures, but only in a still more defective form because of the errors that eluded the redactors; it is impossible that these will not be added to in large number by the copyists. We therefore feel that it is our duty to announce that we are now actively collecting all the papers, notes and remarks left by the late M. Rouelle, and plan to set these in order, with the intention of publishing them ourselves. The family believes it all the more necessary to make this announcement since they fear that the unwise zeal of some disciple of this famous man will bring harm to his reputation while believing that he is enhancing it.[33]

By 1770, however, Rouelle's lectures were already on the path to being outdated; for example, the role of gases in chemical reactions – a topic barely touched upon in his lectures – was beginning to draw the attention of French naturalists.[34] It is likely that the rapid changes that were taking place in the field, the priorities being set by a new generation of chemists, and the vast amount of material that needed to be edited persuaded Diderot to abandon the project in the end.

1.2 Rouelle's Laboratory

Following the presentation of his first paper to the Académie royale des sciences, a groundbreaking study on the classification of neutral salts, Rouelle came to be recognised as one of the most authoritative chemists in

29 Melchior Grimm describes an incident in which Rouelle's awkward hand at chemical experiments led to an explosion in the laboratory. "Un jour, étant abandonné de son frère et de son neveu, et faisant seul l'expérience dont il avait besoin pour la leçon, il dit à ses auditeurs: *Vous voyez bien, Messieurs, ce chaudron sur ce brasier; eh bien, si je cessais de remuer un seul instant, il s'en suivrait une explosion qui nous ferait tous sauter en l'air!* En disant ces paroles, il ne manqua pas d'oublier de remuer, et sa prédiction fut accomplie: l'explosion se fit avec un fracas épouvantable, qui cassa toutes les vitres du laboratoire et, en un instant, deux cents auditeurs se trouvèrent éparpillés dans le jardin: heureusement, personne ne fut blessé, parce que le plus grand effort de l'explosion avait porté par l'ouverture de la cheminée. Le démonstrateur en fut quitte pour cette cheminée et une perruque. C'est un vrai miracle, que Rouelle, faisant ses essais presque toujours seul, parce qu'il voulait dérober *ses arcanes*, même à son frère, homme fort habile, ne se soit pas fait sauter en l'air par ses inadvertances continuelles: mais à force de recevoir sans précaution les exhalaisons les plus pernicieuses, il se rendit perclus de tous ses membres, et il passa les dernières années de sa vie dans des souffrances terribles". Grimm (1812–1813), vol. 1, p. 249.

30 "On ne pourra refuser le titre de fondateur de la chimie en France […] Il créa la chimie […] Quand il parut, on comptait à peine à Paris, trois cabinets; il y en a peut-être deux cents aujourd'hui […] Grand savant, profond théorique, il était manipulateur distrait et maladroit. Je lui ai vu manier le phosphore; le feu dévorant enveloppait ses mains de toutes parts, les pénétrait, les consumait, sans qu'il sût comment la chose était arrivée". Diderot (1969–1972). Diderot had attended Rouelle's public lectures between 1754 and 1757, long before the advent of pneumatic chemistry.

31 Beretta (1998); Pépin (2010).

32 Jacques (1985).

33 "La réputation que feu M. Rouelle s'étoit faite par ses Cours, engageoit ceux qui le suivoient à recueillir tout ce qu'ils pouvoient saisir de ses leçons: plusieurs de ses Disciples ont même pris le soin de rassembler la plupart de ces lambeaux épars, & d'en former des Cahiers suivis, qu'on s'est transmis de main en main, & qu'on a tenté plus d'une fois de donner au grand jour de l'impression. Sa famille, craignant avec raison qu'à la fin quelqu'un ne parvienne à publier ces Leçons d'autant plus difformes, qu'aux fautes échappées aux Rédacteurs, il est impossible qu'il ne s'en soit pas joint une infinité de la part des Copistes, croit devoir faire annoncer qu'elle travaille à recueillir tous les Papiers, Notes & Remarques que feu M. Rouelle a délaissés, & qu'elle se propose de les faire mettre en ordre, pour les publier elle-même. Elle a cru d'autant plus nécessaire de se hâter de donner cet Avis, qu'elle a lieu de craindre que le zele indiscret de quelque disciple de cet homme célèbre ne nuisît à sa réputation, en croyant l'illustrer". Avis (1770).

34 Rouelle was nevertheless one of the first chemists to perfect Hales' apparatus for collecting gases.

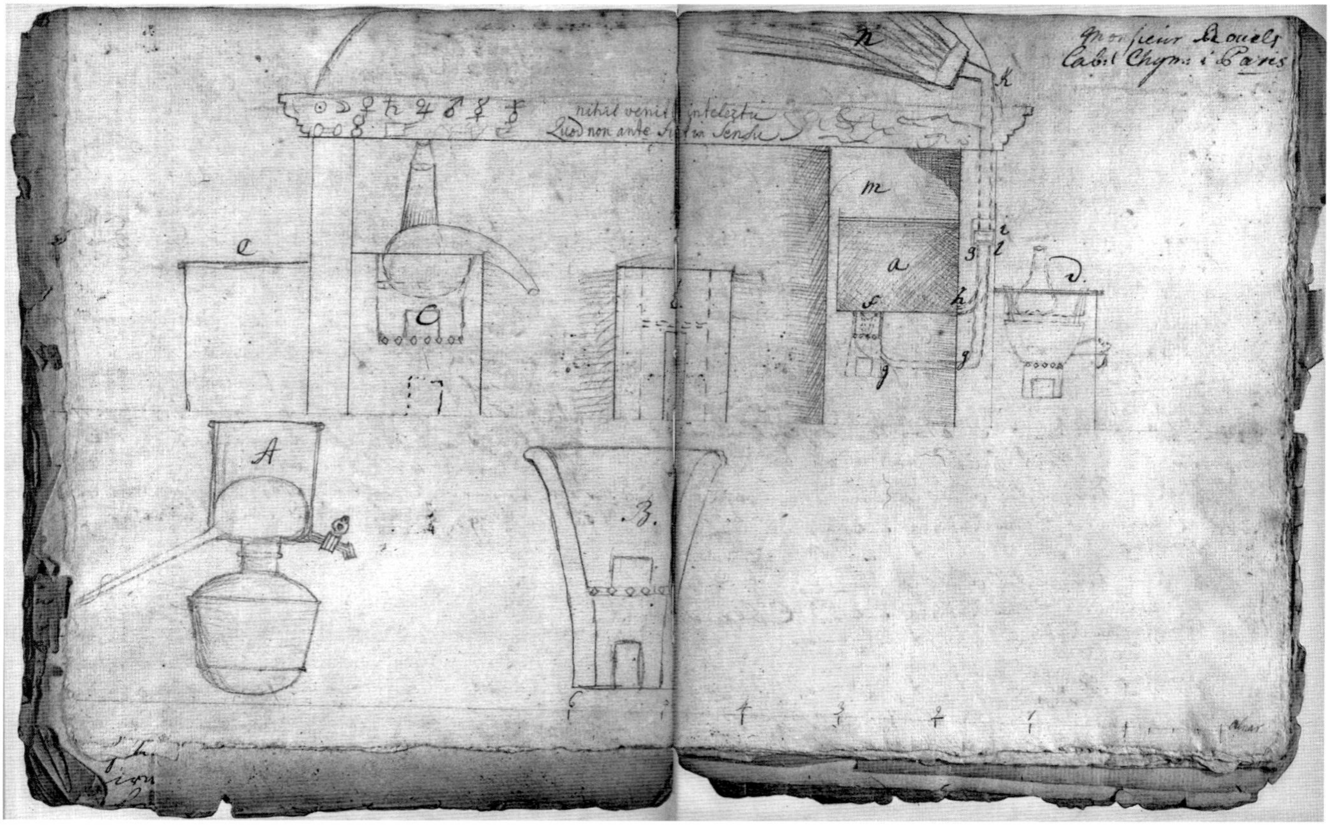

FIGURE 3 Rouelle's laboratory in the Rue Jacob, 1747
FROM BODMAN (1949–1952). COURTESY OF THE TEKNISKA MUSEET – STOCKHOLM

Paris.[35] Not surprisingly, visitors from outside of France who wished to assess the progress of chemistry in Paris asked him to show them his laboratory. Of the visitors to Rouelle's laboratory in the Rue Jacob the Swedish mineralogist Sven Rinman, who visited in 1747, is most remarkable: his is the only description we have of the most important research laboratory in Paris.[36] Rinman's exceptional description and the accompanying sketch – a unique piece of visual evidence – shed valuable light on the architectural features and experimental facilities in this eighteenth-century laboratory (Fig. 3). The drawing includes unusual details, such as the Latin inscription written near the top of the chimney: "[n]ihil venit intelectu [*sic*] quod ante fuit in sensu", a slight variation on the famous Aristotelian maxim, "[n]ihil est in intellectu quod non prius in sensu" ("nothing is in the intellect that was not first in the senses"). These words demonstrate Rouelle's views on the importance of experimentation, but also hint at his possible tendency towards empirical philosophy.

To the left of the inscription are a series of chemical symbols arranged in two rows: at the top, from left to right, are the metals (gold, silver, lead, tin, iron, mercury, and lead), and at the bottom, from left to right, are arsenic and litharge, followed by symbols that cannot be easily identified, perhaps for the common salts and nitric acid. Their significance is unclear, unless they represent the first lines of Geoffroy's table of affinity. Rinman does not comment on either the symbols or the inscription, but focusses instead on Rouelle's furnaces and laboratory apparatus. The description that accompanies his sketch reads (in its entirety):

> [o]n the Chemical Laboratory of M. Rouel, Demonstrator of Chemistry, in Paris on the Rue de Jacob near the Rue des Deux Anges.
>
> The whole building is approximately 50 feet (15m) long and 18 feet (5.5m) wide, consists of three rooms, the middle one being the biggest, 30 feet (9m) long and with a stove against one side, 16 feet (4.7m) wide and 4 feet (1.2m) deep, so that the hood is somewhat above it. Here the following are installed, see drawing *Tab*.

35 Rouelle (1744). The *mémoire* was read at a public session of the Académie on June 3, 1744 and had an immediate influence on French chemists. Rouelle had joined the Académie royale des sciences in May 1744. He had succeeded Boulduc with his appointment to *demonstrateur de chimie* at the *jardin du roi*.

36 The description of the laboratory, together with the original Swedish text, is published in Beretta (2011).

1st: within the walls of the stove a forge hearth 'a' that has its own little stove wall and chimney 'm'. Within it there is a small central furnace 'f'. The plaster 'm' of the furnace roof is above the stove and the air is drawn from there through a pipe 'k-i' to an air chamber 'gi', whence two pipes go out 'ggg' to the air furnace 'F' and beneath that to the hearth 'h', both of these pipes being furnished with sliding lids, allowing air to be admitted in whichever direction is desired to have its effect there, either in the hearth or the air furnace where the air goes into all 4 for [*illegible*]. 'b' is a larger air furnace for big melts, in which the heat becomes quite strong by means of a dome mounted on top.

'C' is a distilling furnace and beyond that there is a sublimation furnace.

Outside the stove 'd' is a square sand heat for digestions, in which the heat rises through narrow pipes at all four corners. Beyond this sand heat there is a reverberatory furnace. At the same [*illegible*] is also installed on the other side of the stove at 'e'- And at both corners in the building there are two distilling boilers, one large and one small with a 4 foot [1.2m] long neck for [*spiritus vini*].

NB It was stated by M. Rouel that the tall neck does not make better [*spiritus vini*] than a short one. The only thing visible above the alembic was the cooling barrel 'A'.

In addition to these large furnaces there were various small portable furnaces, made solely of clay, round and largely conical as in 'B'. The two small rooms were used in one case for quite a large smelting furnace of copper and lead ores, in the other for a small glass furnace for artisanal production of ceramics and enamelling.

Memo. In the furnace 'C' phosphorus was distilled in one day, with a round dome over the furnace.[37]

Rouelle's laboratory was relatively small, and most of the apparatus seems to have been traditional. As Rinman pointed out, however, Rouelle was already improving the construction of his furnaces, some of which some of which would be illustrated in Diderot's and d'Alembert's *Encyclopédie*.[38] It is a pity that Rinman does not describe the glass retort in detail, although the absence of the apparatus for the collection of gases that had only recently been perfected by Hales is not surprising.[39]

During the 1750s and the early 1760s, laboratories, apparatus, and experiments developed considerably, partly due to the huge success of Rouelle's chemical courses. This progress can be seen clearly in the eighteen plates that illustrate the section on chemistry in the twentieth volume of the *Encyclopédie*, published in 1763 by Denis Diderot and Jean d'Alembert.[40] The first plate (Fig. 4) depicts a chemical laboratory; it is likely that Diderot here illustrates the most famous chemical laboratory in Paris, namely Rouelle's, since he had been his pupil and assistant. The scene shows all major elements of an eighteenth-century laboratory; the most innovative feature is a large chemical table on which all chemical experiments and reactions were performed that did not require the use of violent fire.[41] It holds glasses and funnels used to filter liquors, glass jars with different kinds of stoppers, a small portable furnace, and other receivers.

To the left of the table a chemist is shown preparing a metallic solution while conversing with a "*physicien*" (physicist). It is the chemist who manipulates matter while the physicist merely converses with him; this juxtaposition of roles underlines the superiority of chemistry as a science that, due to the central role of experimentation within it, breaks through the surface of matter and reveals

37 "Om Laboratorium Chemicum hos Chymie Demonstratorn Hr. Rouel uti Paris dans la rue de Jacobe pres la rue de Deux AngesHela huset är ungefär 50 fot (15 m) långt och 18 fot (5,5 m) bredt, består af 3:ne rum, varav det mellersta är störst, 30 fot (9 m) långt och har emot ena sidan spis, 16 fot (4,7 m) bred och 4 fot (1,2 m) djup, så att spiskupan står något öfwer. Härwid finnes inrättat, se ritningen Tab.1:o inom spismurarne en smedjehärd 'a' som har sin egen lilla spismur och skorsten 'm'. Theruti är en liten midtugn 'f'. Putsen 'm' är ställd öfwer spisen och ledes wedret thereifrån genom et rör 'k-i'til en wederkista 'gi', hwarffrån gå 2:ne rör ut 'ggg' ut til windtugnen 'F' och thet undre til härden 'h', warandes bägge thesse rör med skuflock försedde, hwarigenom wedret kan insläppas uti hwilketdera man behagar, at giöra ther effect, antigen uti härden el:r [eller]windtugnen ther wedret går in alla 4 för [*illegible*]. 'b' är en större windtugn til growfe smältningar, hwaruti hettan giöres ganska stark förmedest en dom som sättes theruppå.'C' ära n distillerugn och therebortom an sublimerugn.Utom spisen är 'd' en fyrkantig sand capell til digestioner, hwaruti warmen wid alla fyra hörnen gienom små rör upstiger. Bortom samma sand capell är en reverberugn [reverberatory furnace]. På samma [*illegible*] thet ock inrättat wid andra sidan om spisen wid 'e'- Ock wid bägge hörnen uti huset äro 2:ne distillerpannor, en store och en mindre med en 4 fot [1,2 m] hög hals til [spiritus vini]. NB af Hr Rouel berättades at then höga halsen giör intet batter [spiritus vini] än en kort. Alenast observeras öfwer alembiquen woro kylfat 'A'. Utom tesse stora ugnar woro åtskillige små portative giorde af ler alene, runde och merendels coniske såsom 'B'. The 2:ne små rumen amplojerades thet ena til en något stor smältugn för koppar- och blymalmer, thet andra til en liten glasugn [för] slöjders tillwärkning av porcelains samt emailarbeten.P.M. Uti ugnen 'C' distillerades phosphorus på en dag, då en rund dom sattes öfwer ugnen". Beretta (2011).

38 They include his smelting and portable conical furnaces.

39 Hales (1727) was largely ignored by chemists before Joseph Black conducted his pneumatic experiments in the early 1750s; see Guerlac (1951).

40 *Encyclopédie* (1751–1772), vol. 20. Lehman, Pépin (2009).

41 Morris (2015), pp. 52–53.

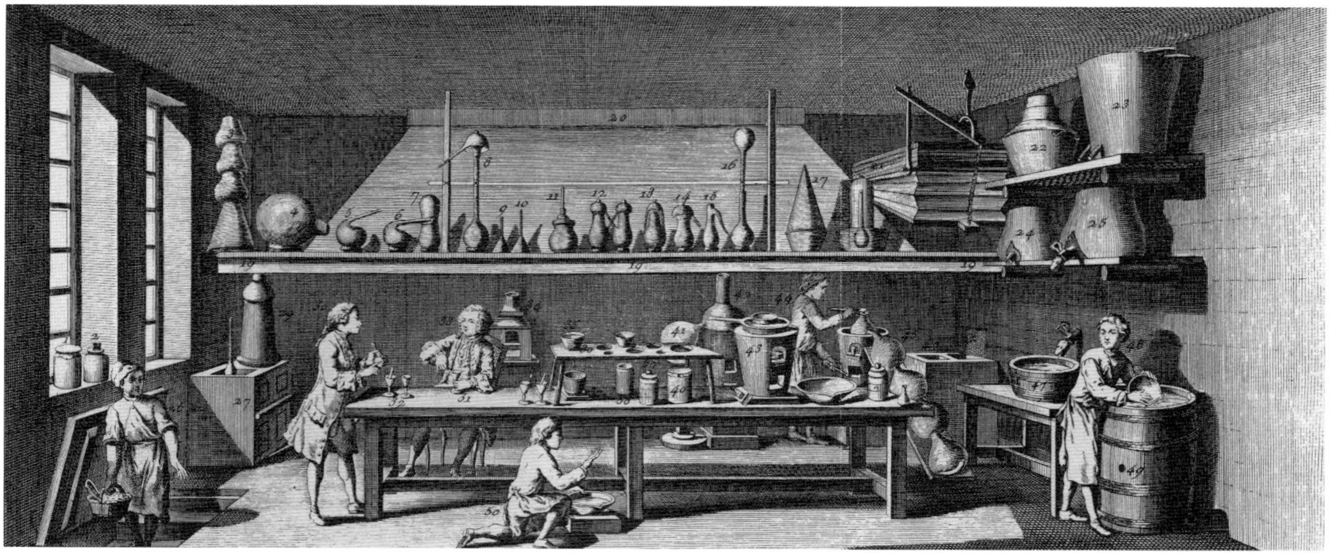

FIGURE 4 Illustration of a chemical laboratory (probably Rouelle's) Diderot and d'Alembert's *Encyclopédie* (1763), s.v. "chimie"
COURTESY OF THE ARTFL ENCYCLOPÉDIE PROJECT, UNIVERSITY OF CHICAGO

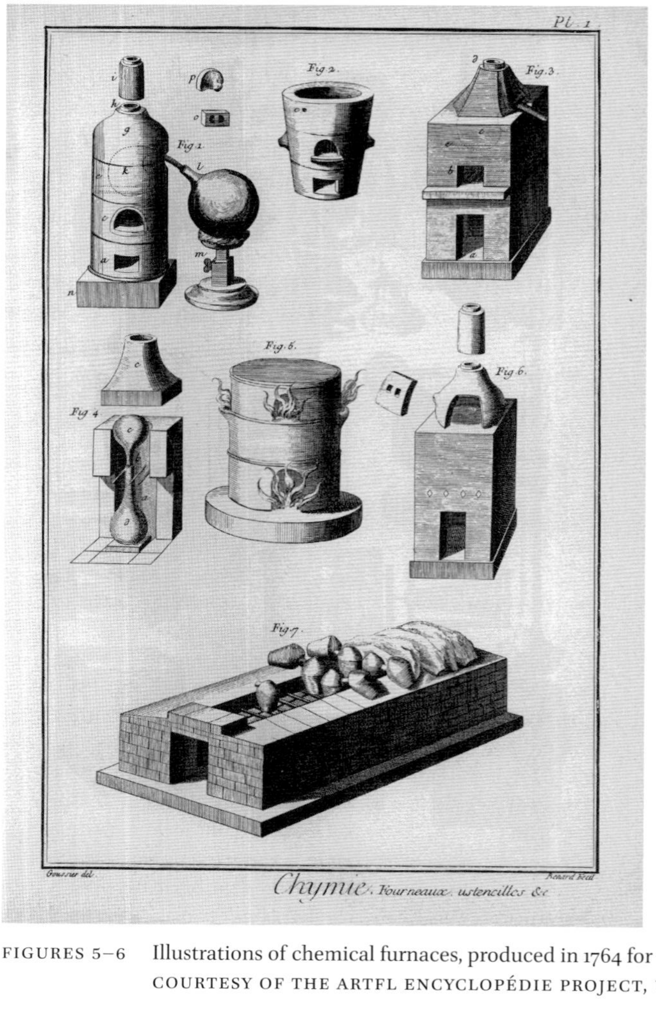

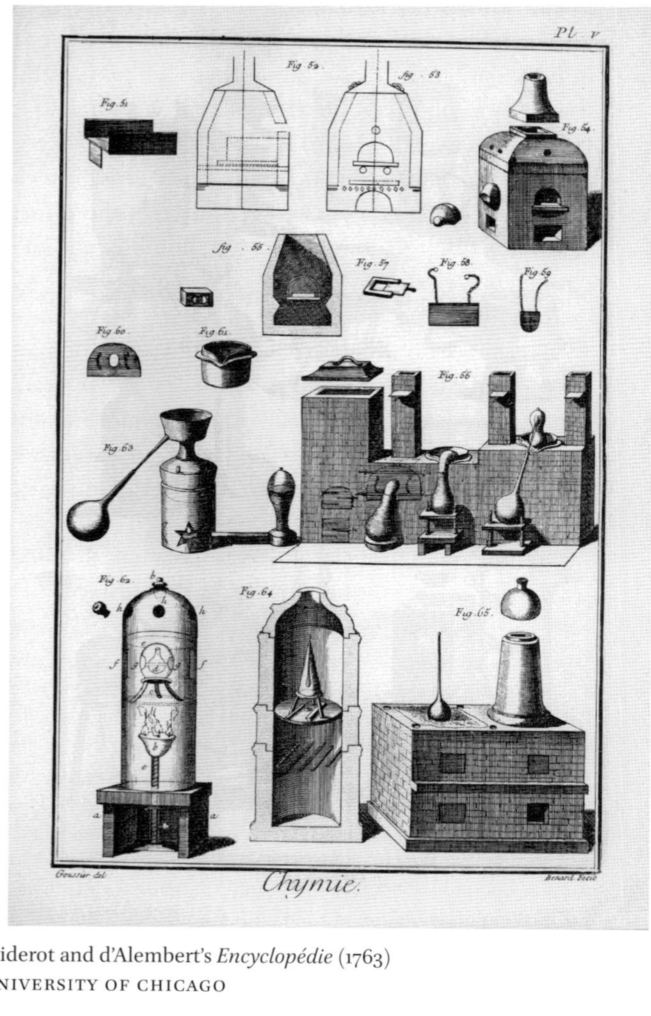

FIGURES 5–6 Illustrations of chemical furnaces, produced in 1764 for Diderot and d'Alembert's *Encyclopédie* (1763)
COURTESY OF THE ARTFL ENCYCLOPÉDIE PROJECT, UNIVERSITY OF CHICAGO

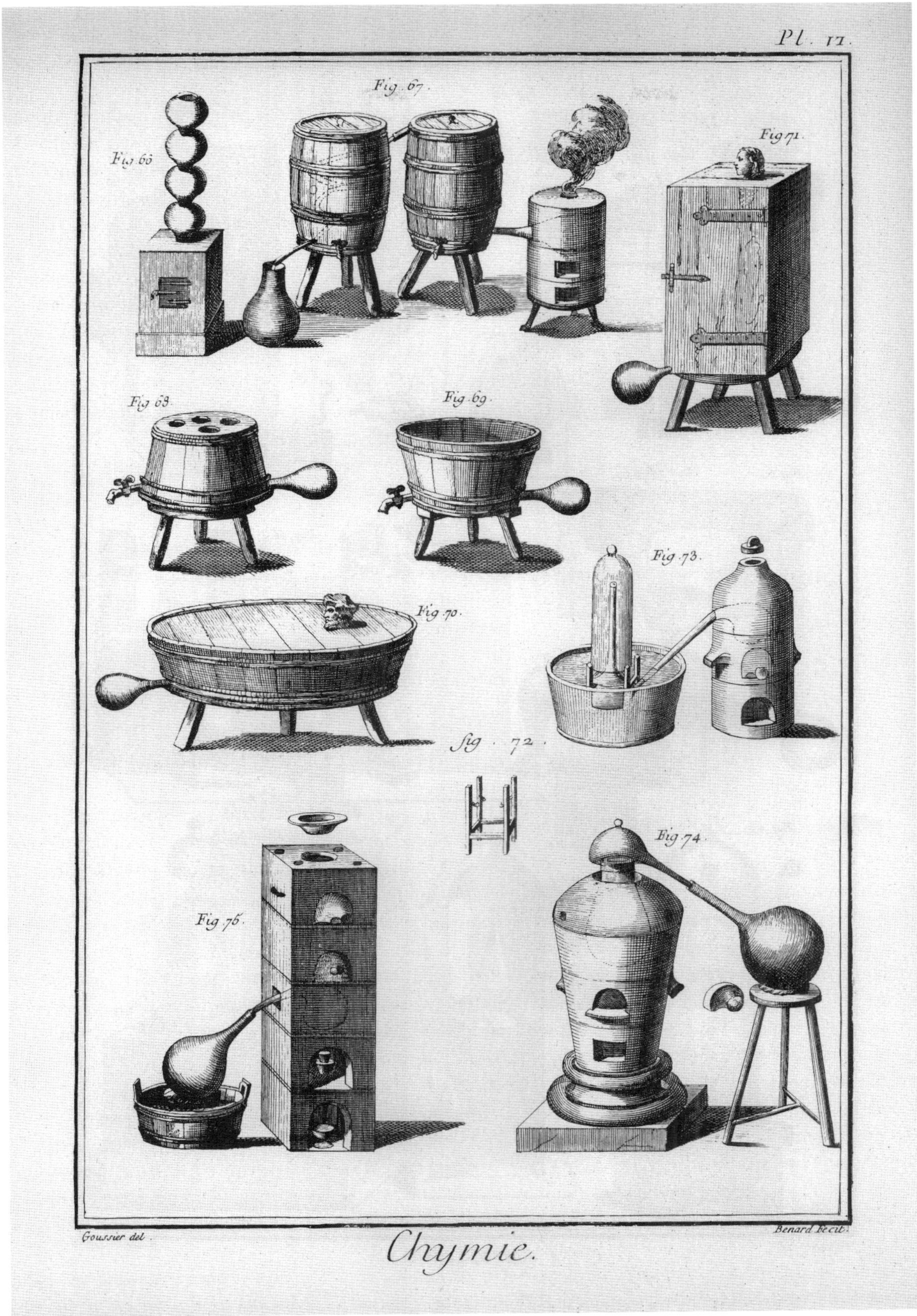

FIGURE 7 Distilling and pneumatic apparatus (1763). Diderot and d'Alembert's *Encyclopédie* (1763)
COURTESY OF THE ARTFL ENCYCLOPÉDIE PROJECT, UNIVERSITY OF CHICAGO

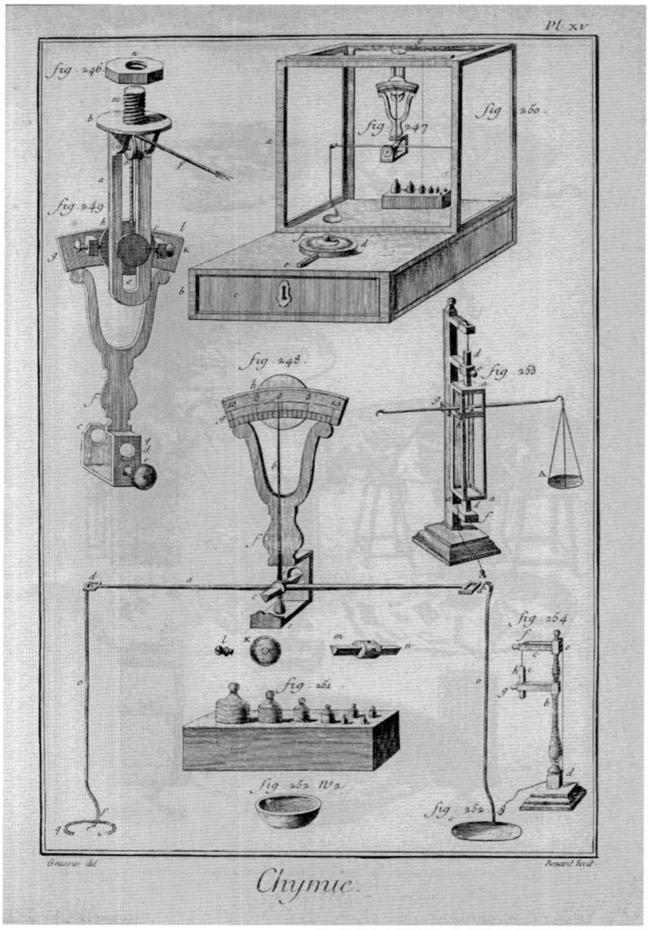

FIGURE 8 An assay balance by Gallonde (1763). Diderot and d'Alembert's *Encyclopédie* (1764)
COURTESY OF THE ARTFL ENCYCLOPÉDIE PROJECT, UNIVERSITY OF CHICAGO

its inner structure. In addition to the chemist and the physicist there are three laboratory assistants and another chemist: the chemist (in the right background) is shown controlling vapours caused by the detonation of nitre; one laboratory assistant (on the left) is carrying up charcoal to feed the furnaces from storage below; another assistant, positioned in front of the table, is collecting water from a basin, and the final assistant (on the right) is cleaning a vessel in a barrel filled with water.

The simple operations performed by the assistants underline the importance of water and fuel to chemical experimentation. Equally important is the ventilation of the room, as shown by a large chimney and bellows. From left to right there are an *athanor* (a furnace that keeps a constant heat), an assaying furnace, and another furnace connected to a large retort. On top of the mantelpiece we find a variety of glass receivers, retorts, alembics, and alchemical vessels used for slow distillation like the pelican, as well as, on the extreme right, an apparatus for measuring the quantity of air escaping from fermenting bodies. Apart from a description of the laboratory Diderot also added several engravings illustrating other types of chemical apparatus, most of which were drawn from Rouelle's own equipment.

In all, Diderot listed 265 pieces of apparatus. Among the most important instruments he included various furnaces for smelting, distillation, decoction, calcination, and glassworking, as well as reverberatory furnaces, ovens of various types, (Figs. 5–6) stoves, *bains marie* and related equipment, apparatus for ascertaining the quantity of air which emanates from a body (Fig. 7), an assay balance made by Gallonde (Fig. 8), as well as all sorts of alembics, flasks, retorts, balloons, glass receivers, mortars, and various smaller utensils. While most of the depicted apparatus came from Rouelle's laboratory, its original design was often that of other chemists, physicists, and alchemists. The pneumatic apparatus, for instance (Fig. 7 n.73), was a modification of that described by Stephen Hales in his *Vegetable Staticks* (1727); and several furnaces were inspired by Johan Rudolf Glauber's *Furni novi philosophici* (1646–1647), which was a comprehensive work on the subject. A considerable number of the instruments were reminiscent of the alchemical tradition, indicating that eighteenth-century chemistry continued to profit from the knowledge of older traditions from outside of academia or the artisanal trades.

1.3 Laboratory Life with Macquer and Baumé

The clarity and structure of the plates published by Diderot in the *Encyclopédie* showed that a detailed description of the laboratory and its apparatus could provide chemists with useful instructions. However, laboratory life was not easy, and experimentation was full of unexpected and often unpleasant disappointments. Moreover, the cost of maintaining a laboratory in good order was considerable: glassware often broke, earthenware vessels were rapidly consumed, chemicals were not always readily at hand, and some of them, such as mercury, were extremely expensive.

In 1766 Pierre-Joseph Macquer, a physician who studied chemistry under Rouelle, published a *Dictionnaire de chymie* which was translated into English, German and Italian, and would be the standard reference work for European chemistry for the following decade.[42] Macquer was a highly experienced and versatile chemist.[43] (Fig. 9) He worked for the *Bureau du commerce* solving technical problems relating to dyestuffs and, from 1757 onwards,

42 Partington (1961–1970), vol. 3, pp. 89–90.
43 On Macquer's laboratories see Lehman (2019).

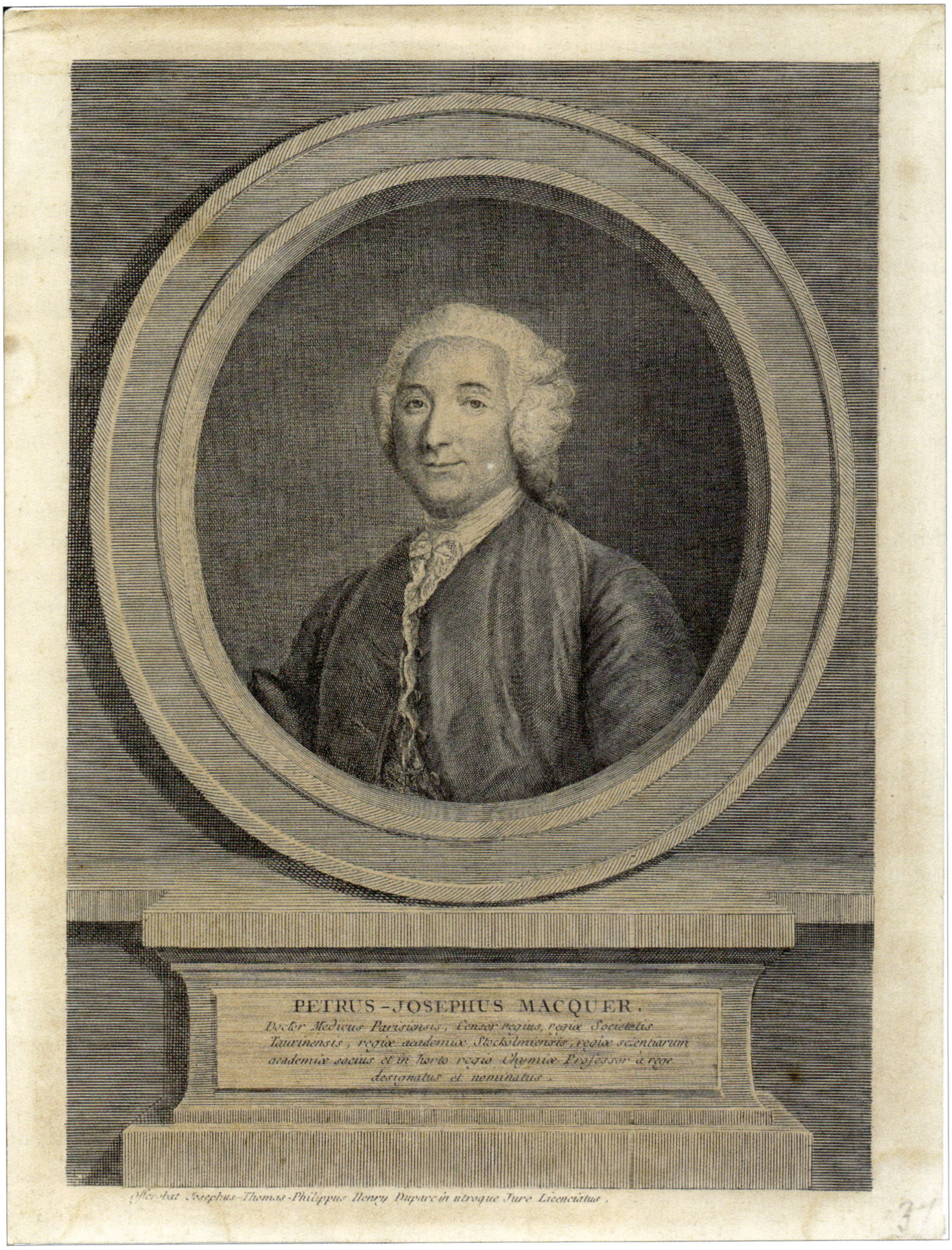

FIGURE 9
Eighteenth-century engraved portrait of Pierre-Joseph Macquer
PRIVATE COLLECTION

was a consultant at the famous porcelain factory at Sèvres. Since the mid-1750s Macquer had also worked in the laboratory of the apothecary Antoine Baumé in the Rue Saint-Denis, where Macquer and Baumé delivered a course in chemistry together for several years. Baumé, who moved his laboratory to the Rue Coquillière in 1762, was one of the most important suppliers of chemicals and chemical apparatus in Paris, and in addition to his shop he had several specialised workshops and laboratories.[44] With his extended experience in chemical manufactories and artisanal, teaching, and research laboratories, Macquer was a distinguished academic chemist with a long record of publications on both practical and theoretical chemistry. Despite its lack of illustrations, the entry headed "laboratory" in Macquer's *Dictionnaire de chymie* was, indeed, far more informative than the one written by Venel for Diderot's *Encyclopédie*.

According to Macquer, chemistry was a science entirely based on experience, and therefore it was impossible to understand it properly without a thorough knowledge of the instruments and apparatus needed to perform it. Macquer's effort to describe the ideal laboratory "proper for a philosophical chemist" without reference to apothecaries' shops, the workshops of the artisans, or the large

44 Davy (1955), pp. 19–26.

manufactories, was an important advance in comparison with the *Encyclopédie*. Although all mentioned sites had many characteristics and apparatus in common, Macquer believed that the chemical laboratory had finally become a scientific site in its own right.

He recommended that laboratories be constructed above ground level. Although the ground level was "most convenient, for the sake of water, pounding, washing, etc.", dampness often compromised the outcome of long experiments. "In such a place, most saline matters become moist in time; the labels fall off, or are effaced; the bellows rot; the metals rust; the furnaces moulder, and every thing almost spoils. A laboratory therefore is more advantageously placed above than below the ground, that it may be as dry as is possible".[45]

The ventilation of the laboratory was also very important, because the exhalations and fumes could be extremely toxic. Macquer was one of the earliest chemists to highlight the dangers of chemical experimentation throughout the entries of his *Dictionnaire*, and to suggest that necessary precautions be taken to avoid accidents; among them, a well-aired laboratory was certainly the first and most important. But even less harmful fumes and dust, such as that caused by charcoal, had to be carefully handled for the laboratory to be kept tidy. Macquer also insisted on the importance of shelves for the storage of glassware and chemicals. "In a laboratory where many experiments [are] made, one [could] not have too many shelves".[46]

The chemical table introduced by Rouelle soon became an essential feature of the chemical laboratory.[47] Macquer placed it in the middle of the room and declared it to serve for all the "preparations for operations, solutions, precipitations, small filtrations; in a word, whatever does not require fire, excepting that of a lamp". Macquer, like Rouelle, placed most of the furnaces beneath the large fireplace. His laboratory, however, contained only some of the furnaces listed in the *Encyclopédie*; those that could be traced back to the alchemical tradition were left out.

In addition to apparatus and utensils, Macquer also included a long list of chemicals: metals, semi-metals, vitriolic and nitric acids, common salt, vinegar, lime water, quicklime, various alkalis, "refined essential oil of turpentine; oil of olives; soap, galls, syrup of violets; tincture of turnesol, or turnesol in rags; fine blue paper; river or distilled rain-water".[48] He also recommended the preparation of neutral salts, which were most commonly used during chemical operations.

The presence of such an amount of equipment and chemicals led Macquer to insist on a cautious and methodical approach to experimentation. Chemists had to be

> well persuaded that method, order, and cleanliness, are essentially necessary in a chemical laboratory. Every vessel and utensil ought to be well cleaned as often as it is used, and put again into its place: labels ought to be fastened upon all the substances. These cares, which seem to be trifling, are however very fatiguing and tedious; but they also are very important, though frequently little observed. When a person is keenly engaged, experiments succeed each other quickly; some seem nearly to decide the matter, and others suggest new ideas: he cannot but proceed to them immediately, and he is led from one to another: he thinks he may easily know again the products of the first experiments, and therefore he does not take time to put them in order: he prosecutes with eagerness the experiments which he has last thought of; and, in the mean time, the vessels employed, the glasses and bottles filled, so accumulate, that he cannot any longer distinguish them; or, at least, he is uncertain concerning many of his former products. This evil is increased if a new series of operations succeed, and occupy all the laboratory; or if he is obliged to quit it for some time: everything then goes into confusion. Thence it frequently happens that he loses the fruits of much labour, and that he must throw away almost all the products of his experiments.[49]

Macquer's detailed description of the chaotic results of the complex and uncertain experiments carried out in the chemical laboratory reveal that the apparatus had not yet achieved the accuracy for which he was aiming.

The time was approaching for a radical reform of both the apparatus and the method of conducting experiments. Before I attend to the discoveries leading to the transformation of chemical laboratories, a brief survey is necessary of the sites created by the Parisian apothecary Antoine Baumé, (Fig. 10) whose early career, as mentioned, was associated with that of Macquer. Baumé represents a key figure in the transition from the traditional chemical laboratory to a new phase of chemical experimentation.

Like Rouelle, Baumé was an ambitious apothecary who was confident in the roles of both the scientist and the

45 Macquer (1777).
46 Macquer (1777).
47 Morris (2015), pp. 52–53.
48 Macquer (1777).
49 Macquer (1777).

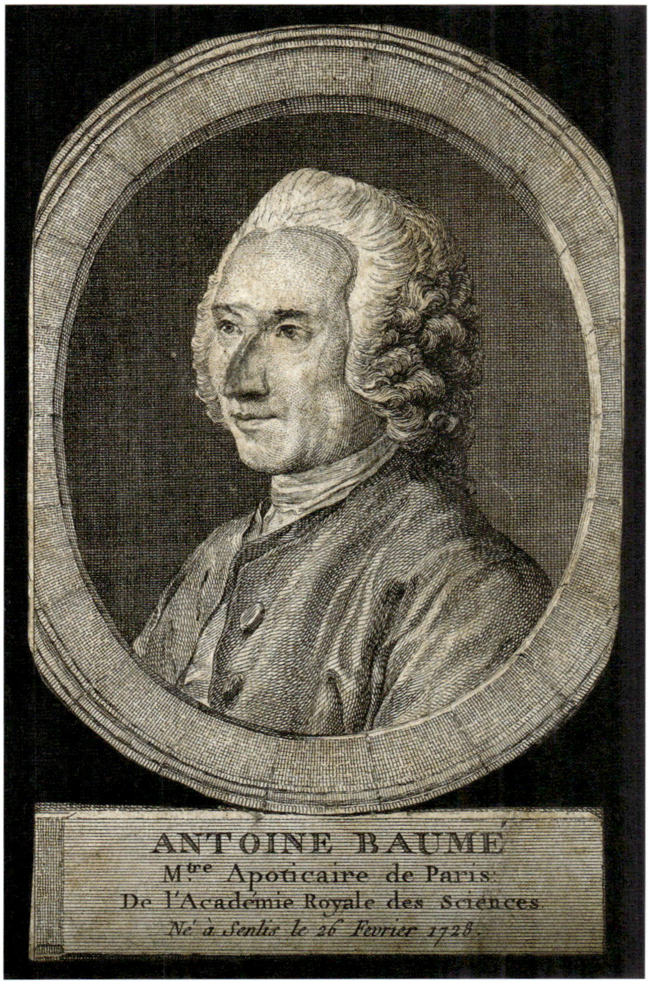

FIGURE 10 Eighteenth-century engraved portrait of Antoine Baumé
COURTESY OF THE WELLCOME COLLECTION

practitioner. He surpassed the low social status of pharmacists at the time by associating himself with Macquer, who was a distinguished member of the Académie des sciences, as well as of other important academic and state institutions. In 1758 Baumé and Macquer signed a contract to deliver an advanced course in the laboratory that Baumé had opened in the Rue Saint-Denis in 1752.[50] The contract covered the acquisition of the necessary equipment, the shared authorship of the discoveries made in the laboratory, and the joint publications. Although, like Rouelle at the Jardin du roi, Baumé acted as the demonstrator, he was Macquer's equal here.

They delivered the courses regularly until 1772, when Baumé was elected a member of the chemistry section of the Académie des sciences. As the number of experiments rapidly grew, Baumé moved his laboratory into a far more spacious building in the Rue Coquillère in 1762, where he was soon able to build five different laboratories: one was devoted to the preparation of drugs, another to the large-scale production of lead acetate, compounds of antimony, mercurial salts, muriate of tin (stannous chloride), sal ammoniac (ammonium chloride), Glauber's salt (sodium sulfate), and volatile oils.[51] Baumé was also a producer and supplier of hundreds of medical remedies, some using ingredients imported from the French colonies in the Americas, and despite the municipal prohibition of the advertisement of the sale of commodities, he published a comprehensive catalogue of the drugs for sale in his shop with their prices in 1775.[52]

In addition to providing his clientele with remedies and chemicals, Baumé was also a major maker and supplier of chemical instruments and apparatus. As early as 1768 he advertised in the *Avantcoureur* the sale of a chemical hydrometer of his own design, an instrument which remained in use up to the twentieth century. He sold all kinds of utensils, glassware, retorts, alembics, cucurbits, and furnaces. Baumé's trade – closely connected with the activity in his own laboratory – is important, because he helped improve and standardise the equipment used in chemical laboratories in Paris as well as the provinces.

The discovery of the active role of gases in chemical reactions radically transformed apparatus and analytical methods. It is, therefore, significant that in the late 1770s and 1780s crucibles, glassware, and the most sophisticated equipment was made by professional instrument makers, natural philosophers or by entrepreneurs with first-hand knowledge of chemical processes, such as Josiah Wedgwood. It is within this context of the rapid and radical transformation of chemical practice that Lavoisier entered the scene.

50 Contant (1952).

51 Davy (1955), pp. 19–21.

52 On the prohibition, see Bouvet (1937), p. 236. Baumé (1775).

CHAPTER 2

Lavoisier's Approach to Chemical Instrument Making

2.1 Lavoisier's Early Education in Science

Unlike most who practiced chemistry in Paris during the second half of the eighteenth century, Lavoisier studied neither medicine nor pharmacy, and for a long time he was regarded more as a physicist than as a chemist. Since his approach to experimentation and to instrument making was influenced by his early education, I shall briefly outline his background.[1]

According to Grimaux, Lavoisier entered the Collège Mazarin in 1754 as a day-student (*externe*) when he was eleven years old.[2] The Collège Mazarin or Collège des Quatre-Nations was founded in 1688 with an endowment from Cardinal Mazarin, in order to provide free education for children from noble families.[3] (Fig. 1) As agreed with the University, the Collège excluded theology, law, and medicine from its education. Students wishing to join the Collège needed to be aged between ten and fifteen, and of Catholic faith. The courses took nine years to complete. They began with six courses in the humanities (*classes d'humanitez*) followed by one on rhetoric. Next were one class in mathematics, one in physics, and one in logic. The final two classes belonged to the general category of *classe de philosophie*. According to Franklin, the Collège Mazarin was particularly popular with day-students for its courses in the natural sciences and mathematics.[4] This is supported by the facts that the library of the Collège owned one of the best collections of books on the history of medicine in Paris, and that a small observatory was built for Nicolas-Louis de La Caille, who became Professor of mathematics at the Collège in 1740 (Fig. 2).

During his stay at the Collège Lavoisier distinguished himself as a brilliant student in the humanities, and he was awarded several prizes in his first six years, the last

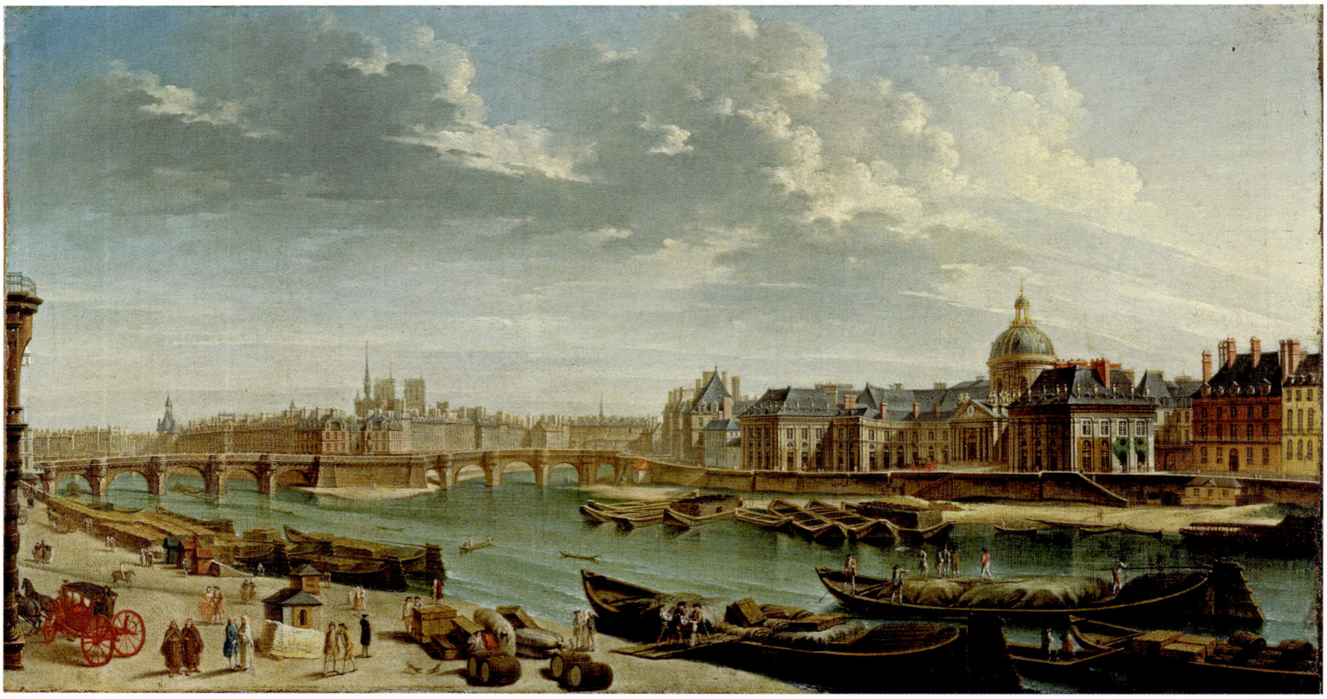

FIGURE 1 View of the Collège Mazarin (1763) by Jean Baptiste Raguenet
COURTESY OF THE GETTY MUSEUM

1 Parts of this chapter are based on Beretta (1994) and Beretta (2014).

2 Grimaux (1888), p. 3.

3 Franklin (1862).

4 "Les écoliers étaient nourris noblement et instruits fondamentalement en toutes sortes de sciences […]. Il devait en grande partie cette préférence à son cours de Mathématiques, car c'était le seul Collège de Paris où les sciences fussent réellement enseignées". Franklin (1862), pp. 108–109.

 | DOI:10.1163/9789004511217_004

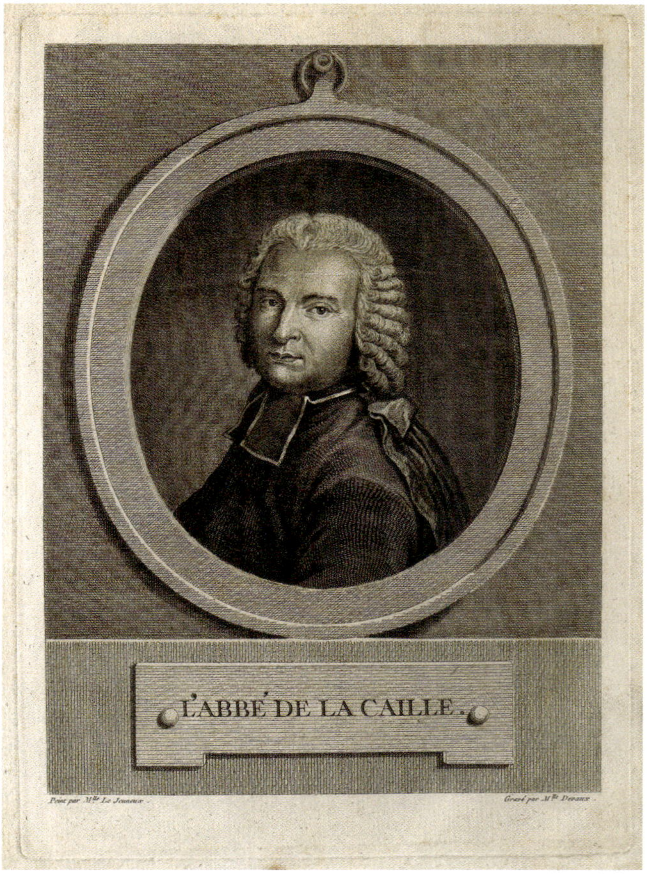

FIGURE 2 Eighteenth-century engraved portrait of Nicolas La Caille
COURTESY OF THE WELLCOME COLLECTION

in 1760 when completing the class of rhetoric. Lavoisier then attended classes in mathematics and philosophy and thereby completed the curriculum of the Collège in 1763.

Between 1760 and 1761 Lavoisier appears to have attended Nicolas La Caille's mathematics and astronomy classes.[5] Given his early interest in astronomy, it is interesting to speculate whether he received instructions in observations of the sky at the Collège's observatory which, therefore, could be regarded as the first site where Lavoisier practised science. Henry Guerlac, however, remarked that from 1761 Lavoisier studied law, and that he gained a bachelor of laws in September 1763, so that "it is difficult to believe that he attended simultaneously the lectures [in philosophy] at Mazarin and the *Faculté de Droit*, while embarking at the same time on his scientific studies".[6] Guerlac concluded that Lavoisier must have left the Collège in 1761 without attending the philosophy classes. But this contradicts an autobiographical note written by Lavoisier in around 1792, in which he claimed that he attended a "bon cours de philosophie" at the beginning of his career in science.[7] Whether Lavoisier stayed at the Collège or not during the years 1762 and 1763 is difficult to confirm, but there is no reason to doubt Lavoisier's own reconstruction.

In the early 1760s Lavoisier became acquainted with the botanist and mineralogist Jean-Etienne Guettard, and as early as 1763 he followed him on his mineralogical excursions at Villers-Cotterets.[8] On these excursions Lavoisier collected mineral specimens and wrote two *mémoires* on natural history. In 1763 he also joined the botanist Bernard de Jussieu "dans ses herborisations".[9] The question of how and when Lavoisier began to study chemistry is more difficult. Grimaux was among the first to state that Lavoisier's initial studies in chemistry occurred in the courses delivered by Guillaume-François Rouelle which he occasionally attended, according to Rhoda Rappaport, from 1762 to 1764.[10]

In 1990 Bernadette Bensaude-Vincent published an extended excerpt of Lavoisier's abovementioned autobiographical from 1792, a part of which had been published by Daumas in 1955, but in spite of being the only direct source on Lavoisier's early education, it did not attract the due attention. This is quite surprising, as in it, Lavoisier provides some previously unknown information about his early forages into chemistry. It is, therefore, worthwhile to reproduce Lavoisier's note here:

> When I began for the first time to attend a course in chemistry I was surprised to see how much obscurity surrounded the first approaches to the science, even though the professor whom I had chosen was regarded as the clearest and most accessible to beginners, and even though he took infinite pains to make himself understood.

5 He also found the time to compete for a prize in eloquence at the Académie de Besaçon; see Bret (2005).

6 Guerlac (1956), p. 213.

7 The autobiographical note by Lavoisier was found by Maurice Daumas (1955), at the beginning of a manuscript entitled *Sur la manière d'enseigner la chimie*, which he dated to 1792, and which is kept at the archives of the Académie des sciences (Paris), Fonds Lavoisier Dossier 1259. Some original comments on the draft published by Daumas can be found in Albury (1972), pp. 112–115. Bernadette Bensaude-Vincent (1990), pp. 456–460, published am extended excerpt of Lavoisier's manuscript without commenting on the autobiographical note. Henceforth I will quote Lavoisier's manuscript from the version published in 1990.

8 On this early studies see Palmer (1998).

9 Grimaux (1888), p. 4.

10 "Lavoisier attended Rouelle's private course in the rue Jacob during the year 1762–63 or 1763–64". Rappaport (1960), p. 77.

I had taken a useful course in physics, I had followed the experiments of the Abbé Nollet, I had also studied elementary mathematics with some success in the works of the Abbé La Caille and had attended his lessons for a year. I was used to the intellectual rigor with which mathematicians do their work, never accepting a proposition until the one preceding it has been proved. Everything is connected, it forms a chain, from the definition of a point, of a line, to the most sublime truths of transcendental geometry.

In chemistry, things are quite different. In the first instance, one starts by assuming instead of proving. [...] So when they were beginning to teach me this science they assumed that I already knew.

What is more, I was not alone in finding matters obscure. For M. de La Planche, under whom I was studying, wanting to have a few quiet words with me in the interval between one class and the next, could not refrain from telling me that I ought to expect to understand nothing for half the course, that a beginner's course could only teach me what he called the dictionary of the science [...].

I found the remainder of the course to be as M. de La Planche had predicted. He assumed from his very first lessons many things that he promised to demonstrate in later lessons, and these lessons passed without the assumptions being demonstrated. From the first day, he talked to us about affinities, which are the most difficult thing to understand in chemistry. [...]

At the same time, it was true that M. de La Planche was the clearest of the teachers then teaching chemistry. He was a pupil of Rouelle, and followed the same divisions and the same method. When the course ended, I wanted to take stock of the knowledge that I had acquired in chemistry, and I realized that I knew quite well everything about the composition of neutral salts, the preparation of mineral acids [...]

I no longer experienced the same difficulty when, in the years that followed, I took Rouelle's course. The eminent professor combined much method in the presentation of his ideas with much confusion in his expression of them. But with the aid of the preliminary knowledge that I had acquired before coming to listen to him by dint of three years' hard work, I succeeded in forming a clear and accurate idea of the state reached by the science of chemistry at this time. But it is no less true that I had spent four years studying a science that was based on only a small number of facts.[11]

In these lines Lavoisier adds substantial new information to our knowledge of his introduction to the scientific world. First of all, we learn that he attended Nollet's public lectures on experimental physics, possibly while attending La Caille's course in mathematics in 1760–1761. Also, contrary to what is commonly believed, Lavoisier asserts that the first chemistry course he attended was taught by

11 "Lorsque j'ay commencé pour la première fois à suivre un cours de chimie, quoique le professeur que j'avais choisi passât pour le plus clair et le plus a portée des commençants, quoiqu'il prit infiniment de peine pour se faire entendre, je fus surpris de voir de combien d'obscurité les abords de la science se trouvaient environnes. J'avais fait un bon cours de philosophie, j'avais suivi les expériences de l'Abbé Nollet, j'avais étudié avec quelque fruit la mathématique élémentaire dans les ouvrages de l'Abbe de la Caille et j'avais suivi pendant un an ses leçons. J'étais accoutumé a cette rigueur de raisonnement que les mathématiciens mettent dans leurs ouvrages jamais ils ne prennent une proposition que celle qui la précède n'ait été découverte. Tout est lié, tout est enchainé, depuis la deffinition du point, de la ligne, jusqu'aux vérités les plus sublimes de la géométrie transcendante. Dans la chimie, c'était une tout autre marche. Dans les premiers pas on commençait par supposer au lieu de prouver. [...] Ainsi en commençant à m'enseigner la science on supposait déjà que je la savais. Cette obscurité au surplus n'était pas pour moi seul. Car M. de la Planche sous lequel j'étudiais et qui voulait bien m'accorder quelques instans de conférence dans l'intervalle d'un cours à l'autre ne laissait de m'avertir que je devais me résoudre à ne rien entendre pendant la moitié du cours, qu'un premier cours ne saurait qu'apprendre ce qu'il appelait le dictionnaire de la science [...] J'éprouvai dans la suite du cours ce que M. de la Planche m'avait annoncé. Il supposait dans ses premiers cours beaucoup de choses qu'il promit de démontrer dans les cours subséquents et les cours se passaient sans que les suppositions fussent démontrées. Dès le premier jour, il nous parlait d'affinités, ce qu'il y a de plus difficile à entendre dans la chimie. [...] Cependant, il était vrai que M. de la Planche était le plus clair des proffesseurs qui enseignaient alors la chimie. Il était élève de Rouelle et suivait la même division et la même méthode. Lorsque le cours fini, je voulus faire l'inventaire des connaissances que j'avais acquises en chimie, je reconnus que je connaissais assez bien tout ce qui concernait la composition des sels neutres, la préparation des acides minéraux [...]. Je n'éprouvais plus les mêmes difficultés lorsque les années suivantes je suivis le cours de Rouelle. Le célèbre proffesseur réunissait à beaucoup de méthode dans la manière de présenter ses idées beaucoup d'obscurité dans la manière de les énoncer. Mais à l'aide des connaissances préliminaires que j'avais acquises avant de venir l'entendre pendant trois années d'assiduité, je parvins à me former une idée nette et précise de l'état ou la science chimique était parvenue à cette époque. Mais il n'en était pas moins vrai que j'avais employé quatre années a étudier une science qui n'était fondée que sur un petit nombre de faits". Text from the edition of Bensaude-Vincent (1990), pp. 456–467.

the Parisian apothecary Laurent-Charles de La Planche, not Rouelle. Even more surprising is the fact that Lavoisier considered La Planche "le plus clair des proffesseurs qui enseignaient alors la chimie", and that he learned about the composition of neutral salts and the preparations of mineral acids in his course. As is well known, Lavoisier was not very generous in his praise for his predecessors, which places even greater emphasis on his positive judgement of La Planche. Furthermore, Lavoisier's high praise of La Planche and the role he played in his own approach to chemistry challenges the common narrative that Rouelle was the only influential figure in Lavoisier's chemical education. Lavoisier actually referred to Rouelle's works mostly in connection with his geological works, and as Rappaport convincingly demonstrated, "it is highly unlikely that Rouelle's lectures stimulated in him a lively interest in chemistry. […] That Lavoisier did not consider Rouelle the major source of his knowledge is further confirmed by his constant readily available published writings of other chemists – notably Stahl, Boerhaave, and Macquer – when he speaks of the history of chemistry".[12]

Before analysing La Planche's work in some detail it is worth noting another interesting piece of information given by Lavoisier's in his autobiographical sketch: he declares that he studied chemistry for four consecutive years, the first under the supervision of La Planche, and three under Rouelle. This changes the commonly accepted chronology of Lavoisier's early experiences in chemistry, according to which he studied chemistry for no more than three years. Considering that Lavoisier attended the humanities classes at the Collège Mazarin until 1760, it is likely that attended La Planche's course in 1761, when he also became interested in other sciences. Lavoisier then attended Rouelle's chemical lectures in 1762, 1763, and 1764. It is, in fact, unlikely that he attended any courses in 1765, as he accompanied Guettard and de Jussieu on their scientific excursions to Normandy and the areas outside Paris.

Furthermore, in 1765 Lavoisier's scientific interest seemed to be focused on mineralogical topics, and his *mémoire* on gypsum, written under the guidance of Guettard, further supports this. Finally, all of Lavoisier's biographers agree that he must have finished his chemical education in 1764. As we read in his note, at the end of his education Lavoisier was dissatisfied with his chemical knowledge and blamed his teachers (both La Planche and Rouelle) for reversing the logical order of chemistry by beginning their courses with the most difficult and complex part of chemistry (the vegetable kingdom) instead of the simplest and easiest to develop (the mineral kingdom). Most of Rouelle's courses began with the chemical analysis of the vegetable kingdom, and they usually provided a poor and superficial survey of mineralogy.[13] Unfortunately, there is no equivalent documentation for La Planche's teaching, and no manuscript notes recording the whole course appear to survive. La Planche's life is similarly poorly recorded. In a rare *mémoire* on La Planche, however, we find the following biographical notes:

> Monsieur de La Planche is a pupil of the celebrated Rouelle. He has attended the lessons of this *grand maître* for three years; and his activity, his precision, his industry in benefitting from it, together with the knowledge which he had already acquired, have soon put him in a position himself to train students who are capable of doing credit to them both. He became *maître apothicaire* in 1748; and since 1746 he has given special courses in chemistry at his own home with the authorisation of the Faculty of Medicine. In 1758 the *corps des apothicaires*, wishing to contribute more to the progress of this art, decided to start a public course at the Jardin de la compagnie.[14]

The public courses in chemistry were inaugurated in 1759, and the Compagnie assigned the post of teachers to La Planche, Bataille, Azema, Guénaud, Couzier, Demoret, Santerne, Laborie, and Jullot. The lectures were delivered in the Jardin de la compagnie, which had been built in 1700 and included a chemical laboratory. The public

12 Rappaport (1961), pp. 97–98.

13 Here I must disagree with Guerlac, who believed that "Rouelle […] was one of the most enthusiastic supporters of the new mineralogical emphasis". Guerlac (1956), p. 215. From the manuscripts recording Rouelle's course of chemistry that I have consulted at the Bibliothèque Nationale, the Bibliothèque de l'Arsenal, the Bibliothèque de la Sorbonne, the Bibliothèque interuniversitaire de pharmacie, and the Bibliothèque du muséum d'histoire naturelle in Paris, it is clear that Rouelle devoted far more time to vegetable chemistry than to mineralogy. The fragment of his course owned by Lavoisier further confirms this; Beretta (2011).

14 "Le Sieur de la Planche est élève du célèbre Rouelle. Il a reçu pendant trois ans les leçons de ce Grand Maître; et son activité, son exactitude, son application à en profiter, jointes aux connoissances qu'il avoit déjà précédemment acquises, l'ont bientôt mis en état de former lui-même des Elèves, capables de faire honneur à l'un et à l'autre. Il a été, reçu Maître Apothicaire en 1748; et de 1746, il faisoit chez lui, sous l'autorisation de la Faculté de Médicine, des Cours particuliers de Chymie. En 1758, le Corps des Apothicaires voulant contribuer de plus en plus au progrès de cet Art, prit le parti d'établir un Cours public, au Jardin de la Compagnie". *Mémoire* (1764), p. 2. On the Jardin see also Lehman (2006).

courses were discontinued in 1765 after a legal controversy involving La Planche in 1764. Lavoisier probably attended La Planche's course at his private laboratory in the Rue de la Monnoie. Unlike for Rouelle, there are no known surviving manuscript copies of La Planche's courses, and apart from Lavoisier's, no contemporary comments on their contents or quality appear to have survived. Nevertheless there are two printed copies of the course plans for 1750 and 1764 which offer a rough idea of the course contents.[15] The plan printed in 1764 is likely to be more closely related to the course Lavoisier attended at the beginning of the 1760s than the other.

La Planche began this course with a series of experiments aimed at demonstrating the existence of phlogiston. After a traditional description of chemical instruments, La Planche continued his course with the chemical analysis of the vegetable, animal, and mineral kingdoms, and with a part devoted to metallurgy. Unlike Rouelle, La Planche apparently devoted more time to mineralogy and metallurgy than to vegetable chemistry, and he included a detailed programme of the investigation of the calcination of lead and the chemical analysis of gypsum, two subjects which would eventually be studied thoroughly by Lavoisier. Undoubtedly, a section entirely devoted to metallurgical matters was important and original. Unlike their German and Swedish equivalents, French naturalists did not study metallurgy in connection with chemistry. During the 1750s and early 1760s d'Holbach (the author of metallurgical articles for d'Alembert's Diderot's *Encyclopédie*)[16] and Jean Hellot (the translator into French of Schlüter's work on mining) successfully created a renewed interest in metallurgy, but their writings were not prompted by a generalised research programme, and their impact on French readers was not comparable with the prominence and authority of the metallurgical literature produced in the Germany and Sweden. Interestingly, La Planche appears to have been familiar with the operations and instruments of metallurgy. However, the prospectus' cumbersome nomenclature and its lack of descriptive details for the experimental procedures make it difficult to assess the quality and content of the course. It is, therefore, difficult to trace a direct and unequivocal influence of La Planche and Rouelle on Lavoisier's early published papers on chemical subjects.

Given Lavoisier's negative impression of both his teachers as well as the state of chemistry we may wonder who inspired him to enter into a career in chemistry rather than experimental physics. The answer to this question certainly lies in his early apprenticeship with Jean-Etienne Guettard. Guettard, who was a distinguished botanist, chemist, geologist, and technologist, and a friend of Lavoisier's father, and therefore acted as a mentor to the young Lavoisier.[17] Although Guerlac perceptively pointed out that "it was Guettard who exerted the greatest influence upon Lavoisier", he did not explore their scientific relations in depth.[18] The references to Guettard's role in Lavoisier's early scientific biography are even less satisfactory in Grimaux's and Poirier's biographies.[19] This is especially disappointing because Grimaux had access to Lavoisier's entire correspondence, and he could not have missed the wealth of documents connecting the two scientists. Unfortunately, for unknown reasons, the rich correspondence between Lavoisier and Guettard between 1760 and 1775 was not published by René Fric, thus leaving a lacuna in the accessible information on Lavoisier's early apprenticeship in science. Their correspondence reveals that Guettard, immediately impressed by Lavoisier's talent for natural history, guided his first steps in science with the greatest care and affection, and that he continued to mentor him intensely up to the late 1760s.[20]

The reason of such a close relationship remains to be explained. Lavoisier was the only member of his family interested in science, and Guettard did not have any connections with the teachers of the Collège Mazarin. An interesting clue is a note published by the mineralogist Antoine-Grimoald Monnet who, in his youth had been closely acquainted with Guettard. Although not an entirely reliable source due to his well-known animosity against Lavoisier, Monnet claims that the friendship between Guettard and Lavoisier's father, Jean-Antoine, was based on their common religious sympathies for Jansenism,[21] and that it was upon Guettard's recommendation that the young Lavoisier attended Nollet's and Rouelle's courses.[22] According to an unpublished sketch of

15 I have reprinted that dating 1764 in Beretta (1994), pp. 79–91.

16 Rappaport (1994).

17 The only biographer of Lavoisier who did not underestimate Guettard's influence on Lavoisier is Guerlac (1956), p. 215 who remarked that "it was owing to Guettard that Lavoisier was introduced to the science of chemistry". Against this claim Palmer has maintained that "[t]here is no evidence in Lavoisier's papers to support these claims of Guettard's alleged influence in matters chemical"; Palmer (1998), p. 240.

18 Guerlac (1975), p. 52.

19 Grimaux (1888) and Poirier (1996).

20 The letters are expected to be published in the forthcoming supplement to Lavoisier's *Correspondance* (LC, vol. 8).

21 "Lavoisier fils d'un procureur au parlement de Paris, janseniste, et en cette qualité ami de Guettard". Monnet (1797), p. 365.

22 "[Guettard] s'appercevant des grandes dispositions du jeune Lavoisier pour les sciences physiques, lui promit de le pousser et de le faire entrer à l'académie des sciences, à quoi il réussit bientôt, c'est-à dire, après que son élève eut suivi les cours de

Guettard's life dated 1786, which was drafted by Lavoisier for Condorcet in preparation for the publication of his obituary, Lavoisier's family, who had recognised his disposition for the natural sciences, asked Guettard to take him under his wings.[23]

Guettard, and to some degree Bernard de Jussieu (who had been Guettard's teacher), encouraged the young Lavoisier to look upon nature as an open-air laboratory.[24] Contrary to previous assumptions, it seems that Lavoisier was already under the guidance of Guettard in 1760 when he began collecting minerals.[25] While Guettard usually is portrayed as a botanist, mineralogist, and geologist, it is noteworthy that he also was an experienced chemist, and that he worked intensely in the chemical laboratory of the Duc d'Orléans between 1748 and 1752;[26] this was located near the Abbey of Sainte Geneviève, and there Guettard worked on the production of porcelain, on the dry analysis of the mineral specimens from the collection of his patron, and on vegetable chemistry.[27] For the production of porcelain the Duc d'Orléans had a second laboratory built in Bagnolet, where Guettard made his breakthrough discovery of the chemical properties of Kaolin. In the early 1780s Guettard and Lavoisier resumed their collaboration and devised several experiments related to the production of porcelain with white steatite.[28]

In the early 1750s Guettard, together with Macquer and Malouin, was appointed as *censeur royal* for chemical literature.[29] After the death of the Duc Louis d'Orléans in 1752 he inherited the *cabinet d'histoire naturelle*, which mostly consisted of minerals, and was granted an apartment in the Palais royal by the Duc's son, Louis Philippe;[30] both the cabinet and the rich library were also moved to the Palais.[31] It appears, however, that Guettard no longer had regular access to a chemical laboratory after 1752, and certainly not to the laboratory near the Abbey.[32] Despite these logistic difficulties Guettard continued his experimental chemical research, and in February 1765 the *Journal des sçavans* reported of his progress on the dissolution of salts.[33] The lack of access to a large laboratory was, in fact, not a serious obstacle to Guettard, who claimed that the most profitable laboratory was that of nature: "it is in the mines and not in the crucible of a chemist that you can learn how the minerals are formed in the depths of the

physique expérimentale de l'abbé Nollet et ensuite ceux du célèbre Rouelle". Ibid.

23 MS Lavoisier (1786), fol. 9v.

24 Condorcet, (1799), p. 182. It is worth noting that Condorcet's obituary was prepared on the basis of material collected by Lavoisier; see MS Lavoisier (1786). The extent of Lavoisier's affection for Guettard is also obvious in his effort to ensure that the library and collection of natural history of his mentor were not dispersed when he was forced to leave the apartment at the Palais Royal in May 1785. On this, see LC, vol. 4, pp. 220–221. Guettard died on January 6, 1786.

25 In 1761 Lavoisier collected the "[a]rgile brun blanc rougeâtre très douce au toucher" now preserved in the Muséum Lecoq of Clermont Ferrand, inventory no. T-La 0252.

26 Louis d'Orléans (1703–1752).

27 "Les terres & les pierres, de même que les plantes, sont des substances qui sont d'une très-grande utilité dans les Arts. La peinture, l'art de l'émailleur, les manufactures de faïance, de porcelaine, de toutes sortes de poteries, les tuileries & les briqueteries en peuvent tirer de grands avantages. Persuadé de cette vérité, feu M. le Duc d'Orléans m'ayant fait l'honneur de venir voir une collection que j'avois faite d'un grand nombre de ces matières, & m'ayant ordonné de les transporter dans le laboratoire de Chymie, qu'il avoit à Sainte-Géneviéve, pour, comme il disoit, les analyser & les brûler toutes, je commençai peu âpres à en soumettre à des expériences qui demandent un feu plus ou moins violent. Le plan que je me proposai de suivre, fut d'examiner d'abord ces différentes substances telles qu'on les tire de la terre, ensuite de les dégager des parties qui peuvent entrer accidentellement dans leur composition. J'étois occupé de ce travail lorsque la découverte des matières à porcelaine se fit. Il ne fut pas discontinué, au contraire, non-seulement on continua les expériences au laboratoire de Chymie, mais encore sous le four à porcelaine, que feu M. le Duc d'Orléans fit construire à Bagnolet. Pour celles-ci, je remettois au sieur le Guay, qui étoit à la tête des ouvriers, les matières que je proposois d'examiner. Le sieur le Guay tenoit registre de ce que je lui donnois, & il étoit obligé de me remettre ces substances passées au feu, & une copie des registres. C'est d'après cette copie que j'ai entre les mains, que j'ai formé des tables des résultats des expériences qui ont été faites à Bagnolet". Guettard (1768–1783), vol. 1, pp. xl–xli.

28 See LO, vol. 2, pp. 238–240.

29 He approved the publication of Poncelet (1755), p. 390. The expert censors were introduced by Louis XV in order to avoid arbitrary censorship.

30 Louis-Philippe d'Orléans (1725–1785).

31 "Enfin Guettard remit de lui-même les collections d'histoire naturelle au prince, les rangea dans des salles du Palais-Royal, et en devint conservateur". Franklin (1867–1873), vol. 1, pp. 81–82.

32 In 1768, while reconstructing his 1748 experimental work with exsiccated plants at the chemical laboratory of the Duc d'Orléans, Guettard remarked: "Je ne me suis pas trouvé à portée de faire commodément les expériences nécessaires pour s'assurer de quelques vérités sur ce sujet depuis que je ne suis plus à Sainte-Géneviéve [*sic*], & je ne sçais si je serai jamais en situation de les faire avec commodité: pour concourir avec ceux qui pourront suivre ce travail, je rapporterai ici une expérience que je trouve sur le Journal d'observations & d'expériences du laboratoire de feu S.A.S". Guettard (1768–1783), vol. 1, p. cx. Guettard also reports on his chemical experiments in vol. 2 of Guettard (1768–1783), pp. xxxv–lviii. In the same volume, on pp. 249 and 267, he highly praises Lavoisier's early work on gypsum.

33 "M. Guettard dans son Mémoire sur les salines, commence par donner en Naturaliste instruit, & versé dans la Chimie, l'histoire du Sel, la manière de le tirer ou l'extraire des différens corps ou substances avec lesquels il se trouve mêlé". *Journal des sçavans* (Feb. 1765), p. 77.

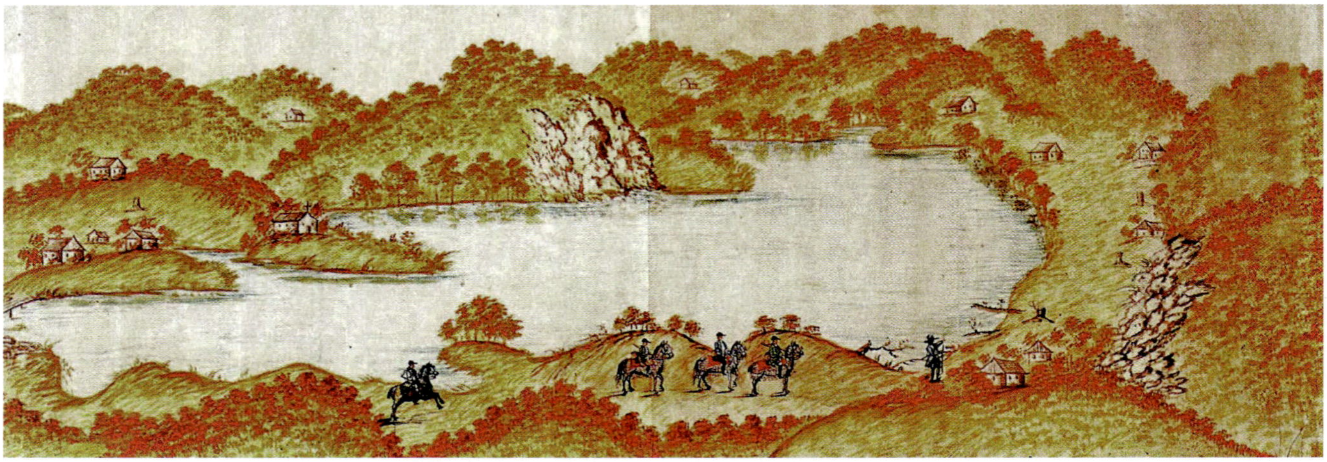

FIGURE 3 Watercolour showing Lavoisier, Guettard, their servants and, on the right, a local guide, during a mineralogical trip to Alsace (summer 1767)
COURTESY OF CORNELL UNIVERSITY LIBRARY – DIVISION OF RARE AND MANUSCRIPTS COLLECTIONS

earth".[34] Although I have not yet been able to trace it, it is likely that Guettard maintained a small laboratory in his new residence at the Palais royal.

From 1762 onwards, Guettard took his young protégé along to his mineralogical excursions in the outskirts of Paris. (Fig. 3) Lavoisier's budding interest in chemical mineralogy is revealed in a note dated 16 August 1763, where he writes about a stone collected at Saint Germain en Laye (together with Guettard?) which, given its violent effervescence upon contact with acids, he identified as "une veritable pierre a chaux".

In September 1764, having been guided a small village near Mézières by Guettard, Lavoisier accurately recorded the mineral composition of the strata of a "coupe" (cross section).[35] With the help of a barometer Lavoisier determined the levelling measurement of the rock layers.[36] Through this technique he eventually developed the idea of studying mineral ores in relation to their stratigraphic positions, and in 1767 he introduced cross-sections of strata to the right-hand margin of the 18 maps engraved by Jean Louis Dupain-Triel under his direction (Fig. 4).[37] Alongside his geological observations, Lavoisier also pursued more ordinary mineralogical surveys, including one examining the *plâtrières* (gypsum quarries) around Paris. Following this survey, Lavoisier's first chemical paper of 1765 was, as mentioned briefly above, devoted to the analysis of gypsum, and most of the mineral specimens he analysed came from Guettard's and the Duc d'Orléans' collections.[38] But where did Lavoisier perform his chemical analyses? He may have done them either in his aunt's apartment in the Rue Du Four or, more likely, in Guettard's *cabinet d'histoire naturelle* at the Palais royal. In addition to his specimens of gypsum, Lavoisier also borrowed the Duc d'Orléans' microscope for his experiments, and it seems unlikely that he would have transferred instruments and specimens to his aunt's residence when he could easily access them at the Palais royal with the help of Guettard.[39] At any rate, following the instructions of his mentor Lavoisier preferred conducting chemical tests, whenever the circumstances allowed this, directly in the

34 "C'est dans les mines, & non dans les creusets d'un Chymiste, qu'on peut apprendre, autant qu'il nous est accordé, comment les mines se forment dans les profondeurs de la terre". Guettard (1768–1783), vol. 1, pp. lv–lvi. On pp. lii–liii of the same work, Guettard also remarked: "Ce n'est point la Philosophie qu'on étudie dans les Collèges, qui peut procurer tous ces avantages [...]. C'est en étudiant la nature dans elle-même qu'on parvient à en dévoiler les secrets. Les voyages sont un des meilleurs moyens qu'on puisse employer pour acquérir les connoissances les plus sûres & les plus certaines sur ces deux objets importans".

35 LO, vol. 5, pp. 4–5. I surveyed Lavoisier's mineralogical approach to chemistry in Beretta (2005).

36 Rhoda Rappaport thought that "this technique was probably learned by Rouelle" (Rappaport (1973), p. 253) but no evidence is given to support this thesis. Since Guettard had been working with Buache, an experienced cartographer, to produce his mineralogical map of 1746, Lavoisier more likely picked up the idea of levelling with the barometer from him. In effect, in his *mémoire* (Guettard (1756), p. 236), Guettard explicitly refers to the operation of precision levelling.

37 Duveen, Klickstein (1954a), pp. 243–244; see also the additions and corrections by Rhoda Rappaport published in Duveen (1965), pp. 129–132. Rappaport discovered that Lavoisier and Guettard prepared ten maps together.

38 MS Lavoisier (1764); MS Lavoisier (1765); MS Lavoisier (1765a).

39 On December 15, 1764 Lavoisier examines a sample of calcined plaster and artificial selenite "au microscope de mr le duc d'orleans": MS Lavoisier (1764), fol. 122.

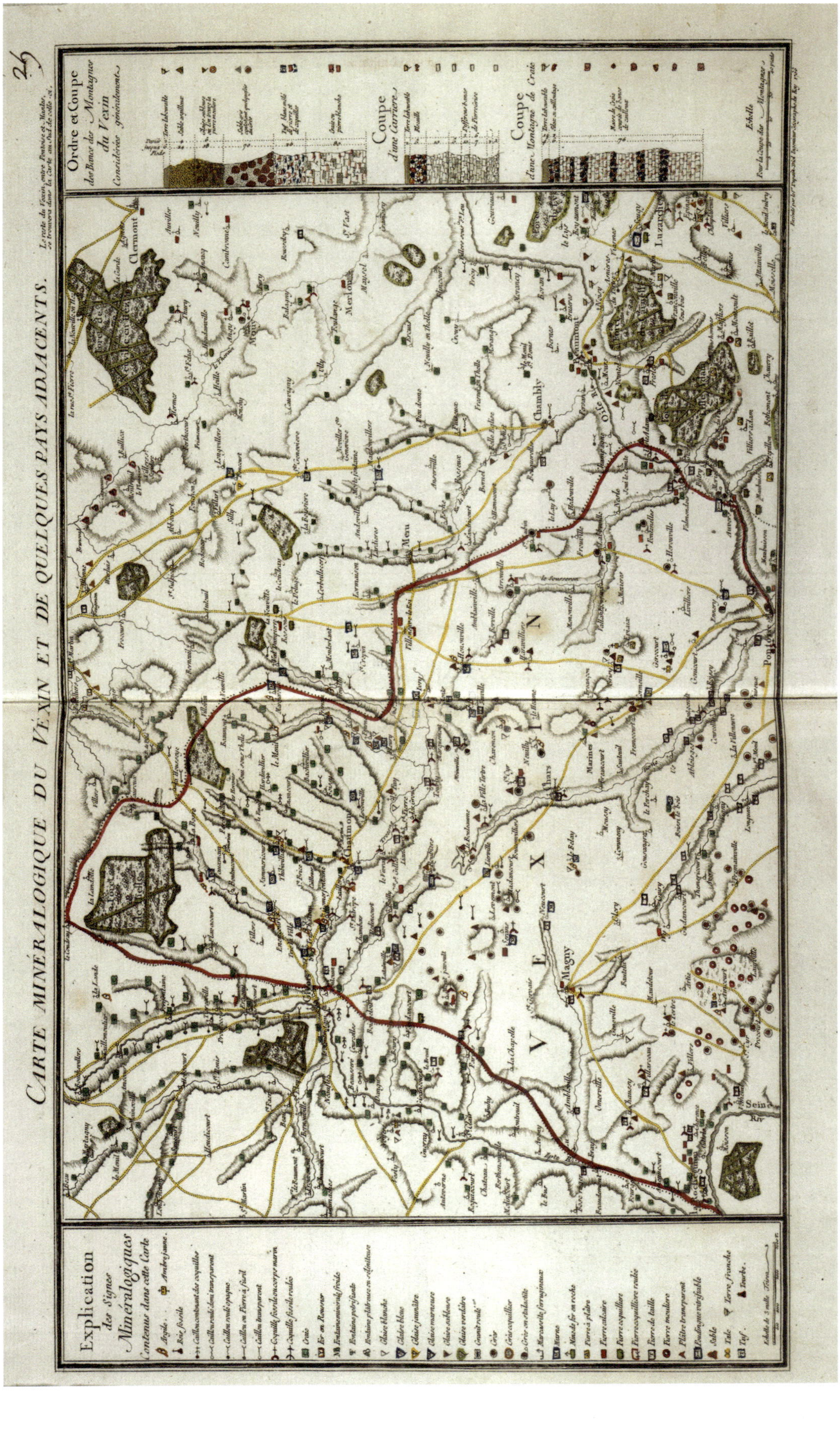

FIGURE 4 Hand-coloured mineralogical map of the Véxin region (1767), from Guettard, Monnet (1780). Clearly visible on the right is Lavoisier's cross-section of geological strata
COURTESY OF CORNELL UNIVERSITY LIBRARY – DIVISION OF RARE AND MANUSCRIPTS COLLECTIONS

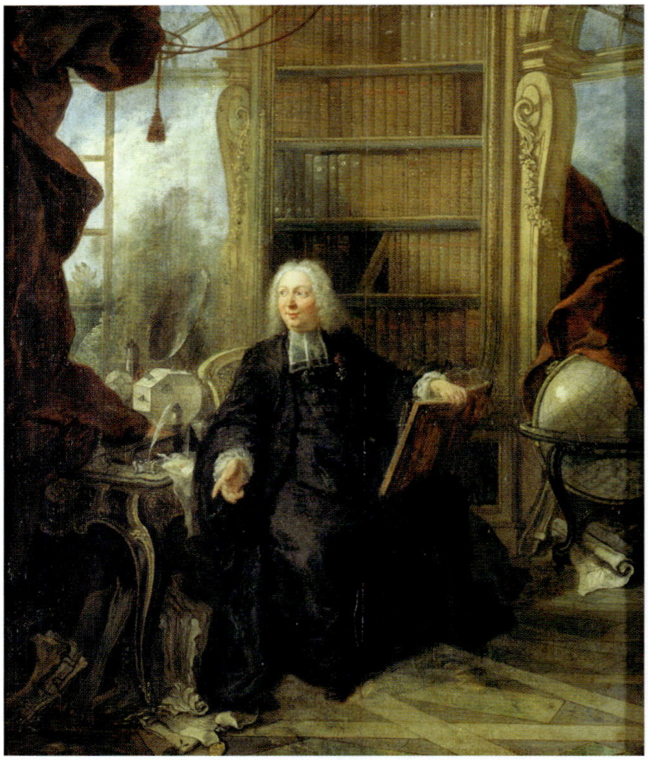

FIGURE 5 Jacques de Lajoüe's presumed portrait of Jean Nollet (ca. 1740)
COURTESY OF PARIS MUSÉES/MUSÉE CARNAVALET

filed: Lavoisier, like Guettard, outlined his approach to the examination of the salts obtained from different specimens of gypsum as follows: "[w]e will first examine the formation of these salts in the laboratory of the chemist, and then we will consider them in the laboratory of nature".[40] The same preference for open air experiments was shown by Lavoisier in his hundreds of analyses of mineral waters.[41]

It was Jean Etienne Guettard who sent his young apprentice Lavoisier to Nollet's course in 1761. Both Guettard and Nollet had studied with Réaumur, who followed a very particular approach to experimental science with detailed attention to apparatus, systematic experimentation, and the constant consideration of the possible applications, both technical and economic, of the scientific endeavour – three principles which Lavoisier took on and later used in the organisation of his laboratory.[42] It worth emphasising that Réaumur declared in his project of reforming the Académie: "chemistry, the investigation of which seems rather frivolous to those who know its true purpose, could become one of the most useful parts of the Academy".[43] With time, Lavoisier embraced Réaumur's vision of chemistry, and most significantly, when he became director of the Académie des sciences in 1785, he resumed the *Description des arts et métiers*, a project which had languished since the death of Duhamel de Monceau in 1782.

Jean Nollet (Fig. 5), exerted an inspiring and long-lasting influence on the young Lavoisier. In his popular course in experimental physics he dealt extensively with chemical topics. In his very first lesson with Nollet, the young Lavoisier learned that chemical laboratories should abandon the chimerical pretensions of alchemy and focus on the decomposition and composition of matter.[44] In many of his lectures Nollet employed instruments from experimental physics in numerous chemical experiments (Fig. 6), and Lavoisier was certainly fascinated by his teacher's instrument cabinet: Lavoisier's surviving instruments include a few purchased from Nollet.[45]

In an extremely interesting reconstruction in his as yet unpublished of the history of French chemistry, Antoine Grimoald Monnet acknowledged Nollet's prominent role in the creation of the popularity of chemistry in Paris which prepared the ground for Rouelle's successful courses. Eventually, Rouelle resented Nollet's success and grew critical of his approach to chemical topics.[46] There is further evidence that shows the importance of Nollet for the Parisian chemical community. It was, in fact,

40 "Nous allons examiner d'abord la formation de ces sels dans le laboratoire du chimiste; nous les considérerons ensuite dans celui de la nature" (1766). LO, vol. 3, p. 129.

41 The first of these is recorded for 25 May 1763, when Lavoisier tested a spring in the vicinity of Villers-Cotterets with the *noix de galle* to determine the presence of iron. Palmer (1998), p. 58. During his travels in the Alsace and Lorraine with Guettard in the summer of 1767, Lavoisier analysed 160 different mineral waters. LO, vol. 3, pp. 180 and 189–205.

42 Terrall (2015).

43 Réaumur (1716–1727), pp. 104–105.

44 Nollet (1764), vol. 1, pp. 13–14.

45 MAM, inv. nos. 20097-0000-, 20100-0000- and 20101-0000-. See catalogue pp. 216–214. On Nollet and chemistry Piot (2019).

46 "Cette science, démontrée avec une clarté, une méthode par l'Abbé Nollet, dont on n'avais pas eu d'exemple jusqu'à lui, faisait aussi une grande sensation. Il n'y avait pas jusqu'au belles dames qui ne voulussent l'étudier et bientôt la même passion les porta vers la chymie, qui leur semblait en être à certains égards une sœur. Il était pourtant fâcheux que ces deux sciences, malgré leur affinité, fussent nettement séparés, que le langage de l'une fut entièrement étranger à l'autre […] Rouelle qu'on pouvait regarder alors comme le chef de la chymie était d'ailleurs si rempli de lui-même et si intolérant pour les idées des autres, qu'il ne pouvait souffrir aucune des explications que l'abbé Nollet donnait des choses qu'il traitait. Il est vrai que le célèbre démonstrateur de physique expérimentale, lors qu'il traitait de quelque objet qui avait du raport avec la chymie proprement dite, tel que le nitre ou le sel marin employés dans ses expériences, se servait d'expressions et en donnait des idées absolument contraires à celles des chymistes". MS Monnet (1785), vol. 2, fols. 156–158.

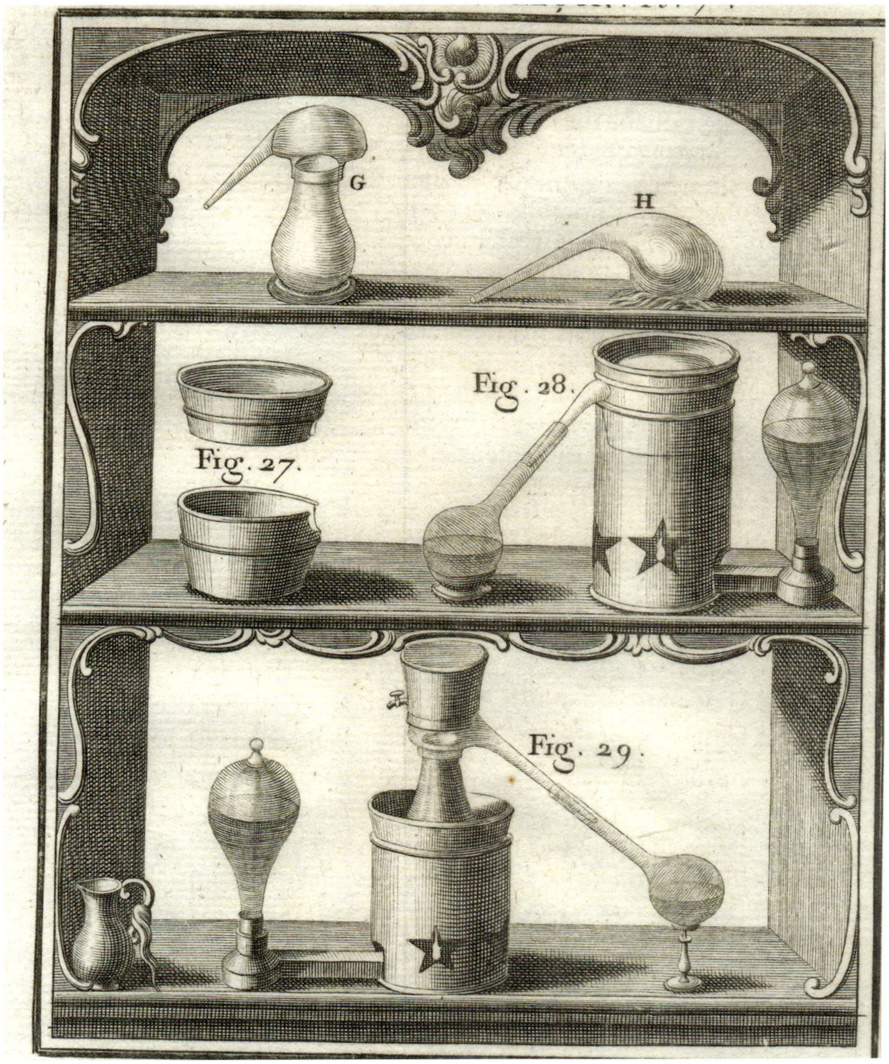

FIGURE 6
Chemical experiments and apparatus presented in Nollet (1764)
PRIVATE COLLECTION

Nollet who introduced Hales' work to the Parisian public in 1745.[47] Interestingly, when investigating the properties of fire, Nollet used chemical equipment he had previously used for experiments on fermentation and distillation.[48] Nollet's inclusion in a course of experimental physics of devices which could be used by chemists was new to France, but probably inspired by Dutch natural philosophers like Petrus van Musschenbroek, who listed instruments like the pyrometer and the thermometer in his works – instruments designed for use in Herman Boerhaave's chemical laboratory. Nollet's success in creating a wide public interest in apparatus created the basis for a steady market in scientific instruments and machines.[49] It is, therefore, plausible that Nollet also sold the chemical apparatus described in his lectures, and thus introduced change to the relatively static environment of Parisian chemical laboratories.[50] Indirect evidence for this is provided by the fact that sometime around during the 1757 Rouelle introduced in his lectures a lengthy description of Hales' pneumatic pedestal, a device which he and Venel significantly transformed.[51] Thus, English pneumatic experiments entered the stage around the time when Lavoisier finished his schooling at the Collège Mazarin. Since Lavoisier attended both Rouelle's and Nollet's lectures we may conclude that he was introduced to Hales' pneumatic chemistry from two different but complementary perspectives. Furthermore, he had the opportunity to experience and, eventually, exploit the favourable

47 Nollet (1764) vol. 3, pp. 243–286.
48 Ibid., vol. 4, pp. 228–236; 272–316; 320ff.
49 See Pyenson, Gauvin (2002).
50 As Paolo Brenni points out "Nollet's action in spreading the culture of scientific instruments prepared the ground for French instrument-makers like Jean Nicolas Fortin and Etienne Fortin". Brenni (2002), p. 27.
51 Holmes (1998), pp. 60–81.

conditions created by the competition between the courses by Rouelle and Nollet.

2.2 Chemical Instrument Making in Paris before 1770

French instrument making did not become competitive until after 1760, and rarely reached the standards set by English instrument makers.[52] During Lavoisier's lifetime Paris became an important centre for innovation, and the precision developed by French instrument makers often equalled those found in English manufacturing.[53] In the second half of the century a relatively large community of Parisian makers of mathematical, optical, and astronomical instruments ran profitable workshops and sold their instruments to an expanding clientele of professional scientists and amateurs. They also worked for institutions such as the Académie royale des sciences, the Observatoire, and the colleges of Paris, which either considerably enlarged workshops and laboratories or created new ones.[54] These instrument makers belonged to different guilds including the founders, *couteliers*, mirror makers, enamellers, and ceramicists, and from the middle of the seventeenth century they were allowed, if not without difficulties, to specialise in the production of scientific instruments.[55] Despite their legal rigidity, guilds offered significant guarantees of survival to their members.[56] In order to overcome the constraints of the guilds' regulations, the Académie des sciences obtained special concessions for instrument makers from the king, which in 1787 led to the creation of a recognised body of engineers specialised in the construction of all sorts of mathematical instruments.[57]

Considering this background, the absence of specialised makers of chemical apparatus is conspicuous. Unlike mathematical instruments, chemical apparatus deteriorated easily and needed to be replaced frequently, which would seem to be a good argument for specialisation. However, makers of chemical instruments were not officially recognised as an independent profession.[58] Who then supplied the Parisian chemical laboratories? Not the makers of the physical and mathematical instruments – their trade catalogues rarely referred to apparatus that could be used in chemical experiments. One exception was a burning lens with a diameter of 122 cm that was advertised by one of the most successful Parisian instrument makers of the 1760s, Claude Passemant, as a device that might be productively used in chemical experiments as well as in architecture and painting.[59] However, I believe it is not a coincidence that this lens was not actually used for chemical experiments, and that the chemistry section of the Académie, which needed a lens in 1772–1773, preferred Tschirnhaus' lens as well as building a new, larger one under the patronage of Trudaine de Montigny.[60]

From the late 1750s to the French Revolution the market for chemical equipment must have been enormous. Chemists were not the only ones in need of such instrumentation: druggists, pharmacists, perfume makers, bakers and all the guilds involved in the preparation and preservation of food, winemakers, and distillers were in constant need of furnaces, alembics, stills, crucibles, mortars, and glassware of all sorts, in addition to other equipment that was commonly found in chemical laboratories. Paradoxically, it was probably because of this widespread market for 'chemical' equipment that specialisation was not needed. During the 1750s, this kind of apparatus was easily available and produced in large quantities by the various guilds that had met the needs of artisans, pharmacists, and cooks long before professional chemists appeared on the stage. This explains why the chemists' laboratories relied on relatively simple apparatus that changed little over a long period of time. Another factor that prevented chemists from making major changes in their equipment was the fact that most of the chemical courses were organised by the guild of Parisian pharmacists: a guild that was subjected to strict regulation and regarded all the changes introduced by its members with suspicion. Only once the Parisian chemical community included a consistent number of practitioners with different educational backgrounds did the resulting competition open new paths of enquiry. However, it took a long time for this to happen, and as late as 1773 Parisian chemists and pharmacists sourced their equipment from non-specialised artisans. Antoine Baumé provides us with

52 Daumas (1989), p. 269.

53 Étienne Lenoir, for example, is among those commonly known for initiating precision instrument making in France; see Turner A. (1989a).

54 Daumas (1989), pp. 258–268.

55 Henri Bouchot wrote that there were 334 *maîtres fondeurs* working in Paris by the end of the eighteenth century; see Bouchot (1887), p. 140.

56 Daumas (1989), pp. 91–100.

57 Daumas (1952).

58 They were either recognised by the king or by means of a patent provided by the Académie des sciences.

59 Passemant (1764), p. 64. The mirror was presented to the king in 1757.

60 Lehman (2013).

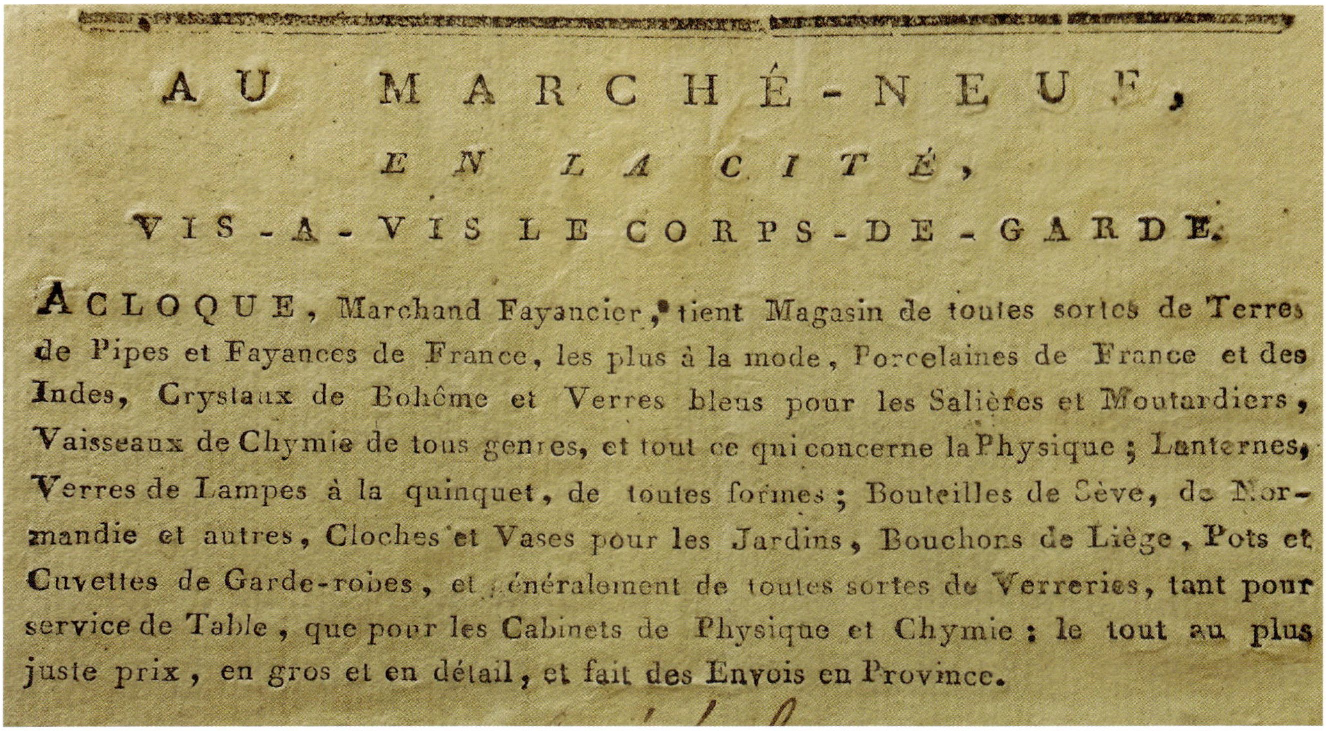

FIGURE 7 Trade card for the Parisian *fayencier* Acloque (1790)
COURTESY OF THE BIBLIOTHÈQUE INTERUNIVERSITAIRE DE SANTÉ – PARIS

a list of the Parisian suppliers to the chemical laboratories of his day:[61] most of the glassware (alembics, retorts, phials, recipients, etc.) could be bought from any *faïencier* (ceramicist) (Fig. 7) or the glazers (*émailleurs*); earthenware from the *marchands potiers de terre*; furnaces and crucibles from the *fournalistes* (furnace makers) or the founders. Many useful pieces of equipment could be obtained from the ironsmiths (*quincailleries*); balances from the balance makers (*balanciers*); and copper recipients from the boilermakers (*chaudronniers*). German crucibles were widely imported into Paris by tradesmen because they were widely regarded as the most durable. The only device made by professional instrument makers was the thermometer, an instrument that, as I shall point out later, was introduced into the chemical laboratory at a relatively late date.[62] It is, therefore, not surprising that the equipment used in Parisian chemical laboratories was rarely innovative. As I have already pointed out in chapter 1, the apparatus in the laboratory of the most successful chemistry teacher of the period, Guillaume François Rouelle, was almost identical to that identified by Nicolas Lémery in the first part of his *Cours de chymie* (Paris, 1675).

Although glassware was of the greatest importance in chemical experimentation, Parisian chemists were unable to find products with the required resistance to both heat and cold. Macquer remarked:

> Vessels intended for chemical operations should, to be perfect, be able to bear without breaking the sudden application of great heat and cold, be impenetrable to every substance and inalterable to any solvent, be unvitrifiable and capable of enduring the most violent fire without fusing. But up to the present no vessels are known which combine all of these qualities.[63]

By 1765 French academic chemists and scientists had developed some interest in glassmaking but did not try to develop a consistent theory of its composition; in fact, glass had not yet become an object of chemical inquiry. By contrast, glassmakers had introduced significant technical innovations, the secrets of which were carefully guarded by the manufacturers and by the guilds.

Pneumatic discoveries prompted a renewed attention to the nature of fire and stimulated active research into new instruments, but this field was developed more

61 Baumé (1773), vol. 1, pp. cxxviii–cxlvi.
62 Powers (2014).
63 Macquer (1753), p. 297.

within the domain of natural philosophy than in that of chemistry. Since Lavoisier had the advantage of a training in both disciplines, his approach to chemical instruments proposed a synthesis of both traditions.

2.3 Lavoisier and Chemical Instrument Making

Lavoisier's approach to instrument making was, in many respects, innovative but not entirely different from that of many of his contemporaries. In fact, the surviving instruments, chemicals and minerals from Lavoisier's collections offer a very insightful overview of the routine of chemical experimentation in a research laboratory typical for the second half of the eighteenth century. From the detailed inventories of his laboratory ordered by the revolutionary authorities in 1794 we know that Lavoisier accumulated some 10,000 pieces of apparatus, mostly glassware, and that the so-called 'precision' and 'physics' instruments accounted for less than 3% of the total.[64] Although a few of these instruments were quite exceptional, most of Lavoisier's apparatus agreed with the standards developed by eighteenth-century chemical research laboratories after the pneumatic discoveries made between 1755 and 1770s attracted the attention of several chemists, apothecaries, physicians, and natural philosophers throughout Europe.[65]

In what follows I will briefly focus on the evolution of Lavoisier's approach to instrument making which very much depended both on the cultural context in which his interest originated and which is exemplarily highlighted in the history of the gasometer.

The first recorded instrument used by Lavoisier is the barometer he began to carry on his person in September 1764 in order to take levelling measurements of rock layers.[66] It was thanks to his use of the barometer that Lavoisier developed the idea of studying mineral ores in connection with their stratigraphic positions, to determine the presence of a rock layer and the causes of its formation, and to reconstruct geological type sections. After exploring the mineralogical and geological resources on the outskirts of Paris, Lavoisier recorded the results of his chemical analysis of the specimens of gypsum he had collected in his first laboratory notebook, which begins in July 1764 and contains the raw data of his numerous experiments.[67] In contrast to Pott, who experimented on his specimens of gypsum with fire', Lavoisier preferred to dissolve his specimens in water, as this was the easier and more natural method of analysis. This method further suggested itself because Lavoisier not only wished to decompose gypsum into its constituent parts, but also wanted to produce an artificial sample by combining vitriolic acid and calcareous earth. To this end he began to use the hydrometer for taking exact gravimetric measurements in August 1764. He subsequently continued to use this instrument to measure the specific gravities of the ingredients of a chemical solution, and in 1768 he perfected the construction of a new type of chemical hydrometer.[68] In 1765, during his analysis of gypsum, Lavoisier also used a microscope, but he did not provide a description of it (Figs. 8 and 9).[69] Thereafter, references to the use of physical instruments were constant in his daily work. During his mineralogical travels with Jean Etienne Guettard in 1766–1767, Lavoisier tried using several instruments in new ways. One example is the barometer that he used in his observations of the geological strata of mountains.[70] Probably inspired by Réaumur's observations on the changes in temperature during reactions with salts, Lavoisier introduced the systematic use of thermometers to chemistry;[71] in the 1760s he used them in the analysis of mineral waters, and had different ones made by Cappy (in 1761), Claude Siméon Passemant, and Gallonde for this purpose.[72] In his 1767 observations Lavoisier analysed mineral waters with three different comparable thermometers, his own hydrometer, and a balance made by Chemin.[73] Thermometers would play a crucial role for the rest of Lavoisier's career, and he employed the best makers to design them for specific chemical experiments. Lavoisier himself perfected the construction of a new type of thermometer (Fig. 10) that he used during his experiments on cold weather, which was placed in the cave of the Observatoire and set the standard for many years to come.[74]

64 For the inventory, see Appendix 3.
65 See chapter 1.
66 LO, vol. 5, pp. 4–5; on this topic see Bourguet (2002).
67 MSS Lavoisier (1764), (1765), and (1765a).

68 MAM inv. nos. 07508-0001, -0006-. See catalogue pp. 319–320.
69 LO, vol. 3, p. 141.
70 LO, vol. 5, p. 109.
71 "Sans les Thermomètres nous n'aurions jamais découvert que certains Sels, en se fondant dans l'eau, la refroidissent"; cited in Birembaut (1958), p. 311.
72 On Cappy, see LC, vol. 1, pp. 44, 65; MAM, inv. n. 19934-0000-. On Passemant and Gallonde, see MAM, inv. n. 19935-0000-. Lavoisier used this thermometer in his examination of natural waters in 1767 and for experiments on distilled water ("Pesanteur absolue de l'eau distillé", LO vol. 3, pp. 451–454). A comparison of Gallonde's thermometer with two other thermometers may be found in MS Lavoisier (1767–1768).
73 In his manuscript notes, Lavoisier mentions a "bleue" [*sic*] thermometer and one "corrigé" that I was not able to identify. On the balance, see MS Lavoisier (1767).
74 Lavoisier (1785).

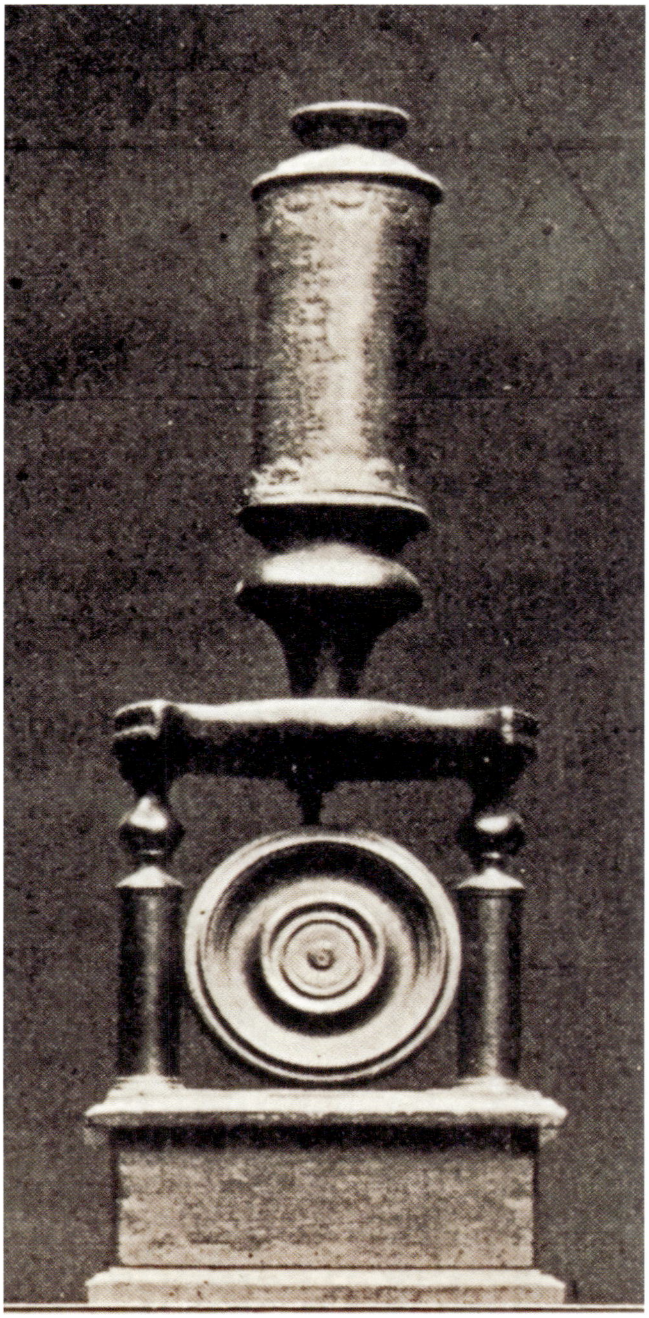

FIGURE 8 Early microscope from Lavoisier's collection exhibited in 1900. Its present location is unknown. *Musée Centennal* (1900)

Once Lavoisier was interested in a particular instrument he employed different makers to create several copies of the same, and a series of comparisons then enabled him to assess their accuracy as well as to understand how they might be improved. Lavoisier used the same approach in the construction and perfection of barometers, hydrometers, balances, burning lenses, Papin digesters, and pyrometers.

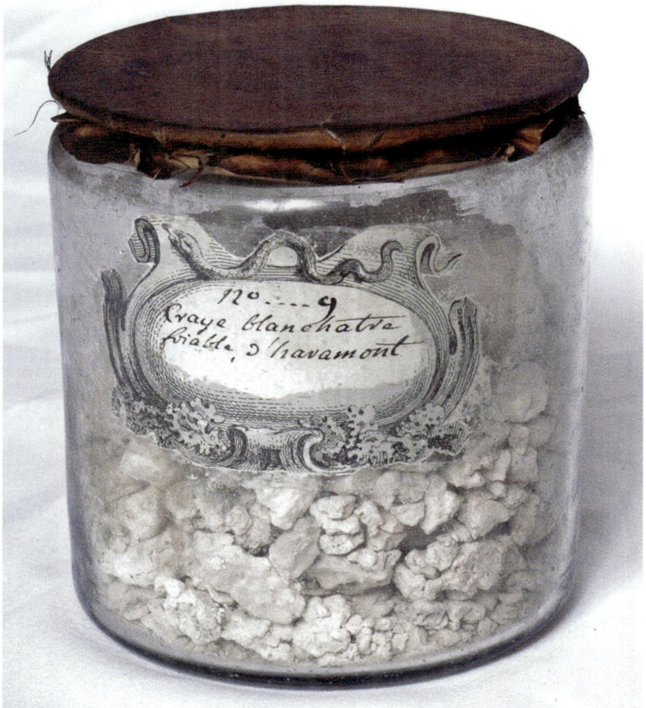

FIGURE 9 Glass jar containing specimens of gypsum collected by Lavoisier during his mineralogical trip with Guettard COURTESY OF THE MUSÉE LOCOQ – CLERMONT FERRAND

Like Nollet before him, Lavoisier combined several instruments in use and adapted them to specific areas of investigation. This particular approach not only required different professional skills, but also creativity in exploiting the interdisciplinary potential of different types of instruments which had been used in a relatively static way for a long time.

In 2000 Larry Holmes depicted Lavoisier as a *bricoleur* skilled in assembling pieces of apparatus he had in hand.[75] Though Holmes used the term in a derogatory way, I nevertheless think that this definition is very helpful to portray Lavoisier's creative views on chemical instrumentation.[76] Lavoisier's creativity, combined with his well-known drive for accuracy, is particularly evidenced by his early efforts to introduce precision measurement to pneumatic chemistry. This important field of investigation revolved around several types of apparatus and materials which became central in chemical investigation. Historians have emphasised the role of the precision balance, Priestley's pneumatic trough, and Hales' pneumatic apparatus, but

75 Holmes (2000), pp. 149–150.

76 Holmes (2000) wrongly claims that Lavoisier did not design his own apparatus.

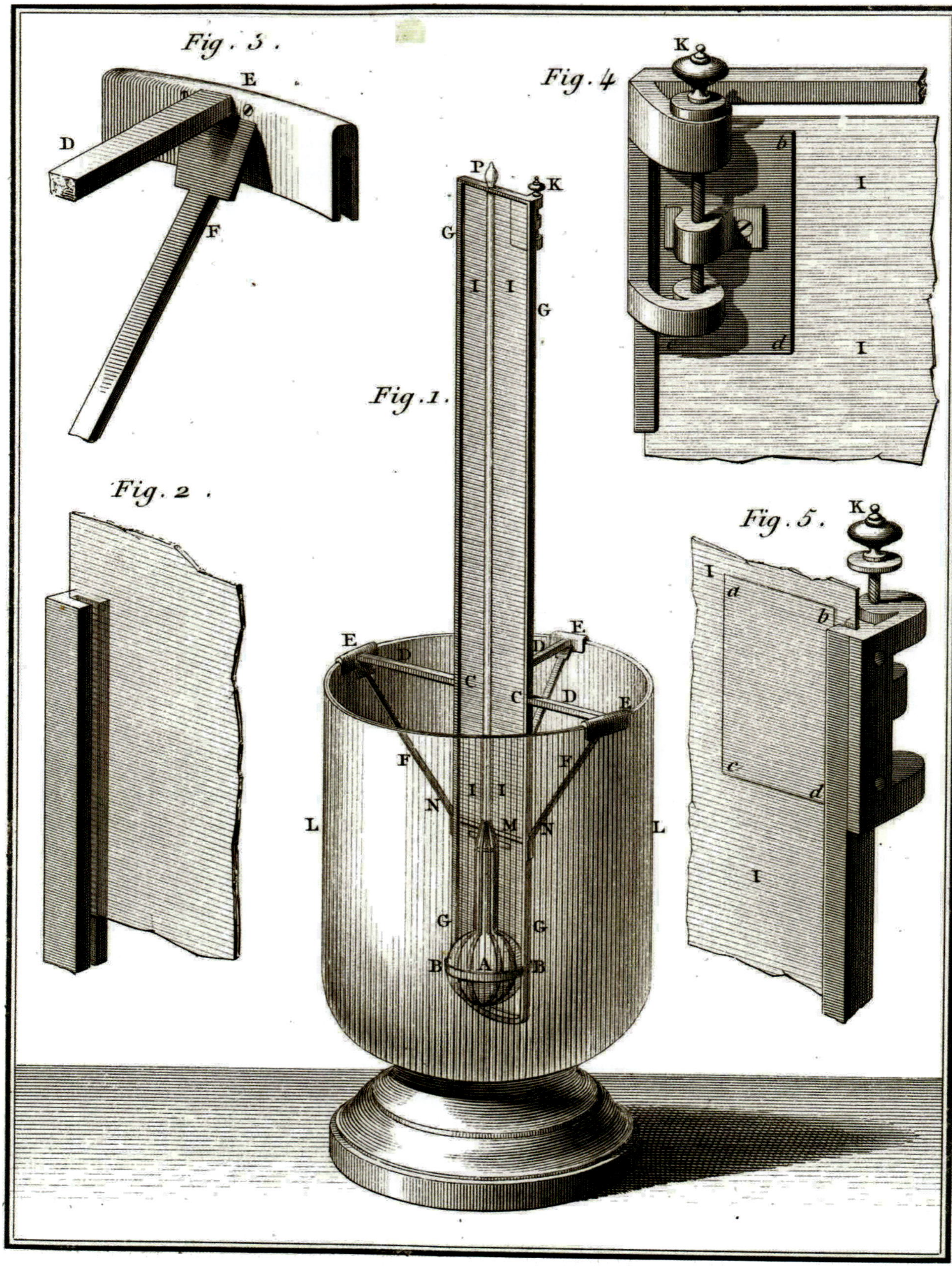

FIGURE 10 Lavoisier's thermometer from the Observatoire. LO, vol. 3, plate XI

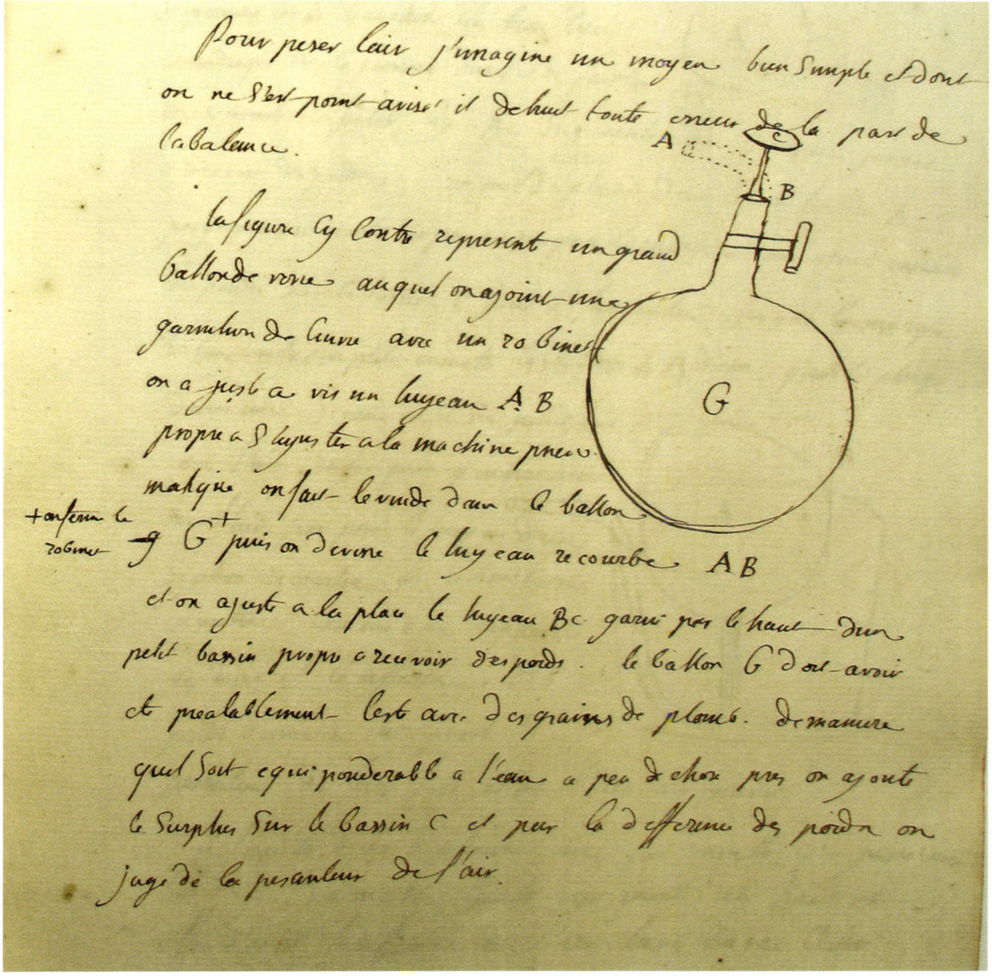

FIGURE 11
Lavoisier's balloon for weighing gases. MS Lavoisier (1772–1788), vol. 1, fol. 14r
COURTESY OF THE ARCHIVES DE L'ACADÉMIE DES SCIENCE – PARIS

somehow neglected the innovations in glass apparatus and its pervasive presence in most chemical experiments involving gas. Moreover, chemists were in constant, and at times sudden, need of glassware.

The shape of glass recipients changed radically, and chemists were extremely creative in testing the potential created by the malleable properties of glass.[77] Lavoisier was no exception, and once he decided to work systematically on the nature of gases he began to both study the properties of glass and to conceive new glass apparatus.[78] As early as March 1773, Lavoisier conceived a recipient to weigh gases by plunging it into water (Fig. 11), but it was not easy to have this kind of instrument made.[79] The delivery of his apparatus for the calcination of lead by means of the burning lens (also dated March 1773) was delayed due to the workmen's slow pace.[80] It is interesting that Lavoisier refers to the makers as *ouvriers*, thus confirming that in 1773 there were no specialised makers of this kind of apparatus. At this point he was still working in his residence in the Rue des Bons Enfants, which was unsuitable for a large laboratory. After waiting for his balloon for two weeks Lavoisier was still not able to perform the experiments he had in mind, and the entries in his laboratory notebooks reveal a growing sense of frustration. With this in mind, and given the quantity and variety of glassware he needed, it is likely that when Lavoisier moved to the larger residence at the Arsenal in Spring 1776 he used the opportunity to set up a small glassmaking workshop as well.

In Lavoisier's 1794 laboratory inventory for his residence at Boulevard de la Madeleine two tables are listed, one for glassblowing and one for glass grinding, and among the surviving items now at the Musée des arts et métiers there are one glassblower's chair and two glass cutters which

77 Macquer, for instance, remarked how important the making of balloons was after the discovery of gases: Macquer (1778), vol. 1, pp. 223–225.

78 Beretta (2014a).

79 MS Lavoisier (1772–1788), vol. 1, fol. 14r.

80 Ibid., fol. 18r.

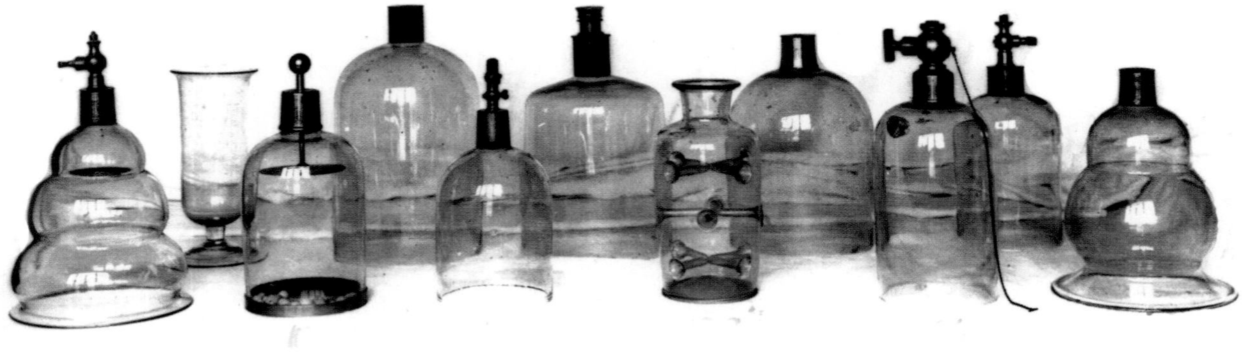

FIGURE 12 Chemical glassware from Lavoisier's chemical laboratory. Lavoisier scientific instruments and card index from the Chazelles Collection (Accession no. 1983.218)
COURTESY OF THE HAGLEY MUSEUM AND LIBRARY

conclusively show that some of Lavoisier's glassware was made at the Arsenal.[81]

These important findings are not exceptional; many of Lavoisier's contemporaries also produced their own chemical glassware. This was cheaper and more effective than glass produced by professional glassmakers, who mostly made thermometers and barometers and were not familiar with the needs of a chemist.[82]

Lavoisier owned a vast quantity and variety of glassware (Fig. 12). During the 1760s and 1770s he purchased small glass recipients from glazers working with the enameller's lamp to produce test tubes and other refined works; large recipients were provided by ceramicists, while crystal glass vases were imported from England.[83] The difficulty to obtain more elaborate apparatus from others prompted Lavoisier to research on his own extensively.

The significant role of glassmaking technique is also evidenced in the complex origin of the gasometer, the best known of Lavoisier's instruments. This apparatus was, in fact, built over the course of several years, and, at least initially, its main parts were not intended for use in the famous experiments on the synthesis and analysis of water of 1783 and 1785. As Lavoisier acknowledged in 1789, the experimental context of the first piece of the equipment was, in fact, quite different, as it was used to "furnish a uniform and continued stream of oxygen gas in experiments of fusion".[84] This early apparatus, conceived in 1781, was not particularly revolutionary, and represents a relatively traditional technique in Lavoisier's efforts to quantify the elements of oxygen . He aimed to keep rigid control over the conditions of chemical experimentation, which was quite a difficult task in a science dominated by qualitative assessments. Most frustrating was the difficulty to control heat when submitting substances to the action of fire. Up to the 1780s it was nearly impossible to know even by approximation what temperatures furnaces would produce.[85] This is why Lavoisier preferred either wet analysis or used instruments like the burning lens, which allowed chemists to observe the combustion. In the early 1770s, Lavoisier and his associates at the Académie's chemistry section conducted several large-scale experiments on the combustion of diamonds using several large burning lenses, from Tschirnhaus' to the one built by Bernière for Trudaine de Montigny.[86] While the heat produced with these lenses was, indeed, considerable, the difficulty to control them, combined with their cost, made this method impractical.

In 1774, when oxygen was discovered as the gas that caused the combustion and calcination of substances, Lavoisier began to identify different experimental methods of controlling oxygen's action and effects with the

81 MAM inv. no. 20057-0000- (chair). MAM inv. nos. 20048-0000- and 20069-0000- (glass cutters).

82 On this see Danger (1829).

83 Barrelet (1953), pp. 116–120.

84 Lavoisier (1789), vol. 2, pp. 346–347.

85 Most efforts from before the mid-1780s resulted in qualitative assessments; see the entry on "fourneaux" in Macquer (1778), vol. 2, pp. 253–263.

86 Lehman (2013); see also MAM inv. n. 19884-0000- on p. 400.

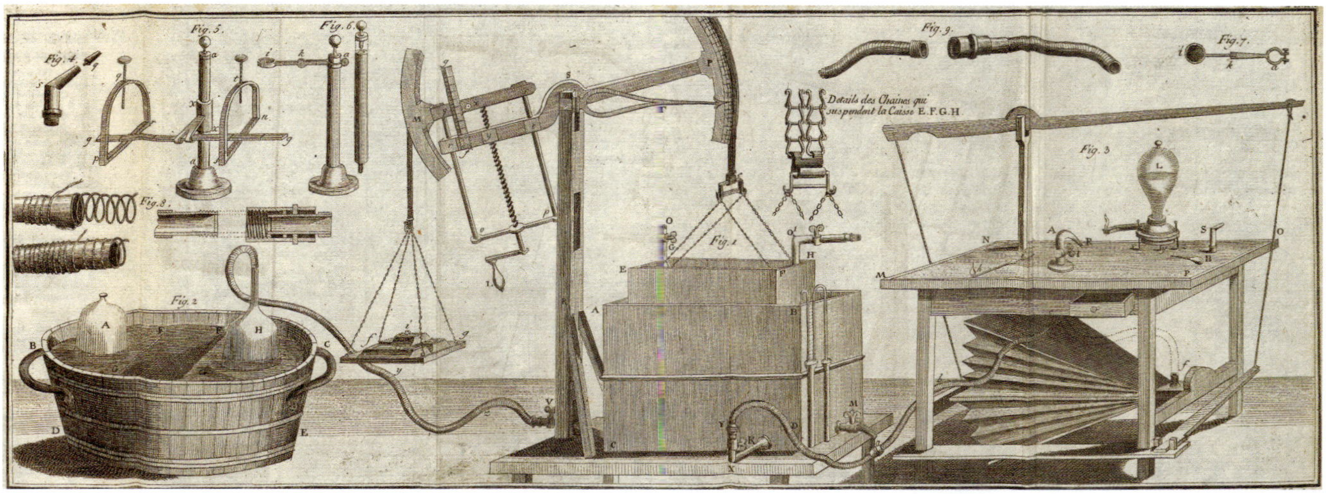

FIGURE 13 Lavoisier's hydrostatic bellows. Ehrmann (1787), Plate II
COURTESY ETH-BIBLIOTHEK ZÜRICH, RAR 1394

highest possible precision.[87] In the early 1780s, when he knew a lot more about oxygen, he began to work on a device with which he could exploit its combustive properties. This so-called 'hydrostatic bellow' (Fig. 13), of which only a few fragments have survived at the Musée des arts et métiers, consisted of a pneumatic trough in which the oxygen was collected and, through copper pipes, let into a large iron case in which it could be weighed with precision.[88] The gas was eventually collected in an enameller's desk with a lamp through which given quantities of oxygen could be set alight and, via a glass blowpipe, directed onto the object or material to be fused. The construction of this apparatus resulted from the collaborative effort of several skilful technicians. According to Daumas, at the end of 1781 Lavoisier was assisted by Philippe Joachim Gengembre, who was joined by the engineer Jean Baptiste Meusnier de la Place in spring 1782, and by the instrument maker Pierre Bernard Mégnié sometime in 1783.[89] Both Lavoisier's tinsmith, Naudier, who had worked in his laboratory since 1774, and Nicolas Fortin were probably involved in the making of the pneumatic tank.

The hydrostatic bellows were made from several pieces and devices, none of which were entirely new. Its making shows Lavoisier's method of working with instruments, and although very little documentation survives, a brief survey of its components and their evolution offers both historical background for the construction of the gasometer and an idea of what laboratory life was like at the Arsenal.

Let us begin with the glass blowpipe.[90] The use of the blowpipe is quite surprising, because Swedish chemists are usually credited as the exclusive inventors and users of this instrument (Figs. 14 and 15).[91] In fact, this handy device, which was made from either glass or iron, had been used since antiquity and in several chemical trades. In the first half of the eighteenth century, Swedish chemists were the first to invent a portable burner connected to a blowpipe, and this became an extremely important instrument for the chemical analysis of mineral ores. Though simple, it required much skill to use it effectively. Through the use of the blowpipe, the Swedish chemists discovered several important elements including nickel (1751), manganese (1774), molybdenum (1781) and tungsten (1783). French chemists used this device as early as the 1760s, but without significant results. In a letter from January 1769 to Torbern Bergman, Macquer lamented about the practical difficulties he had encountered in keeping the heat constant.[92] Some of these difficulties were solved in 1770 when a newly published English translation of Cronstedt's work on mineralogy presented a simpler device, a 'pocket-laboratory', which soon became very successful.[93] This

87 As Lavoisier admitted (Lavoisier (1782), vol. 2, p. 424), he discovered oxygen after Priestley had discovered it in an experimental campaign organised in Trudaine's laboratory at Montigny.

88 The machine was presented by Lavoisier before the Académie des sciences in April 1782. Lavoisier (1782).

89 Daumas (1955), p. 143.

90 MAM, inv. n. 20025-0000-, see p. 335.

91 See, for instance, Jensen (1986) and Abney Salomon (2019).

92 "Je connais bien l'usage du chalumeau dont vous me parlés; je m'en suis servi quelques fois, c'est un instrument fort commode pour certaines experiences, ou essais en petit aux quels le contact immediat du principe inflammable n'est pas contraire, mais je vous avoue que je n'ai pas le même talent que vous pour le faire aller sans interruption". Bergman (1965), p. 236.

93 Cronstedt (1770); see also Bergman (1781) and Bergman (1784).

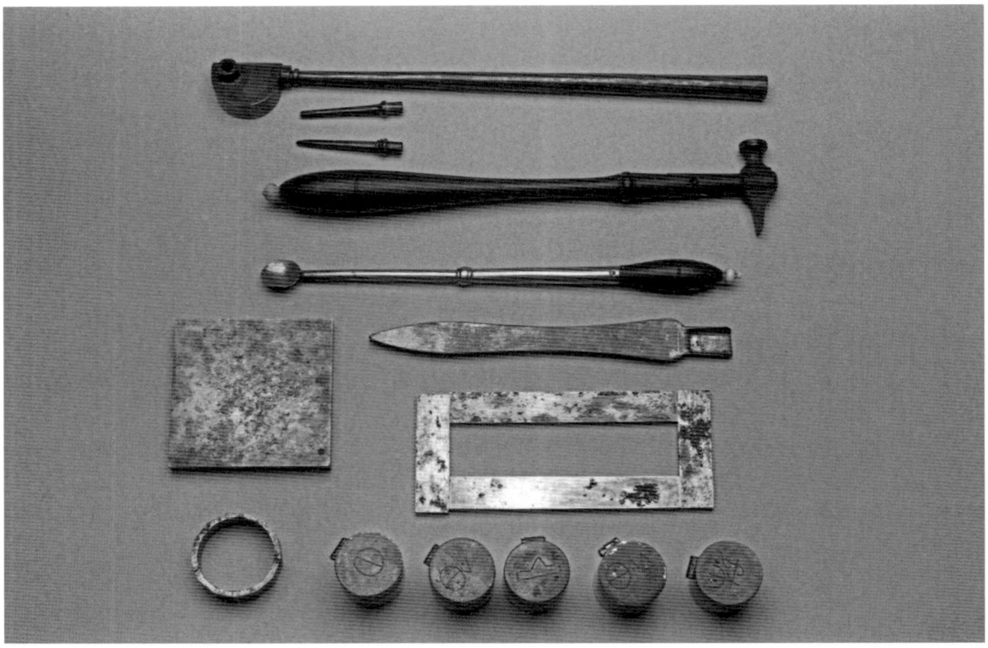

FIGURE 14
Torbern Bergman's blowpipe (Bergman, 1780)
COURTESY OF UPPSALA UNIVERSITY – MUSEUM GUSTAVIANUM

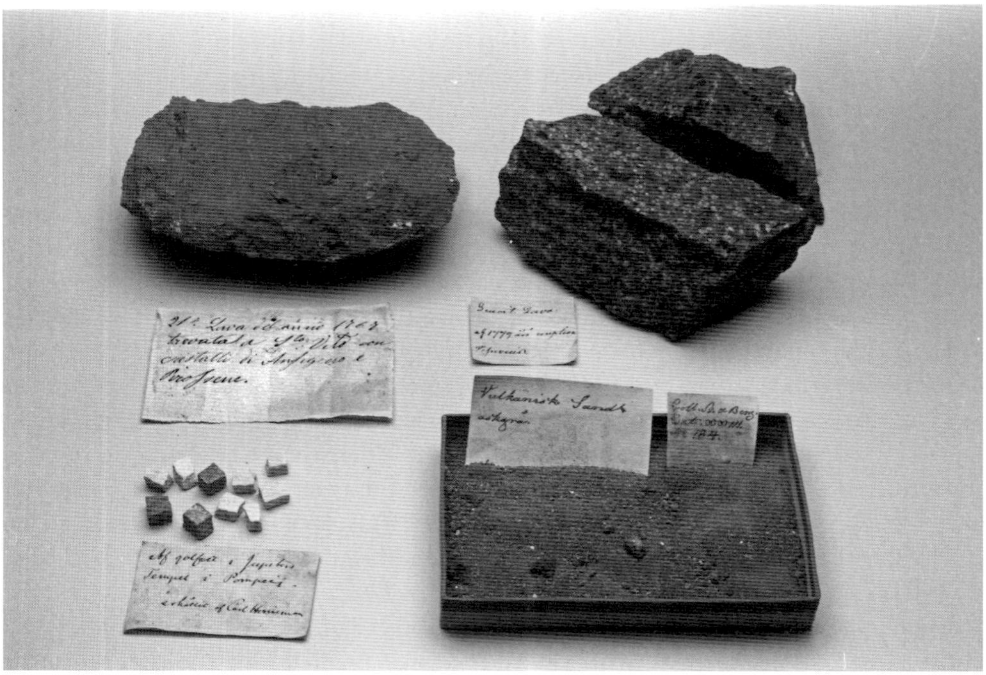

FIGURE 15
Volcanic rocks and fragments of a Pompeian mosaic, from the mineralogical collection of Torbern Bergman and submitted to analysis by blowpipe
COURTESY OF UPPSALA UNIVERSITY – MUSEUM OF EVOLUTION

portable laboratory consisted of a burner, a few reagents, and an iron blowpipe. The Portuguese inventor Jean Hyachinte Magellan, who resided at the time in London, was a key figure in the trade of scientific instruments, and in March 1771 he sent two pocket laboratories to Jean Charles Philibert Trudaine de Montigny, the *intendant des finances*, and an amateur chemist and patron of Lavoisier.[94] Since Lavoisier often worked in Trudaine's laboratory at Montigny in the early 1770s, it is highly probable that he saw the pocket laboratory there, and that one of the two sets sent by Magellan was actually intended for him.[95] Lavoisier's works do not record this instrument,

94 Home *et al.* (2017), vol. 1, p. 161.

95 During the 1770s Magellan became Lavoisier's main supplier of English instruments and books. Lavoisier was acquainted with Magellan via Trudaine. See the portable burner in the catalogue MAM inv. no. 20107-0000-, on pp. 305–306. I provide more details on Trudaine and Lavoisier's collaboration in the next chapter.

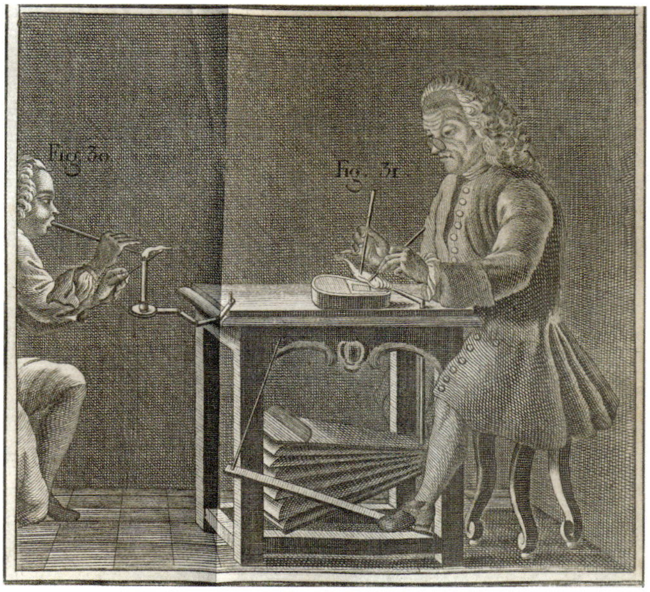

FIGURE 16 Jean Raux demonstrating the use of the blowpipe in 1739. Plate from Nollet (1764)
PRIVATE COLLECTION

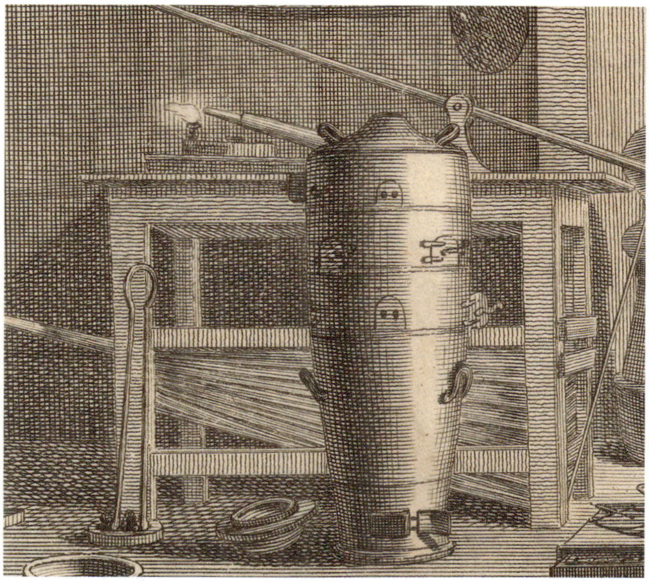

FIGURE 17 William Lewis' enameller's desk. Detail of a plate published in Lewis (1763)
COURTESY OF THE SMITHSONIAN INSTITUTION

and he probably faced the same difficulties that Macquer had encountered a few years before. Other reasons why he might have not been too impressed with the device were that its functionality depended in large measure on the dexterity of the experimenter, and that the analytical results obtained were mostly qualitative. The blowpipe, by contrast, was a handy and extremely useful instrument for a chemist.

Lavoisier solved his difficulties with an ancient technology which he probably learned about from Jean Nollet, either while attending his lectures or from his own copy of the *Léçons de physique expérimentale*. In this work Nollet describes at length Jean Raux' (Fig. 16), the royal enameller's public demonstration before the officers of the court in 1739, including the apparatus he used to control fire and to heat the small objects he wished to manipulate.[96] As we can see from the accompanying plate in Nollet, the apparatus is very similar to Lavoisier's and places the use of the blowpipe in an entirely new material context. The adoption of this enamelling technology in chemistry was already known from a 1763 illustration of William Lewis' London laboratory (Fig. 17), which shows an enameller's desk.[97] Since I doubt that Lavoisier had access to Lewis' book, his employment of this technique seems to show how natural it was for chemists engaged in research to adapt devices commonly used in the chemical trades for their own purposes.[98]

Lavoisier fed the blowpipe from the enameller's desk with a specific amount of oxygen from a previously filled leather bag,[99] but this method did not enable him to measure precisely the volume of oxygen employed during the fusion.[100] Lavoisier therefore asked the engineer Jean Batiste Meusnier de la Place for assistance in perfecting the device by introducing pneumatic chests or gasholders (*caisses pneumatiques*) and a large balance beam which could establish both the quantity and the volume of oxygen.[101] Interestingly, the balance beam was similar, if not identical, to that of Watt's steam engine. In 1781 the brothers Périer, Lavoisier's suppliers of copper pipes and laminated iron, installed the first steam engine in Paris at Chaillot, and it is more than likely that Lavoisier, always interested in the latest technological advancement, saw the machine in action.[102] Meusnier de la Place's and Lavoisier's balance beam was an adaptation from the Périers', and was the forerunner of that which would later

96 Nollet (1764), vol. 4, pp. 502–512.
97 Lewis (1763); the engraving is bound in before the frontispiece.
98 The title is not present in Lavoisier's library: Beretta (1995).
99 Six of the bags have survived in the collection: MAM inv. nos. 20160-0001-, 20160-0002-, 20161-0000-, 20162-0000-, 20163-0001-, 20163-0002- and 20129-0003-.
100 LO, vol. 2, pp. 425–426.
101 The chests, made in *fer blanc* (tin), were inserted one into the other. The smaller tank contained water and the larger one oxygen.
102 Payen (1969).

form part of Mégnié's and Fortin's gasometers, which were made in 1785 and 1788 respectively.[103]

Lavoisier tested the hydrostatic bellows on several occasions in 1782, when he subjected a number of precious stones, minerals, and platinum to oxygenated heat.[104] On the 5th of June he showed a spectacular experiment at the Académie des sciences in which he was able to fuse a specimen of platinum at an unprecedented speed.[105] Since platinum was difficult to handle, this news reached the press. Benjamin Franklin was so impressed with that he immediately reported the results to Joseph Priestley:

> Yesterday the Count du Nord was at the Academy of Sciences, when sundry Experiments were exhibited for his Entertainment; among them one by M. Lavoisier, to show that the strongest Fire we yet know, is made in a Charcoal blown upon with dephlogisticated Air. In a Heat so produc'd he melted Platina presently, the Fire being much more powerful than that of the strongest burning Mirror.[106]

The experiment was repeated by Lavoisier with the assistance of the master goldsmith and instrument maker Marc Etienne Janety, who produced before the Académie des sciences several platinum objects as well as a *nécessaire* which was to be sent to the King of Spain. (Fig. 18)

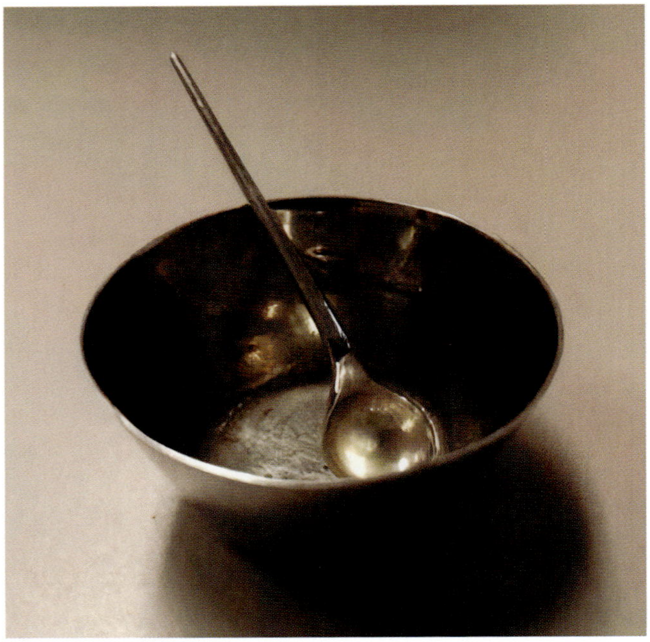

FIGURE 18 Janety's platinum salt cellar with spoon
COURTESY OF CORNELL UNIVERSITY LIBRARY – DIVISION OF RARE AND MANUSCRIPTS COLLECTIONS

At the Arsenal Lavoisier continued his experiments with the bellows by submitting a great number of precious stones to the action of heat, many of them lent by jewellers and academicians. He also conducted these experiments before a large audience of members of the Académies, artisans, and laymen. Although the results were not as spectacular as those for platinum, Lavoisier was able to assess the nature of precious stones with this new evidence.[107] It was certainly because of the publicity and interest attracted by the experiments that, in July 1782, the astronomer Bochart de Saron suggested that Lavoisier fill the blowpipe with a mixture of oxygen and hydrogen, and assess whether the fusion of platinum could be further accelerated.[108] It is not known if Lavoisier tried Bochart de Saron's suggestion immediately, but it is reasonable to believe that it inspired him to manipulate hydrogen and oxygen together. At this stage, at the end of 1782, he probably did not yet have a clear idea about the composition of water, but Bochart's suggestion would have acquired a new meaning in Lavoisier's thoughts about his experiments with Alessandro Volta of earlier that Spring, as well as the instrument he presented to the Académie des sciences.

Even if there is only indirect evidence available for it, the collaboration between Volta and Lavoisier brought about some important and fruitful results.[109] The experiments on the vaporisation of water and the use of several electrical instruments suggest that, during his stay in Paris Volta, had demonstrated his electrical eudiometer, or as he called it, the *pistola elettrica* (electrical gun) (Fig. 19), for Lavoisier and other French scientists. Volta first used this instrument for his experiments on the inflammable airs in spring 1777, when he used electrical discharge in a closed recipient to measure the degrees of the flammability of airs.[110] During one of these experiments Volta observed that inflammable air (hydrogen), when combusted in the presence of dephlogisticated air (oxygen), ceased to be

103 See the description of the gasometers MAM inv. nos. 07547-0001-001- and 07547-0002-001-, in the catalogue on pp. 340–346.
104 See catalogue MAM inv. no. 20185-0000-, on pp. 377–381. and Lavoisier (1782a).
105 It was on this occasion that he showed that a stream of dephlogisticated air (oxygen) could melt platinum. MS Procès-verbaux de l'Académie des sciences, 1782, fol. 103.
106 Franklin (1959–2018), vol. 37, p. 446.
107 The publication of these results was delayed until 1785.
108 Their combination would have helped raise the flame's temperature.
109 Beretta (2001b).
110 On Volta's contributions to pneumatic chemistry and the debate on the composition of water, see Abbri (1984), pp. 275–289. On Volta's eudiometer see my introduction to Landriani (1995), pp. 36–39 and 109–124, and Beretta (2000).

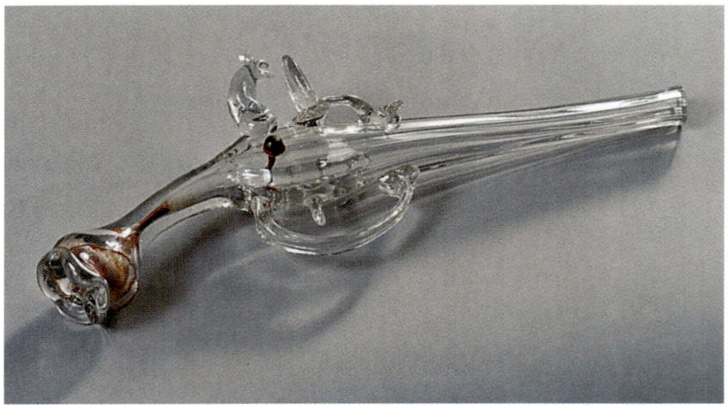

FIGURE 19 Alessandro Volta's electrical gun coming from Charles' Cabinet (inv. n. 01651-0001-)
© MAM/PHOTO SYLVAIN PELLY

a gas, and a dew would be left in the recipient – a completely unexpected result which he did not identify as water. Interestingly, it was not until 1784, after the publication of Cavendish's and Lavoisier's crucial papers , that Volta understood the meaning of the surprising results of his experiment.[111]

Sometime between 1782 and 1784 Lavoisier used Volta's electrical eudiometer, and in August 1784 the chemist Jean Darcet sent Volta a report on behalf of Lavoisier, on Lavoisier's and Jean Baptiste Meusnier de la Place's latest experiments on the decomposition of water.[112] In the accompanying letter Darcet showed that he was not only aware of Volta's eudiometer, but also of its possible consequences for Lavoisier's work on the synthesis of water:

> I recall very well the cloudy vapour that results from the total or near-total destruction of the two airs, when they are set aflame with the aid of the electric spark in your new apparatus, which you kindly had delivered to me.[113]

It is clear from Darcet's testimony that while in Paris, Volta had shown his eudiometer and conducted the experiments that allowed him to synthesise water. It is rather unlikely that Volta would have shown these experiments to Darcet but not to Lavoisier, but even if that was the case, it is extremely probable that Darcet would have told Lavoisier, his colleague at the Académie, about them. Moreover, we know from the revolutionary inventory of Lavoisier's laboratory, which was compiled in 1794, that he owned an "eudiomètre de Volta", one "pistolet de Volta en cuivre" and one "pistolet de Volta en fer blanc".[114] Considering that Lavoisier and Laplace had been engaged in a series of experiments on the vaporisation of water since Volta's arrival in Paris, it is likely that Lavoisier was inspired to investigate the nature of water from another perspective when he saw Volta's electrical demonstration on what he called the decomposition of inflammable air. Although there is no conclusive evidence for this, the numerous coincidences listed above, combined with Lavoisier's intense programme of experiments after 1782, suggests that we should look at the investigations into the nature of water from a wider perspective and within a richer context.

No matter whether Lavoisier took his inspiration from Bochart de Saron or from Volta, he decided to work on water on a large scale in 1783.[115] He first attended to this in April 1783, repeating Priestley's experiments on the presumed conversion of air into water.[116] On 24th June of the same year he organized a set of experiments devoted to the origin of water ("Eau, sa formation") at the Arsenal, for which Blagden, Dionis du Séjours, Laplace, Vandermonde, Fourcroy, and Meusnier were present. In a glass balloon Lavoisier burned specific quantities of hydrogen and

111 In 1798, however, in a letter to Van Marum, Volta vindicated the priority of the discovery with the following words: "I am very fond of chemistry, and particularly pneumatic chemistry which I was the first to cultivate in Italy. You will probably know my discoveries and my experiments on inflammable air. Was not I the first to discover by a series of experiments with my pistol and my inflammable air lamp, and those with an eudiometer or apparatus to burn this air in closed vessels, mixed in different proportions of common air, or vital air, likewise my invention – who had discovered, I say, and found before 1781 that all inflammable air disappears and takes with it the destruction of a volume of vital air which is approximately half of its own volume? […] This is the point at which I had arrived with my experiments several years before Lavoisier […]. He only verified or completed therefore my discovery". In Forbes (1969–1976), vol. 6, p. 360. Since he did not recognise the composed nature of water until the late 1780s, none of Volta's claims are historically grounded.

112 Volta (1949–1955), vol. 2, pp. 236–239.

113 "Je me suis rappellé for bien de cette vapeur nebuleux qui resulte de la destruction entiere ou presque entiere, des deux airs, en les enflammant à l'aide de l'etincelle electrique, dans votre nouvel appareil, que vous aviez eu la complaisance de faire porter chez moi". Letter by Darcet to Volta of August 16, 1784 Volta (1949–1955), vol. 2, p. 237.

114 MS Charles, Lenoir, Fortin (1794), fols 2–3. Transcribed in Appendix 3, p. 154.

115 Lavoisier had worked extensively on water since the late 1760s, when he began to experiment on the presumed conversion of water into earth.

116 "M. Lavoisier a lu [before the Académie des sciences] une note sur l'expérience par laquelle M. Priestley a prétendu que l'eau se convertissait en air et de laquelle il résulte que l'air dans cette expérience passait à travers les vaisseaux". Cit. in Daumas (1955), p. 48. See also LC, vol. 3, p. 735.

oxygen, and obtained pure water. The balloon was connected to two pneumatic chests, one of which, according to Daumas, had been previously made for the hydrostatic bellows.[117] The following day, Lavoisier and Laplace declared before the Académie des sciences:

> Water is not a simple substance; it is composed on a weight-for-weight basis of inflammable air [hydrogen] and vital air [oxygen].[118]

The first version of the gasometer, which was an evolution of the hydrostatic bellows, was not sufficiently accurate. Before Lavoisier could plan its improvement, a fortunate circumstance helped him to gather both new data and funds for his project. In the summer and autumn of 1783 the Montgolfier brothers demonstrated their latest invention, the hot air balloon, several times. This sensation, first in Paris and then throughout Europe, prompted the Académie des sciences immediately to create a commission (with Lavoisier among its members) to investigate how such an important invention could be improved.[119]

Once hydrogen proved to be a more effective gas to lift the balloon than that used by the Montogolfier brothers, the commission's priority was finding a new way to produce hydrogen on a large scale. Unsurprisingly the Académie asked Lavoisier to solve the problem, and provided him with financial support for new experiments. Upon Lavoisier's request, Meusiner de la Place was admitted to the commission in early January 1784, and as early as in March they were able to make a new apparatus: a rifle barrel (of incandescent iron) with which water was decomposed and 82 pints of hydrogen were produced. The experiment was successfully repeated on 29th March in the presence of the entire aerostatic commission. This experiment not merely showed that it was possible to produce hydrogen on a large scale, but it also demonstrated that water was not a simple substance. From spring 1784 onwards Lavoisier put much effort into perfecting the apparatus he had built in 1781, and recruited a team of assistants including, in addition to Meusnier, Gengembre, Fortin, Mégnié, Fallot, and Naudier.[120] Moreover, he ordered a long copper pipe from the Périer brothers.[121] Throughout the next month Lavoisier's team worked on perfecting the gasometer (the evolution of the hydrostatic bellow) and the apparatus for the decomposition of water (the rifle barrel). Their correspondence shows in some detail the technical difficulties they had to face and, on occasion, Meusnier's and Naudier's, the engineer's and the tinsmith's, diverging views on specific technical solutions.[122] (Fig. 20)

After several tests on the gasometer calibration, the whole apparatus was ready in February 1785, and on the 16th Lavoisier asked the Académie to form a commission which included the chemistry section as well as Bochart de Saron, Brisson, Bailly, Laplace, Monge and the Duc de la Rochefoucauld.[123] Lavoisier invited several other scientists and lay people to assist the large-scale experiment, which took place at the Arsenal on the 27th and 28th of February, and again on 1st March. Although the experiments were extremely successful, the quantitative data produced was far from accurate, and in a letter to Lavoisier of March 1785, Meusnier suggested some improvements to the pneumatic chests and the balance beam which led, in the following two years, to the construction of new gasometers which Mégnié would not complete until March 1788.[124] The gasometers were, thus, constructed from several pieces of apparatus originally intended for experiments without direct connection to the analysis and synthesis of water. Moreover, some of their parts originated in already existing instruments and techniques (the enameller's desk and the blowpipe).

As I have pointed out above, Lavoisier was very capable in utilising technical suggestions from those he employed as assistants, but he was also very keen to give credit to Meusnier for his important improvement. Although it is difficult to know if Lavoisier deliberately exploited the opportunity presented by of the Académie's balloon commission for his own purposes, he was certainly extremely successful in highlighting the revolutionary value of his results.

As the history of the gasometer shows, from the late 1770s Lavoisier used his laboratory at the Arsenal as a collaborative experimentation site where, with the assistance of instruments makers, artisans, and colleagues from the Académie, he assembled and dissembled apparatus so

117 Daumas (1955), p. 143.

118 "L'eau n'est pas une substance simple, elle est composée poids par poids d'air inflammable et d'air vital". Cit. in Daumas (1955), p. 49.

119 On this see Gillispie (1983) and Thébaud-Sorger (2007).

120 Fallot, who prior to working for Lavoisier was Macquer's laboratory assistant, was eventually employed by Lavoisier at the refinery of the Arsenal. See the next chapter.

121 LC, vol. 4, p. 17.

122 LC, vol. 4, pp. 32–33.

123 LC, vol. 4, p. 67.

124 LC, vol. 4, pp. 87–88. MAM, inv. nos. 7544-0000- and 7545-0000-.

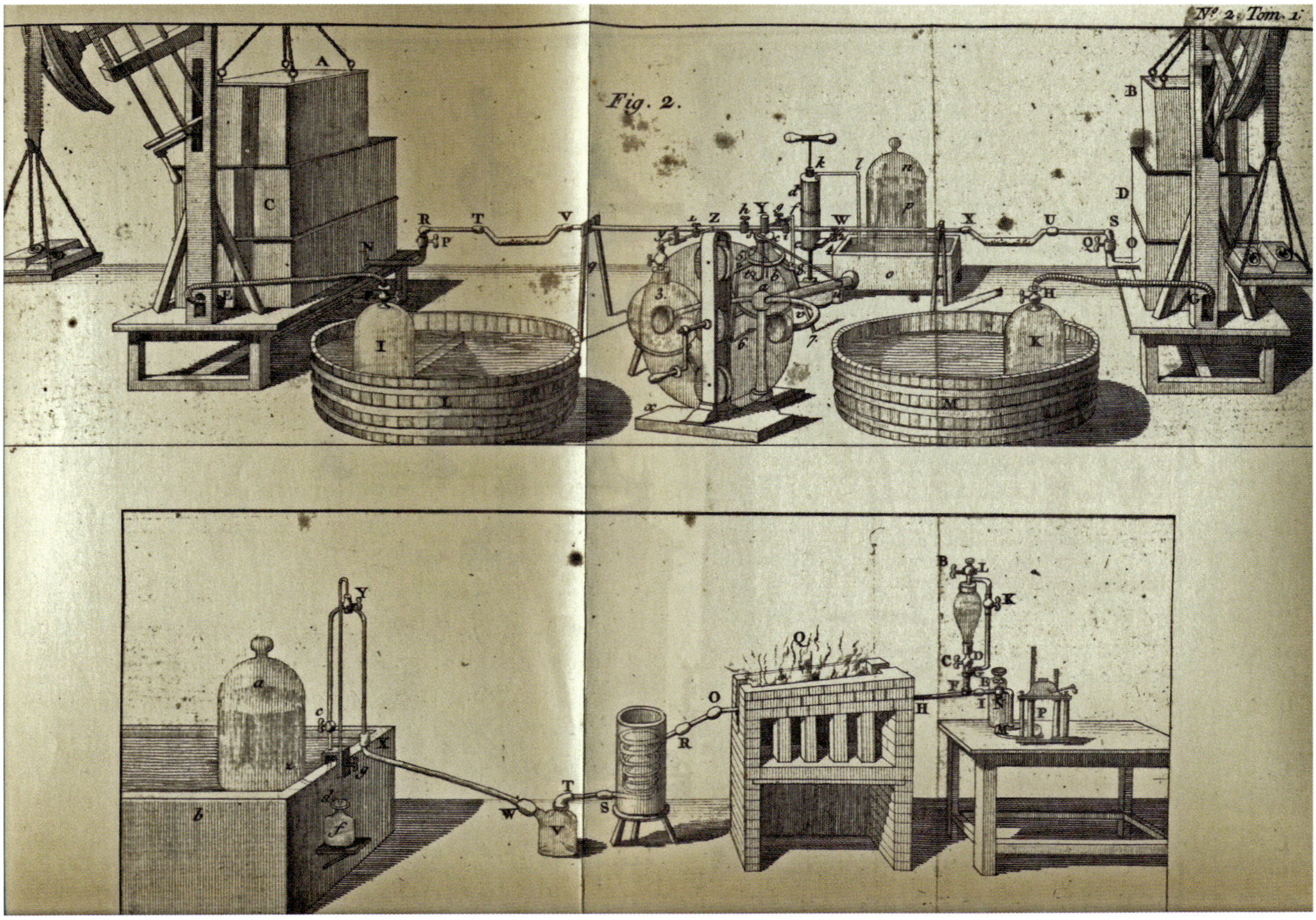

FIGURE 20 Lavoisier's and Meusnier de la Place's apparatus used in the spring of 1785 in experiments on the analysis and synthesis of water *JOURNAL POLYTYPE* (27 FEBRUARY 1786). COURTESY OF CORNELL UNIVERSITY LIBRARY – DIVISION OF RARE AND MANUSCRIPTS COLLECTIONS

that their growing complexity could be adapted for a variety of public experiments.

2.4 Parisian Instrument Makers at the Arsenal

Lavoisier moved to the Arsenal in Spring 1776, where he had access to a large building and soon built up the most elaborate chemical laboratory of the French capital. In his eulogy for Lavoisier, written following an interview with Madame Lavoisier, Georges Cuvier recalled the Lavoisier's organisation of his laboratory team thus:

> In the morning, several enlightened friends whose cooperation he had requested would gather in his laboratory. He even accepted very young people whose shrewdness he had recognised, and certain workers who were especially skillful in producing accurate instruments. He always announced his plans to the assistants with great clarity. Each one made suggestions about their execution; and everything that was considered plausible was soon put to test.[125]

This statement needs further clarification. Cuvier seems to be referring to two different groups: skilled artisans and young scientists. Nicolas Fortin, a talented artisan who had trained in Lavoisier's laboratory and specialised in making chemical instruments, is the most prominent of the former group. Lavoisier first recognised Fortin's abilities in June 1779, when the latter reported on his vacuum pump at the Académie.[126] Before this presentation Fortin had been unknown (Fig. 21). Lavoisier must have been rather impressed with Fortin's abilities because he soon engaged him for the construction of various sophisticated

125 Cuvier (1819), p. 462.
126 LO, vol. 4, p. 327.

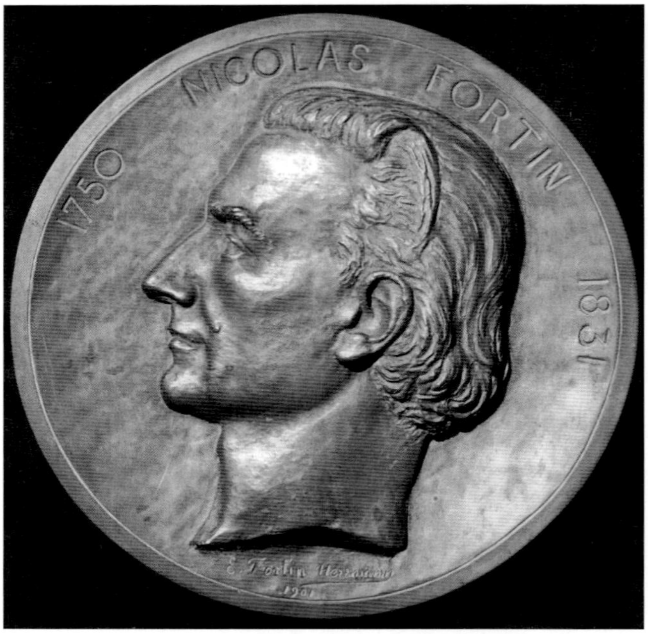

FIGURE 21 Medallion with a portrait of Nicolas Fortin (1906) made after a lithography by Godefroy Engelmann (MAM inv. no. 13946-0000-)
© MAM

instruments, for example, a pneumatic piece of apparatus, a gasometer, a large precision balance, and an apparatus for combusting oils and for fermentation. Fortin also provided Lavoisier and Meusnier de la Place with technical solutions for assembling the apparatus used during the famous experiments on the decomposition and composition of water.[127] In a letter to Achard dated 6 December 1789, Lavoisier declared that Fortin was the artisan he employed most commonly.[128] It is likely that Fortin worked regularly in Lavoisier's laboratory, at least early on, but sometime during the 1780s he opened his own workshop in the Place de la Sorbonne, where he had only two employees, a strikingly small figure in contrast with the forty workmen employed by Jesse Ramsden in London. Thanks primarily to Lavoisier, Fortin's reputation rapidly grew, and he began to work for other scientists and, above all, for the Académie des sciences. Fortin became one of the most authoritative instrument makers used by the Commission des poids et mesures in 1793. Eventually, Fortin was officially recognised, with others, to have "revolutionise[d] the science of physics and created modern chemistry".[129]

Another Parisian instrument maker used regularly by Lavoisier who, in time, specialised in the construction of chemical instruments was Pierre Bernard Mégnié. As *ingénieur en instruments mathématiques*, Mégnié was a specialist before he met Lavoisier, and owned a shop at the Cour de commerce, Faubourg St. Germain in Paris, where he made mathematical and astronomical instruments. Like Fortin, he was very young when Lavoisier recruited him. His youth probably made him more open to the challenge of exploring a new field and to apply his experience with precision instruments to chemical experimentation. He was employed extensively by Lavoisier, whom he called his patron (*protecteur*) and benefactor (*bienfaiteur*).[130] The first recorded instrument he made for Lavoisier was a barometer used for the meteorological campaign the latter promoted in 1778, and the last a pair of gasometers in 1787.

The examples of Fortin and Mégnié show that Lavoisier was especially fond of makers he could trust and guide, and who could be actively engaged in his laboratory at the Arsenal. This also explains why most of the makers he used were Parisian, despite his awareness of the superiority of English craftsmanship.[131]

Situated somewhere between Lavoisier the chemist and the instrument makers such as Fortin were the engineers, including Jean Baptiste Meusnier de La Place, military engineer and, since 1776, a corresponding member of the Académie des sciences de Paris. As I already pointed out above, Lavoisier employed him after he had seen his abilities during the ballooning experiments promoted by the Académie des sciences.

Another group working on the construction of equipment in Lavoisier's laboratory were a few young students who, because of their positive attitude and intelligence, were involved in experiments as laboratory assistants. Pierre Adet, Jean-Henri Hassenfratz, Subrin, Philippe Gengembre, Eleuthère Iréné Dupont, and Armand Séguin were among Lavoisier's assistants:[132] all began their apprenticeship at the Arsenal, and on several occasions they were asked to propose improvements for the existing instruments or to invent new ones. None of them

127 See the invoice sent by Fortin Lavoisier on July 20, 1785 which shows that his role in the construction of the perfected gasometer has been underestimated: LC, vol. 4, pp. 137–140.

128 LC, vol. 6, p. 90.

129 "Fortin. Ce dernier, plus spécialement adonné à la construction des instrumens de physique, a secondé, par son talent, les travaux des physiciens français qui ont changé la face de la physique et créé la chimie moderne". Rapport (1819), p. 256.

130 LC, vol. 4, p. 67.

131 See the biographical dictionary of Lavoisier's instrument makers and suppliers on pp. 162–180.

132 On Hassenfratz, see Grison (1996), pp. 45–122. Subrin, a student at the École des mines, was introduced to Lavoisier by Hassenfratz in 1786; LC, vol. 4, p. 209. Philippe Gengembre (1764–1838) had been Lavoisier's pupil as part of the activities of the Régies des poudres since 1783; see Bret (1994), pp. 70–72. On Dupont, see Dujarric de la Rivière (1954), Bret (1994), pp. 68–69; Bret (1995), Bret (1996).

became prominent chemists, but almost all of them had important careers: Adet became an ambassador in America, Hassenfratz was professor of physics at the École polytechnique, Dupont became the founder of the factory 'Lavoisier's Mills' (eventually renamed Dupont) in Wilmington, Delaware; Gengembre would be the "inspecteur général" at La Monnaie (1803–1815), and Séguin the director of a very successful tannery.

Séguin may serve as a significant example for their work as instrument makers.[133] He first joined Lavoisier's laboratory in this capacity during the second campaign of experiments on the decomposition of water in 1785. Apparently impressed by his talent, Lavoisier soon employed him on a permanent basis, and by 1791 Séguin and his brother lived "in the faubourg St Antoine beyond M: Lavoisiers".[134] By then he had established a career, and the Scottish naturalist James Hall reported during one of his numerous visits to the Arsenal that Lavoisier entrusted Séguin with showing him the machines "for burning charcoal and for oils" as well as an air pump.[135] Séguin is well known for his contributions to the experiments on human breathing of 1790 to 1792.[136] In 1791 he had devised a new eudiometric method for measuring the salubrity of the air, thus providing Lavoisier with an extremely effective instrument that challenged those used by most of his contemporaries.[137] In the same year Séguin and Lavoisier conducted a new series of experiments on human perspiration and devised new apparatus for this purpose. They reported their results to the Académie on 12 February 1792, to demonstrate the fallacy in Albrecht von Haller's claim that the human body increased in weight when immersed in water (because of the absorption of water).[138] Since the combustion of oxygen in the lungs and during digestion generated heat in the human body, Séguin and Lavoisier wished to measure the dissipation of heat that kept the body temperature constant. Séguin thought that the excess caloric escaped through the pores of the skin in the form of sweat, and to study this phenomenon, Séguin devised a complex experimental procedure (Figs. 22–23). On 21 May 1791, Hall accompanied Séguin to Fortin's workshop to order an air pump for Lavoisier. A few weeks later Hall reported about a meeting at Charles' cabinet at the Louvre, where Séguin, Lavoisier, and Charles used a large electrical machine made by Charles to repeat van Troostwijk's and Deiman's experiment on oxygen and hydrogen. Séguin also had a prominent role in the fusion of platinum, which was the experimental challenge of the day. After the disappointing results achieved with the burning lens and other naturalists' methods, Lavoisier asked Séguin "to construct a very simple furnace […] of very refractory earth […] to produce a heat greatly more intense than any hitherto known".[139] In order to increase the temperature, Lavoisier suggested that Séguin use his previously mentioned hydrostatic bellows and nourish the fire with oxygen only. Lavoisier also tried an "oxy-hydrogen blowpipe" inspired (as mentioned above) by Bochart de Saron.[140] The refractory material of Séguin's furnace was supplied by Josiah Wedgwood, who also sent Lavoisier two thermometers for measuring the heat capacity of his clay. Thus, Séguin played a very prominent role in Lavoisier's laboratory as a demonstrator, as a maker and designer of instruments, as a supervisor of other instrument makers (both Parisian and foreign), and, last but not least, as a young scientist. Unfortunately, Lavoisier's early death and Séguin's great success as an entrepreneur later distracted the latter from his scientific pursuits.

From the all of the above it is clear that Lavoisier realised that he could only reform chemistry by radically changing experimental practice, and that his laboratory needed to accommodate a new apprenticeship model. The involvement of so many different instrument makers, young apprentices, and experienced scientists from other disciplines indicates a deliberate and consistent effort to change the laboratory practice of a science that, in Lavoisier's view, was too rudimentary to enable substantial progress. Unlike experienced chemists, young apprentices and instrument makers had, for different reasons, no preconceived ideas, and they considered Lavoisier's laboratory and patronage a formidable opportunity to acquire a public reputation rapidly. Many of them had the chance to present their ideas and results before the Académie – an exclusive and selective institution – at a remarkably young age.

Thus, the eighteenth-century transformation of chemical instrument making to a large extent took place in the laboratory of the Arsenal. However, the pioneering changes introduced by Lavoisier prioritised the creation of an innovative site of collegial experimentation over the construction of costly and sophisticated equipment. The new group of professionals collaboratively produced a variety of instruments at the Arsenal, and thereby demonstrated the crucial importance of active instrument makers to chemical research. Over the years Meusnier, Fortin, Mégnié, Gengembre, Hassenfratz, and Séguin came up

133 On Séguin, see Mercier (1976); Beretta (2001a).
134 Diary by James Hall of 1791 cited in Chaldecott (1968), p. 32.
135 Cited in Chaldecott (1968), p. 25.
136 Beretta (2012).
137 On Séguin's eudiometric method, see Lavoisier (1805), vol. 2, pp. 143–153.
138 LO, vol. 5, pp. 379–390, especially pp. 383–384.
139 Cited in Chaldecott (1968), p. 30.
140 Chaldecott (1968), p. 34.

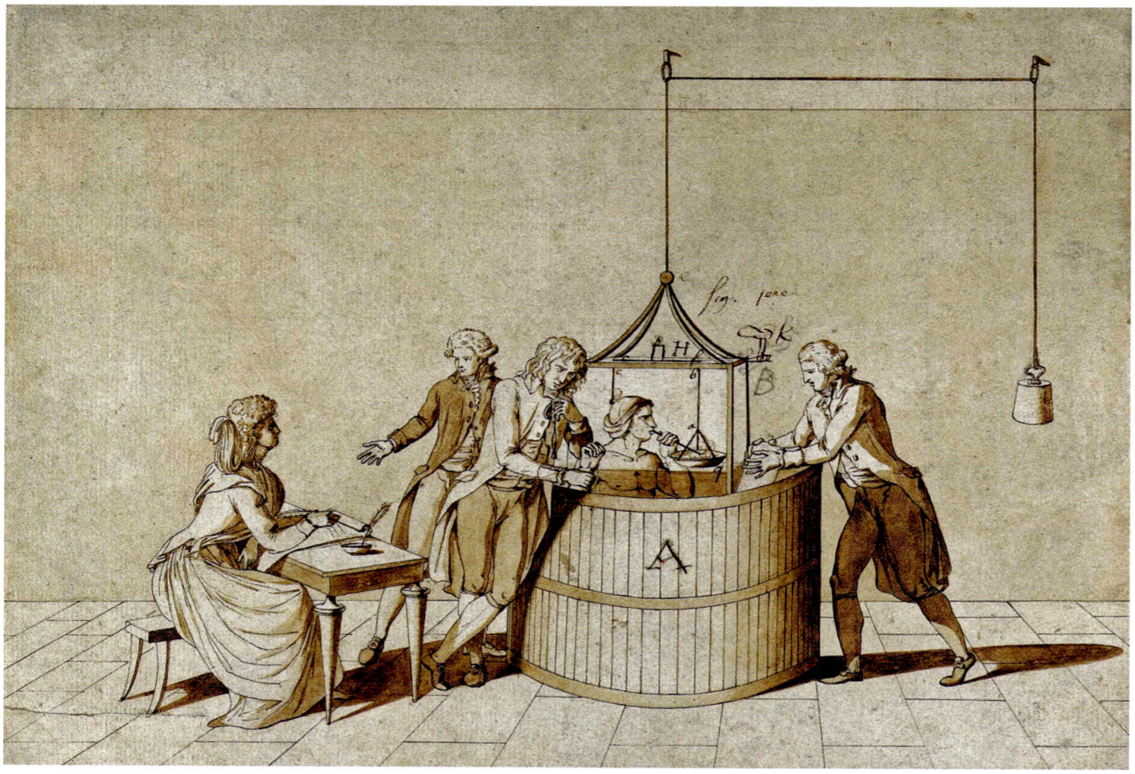

FIGURE 22 Séguin, seated, with his head under a glass canopy that is illuminated by a candle (ca. 1790)
COURTESY OF THE WELLCOME LIBRARY

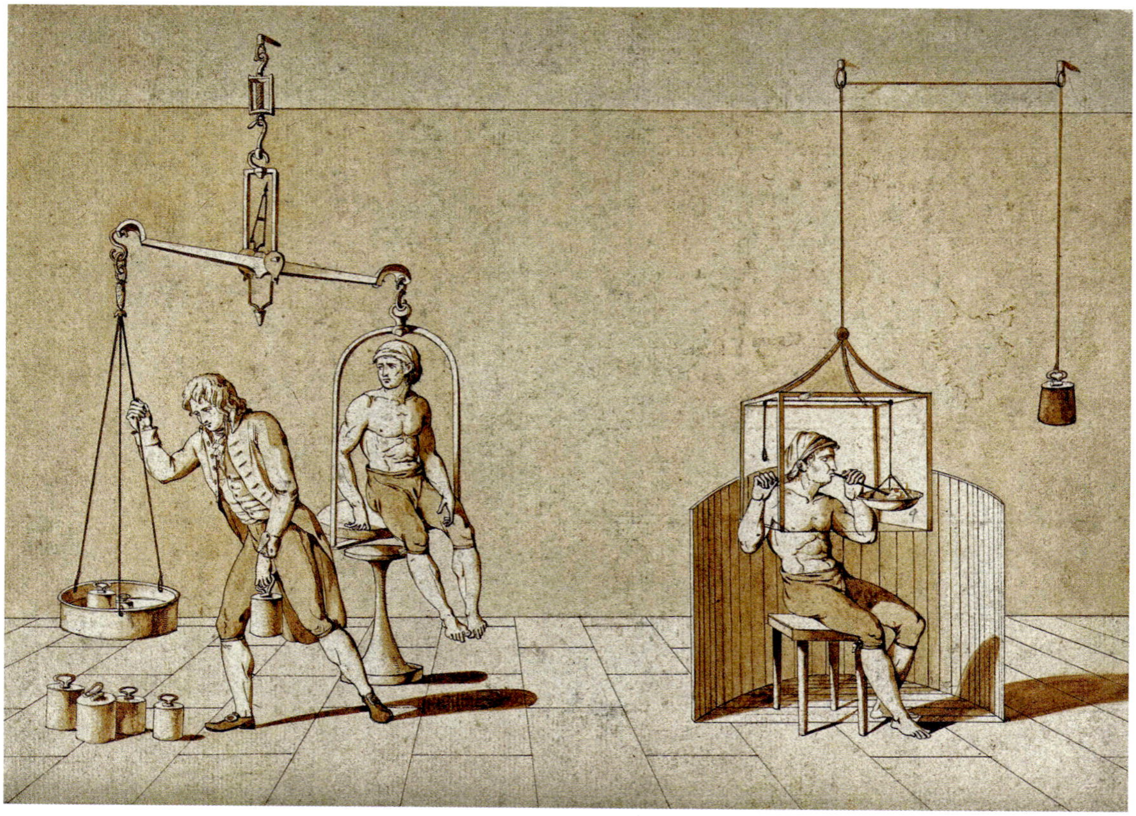

FIGURE 23 Séguin being weighed on a large set of scales (ca. 1790)
COURTESY OF THE WELLCOME LIBRARY

with technical solutions and suggestions that provided solid experimental support for Lavoisier's theories; at the same time, their technical contributions also increased their confidence and, for some, turned them into scientific authors. This collaborative network opened up a market for precision instruments that would soon become commonplace in the major Parisian research laboratories. It is not a coincidence that Balthasard Georges Sage, a fierce opponent of Lavoisier's theory, kept several precision instruments in his laboratory at La Monnaie, including a balance made by Mégnié. Fortin's devices became relatively common and were used by all chemists engaged in research, from Berthollet to Gay-Lussac. The Arsenal, as is well known, was also the laboratory for a new generation of chemists such as Fourcroy, Berthollet, and, in the late 1780s, Guyton de Morveau, all of whom became increasingly engaged in the development of the chemical arts and, in a few cases, the introduction of new chemical instruments.

Lavoisier's long-term collaboration with different instrument makers and chemists made him appreciate the importance of the chemical trades and technology toward the end of his career. This increasing awareness led him to undertake several projects aimed at the education of chemical artisans and instruments makers. In early 1787 he founded the Société pour le progrès des sciences et des arts, and he had an active role at the Bureau des consultation des arts et métiers as well as at the Lycée des arts during the French revolution. All this was aimed at bridging the gap between theoretical chemistry, the chemical trades, and instrument making, and inspired Lavoisier to draft a general reform of education.[141] The Arsenal was thus intended to be a model for public institutions, and it is probably not a coincidence that the École Polytechnique incorporated most of Lavoisier's ideas into the organisation and composition of its chemical laboratories.[142]

The Arsenal, then, cannot be regarded as a laboratory in which Lavoisier displayed his lavish wealth through the acquisition of exceedingly expensive devices, but as a pioneering site where he advocated the specialised manufacture of chemical instruments. At the same time, Lavoisier encouraged the development of the scientific skills of instrument makers by involving them in sophisticated scientific and experimental research programs.

141 LO, vol. 4, pp. 649–668; LO, vol. 6, pp. 516–532; Bret (2012).

142 Thomas Bugge thus wrote in 1798: "The Polytechnic School has two very large and fine chemical laboratories, besides two of inferior extent, and some mechanical workshops". Cited in Crosland (1969), p. 39. Interestingly, Hassenfratz was responsible for the mechanical workshops.

2.5 Second-Hand Instruments

Lavoisier not only borrowed, bought, conceived and contributed to the making of instruments, but he also purchased a few instruments from other collections. Although it is impossible to reconstruct the size of the second-hand instrument market, an interesting episode told by Jean Michel Raymond-Latour suggests that Lavoisier bought an instrument from the collection of the late Duc de Chaulnes (Joseph d'Albert d'Ailly) sometime in 1792. Since this story sheds light on Parisian laboratory life it is worth quoting in full:

> One of the fads of the wealthy lords of the *ancien régime* was to own a *cabinet de physique* embellished with beautiful equipment and a chemical laboratory well furnished with apparatus and reagents stored in pretty jars. These objects were never handled; their possessors imagined that mere ownership would grant them, in the eyes of the world at large, the title of *savant*.
>
> The Duc de Chaulnes shared this characteristic, although to a somewhat lesser degree, with his noble colleagues and devoted himself to a slight extent to alchemy. Like many other idiots he had melted and evaporated away a large part of his gold in his crucibles in a vain attempt to find the philosopher's stone which he sought. This search he pursued with the same passionate zeal as those adepts who had preceded him; his reserve in this hunt was no less ruinous to him for being slower.
>
> After his death the fine scientific equipment was put up for sale. I made my way to his house with no intention of doing much harm to the bidders nor contributing substantially to the receipts of the sale for I was only carrying on my person the modest sum of fifteen francs. This sum was intended for the acquisition of some instrument without any previous idea as to which it should be. As I entered the sale room the auctioneer was offering a small instrument made of metal, the use of which I was completely ignorant. A bid was immediately made for this lot by a small man with a pale, thin, oval, and not particularly gracious looking face; his obvious keenness made me bid for the same object; he stared at me, amazed to see a raw youth battling against him in order to deprive him of or at least cause him to overpay for a gadget the use of which he alone understood. Anyhow, he refused to give up and topped my bid; I immediately raised my own bid. At this second slap in the face I see the little man stamp his foot and

obviously become furious; once more he tops my bid and I reply at once. This disconcerted my opponent who evidently saw in me a competitor determined to make him pay an exorbitant price for an object whose real value had already been exceeded. My last bid had absorbed my fifteen francs and if my antagonist had only advanced the bidding by a penny I should have been forced out of the running. After I had foolishly exchanged my good money for an instrument which was an embarrassment to me, my rival sweetly approached me and asked, with a mocking smile, to let him know the use of a piece of equipment to which I had apparently attached so great a value. In reply I said that I had intended asking him exactly the same question when I had seen how keen he had been to prevent me from possessing it. He smiled again with a smug air at seeing me caught with my acquisition. He then walked away from me with an ironical laugh and a shrug of the shoulders as if wishing to say "well, of all the fools" to me.

Two days after this adventure went to show my new acquisition to Fourcroy and to inquire of him the use to which I could apply it. On seeing the small gadget he burst into a laugh and said "You have certainly achieved something, by golly, with your purchase. You have succeeded in getting one of Europe's leading scientist's back up; it is Lavoisier whom you have been unfortunate enough to cross; go quickly to him, taking your new purchase, and obtain his forgiveness". I did not wait to be told twice.

When I reached Lavoisier's house I was taken to see him; he recognized me immediately and turned his back at the sight of the controversial instrument. I did not allow this far from courteous reception to discompose me and told him that I had hastened to bring him the object that he had missed buying. I said I regretted having deprived him of it and that if I had only previously enjoyed the honor of his acquaintanceship I would never have topped his bid. This honorable amend produced a good effect and the *savant* turned around, carne toward me and said that in view of my action he would be glad to relieve me of something with which I had got myself stuck and pay me the full price that it had cost me. He then asked me about my work and I told him that I devoted myself especially to the study of chemistry, a taste for which subject had taken a hold of me after reading his *Traité* on this science. Praise, from whatever source it may come, is always gratifying to hear; our savant unbent and offered to show me some of the more important equipment he had devised, in particular that for the decomposition and recomposition of water. I accepted with alacrity because I had already heard about it and was keen to see it. I was careful not to impose by staying too long in his scientific museum and soon left, proud and overwhelmed by his kindness.[143]

143 "L'une des manies des seigneurs opulens de l'ancien régime était d'avoir un cabinet de physique décoré de belles machines, et un laboratoire de chimie bien fourni d'ustensiles et de réactifs renfermés dans de beaux flacons, auxquels ils ne touchaient pas, se persuadant qu'il suffisait de tout ce matériel de la science pour leur procurer, dans le monde, quelqu'importance, à titre de savans. Le duc de Chaulnes partageait aussi ce travers, mais moins que ses confrères titrés, et s'adonnait un peu à l'alchimie; il avait, comme tant d'autres fous, fondu et évaporé dans ses creusets une grande partie de son or, sans avoir réussi à faire sortir de ses fourneaux: cette pierre philosophale à la recherche de laquelle il se livrait, non pas avec la même ardeur que les adeptes ses devanciers, mais avec une réserve qui, pour être plus lente, n'en devenait pas moins ruineuse. Après sa mort, son riche mobilier scientifique fut mis en vente. Je me rendis à son hôtel; je n'y allais pas avec des intentions bien hostiles contre les enchérisseurs, ni pour grossir beaucoup la recette, puisque je n'emportais sur moi que la modique somme de quinze francs, destinée à faire l'emplette d'un instrument, sans m'être précisément décidé sur le choix. Au moment où j'entrai dans la salle de vente, l'huissier priseur mit à l'enchère un petit instrument de métal dont j'ignorais alors complètement l'usage; elle fut aussitôt couverte par un petit homme, d'une figure blême, maigre, ovale et peu gracieuse; son empressement me décide à convoiter le même objet; il me regarde, tout étonné de voir un blanc-bec lutter contre lui, pour le priver ou lui faire payer trop cher un instrument dont il était le seul qui en connût l'emploi. Toutefois, il y tenait, et couvrit mon enchère, et moi aussitôt d'aller sur ses brisées; à cette seconde incartade, je vois mon petit homme qui trépignait et devenait de plus en plus furibond; il couvre de nouveau ma mise; je riposte à l'instant, et je déconcerte mon homme, qui ne voit décidément en moi qu'un concurrent obstiné à lui faire payer un prix exorbitant un objet qui avait déjà dépassé sa valeur. Ma dernière enchère absorbait mes quinze francs, et si mon antagoniste l'avait couverte d'une obole, j'etais mis hors de combat. Après que j'eus niaisement troqué mon argent contre une machine qui m'était déjà à charge, mon concurrent s'approcha mielleusement de moi, et me demanda, avec un sourire moqueur, de vouloir bien lui faire connaître l'emploi de l'instrument auquel j'avais paru attacher tant de prix. Je lui répondis que je m'étais proposé de lui adresser la même question, en voyant son empressement à me le disputer. Il sourit de nouveau, d'un air de satisfaction, de me voir tout attrapé de mon emplette. Il s'éloigna de moi, encore courroucé, avec un rire de satire accompagné d'un haussement d'épaules, comme s'il avait voulu me dire asinus asinorum! Deux jours après cette aventure, j'allai trouver Fourcroy, pour lui montrer mon acquisition, et le prier de m'apprendre à quel usage je pourrais l'employer; il se mit à rire aux éclats en voyant la petite machine, et me dit: 'Tu as fait, parbleu, de belles affaires, avec ton achat; tu t'es mis à dos l'un des premiers savans de l'Europe: c'est Lavoisier que tu as eu le malheur de contrarier; cours vîte obtenir ton pardon, en lui portant ton emplette.' Je ne me le fis pas dire deux fois. Arrivé à son hôtel, je fus introduit auprès, de Lavoisier, qui me reconnut aussitôt, et me tourna le dos en

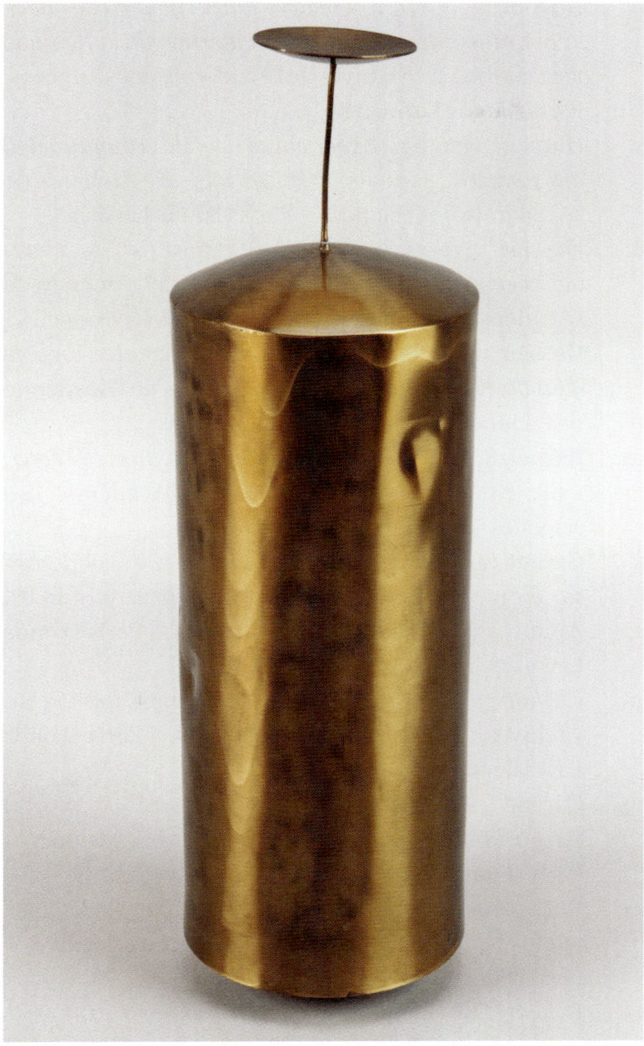

FIGURE 24 This slightly battered hydrometer could be the one lent by Lavoisier to Rouelle in 1767. (MAM inv. no. 07508-0004)

This anecdote is quite interesting, because it shows instruments from the collections of deceased scientists were offered at auction exactly like books and natural history collections. Unfortunately, as far as I know, there are no auction catalogues of instruments that allow us to assess the size of this market.[144] However, this must have been a relatively common way to enlarge existing collections. It is well known that in 1786, after the death of his master, Lavoisier purchased part of Guettard's collection of books, manuscripts, and minerals.[145] After Lavoisier's execution a few of his instruments entered into the collection of Charles and Rumford and, as I will illustrate in chapter 5, many of them continued to be used in Marie Anne Lavoisier's cabinet.

Lavoisier also frequently borrowed and lent instruments. As early as summer 1767 he lent one of his hydrometers to Rouelle who didn't return it in perfect condition.[146] (Fig. 24)

2.6 The Laboratory Notebooks

Lavoisier's laboratory notebooks, most of which are preserved at the Archives de l'Académie des sciences in Paris, contain extremely interesting information on the evolution of instrument making, on the collegial nature of the team set up by Lavoisier from the late 1760s onwards, and on the complex organisation of the information collected during laboratory practice.

In a famous and still-used appendix to his biography of Lavoisier, Marcellin Berthelot published a summary of what he believed to be all of Lavoisier's *registres de laboratoire*.[147] These were 14 folio volumes, which were subsequently the object of systematic interpretation and small emendations.[148] These historiographic interpretations aside, I do not know of more recent studies recording other notebooks than those listed by Berthelot in 1890.[149]

---

voyant le malencontreux instrument dans mes mains. Je ne me laissai pas décourager par cette réception peu civile; je lui dis que je m'empressais de lui apporter l'objet qu'il avait paru regretter; que j'étais désolé de l'en avoir privé, en ajoutant que, si j'avais eu l'honneur de le connaître, je me serais abstenu de couvrir son enchère. Cette espèce d'amende honorable produisit un bon effet. Le savant se retourna et s'approcha de moi, en me disant que, par égard pour mon procédé, il voulait bien me débarrasser d'une chose qui m'était à charge et me rembourser le prix que je l'avais payée, Il m'adressa ensuite quelques questions sur la nature de mes occupations; je lui dis que je me livrais surtout à l'étude de la chimie, dont j'avais puisé le goût dans la lecture de son Traité sur cette science. La louange, de quelque part qu'elle arrive, chatouille toujours un peu l'oreille; notre savant se dérida tout de bon, et m'offrit de me montrer quelques-uns des principaux appareils qu'il avait imaginés; celui surtout pour la décomposition et recomposition de l'eau. J'acceptai avec d'autant plus d'empressement, que j'en avais beaucoup entendu parler et que je désirais le connaître. Je me gardais d'être importun par un trop long séjour dans son musée scientifique, et j'ensortis fier et confus de sa complaisance". Raymond-Latour (1836), pp. 222–226, translated into English by Duveen (1958), p. 470.

144 By contrast, book auction catalogues were published quite frequently: see Beretta (1995).

145 LC, vol. 5, pp.

146 "J'ai preté mon areometre de cuivre le grand dont je me suis servi pour toutes mes experiences et mes tables à M. Rouelle on me l'a rendu un peu bosselé". MS Lavoisier (1767–1788), vol. 9, fol. 29r. I thank Francesca Antonelli for bringing me this note to my attention. The hydrometer mentioned in this note is probably one of MAM inv. nos. 07508-0001-, -0006-; see catalogue pp. 319–321.

147 Berthelot (1890), pp. 209–310.

148 The notebooks were compiled in different hands, and Berthelot had difficulties in identifying them. Antonelli (2021) shows the pervasive presence of Marie Anne Lavoisier's hand in many of them. *Registre* 5 was almost entirely compiled by Madame Lavoisier with the title *Année 1777. Produit du cours de M. Bucquet.* MS Lavoisier (1772–1788), vol. 5.

149 Antonelli (2021) provides a comprehensive study and a thorough reassessment of their cultural value.

However, this is only one part, and not necessarily the most important one, of Lavoisier's laboratory's activities. In fact, the laboratory notebooks number 32 in total, excluding the notebooks concerning his experiments on agriculture which, in comparison with the chemical notebooks from a methodological point of view at least, could provide some useful information.

As we can see from the chronological list of laboratory notebooks in the following table, Lavoisier kept constant and extensive records of his experiments from the very beginning of his career when, following a routine he had learned from Guettard, he kept records of both his field observations and his experiments.[150] In the early 1770s these notebooks recorded his collaborative experiments with other scientists, and in many cases it is rather difficult to establish the primary author. Significantly, we know that the laboratory notebook recording the crucial 1785 experiments on the decomposition and composition of water was left to Meusnier de la Place, and only a few leaves have survived at the Kroch Library of Cornell University.[151] Lavoisier's final records of experiments on respiration, now lost, were kept by Armand Séguin, the principal author of the *mémoires* on human respiration, and were only partially published between 1791 and 1793. The fate of these *registres* shows that in the 1780s laboratory life at the Arsenal became a collective endeavour.

Table of Lavoisier's Laboratory Notebooks

1. Dossier 1388-03. *Journal d'experiences commencé le 2 juillet 1764 (sur le gypse)*. Ends on 1st September. 32 fols. 4°. Archives de l'Académie des sciences – Paris (Fonds Lavoisier).
2. Dossier 1388-04. *Journal d'experiences commencé le 15 janvier 1765 (sur le gypse)*. 20 fols. 4°. Archives de l'Académie des sciences – Paris (Fonds Lavoisier).
3. Dossier 1388–05. *Journal d'experiences commencé le 1er mars 1765 (sur le gypse)*. 20 fols. 4°. Archives de l'Académie des sciences – Paris (Fonds Lavoisier).
4. Dossier 1371–01. *Experiences sur les meches plattes*. April 1765. 25 fols. 4°. Archives de l'Académie des sciences – Paris (Fonds Lavoisier).
5. Carnet 9. *Expérience sur le gypse. 1766–1768*. Also experiments made in 1792. 75 fols. 8°. Archives de l'Académie des sciences – Paris (Fonds Lavoisier).
6. Dossier 1371–05. *Experiences faittes pour determiner la quantitè d'air necessaire a l'entretien de la flamme*. 1767. 22 fols. 4°. Archives de l'Académie des sciences – Paris (Fonds Lavoisier).
7. Dossier 1371–07. *Experiences sur la comparaison des lumières*. January 1767. 46 fols. 4°. Archives de l'Académie des sciences – Paris (Fonds Lavoisier).
8. Dossier 1371–10. *Experiences faittes sur les chandelles en janvier et fevrier 1767*. 22 fols. 4°. Archives de l'Académie des sciences – Paris (Fonds Lavoisier).
9. Dossier 1371–06. *Experiences sur differentes especes d'huille*. April 1767. 32 fols. 4°. Archives de l'Académie des sciences – Paris (Fonds Lavoisier).
10. Carnet 8. *Expériences et analyses de diverses eaux*. 1768. 78 fols. In 8°. Archives de l'Académie des sciences – Paris (Fonds Lavoisier).
11. Carnet 10. *Registre contenant une suite d'expériences sur les eau de vie et l'esprit de vin*. 1769. 22 fols. In 8°. Archives de l'Académie des sciences – Paris (Fonds Lavoisier).
12. Carnet 13. *Phisique et géometrie, 1771*. 81 fols. In 8°. Archives de l'Académie des sciences – Paris (Fonds Lavoisier).
13. Carnet. *Analyse de différentes eaux. 1771–1772*. 80 pp. 8°. Archives de l'Académie des sciences – Paris (Fonds Lavoisier).
14. Registre de Laboratoire 1. *Du 20 février au 28 aoust 1773*. 139 fols. Folio. Archives de l'Académie des sciences – Paris (Fonds Lavoisier).
15. Registre de Laboratoire 2. *Du 28 août 1773 à le 23 mars 1774*. 122 fols. Folio. Bibliothèque Municipale de Perpignan. Mf MS 61–61 bis.[152]
16. Registre de Laboratoire 3. *Du 23 mars 1774 au 13 février 1776*. 138 fols. Folio. Archives de l'Académie des sciences – Paris (Fonds Lavoisier).
17. Copie du registre tenue par les Commissaires nommées par l'Académie Royale des Sciences en conséquence des ordres de sa Majesté pour le Jugement du prix et les recherches sur la fabrication du salpêtre. Dossier Lavoisier 837. Archives de l'Académie des sciences – Paris (Fonds Lavoisier).
18. Registre de Laboratoire 14. *Recueil d'expériences chimiques. Par MM Macquer & Lavoisier 1775–1776–1777*. 14 fols. Folio. Archives de l'Académie des sciences – Paris (Fonds Lavoisier).
19. Registre de Laboratoire 6bis. *Sur le salpetre* [1775–1777]. 162 fols. Folio. Archives de l'Académie des sciences – Paris (Fonds Lavoisier).

150 The Archives de l'Académie des sciences in Paris hold several travel diaries by Guettard in which scientific information is arranged in a similar way.

151 Daumas, Duveen (1950).

152 A xerox copy of this notebook is preserved at the Archives de l'Académie des sciences – Paris (Dossier Lavoisier).

20. Registre de Laboratoire 4. *Du 13 février 1776 au 3 mars 1778. Il y a une expérience de 1779*. 235 fols. Folio. Archives de l'Académie des sciences – Paris (Fonds Lavoisier).
21. Registre de Laboratoire 6. *Tome sixième, depuis aoust 1778 jusqu'au 7 septembre 1782*. 248 fols. Folio. Archives de l'Académie des sciences – Paris (Fonds Lavoisier).
22. Dossier 1323–04. *Journal d'experiences faittes avec M. de La Place à l'arsenal et pour les quelles il a eté pris datte a l'academie le 19 decembre 1781* (*26–12–1781–23.02.1782*). 24 fols. 4°. Archives de l'Académie des sciences – Paris (Fonds Lavoisier).
23. Registre de Laboratoire 8. *25 Mars 1783 au février 1784*. 148 fols. Folio. Archives de l'Académie des sciences – Paris (Fonds Lavoisier).
24. Registre de Laboratoire 7. *Expériences sur la chaleur et autres, du 16 décembre 1782 au 14 avril 1784; premières expériences sur la décomposition de l'eau*. 164 fols. Folio. Archives de l'Académie des sciences – Paris (Fonds Lavoisier).
25. Registre de Laboratoire 9. *Tome neuvième du 26 avril 1784 jusqu'au dernier de décembre 1784*. 125 fols. Folio. Archives de l'Académie des sciences – Paris (Fonds Lavoisier).
26. Registre de Laboratoire 10. *Année 1785. Machines pour manoeuvrer le gaz. Expériences en grand sur la composition de l'eau, avec procès-verbaux signés des commissaires de l'Académie*. 121 fols. Folio. Archives de l'Académie des sciences – Paris (Fonds Lavoisier).
27. Registre de Laboratoire 11. *Du 24 avril 1785 au*. 256 fols. Folio. Archives de l'Académie des sciences – Paris (Fonds Lavoisier).
28. Registre de Laboratoire 12. *De septembre 1786 à la fin de 1787*. 164 fols. Archives de l'Académie des sciences – Paris (Fonds Lavoisier).
29. Dossier 26–13. *Journal d'experiences faittes par les commissaires nommés par l'academie des sciences*. 1787?. 2 fols. 4°. Archives de l'Académie des sciences – Paris (Fonds Lavoisier).
30. Carnet 16. Experiments at Trianon Garden and *Voyage en Bourgogne* 1786–1788. 90 pp. In 8°. Archives de l'Académie des sciences – Paris (Fonds Lavoisier).
31. Registre de Laboratoire 12. *Du 20 mars 1788 au*. 100 fols. Folio. Archives de l'Académie des sciences – Paris (Fonds Lavoisier).
32. Dossier Chabrol 05–08. *Fragments concernant les expériences de laboratoire*. Undated. 9 fols. 4°. Archives de l'Académie des sciences – Paris (Fonds Lavoisier).

Lavoisier attached particular importance to the precision of notes in laboratory notebooks, while being aware that relying entirely on these records would be dangerous. As early as in the introduction to the projected second volume of the *Opuscules physiques et chymiques* (ca. 1778), he declared that chemistry was not a mere accumulation of disorganised facts, because this would merely have required him "to copy […] the laboratory notebooks".[153] "Facts", he wrote at about the same time in the famous "Mémoire sur la combustion", "observations, experiments, are the materials of a great edifice. But in assembling them, we must not encumber our science. We must, on the contrary, devote ourselves to classifying them, to distinguish which belong to each other, to each part of the whole to which they pertain".[154] Lavoisier here criticised Priestley's *Experiments and Observations on Different Kinds of Air* (London, 1774), which provided a detailed history of his experiments without any order, exactly as they were performed and recorded in the laboratory. The shortcomings of the mere accumulation of facts, aggravated by the typical confusion in a chemical laboratory, prompted Lavoisier to develop new techniques for recording his experiments. In his notebooks we see that he regularly described the initial conditions (temperature and atmosphere) in his laboratory and further monitored any changes as the experiment went on. The methodical entries to the notebooks are strikingly consistent throughout Lavoisier's scientific career and emphasise their increasing relevance in the construction of published results. This record of a variety of data, many of which were directly relevant to the experiments, provided Lavoisier with ample margins for the interpretation of errors and discrepancies. This procedure, although common in modern experimentation, has often raised doubts among historians about the accuracy of Lavoisier's experimental method, to the extent that it was recently called a "mildly deceptive practice".[155] In fact, the result of an experiment depended on the construction and classification of data rather than its mere description.

153 "Si j'eusse suivi cette méthode, il ne m'aurait pas été difficile de faire un grand ouvrage: *j'aurais pu me contenter de copier les registres de mon laboratoire*, de les accompagner de réflexions succinctes, et j'aurais vu en peu de temps les volumes se multiplier sous ma plume" (my italics). Lavoisier (1778), pp. 267–268.

154 "Les faits, les observations, les expériences, sont les matériaux d'un grand édifice; mais il faut éviter, en les rassemblant, de former encombrement dans la science; il faut, au contraire, s'attacher à les classer, à distinguer ce qui appartient à chaque ordre, à chaque partie du tout auquel ils appartiennent". Lavoisier (1779a), p. 225.

155 In commenting on Lavoisier's pneumatic experiments of July 1773, Holmes points out: "There are several remarkable features of Lavoisier's reasoning. The most familiar, because it became a stable feature of his quantitative style, was that in presenting his results publicly, he ignored the sources of error he had observed, and the approximations that he had made of his experiments, and presented the numbers resulting from his calculations as though they were exact". Holmes (1998), p. 91, see also p. 116.

CHAPTER 3

Lavoisier's Sites of Experimental Practice: From the Field to the Laboratory (1764–1794)

3.1 Sites of Experiments

Throughout his career Lavoisier performed his experiments in numerous laboratories, and although the Arsenal laboratory has attracted much attention, many of his most important discoveries were, in fact, made in other laboratories scattered across greater Paris. Between 1761 and 1794 Lavoisier performed experiments and observations on ca. 50 different sites in Paris and closer outskirts.[1] If we extend the survey to the whole of France, the number of sites is significantly higher. The variety of the sites is largely due to Lavoisier's education which, as I have pointed out in chapter 2, introduced him to field research and experimentation. As is well known, most of Lavoisier's residences and sites of experimentation no longer exist, and the evidence for this survey is inevitably fragmentary and, in a few cases, conjectural.[2]

In addition to the experiments he conducted while travelling with Guettard (on behalf of the Ferme générale and of the Académie des sciences), Lavoisier undertook a chemical analysis of mineral waters in the surroundings of his father's residence at Le Bourget in 1773, at the "Puits de l'Hotel des poudres" at the Arsenal, and by the Seine in October 1778.[3] On the topic of extreme cold weather , Lavoisier performed experiments in January 1776 in both open and closed environments, namely in the caves of the Observatoire royal, at the Jardin de Luxembourg, at the École militaire, and at the Palais royal.[4]

Equipped with portable or pocket laboratories Lavoisier was able to perform experimental work outside the chemical laboratory. It is interesting to note here that Lavoisier conceived a new kind of portable aerometer for his analysis of mineral waters. Lavoisier also commissioned other portable instruments such as barometers and thermometers, which he took with him on his mineralogical travels.

In April 1765 Lavoisier began to record his observations on the best methods and equipment for illuminating the approximately 800 streets of Paris in a new laboratory notebook. This material served as preparation for an essay presented at the end of that year, with which he would enter the prize competition that had been launched by the Académie royale des sciences in 1764.[5] Lavoisier's approach was multidisciplinary. He applied his chemical skills to the analysis of the combustive properties of different kinds of oil and their consumption; he studied the best ways to fuel light in detail and, using a thermometer, the temperature of lamps; and he examined the optical performance of his reverberating lamp, both indoors and, most importantly, outside, where he substituted the old lamps with its own in a number of streets near the Rue du Four, where he lived (Fig. 1).[6] Lavoisier reported the results of his experiments to the Académie in 1765.[7] These results were recorded in a laboratory notebook which set the standard for Lavoisier's later written records of experiments.[8] Among these experiments there is also a silver model of a street lamp holder with a wrought-iron cage for the lamp, and it is quite surprising that, given this availability of evidence and primary sources, the primary biographies of Lavoisier mention this prize competition and the related writings only in passing.[9]

After the death of Théodore Baron d'Hénouville in 1768 Lavoisier joined both the Académie royale des sciences

1 On Lavoisier's laboratories see Beretta (2003), pp. 313–334; Beretta (2004); Beretta (2009); and Beretta (2014).

2 The only exception being the château at Freschines. See Lemay (1934), Duveen (1951) and Appendix 1, pp. 115–148.

3 Le Bourget is a small village just outside Paris. At the end of 1773 Lavoisier recorded in his *registre de laboratoire* the analysis of "eau de la marre". MS Lavoisier (1772–1788), vol. 3, fol. 28. Lavoisier had inherited the property at Le Bourget from his mother. On 20 November, 1772, in a wood near Le Bourget, Lavoisier also conducted experiments on the supposed properties of the divining rod. LO, vol. 6, pp. 90–93. On the experiments at the Arsenal and by the Seine, see MS Lavoisier (1772–1788), vol. 6, fols. 54–56, 38, 58–61 and 64.

4 LO, vol. 3, pp. 355–386, at p. 366.

5 Lavoisier was awarded a gold medal for his contribution. His "Mémoire sur les différents moyens qu'on peut employer pour éclairer une grande ville" was not published until 1868, in LO, vol. 3, pp. 1–77. On this see Bothereau (2018).

6 Madame Lavoisier declared that her husband "fait tendre sa chambre en noir, passé six semaines sans voir le jour pour accoutumer ses yeux à l'obscurité et pour mieux juger de l'effet de ses expériences". Cited in Gillispie (1956), p. 57.

7 LC, vol. 1, pp. 5–6.

8 MS Lavoisier (1765b); MS Lavoisier (1767a); MS Lavoisier (1767b); MS Lavoisier (1767c); MS Lavoisier (1767d). It is a pity that, in his otherwise excellent study, Botherau (2019) has not examined these documents.

9 MAM inv. no. 20235-0001-.

 | DOI:10.1163/9789004511217_005

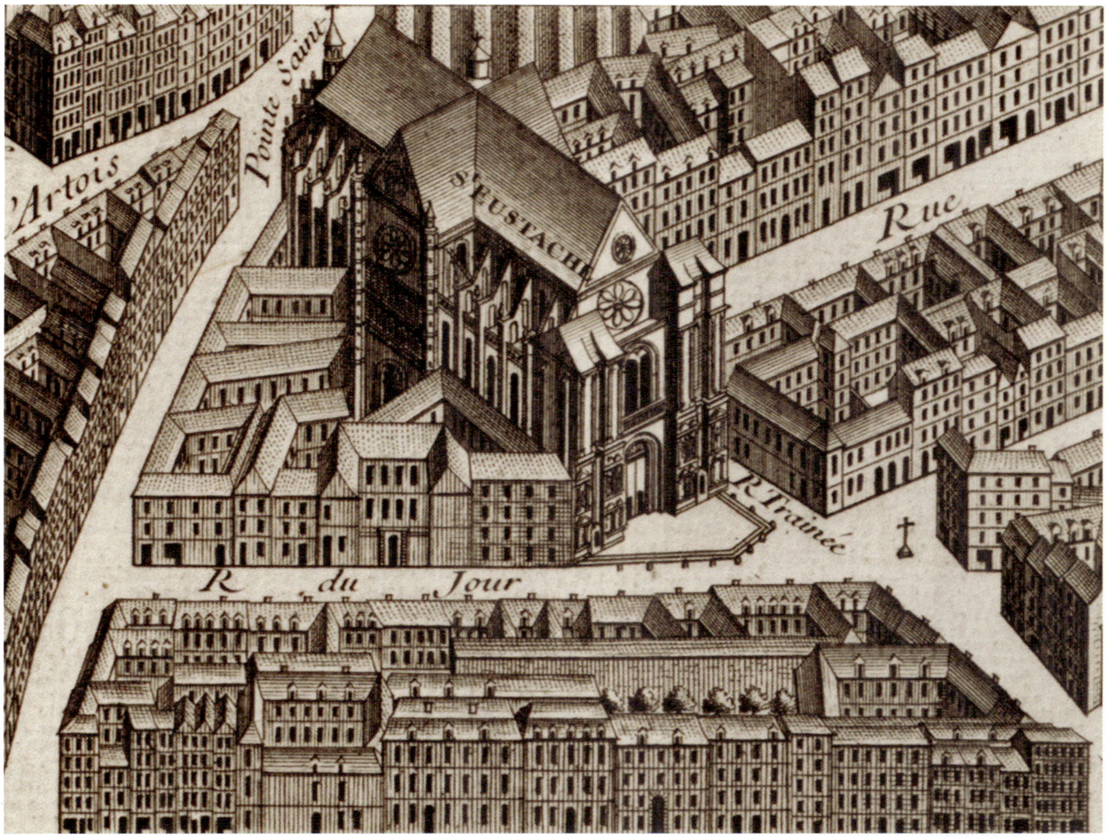

FIGURE 1 View of the Rue du Four, close to the church of Saint-Eustache
FROM TURGOT (1739). COURTESY OF HTTPS://FR.WIKIPEDIA.ORG/WIKI/PLAN_DE_TURGOT

and the Ferme générale as *adjoint*.[10] During the summer he was asked by the Ferme, which controlled (among other things) the tobacco monopoly, to visit several tobacco manufacturers and to analyse the tobacco quality chemically.[11] In December 1771, after he married Marie Anne Pierrette née Paulze, Lavoisier moved not far from the Rue du Four to a newly built house at the corner of the Rue des Bons Enfans et Rue Neuve des Bons Enfans in the proximity of the Palais royal (Figs. 2–3). From a manuscript description of April 1804, when the house was put up for sale by Marie Anne Lavoisier, we know that it consisted of three floors and a loft.[12] According to Bedel, Lavoisier moved in with a chemical laboratory of which we have no further details.[13] In June 1772 Lavoisier dug a hole of 17 feet in his courtyard in order to measure the stratigraphic composition of the soil.[14] At the end of September 1774 he repeated some pneumatic experiments together with Joseph Priestley, who reported the following:

> The experiments, of which an account will be given in this section, were occasioned, in, part, by a hint thrown out by Mr. Bewley, in his letter to me, printed in the Appendix of my former volume; but more immediately by an experiment which I had the pleasure to see at Paris, in the laboratory of Mr. Lavoisier, my excellent fellow-labourer in these inquiries, and to whom, in a variety of respects, the philosophical part of the world has very great obligations.

10 His entering the Académie royale des sciences is an interesting coincidence, because Guettard owned a substantial part of Baron's papers. I recently found among them a manuscript devoted to the increase of weight in calcined metals and a conjecture about the possible cause. At this stage it is impossible to say if and when Lavoisier saw the manuscript, but given that Guettard was his closest acquaintance among the Parisian scientists at the time, it seems a plausible conjecture. See MS Baron (1750–1760).

11 LC, vol. 1, p. 233. Similar analyses were made by Lavoisier about a year later at Charleville and Stenay; LC, vol. 2, pp. 287–289.

12 MS Gabriel (1804); I thank Francesca Antonelli for providing me with this document.

13 "Pour réaliser ses experiences personelles, il fit aménager un laboratoire rue des Bons-Enfants, à Paris". Bedel (1986), p. 651.

14 LO, vol. 5, p. 173.

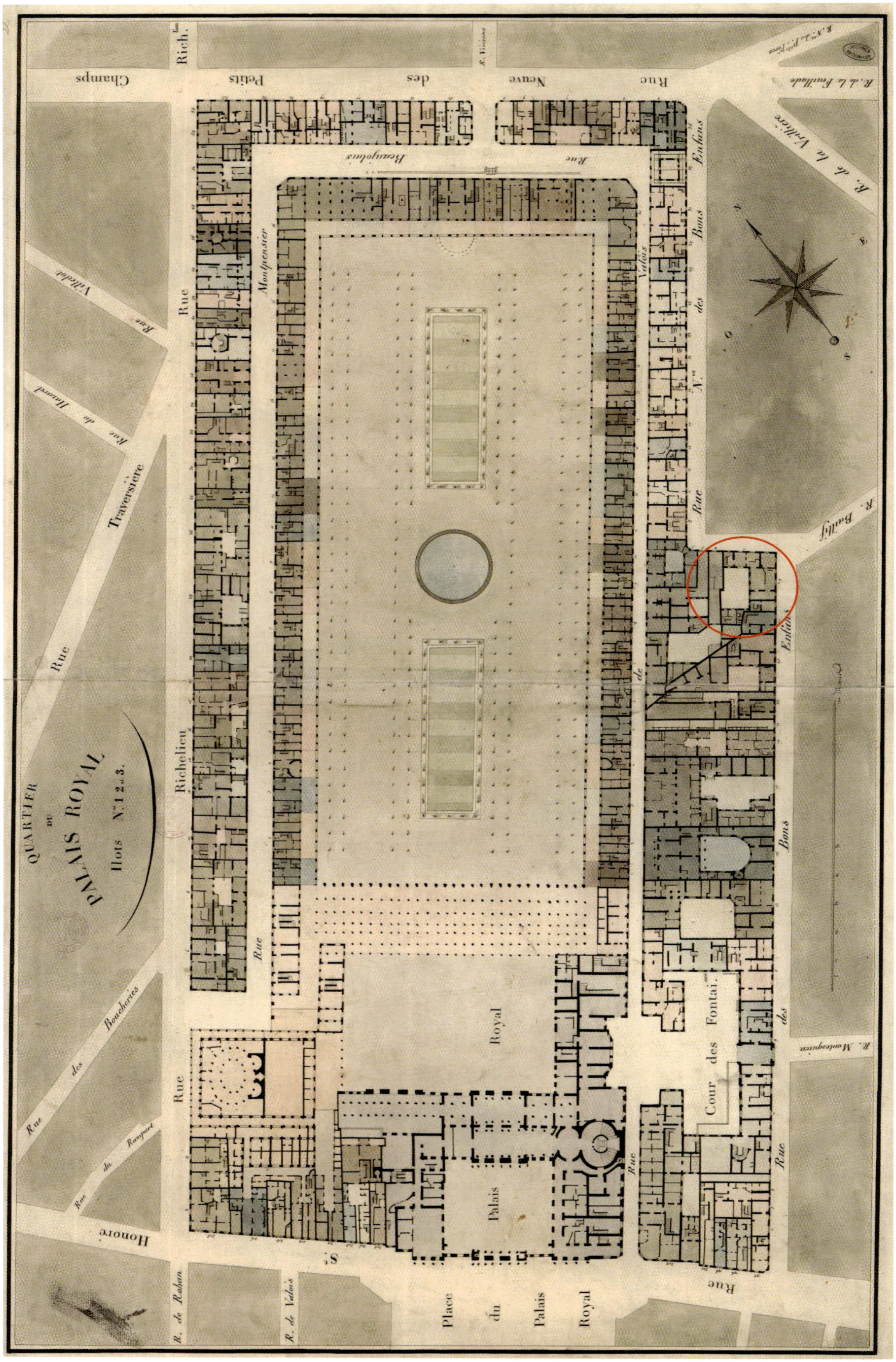

FIGURE 2 Plan of the Rue des Bons Enfants and the Palais royal. The Lavoisiers' residence is indicated with a red dot. Le cadastre de Paris par îlot, dit Atlas Vasserot (1810–1836)
COURTESY OF THE ARCHIVES DE PARIS

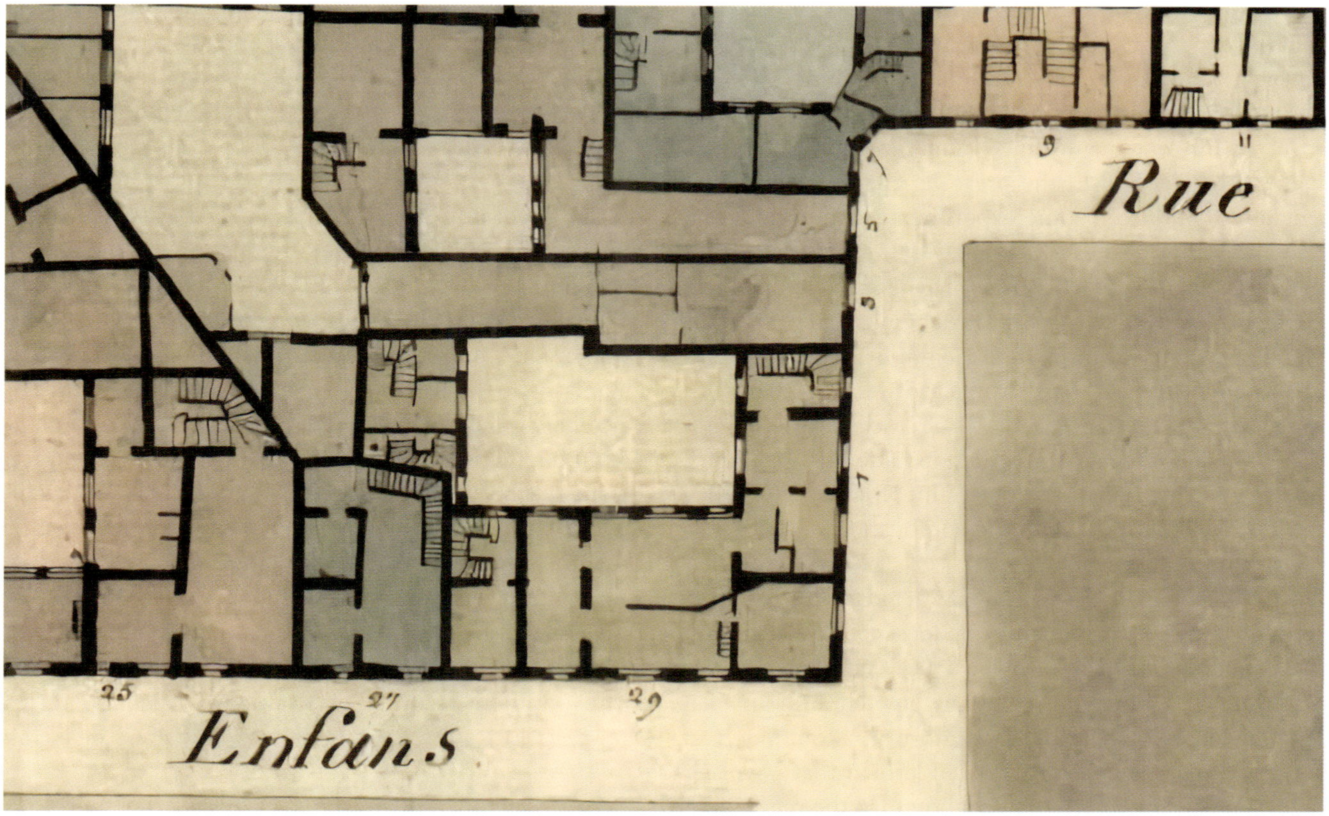

FIGURE 3 (detail of fig. 2). The Lavoisiers' apartment (spread across three floors and a loft) between 1771 and 1776. Le cadastre de Paris par îlot, dit Atlas Vasserot (1810–1836)
COURTESY OF THE ARCHIVES DE PARIS

> Mr. Bewley says, that he had always taken it for granted, that the elastic fluid, generated in the preparation of nitrous ether, without distillation, was fixed air; but that, after seeing the first publication of my papers relating to air, he found, on examination, that it had the general properties of nitrous air.
>
> At Mr. Lavoisier's I saw, with great astonishment, the rapid production of, I believe, near two gallons of air, from a mixture of spirit of nitre and spirit of wine, heated with a pan of charcoal; and when that ingenious philosopher drew this air out of the receiver with a pump, and applied the flame of a candle to the orifice of the tube through which it was conveyed into the open air, it burned with a blue flame; and working the pump pretty vigorously, he made the streams of blue flame extend to a considerable distance. Being very much struck with this experiment, I determined with myself to give particular attention to it, and pursue it after my return to England.[15]

This experiment probably ended up in the dinner party on which Priestley remarked:

> Having made the discovery [of oxygen] sometime before I was in Paris in 1774, I mentioned it at the table to Mr. Lavoisier, when most of the philosophical people of the city were present; saying that it was a kind of air in which a candle burned much better than in common air, but I had not given it any name. At this all the company, Mr. and Madame Lavoisier as much as many, expressed great surprise. I told them that I gotten it from *precipitate per se*, and also from *red lead*. Speaking French very imperfectly, and being little acquainted with the terms of chemistry, I said *plomb rouge*, which was not understood until Mr. Macquer said, I must mean *minium*. Mr. Scheele's discovery was certainly independent of mine, tho' I believe not made quite so early.[16]

15 Priestley (1774–1777), vol. 2, pp. 121–122.

16 Priestley (1803), p. 116.

Priestley's important testimonies are the only known references to chemical experiments performed in Lavoisier's laboratory in the Rue des Bons Enfans.

Since Guettard's cabinet was situated at the Palais royal, within walking distance from Lavoisier's residence, it is (as I suggested in the previous chapter) likely that the latter conducted some experiments in mineralogical chemistry there. However, since Guettard was not interested in pneumatic chemistry, Lavoisier is unlikely to have profited from this traditional chemical laboratory for very long. More sophisticated apparatus was to be found elsewhere. On 25th April 1772 Lavoisier, together with Macquer and the Duc de Croÿ-Solre, joined the laboratory of the apothecary Louis-Claude Cadet de Gassicourt, which was situated in the Rue Saint-Honoré; there they performed experiments on the combustion of diamonds with the aid of a burning lens.[17]

The place where Lavoisier conducted his early experiments in pneumatic chemistry does not seem to have been recorded prior to 14th July 1772, when Jean Charles Philipbert Trudaine de Montigny invited him to visit his newly built chemical laboratory at Montigny, just outside Paris.[18] Although it is well known that Lavoisier spent much time in Trudaine's laboratory, the reasons for their close acquaintance have not usually been explored. Trudaine, the *intendant des finances*, was little more than an amateur in chemistry, but during the early 1770s he was constantly informed of the progress in British pneumatic chemistry by Magellan, who also provided Trudaine's secretary Louis Henri Duchesne with instruments, books, and information.[19] Driven either by curiosity or by perceptiveness about progress in the field, Trudaine invested his resources without hesitation, and his laboratory became a site of experimentation for both Lavoisier and Macquer. Trudaine also invited foreign guests, as once more reported by Joseph Priestley:

> [In October 1774] I was in company with Lord Shelburne at the seat of Mons. Trudaine, at Montigny in France; where, with that generous and liberal spirit by which that nobleman is distinguished, he has a complete apparatus of philosophical instruments, with every other convenience and assistance for pursuing such philosophical inquiries as any of his numerous guests shall chuse to entertain themselves with.[20]

In this laboratory Lavoisier performed crucial experiments on the calcination of metals, on fictitious airs, and on animal respiration, as well as many other different experiments, mostly in pneumatic chemistry. It is, therefore, not surprising that Lavoisier dedicated his first chemistry book, the *Opuscules physiques et Chymiques* of December 1773, to Trudaine de Montigny, (Figs. 4 and 5) whom he portrayed in the preface as the true inspiration for his recent interest in pneumatic chemistry.[21] It is relatively well known that Lavoisier was indebted towards Trudaine in his early career, but it may be more surprising that he continued his important chemical experiments at the laboratory at Montigny, such as those on animal respiration, even after he moved to the Arsenal.[22] In fact, the last experiment recorded at Montigny was conducted in May 1777, just a few months before Trudaine's death on 5th August.

From August 1772 onwards Lavoisier, together with Macquer, Brisson, and Cadet de Gassicourt, performed some 220 experiments on diamonds and other precious stones using Tschirnhaus' burning lens.[23] The lens was kept "dans une salle basse du Louvre" and the experiments were made at the Jardin de l'infante near the Louvre.[24] Between July and October 1774, thanks to the generosity of Trudaine de Montigny, a new and larger burning lens was used for experiments at the Jardin de l'infante, and a new shelter was built to host the instrument (Figs. 6–7).[25] (See also Fig. 33 of the Catalogue on p. 405.)

17 Grouchy, Cottin (1906–1907), vol. 3, p. 19;LO, vol. 2, pp. 38–64.

18 "Je scais votre exactitude sur les details de physique et de chimie […]. Je travail actuellement a devenir votre confrere chimiste. J'ay un laboratoire dont je ne sors pas depuis trois jours qu'il est fini. Si je suivois mon gout je m'y livrerois davantage mais vous êtes jeune et qui avez du tems a vous, je vous exorte a l'employer aussi utilement". Letter from Trudaine to Lavoisier, reproduced in LC, vol. 2, p. 369. The castle was demolished in 1852, and I have been unable to find any records relating to its laboratory.

19 On Trudaine, see Delorme (1950). The correspondence is published in the first volume of Home *et al.* (2017).

20 Priestley (1790), vol. 2, pp. 296–297.

21 Interestingly, the book was published by Prault, who were also Guettard's publisher. The name of the publisher is recorded at the end of the volume (on p. 432) and not on the frontispiece.

22 It was Trudaine who, via his correspondence with Magellan, put Lavoisier in contact with those British naturalists who were engaged with experiments on gas. LO, vol. 2, pp. 174–183.

23 LO, vol. 2, pp. 64–88; LC, vol. 2 , pp. 374–384, and LO, vol. 3, p. 284. Most of the experiments were made between August and October 1772. On this, see Lehman (2013).

24 LO, vol. 2, p. 75.

25 The *Mémoires secrets* report the event as follows: "*10 Juillet*. On construit au jardin de l'infante, au bas de la galerie du Louvre un petit bâtiment pour faire de nouvelles expériences sur la fusion & l'évaporation du diamant. Ce sont messieurs Cadet & Lavoisier, commissaires de l'académie des sciences, qui continueront à y présider. On doit se servir du fameux miroir ardent de M. le régent". "*25 octobre 1774*. M. de Bernieres, ancien contrôleur-général des ponts & chaussées, connu par diverses

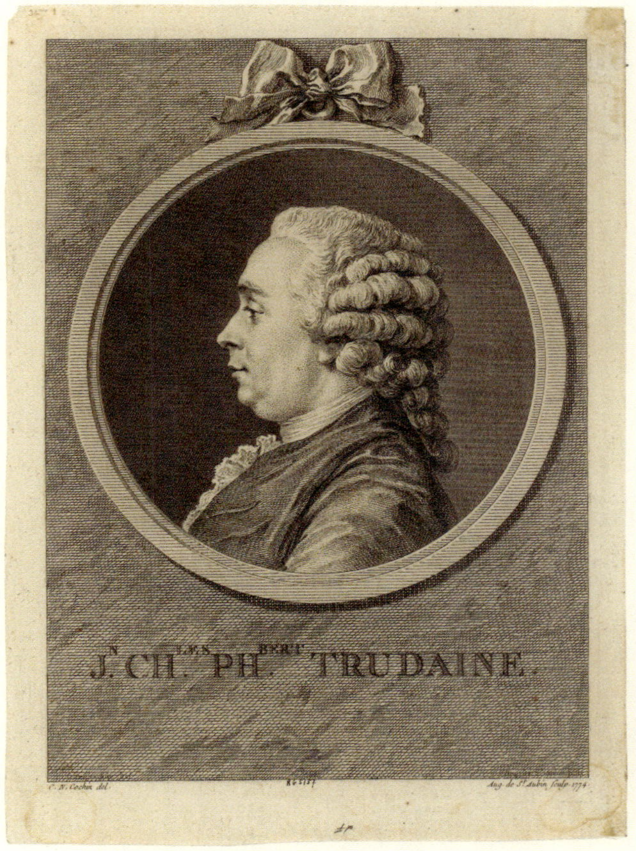

FIGURE 4 Augustin de Saint-Aubin, engraved portrait of Jean Charles Philibert Trudaine de Montigny
COURTESY OF PARIS MUSÉES / MUSÉE CARNAVALET

With his appointment at the Régie des poudres at the beginning of July in 1775 Lavoisier embarked on a new series of experiments on saltpetre. The first recorded experiment in this series was done at the saltpetre manufacture at Essones, outside Paris, but it was only in spring 1776 that he was able to move to his new residence at the Petit Arsenal.[26] From this moment onwards, Lavoisier began to build up a chemical laboratory that would pass for the one of the best equipped in Paris about a decade later. The site of the Arsenal was particularly apt for the creation of a chemical laboratory, because there were deposits of saltpetre and two foundries already situated

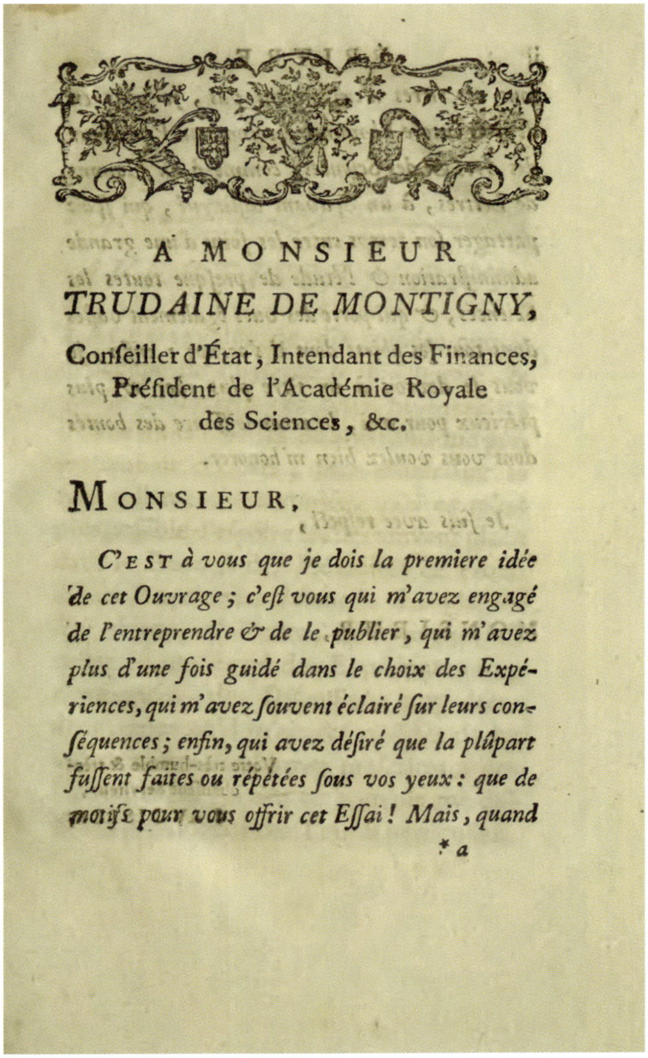

A MONSIEUR
TRUDAINE DE MONTIGNY,
Conſeiller d'État, Intendant des Finances, Préſident de l'Académie Royale des Sciences, &c.

MONSIEUR,

C'EST à vous que je dois la premiere idée de cet Ouvrage; c'eſt vous qui m'avez engagé de l'entreprendre & de le publier, qui m'avez plus d'une fois guidé dans le choix des Expériences, qui m'avez ſouvent éclairé ſur leurs conſéquences; enfin, qui avez déſiré que la plûpart fuſſent faites ou répétées ſous vos yeux: que de motifs pour vous offrir cet Eſſai! Mais, quand

* a

FIGURE 5 Lavoisier's dedicatory epistle to Trudaine de Montigny
FROM LAVOISIER (1774). PRIVATE COLLECTION

there.[27] With the foundation of the Régie in 1775 a refinery and an atelier for the production of saltpetre were also

machines qu'il a imaginées, s'est: propose de faire un miroir ardent, tel que celui qu'on impute à Archimede. Il en a fait l'essai, il y a quelques jours, dans le jardin de l'infante, donnant sur la riviere, & il paroît que les soins n'ont point été infructueux. Il doit incessamment en réitérer les opérations en présence de MM. de l'académie des sciences, auxquels il sera voir la machine". *Mémoires* (1777–1789). See also LO, vol. 3, pp. 274–283.

26 LC, vol. 6, p. 14.

27 "Le petit Arsenal commence par le magazin des armes, rue St. Antoine, & finit à une sort de jardin près de la rivière. Il a, en y comprenant le jardin, 338 toises de long sur 42 de large. Ces Arsenaux, l'un & l'autre voisins de la Bastille, ont été commencés sous le règne & par les ordres de Charles V, dit le sage. C'est un composé de 7 grandes Cours, où il y avoit peu de Bâtiments. On y a fait depuis peu deux belles maisons; des ouvriers travaillent encore aujourd'hui pour en augmenter le nombre. Il y a un magazin à poudre, & 2 fonderies, où l'on travailloit autrefois l'Artillerie; mais depuis que Louis XIV en a transféré la fabrique sur les frontières, elles ne servent que pour la fonte des statues de bronze […]. Le jardin de l'Arsenal est ouvert au public pour la promenade. L'air en est bon, & la vue des plus belles. Il y a pour l'enclos de l'arsenal une Jurisdiction qui connoit des fontes de Canon, des poudres, & de leur façon &c. On rappelle le *Baillage de l'arsenal*". Büsching (1768–1779), vol. 4, pp. 146–147.

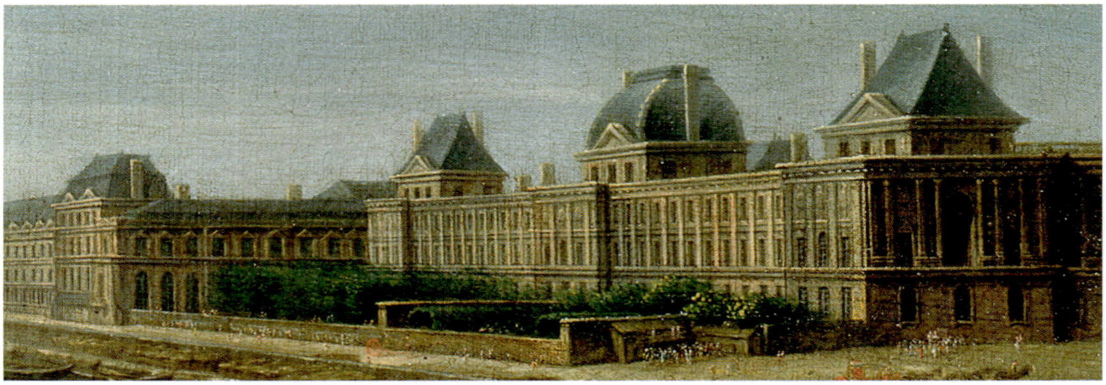

FIGURE 6 The Jardin de l'infante and the Louvre viewed from Pont Neuf. Jean Baptiste Raguenet (1763)
COURTESY OF THE GETTY MUSEUM

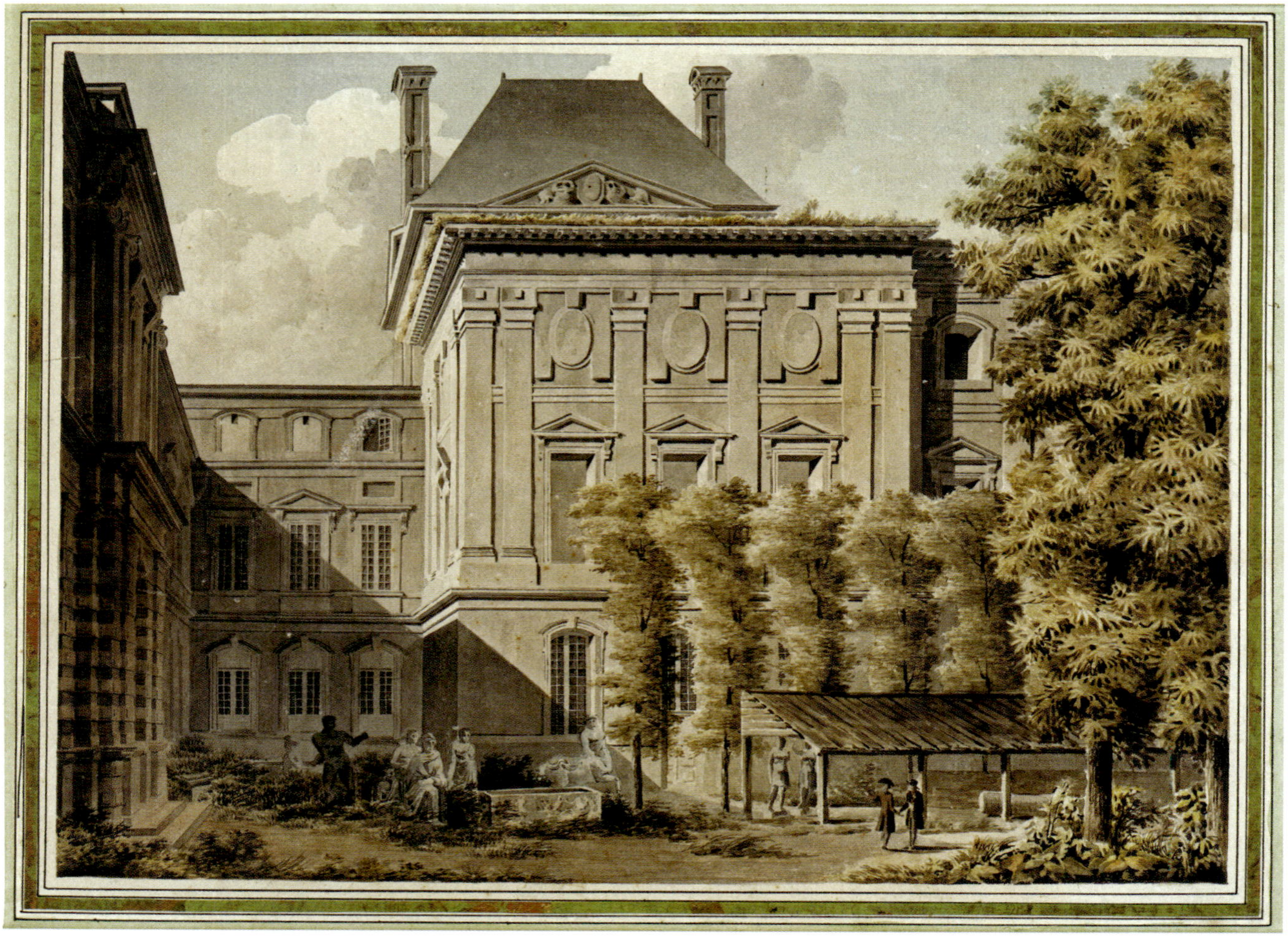

FIGURE 7 The shelter for the large burning mirror at the Jardin the l'Infante was apparently still in its place at the end of the 18th century when the architect Louis-Pierre Baltard made this drawing entitled "Vue de la façade dédoublant la façade de Le Vau au Louvre, côté du jardin de l'Infante."
COURTESY OF PARIS MUSÉES / MUSÉE CARNAVALET

created.[28] Upon a suggestion by Turgot in July 1775 a prize for the best production of saltpetre was created at the Académie royale des sciences. To facilitate the activities of this strategic initiative Turgot also promised to provide the commissioners, who were presided over by Lavoisier, with a large facility where they would be able to repeat the competition participants' experiments. Turgot suggested that the site should either be the Régie des poudres or another establishment yet to be determined. The newly installed Régie des poudres was not available because the whole site was now under renovation, and the matter was too urgent to wait for the completion of the works. In the end, the choice of venue for the commission meetings from October 1775 fell upon a *maison du salpêtre* in faubourg Saint-Denis.[29] This comprised a house, a garden, and a large shelter. The Académie supplied the commission with 1,200 *livres* for the rent and with 1,200 *livres* to cover the costs of experiments.[30] Since the commission's principal aim was the reconstruction of experiments proposed by the prize competition participants, the activities at Saint-Denis were not particularly innovative. This does not mean that they were not important, and it is likely that Lavoisier and other chemists who became actively involved in this innovative research programme, such as Baumé, were inspired by this routine work.

Before I describe the Arsenal in more detail, I briefly want to focus on other sites of experiments used by Lavoisier from the late 1775s onwards. As a supporter of the *physiocrates*, Lavoisier actively engaged in the reform of agriculture.[31] In 1775, after his father's death, he exploited the property he inherited at Bourget for agricultural experiments. Interestingly, on 1st September 1785, this property was adopted, on Lavoisier's suggestion, as the laboratory for the Comité d'administration d'agriculture.[32] In 1778 Lavoisier purchased a large piece of land at Fréchines, between Blois and Vendôme, where he conducted large-scale experiments in different branches of agriculture. He went to Freschines for two or three weeks every year, and thanks to a sizeable investment he managed to double the productivity of his 360 hectares in little more than 10 years.[33] During this period Lavoisier kept accurate records in the *registres de laboratoire*, some of which have survived in the dossiers of the Archives de l'Académie des sciences.[34] We do have record documenting the use of instruments and apparatus at Freschines[35] and it is probable that one of the farm buildings was used as a laboratory (Figs. 8–9).

In the late 1780s and early 1790s these agricultural experiments inspired Lavoisier to explore topics related

28 "Dans le petit Arsenal sont l'hôtel de la Régie des Poudres & Salpêtre, & les bâtimens où se font la fabrication & la raffinerie du salpêtre". Thiéry (1784), p. 95.

29 The commission was formed of Macquer, Sage, Baumé, Trudaine de Montigny, d'Arcy, and Lavoisier. After the death of Trudaine in 1777 Tillet took his place.

30 "L'intention du Roi, suivant la lettre de M. Turgot, du 17 août, étant de procurer aux commissaires de l'Académie un emplacement commode, soit pour y répéter les expériences qui paraîtraient propres à procurer des lumières sur l'origine et sur la composition du salpêtre, les commissaires, louèrent au mois d'octobre suivant, une maison, un jardin et un grand hangar dans le faubourg Saint-Denis; et non seulement M. le Contrôleur général voulut bien approuver la location de 1,200 livres, mais il y joignit encore une somme de 1,200 livres pour subvenir aux frais des expériences". LO, vol. 5, pp. 472–473; on this site see Lehman (2019), pp. 247–252.

31 On this topic I have relied on the valuable study by Lenglen (1936).

32 Ibid., pp. 21–22.

33 A report on this success was presented by Lavoisier before the Comité d'agriculture on 18th September 1787; Lavoisier (1787a). Marie Anne Lavoisier added that Lavoisier "introduisit les pommes de terres, inconnues avant lui. Les prairies artificielles furent etablies ou elles ne l'avaient jamais été. Des troupeaux de belles especes meublerent la ferme. Des registres de la culture de chaque piece de terre, et de leur produit, furent tenus exactement". Gillispie (1958), p. 59.

34 MS Lavoisier (1781–1782). These books only cover the period from 1781 to 1787.

35 A revolutionary inventory compiled on September 5, 1795 we read: "Quatre baromètres : le premier portant le n° 14, donné par Magnié [Mégnié] à Lavoisier, monté partie en cuivre, partie en bois d'acajou, encastré dans 4 pieds de bois dont la hauteur est pareillement garnie de cuivre ; le second en cadran, au-dessus, un thermomètre à l'esprit-de-vin ; le troisième de Réaumur, à côté un thermomètre à l'esprit-de vin, l'un encastré dans l'autre à l'aide de charnières ; le quatrième, de Réaumur aussi, endommagé; ces deux derniers n'ont rien de précieux. 4° Un trébuchet renfermant des poids propres à peser l'or (quelques pièces manquent). 5° Des petites balances de cuivre. 6° Deux romaines rondes travaillées en Angleterre : une grande, l'autre petite, dont le contour est en cuivre. 7° 7 cahiers de papier ordinaire et 6 à lettre. 8° 4 flacons vides : deux grands et deux petits, 6 vases ronds en verre ordinaire vides : trois grands et trois fort petits. 9° 6 bocaux dont 2 montés sur pied. 10° Un flacon d'alcali fixe caustique. 11° 2 flacons d'acide vitriolique concentré. 12° Deux flacons d'eau-forte. 13° Un flacon d'une liqueur qui nous est inconnue. 14° Un petit mortier en métal. 15° Un alambic de métal, le dessous seulement, avec son serpentin et un vase de métal à deux anses. 17° Une fontaine très curieuse en bois, piédestal peint en couleur de marbre gris uni, le corps peint couleur de maron, le vase au-dessus peint en marbre gris, orné dans le milieu d'une tête dans un cadre sculpté et peinte en bronze, le reste est sculpté en gouttes d'eau. 18° Une pompe à incendie sans tuyau. 19° Deux petits moufles en cuivre. 20° Un pied de niveau garni en cuivre, incomplet." Cauchie (1900), pp. 64–65.

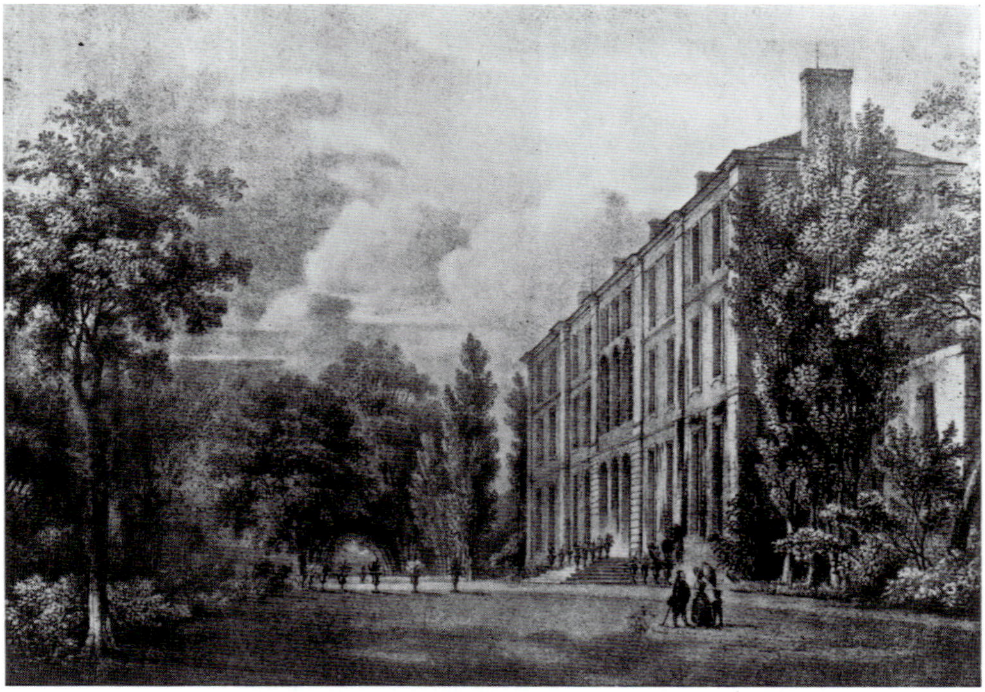

FIGURE 8
Lavoisier's château at Freschines
FROM DUVEEN (1965)

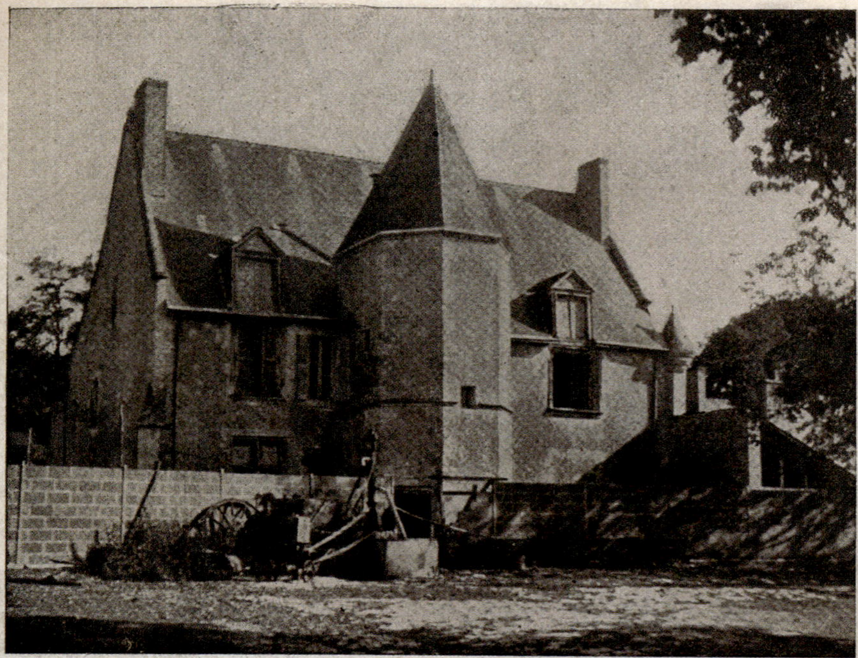

Le vieux château de Freschines, où Lavoisier avait établi son laboratoire.

FIGURE 9
Lavoisier's chemical laboratory at Freschines
FROM WEELEN (1943)

to vegetable and animal chemistry. In the *Programme du prix proposé par l'Académie sur la nutrition, pour 1794*, which was published in 1792, Lavoisier announced that a commission of five members of the Académie, assisted by Hassenfratz and Séguin, was ready for a series of experiments on vegetable chemistry in a room at the Jardin des plantes.[36] Unfortunately, the fate of these experiments is unknown.

36 The Académie "a nommé une commission de 5 de ses membres, auxquels ont été adjoints MM. Hassenfratz et Séguin. Un local est disposé au Jardin des plantes et les commissaires vont

Lavoisier carried out several experiments as a referee or inspector on behalf of the Académie des sciences. In June 1778 he visited the Mines de Poullaouen et d'Huelgoat, where he assisted with metallurgical experiments, and he conducted experiments on vegetable ashes in Baumé's and Sage's laboratories in August 1779.[37] Sometime in 1780, together with Dietrich, he collected inflammable air in the Fossés de la Bastille with a special apparatus; in March 1780 he visited and measured the quality of air in the prisons of the Grand et Petit Châtelet, For-l'Evêque, and the Conciergerie; in 1781 he experimented with a new method of lighting in the Salon des tableaux at the Louvre, and in September 1783 he organised a demonstration of Montgolfier's balloon in the Place royale;[38] on 27 March 1784 he conducted experiments on gold and silver at the Hôtel des Monnaies, and on animal magnetism at Benjamin Franklin's residence in Passy.[39] In April of the same year, with Séguin, he collected eudiometric data at the Comédie française, at the Tuileries, and in the Salle de l'Académie française at the Louvre;[40] at the beginning of 1785 he experimented on cider at the Château de Fernes; in December of the same year he conducted eudiometric and pneumatic experiments in the Hôpital de la Charité, at the hospice de Saint-Sulpice, and in the infirmaries de la Salpêtrière and of the Invalides;[41] at about the same time, together with Monge, he replicated his experiments on the decomposition of water in the Président du parlament Jean Baptiste Gaspard Bochart de Saron's chemical laboratory; another replication of the experiment took place in Lefebvre de Gineau's laboratory at the Collège royal in May 1788, and a third one, following the experiments proposed by the Dutch natural philosopher Paets van Troostwijk, in the *cabinet* of Charles, which was situated in the beautiful Apollon gallery of the Louvre, in May and June 1791.[42] Lavoisier assisted as an expert with the attempt to convert peat into coal at the Trianon in 1786;[43] in March of the same year he was at Courbevoie, near Paris, to experiment on a fluid supposedly effective in extinguishing fires, and in November at the manufacture of Saint-Gobain, to experiment on the cupellation of platinum.[44] Together with Vandermonde, Monge, Berthollet, and Fourcroy, Lavoisier inspected the Creusot coke furnaces and glassworks at Moncenis (Figs. 10–11).[45] In December 1787 he was in the Rue des Anglais to test the so-called "pompes anti-méphitique".[46] In August 1788 he was at the opera to try a substitution of gunpowder with hydrogen in fireworks.[47] In April 1789 he surveyed Paris' slaughterhouses.[48] In 1790 he conducted experiments on the dilatation of metals at the Jardin de l'Arsenal.[49] The same set of experiments was resumed in May 1793 in the garden of Lavoisier's new residence on the Boulevard de la Madeleine.[50] The intense activities of the chemistry section at the Académie prompted Lavoisier to write a letter to the Ministry of Interior Roland on 30th December 1792, requesting funds for the construction of a chemical laboratory.[51] Even without a laboratory the members of the chemistry sessions did not confine themselves to presenting their experimental work during the *séances publiques*. Lavoisier presented numerous experiments during

entreprendre une suite d'expériences sur la végétation". Cited in Lenglen (1936), p. 98. The text of the prix published in Lavoisier's *Œuvres*, does not correspond to that published by Lenglen.

37 On the 1778 experiments at the Mines de Poullaouen et d'Huelgoat, see LO, vol. 5, pp. 179–85. On the 1779 experiments at Baumé's and Sage's laboratories, see LO, vol. 4, pp. 328–338.

38 For the experiments at the Louvre, see LO, vol. 3, pp. 91–102. On the demonstration of Montgolfier's balloon, see LC, vol. 3, p. 742.

39 The experiments on gold and silver can be found in MS Lavoisier (1772–1788), vol. 7, fol. 82. The house in Passy no longer exists.

40 LO, vol. 2, pp. 676–687.

41 LO, vol. 3, pp. 536–561 (Hôpital de la Charité). LO, vol. 3, pp. 602–668 (infirmary de la Salpêtrière and infirmary of the Invalids).

42 On the 1788 experiments, see LC, vol. 5, p. 250. For 1791, see Chaldecott (1968), p. 27.

43 LO, vol. 4, pp. 462–467. The botanical garden at Trianon had been created by Bernard de Jussieu.

44 See LO, vol. 4, pp. 426–430 and LO, vol. 5, pp. 340–345, respectively.

45 LC, vol. 5, pp. 87–92.

46 LC, vol. 5, p. 8.

47 LC, vol. 5, pp. 203–205.

48 LO, vol. 3, pp. 579–601.

49 "Le grand et important travail des nouvelles mesures à établir dans toute la république, occupa l'académie dès 1790 mais personne n'y a pris plus de peine, et n'y a été plus utile que Lavoisier. La dilatation des métaux, par la chaleur, étoit une chose importante et qui n'étoit point assez connue. Il fit construire dans le jardin de l'Arsenal, un appareil où les règles de métal, plongées dans l'eau et soumises à divers degrés de chaleur, faisoient mouvoir une lunette qui marquoit sur un objet éloigné, les moindres degrés de dilatation". Lalande (1795), p. 185.

50 "Les expériences pour la dilatation relative du cuivre et du platine se firent l'année suivante, du 24 mai au 5 juin, dans le jardin de la maison que M. Lavoisier occupoit alors sur le boulevard de la Nouvelle-Madeleine, et les bornes qu'on y avoit solidement établies pour cet objet ont subsisté tant qu'on a cru que leur conservation pourroit être utile. On verra tous les détails de ces diverses expériences dans deux mémoires de Borda". Delambre (1806–1810), vol. 1, p. 21.

51 "Un laboratoire de chimie est une pièce essentielle pour l'Académie parce qu'elle a journellement des expériences à répéter, souvent même pour satisfaire à des décrets de l'Assemblée n[ation]ale et qu'il est important qu'elle puisse s'en occuper dans un local voisin du lieu de ses séances". LC, vol. 7, pp. 181–182. Roland replied to Lavoisier on January 14, 1793 the following: "Quant à la lettre que vous m'avez adressée dans le même temps au sujet du local que vous m'annoncez avoir été accordé à l'Académie pour y établir son laboratoire de chymie auprès des salles destinées à ses assemblées, j'ai engagé le C. Heurtier inspecteur des Maisons nationales à le voir & à m'en rendre compte pour que je puisse statuer sur la demande de l'Académie". LC, vol. 7, pp. 195–196.

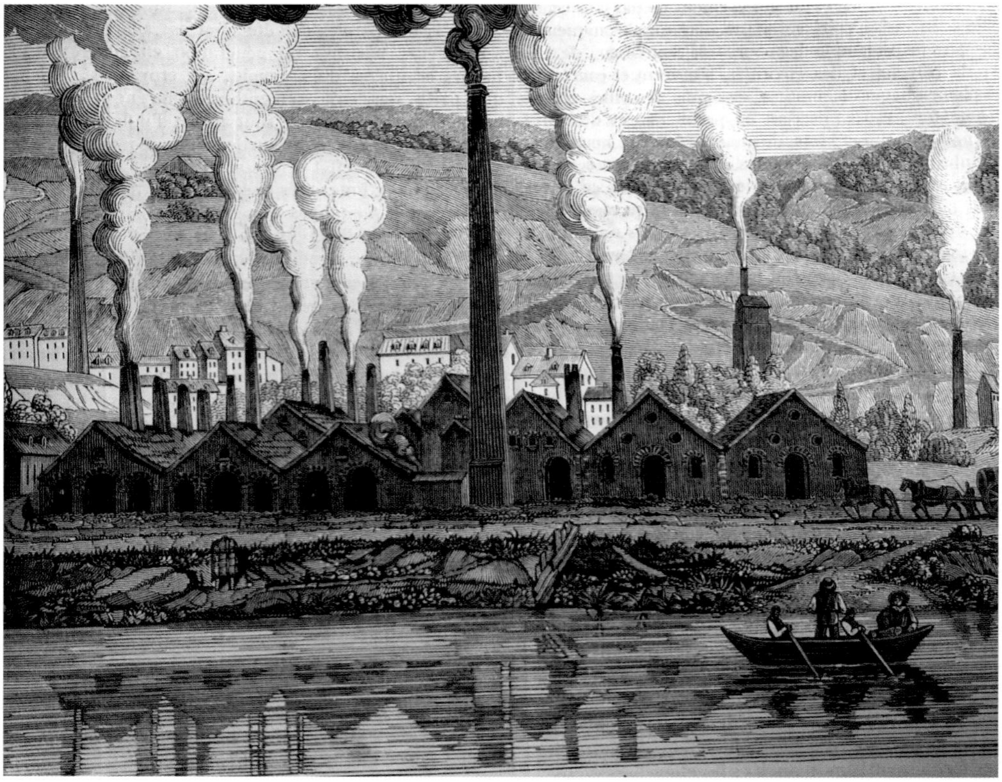

FIGURE 10
The chemical works at Le Creusot in 1827
FROM THE *MAGAZIN PITTORESQUE* (1834) P. 229

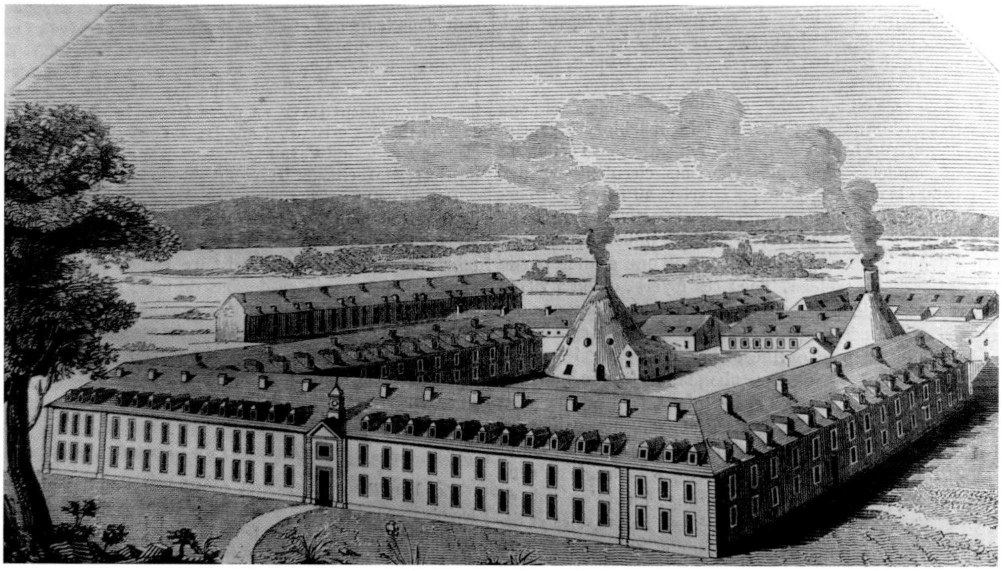

FIGURE 11
The glass and crystal work at Le Creusot
FROM THE *MAGAZIN PITTORESQUE* (1834) P. 228

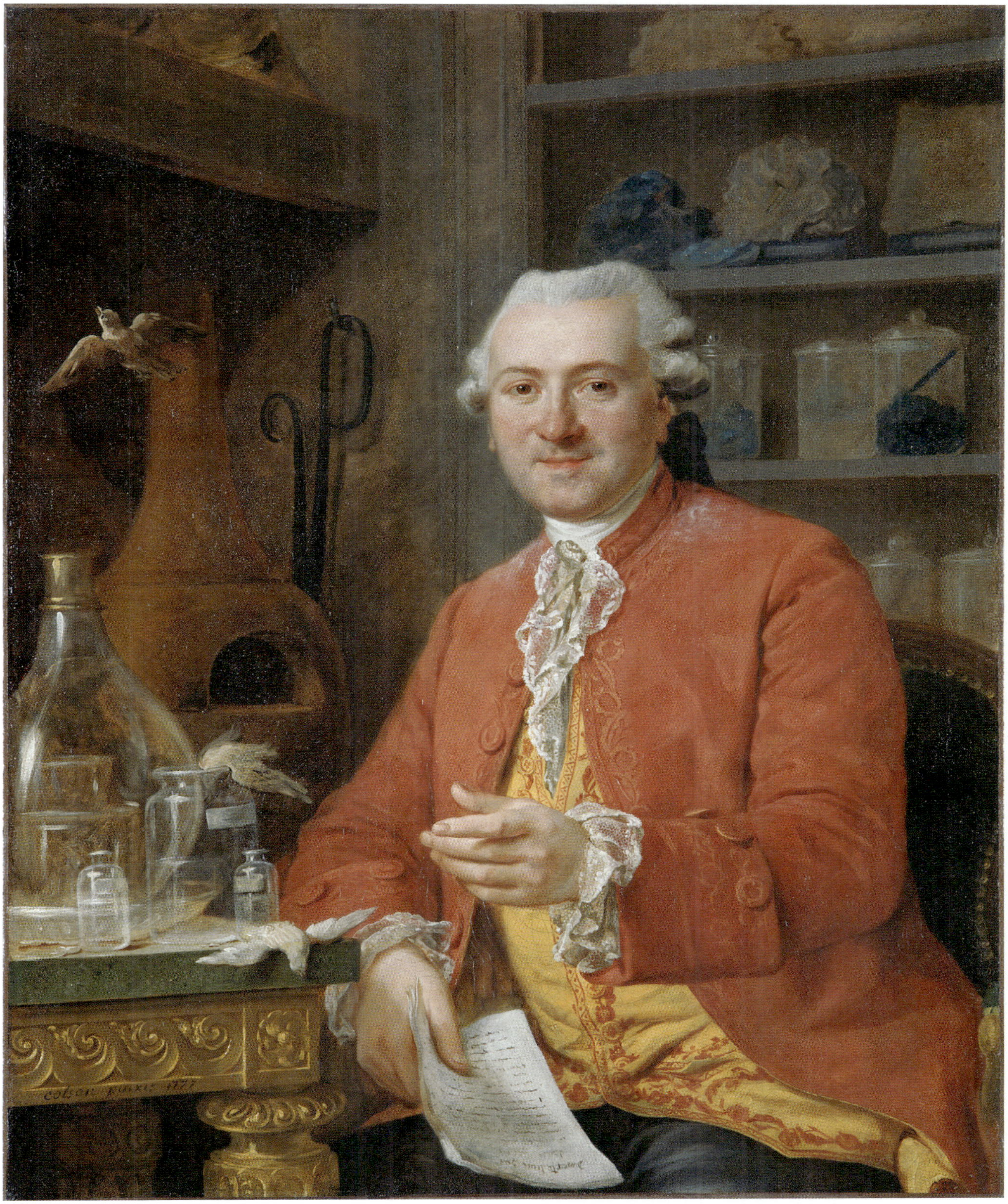

FIGURE 12 Portrait of Balthasard Sage made by Jean-François Colson in 1777. This is an exceptionally interesting painting, because commemorates Lavoisier's public experiment on the respiration of birds and fixed air, performed before the Académie des sciences in May 1777, and their 'presumed resuscitation', which was made possible by Sage. One bird is breathing air from a jar, presumably filled with fixed air, while a second, apparently dead, is lying on the desk, and a third is flying in the air. This was the unexpected outcome, according to Sage, of Lavoisier's public demonstration.
COURTESY OF THE COLLECTIONS DU MUSÉE DES BEAUX-ARTS DE DIJON/FRANÇOIS JAY

the public sessions of the Académie. On 10 May 1777, in the presence of the Comte de Falkenstein *alias* Emperor Joseph II, Lavoisier experimented on the respiration of animals and fixed air (Fig. 12).[52] On 5 June 1782 Lavoisier conducted experiments on the fusion of platinum in the presence of the Grand Duchess of Russia;[53] on 24 January 1784 he presented before the Académie an apparatus of his own invention for the collection of gases during the experiments on the composition of water;[54] and this was followed on 8 March 1788 Meusnier de la Place's the presentation on the two gasometers.[55] But the most important public experiments were probably those on human respiration, for which Lavoisier collaborated with Séguin, and which were presented before the Académie on 7 November 1790.[56]

As a manager of the Régie Lavoisier visited several parts of France where he performed experiments on the nature of saltpetre.[57] Sometimes Lavoisier was inspired by external circumstances to conduct unplanned experiments. This was the case with the electrical experiments on the torpedo fish. While stopping at La Rochelle during a trip commissioned by the Régie des poudres, Lavoisier reported on the different intensity of the shocks emanated by different parts of the body of the torpedo. While holding one hand to the lower surface of the fish, Lavoisier touched the upper part with a piece of iron that he held in the other hand, hoping to cause an electrical shock via this physical contact.[58] Between 1783 and 1791 Lavoisier used the refinery of the Arsenal rather than his own laboratory for the assay of saltpetre.[59] This refinery became quite important, because it provided the manufactory of Essonnes – where on October 27, 1788 Lavoisier and Berthollet conducted their famous and tragic experiment on the explosive effects of a new powder with potassium chlorate.[60]

From the early 1780s onwards Lavoisier had, in addition to the laboratory at the Arsenal, another laboratory situated in his cousin's apartment at the Palais royal.[61] No details concerning the kind of experiments conducted there survive, but given its vicinity to the Académie and to Baumé's and Cadet de Gassicourt's chemical laboratories it was probably used by Lavoisier as a more convenient place for his academic meetings.

3.2 The Arsenal

Lavoisier's apprenticeship with Guettard and Nollet must have played an important role in making him at home in so many and different sites for his experiments. Amidst

52 This is the interesting report given by Sage: "La fermentation vineuse produit un acide méphitique, sur lequel M. Sage a fait des expériences qui méritent d'être rapportées dans ce dictionnaire. Voici ce qu'il raconte lui-même. Le 10 mai 1777, M. le comte de Falkenstein (l'empereur) s'étant rendu â l'académie des sciences, M. Lavoisier répéta en sa présence quelques-unes des expériences du docteur Priestley sur l'air fixe. Il mit un moineau dans un bocal, où à peine eut-il versé de l'air fixe, qu'on vit l'oiseau s'agiter, & un instant après tomber sur le côté. M. Lavoisier le retira du bocal, & le présenta pour mort à M. le comte de Falkenstein. Ayant demandé cet oiseau, je versai dans le creux de ma main environ un gros d'alkali volatil-fluor, & j'y posai le bec de l'animal: je le mis sur la table au premier signe de mouvement qu'il me donna; mais à peine eut-il étendu ses ailes, qu'il retomba: je le présentai de nouveau & de la même manière à l'alkali volatil, qui acheva de produire son effet. L'animal eut alors assez de force pour se tenir sur ses pattes; il marcha, battit des ailes & s'envola; on fit ouvrir les fenêtres, & le petit ressuscité partit à tire d'ailes". Article "vin" by Balthazard Sage in *Encyclopédie* (1778), vol. 35, pp. 487–491 on p. 487.

53 LO, vol. 2, pp. 413–431, and LC, vol. 4, p. 266.

54 At the Académie "M. Lavoisier a fait voir une machine destinée à employer l'air déphlogistiqué dans les expériences de chimie et à mesurer la quantité d'air qu'elle dépense et à la graduer à volonté". MS Procès-verbaux, 1784, fol. 17.

55 Presentation by Meusnier of "deux machines pour les expériences sur les airs". MS Procès-verbaux, 1788, fol. 62.

56 At the Académie "MM. Lavoisier et Seguin ont fait des expériences sur la respiration humaine et celle des animaux [...]. M. Lavoisier a lu un Mémoire sur la respiration des animaux". Continued on December 7 and 11. MS Procès-verbaux, 1790, fol. 235. The presentations continued at least until 4 May 1791, when Lavoisier presented "a machine of oiled silk to contain a man. A mask for breathing joined by tubes to a system of glass vessels". Chadelcott (1968), p. 24.

57 In 1777 he visited the manufactories of Languedoc, where he experimented on the ashes used in the fabrication of saltpetre. LO, vol. 2, pp. 166–173. He travelled to Roche-Guyon, the Montagnes de Saint-Chamas, and Miramas in spring 1778 and there analysed different kinds of native saltpetre. LO, vol. 5, p. 563, and MS Lavoisier (1772–1788), vol. 6, fols. 50–51.

58 On this episode see Beretta (2001b), pp. 39–40.

59 LO, vol. 5, pp. 648–49.

60 "La raffinerie de l'A. de Paris fournit de salpetre la fabrique d'Essonne"; LC, vol. 6, p. 83. On the tragic experiment, see LO, vol. 5, pp. 741–745.

61 This is first mentioned in March 1782, when Lavoisier met with Laplace: LC, vol. 3, p. 714. Lavoisier's cousin's apartment belonged to Clément Augez de Villers, who lived in the Rue Croix des Petit-Champs, parallel to the Rue des Bons Enfants and close to the Palais royal; an interesting note by Lavoisier's colleague at the Académie Auguste Denis Fougeroux de Bondaroy dating August 6, 1789, reports the following "M. Lavoisier ayant été déclaré non couplable s'étoit retiré, avoit passé sans être connu par la porte sous l'arcade Saint-Jean et avoit gagné un appartement près le Palais-Royal, qu'il a loué pour recevoir les mercredi et samedi ses confrères de l'Académie, y parler science et y faire des expériences relatives à la chimie ou physique". Scheler (1960), vol. 2, p. 127. See also Viel (1995a) and LC, vol. 6, pp. 59–61.

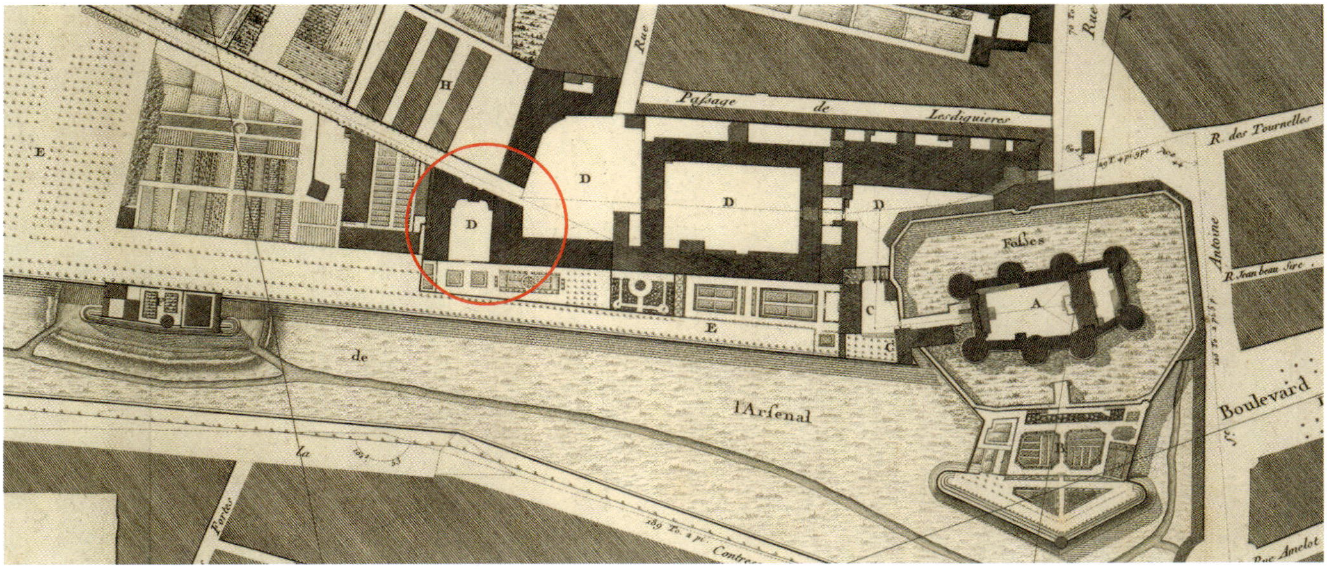

FIGURE 13 Map of the Arsenal and (on the right) the Bastille. Ca. 1789
COURTESY OF THE BIBLIOTHÈQUE HISTORIQUE DE LA VILLE DE PARIS

this remarkable versatility, the laboratory at the Arsenal became Lavoisier's centre of chemical experimentation.

The Arsenal in Paris consisted of a large number of buildings close to the Bastille with a long history. As early as the fourteenth century several warehouses were used as a depot for weapons; in 1533, by order of King François I, one of these warehouses was transformed into a forge for the production of cannons; during the seventeenth century the area expanded considerably, and during the eighteenth century it became an important centre for the production, refinery, and storage of saltpetre.[62]

Since Lavoisier's residence was not a private property, all efforts to date to find the building plan for the period from 1776 to 1791 in French archives have been fruitless. As early as 1788 some of the buildings of the Petit Arsenal were already abandoned by the Régie, and by 1806 the whole headquarter was renewed (Fig. 13);[63] by 1809 the only remaining buildings were the general headquarters of the Régie – presumably Lavoisier's residence – and the refinery.[64] The original buildings of the Petit Arsenal were destroyed during the Franco-Prussian war of 1870, and no information has survived concerning the size and location of Lavoisier's residence and laboratory (Fig. 14). It is even difficult to determine with absolute certainly which building constituted Lavoisier's residence. At the Bibliothèque historique de la ville de Paris a manuscript map of the Arsenal, probably dating from around 1780, indicates with precision the building of the Régie des poudres (Fig. 15);[65]

62 Several interesting histories of the Arsenal have been written, but references to the Régie des poudres and its period of activity are quite scarce: see De la Cité (1902); Batiffol (1929); Babelon (1970); Faure (2002).

63 "L'Arsenal a été supprimé en 1788. D'après l'ordonnance de suppression, on devait l'abattre pour élever sur son emplacement un nouveau quartier; mais ce projet ne reçut point d'exécution. Seulement, en 1806, le boulevard Bourdon remplaça le jardin; on forma de l'esplanade (l'ancien Mail), qui suivait le bord de la rivière, le quai du Mail, aujourd'hui quai Morland; une grande de partie de ce qu'on nommait le Petit Arsenal servit à la construction de la rue Neuve de la Cerisaie. Enfin, en 1807, on éleva sur ce qui restait du jardin, et le long du boulevard, un vaste édifice connu sous le nom de Grenier d'Abondance, Grenier de Réserve. Comme grand-maître de l'artillerie, Sully logeait à l'Arsenal. C'est en mémoire de ce grand homme que la rue où se trouve l'entrée du bâtiment principal a reçu le nom de rue de Sully. Ce qui reste des anciens bâtiments est occupé aujourd'hui par la belle bibliothèque publique appelée Bibliothèque de l'Arsenal". De Gaulle (1839), p. 429.

64 "Le petit Arsenal a été démoli en grande partie, pour l'ouverture d'une autre rue qui donne également sur le boulevard, et qu'on nomme rue Neuve de-la-Cerisaie. Les deux pavillons encore existants sont occupés, l'un par l'administration générale, l'autre par la raffinerie des salpêtres". Bins (1809) p. 527.

65 MS Plan Plan de l'Arsenal (1780); the map was certainly drawn sometime between spring 1776, when the time Régie des poudres moved into its headquarters, and the fall of the Bastille in 1789. The bibliographical description of the map reads: "Plan du secteur compris entre la forteresse de la Bastille, la rue Saint-Antoine, celle du Petit-Musc, l'île Louviers et le grand fossé de l'Arsenal (actuel boulevard Bourdon). On remarque les bâtiments du Petit-Arsenal, occupé par la Régie des poudres. On y voit la cour du salpêtre qui communique avec la rue du Petit-Musc par la rue de Cerisaie, et avec la rue Saint-Antoine, par celle de Lesdiguières. Une allée plantée de deux rangées d'arbres, parallèle au fossé de l'Arsenal, relie cet ensemble au Grand-Arsenal, situé le long de la Seine et comprenant principalement l'hôtel

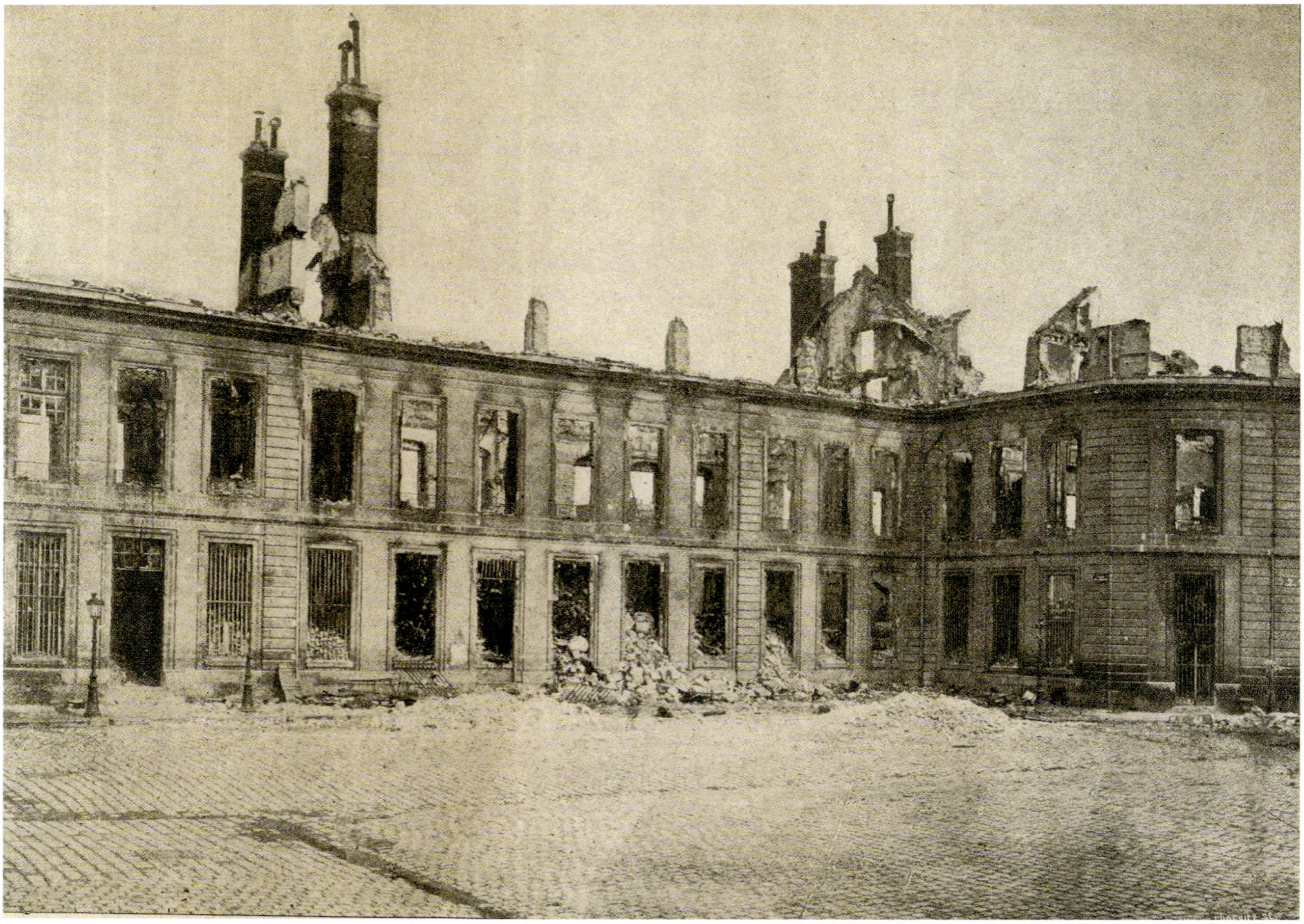

FIGURE 14 The building of the Arsenal after the Franco-Prussian war in 1870
PRIVATE COLLECTION

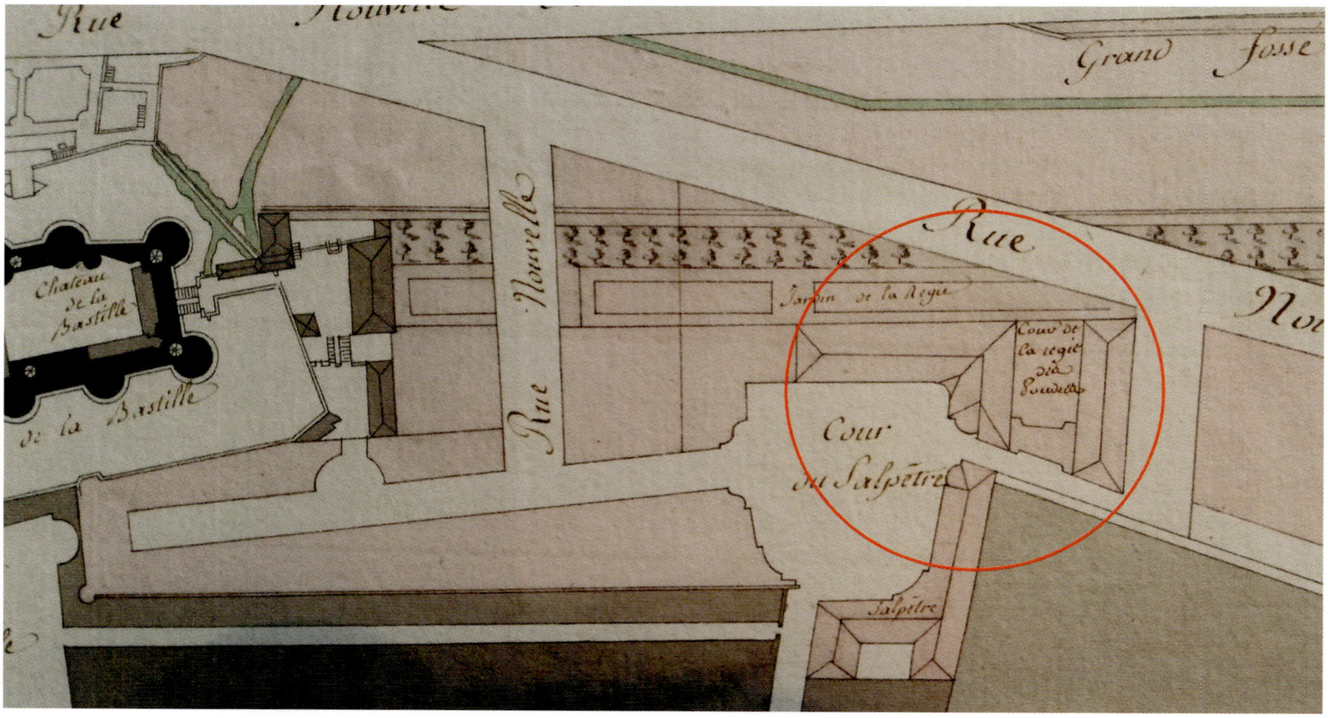

FIGURE 15 Map of the Arsenal. Ca. 1780
COURTESY OF THE BIBLIOTHÈQUE HISTORIQUE DE LA VILLE DE PARIS

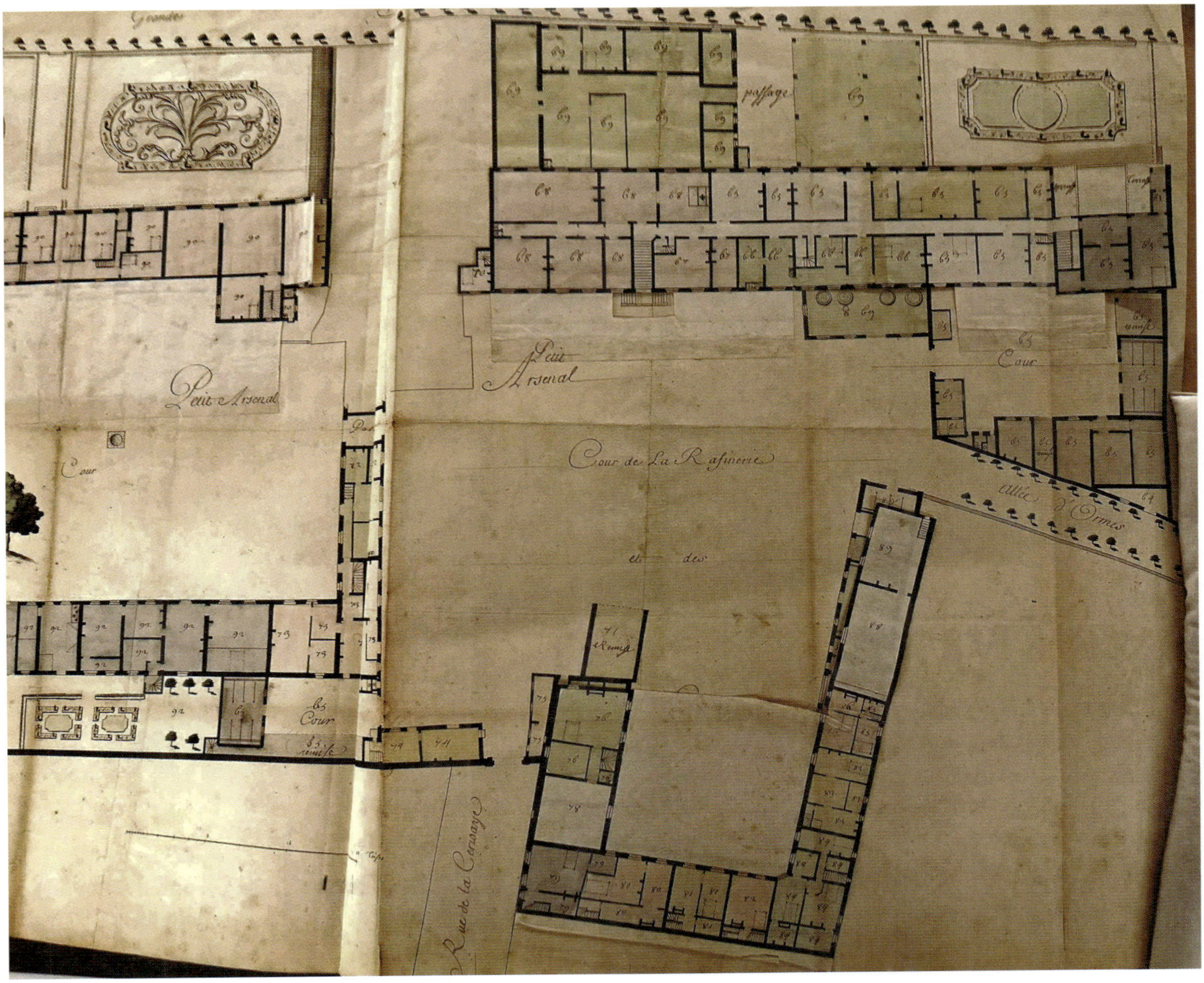

FIGURE 16 The Petit Arsenal before the renovation ordered by Turgot in 1775
COURTESY OF THE BIBLIOTHÈQUE NATIONALE DE FRANCE

but while it is possible, as suggested by Belhoste, that this was the location of Lavoisier's residence, the fact that some of the other *régisseurs* appointed by Turgot made the Petit Arsenal their home leaves this question open for speculation.[66]

The Bibliothéque Nationale de France holds a detailed plan of the Petit Arsenal, almost certainly dating from before 1776, but while this document is a very useful map detailing the boundaries of the whole area, it neither shows the location of the laboratories nor of Lavoisier's later residence.[67] (Fig. 16) The Comte d'Argenson commissioned a manuscript map of the area surrounding the Arsenal (including the Petit Arsenal) in 1756 (now at the Bibliothéque de l'Arsenal), and this provides useful

et les jardins du ministre de la Guerre (prince de Montbarrey), ainsi que l'hôtel occupé par la comtesse d'Amblimont et le marquis de Paulmy. Ce secteur est traversé par plusieurs 'rues nouvelles': il s'agit d'un projet d'aménagement, dans lequel on peut retrouver le tracé du boulevard Bourdon actuel, celui de la rue de Sully et d'une partie de la rue de l'Arsenal".

66 Belhoste (2011), p. 49. Lavoisier's colleagues were Le Faucheux (ca. 1720–1792), a member of the old administration; Barbaut de Glatigny (1726–1783), who was responsible for the finance and administration, and replaced in the 1780s by Le Tors (ca. 1748–1788); and Jean-Baptiste-Paul-Antoine Clouet (1739–1816), whose expertise was both technical and administrative. See Bret (2004). According to the *Almanach royal* all but one had their residence at the Arsenal: Barbault de Glatigny lived in the Rue des Jeûneurs. Clouet was (like Champy later on) probably housed at the refinery, Cour des salpêtres. Le Faucheux and his son Le Faucheux des Aunois lived, like Lavoisier, at the Hôtel de la Régie. The residence of Le Tors, who replaced Glatigny until his own accidental death at Essonnes in 1788, is unknown. I thank Patrice Bret for providing me the with this information.

67 MS Plan du Petit Arsenal (ca. 1750).

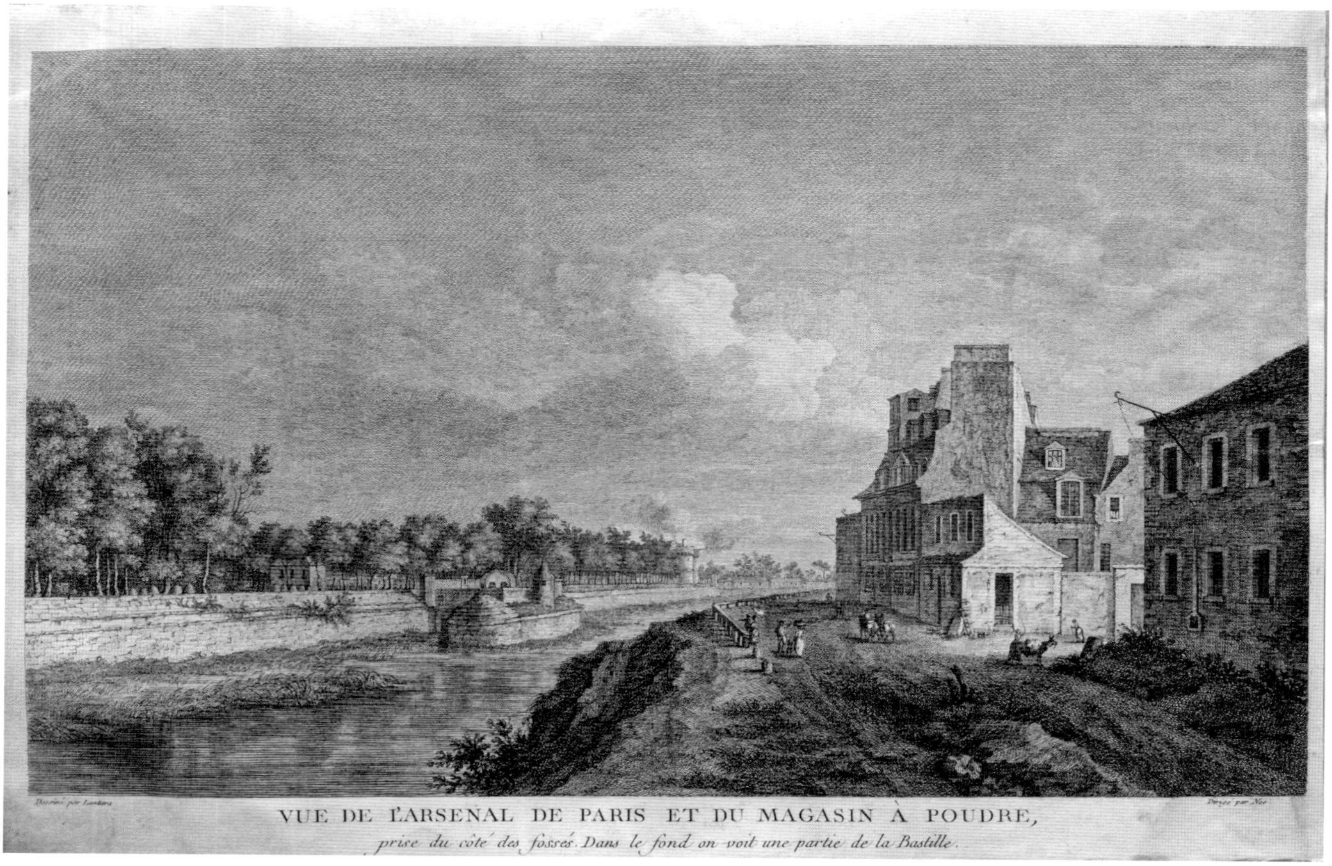

FIGURE 17 View of the Fossés de l'Arsenal and, on the left, a gunpowder warehouse. Ca. 1770
PRIVATE COLLECTION

visual information on the activities before the Turgot's re-organisation as well as the "logement du commissaire general des poudres", which was eventually renovated in 1776 in preparation for its becoming the residence of the Régisseurs des poudres.[68] The King wished that the Régisseurs lived in the proximity of the refinery in order to easily inspect its activities.[69] As the Hôtel de la Régie was already the residence of Micault de Courbeton, Commissaire général et fermier des poudres et salpêtres, Turgot hoped that by renovating it, the building could host not only de Courbeton but also Lavoisier and Le Faucheux.[70] However, he disagreed with this new arrangement, complaining that the building was neither comfortable nor sufficiently large to host the régisseurs.[71] In the end, de Courbeton was persuaded to move elsewhere at his own expense.

Lavoisier moved to the Petit Arsenal in spring 1776. The area consisted of l'Hôtel de la régie des poudres & salpêtre in the Rue des Ormes, a building facing the Cour du saltpêtre, probably occupied by Lavoisier, and the ateliers for the production and refinery of saltpetre. Additionally there was a public garden with a nearby access to water at the Fossées de l'Arsenal, where a gunpowder warehouse was located. (Fig. 17)

On the basis of his wife's descriptions in word and image, as well as those of his guests and his associates, we know that Lavoisier's residence consisted of two floors, and that his laboratory occupied several rooms. The English agriculturist Arthur Young, who presented a recommendation letter from Priestley, was a guest at the Arsenal on 16 September 1787, and he reports of the visit:

68 MS Détail (1756).

69 "L'Intention du Roy Monsieur, Etant que vous soyés logé dans les batiments affectés au service des poudres afin d'être plus à portée de surveiller les operations de la raffinerie à salpêtre, d'examiner la qualité de Celui qui est fourni par les salpetriers et de suivre toutes les operations de la Regie dont vous ètes chargé." Letter by Turgot to Lavoisier dating July 26, 1775; LC, vol. 2, p. 498.

70 "Vous voudrés bien vous y transporter aussitòt que la distribution des logements occupés aujourd'huy par M. De Courbeton sera faitte et arretée par M. D'ormesson, entre vous, Le Sr Lefaucheux et le S^r De Courbeton a qui le Roy veut bien en laisser un en sa qualité de Commissaire general." Ibid.

71 Cochu (1777), p. 35.

To Mons. Lavoisier, by appointment. Madame Lavoisier, a lively, sensible, scientific lady, had prepared a *dejeuné Anglois* of tea and coffee, but her conversation on Mr. Kirwan's Essay on Phlogiston, which she is translating from the English, and on other subjects, which a woman of understanding, that works with her husband in his laboratory, knows how to adorn, was the best repast. That apartment, the operations of which have been rendered so interesting to the philosophical world, I had pleasure in viewing. In the apparatus for aerial experiments, nothing makes so great a figure as the machine for burning inflammable and vital air, to make, or deposit water; it is a splendid machine. Three vessels are held in suspension with indexes for marking the immediate variations of their weights; two that are as large as half hogsheads, contain the one inflammable, the other the vital air, and a tube of communication passes to the third, where the two airs unite and burn; by contrivances, too complex to describe without plates, the loss of weight of the two airs, as indicated by their respective balances, equal at every moment to the gain in the third vessel from the formation or deposition of the water, it not being yet ascertained whether the water be actually made or deposited. If accurate (of which I must confess I have little conception), it is a noble machine. Mons. Lavoisier, when the structure of it was commended, said, *Mais oui monsieur, & meme par un artiste Francois*! with an accent of voice that admitted their general inferiority to ours. It is well known that we have a considerable exportation of mathematical and other curious instruments to every part of Europe, and to France amongst the rest. Nor is this new, for the apparatus with which the French academicians measured a degree in the polar circle was made by Mr. George Graham. Another engine Mons. Lavoisier shewed us was an electrical apparatus inclosed in a balloon, for trying electrical experiments in any sort of air. His pond of quicksilver is considerable, containing 250lb. and his water apparatus very great, but his furnaces did not seem so well calculated for the higher degrees of heat as some others I have seen. I was glad to find this gentleman splendidly lodged, and with every appearance of a man of considerable fortune.[72]

Bedel pointed out, Lavoisier's laboratory was organised across several rooms:

In his laboratory at the Arsenal [...] [Lavoisier] had all the space one could wish for to set up a fine laboratory. There were special rooms for the scales and for the equipment in general. The most commonly used apparatuses and instruments were tidily arranged on shelves along the walls. On a table were the laboratory notebooks in which details of the experiments were recorded.[73]

Lavoisier used every corner of the Arsenal for his experiments. In June 1781 his experiments with Laplace on the dilatation of bodies took place in the garden of the Arsenal and used a new optical pyrometer, which they then presented before the Académie on 22 December of the same year.[74] In March 1782 the garden was used again, this time by Alessandro Volta, Laplace, and Lavoisier, for experiments on the vaporisation of fluids, in which they used Volta's condenser.[75] The garden, as I already mentioned

72 Young (1792), pp. 64–65.

73 "Dans son laboratoire de l'Arsenal [...] [Lavoisier] disposait de toute la place que l'on peut souhaiter pour aménager un beau laboratoire. Il y avait des pièces spéciales pour les balances et le matériel en général. Aux murs, les appareils ou instruments d'un usage courant étaient bien rangés sur des rayons. Enfin, sur une table, se trouvaient les cahiers de laboratoire sur lesquels étaient notés les détails des expériences". Bedel (1986), pp. 648–649. Unfortunately Bedel does not quote the source of his claim.

74 On the pyrometer, see LO, vol. 2, pp. 739–764 and p. 776. "M.M. Lavoisier et Laplace ont retenu date pour la description d'un nouveau hygromètre [sic], au moyen duquel on peut mesurer la précision au moins d'un centième de ligne, les allongements des corps solides par la chaleur, avec une suite d'expériences sur la dilatation du verre et des métaux. L'hygromètre est exécuté". MS Procès-verbaux de l'Académie des sciences, 1781, fol. 254.

75 "All'incontro quelli che ripeterono l'istesso Sig. de la Place e Sig. Lavoisier ad una campagna di quest'ultimo ebbero buon riuscimento. La qual cosa c'invogliò a ripetere e moltiplicar le sperienze, e il successo fu completo, avendo ottenuto segni chiarissimi dì elettricità dall'evaporazione dell'acqua, dalla semplice combustione dei carboni, e dall'effervescenza delle limature di ferro nell'acido vitriolico diluto. Ciò avvenne il giorno 13 Aprile e la maniera di far l'esperienza fu questa: si isolò in un aperto giardino una gran lastra di metallo, alla quale era attaccato un lungo filo di ferro che veniva a terminare in contatto dello scudo o disco posato sul piano di marmo, e questo tenevasi continuamente asciutto e caldo da alquanti carboni sottoposti. Ciò fatto posimo su la detta lastra isolata alcuni scaldini ripieni di carboni mezzo accesi, e lasciammo che la combustione ajutata da un gentil vento che spirava andasse rinforzandosi per alcuni minuti: allora rimovendo lo scudo dal contatto del filo metallico e quindi da quello del marmo con alzarlo al consueto modo vi comparvero i segni aspettati di elettricità, mentre accostato al nuovo elettrometro del Sig. Cavallo, fece che s'aprissero i due fili colle pallottoline: esaminata questa elettricità si trovò essere *negativa*". Volta (1784), p. 159. The experiments were repeated also indoor in "a grande stanza".

above, became a venue for the experiments commissioned by the Académie des sciences on the new weights and measures. Lavoisier explored the nearby *fossés* both to analyse water and to collect inflammable gases.

Lavoisier's laboratory underwent some architectural changes. In August 1788 Hassenfratz reported the progress made by Lavoisier's carpenter in what he calls "notre laboratoire".[76] The Arsenal also included Séguin's small workshop nearby, where he and Lavoisier conducted experiments like those on the fusion of platinum in July 1791.[77]

Some of the visual features of Lavoisier's laboratory can be inferred from the drawings prepared by Madame Lavoisier of Lavoisier's and Séguin's experiments on human respiration and transpiration (Figs. 18–19). Two of these drawings are well known, since Grimaux reproduced them in his 1888 biography of Lavoisier.[78] They show a spacious room which, given the inclination of the walls, was most probably located in the attic. The same room seems to be shown from opposite angles. Two additional, recently discovered drawings that show Lavoisier and Séguin engaged in experiments on transpiration.[79] It seems that here, too, the same room is illustrated. In my view this room only occasionally served as a site for experiments rather than being a part of the laboratory.[80] The chemical containers shown on the shelves seem to indicate that the room was also used as storage for chemicals. Otherwise the room is quite empty, and the apparatus shown in these pictures appears to be a provisional assemblage of instruments and equipment. It is in stark contrast with Lavoisier's vast collection of instruments, apparatus, and chemicals, which amounted to some 10,000 items.

If this is correct, where was the core laboratory? And what did it look like?

Young tells us that Lavoisier kept his large apparatus in one large room, and the records taken in 1794, when Lavoisier's laboratory at the Boulevard de la Madeleine was confiscated, attest that the laboratory consisted of three rooms, two of which dedicated to the chemical laboratory and one to the *cabinet de physique* (all on the ground floor of the house).[81] Interestingly, the mineralogy cabinet remained packed, which seems to indicate the existence of a special room at the Arsenal where Lavoisier kept his mineralogical collection. Lavoisier had created a "magasin de verrerie"[82] (glassware warehouse) in the cellar where he also kept, across several rooms, all his chemicals which, if the above assumption is correct, were kept in the attic at the Arsenal. It is naturally dangerous to compare the Arsenal laboratory to the one Lavoisier created in haste at the Boulevard de la Madeleine, but I nevertheless think it likely that he kept his *cabinet de physique*, the chemicals and glassware, and the chemical laboratories in separate room in both places. With the accessibility of books of great importance, it is likely that the library was not far from the laboratory. Lavoisier's *registres de laboratoire* often contain precise references to printed works that he clearly had consulted during an experiment. The combination of the laboratory and the library was extremely common among naturalists and natural philosophers. As reported Young in detail from his visit to Dijon, Guyton de Morveau had a laboratory filled with books and instruments, a practice that was also very common among Swedish eighteenth-century chemists.[83]

An insightful description of Lavoisier's laboratory which matches the abovementioned features appears in Antoine-François Fourcroy's publications. Fourcroy was working in Lavoisier's laboratory for years, and his obituary of Lavoisier, published in 1796, provides an extremely informed description of the experimental routines followed at the Arsenal:

> He devoted his fortune to the expansion of science; his house became a vast laboratory where nothing was lacking. The ablest engineers were busy constructing instruments infinitely better than those that had been used before him, new apparatuses that were valuable for their sensitivity and accuracy. He spared no expense for such a fine and useful pursuit. Besides this initial advantage of fortune – which so few men know how to apply to the benefit of their peers' well-being – Lavoisier combined several others, which he also exploited well. Twice a week, he held gatherings to which the men with the greatest distinction in geometry, physics, and chemistry were invited. Instructive conversations – exchanges similar to those that had preceded the foundation of the academies – became the focus of all enlightenment. The opinions of the best informed men in Europe were discussed there; the most striking and novel passages of books published abroad were read there; theories were compared with experiments; scientists

76 LC, vol. 5, p. 199.
77 Chadelcott (1968), p. 32.
78 Grimaux (1888), pp. 118 and 128.
79 Beretta (2012). See Figs. 22 and 23 in Chapter 2, p. 44.
80 This is not the opinion of Bret, Lanoë (2006), p. 142, who believe that Lavoisier's laboratory was in the attic.
81 See appendix 1, p. 132.
82 See appendix 1, p. 124.
83 Young (1792), p. 153. On Swedish laboratories, see Beretta (1995), p. 80. Torbern Bergman's rich library was also near his laboratory.

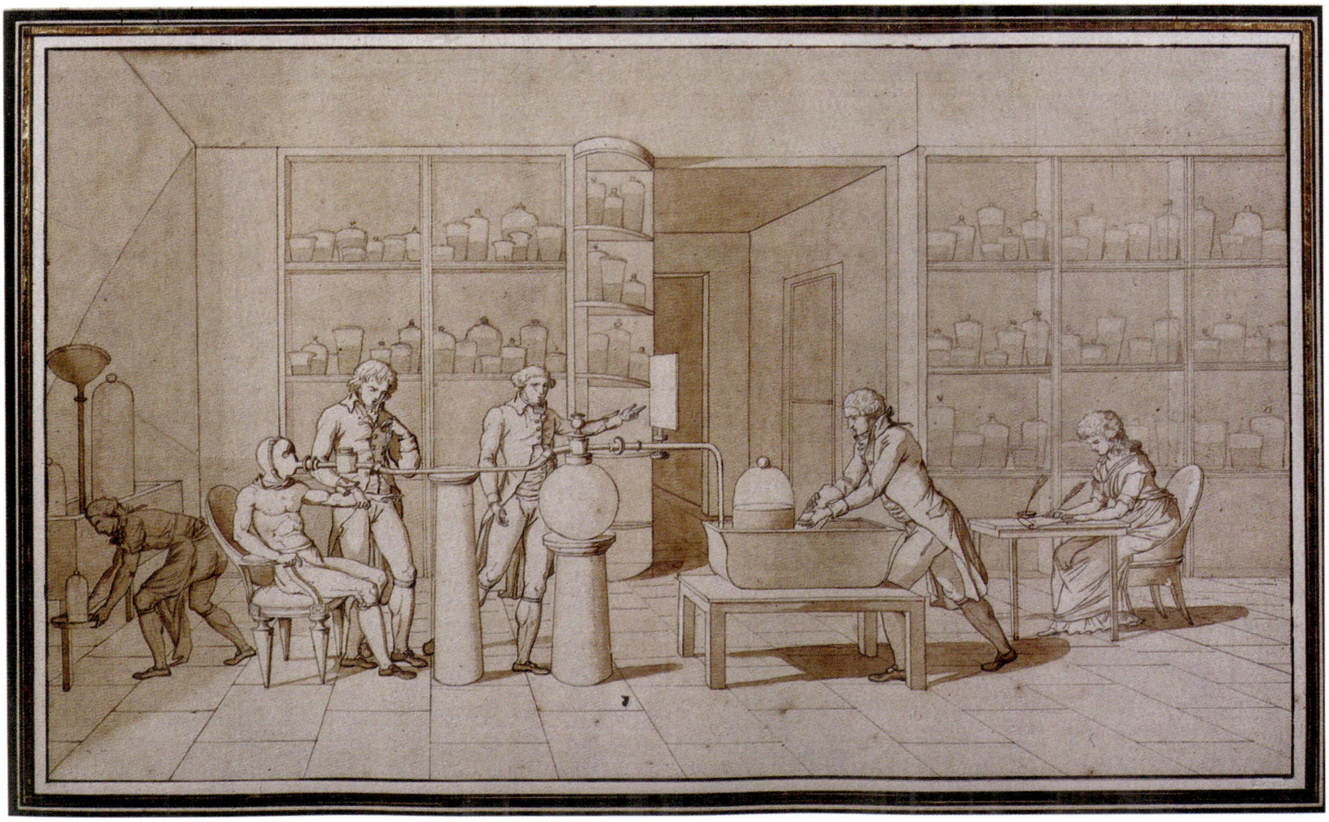

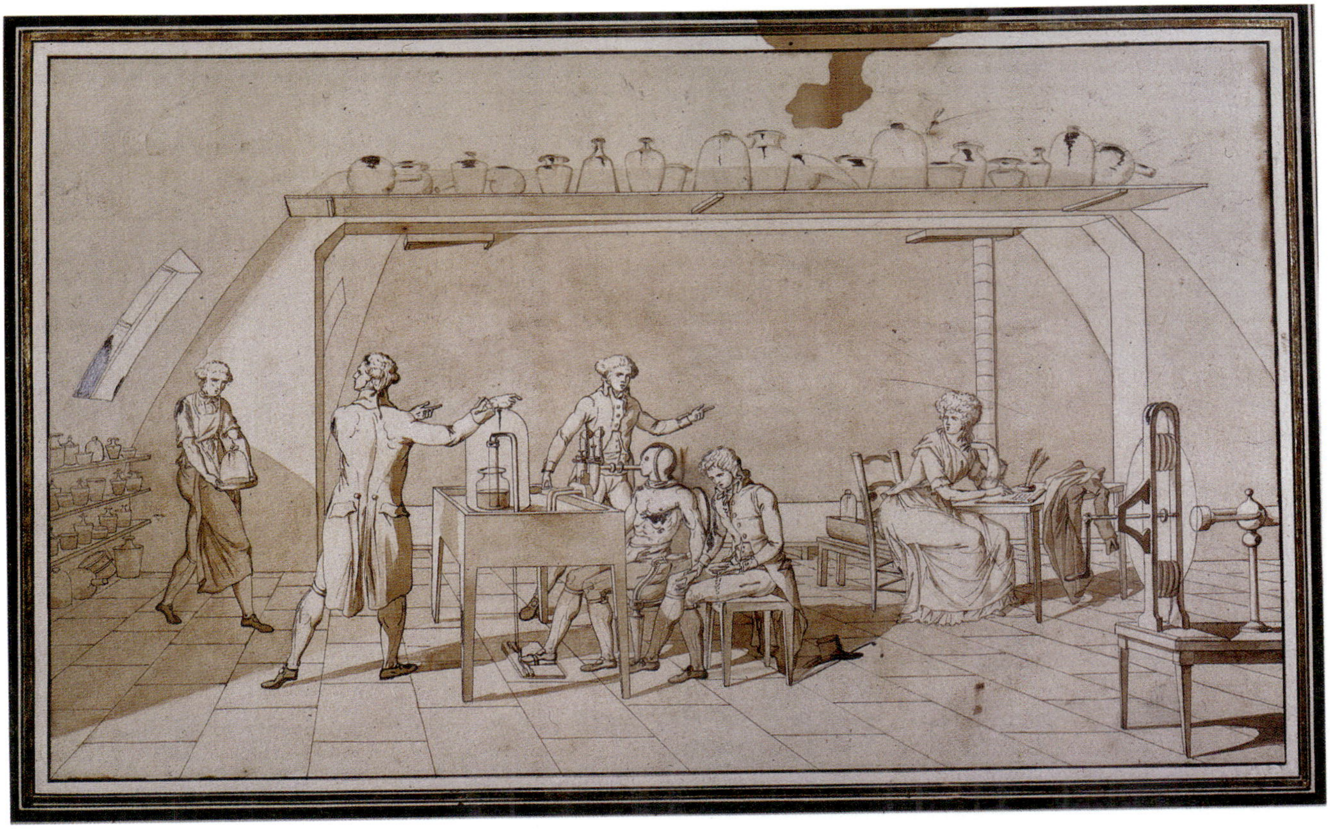

FIGURES 18–19 Marie Anne Lavoisier's drawings of Lavoisier's experiments on human respiration made at the Arsenal in 1790
PRIVATE COLLECTION

of all nations were admitted: Priestley, Fontana, Blagden, Ingenhousz [sic], Landriani, Jacquin the younger, Wath [sic], Bolton and other illustrious physicists and chemists from England, Germany and Italy found themselves in the company of Laplace, Lagrange, Borda, Cousin, Meunier, Vandermonde, Monge, Guyton and Berthollet. I shall never forget the wonderful hours that I spent in these learned conversations; all that I heard and collected there that was useful for the progress of science and the happiness of mankind will never leave my memory. Among the great benefits of these meetings, the one that most struck me – and the one whose immeasurable influence was soon felt in the Académie des sciences, and later in all works of physics and chemistry published in France in the past twenty years – is the agreement that developed between the manner of reasoning of geometers and that of physicists.[84]

Fourcroy published another, more evocative description was published in the entry titled *laboratoire* in the *Encyclopédie méthodique*. Although published in 1805, it seems to me that this detailed description of the ideal chemical laboratory recalls with remarkable precision the features of the Arsenal laboratory. Fourcroy's description is certainly not of the laboratories set up at the École polytechnique, which were designed for teaching rather than research.[85] Moreover, at the beginning of his article, Fourcroy makes clear that he is going to describe the best research laboratory of his time, and explicitly mentions those that had been built and that flourished before the outbreak of the Revolution, including Lavoisier's:

> In order to properly grasp the spirit that will guide me in this important article, one must begin by assuming that I envisage the case where the chemist controls the choice of premises, the construction, the expenditures, etc., and that I consider the point closest to perfection in this type of facility. In a manner of speaking, I shall state, I shall describe the utmost good, greatness and completeness [...].
>
> I have seen great figures, the Orléans, the Ayens, the Larochefoucaulds, the Lauraguais, the Courtivrons, the Chaulnes, the Sarons, the Malesherbes, the Poulletier de la Salle, etc. made happier by the chemical researches that they conducted with the Rouelles, Darcets, Macquers and Bucquets – chemists who made the fame of their age and of their country [...]. What a difference between the reputation of Lavoisier, who employed a portion of his fortune for the chemical discoveries that gave lustre to his name, and the vain renown of the world's grandees and rich, the sole legacy of whose wealth are piles of stones, showy monuments in which posterity sees only the titles of pride and vanity.[86]

84 "Sa fortune fut consacrée à l'agrandissement de la science; sa maison devint un vaste laboratoire où rien ne manquoit. Les plus habiles ingénieurs furent occupés à lui construire des instrumens infiniment meilleurs que ceux qu'on avoit employés avant lui, des appareils nouveaux et précieux par leur délicatesse et leur exactitude: rien ne lui coûtoit pour une si belle et si utile occupation. A ce premier avantage de la fortune, dont si peu d'hommes savent profiter pour le bonheur de leurs semblables, Lavoisier en réunit plusieurs autres dont il sut également tirer parti. Il tenoit chez lui, deux fois la semaine, des assemblées auxquelles étoient appellés les hommes les plus distingués dans la géométrie, la physique et la chimie; des conversations instructives, des entretiens semblables à ceux qui avoient précédé l'établissement des académies, y devenoient le centre de toutes les lumières. On y discutoit les opinions des hommes les plus éclairés de l'Europe; on y lisoit les passages les plus frappans et les plus neufs des ouvrages publiés chez l'étranger; on y comparait les théories avec les expériences; les savans des toutes les nations y étoient admis: Priestley, Fontana, Blagden, Ingenhoutz, Landriani, Jacquin le fils, Wath, Bolton et d'autres physiciens et chimistes illustres d'Angleterre, d'Allemagne, d'Italie, s'y trouvoient réunis avec Laplace, Lagrange, Borda, Cousin, Meunier, Vandermonde, Monge, Guyton, Berthollet. Je n'oublierai jamais les heures fortunées que j'ai passées dans ces doctes entretiens; tout ce que j'y ai entendu et recueilli d'utile pour les progrès des sciences et pour le bonheur des hommes, ne sortira jamais de ma mémoire. Parmi les grands avantages de ces réunions, celui de tous qui m'a le plus frappé, et dont l'inappréciable influence s'est bientôt fait sentir dans le sein de l'Académie des Sciences, et par suite dans tous les ouvrages de physique et de chimie, publiés depuis vingt ans en France, c'est l'accord qui s'est établi entre la manière de raisonner des géomètres et celle des physiciens". Fourcroy (1796), pp. 32–33.

85 On the laboratories at the École, see Smeaton (1975), Langins (1987), pp. 64–67 and Tron (1996).

86 "Il faut d'abord supposer, pour bien saisir l'esprit qui va me guider dans cet article important, que j'admets le cas ou le choix d'un local, la construction, les dépenses &c. sont au pouvoir du chimiste, & que je considère le point le plus rapproché de la perfection dans ce genre d'établissement. C'est en quelque sorte le maximum du bien, du grand, du complet que j'énoncerai, que je décrirai [...]

J'ai vu des grands personnages, les Orléans, les Ayen, les Larochefoucauld, les Lauraguais, les Courtrivon, les Chaulnes, les Saron, les Malesherbes, les Poulletier de la Salle, &c. plus heureux par les recherches chimiques qu'il suivoient avec les Rouelle, les Darcet, les Macquer, les Bucquet, chimistes qui ont fait la gloire de leurs temps & de leur pays [...]. Quelle différence entre la réputation de Lavoisier, employant une portion de sa fortune aux découvertes chimiques qui ont illustré son nom, & la vaine renommée des grands & des riches de la terre, qui n'ont laissé de leur richesse que des monceaux de pierre, monumens

Fourcroy remarked that the ideal research laboratory consisted of a building of two floors, a courtyard, warehouses, a garden, and a water source.[87] On the ground floor the laboratory should consist of three rooms. At the centre was the room for chemical experiments,[88] which was adjacent to two further rooms, one "devoted to housing the fountains, and to conducting all operations requiring much water"; the other was used as a store room for all "balances, pneumatic, compression and electric machines, and in general all metal utensils that could be attacked by contact with acid vapours, which would rapidly alter their accuracy and precision".[89]

According to Fourcroy, it was useful to have another room on an upper level where all the chemicals were kept in order, which, as I have already pointed out, was the case for Lavoisier's experiment room in the attic portrayed by Madame Lavoisier.[90] Interestingly, when it came to describing the contents of the special room for keeping precision instruments, Fourcroy listed the gasometer, eudiometers, electrical machines, pneumatic pumps, burning mirrors, barometers, thermometers, balances, the calorimeter and the hydrometer which, as we know for certain, was the apparatus featured in Lavoisier's *Traité élémentaire de chimie*.

While most of the witnesses praised the generous proportions of Lavoisier's laboratory, it should be noticed that these were not exceptional. While passing Dijon and visiting Guyton de Morveau's chemical laboratory in 1789, Arthur Young claimed that he had not seen elsewhere a "similar variety and extent of apparatus", and highly praised the admirable arrangement of the furniture in the two rooms which were, in fact, probably similar to Lavoisier's.[91] It is interesting that a neutral observer like Young was more impressed with Guyton's laboratory than with Lavoisier's – the latter was apparently not as exceptional as historians have claimed.

With his resignation from the Régie des poudres on 15 August 1792, Lavoisier was forced to move from the Arsenal, and it was only at the end of September that he was able to move into a new residence at 243 Boulevard de la Madeleine, a house belonging to the banker Lecoulteux de La Noraye.[92] (Fig. 20) The inventories taken by the Revolutionary authorities in 1794 provide a detailed description of the house and of the make-up of his laboratories.[93]

From my brief survey Lavoisier emerges as an extremely assiduous experimenter, and his keen interest in the production of ever more accurate instruments and apparatus are the results of the daily routine he followed on different sites of chemical practice. Guettard's fundamental lesson of the importance of investigating nature in the field was yet another constant feature of Lavoisier's approach to chemical experimentation. Throughout his career he conducted experiments outdoors, always trying to go beyond the architectural limits of his own laboratory. Both types of venues were populated with a community of assistants, relatives, instrument makers, friends, peers, officials, and colleagues, which made them exceptional scenes for collegial work. This complex geography of relations and sites merged in his published work in a discreet, albeit not invisible, way, and the traces which I have so far been able to detect seem to reveal conventional new profile of Lavoisier's experimental background.

fastueux où la postérité ne voit que les titres d'orgueil & de la vanité". *Encyclopédie méthodique – Chymie* (1786–1815), vol. 4, p. 566.

87 "Le grand côté du laboratoire opposé à la cour peut être un jardin planté d'arbres assez écartés du mur de face, pour laisser pénétrer le soleil du midi. Il est très avantageux d'avoir à porté du laboratoire un vaste réservoir d'eau de source ou d'eau de pluie, & une glacerie, la glace étant aujourd'hui un des premiers besoins de beaucoup d'expériences. Quand le local, composé d'une cour, des hangars, d'une glacière, d'un bâtiment à deux étages, est construit […], on s'occupe d'arranger & de meubler l'intérieur du laboratoire & des pièces adjacentes, de manière à les approprier aux recherches de chimie". *Encyclopédie méthodique – Chymie* (1786–1815), vol. 4, p. 567.

88 "La grande pièce du centre, ou le laboratoire proprement dit, doit être dallée sur son sol en pierres bien jointes & dures; il faut y pratiquer un assez grand nombres de fenêtres opposées, pour y recevoir une belle lumière, & y exciter au besoin les vapeurs des courans rapides pour entraîner les vapeurs dangereuses qui se dégagent souvent dans les opérations." Ibid.

89 "[J]'ai dit qu'on devoit avoir un pièce particulière destinée à renfermer les instrumens de physique qu'on doit écarter des vapeurs salines & acides". Ibid. p. 568.

90 "Il est très utile d'avoir au dessus du laboratoire un premier étage, dans lequel on porte l'ouvrage par ordre, les produits des opérations". Ibid., p. 567.

91 Young (1792), p. 153.

92 Duveen (1951).

93 It is only thanks to these inventories, which are published in the appendix, that we have information about the quantitative and qualitative composition of Lavoisier's laboratory.

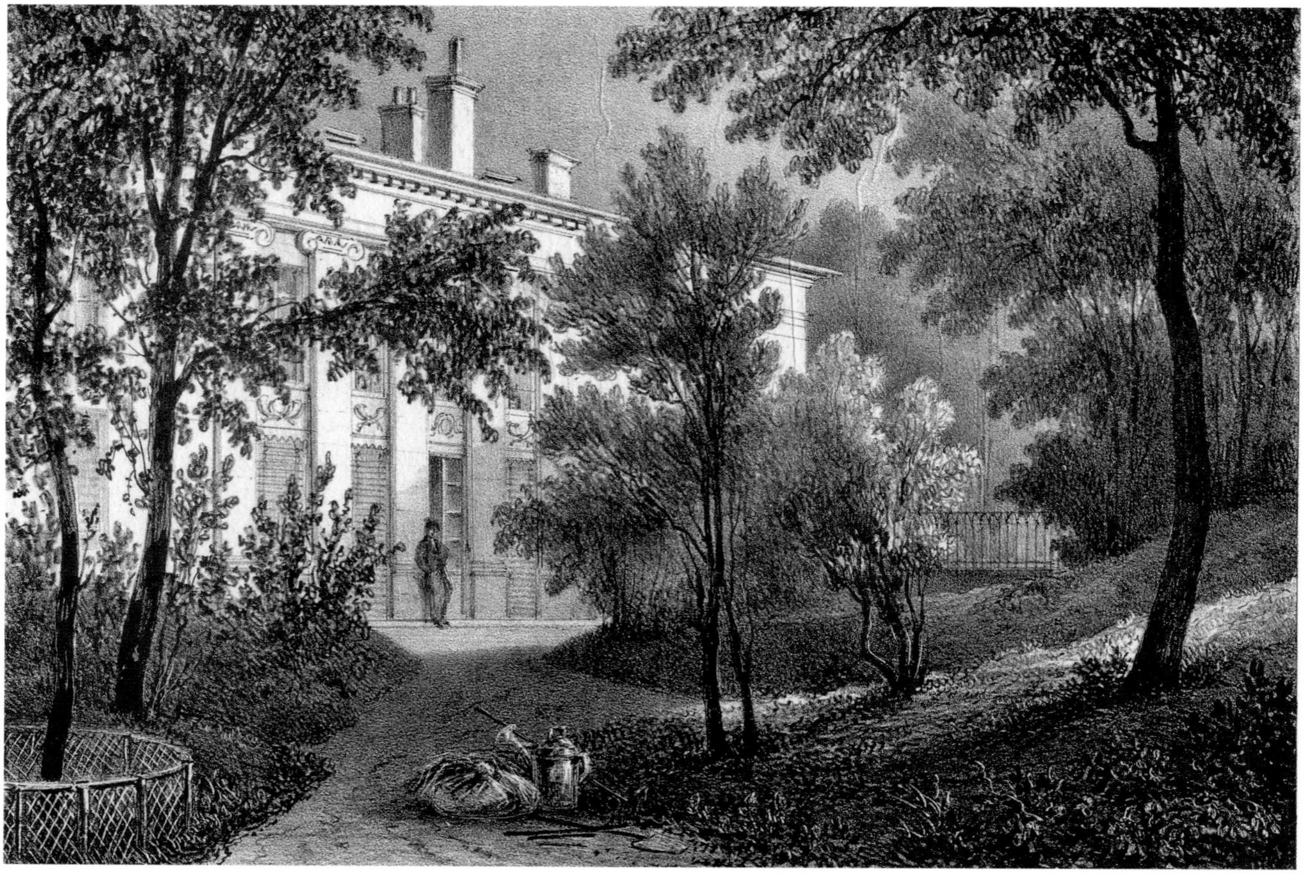

FIGURE 20 Lavoisier's residence in the Boulevard de la Madeleine in 1792–1793, shown in an engraving from 1835
PRIVATE COLLECTION

CHAPTER 4

The Cost of Lavoisier's Laboratory

4.1 Was Chemistry Cheap or Expensive?

Lavoisier's laboratory has often been regarded as an extravagant and expensive site of experimentation. However, the sources – some of which have only recently become accessible – call for a comprehensive reassessment of this impression. Lavoisier's letters and manuscripts not only offer an insightful overview of the trade of Parisian instrument makers, including their costs, but also provide interesting information about others working at the Arsenal at the time, such as the *garçons de laboratoire* and more skilled laboratory assistants. The first part of this chapter will survey the costs of instrument making and Lavoisier's laboratory as a whole, while the second part will investigate the organisation of salaried labour at the Arsenal.

Those who have most emphasised on the lavish costs of Lavoisier's laboratory have suggested that the results of the large-scale experiments at the Arsenal became famous due to the impact of the costly apparatus rather than the actual experimental results.[1] Jean-Luc Chappey went so far as to claim that Lavoisier's laboratory practice transformed chemistry into a "science sévère", preventing the laymen, artisans, and women (who, prior to his revolution, were able to access an open "science mondaine") from participating.[2] These positions are mainly rooted in the conviction that eighteenth-century chemistry before Lavoisier was relatively simple in its experimental features, and that the most important discoveries on the nature of gases were made with rudimentary instruments. According to this view, the extraordinary experimental skills of the Swedish apothecary Carl Wilhelm Scheele, who is usually compared with Lavoisier and credited with the discovery of oxygen and several other elements, were enhanced by his use of remarkably simple apparatus.[3] In fact, Scheele made his main discoveries in Torbern Bergman's lavishly equipped Uppsala laboratory (Fig. 1), where he found a great variety of new instruments and expensive apparatus for his pneumatic experiments. In addition to minerals, models, and a "cupboard of noteworthy chemicals" such as salts, metals, and acids, Bergman's laboratory had a forge with good double bellows, both reverberatory and glassworking ovens, one *athanor* (constant-temperature oven) for digestion, both open and closed boilers for distillations, an oven for colouring and salt boiling with a copper cauldron was bricked-in as a container, and two cast-iron cauldrons working as sand baths, into which two loose cylindrical pots (one made of fine tinplate, the other of lead) could be put when needed for colours and various evaporations. There were several contraptions for hanging distillation vessels over an open fire within small, open furnaces so that, as Bergman remarked, one could "conveniently see all that [took] place in a vessel from beginning to end".[4]

Bergman also designed different kinds of blowpipes and owned a remarkably rich collection of vessels and receivers (made of glass, refractory clay, and metals), mortars, and other common equipment. Like Priestley, Bergman kept a special room in his laboratory for mathematical instruments such as hydrometers, an electrical machine, two electrophores (electrostatic generators), a hygrometer, an air pump, various balances, a compound microscope, various kinds of barometers and thermometers, and special glassware. A comprehensive list of both the fine instruments and the chemical apparatus was compiled soon after Bergman's death by his successor in the chair of chemistry, Johan Afzelius.[5]

Bergman justified the quantitative and qualitative expansion of his laboratory equipment with the growing complexity of experimentation and the fact that accuracy was of utmost importance if useless disputes were to be avoided.

Bergman's equipment could, in many ways, be compared to Lavoisier's and that of other natural philosophers such as Henry Cavendish, Felice Fontana, and Joseph Priestley.[6] In the chemistry of gases without the use of new apparatus (most of which was derived from natural philosophy) it was practically impossible to conduct original experiments. Such apparatus was quite expensive, and had been so long before Lavoisier entered the scene. Having said this, there is no doubt that, towards the end of the century, Lavoisier contributed to the development of instrument making in chemical laboratories to a higher

1 See, for instance, Roberts (1991); Golinski (1994, 1994a, 1995), and Holmes (1998 and 2000). An earlier version of this chapter has been published in Beretta (2022).

2 Chappey (2004), p. 23; Lilti (2005), pp. 262–263. I shall discuss this contrast and the social aspects of Lavoisier's laboratory life in more detail in the next chapter.

3 This thesis has recently been emphatically supported in Fors (2003).

4 Bergman (1985), p. 471.

5 MS Afzelius (1785), fols. 133–196.

6 On this see Beretta (2021).

 | DOI:10.1163/9789004511217_006

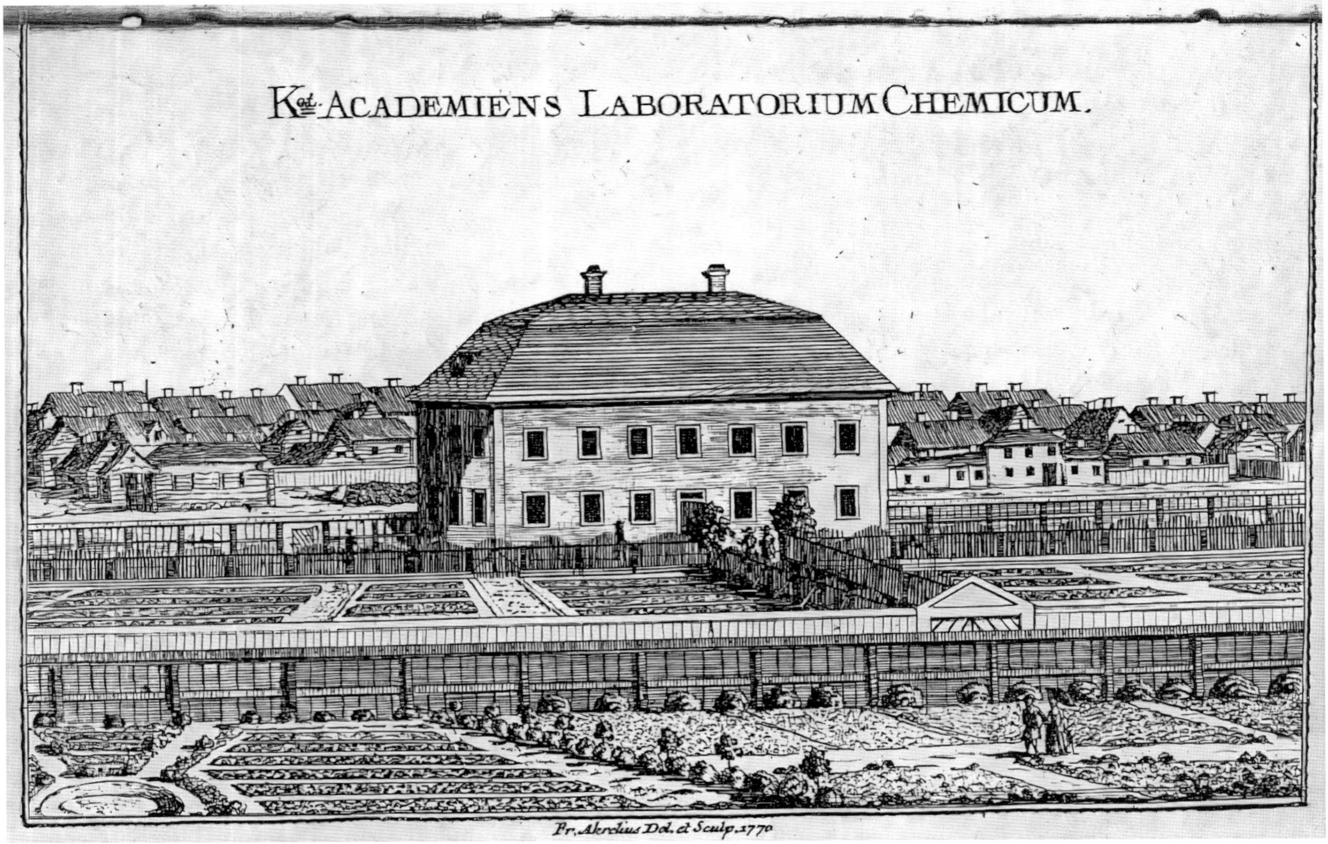

FIGURE 1 Torbern Bergman's chemical laboratory in Uppsala
FROM BUSSER (1769–1773). PRIVATE COLLECTION

standard, and thereby became a key protagonist in making the science more precise and, consequently, more expensive. The process by which he achieved this reputation is quite complex and it involves several aspects that should be taken into consideration.

When it comes to accurate figures and their modern equivalents for the cost of Lavoisier's apparatus, the information available is both poor and contradictory. This is not surprising given that pre-revolutionary currency is not easily converted into modern terms. In 1996, in his biography of Lavoisier, Jean Pierre Poirier claimed that in 1788 one *livre* was equivalent of 40 USD (ca. 36 Euro). This calculation was, according to Poirier, based exclusively on the average daily salary of a skilled worker (between 1 and 2.5 *livres*) without taking into account the cost of living and the highly variable cost of luxury commodities.[7] While salaries in the eighteenth century were uniformly low, the cost of living fluctuated considerably, and luxury commodities could reach prices which would be extremely difficult to compare to today's standards. This is shown in the following table:

- Daily average salary 1750–1780: 1.3 *livre*
- Cost of one kg of bread in Paris in 1782: 0.2586 *livre*
- Cost of a mirror of $4m^2$ in 1702: 2.450 *livres*[8]

Assuming the validity of Poirier's estimation, one kg of bread would cost more than 10 USD in today's value, a price that most of the population in Paris would have not been able to afford.

If we take inflation instead of a worker's average daily salary as the main parameter for assessing the value of money, one *livre* in 1786 would equate to 11.5 Euro in 2019. All of a sudden average salaries, the cost of common goods and luxury commodities begin to make sense.

David's famous double portrait of the Lavoisiers cost 7000 *livres* in 1788, and it seems to me that the sum of 71,500 EUR following the inflation index is a far more realistic figure than that of ca. 300,000 Euro suggested by

7 Poirier (1996), pp. 413–414. Poirier insisted on taking the average salary as exclusive parameter.

8 This table is drawn from Wikipedia, article *Livre tournois*. https://fr.wikipedia.org/wiki/Livre_tournois (accessed October 2020).

Poirier.[9] We shall later see that my estimation is also consistent with the historic exchange rate between French *livres* and the English pound.

Following this preamble and independently of our assumptions, it is clear that an assessment of the cost of Lavoisier's instruments is not as easy as some historians have so far believed. This is further complicated by the fact that some of Lavoisier's contemporaries deliberately exaggerated the overall cost of his laboratory equipment. In March 1787 Gaspard Monge wrote a letter to Martinus Van Marum, in which he remarked:

> The unfortunate part of the system of the composition and decomposition of water stems from the fact that Mr. Lavoisier has carried out his experiments with apparatus that was too expensive, and that all the natural philosophers think that to be sure of the fact it is necessary to spend five to six thousand *livres*. I first carried out the experiment on the composition and it only costed two *Louis* [48 *livres*]...[10]

However, as Monge openly admitted, nobody was really persuaded by the results he had obtained from using simpler apparatus, and as far as I know, no other scientist tried to replicate Monge's experiments. Moreover, Monge was a mathematician, and most of the members of mathematics section at the Académie des sciences immediately supported Lavoisier's theory and experiments. The mathematicians and natural philosophers of the Académie, who had little or no background at all in chemistry, did not have any reason to be attached to the phlogiston theory, and they clearly applauded Lavoisier's effort to apply accuracy and precision to chemical experimentation. Although Lavoisier sought the support of physicists and mathematicians, his large-scale experiments on the analysis and synthesis of water were principally addressed to chemists, and simpler apparatus like that conceived by Alessandro Volta (1782), Henry Cavendish (1783), and Gaspard Monge (1783) did not produce any compelling result to that effect. This is why Lavoisier planned the construction of gasometers in collaboration with Jean Baptiste Meusnier de la Place between 1783 and 1787.

The cost of Lavoisier's apparatus was also discussed by Parisian chemists. As early as 1790 the apothecary Jean François Demachy strongly criticised Lavoisier's use of complex apparatus:

> what would we say of our current physics, once so simply pursued with the aid of practical instruments with so few complications? Do we not see it subservient to machines, countless gears, gaudy embellishments that, with the excuse of offering higher precision, lead an honest observer to harbour greater doubts about the real success of the experiment?[11]

In 1793 Demachy wrote an extremely long and critical review of Lavoisier's *Traité élémentaire de chimie*, in which he concluded that the geometric precision of the costly apparatus was impossible.[12] In 1796 Philippe Quénard, *représentant de la commune*, who had become acquainted with Lavoisier during the Revolution, wrote the following malicious statement:

> Lavoisier was still only fifty years old. Study had left him all his vigour: all difficulties had been overcome; he had created the science; and, as he enjoyed a brilliant fortune, and had a passion for his art, he spared nothing to discover a truth. He spared no effort and no expense when the goal was to move one step ahead. The experiment of the decomposition of water, which was so simple, had alone cost him more than five hundred thousand *livres* [approximately 5,500,000 Euro]. It was no doubt this passion for research that had made him ungenerous, even insensitive to everything that came close to him.[13]

9 On the double portrait, see LC, vol. 5, p. 238. Poirier's estimation is uncritically supported by Baetjer (2019), pp. 317–324.

10 Forbes (1969–1976), vol. 6, p. 257.

11 "que dirions-nous de notre physique actuelle, autrefois si simplement cultivée à l'aide d'instruments commodes et si peu compliqués? Ne la voyons-nous pas subordonnée à des machines, à des rouages sans nombre, à des embellissements fastueux qui, sous prétexte d'une plus exacte précision, portent dans l'esprit d'un observateur de bonne foi plus de doutes sur le succès réel de l'expérience?" Demachy (1790), cited in Chappey (2004), p. 26.

12 "La précision géometrique est impossible [...] dans le fait; les appareils des pneumatistes, trop compliqués, y apportent eux-mêmes des obstacles incalculables [...]. Le rigorisme du calcul est incompatible avec les expériences, quelqu'exactes soient'elles d'ailleurs. Donc les pneumatistes ont retardé la marche de la chymie, en excédant les bornes de l'exactitude". Demachy (1794), pp. 299–300.

13 "Lavoisier n'avait encore que cinquante ans. L'étude lui avait laissé toute sa vigueur: toutes les difficultés étaient vaincus; il avait créé la science; et, comme il jouissait d'une fortune brillante, et il avait la passion de son art, il n'épargnait rien pour découvrir une vérité: il ne calculait ni la peine, ni l'argent, quand il s'agissait d'avancer d'un pas. La seule expérience de la décomposition si simple de l'eau lui avait coûté plus de cinq cent mille livres. C'est sans doute cette passion des recherches qui l'avait rendu avare, insensible même à l'égard de tout ce qui l'approchait". Quenard (1796).

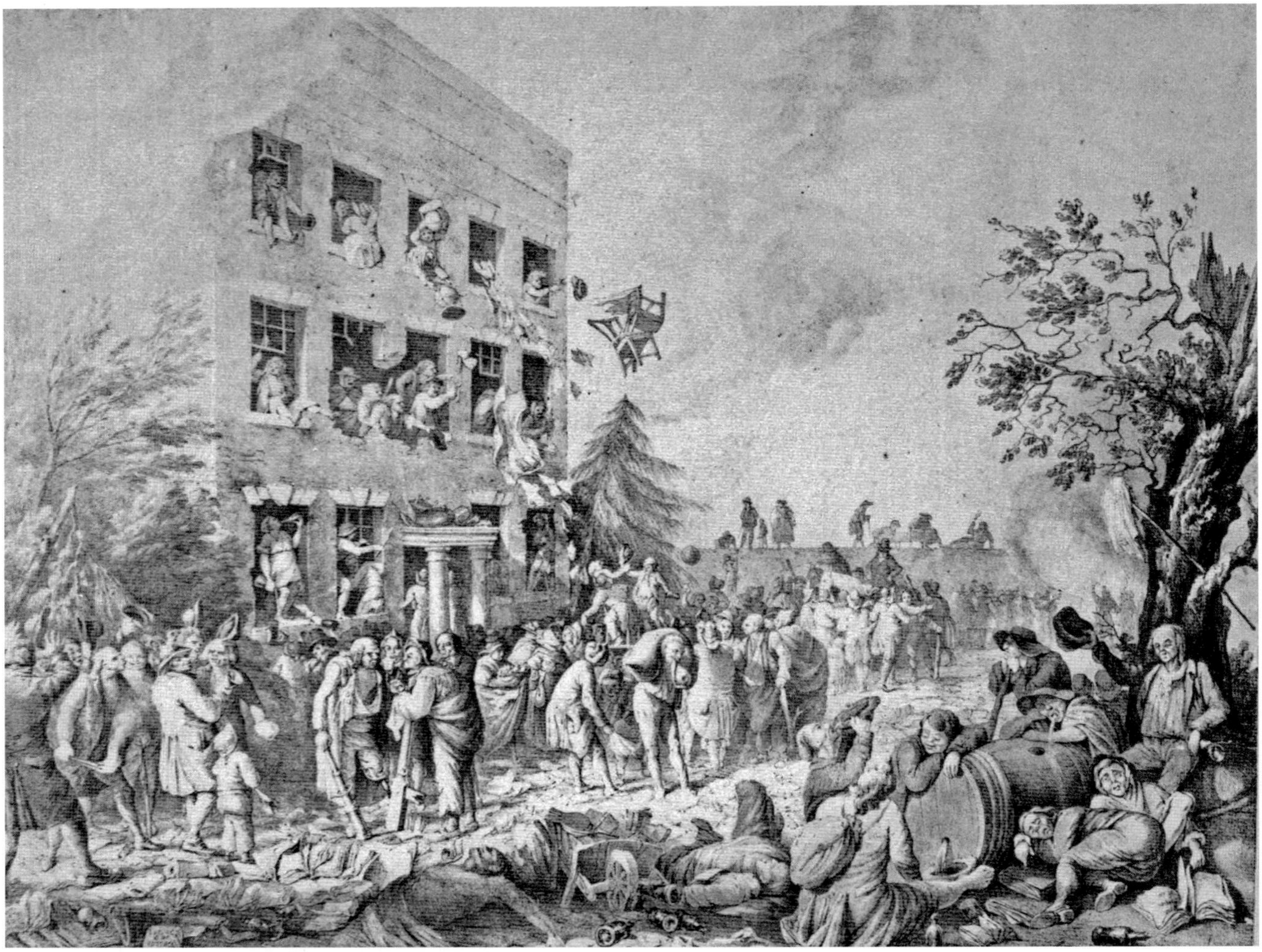

FIGURE 2 The destruction of Priestley's laboratory and library during the Birmingham Riots of 1791
COURTESY OF THE EDGAR FAHS SMITH COLLECTION, UNIVERSITY OF PENNSYLVANIA

Other, similarly imaginative figures have been proposed for these experiments, but from the correspondence and all the invoices of the instrument makers involved in the construction of most of the instruments used in these famous experiments, it is difficult to get an estimate going beyond 10,000 *livres*. The controversy, triggered by the large-scale experiments on the composition of water, fell victim to rhetoric and overemphatic arguments, and unsurprisingly a controversial figure like Lavoisier was subjected to similar accusations by many of his contemporary critics. Although Joseph Priestley was certainly among Lavoisier's less prejudiced acquaintances he, too, criticised the "difficult and expensive apparatus" used by Lavoisier and Meusnier de la Place for the experiments on the decomposition and composition of water.[14] However, these much-quoted lines were written in 1800, after Priestley had moved to US and was extremely frustrated about his lack of access to suppliers of scientific instruments. As is well known, the Birmingham riots of 1791 were at the heart of the complete destruction of Priestley's laboratory, which was very well equipped with a large number of expensive instruments (Fig. 2). Furthermore, Priestley's early pneumatic apparatus was not as simple as he claimed it to be in the prefaces for his books.[15] For manipulating gas he perfected Hales' earthenware pneumatic trough, which would later be made of wood; it was filled with water and used for collecting and manipulating different gases with the help of cylindrical glass jars. The glass jars, receivers, tubes, and balloons were adapted in shape to the purpose of storing gases, and special attention was given to the construction of the joints, stoppers, and taps. Significantly, all this glassware supplied by none other than William Parker, one of the most renowned and expensive instrument makers in London (Figs. 3–4). Priestley also presented in his works

14 Priestley (1800), p. 77.

15 There are no recent studies on Priestley laboratory, but McKie (1956) and Badash (1964) are still extremely useful.

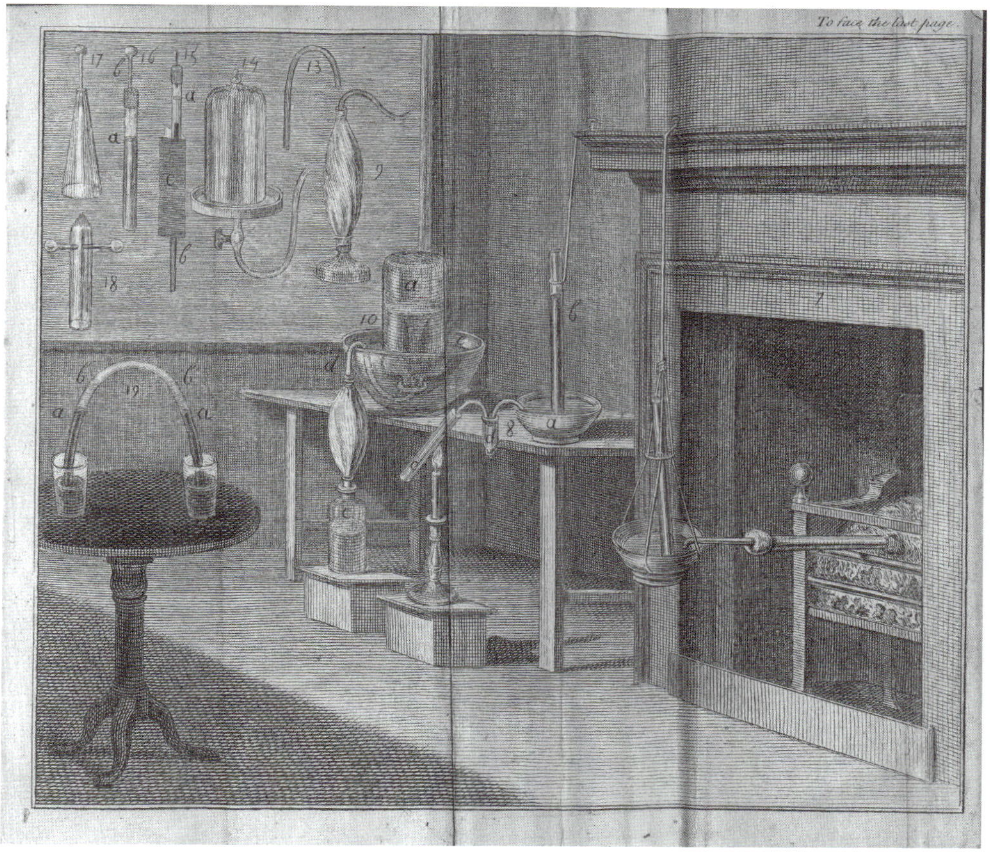

FIGURE 3
Priestley's laboratory and apparatus. From Priestley (1774–1777)
COURTESY OF THE MUSEO GALILEO

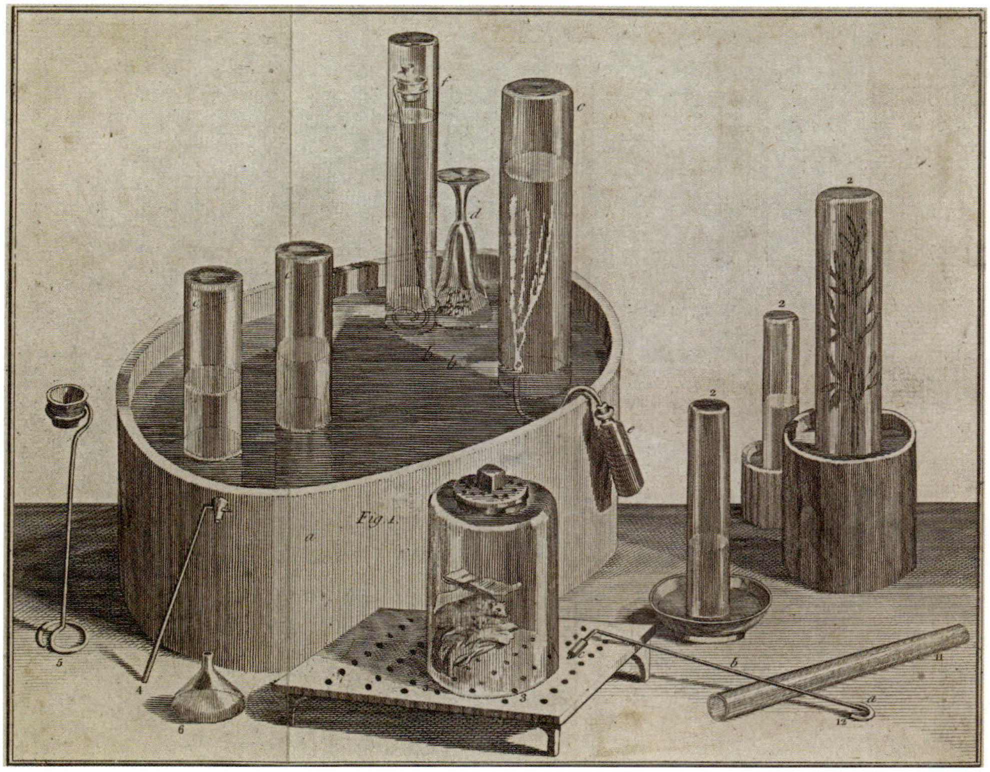

FIGURE 4
Priestley's pneumatic apparatus. From Priestley (1774–1777)
COURTESY OF THE WELLCOME COLLECTION. PUBLIC DOMAIN MARK

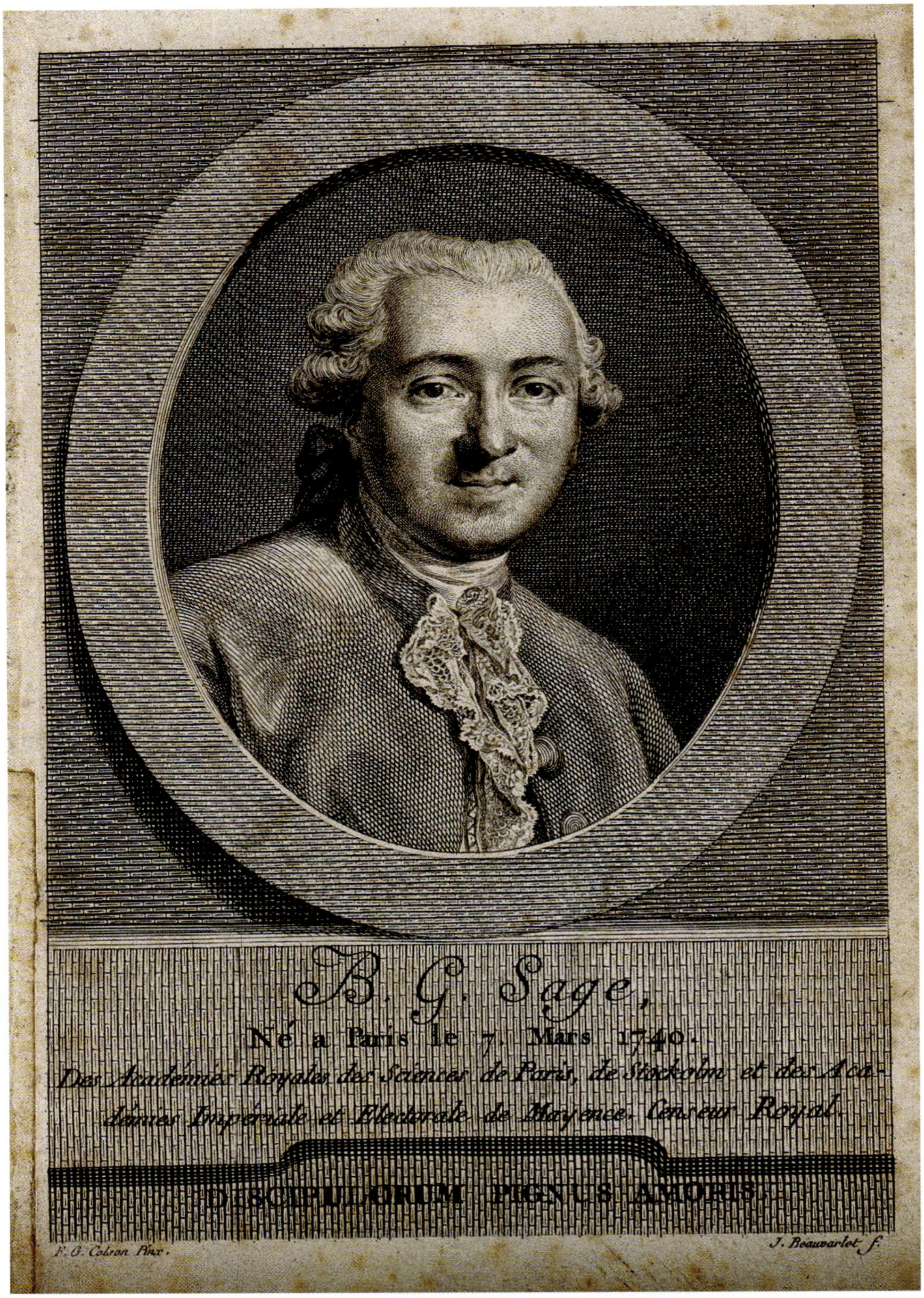

FIGURE 5
Engraved portrait of Sage
COURTESY OF THE WELLCOME COLLECTION

a sophisticated apparatus for expelling gas from solids and for impregnating a fluid with gas, an eudiometer, and apparatus for taking the electric spark in any kind of gas. Inspired by his background in natural philosophy, Priestley used a powerful burning lens made by William Parker to heat *mercurius calcinatus per se* (mercuric oxide), which led to the discovery of oxygen or, as Priestley called it, 'dephlogisticated air'. But I shall return to Priestley's laboratory towards the end of this chapter.

Back in Paris, the apparatus adopted at the Arsenal was, notably, not at entirely unique, and other laboratories matched Lavoisier's both in their size and their expenses. The most interesting case is certainly that of the apothecary Balthazard Georges Sage (Fig. 5), whose education was surprisingly similar to Lavoisier's:[16] born in 1740, he studied at the Collège Mazarin and began his training as a scientist in Rouelle's courses on chemistry, a course in experimental physics taught by Jean Nollet, and some

16 The most comprehensive accounts of Sage's life, based on his many autobiographical notes, are contained in Dorveaux (1935) and (1935a). See also Sage (1818) and (1820).

practical instruction in botany with the naturalist Bernard de Jussieu. Interestingly, like Lavoisier, Sage exhibited an early interest in mineralogical chemistry, and in 1760, at the age of twenty, he opened a *boutique* on the corner of the Rue des Mauvais Garçons, where he delivered free lectures on *minéralogie docimastique*.

In the 1760s Sage decided to specialise in mineralogical chemistry and began to put together a considerable collection of natural history specimens. He managed to attract the attention of Louis XV, was elected a member of the Académie in 1770, and from then on enjoyed the protection of the king. With his royal patronage he was able to publish his chemical books at the Imprimerie de Monsieur (the royal printing house) without having to ask the permission of the Académie des sciences. Moreover, he gained a fortune, which enabled him to take up residence in an elegant *hôtel particulier*, and was granted the astronomic sum of 30,000 *livres* for setting up galleries for his mineralogical collection. In 1778 Sage was appointed professor of chemistry at the newly established École royale des mines at La Monnaie, with a yearly allowance of 6,000 *livres* (Fig. 6).[17] His palatial townhouse allowed Sage not only to keep his collection there, but also to construct a laboratory that must have been one of the most impressive in the capital at the time. (Fig. 7) Like Lavoisier he owned several precision instruments, including one of Mégnié's balances. Sage's brief description of his own laboratory is worth to quote:

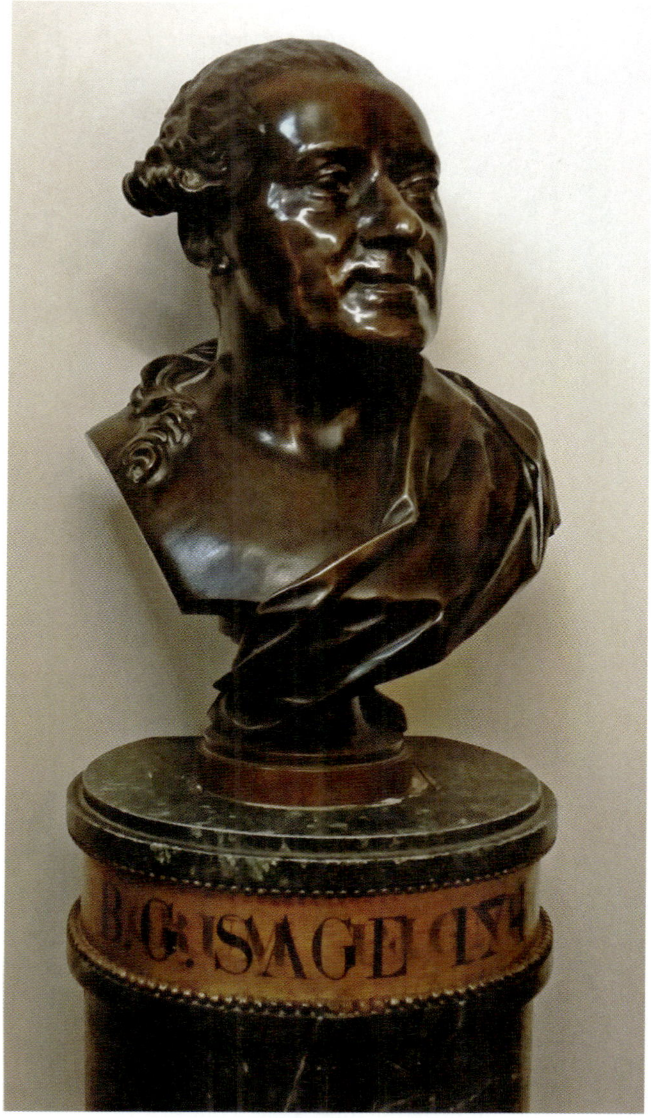

FIGURE 6 Sage's bust, commissioned by his students
COURTESY OF THE ÉCOLE DES MINES – PARIS

> When the King created the Chair of Metallurgical Chemistry in 1777, which M. Valdec de Lessart had enjoined M. Necker to propose to His Majesty, it was established at the Royal Mint by a decision of the Council. This location was the preferred choice, because there is a direct connection between mines and coins in all countries. The laboratory and the office are installed there in the most beautiful premises of Paris, and capture the attention of Foreigners.
>
> The laboratory is equipped with the ovens and utensils specific to experiments in Chemistry, whose details we shall not describe here; but as Docimasy requires special instruments, which are precise and very delicate, I felt it necessary to describe some of those to be found in the laboratory of the École royale des Mines.
>
> A rolling mill, with five-inch cylinders.
>
> Three assay scales, enclosed in cages or lanterns; these instruments are perfect examples of their kind.
>
> An assay scale, weighing from the *marc* to one-twenty-fourth of a grain; it is fitted with a *poids de semelle*, and all weights calibrated up to the *marc*. This fine scale belonged to the Marquis de Courtanvaux.
>
> An assay scale, whose centre of gravity lies below the centre of suspension. This ensures its constant balance, and does not keep it from tilting at the one-hundred-forty-fourth part of a grain. The scale contains a series of assay weights, in which the grain is divided into thirty-six parts. This shortens docimastic calculations, for, as one hundred grains yield

17 Soon after, his students dedicated to him a bust (see fig. 6) with a latin inscription: "J'ai créé le corps des mines, et je devais par droit être compris avant mes élèves dans la nouvelle organisation, où l'on n'a pas même fait mention de moi. Il fut un temps où mes élèves me marquèrent leur reconnaissance en faisant faire mon buste en bronze, et en mettant pour épigraphe sur le cippe: *Discipulorum pignus amoris*. Mais, depuis la révolution, je n'ai plus entendu parler d'eux". Sage (1813a), pp. 28–29.

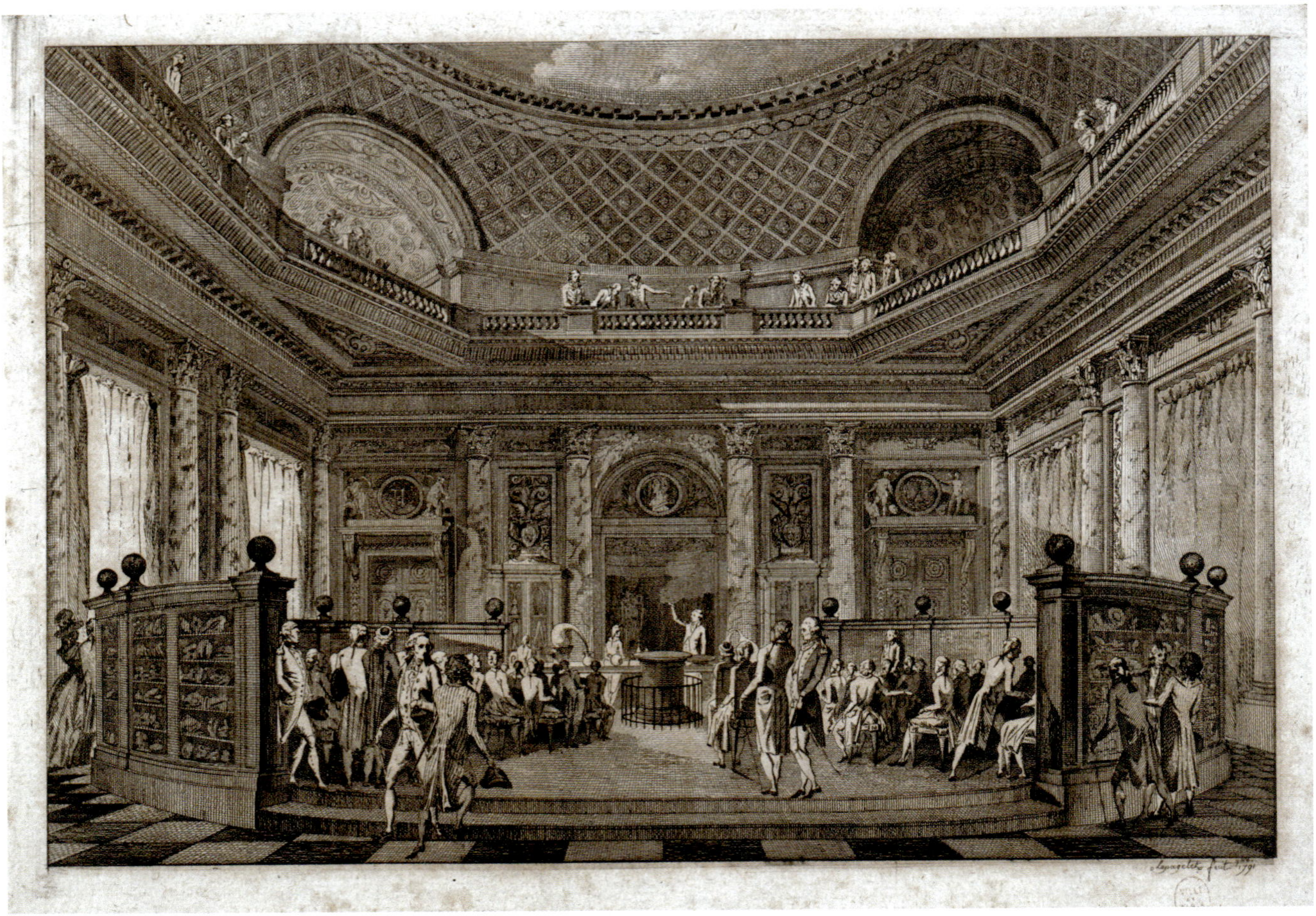

FIGURE 7 Le Pagelet's engraving of a chemical course delivered at La Monnaie (1791)
COURTESY OF THE PARIS MUSÉES/MUSÉE CARNAVALET

one-thirty-sixth of a grain of fine metal, one quintal will yield three gross and forty grains.

The lantern of this Scale contains a series of cupels, offering a picture of the remarkable results that this operation can produce.

The drawer of this scale contains the gilder's brush, tweezers, files, a silver goniometer, and a complete pocket laboratory.

An assay scale, made by M. Mégnié, which cannot remain in balance except when the weights are absolutely equal, because its centre of gravity is slightly higher than its centre of suspension. This scale tips at the two-thousandth part of a grain. It has been made suitable for hydrostatic experiments.

The cabinet is equipped with Dellebarre's microscope,[18] and pneumatic, electric, hydropneumatic and other machines.[19]

At the request of the king, the minister of finance, Calonne, gave 40,000 francs to Sage to help with the cost

18 As was Lavoisier's laboratory.

19 "Lorsque le Roi créa en 1777, la Chaire de Chimie métallurgique, que M. Valdec de Lessart avoit engagé M. Necker à proposer à Sa Majesté, elle fut fondée à l'Hôtel royal des Monnoies par arrêt du Conseil: ce lieu fut choisi de préférence, parce qu'il y a dans tous les pays un rapport immédiat entre les mines & les monnoies. Le laboratoire & le cabinet y sont placés dans le plus beau local de Paris, & fixent l'attention des Étrangers.

Le laboratoire est muni des fourneaux &. ustensiles propres aux expériences de Chimie, dont on n'expose point ici le détail; mais la Docimasie exigeant des instrumens particuliers, exacts & très-délicats, j'ai cru devoir décrire quelques-uns de ceux qui sont dans le laboratoire de l'École royale des Mines.

Un laminoir, dont les cylindres ont cinq pouces.

Trois balances d'essai, renfermées dans des cages ou lanternes; ces instrumens sont parfaits dans leur genre.

Une balance d'essai, pesant depuis le marc jusqu'a la vingt-quatrième partie du grain; elle est munie des poids de semelle, & de tous les poids étalonnés jusqu'au marc: cette belle balance a appartenu à M. le marquis de Courtanvaux.

Une balance d'essai, dont le centre de gravité est au-dessous de celui de suspension, ce qui la porte à un équilibre constant, & ne l'empêche point de trébucher à la cent quarante-quatrième partie d'un grain: cette balance renferme une suite de poids d'essai, où le grain est divisé en trente-six parties, ce qui abrège

of adapting his residence for his laboratory and galleries. Sage's lectures were attended by distinguished scientists such as Romé de l'Isle and Jean-Antoine Claude Chaptal. By the mid-1770s Sage had achieved a great academic reputation, enjoyed the generous protection of the king, and exerted considerable power and influence.

In 1778, however, the tide began to turn against him. In April and May of that year Sage announced to the Académie that he had discovered a method to draw pure gold from 'vegetable earth', which was refuted by a member of the Académie des sciences' chemistry section via experimental tests. Sage was involved in other controversial experiments which cost him his scientific reputation both in Paris and abroad, but his position did not suffer, and he continued to benefit from the unconditional protection of the king until the outbreak of the French revolution in 1789.[20] In the 1780s Lavoisier conducted experiments with Sage on several occasions, and Sage granted Lavoisier access to minerals and stones that could only be found in his cabinet. Sage's case is interesting in many respects, but the patronage he enjoyed was quite exceptional.

The cost of chemical laboratories, however, increased exponentially regardless of royal subsidies or individual fortunes. A striking example is that of Antoine Baumé who, as mentioned in the preceding chapters, was one of the most influential experimental and industrial chemists in Paris. Although they refer to a completely different social and economic background, Antoine Baumé's pharmaceutical and chemical laboratories in the Rue Coquillière became so important that when he was able to sell them in February 1780, for 126,000 *livres* – much more than the total value of Lavoisier's laboratory according to the inventory and assessment of 1794.[21] The reason why Baumé realised such a high price for his laboratory is not known, although he was certainly among the Parisian chemists most actively seeking to improve the existing apparatus, and, with his shop, had become the most important supplier of apparatus in Paris. Moreover, Baumé traded in all sorts of chemicals, including rare and expensive commodities such as platinum and mercury. (Figs. 8–9)

The private laboratory built by the aristocrat, instrument maker, and chemist Louis Joseph d'Albert d'Ailly, Duc de Chaulnes, also included sophisticated chemical apparatus, for which it acquired great fame and prestige. Aware of the superiority of English instrument making, the Duc de Chaulnes, like Trudaine de Montigny, successfully introduced new experimental techniques in pneumatic chemistry to Paris.[22] In all these laboratories public experiments were routinely performed, and they also contained cabinets which displayed the mineral collections. They were occasionally used as venues for social gatherings for academicians, apothecaries, laymen, and women. The financial turnover of these laboratories was considerable and in many comparable ways to Lavoisier's.

It is not surprising that chemical experiments of any kind required a considerable amount of money, because chemistry was not at all, as is commonly believed, an economic science. While it is true that prior to the discovery of gases the apparatus was simple and relatively inexpensive, its rapid deterioration, due both to frequent use and the destructive action of heat and fire, required the crucibles, alembics and other common devices to be constantly replaced. Chemicals and reagents were also relatively expensive, and one might reasonably argue that the waste of apparatus, chemicals, and fuel was considerable before the introduction of precision instruments for the control and measurement of heat. Indeed, depictions of early-eighteenth-century chemical laboratories show them overcrowded with all sorts of objects and instruments. Even given that these were relatively inexpensive, their frequent turnover necessarily increased the cost in the long run.

---

les calculs docimastiques, car cent grains produisant un trente-sixième de grain de fin, un quintal produira trois gros quarante grains.

La lanterne de cette Balance renferme une suite de coupelles, qui offre le tableau de ce que cette opération peut produire de remarquable.

Le tiroir de cette balance renferme le grattebosse, les pinces, les limes, un goniomètre en argent & un laboratoire de poche complet.

Une balance d'essai, faite par M. Mégnié, qui ne peut conserver l'équilibre que quand l'égalité est absolue, parce que le centre de gravité est un peu plus haut que celui de suspension; cette balance trébuche à la deux millième partie d'un grain; elle est rendue propre aux expériences hydrostatiques.

Le cabinet est muni du microscope de Dellebarre, de machines pneumatique, électrique, hydropneumatique, &c." "État du laboratoire de l'école royale des mines"; Sage (1784), pp. 483–487.

20 Several harsh statements against Sage may be found in Bergman's correspondence; Bergman (1965); Beretta (2014b).

21 Baumé's pharmacy was purchased by two apothecaries, Pierre François Dreux and François Fourcy. Davy (1955), pp. 24–26.

22 For a comparison between the English and French contexts, see Hilaire-Pérez (2000).

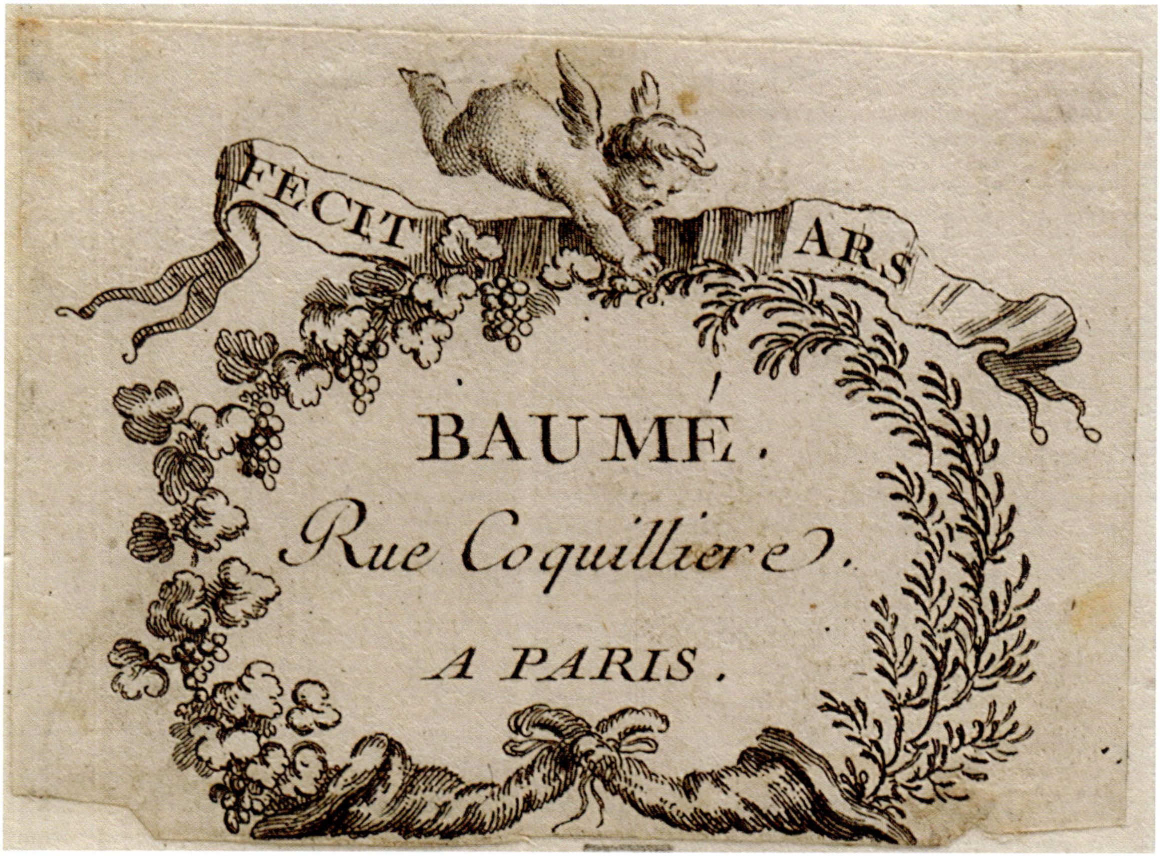

FIGURE 8 Trade card of Antoine Baumé's pharmacy, *c.*1760
COURTESY OF THE WADDESDON (NATIONAL TRUST) BEQUEST OF JAMES DE ROTHSCHILD, 1957. ACC. NO: 3686.2.55.140. PHOTO: WADDESDON IMAGE LIBRARY, UNIVERSITY OF CENTRAL ENGLAND DIGITAL SERVICES

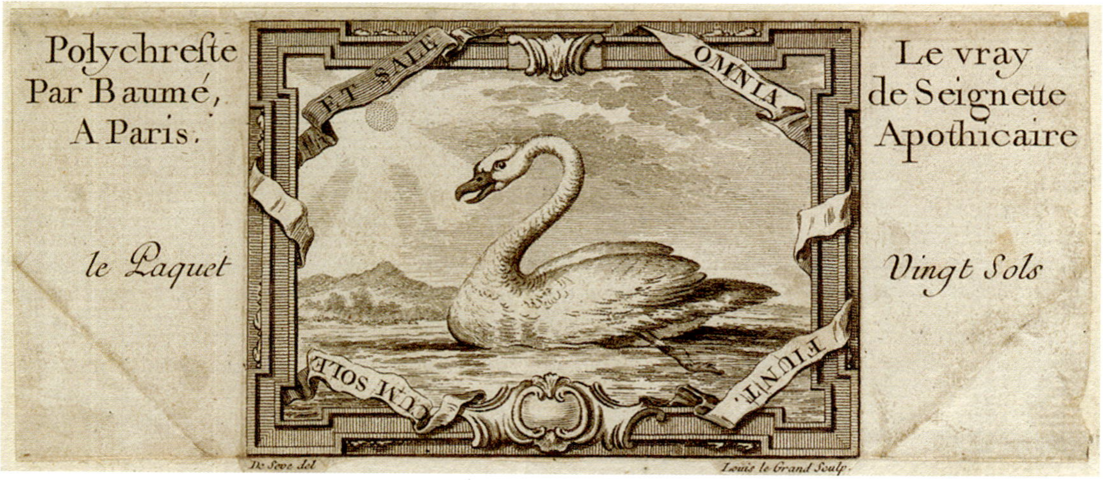

FIGURE 9 Jacques de Sève, Wrapper and Label for Seignette Salt, sold by Antoine Baumé, 1750–1770
COURTESY OF THE WADDESDON (NATIONAL TRUST) BEQUEST OF JAMES DE ROTHSCHILD, 1957. ACC. NO: 3686.2.35.89. PHOTO: WADDESDON IMAGE LIBRARY, UNIVERSITY OF CENTRAL ENGLAND DIGITAL SERVICES

4.2 The Cost of Lavoisier's Laboratory

Let us now approach Lavoisier's laboratory and its cost in more detail. Lavoisier was a very rich man, but he limited his investment in the chemical laboratory to what he considered strictly indispensable to the success of his experiments. In 1791 one of his closest and most talented assistants, Armand Séguin, reported to James Hall that Lavoisier "spent not above 250£ on chemistry – that his revenue is 16,000£ a year".[23] The approximate modern equivalent of 250 pounds is about 50,000 Euro, which is a considerable sum but not extravagant, and even less so once we consider his annual income of more than 2,5000,000 Euro. It should also be noted that the cost of Lavoisier's apparatus reached its peak in the 1780s and 1790s, i.e. exactly at the time of Séguin's mentioned report.

Earlier in his career, probably in late 1774 or early 1775, Lavoisier wrote in detail on the necessary equipment for a standard chemical laboratory.[24] There he listed, among other things, one large fusion furnace (500 *livres*) eleven furnaces (fusion, cupellation, distillation, and reverberatory), mostly of Baumé's and Macquer's design (170 *livres*), two furnaces for gases after Priestley (9 *livres*), more than 300 crucibles of different sizes and materials (60 *livres*), 50 retorts made of refractory clay (50 *livres*), several types of glassware such as alembics, cucurbits, pelicans, 350 glass retorts of different sizes, 40 recipients made of crystal, four glass balloons made of glass or crystal, five Woulfe's bottles for the distillation of mineral acids, 24 mattrasses for distilling, 24 glass funnels, 150 recipients of different capacities, 400 jars, 500 bottles (for a total amount of more than 700 *livres*), four balances (for a total amount of 240 *livres*), four mortars of different sizes and materials (glass, marble, and iron), a copper alembic and serpentine (348 *livres*), and a pneumatic trough and 72 kilos of mercury for the reconstruction of Priestley's experiments (for a total amount of 600 *livres*). In addition to these pieces of apparatus and chemicals Lavoisier listed several kinds of utensils, recipients, and other minor devices. Lavoisier excluded the costs of the rooms and the tables on which the listed experiments were to be performed in part. The total cost of the laboratory was, according to Lavoisier, 3,600 *livres*, a sum lower but not too different from Séguin's estimate of 1790. However, from Lavoisier's conclusion it is clear that he had been asked to produce a budget for a relatively economic laboratory. Surprisingly, there are no references to instruments of natural philosophy like hydrometers, barometers, and thermometers, which Lavoisier had regularly used in his experiments from the mid 1760s onwards. However, it should be noted that, while precision instruments played an important role in Lavoisier's laboratory, they were kept separately from the chemical apparatus, and were allocated to a *cabinet de physique*. As shown in the Appendix 3, the precision instruments were similarly separately listed in the inventory of the commissioners appointed by the revolutionary authorities in 1794.

It is, in fact, the apparatus used in experimental physics for which Lavoisier's correspondence provides the most detailed information about the costs and the people involved in its construction. Even in the early stages of his scientific career Lavoisier wished to adapt the methods of experimental physics to chemistry, and to this purpose he had several instruments made which helped him with the quantitative measurement of the phenomena and reactions he dealt with. This approach was probably influenced by Jean Nollet who not only showed in his *Leçons de physique expérimentale* the central importance of perfecting the design of the apparatus for experiments, but also delivered several lectures on chemical topics.

The first instrument whose acquisition was recorded in Lavoisier's correspondence is a modified version of Papin's digester made in 1776 by François Philippe Charpentier (1734–1819), the *mécanicien du roi*. Lavoisier paid 780 *livres*, a very high price. However, this device, originally conceived by Papin in 1681 for domestic purposes, became a rather complex scientific apparatus. Lavoisier used it in experiments on the expansion of air both under pressure and in heat. He used the new digester in combination with a balance and De Luc's barometer. A digester, probably a different one, was also used in 1783 during the first experimental campaign on the decomposition of water.

In March 1776 Lavoisier purchased a number of items from Antoine Baumé, who owned the largest apothecary shop in Paris, including several chemicals such as red precipitate of mercury, phosphorus, nitre, ammoniac, and marine salts; 50 crystal jars, two chemical bottles, four barrels of different sizes, and other utensils; all for the total sum of 474 *livres* and 14 *sous*. In less than a month then he spent more than 1,100 *livres*. In the same year, on 26th October, Lavoisier purchased further items from Jean-Baptiste Bucquet, an apothecary with whom he had collaborated on an important experimental campaign on various chemical topics. These included one pneumatic trough, two bell jars, four retorts, and almost two kilos of mercury and red precipitate. No price is stated, but it

23 The amount spent was little more than 5,000 *livres*. In 1781 one pound sterling was equivalent to 22 *livres* 8 sols (20 sols made 1 *livre*); on this see Kindleberger (1984), p. 475; Selig (2013). Denton (Paris, 2003), p. 549. I thank Elizabeth Denton for providing me with a copy of her thesis containing the edition of Hall's diary.

24 Lavoisier (1775a).

seems that Bucquet supplied Lavoisier with higher quality apparatus than Baumé. Of course, all these purchases were unusual, but they were justified by Lavoisier's move to the Arsenal, where he had plenty of space for a large laboratory, in the spring of 1776.

Lavoisier's interests were not exclusively chemical, and sometime in 1777 he had the idea to create an international meteorological network using standardised and comparable barometers. He initially asked the Paris engineer, clock maker, and instrument maker Alexis Magny (1712–1779) to make a new barometer. Lavoisier used this barometer at the beginning of his meteorological campaign of 1778, but following Magny's death in 1779 he asked Pierre Bernard Mégnié (1751–1807) to make at least twelve copies a new kind of barometer, which he distributed to several colleagues in France as well as abroad.[25] Marie Anne Lavoisier reported that as late as 1805 meteorological data were still sent in from Baghdad.[26] Mégnié was an *ingénieur en instruments mathématiques*, and owned a shop at the Cour de Commerce, in the Faubourg Saint Germain in Paris. In May 1775 the Portuguese inventor and industrial spy Jean Hyacinthe de Magellan claimed that Mégnié, despite his young age, was already a competent maker of sophisticated astronomical instruments.[27] From 1779 onwards he was extensively employed by Lavoisier, whom he called his patron (*protecteur*) and benefactor (*bienfaiteur*).[28] During the winter of 1781/82 he contributed to the construction of Ramsden's pyrometer which was used by Lavoisier and Laplace in experiments at the Jardin de l'Arsenal.[29] Lavoisier also owned a small assay balance (which had been made in Mégnié's workshop), a mercury thermometer, and a glass recipient used for the experiments on the composition of water (1786).[30] The most important and expensive piece of apparatus made by Mégnié for Lavoisier was a pair of gasometers (1787) designed by Jean Baptiste Meusnier de la Place, which cost 7,554 *livres*.[31] It is interesting to note that, during their evaluation of 1794, Fortin, Lenoir, and Charles still valued the gasometer at 8,000 *livres*, while the notary who inventoried Madame Lavoisier's belongings after her death in 1836 gave a valuation of only 60 francs.[32]

Mégnié was an instrument maker who was highly regarded by many members of the Académie des sciences, but from his correspondence with Lavoisier it is clear that his profession was not particularly financially rewarding, and that the construction of complex apparatus such as the gasometer would fully occupy his workshop for more than 100 working days. It was amidst of these difficulties, Lavoisier's patronage greatly helped Mégnié's reputation.

Another instrument maker who owed much of his fame to Lavoisier's patronage was Nicolas Fortin. He owned a shop on the Place de la Sorbonne in Paris and employed only two assistants: Jean Louis Cheron and Louis Marie Frédérique Pinson.[33] According to Thomas Bugge's travel diary of 1798, Fortin was a "very good craftsman" but he required "better payment than Dumotiez, because his workshop [was] not so well staffed and equipped".[34] Fortin became one of the most important suppliers for Lavoisier. In a letter to Achard dated 6 December 1789 Lavoisier declared that Fortin was the *ouvrier* he employed "le plus habituellement".[35] Fortin's first invoice addressed to Lavoisier dates to 1st March 1784, and is for one of his famous and very expensive air pumps.[36] From an invoice sent on 20 July 1785 it is clear that he provided Lavoisier with most of the equipment used in the experiments on

25 On Lavoisier's use of the barometer, see LC, vol. 3, p. 626. On his employment of Mégnié, see LC, vol. 3, p. 623 ff. Daumas (1955), p. 128. Two barometers are in the collection of the Musée des arts et metiers (MAM), inventory nos. 08761-0000- and 19949-0000-.

26 "Il fit construire par le meilleur ouvrier douze baromètres tous d'accord. Il en plaça un à l'observatoire de Paris, un en auvergne, il en confia un aux soins du consul de bagdad [the Abbé Beauchamp who was in fact astronomer and 'vicaire-général' to the Bishop of Babylone *in partibus*], un fut envoyé en Russie, un en Suéde. J'ignore quel fut le succès de cette belle idee. Le consul de bagdad seul envoya quelques observations". Quoted in Gillispie (1956), p. 59; on this interesting episode see also Bret (2019).

27 In a letter to Jacques-André Mallet, Magellan reported that Mégnié was about to finish a "machine parallatique", and that he was very happy with his work. Considering that Magellan was living in London and was in daily contact with English instrument makers, this compliment, addressed to a young Parisian artisan, is most unusual. This letter is also important in another respect, because it shows that Lavoisier relied on well-known makers of mathematical instruments and persuaded them to adapt their skills to the creation of chemical apparatus. The Magellan letter is published in Home *et al.* (2017), vol. 1, pp. 463–464.

28 LC, vol. 4, pp. 66–67.

29 LO, vol. 2, pp. 739–764.

30 MAM inv. no. 19885-0000- (assay balance); this might have been similar to the one listed in Sage's laboratory at la Monnaie. MAM inv. no. 19914-0000- (mercury thermometer). Musée d'histoire national naturelle Paris, inv. n. OA 289 (glass recipient).

31 LC, vol. 5, pp. 52–55; MAM, inv. nos. 07547-0001-001- and 002-.

32 On Fortin, Lenoir, and Charles' estimation, see the appendix 3. "Deux grands gazomètres de Lavoisier avec tous leurs accessoires, prisés ensemble la somme de soixante francs". Paulze-Lavoisier (1836).

33 LC, vol. 7, p. 391.

34 Crosland (1969), p. 171.

35 LC, vol. 6, p. 90.

36 LC, vol. 4, p. 11 (invoice). The MAM owns an air pump made in 1792 (inv. n. 07517-).

water.[37] The total cost of the equipment amounted to 483 *livres*. In April 1786 Fortin provided Lavoisier with an electrical machine (172 *livres*), two pipes, and an apparatus for collecting nitrous air.[38] In June 1788 he sent another invoice for apparatus used in the experiments on the synthesis of water at the Collège royal, which took place on 27 May and 7 June.[39] In August 1788 Fortin finished the large-scale apparatus for the experiments on oil combustion (1,109 *livres*), a great balance (600 *livres*), the apparatus for the ether (520 *livres*) and an apparatus for the combustion of carbons (110 *livres*).[40] Between 1788 and August 1789 he supplied Lavoisier with two apparatuses for wine fermentation.[41] On 9 May 1790 he sent an invoice of 480 *livres* for a few new pieces of equipment and the readjustment of previously purchased ones.[42] On 1st December 1792 he sent an invoice of 300 *livres* for the construction of several instruments, which are not further specified. Lavoisier also owned a compass and a gasometer of his making.[43]

Mégnié and Fortin were Lavoisier's favourite suppliers of equipment, but they were certainly not the only ones. Lavoisier in fact purchased instruments and apparatus from more than 80 makers, mostly Parisians.[44] Of these many supplied him with relatively standardised items, e.g. the microscope and case made by Louis François Dellebarre (1726–1805) in December 1778, which cost only 18 *livres*.[45]

In addition to professional instrument makers Lavoisier relied on the assistance of others whose important roles are more difficult to recognise in his correspondence.

4.3 The Cost of Labour

Even in the eighteenth century chemistry was a science which could not be practised by one person alone. Throughout his career Lavoisier employed assistants, some of whom later established an independent career in science and industry. It is quite difficult to list these assistants comprehensively, as it is likely that when Lavoisier moved there many already worked at the Arsenal's refinery and were therefore salaried artisans with whom he had daily contact. However, Lavoisier's correspondence and indirectly related documents provide us with some interesting information. As early as 1774, when his laboratory was still situated in the Rue des Bons Enfants, Lavoisier employed a tinsmith named Pierre Naudier (or Nodier); Naudier later had an important role in the making of the calorimeter and worked on the development of the apparatus used in Lavoisier's and Meusnier de La Place's experiments on the decomposition of water of April 1784.[46] He would later be employed by the Commission des poids et mesures.[47] After Macquer's death in 1784, Lavoisier employed Macquer's former *garçon de laboratoire*, Fallot, who worked "almost every day at the Arsenal" and prepared the experiments.[48] Since Fallot was already employed by the Régie des poudres, Lavoisier did not need to pay for his assistance.

Lavoisier also managed to recruit professional scientists, including Jean Baptiste Bucquet, who supplied him with chemicals and apparatus, and Jean-Baptiste Meusnier de la Place, who played a key role in the construction of the gasometers, and again during the campaign devoted to the experiments on water (1783–1785).[49] Since this campaign ran parallel to the work of the academic commission which had been appointed in 1783–1784 to examine the technical and scientific details behind Montgolfier's invention of the balloon, Lavoisier managed to have Meusnier paid for his services by the Académie.

Other qualified assistants, who were all of a humble social background, were Philippe-Joachim Gengembre (1764–1838), Jean Henri Hassenfratz (1755–1827), Jean Joseph Welter (1763–1852), Armand Séguin (1767–1835) and the amateur chemist Charles André Hector Grossart de Virly (1754–1805). The need of this kind of assistant clearly emerged during the mid-1780s, when Lavoisier began to think about large-scale chemical experiments and the conception of certain pieces of apparatus, such as the calorimeter and gasometers, which required a

37 LC, vol. 4, pp. 137–140.
38 Used in March 1786; Berthelot, p. 304 (electrical machine). LC, vol. 4, pp. 212–213 (apparatus for collecting nitrous air).
39 LC, vol. 5, pp. 176–177.
40 MAM, inv. no. 07549-0000- (oil combustion apparatus). MAM inv. no. 19887-0000- (great balance). LC, vol. 5, pp. 209–210 (apparatus for the combustion of carbons).
41 LC, vol. 6, p. 64 and MAM, inv. nos. 07550-0000- and 07551-0000-.
42 LC, vol. 6, pp. 141–142.
43 See LC, vol. 7, p. 169 (invoice); Beretta (2003), pp. 313–334 (compass). "Ancien gazometre de Fortin", ibid.
44 See Lavoisier's list of instruments makers and suppliers; Appendix 6 pp. 162–180.
45 LC, vol. 3, pp. 631–632.
46 LC, vol. 2, p. 458 (1774); MAM, inv. no. 07547-0002-001- (calorimeter); LC, vol. 4, pp. 18, 26–29, 33 (experiments on the decomposition of water).
47 LO, vol. 6, p. 693.
48 Fallot is first mentioned in August 1784 in LC, vol. 4, p. 33. "Le Sieur Fallot, qui, après avoir été pendant long-temps garçon de laboratoire du célèbre Macquer, était employé comme directeur des ouvriers à la raffinerie des salpêtres de l'Arsenal, et travaillait presque tous les jours au laboratoire de notre illustre Lavoisier". Cited in Bottée, Riffault (1811), p. 332.
49 LC, vol. 4, pp. 299–304.

complex division and organisation of the experimental work.

Gengembre was the son of a concierge at the Louvre.[50] He studied with Monge and Sage before entering the École des poudres at the Régie des poudres, and from 1785 he taught courses in chemistry which combined fundamental research with practical chemistry. From 1784 he assisted Lavoisier and Meusnier de la Place in their campaign of experiments on the composed nature of water, and together with Naudier and Fortin set up the apparatus for the experiments. In the same year he prepared the experiments on guinea pigs.[51] From 1783 to 1785 and probably beyond, he was paid by the Régie. In 1787 Lavoisier involved him in the preparation of the sequel to the *Description des arts et métiers*.[52] In a letter of recommendation dated 19 July 1790 Lavoisier writes that Gengembre "a poussé for loin les mathématiques" and that he had good knowledge of practical geometry.[53] In November 1795 he was appointed "artiste-mécanicien" at La Monnaie, where he designed new mechanical devices prior to become "Inspecteur general" (1803–15) and then director of the steam engine factory of the Navy at Indret (1828–38).

Hassenfratz began his training as a carpenter with Nicolas Fourneau, and this practical background influences his later interest in solving technical problems.[54] After being taught mathematics by Monge and chemistry by Sage, he became a mining inspector and began to work regularly in Lavoisier's Arsenal laboratory in the early 1780s. On 14 May 1782 he presented a series of observations made with the eudiometer before the Société royale de médecine, and this seems to be his earliest work dealing with chemical subjects. In the summer of 1783, when he worked in Vienna at the laboratory of Jan Ingen-Housz, he wrote to Lavoisier about the eudiometer he saw there.[55] After his return from Austria, Lavoisier recruited Hassenfratz in his laboratory, who soon became his favourite assistant.[56] In April 1786 he supplied Lavoisier with ca. 100 *livres* of heavy spar (barytine), a mineral of barium sulphate usually used in paint.[57] Around the same time he published the first comprehensive work on the chemical use of the blowpipe to appear in France, in which he also proposed his own technical innovations.[58] In August 1788 he appears to have worked on Madame Lavoisier's drawings of Lavoisier's apparatus, which were soon to be published in the *Traité élémentaire de chimie*.[59] Lavoisier regularly helped Hassenfratz financially while the latter was working at the Arsenal.[60] Lavoisier was so impressed by his skills that, one year later, he proposed him for election as a member of the Académie, albeit unsuccessfully. Nevertheless, Lavoisier not only included Hassenfratz in the circle of the Arsenal, but also appointed him to director of the laboratory, which secured him a salary of 1,200 *livres*.[61] While Hassenfratz showed great ability in marketing his skills, it soon became clear that his role in the highly competitive environment of the Arsenal necessarily needed to be reappraised. Sometime before 1788, when Armand Séguin was recruited to join the Arsenal, the competition became especially fierce.

In the early 1790s Hassenfratz became less assiduous in Lavoisier's laboratory and rented a flat in Fourcroy's residence *à la Couronne d'Or*, in the Rue des Bourdonnes. The rent was probably paid by Lavoisier and it was here that Hassenfratz made his experiments on nutrition on behalf of the French chemist.[62]

First introduced to Lavoisier by Fourcroy in the mid 1780s, Séguin initially worked in the laboratory of the Arsenal as a *garçon de laboratoire* but soon became one of Lavoisier's favourite assistants, and from 1789 onwards he was involved in several crucial experiments.[63] On 20 September 1789 he was in charge of assembling Fortin's apparatus for combustion.[64] In November 1790 he conceived and assembled apparatus for the experiments on human respiration, and he was granted the honour of presenting it, alone, before the Académie des sciences.[65] Significantly, in his famous essay "Détails historiques, sur la cause de l'augmentation de poids qu'acquièrent les substances métalliques, lorsqu'on les chauffe pendant leur exposition à l'air", Lavoisier gave Séguin full credit for his collaboration in laying the foundations for the theory of respiration.[66] Séguin had a prominent role in perfecting the accuracy of the scale invented by Santorio Santorio for

50 On Gengembre, see Bret (1994), pp. 231–250.
51 LC, vol. 4, p. 6.
52 On this, see LC, vol. 5, pp. 277–286.
53 LC, vol. 6, p. 153.
54 Grison (1996).
55 LC, forthcoming in the supplementary volume of Lavoisier's correspondence.
56 At least until Armand Séguin contended this privilege in the late 1780s.
57 LC, vol. 4, p. 206.
58 Hassenfratz (1786).
59 LC, vol. 5, pp. 199–200.
60 Grison (1996), pp. 84–85.
61 Grison (1966), pp. 59–60 and pp. 83–85.
62 Ibid., p. 109.
63 On Séguin, see Mercier (1976). On the collaboration between Lavoisier and Séguin, see Beretta (2001).
64 LC, vol. 6, pp. 70–71.
65 On the apparatus, see LC, vol. 6, pp. 192–193.
66 Lavoisier (1805), vol. 2, pp. 78–87.

weighing the loss of fluids due to transpiration.[67] In 1791 he followed Lavoisier's experiments on the fusion of platinum and, together with Meusnier and James Hall, conceived a furnace that could produce high temperatures.[68] The furnace was built in his workshop in the Faubourg Saint Antoine, close to the Arsenal, where he worked with his brother, and it was used by Lavoisier for the experiments on transpiration. Séguin's furnace would eventually also be highly appreciated by the Commission des poids et mesures;[69] and he further developed a new type of eudiometer in 1792.[70] In 1795 his new method of preparing leather within a few days – "which formerly underwent a preparation of two years" – became vastly successful.[71]

One 'Velter', probably Jean Joseph Welter, is mentioned in Lavoisier's *registre de laboratoire* in connection with the experiments in March 1788 to convert Marsilio Landriani to Lavoisier's new theory of combustion.[72] Welter was a maker of chemical apparatus and an industrial entrepreneur who produced large quantities of gunpowder at Meudon towards the end of the eighteenth century.[73] He eventually founded a chemical factory near Valenciennes, but was also involved in the successful improvement of the manufacture of textiles.[74] He was an occasional collaborator of Berthollet, Laplace, and Guy Lussac at the beginning of the nineteenth century.

With the exception of Hassenfratz, who was slightly older, the assistants were born during the 1760s, and were probably both trained and employed by Lavoisier. The generation gap is important, because apparently Lavoisier sought for artisans who would not hold any prejudice against the reform he wished to introduce to the art of chemical experimentation.

4.4 Deconstructing the Legend

But what was, then, the overall cost of Lavoisier's laboratory? In the spring and autumn of 1794 Lavoisier's laboratory was confiscated, and the apparatus and glassware were carefully inventoried and valued by experts appointed by the Commission temporaire des arts and the Commission d'instruction publique, in the course of several visits.[75] The task was complicated by the fact that Lavoisier's new residence in the Boulevard de la Madeleine was not a particularly suitable for a large laboratory, and the instruments filled several rooms and the cellar. The first inventory was made on 30th May 1794 by the apothecary and inventor Antoine Arnoult Quinquet, who was charged by the commission to select all pharmaceutical items. The list compiled by Quinquet was quite comprehensive, and he often included items without a direct connection to pharmacy, e.g. an electrical machine. The most expensive item he listed was a large quantity of mercury (more than 47 kg), which was estimated at 1,584 *livres*. The overall valuation of the pharmaceutical collection was 5,294 *livres* and 19 *sous*. The collection was moved to the hospice of the Tribunal révolutionnaire, and eventually dispersed.

Nicolas Leblanc and Claude Louis Berthollet compiled a preliminary inventory of the chemical apparatus on 5 November 1794, and in the following days Leblanc inventoried some 7,000 items of instruments and glassware, which he estimated to have cost 7,237 *livres* and 10 *sous*. It is worth noting that the sum nearly equals the amount Lavoisier spent for David's portrait in 1788. However, Leblanc's inventory did not account for the precision instruments, which were catalogued separately by experimental scientists and instrument makers including Jacques Alexandre César Charles, Etienne Lenoir, and Nicolas Fortin; all had collaborated with Lavoisier during the 1780s. This separate list identified 122 items, including Mégnié's gasometers, Fortin's air pump, the precision balance and the apparatus for the combustion of oils, two more precision balances, the calorimeters, and less valuable instruments *de physique* such as hydrometers, thermometers, barometers, Volta's pistols, and Nollet's mechanical models. Although the inspectors did place an overall value on the 122 items, they established that the cost of the seven most valuable items exceeded 14,000 *livres*, a considerable sum that alone doubled Leblanc's estimate for 7,000 ordinary pieces of apparatus.

The estimated cost of the apparatus found in Lavoisier's laboratory exceeded 27,000 *livres* which, if we take inflation as a parameter, would be comparable to roughly 300,000 Euro today. Although a few instruments were left out of the assessment, the revolutionary inventories, compiled by professionals, instruments makers, and chemists, provide a good overview of the real cost of Lavoisier's

67 Beretta (2012).

68 On Meusnier and Hall, see Chaldecott (1968), p. 33. On the furnace, see LC, vol. 6, pp. 327–328.

69 Chaldecott (1968), pp. 50–51.

70 Séguin (1814).

71 Crosland (1969), p. 150.

72 On Welter, see LC, vol. 5, pp. 141 and 220. "On s'est rassemblé au laboratoire: MM. Landriani. Assenfratz(sic). Velter. Lavoisier. M. Landriani a proposé les expériences suivantes". Quoted in Berthelot (1890), p. 307.

73 Bottée, Riffault (1811), p. 333 and Bret (1992).

74 Chaptal (1819), p. 9.

75 Transcriptions of the inventories are published in the appendixes 1–3.

laboratory. With this estimate in hand we are in a better position to assess the criticism of the cost of his laboratory which was voiced by many of Lavoisier's contemporaries. A comparison of the estimate with the cost of other contemporary laboratories may lead to an even more precise picture.

I have already pointed out that Baumé's laboratories were sold for 126,000 *livres* in 1780. Since Baumé is always portrayed as traditional chemist who, like most Parisian pharmacists and apothecaries, performed chemical experiments with apparatus that was cheap to acquire and economic to run, the cost of his shop and workshops exceeds by far the overall cost of Lavoisier's laboratory.

Outside Paris we find a similar scenario. During the Birmingham riots of July 1791, Joseph Priestley lamented the almost complete destruction of his laboratory and library. After careful assessment he claimed damages to the amount of 4083 pounds one year later. The cost of the instruments, apparatus, and chemicals alone amounted to almost 700 pounds, in today's value 102,000 Euro, and about one third of the cost of Lavoisier's laboratory. Since Joseph Priestley has often been compared to Lavoisier to show the presumed simplicity of his experimental approach, this comparison indicates to me that accurate comparative studies on eighteenth-century chemical laboratories and their costs are badly needed.

Martinus Van Marum, who is often remembered for his successful attempt to produce a cheaper version of Lavoisier's gasometers, deserves careful attention. Van Marum was, in fact, converted to the new chemistry by Lavoisier's and Meusnier's large-scale experiments on water, so much so that in January 1792 he admitted that the gasometer was "an indispensable instrument".[76] The challenge was to construct a gasometer that, while maintaining the precision of Lavoisier's, was considerably simpler and less expensive to make. Lavoisier himself invited attempts to simplify the apparatus wherever possible. This was, after all, the aim of most eighteenth-century instrument makers. However, simplicity came only with time, and it would be quite difficult to find any important precision instrument was already cheap and simple at the time of its invention! This is also clear from the instruments and machines that Van Marum constructed or had made for the Teyler Museum in Haarlem.[77] It will suffice to mention here that for the completion of his large electrostatic generator in 1785, Van Marum planned to purchase the largest disks ever made (1.9 m diameter) for 3,000 *livres* in Paris, at the Saint-Gobain glass manufacture.[78] A few years later, in 1790, he was given the grand credit of 500 pounds for the purchase of English instruments. His physics and chemistry cabinets were filled with expensive and highly innovative instruments. Similarly, the laboratories built by Torbern Bergman in Uppsala, by Felice Fontana in Florence, by Alessandro Volta in Pavia, and those of many other natural philosophers and chemists throughout Europe were comparable to Lavoisier's. It is also worth to remember that in astronomy the cost of instruments was definitely higher: "a 8-foot mural quadrant by John Bird [in 1771] cost 8,000 *livres*".[79]

Lavoisier undoubtedly had one of the best-equipped chemical laboratories of eighteenth-century Paris, but a examination of its cost shows that it was not incommensurably different from those of his contemporaries who were engaged in experimentation on the nature of gases. The relatively high cost of Lavoisier's apparatus is due to his approach to experimentation which, from the very beginning of his career, aimed at adapting mathematical apparatus (known to be expensive) to chemistry. This approach, which was also followed by other chemists and natural philosophers, was justified by the problematic nature of gases that had significantly altered the scientific status of eighteenth-century chemistry, and which encouraged the introduction of new pieces of apparatus.

Lavoisier's social status and his financial situation, which made it possible for him to supply the laboratory with exceptional resources, played a secondary role, and it is impossible, perhaps even misleading, to understand his investigative research programme within such a limited historiographical framework. Lavoisier's financial investment in his chemical laboratory did not differ substantially from other Parisian laboratories, e.g. that of Balthasard Sage, but the impact Lavoisier's organisation of chemical experimentation, regardless of the results, was far more innovative and long-lasting in its influence.

Lavoisier's novel approach had not only financial consequences, but it also implied a social rise of new professional figures, such as the instrument makers and laboratory assistants who, thanks to his patronage and support, were able to establish new roles as makers of precision instruments and thereby make themselves indispensable.

76 Forbes (1969–1976), vol. 6, p. 223
77 Turner, Levere (1973).
78 Turner, Levere (1973), p. 313.
79 Turner A. (1989), p. 1.

CHAPTER 5

The Chemical Revolution on Stage: Lavoisier's Collection of Instruments (1789–2020)

5.1 The Chemical Revolution on Show (1789–1836)

The historical collections we admire in museums are often the survivors of a long story of progressive loss and destruction. What has survived is the result of a 'natural' selection through random circumstances.[1] The collection of Lavoisier's instruments currently on display in one room of the Musée des arts et métiers tells one of these stories, and it reminds us of the complex vicissitudes that surround the construction and design of a museum.[2] It also reminds us that over time, collections interact not only with the spaces for they are destined, but also with the cultural and social milieus that adopt and use them for ever-changing purposes. This chapter highlights the crucial moments in the history of Lavoisier's instrument collection and illustrates the changing perspectives on the rationale for their display.

The previous chapters discussed Lavoisier's experimental practice within the chemical and scientific context of his time. However, his laboratory was also extensively used for display and social purposes, with scientific demonstration not being the main occupation of the guests invited to assist. Distinguished guests invited for dinner or other social events were introduced to the laboratory, which occasionally became a place for conversation, spectacle, and proselytism. These kinds of activities, organised by Lavoisier in collaboration with his wife, were inspired by a culture *salonniere* that had become relatively common within the Académie des sciences. The social tradition had been introduced in the early 1770s by Bochart de Saron, who invited a selection of friends for tea or ice cream every Wednesday after the sessions of the Académie des sciences, and engaged in scientific conversations with them as well as organising several experiments (many of which in the field of pneumatic chemistry). Lavoisier entertained along similar lines, at the Arsenal and occasionally at his residence in the Rue des Bons Enfans, which was situated very close to the Académie.[3] A contemporary description of Bochart's scientific salon by Van Marum is a useful record of the atmosphere at events of this type:

> M. Bernard de Scaron [Bochart de Saron], *Président du Parlement, membre honoraire de l'Académie*, invited me, together with several members of the assembly, to have a glass of lemonade at his home. An apparatus, shown at the assembly, for controlling a balloon by means of bellows, caused M. Lavoisier and M. de Laplace to engage in a dispute with M. de Scaron's on the possibility of controlling balloons. M. de Lavoisier maintained the possibility of this and even had favourable expectations about the attempts of M. Meusnier [...]. M. de Laplace on the other hand held that it was altogether absurd to regard the control of a balloon as possible. After this the operation of M. Nairne's Patent electrical machine was tested. So far as I could judge, the power of this machine was almost equal to that of Cuthbertsons's 18-inch machine with 2 discs.[4]

The experiments at the Arsenal were usually organised around dinners and social events, and the lines between domestic sociability and experimental science were often quite blurred. In a short biographical note on her late husband, written for Georges Cuvier's *Eloge* (which was published in 1819), Marie Anne Lavoisier provides a vivid picture of the meetings at the Arsenal:

1 Lacour (2019).

2 In fact, only one fifth of the collection is on display, while the rest is preserved in the warehouse of the Musée in Saint-Denis.

3 "Tous les mercredi, au sortir de l'académie, Bochart invitoit plusieurs de ses confrères à se rendre chez lui. Le prétexte étoit de prendre du thé ou des glaces: on en prenoit en effet; mais bientôt l'entretien rouloit sur les objets le plus relevés; c'étoit une nouvelle séance académique où chacun apportoit le tribut de ses veilles; on s'y occupoit beaucoup des découvertes des autres savans; on ne manquoit jamais d'y faire des expériences sur les airs fixes, le platine, et d'autres infiniment importantes. On ne peut se faire une idée de l'utilité dont ces sortes de comités étoient, et pour ceux qui les composoient, et pour la science en général. Le célèbre Lavoisier qui avoit une estime particulière pour Bochart, suivoit son example: il fasoit, tous les samedis, une semblable réunion, quelquefois chez lui, le plus souvent chez un de ses parens, qui étoit plus voisin de l'académie. On présume bien que là le tems étoit aussi utilement employé que chez Bochart". Montjoye (1800), pp. 84–85.

4 Forbes (1969–1976), vol. 2, p. 227. The report is dated 20 July 1785, and reveals, among other things, how the commission on ballooning, which was promoted by the Académie des sciences, held informal meetings in domestic settings.

 | DOI:10.1163/9789004511217_007

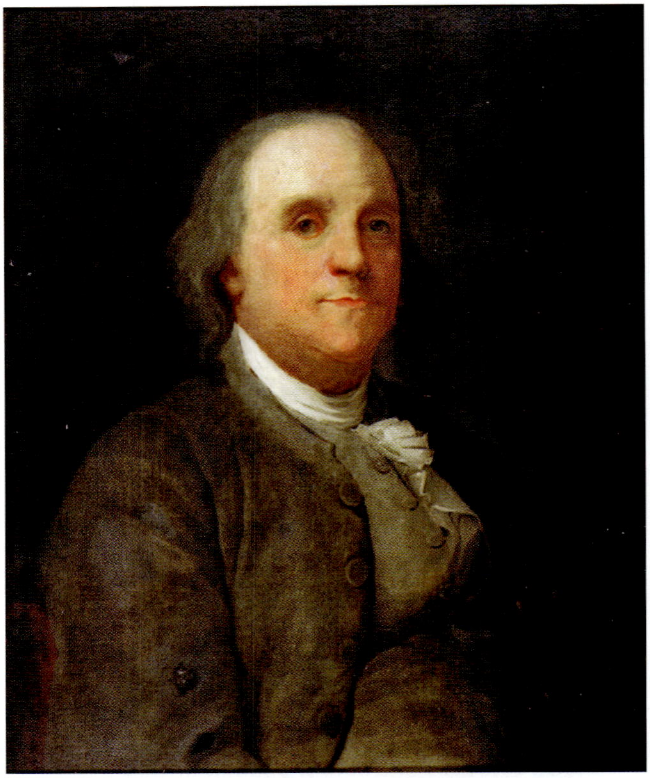

FIGURE 1 Marie Anne Lavoisier's portrait of Benjamin Franklin, made in 1787. Reproduced with kind permission of a relative of Dr Benjamin Franklin's

> an entire day was devoted every week to experiments. It was, said Lavoisier, his happy day. Some enlightened friends – young people proud of having been granted the honour of collaborating in his experiments – would meet in the laboratory in the morning. That is where they breakfasted, discussed, conducted experiments, and where the fine theory that immortalized its author was born.[5]

Unfortunately, just from the few surviving hints and documents, it is extremely difficult to reconstruct in detail the social dynamic of these meetings.

During the late 1770s and early 1780s, Benjamin Franklin, a frequent guest of the Lavoisiers', was often entertained with an experimental performance, political conversation, and social dinner on the same day (Fig. 1). It was in autumn 1774, during one of the dinners the Lavoisiers had organised in their apartment in the Rue des Bons Enfans that Joseph Priestley communicated his discovery of oxygen.[6]

As important as dinners and conversations were, the Lavoisiers planned their social programme carefully following a schedule of events. As I have previously pointed in connection with Arthur Young's visit to Arsenal in 1787 (see the previous chapter), Lavoisier's collection was often displayed for visitors who had little or no knowledge of pneumatic chemistry, and this usually took place after dinner. By showing his precision instruments to non-specialists, Lavoisier wished to demonstrate the progress that Parisian instrument makers had achieved thanks to his patronage.

The display of instruments also offered opportunities for more entertaining social gatherings. Balthasard Sage, the director of the École des mines, reported that in the early 1780s (probably in the spring of 1782) Alessandro Volta had been invited to the Arsenal to assist in the following show:

> I was told of the following by the famous Volta, my friend, who was an eyewitness to the event, one day when he was dining at Lavoisier's. In the drawing-room, after the meal, we found a magic lantern, whose operator said: In these scenes, you will see the means I have imagined to ridicule the Stahlians.
>
> Ah! do you see, in this first scene, the character representing the phlogisticon? He's wearing a crown of thorns on his head, and the flame of genius on his behind. His posture is that of a supplicant; his hands are joined. He's followed by his acolytes, all in mourning.
>
> In this other scene, you see this antiphlogistician, his head on an anvil, on which they want to reforge it. But it is so dense with iron [*sidérotée*][7] that it resists under the hammer.
>
> Ah! do you see, in this third scene, a pit in which the phlogisticon is about to be buried alive? He's followed by his partisans, who are holding lachrymatory vessels.
>
> Ah! do you see this fourth scene? It presents the apotheosis of aerial chemistry, and a concert of parrots.

5 "un jour tout entier dans chaque semaine était consacré aux experiences; c'était disait Lavoisier son jour de bonheur, quelques amis éclairés, quelques jeunes gens fiers d'être admis a l'honneur de coopérer a ses experiences se reunissaient des le matin, dans le laboratoire; c'était là que l'on déjeunait, que l'on dissertait, que l'on travaillait, que l'on faisait des experiences, que naissait cette belle théorie qui a immortalisé son auteur". Quoted in Gillispie (1956).

6 Priestley (1803), p. 116 (see Chapter 3).

7 "Tête de fer" [iron head]. Note by Sage.

> All the oxyphiles exclaimed: Your invention, M. Ludius,[8] is bound to be a success.[9]

These performances seem to have been relatively frequent; in 1788 Marsilio Landriani told Priestley how amused he was to watch a battle between *sulphites* and *sulphates* (Fig. 2).[10] Priestley was also amused but not persuaded of the end of the phlogistic era.

We know from the thirteenth of Lavoisier's *Registres de laboratoire*, preserved in the Archives of the Académie des sciences of Paris, that he, Guyton de Morveau, Meusnier de La Place, and Madame Lavoisier were trying very hard to convert the Italian naturalist and industrial spy Marsilio Landriani to the new theory, and apparently they used every possible means to this end.

Around the same time, according to an anonymous report that was published in Lorenz Crell's *Chemische Annalen* in 1789, Lavoisier – or more likely his wife – organised a performance at the Arsenal, staging a phlogiston inquisition. Marie Anne Lavoisier, with her "great knowledge in chemistry", played the part of a priestess at a sacrifice, and Stahl the victim. The anonymous witness reports the scene as follows:

> I have witnessed a most remarkable performance here [in Paris], which to me, as a German, was very unexpected, and which I noted quite strongly. I saw the famous M. Lavoisier hold a positive *auto-da-fé* of phlogiston in the Arsenal. His wife (who has a really great knowledge of chemistry and has translated several chemical writings) served as the sacrificial priestess, and Stahl appeared as the *advocatus diaboli* to defend phlogiston. In the end, poor phlogiston was burned upon the accusation of oxygen. Do not think this a humorous invention of mine! Everything is literally true. Do not ask whether the cause of phlogiston is irretrievably lost, or what I think about the issue. But I will readily admit that I am glad that this spectacle did not occur in my fatherland.[11]

A similar show, at the presence of Landriani, was apparently put on with a magic lantern in 1788.[12] Sometime between 1788 and 1790 another witness recalled a party given at the Arsenal in celebration of the triumph of the chemical revolution:

> I have seen several of these scientists up close, and I have always recalled, with a pleasure and tenderness since commingled with the most painful feeling, a truly interesting party given at Lavoisier's. The purpose was, no doubt, important. Its performance was as joyous as it was varied. The event celebrated the revolution brought about in chemistry by the new discoveries or, if you will, the triumph of pneumatic chemistry and the adoption of the modern nomenclature. There, I saw gathered nearly all the scientists rightly celebrated in France; and I confess that I have found few meetings to be so pleasant and genuinely friendly. There were various games, which livened the party and surprised several guests, to whom a lottery awarded small prizes reflecting the winners' character or merit; their distribution was accompanied by the recital of verses. After that, the children had their show: it was a magic lantern, depicting the battles between the phlogiston and oxygen, with all their scientific troops. Elsewhere, this small performance would have been seasoned

8 "This word was used by Cicero to denote a buffoon, a jester, a juggler". Note by Sage.

9 "Je tiens le fait suivant du célèbre Volta, mon ami, qui en a été témoin oculaire, un jour qu'il dînait chez Lavoisier. On trouva dans le salon, après le repas, une lanterne magique, dont le compère dit: Vous allez voir dans ces tableaux le moyen que j'ai imaginé pour jeter du ridicule sur les stahliens.

Ah! voyez-vous, dans ce premier tableau, le personnage qui représente le phlogistique? il a sur la tète une couronne d'épines, et au derrière la flamme du génie: son attitude est celle d'un suppliant; il a les mains jointes. Il est suivi de ses sectaires, qui sont tous en deuil.

Dans cet autre tableau, vous voyez cet antiphlogisticien, la tête sur une enclume, où on veut la reforger; mais elle est tant sidérotée, qu'elle résiste sous le marteau.

Ah! voyez-vous, dans ce troisième tableau, une fosse dans laquelle on va enterrer, tout vif, le phlogistique? il est suivi de ses partisans, qui tiennent en main des lacrymatoires.

Ah! voyez-vous ce quatrième tableau? il offre l'apothéose de la chimie aérienne, et un concert de perroquets.

Tous les oxiphiles de s'écrier: Votre invention, M. Ludius, ne manquera pas de prospérer". Sage (1813), pp. 129–130.

10 LC, vol. 5, p. 220.

11 "Vom H[er]rn von E.** in Paris. – Ich habe hier ein ganz besonderes Schauspiel erlebt, was mir, als einem Deutschen, sehr unerwartet war, und ungemein auffiel. Ich sahe den berühmten H[er]rn Lavoisier im Arsenal ein förmliches Auto – da-fé über das Phlogiston halten, worinn seine Gattinn, (die wirklich viele Kenntnisse in der Chemie hat, und verschiedene chemische Schriften übersetzte,) die Opferpriesterin machte, Stahl zur Vertheidigung desselben, als Advocatus diaboli erschien; wo bei allem dem aber doch das arme Phlogiston, auf Anklage des Oxygens, zuletzt verbrannt wurde. Halten Sie dies nicht für eine scherzhafte Erfindung von mir; alles ist buchstäblich war. Ob nunmehr denn die Sache des Phlogistons unwiderruflich verloren sey, oder was ich sonst dabey dachte, darum fragen Sie mich nicht: aber das gestehe ich gern, ich freue mich, daß die Scene nicht in meinem Vaterlande war". Crell (1789). See also LC, vol. 5, p. 136.

12 LC, vol. 5, p. 220.

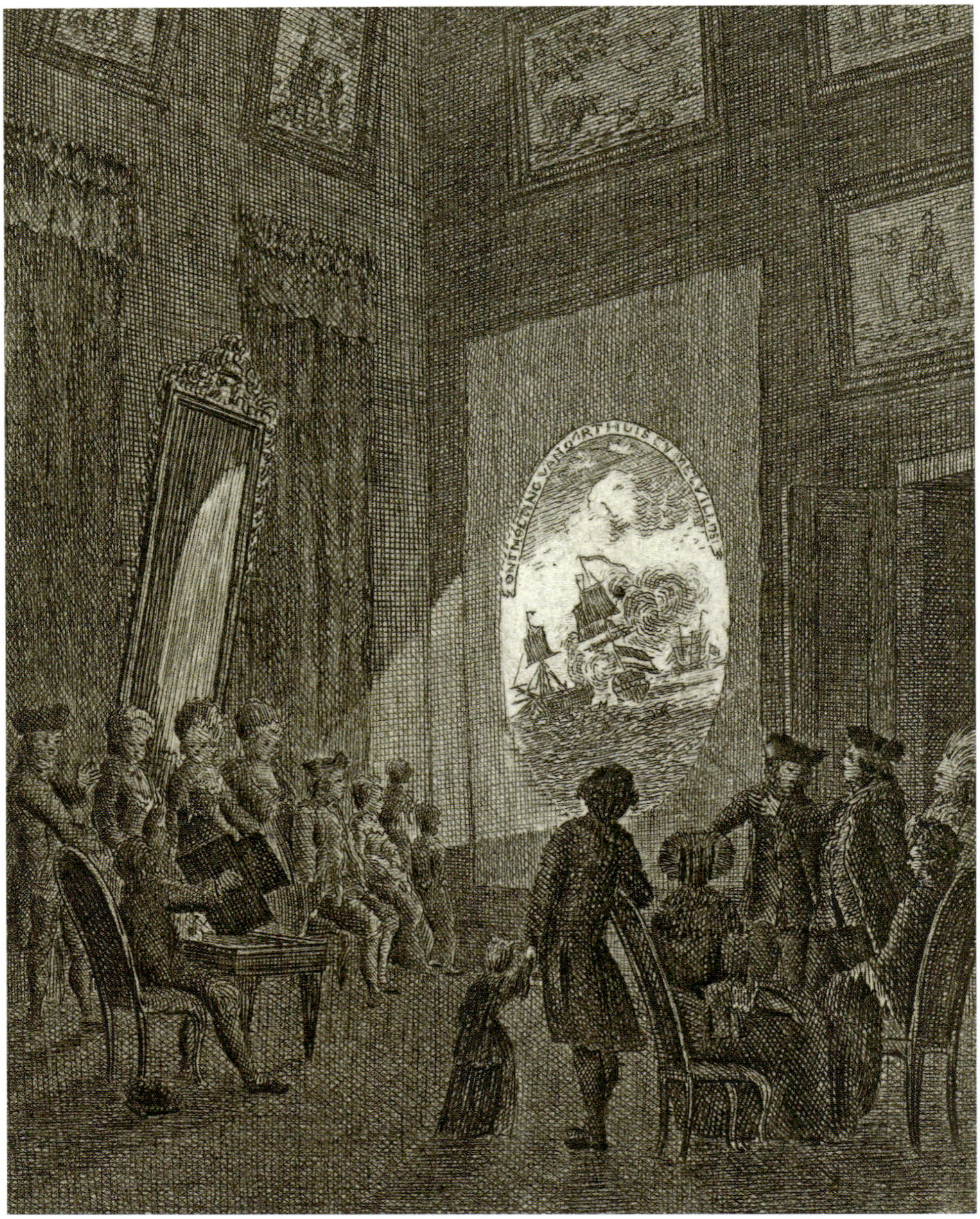

FIGURE 2
Eighteenth-century engraving of a spectacle with a *lanterna magica*
PRIVATE COLLECTION

with epigrams. Here, one could smile without fear of hurting the then still numerous opponents of the new chemical language.[13]

We do not know exactly where these shows took place, but it is not unlikely that some of the instruments, such as Lavoisier's gasometers, were used as props. After all, Lavoisier's instruments, carefully arranged in a chronological order, constitute an important part of the visual

13 "J'ai vu de près plusieurs de ces savans, et je me suis toujours rappelé, avec un plaisir, un attendrissement au quel s'est mêle depuis le sentiment le plus douloureux, une fête, vraiment intéressante, donnée chez Lavoisier. L'objet en était important sans doute; l'exécution en fut aussi gaie que variée. On y célébrait la révolution opérée dans la chimie par les nouvelles découvertes, ou, si l'on veut, le triomphe de la chimie pneumatique et l'adoption de la nomenclature moderne. Là, je vis rassemblé presque tout ce qu'il y avait de savans justement célèbres en France; et j'avoue que peu de réunions m'ont offert autant d'aménité, de véritable amabilité. Après différens jeux, qui animèrent la fête et étonnèrent quelques-uns des convives auxquels on fit échoir à une loterie de petits lots ayant des rapports au caractère ou au mérite des gagnans, et dont la distribution fut accompagnée de couplets, les enfans eurent leur spectacle: ce fut une lanterne magique, représentant les combats du *phlogistique* et de l'*oxigène* avec toutes leurs troupes scientifiques: ailleurs, cette petite représentation eut été assaisonnée d'épigrammes; ici on put sourire sans crainte d'affecter les antagonistes, alors assez nombreux, de la nouvelle langue chimique". Pujoulx (1801), pp. 368–369.

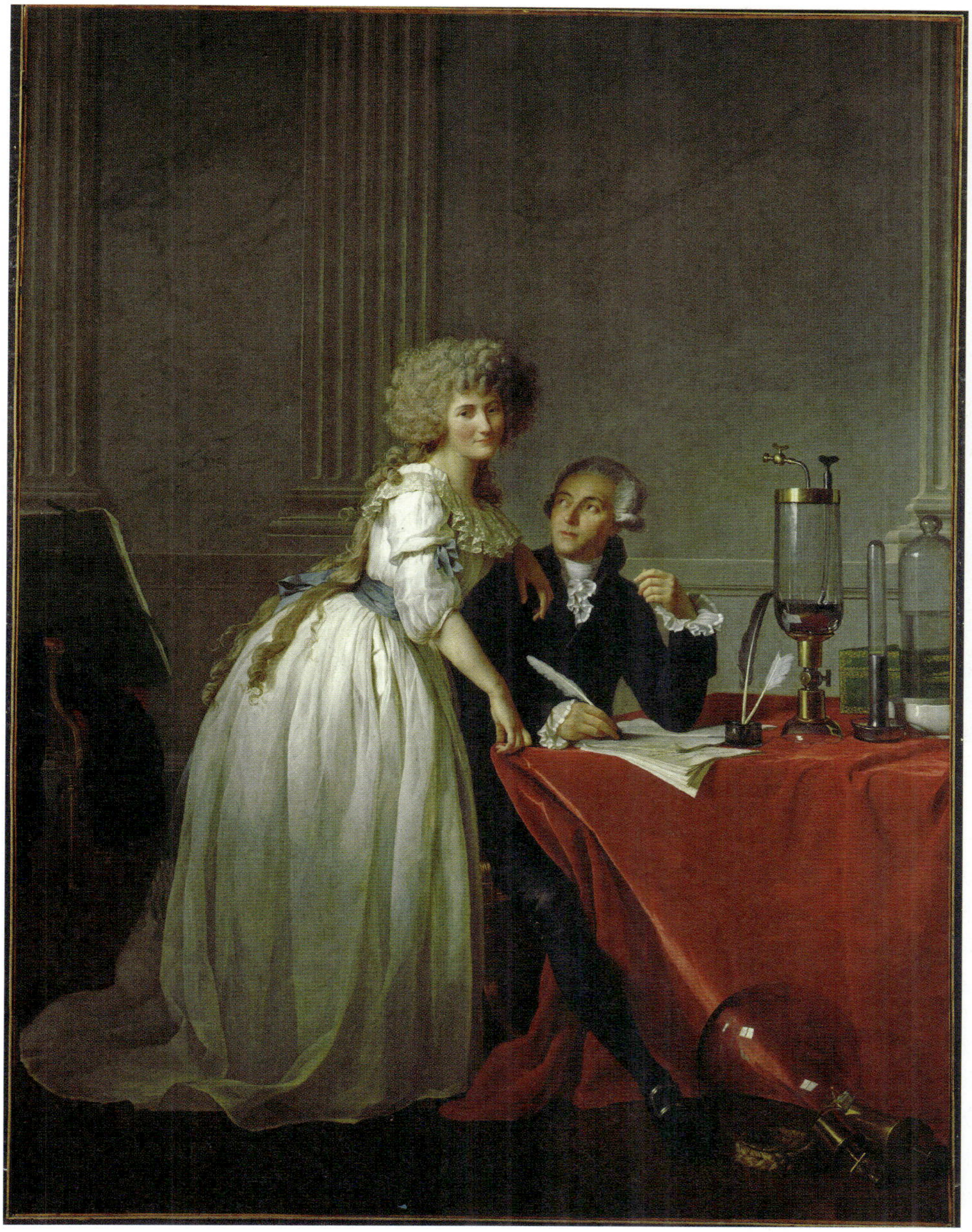

FIGURE 3
Jacques Louis David's portrait of the Lavoisiers (1788)
COURTESY OF THE METROPOLITAN MUSEUM – NEW YORK

narrative of Jacques Louis David's 1788 double portrait of Lavoisier and his wife.[14] The precise meaning of the mise-en-scène of the instruments, the portfolio with Madame Lavoisier's drawings, and the two protagonists is symbolic in a picture epitomising a successful career in science (Fig. 3). This important portrait of the chemical revolution also underlined, among other things, the importance of the domestic setting in eighteenth-century chemistry. Furthermore, the dominant figure of Madame Lavoisier evoked the central role still played by sociability in scientific discourse.

The presumed juxtaposition of the public features of science, represented by spectacles and entertainment, and Lavoisier's more professional and elitist approach to chemistry, which has been emphasised in recent historiography, provides us with a picture both distorted and no longer tenable.[15] As the examples briefly mentioned here show, Lavoisier attended intensely to adapting his scientific practice to the fashion of the day, and the spectacles

14 On this, see Beretta (2001). Recent technical examination of the portrait has revealed significant and previously unknown alterations including the representation of a celestial sphere and, possibly, a case with the laboratory notebooks; Pullins, Mahon, Centeno (2021).

15 See, for instance, Golinski (1992); Chappey (2004); Lilti (2005), pp. 263–266; Bensaude-Vincent, Blondel (2008).

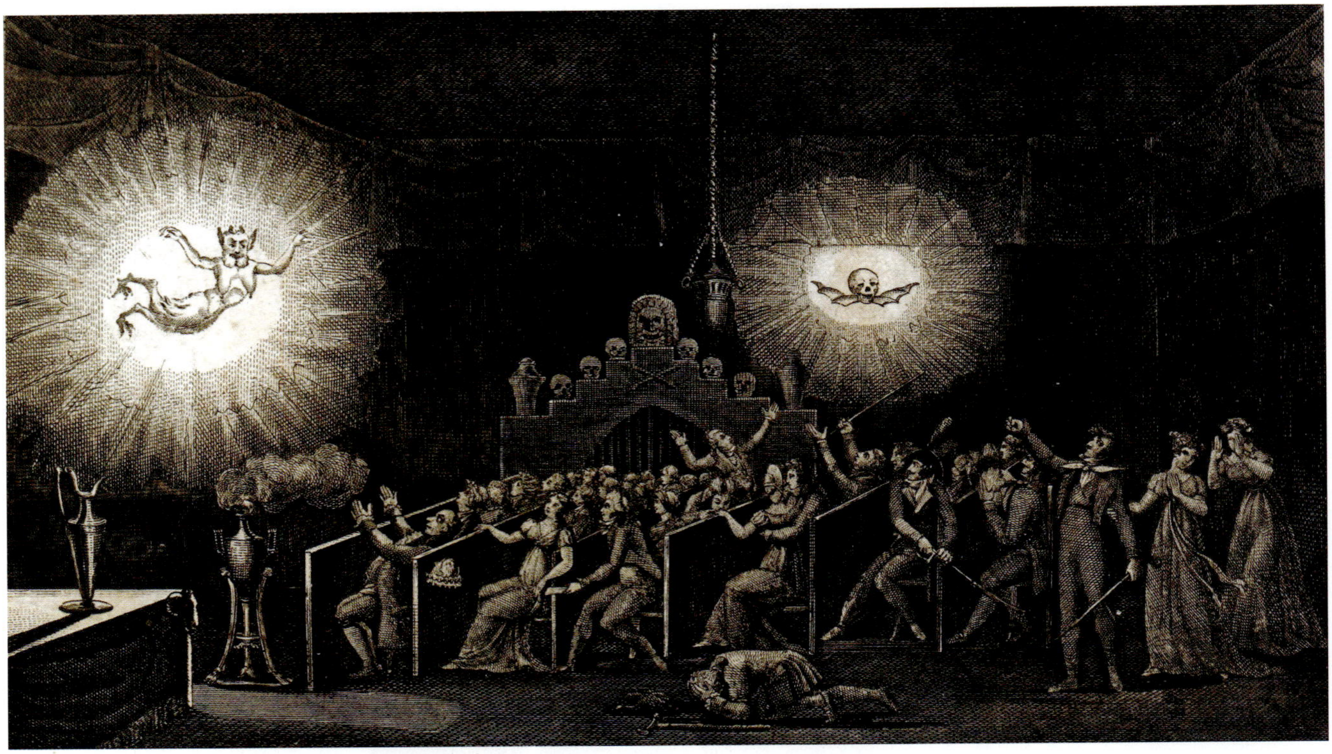

FIGURE 4 Robertson's phantasmagoric show (1797)
FROM ROBERTSON (1831)

and social events his wife organised were an important part of his campaign of persuasion.[16]

Phantasmagoric shows with the magic lantern were part of a growing fashion among both the official and the unofficial Parisian scientific communities.[17] Interestingly, the most popular shows, which were organised by the Étienne-Gaspard Robertson towards the end of the century (Fig. 4), were attended by Marie Anne Lavoisier; and in one of them Lavoisier was represented as the victim of Robespierre:

> Robespierre emerges from his grave, wants to stand up, lightning strikes and turns the monster and his grave into dust. Beloved shadows temper the scene: Voltaire, Lavoisier, J.-J. Rousseau, appear in turn; Diogenes, lamp in hand, seeks a man; to find him, he walks, so to speak, across the rows of spectators, impolitely causing a fright among the ladies, which they all find amusing. Such are the effects of optics that everyone is convinced their hands are touching these approaching objects.[18]

As much as Lavoisier and his wife liked this kind of scientific show, the events following the outbreak of the French Revolution deprived them of the opportunity to pursue them regularly. Due to his prominent role within the Ferme générale, Lavoisier was investigated from 1791 onwards, then put on trial in 1793, and executed on 8 May 1794.

Lavoisier's collection of instruments was confiscated, the Revolutionary authorities began to inventory it in spring 1794, and it was gradually returned to Lavoisier's widow two years later.[19] When Lavoisier was guillotined, he had been preparing an edition of his new and old works, and after his death Madame Lavoisier felt it was

16 These themes are thoroughly discussed in the doctoral thesis of Antonelli (2021). The expression "campaign of persuasion" is taken from Perrin (1981).

17 *Lanterne magique* (1990).

18 "Robespierre sort de son tombeau veut se relever la foudre tombe et met en poudre le monstre et son tombeau. Des ombres chéries viennent adoucir le tableau: Voltaire, Lavoisier, J. J. Rousseau, paraissent tour à tour; Diogène, sa lanterne à la main, cherche un homme, et, pour le trouver, traverse pour ainsi dire les rangs, et cause impoliment aux dames une frayeur dont chacune se divertit. Tels sont les effets de l'optique, que chacun croit toucher avec la main ces objets qui s'approchent". Robertson (1831), vol. 1, p. 283.

19 Beretta (2003).

her duty to bring this to completion while also restoring the original order of the collections at the same time.[20] Her determination to retrieve all the seized items is evidenced in her negotiation with the École centrale des travaux publics, which had been entrusted (among other things) with Lavoisier's collections of minerals and fossils of some 4,000 items.[21] The negotiations were led by the apothecary Jean Charles Pluvinet, who had been one of Lavoisier's collaborators and was one of the few who tried to save him from the guillotine. At the end of 1795, Marie Anne Lavoisier was entitled to get back the whole collection, but Pluvinet, who was aware how difficult it would be to build a similar one from scratch, offered to estimate its value and to purchase it for the École. However, Marie Anne Lavoisier was eager to have the collection back, and this negotiation was not even started.[22]

Even if a few instruments were probably dispersed between 1794 and 1796, the bulk of Lavoisier's collection was returned to Marie Anne Lavoisier, and she immediately set to preserving the collection intact, as a memorial to her husband and the chemical revolution. On 21 November 1801 Benjamin Thompson (Fig. 5) visited Marie Anne Lavoisier during a brief stay in Paris, and reported the following:

> [f]ound Madame Lavoisier at home and alone. Sat with her for an hour, and found her very lively, witty and pleasing in conversation. She received me in an elegant room which had every appearance of being a *Cabinet de Physique*. It was quite filled and even crowded with Philosophical and Chemical apparatus. The various instruments were constructed on the largest scale with great care and accuracy. It was evident that they were the apparatus employed by the late unfortunate M. de Lavoisier – but I did not ask the question.[23]

From 1804 onwards, Marie Anne Lavoisier held a famous salon at her residence, N° 39 Rue d'Anjou-Saint-Honoré – (Figs. 6–7) not far from where both her father and husband were buried[24] – to which many young scientists were invited. While other salons of the time were primarily evenings of animated political, philosophical, and literary discussions, Marie Anne Lavoisier's also provided the setting for scientific research and chemical experiments, and Lavoisier's laboratory was used on a number of occasions.

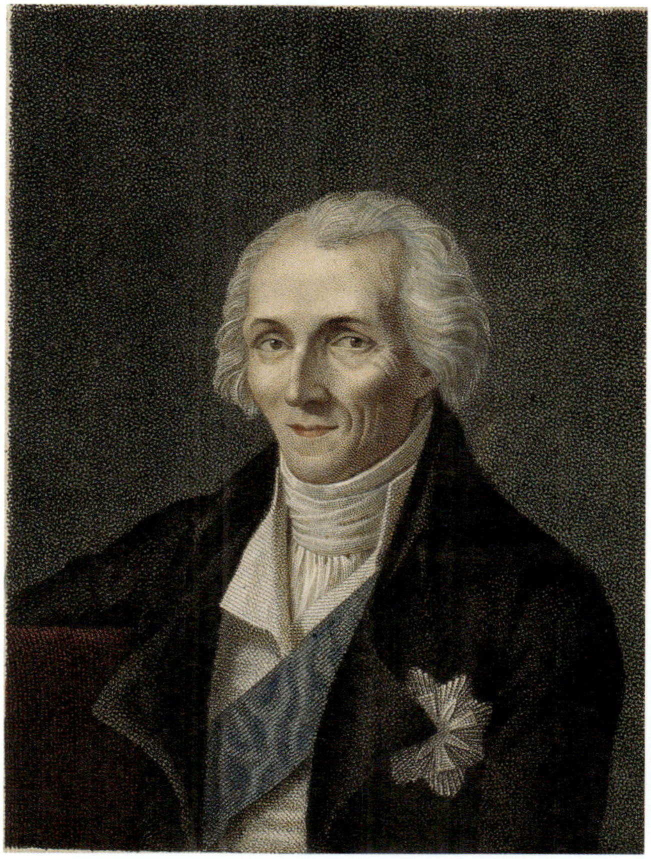

FIGURE 5 Engraved portrait of Benjamin Thompson, Count Rumford
COURTESY OF THE WELLCOME COLLECTION

As stated by Rumford, Lavoisier's *cabinet de physique* was considered suitable for a social visit which, by all evidence, was not directly connected with science or experiments. Given the size of Marie Anne Lavoisier's residence, this was certainly deliberate. The collection of instruments created an atmosphere evoking the genius of Lavoisier while demonstrating the expertise of its present curator. It is not surprising, then, that a few days later Marie Anne Lavoisier offered Rumford to assist him during his experiments; she married him just a few years later, on 24 October 1805. There is evidence that Rumford used

20 On Marie Anne Lavoisier's work on the edition, see Beretta (2001).
21 On this, see Tron (1996).
22 Ibid., pp. 12 and 42.
23 Sparrow (1958), p. 20. In 1801 Marie Anne Lavoisier probably still lived in Boulevard de la Madeleine.
24 In a letter to Charles Blagden dating 15 octobre 1804, one of his French friends (the name is not mentioned) reported: "Mad. Lavoisier vient d'acheter une maison, Rue d'Anjou, très jolie, avec un grand jardin. Ils vont s'occuper à l'arranger, et surement parfaitement bien. Il y a aura une Galerie pour les Intrumens de Physique. Je pense qu'il se marieront bientôt, je le suppose; il est possible qu'ils fassent un voyage cet hyver, mais c'est incertain". Banks (2007), vol. 5, p. 384. Liste (1814), p. 34. Lavoisier's remains were later moved to the catacombes and their present location is unknown.

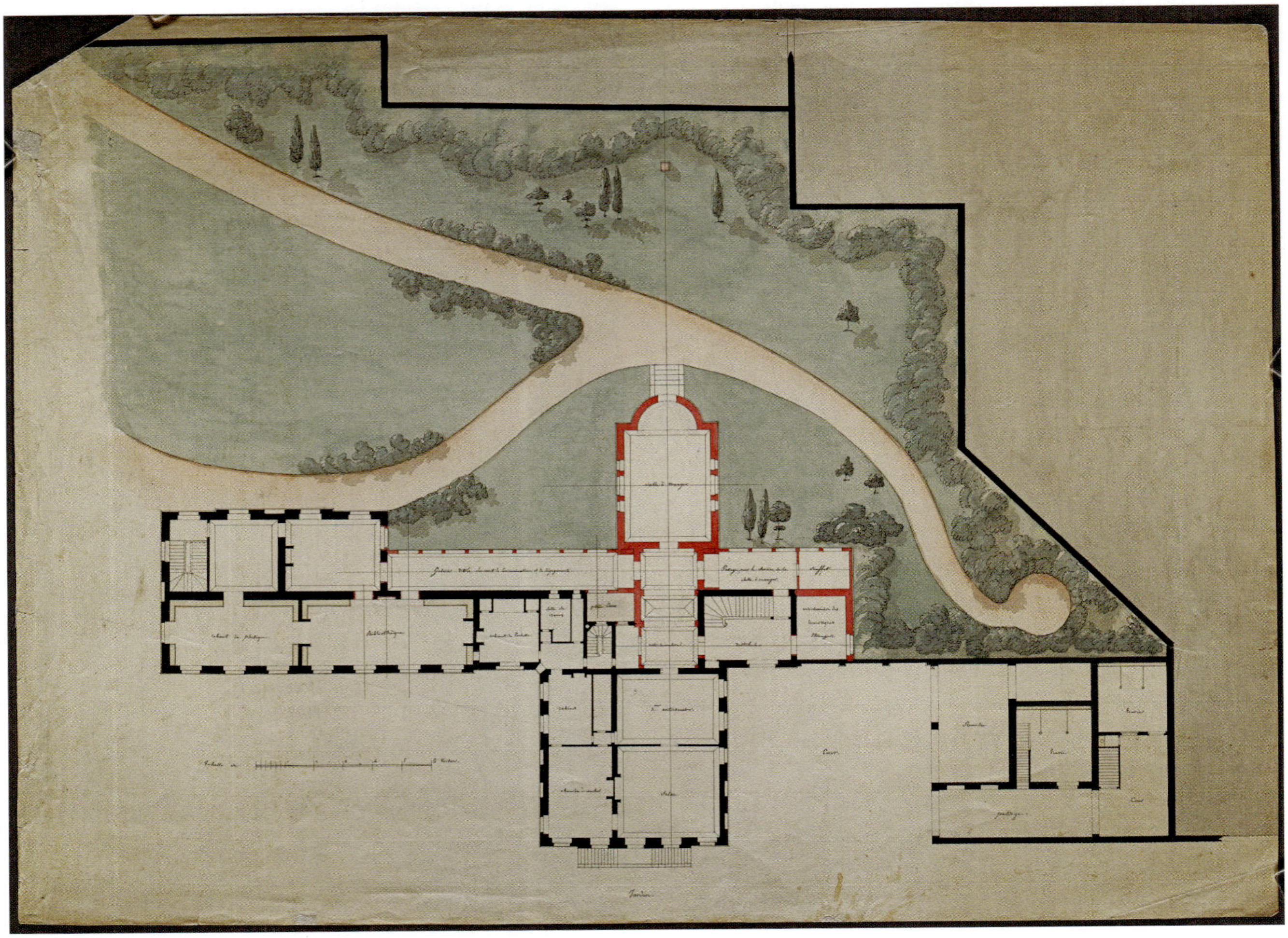

FIGURE 6 *Plan générale du jardin de Madame de Rumford*, with a floor plan of her residence
COURTESY OF CORNELL UNIVERSITY LIBRARY – DIVISION OF RARE AND MANUSCRIPTS COLLECTIONS. LAVOISIER 4712_B24

Lavoisier's apparatus for his own investigations into the nature of heat, and a few of Rumford's instruments ended up in Lavoisier's collection after his separation from Marie Anne Lavoisier in 1809.[25] During this period Lavoisier's instruments were displayed both as a memorial to the chemical revolution and its principal author, and actively used by a younger generation of scientists; in fact, the instruments' enduring utility underlined Lavoisier's anticipative genius.[26] Guests attending the salon must have had the impression that Lavoisier's spirit was manifested in his instruments, and that when these were used, he was resuscitated.

As early as July 1803 the English chemist Alexandre Marcet, who visited Paris at the time, declared:

> Eight days ago, at Fourcroy's, I repeated the experiments on the respiration of nitrogen oxide gas. Messrs. Underwood and Vauquelin both passed out; today, I am repeating them at the home of Madame Lavoisier, with whom I spent four hours in private yesterday, amid her husband's instruments, and as if surrounded by his shadow.[27]

25 On Rumford's experiments on heat, see Ellis (1871).

26 In a letter dated July 1805 and addressed to Marc-Auguste Pictet, Marie Anne Lavoisier described the architectural design of her new residence as follows: "Comme je désire extrêmement Monsieur que vous n'ayez aucun droit de vous plaindre de votre pupille, je vous dirai donc, puisque vous le désirez, que je suis occupée dans ce moment à faire bâtir ma galerie, qui sera de plain pied à la suite de mon appartement. Elle sera composée d'une première pièce, qui fera [*sic*] de bibliothèque et d'une seconde où sera placé mon Cabinet. L'exposition sera au midi. Au-dessous du Cabinet sera un laboratoire qui communiquera avec le Cabinet par un petit escalier. À côté du laboratoire et sous la bibliothèque sera une serre où j'aurai des fleurs l'hiver pour mes amies." Pictet (1996–2004), vol. 4, p. 901 (see Figs. 6–7).

27 "J'ai répété il y huit jours chez Fourcroy les expériences de respiration du gaz oxide d'azote. Mr Underwood et Vauquelin furent l'un et l'autre sur le carreau; aujourd'hui je les répété chez

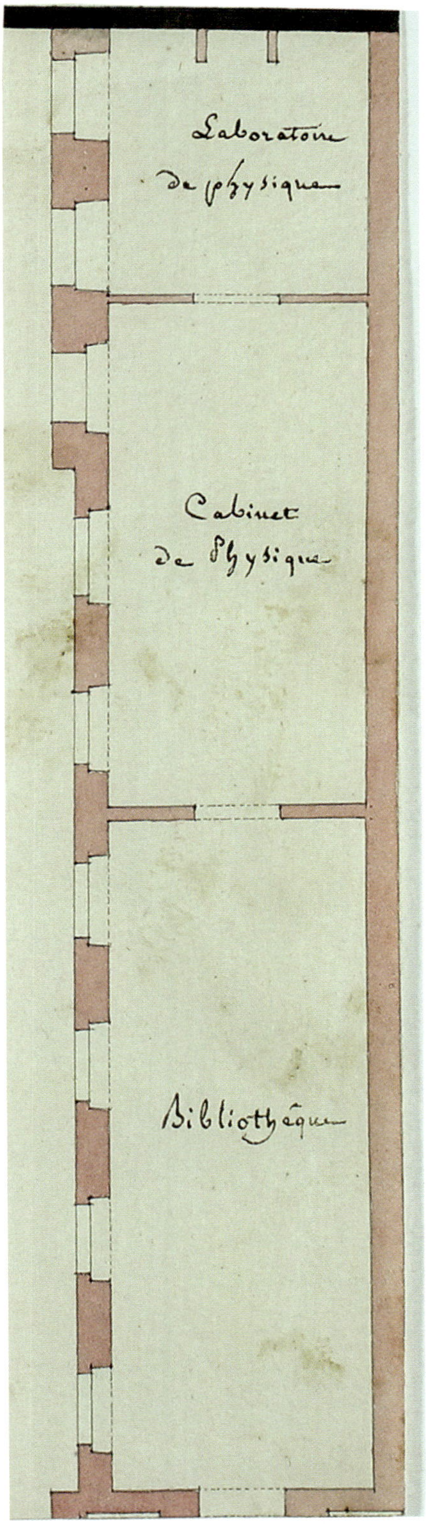

FIGURE 7
Plan of the physics laboratory, the cabinet de physique, and the library in Marie Anne Lavoisier's residence at the Rue d'Anjou
COURTESY OF CORNELL UNIVERSITY LIBRARY – DIVISION OF RARE AND MANUSCRIPTS COLLECTIONS. LAVOISIER 4712_B24

Other eminent scientists were invited to Madame Lavoisier's salon, too, and there had the opportunity to admire, and occasionally use, her late husband's magnificent collection of instruments. Instruments were not the only attraction that visitors were shown: Georges Cuvier, one of the most assiduous guest at Marie Anne Lavoisier's receptions, was encouraged to inspect the collection of minerals and fossils, and she donated one fossil bone to him, on which he (a young naturalist) subsequently published a report.[28] Unsurprisingly, Count Rumford found these frequent events and the persistent shadow cast by Lavoisier quite distressing.

On 17 July 1835, a few months before passing away, Marie Anne Lavoisier donated her late husband's laboratory notebooks, a few other manuscripts, and some printed books to François Arago, the secretary of the Académie des sciences.[29] This donation during her lifetime was an early indication of her wish the Académie should be the institution to take care of Lavoisier's memory after her death.

5.2 Instruments Enter French Politics: The Private and Public Fate of Lavoisier's Collection (1836–1900)

Marie Anne Lavoisier died on 10 February 1836, and the obituary by Jean Baptiste Huzard, a celebrated veterinarian, commemorated her as follows:

> she had set up a charming house in Paris, a town and country retreat, placed in the middle of her garden, far from all her neighbours, and in which she combined the useful and the pleasurable in every way: her friends, the physics cabinet, the chemistry

Madame Lavoisier, avec qui j'ai passé hier quatre heures en tête a tête, au milieu des instrumens de son mari, et comme environnée de son ombre". Letter from Marcet to Marc-August Pictet, published in Pictet (1996–2004), vol. 3, p. 353.

28 Cuvier (1812), vol. 3, "sixième mémoire", pp. 37–38.

29 "Note manuscrite d'un envoi des manuscrits par Mad.e de Rumford à M Arago.

Le 17 juillet 1835/Envoyé à monsieur Arago secrétaire perpetuel de l'Académie des sciences./Trois cartons/Quatorze registres petit in folio reliés manuscrits de M Lavoisier/

Douze petits registres reliés même objet [cancelled]/ Un registre cartonné. [cancelled]/ Huit volumes et quelques brochures ouvrages imprimés de Mr Lavoisier/ Cinq petits registres reliés manuscrits. Notes de Mr Lavoisier/ Total/ 3 cartons/19 registres grands et petits manuscrits/ 8 vol. imprimés ouvrages publiés par Mr. Lavoisier". MS Paulze-Lavoisier (1835). Henry Guerlac mistakenly thought that the donation to Arago was done by Léon de Chazelles around 1846; Guerlac (1979), p. 96.

laboratory, and Lavoisier's library, all of which she had kept.[30]

On 7 May 1836, roughly three months after Marie Anne Lavoisier's death (Fig. 8), one of her acquaintances, the chemist Jean-Batiste Dumas, delivered a long and passionate lecture on Lavoisier at the Collège de France, in which he announced his plans to monumentalise Lavoisier with a complete edition of his works.[31]

Marie Anne Lavoisier left her estate to Gabrielle Ramey de Sugny, the niece of her brother Christian Paulze (1755–1793); Gabrielle married Léon Bérard de Chazelles (1804–1876) (Fig. 9), who was a wealthy landowner at Puy de Dome, and would enter a successful political career at the anti-republican conservative front.[32] De Chazelles would also play a key role in the fate of Lavoisier collection which, thanks to his initiative, assumed both political and public relevance. It is likely that parts of the collection were still kept by Léon de Chazelles in his Parisian residence in the late 1830s;[33] or, more likely, that they remained in Madame Lavoisier's house, but it is unclear whether de Chazelles kept the entire collection of instruments, most of which was simple glassware.[34]

Léon de Chazelles was a very energetic man, and he immediately decided to pull down the residence in the Rue d'Anjou.[35] Following Marie Anne Lavoisier's will, he

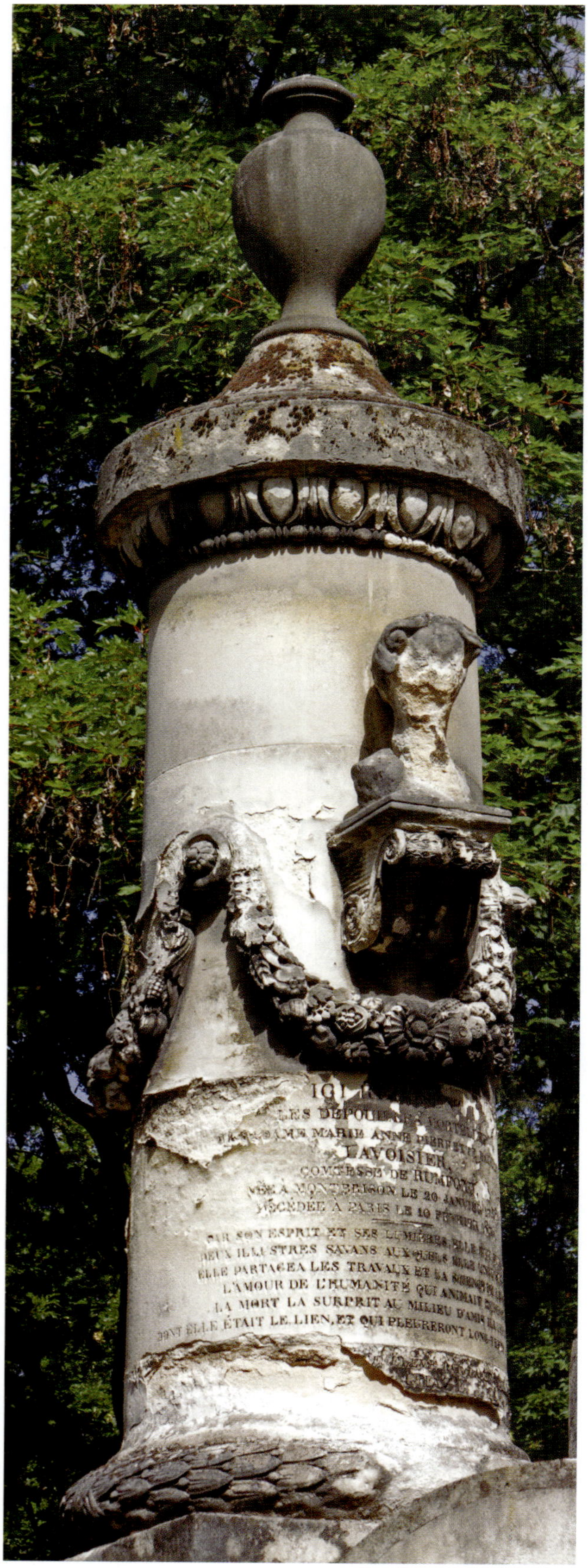

FIGURE 8 Marie Anne Lavoisier's funerary monument at the cemetery of Père Lachaise – Paris (1836)

30 "Elle s'était formé dans Paris une maison charmante, une retraite de ville et de campagne, placée au milieu de son jardin, éloignée de tous voisins, et dans lesquels elle réunissait sous tous les rapports, l'utile et l'agréable; ses amis, le cabinet de physique, le laboratoire de chimie et la bibliothèque de Lavoisier qu'elle avait conservés". Huzard (1836).

31 "Après une vie si honorable, après une mort si cruelle, qu'avons-nous fait pour Lavoisier? Qu'a fait la France pour Lavoisier? Où trouver un monument qui rappelle sa mémoire, un simple buste qui lui soit consacré? La France, hélas! semble l'avoir oublié. [...] Aussi, Messieurs, si l'on vous demande quel est le monument que la cendre de Lavoisier réclame, répondez sans crainte: c'est une édition complète de ses œuvres". Dumas (1837), pp. 184–185.

32 The wedding took place on 29 April 1834. De Chazelles was elected at the Assemblée legislative in 1849 and was mayor of Clermont Ferrand from 1850 until 1860.

33 Throughout the year 1837 Léon de Chazelles lived in Paris in the Rue du Bac.

34 From the "inventaire après dèces" of Marie Anne Lavoisier made by Alexis Vavin on March 1836 it is clear that a large part of Lavoisier's laboratory equipment, including glassware, was still kept in the residence. See MS Paulze-Lavoisier (1836). See Appendix 5.

35 Madame Lavoisier's residence and garden were demolished by Léon de Chazelles and in 1840 with the authorization of the municipality of Paris a Rue Lavoisier was inaugurated. According to Hillaret the Rue Lavoisier "a été ouverte en 1838, sur les jardins de l'Hôtel du comte de Chazelles [...]. On ouvrit, en même temps

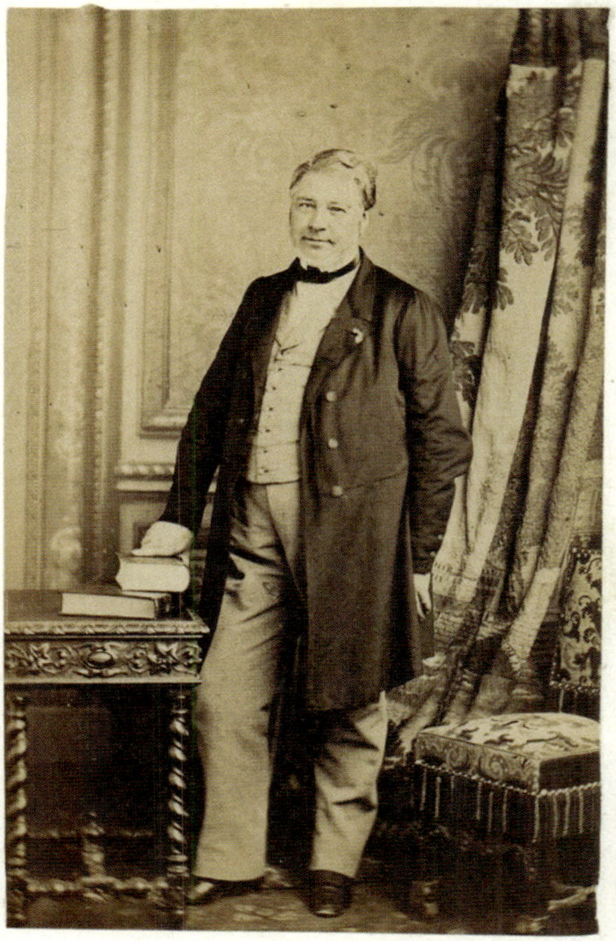

FIGURE 9 Photographic portrait of Léon de Chazelles
PRIVATE COLLECTION

took immediate action to ensure that Lavoisier's collection would go to public institutions. Between June and July 1836 he donated some important instruments to the Académie des sciences.[36] This donation consisted of Mégnié's two famous gasometers made in 1787, the glass vessel with electrodes used by Lavoisier and Meusnier de la Place for studying the combustion of hydrogen with oxygen in a constant flow of the reacting gases, 6 hydrometers, Fortin's air pump of 1792, Naudier's calorimeter of 1782, 30 standard measures of volume probably used during the reform of weights and measures of 1789, the balance Chemin made in 1770, a balance made in 1775, Lavoisier's and Laplace's ice calorimeter, a few tubes for use in the construction of thermometers, Fortin's apparatus for the combustion of oil as well as his apparatus for studying wine fermentation, both made in 1789, Mégnié's barometers of 1778, and a burning lens.[37] Where these instruments were kept and whether they were actually exhibited is not known. It is significant in this regard that, in 1854, Madame Lavoisier's niece donated one balloon used by Lavoisier in his experiments on the composition of water to the Muséum d'histoire naturelle, perhaps expressing her dissatisfaction with the handling of the collection at the Académie des sciences which was not exhibited.[38] Indeed, it does not seem that Lavoisier's instruments played an important public role in this period: on 6 June 1864, in a meeting of the Comité secret of the Académie, a report by the administrative commission, approved by Becquerel, suggested that the instruments should be moved to the Conservatoire des arts et métiers, the Muséum d'histoire naturelle, the École des mines, and the Faculty of Medicine.[39] In December of the same year the Académie, with the approval of the Ministry of Public Instruction, donated Lavoisier's instruments to the Conservatoire des arts et métiers. The possibility cannot be ruled out that, as originally planned by the Comité, other scientific institutions in Paris received

que la rue Lavoisier et également sur les terrains de l'hôtel de Chazelles la rue de Rumford reliant la rue Lavoisier à celle de la Pépinière: cette rue a disparu, en 1854, lors de l'ouverture du boulevard Malesherbes". Hillaret (1964), vol. 2, p. 26.

36 "M. Charles Dupin annonce qu'il est chargé d'offrir pour le cabinet de l'Académie, de la part de M. de Chazelles, héritier de madame la comtesse de Rumford, veuve de Lavoisier, quelques beaux appareils qui ont servi aux expériences de cet illustre chimiste, ainsi qu'une série de cylindres qui durent être employés pour déterminer les étalons des poids et mesures du système métrique. L'Académie accepte cette offre avec reconnaissance". *Comtes Rendus hebdomadaires des séances de l'Académie des sciences* (Paris: Bachelier, 1836), vol. 2, p. 613. For the correspondence between Léon de Chazelles and Dupin, see MS Chazelles, Dupin (1836).

37 MAM inv. nos. 07547-0001-001-, 07547-0002-001- (gasometers); 07548-0000- (glass vessel with electrodes); 07508-0000- (hydrometers; these were conceived by Lavoisier in 1768); 07517-0000- (Fortin's air pump); 07520-0000- (Naudier's calorimeter; this instrument was also used by Laplace in the experiments on heat); 07542-0001- and 07542-0002- (standard measures of volume); 07544-0000- (Chemin's balance); 07545-0000- (balance); 07547-0004-001- (ice calorimeter); 07547-0005-001-, 002- and 003- (thermometer tubes); 07549-0000- (oil combustion apparatus); 07550-0000- and 07551-0000- (wine fermentation apparatus); 07658-0000- and 08761-0000- (Megnié's barometers; the latter item was donated by Léon de Chazelles in 1867); 08229-0001- (burning lens, donated by Léon de Chazelles in 1867).

38 Muséum d'histoire naturelle – Paris N° OA.289. The Muséum's list of acquisitions from March 1854 contains the following entry: "55. Appareil de Lavoisier pour la composition de l'eau. Donné par Mme nièce de Mr de Rumfort [sic]". I would like to thank Pierre-Jacques Chiappero for bringing this important piece of apparatus to my attention.

39 MS Register (1864), pp. 90–91. I would like to thank Florence Greffe for providing me with copies of these documents.

other instruments, but I have not been able to find any evidence for other donations.

The arrival of the instruments at the Conservatoire marked a new life for the collection. Furthermore, in 1864 Jean-Baptiste Dumas published the first volume of Lavoisier's *Oeuvres*, thus attracting a renewed interest in his work. Léon de Chazelles also aided the preparation of this publication, which had been begun in 1836: in May 1846 he had donated 51 boxes containing thousands of Lavoisier's manuscripts to the Académie.[40] While sizeable, this was only a part of the collection in his possession. In fact as early as 1838 he had already donated several manuscripts and letters to the Bibliothéque municipale d'Orléans and to the Bibliothéque municipale de Clermont-Ferrand;[41] in the same year he had contributed to his friend naturalist Henry Lecoq Lavoisier's extremely large collection of minerals and fossils.[42]

After reaching the Conservatoire des arts et métiers, the collection of instruments was only partially exhibited. At a scientific soirée organised at the Conservatoire on 29 October 1864, Henri Édouard Tresca accompanied a selected group of "friends of science" on a tour of the various Conservatoire halls where shows devoted to scientific experiments were combined to the display of the highlights of the collection of historical instruments. About the library, which was located in the former refectory of the Benedictine friars (Fig. 10), he wrote:

> The actual venue for the festival was the library, modelled after the Sainte-Chapelle, splendidly illuminated with gas lamps thanks to M. Vaudoré. On each of its tables, the most varied and most curious precision instruments had been placed. Never, in this respect, had such an abundance of riches been assembled; they were further enhanced by Lavoisier's original apparatus, recently donated to the Conservatoire by the Académie des sciences.[43]

And a little later he added:

> To preside, as it were, over this truly splendid collection, of which it did not want to remain a mere spectator, the Conservatoire had placed at the two ends of the room a series of models of great historical value, recently donated by the Académie des sciences and the Société d'encouragement.
>
> These relics were religiously saluted by the friends of science, gathered that evening, and, like them, we shall pay our tribute to the glories thus represented in our exhibition. Lavoisier's original apparatus, recently donated to the Conservatoire by the Académie des sciences, is the one employed by the founder of modern chemistry in his research on the formation of water by the union of its elements.[44]

Naturally, the precious relics that attracted the attention of the audience were the two gasometers and the calorimeter; Tresca does not indicate whether the other instruments donated by the Académie were also exhibited in the library. Together with Vaucanson's machine, Lavoisier's instruments seem to have been the only historical objects in a museum whose primary goal was the celebration of the progress of recent French science through practical demonstrations. It is significant here that Tresca did not even mention the eighteenth-century cabinet of Jacques Alexandre César Charles' instruments, one of the most expansive surviving. This is consistent with Tresca's later

40 MS Chazelles (1846). Manuscript list of 11 pages, annotated by De Chazelles, describing the contents of each 'carton'. The order of listing was eventually changed entirely to the actual one.

41 Dumas (1862) (Bibliothéque municipale d'Orléans); Demeulenaere-Douyère (1994) (Bibliothéque municipale de Clermont-Ferrand).

42 Pelucchi (2009a). De Chazelles further donated a collection of small Egyptian statues ("série de statuettes égyptiennes en bois et en biscuit") to the city, which also came from Lavoisier's collection. In reply to an enquiry at the Musée Bargoin of Clermont-Ferrand, Madame Camille Gaumant kindly replied that the small statues "ont été ramenés d'Égypte par le baron François de Tott (1733–1793) lors de ses expéditions de 1777–1778. Ils ont ensuite été donnés à Antoine-Laurent de Lavoisier. Sa collection fut donnée à la ville en 1836 par Léon de Chazelles".

43 "Le vrai salon de la fête, c'était la bibliothèque, cette émule de la Sainte-Chapelle, qui, splendidement éclairée au gaz par les soins de M. Vaudoré, avait reçu sur chacune de ses tables les intruments de précision les plus variés et les plus curieux. Jamais on n'avait réuni, sous ce rapport, d'aussi grandes richesses, qui se trouvaient encore rehaussées d'une part par l'appareil original de Lavoisier, récemment donné au Conservatoire par l'Académie des sciences". Tresca (1864), p. 207.

44 "Pour présider en quelque sorte à cette collection vraiment splendide dont il ne voulait pas rester simple spectateur, le Conservatoire avait garni les deux extrémités de la salle avec des modèles d'une grande valeur historique, qui lui ont été donnés récemment par l'Académie des sciences et la Société d'encouragement. Ces reliques ont été religieusement saluées par les amis de la science, réunis à cette soirée, et nous payerons, comme eux, notre tribut aux gloires ainsi représentées à notre exposition. L'appareil de Lavoisier, récemment donné au Conservatoire par l'Académie des sciences, est celui qui a été employé par ce fondateur de la chimie moderne, dans ses recherches sur la formation de l'eau par l'union de ses éléments". Ibid., p. 241.

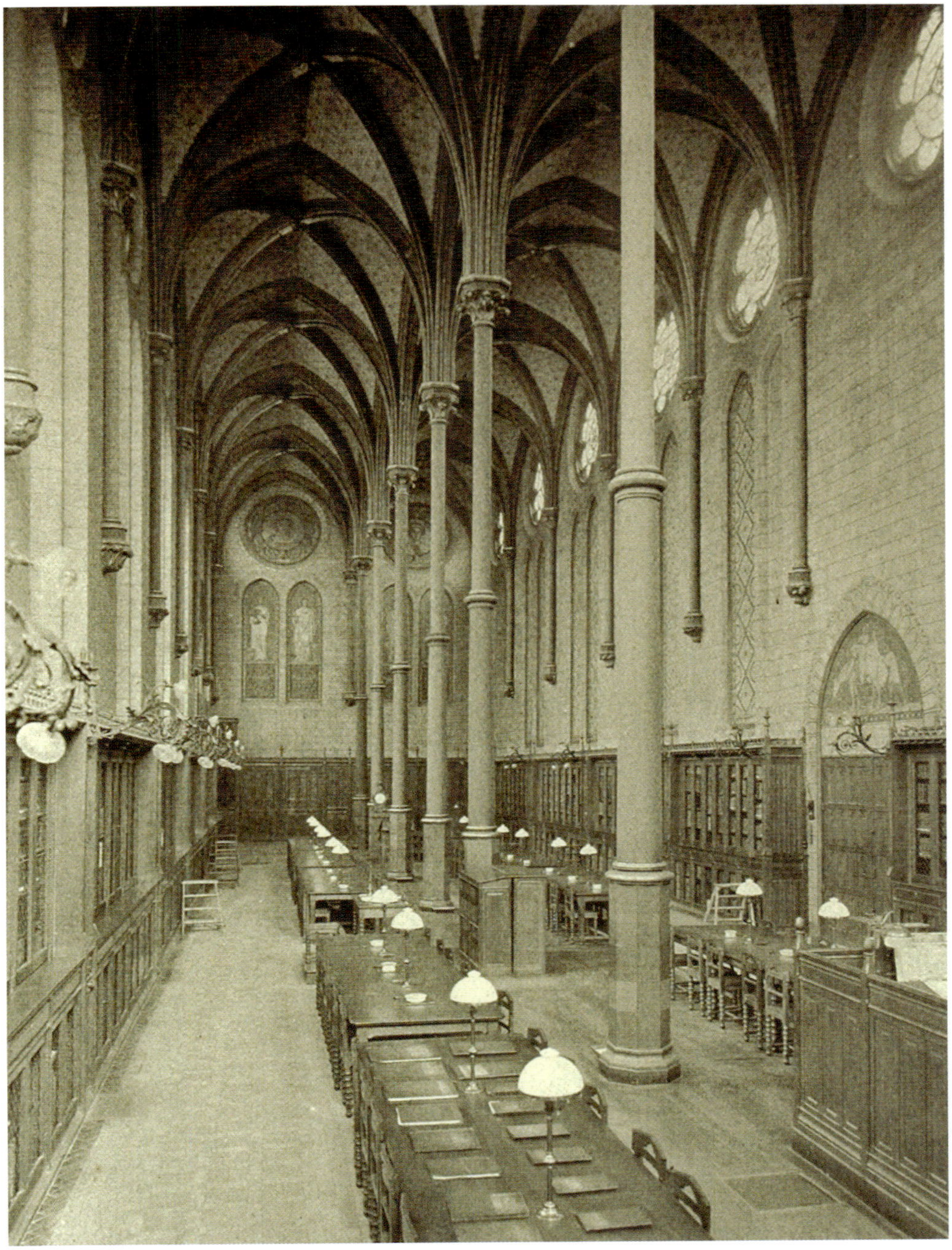

FIGURE 10
The old Benedictine friars' refectory, transformed in the library of the Conservatoire des arts et métiers where, in 1864, Lavoisier's gasometer and calorimeter were exhibited.
COURTESY OF MAM. © MAM

account of the museum's holdings, written in 1869, which omitted even Lavoisier's instruments.[45] The great universal exhibition of 1867 seemed adopted the same vision of demonstrating the utility of science and technology in the present and in the future rather than celebrating their past successes.

Despite the early neglect of the Lavoisier instruments at the Conservatoire, this was the first collection gathered by a chemist to be exhibited in a public museum. Lavoisier's fame increased with the publication of the first few volumes of his collected works in the late 1860s; and the outbreak of the Franco-Prussian war in 1870 provoked a fierce controversy between French and German chemists over the role played by Lavoisier in the history of chemistry, which allocated a more prominent place – and even a political role – to his instruments.[46] The war, in fact, epitomised a lasting conflict between German and French chemists that had begun long before 1870 and that,

45 Tresca (1869).

46 Regarding the outbreak of the Franco-Prussian war, see Béchamp (1871). On the controversy, see Daumas (1955), pp. 66–90; Bensaude-Vincent (1993), pp. 363–392.

throughout the nineteenth century, had accentuated its dramatic tones in all possible variations.[47]

Charles-Adolphe Wurtz's famous claim (which introduces his work on the evolution of chemical concepts) that chemistry was a French science because it had been established by Lavoisier was rebutted by the German chemical community, and the history of chemistry suddenly became an important battle field which attracted the attention of the most influential chemists of the time.[48] Within this context, artefacts also began to play an important role. At the grand exhibition of scientific instruments and apparatus at the South Kensington Museum in March 1876, scientists from all over the world competed against each other by presenting the recent scientific achievements of their nations. 'Relics' and historic collections were extensively used to this purpose. Galileo's and Newton's instruments attracted the attention of tens of thousands of visitors for the first time. While the British, German, and Italian exhibitions were organised with the greatest care in order to emphasise the prominent role of these nations in the history of science via displays of historic artefacts, France did not match them, especially in chemistry. Lavoisier's calorimeter, a landmark in the chemical revolution and the only piece from his collection brought to the London exhibition, was displayed in a section on heat, between sections devoted to thermometry and conduction. The resounding success of the London exhibition must have inspired the French authorities to reassess their policies concerning the public display and function of scientific collections.

Despite the French failure at the 1878 exhibition, French scientists had not forgotten Lavoisier, and following Jean-Baptiste Dumas' death in 1884, the chemist Edouard Grimaux approached the completion of the publication of Lavoisier's collected works with renewed energy. In this crucial period the celebration of the first centenary of the French Revolution was fast approaching, and this also, fortuitously, coincided with the centenary of the publication of Lavoisier's most important work, the *Traité élémentaire de chimie*. It is, therefore, not surprising that there was a desire to organize a series of events that would highlight Lavoisier's prominence in history. Edouard Grimaux, who had been granted access to the manuscripts and documents which remained with Lavoisier's heirs, published the hitherto most authoritative biography in 1888, and only two years later Marcellin Berthelot published his, which contained a detailed description of Lavoisier's laboratory notebooks.[49]

Sometime before 1882 Lavoisier's collection of instruments was finally exhibited on the ground floor of the Conservatoire national des arts et métiers, in the so-called 'salle Lavoisier', between the collection of physics instruments and the *salle de l'écho* (Figs. 11 and 12).[50] This added greatly to the visibility of the collection, and in the Parisian Exposition universelle of 1889, Lavoisier's laboratory, together with Madame Lavoisier's portrait and her drawings of the experiments on respiration, filled an entire room.[51] Lavoisier's name was also added to the 72 inscriptions adorning the Eiffel tower.

In 1894, at the commemoration of the centenary of Lavoisier's death, the public instrument exhibition attracted such widespread attention that it was exhibited in a part of the Vaucanson gallery of the Conservatoire.[52] The exhibition was "a part of the national reparation to Lavoisier proposed to the Academy of Sciences by M. Grimaux, member of the chemistry section, and joint editor of Lavoisier's works with the late Dumas".[53]

For the Universal Exhibition in Paris in 1900, Berthelot organised a grand celebration of Lavoisier as the founder of modern chemistry. Under the auspices of the Académie des sciences and guided by Berthelot, the French chemical community decided to celebrate Lavoisier's life work with a monumental statue. Between 1894 and 1900 over 100,000 francs were collected in France, Russia, the USA, and further European countries. The statue installed in front the church of La Madeleine was completed in July 1900 (Fig. 13).

The date of the statue's unveiling coincided with the opening of the International Congress of Pure and Applied Chemistry and anticipated the opening of the *Exposition universelle internationale de 1900*, which would take place a few months later, and for which Berthelot and Louis Troost organised an exhibition celebrating Lavoisier and French chemistry, the "Musée centennal de la classe 87".[54] The congress, the exhibition, and the symbolic unveiling of the statue were all the result of Berthelot's energetic campaign intended to vindicate the primacy of Lavoisier's

47 Fell (2000).

48 "La chimie est une science française: elle fut constituée par Lavoisier, d'immortelle mémoire". Wurtz (1869), p. 1. Beretta (2011a).

49 Grimaux (1888). Berthelot (1890).

50 *Conservatoire* (1882), p. lxxii.

51 On the experiments, see the *Gazette des beaux arts* (1889), p. 538. On the room, see *Catalogue* (1889), p. 10.

52 "The above-named apparatus, with a quantity of other scientific instruments, are now being exhibited in a portion of the Vaucanson Gallery, which is to be called Salle Lavoisier". Anonymous (1894).

53 Ibid.

54 On this exhibition, see Beretta (2019).

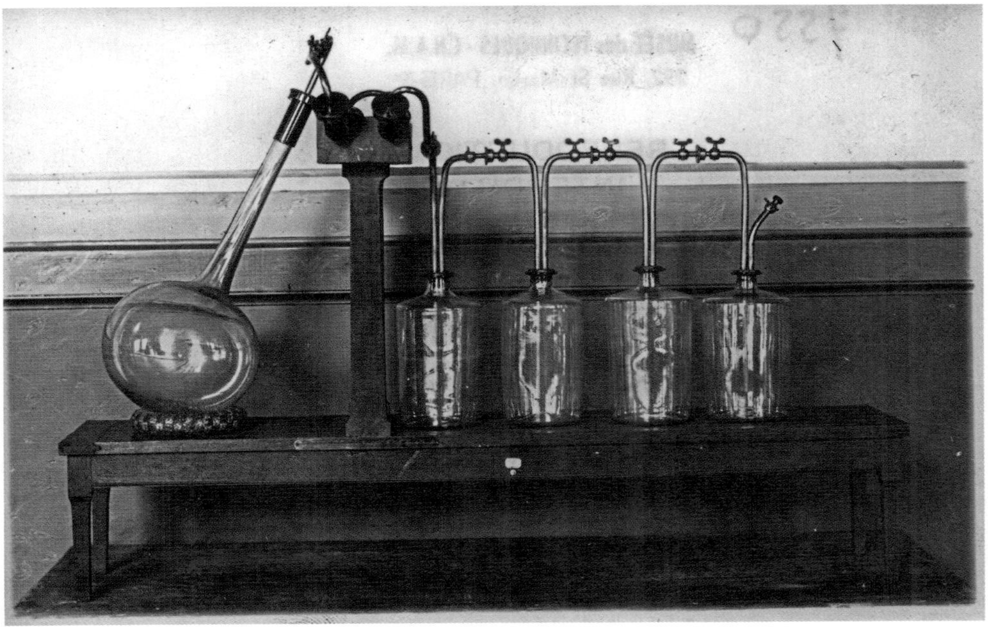

FIGURE 11
Lavoisier's apparatus for studying fermentation in the exhibition of 1882
COURTESY OF MAM. © MAM

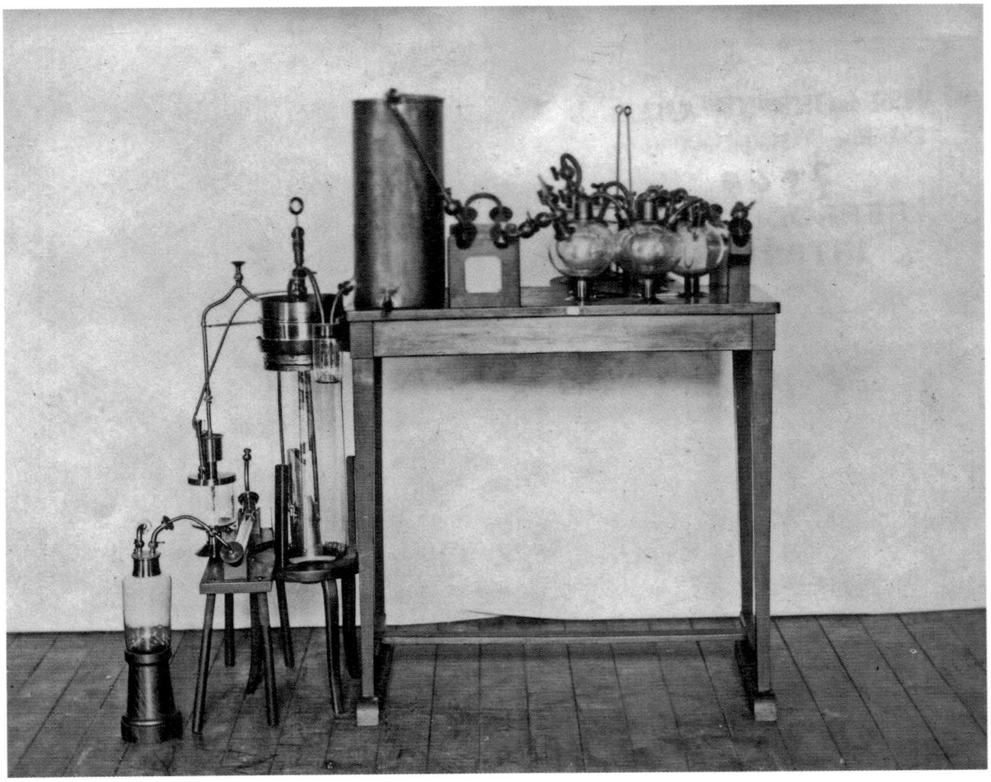

FIGURE 12
Lavoisier's apparatus for the combustion of oils in the exhibition of 1882
COURTESY OF MAM. © MAM

chemistry, to the French nation, and to some degree to himself.[55]

The Musée with the chemical exhibition was one of many devoted to the arts and crafts, but arguably the most important one, since it exclusively focused on the history

55 Significantly, one year later, on 24 November 1901, the international scientific community celebrated Berthelot's 50 years of scientific activities at a meeting at the Sorbonne. The transactions were published in Anonymous (1902). In 1927 the Berthelot's centennial was celebrated with the erection of a statue and his commemoration in the Pantheon, and much of the rhetoric of celebrating Lavoisier as the champion of French chemistry in 1900 was re-used to commemorate Berthelot. On this, see Anonymous (1929) and the informative essay by Fox (2016).

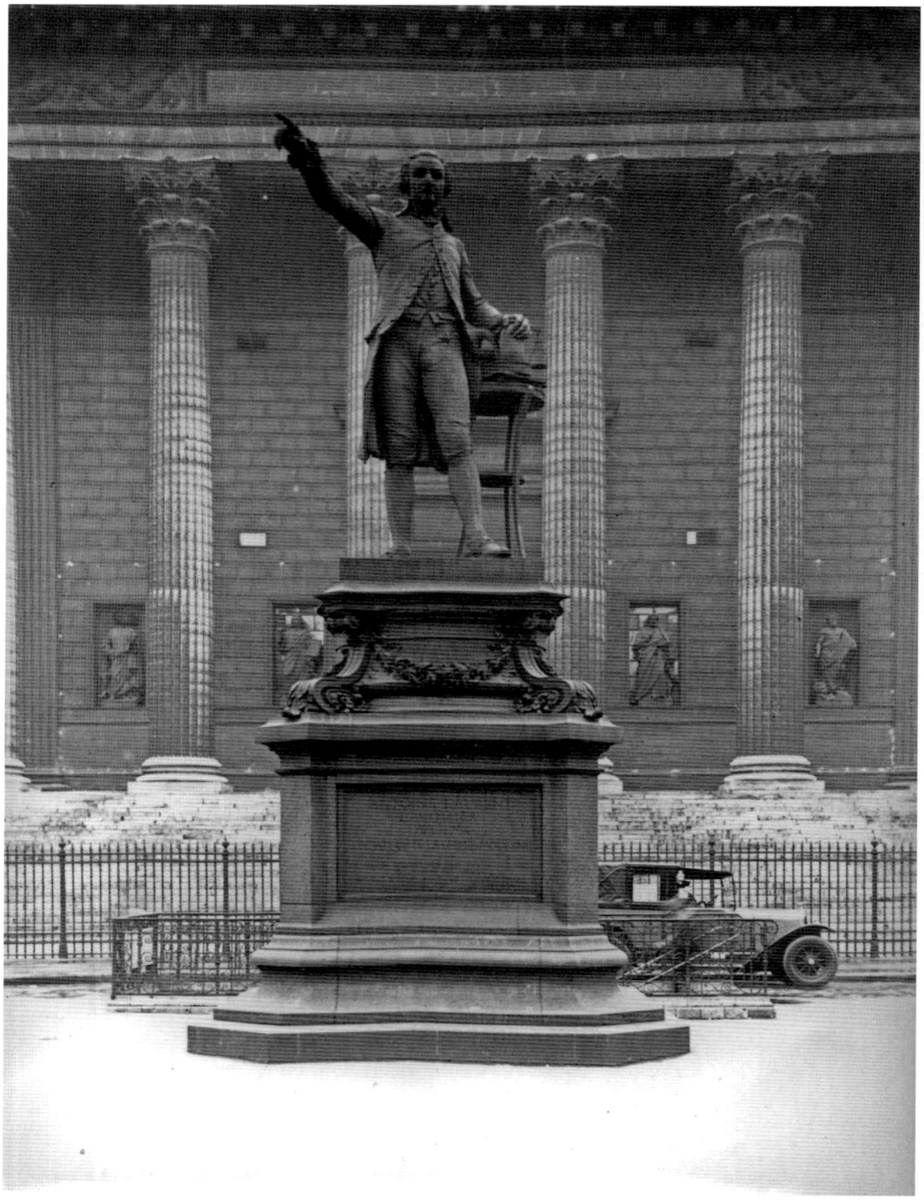

FIGURE 13
Barrias' statue of Lavoisier, in front of La Madeleine (1900)
PRIVATE COLLECTION

and achievements of a specific French science. This was the first time that French history was presented through the artefacts, instruments, machines, chemicals, and (to a lesser degree) written documents left by its protagonists. It is significant that the exhibition, titled "Les chimistes français du XIXe siècle", was limited to French chemists. "Science", admitted the anonymous author of the introduction to the catalogue, had "no fatherland", but for the sake of consistency it was necessary to focus on the French context. As Lavoisier was the "founder of modern chemistry", only the French lineage of chemists, from Berthollet to Berthelot, could reveal its rapid progress in the purest form. Apart from the ideology that had inspired these initiatives, the Musée centennal of chemistry was a very fine exhibition where, among other things, various instruments, manuscripts, and portraits of Lavoisier were displayed for the first time (Fig. 14). Telling the nationalistic story of the history of French chemistry was not the only purpose of the exhibition; rather, in fact, the organisers intended to build up a permanent museum of French chemistry, but for various reasons this project was never realized, and the collection was, once more, dispersed. Those of Lavoisier's items that had been lent by his heirs were returned to them, and the remaining instruments went back to the Conservatoire. It was probably in 1900 (and certainly not later than 1904) that the 'salle Lavoisier' was rearranged, and the most important instruments and apparatuses were placed into a large glass case and

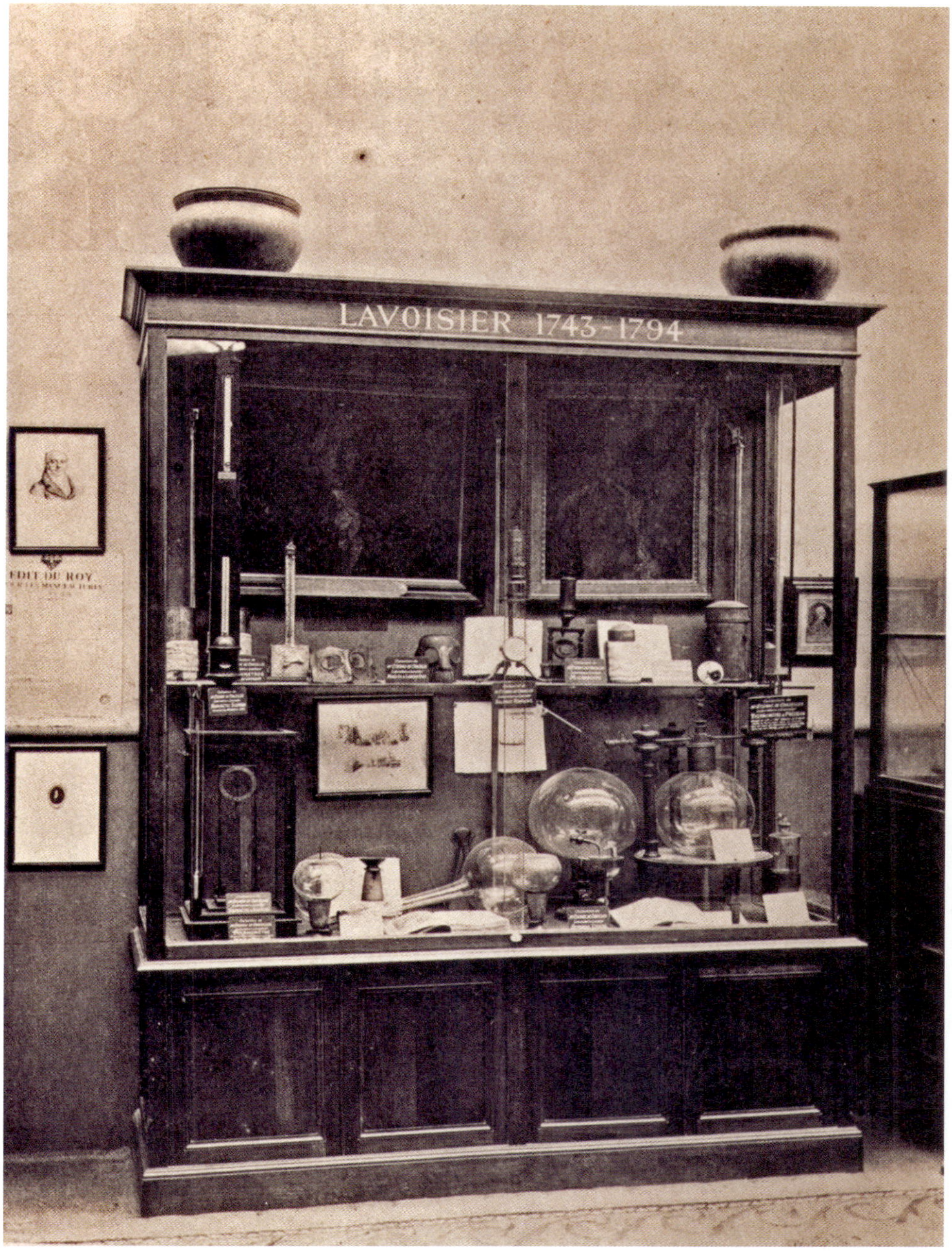

FIGURE 14 The case devoted to Lavoisier at the Musée Centennal, displaying several instruments and iconographic items which were eventually dispersed
MUSÉE CENTENNAL DE LA CLASSE 87. ARTS CHIMIQUES ET PHARMACIE, 1900. PRIVATE COLLECTION

exhibited in the *salle carrée*, the first hall at the top of the main stairs, on the first floor (Figs. 15 and 16).[56]

56 The photo was taken and published in a commemorative volume by the Conservatoire des arts et métiers: *Conservatoire* (1904).

Culturally and politically, the grand commemoration of the centenary of Lavoisier's death undoubtedly produced a better knowledge of his work and his chemical heritage, but at the same time, the deliberate effort to use and display the objects in order to create a cult of Lavoisier as the founder of modern chemistry proved to be shortsighted: it transformed Lavoisier into a myth.

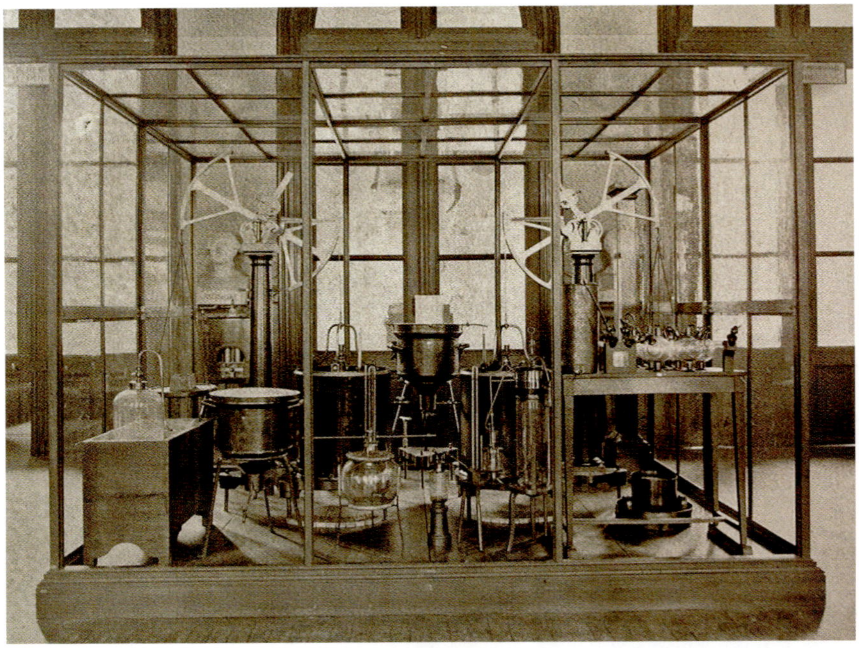

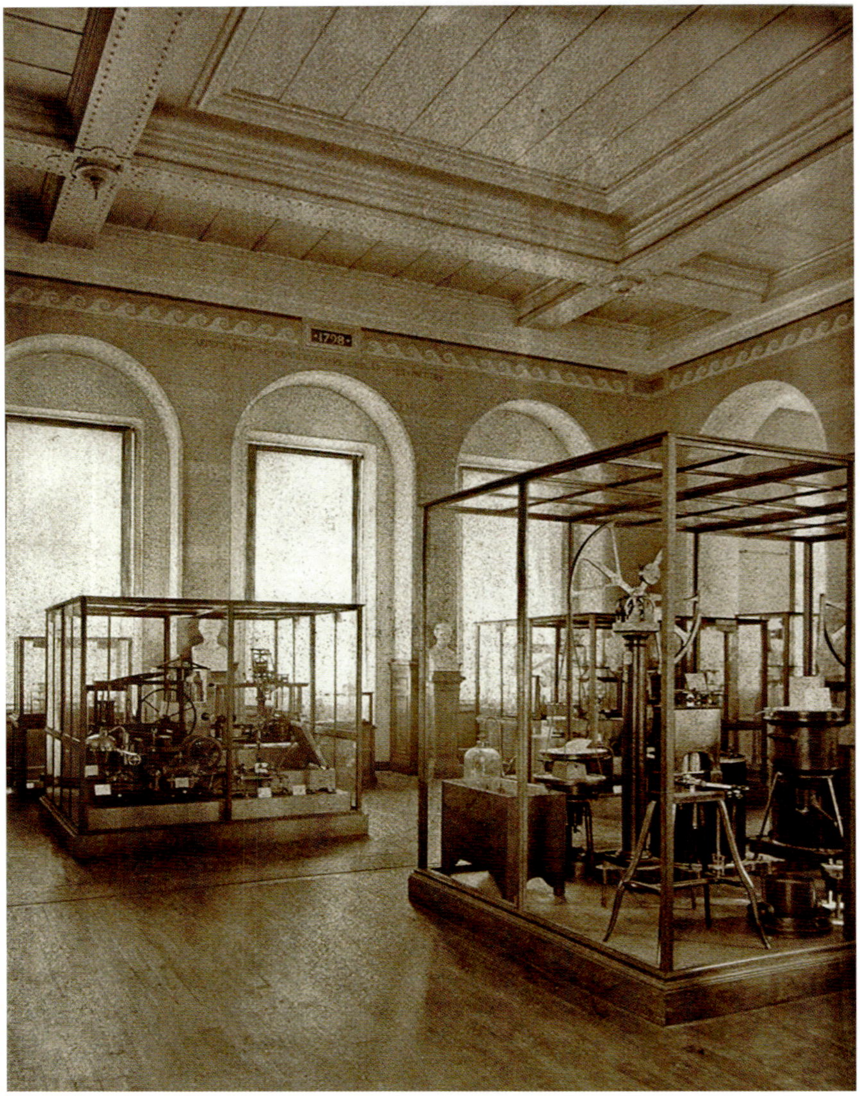

FIGURES 15–16
Photographic views of Lavoisier's gasometers and calorimeters displayed in a glass case of the 'salle carrée' (1900–1904)
COURTESY OF MAM. © MAM

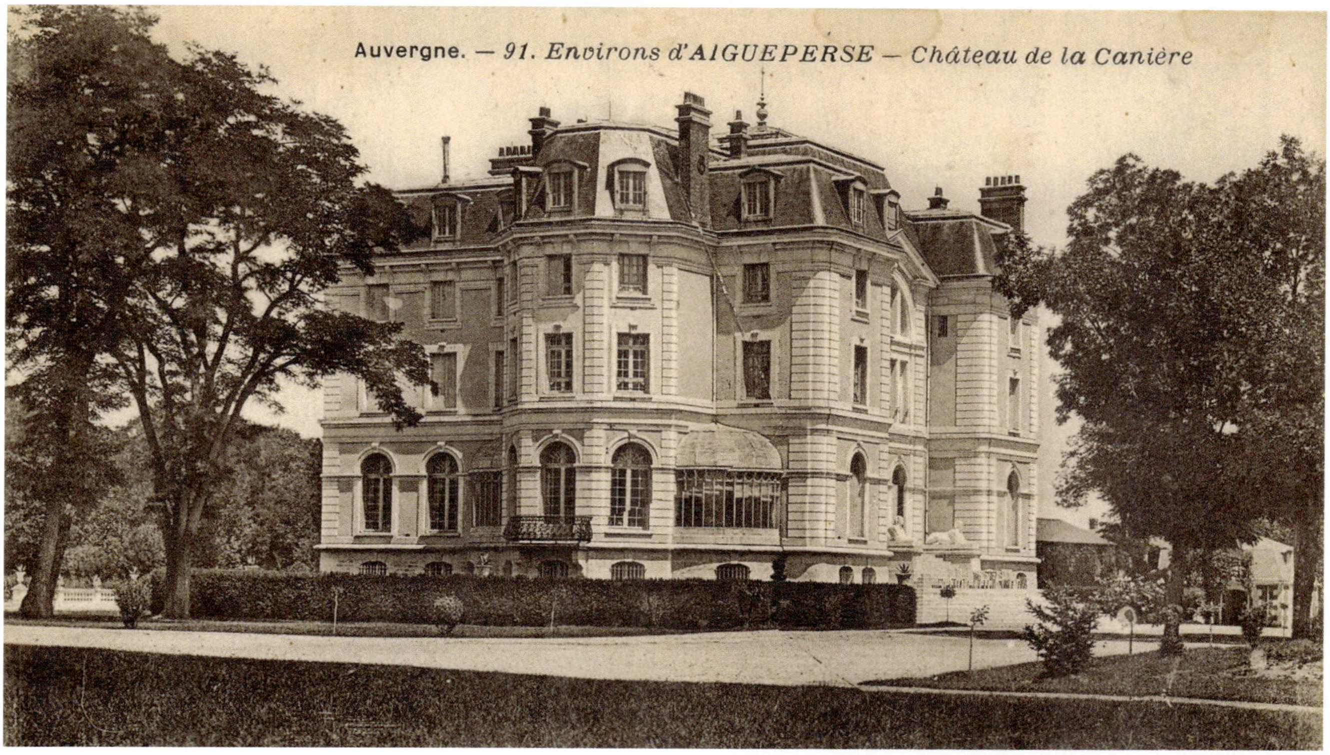

FIGURE 17 The château of La Canière in the early twentieth century
PRIVATE COLLECTION

5.3 The 1943 Exhibition: Lavoisier vs Nazi Germany

The story of Lavoisier's collection of instruments between the two World Wars is intertwined with the complex vicissitudes of what was still left in the hand of his heirs. When Léon de Chazelles died in 1876, all of Lavoisier's collection – instruments, books, and a significant portion of his manuscripts – was moved by his son, Étienne de Chazelles, to a château in La Canière in Aigueperse, near Riom (Puy-de-Dôme) (Fig. 17). In 1879 Pierre Truchot, a professor of chemistry of the University of Clermont-Ferrand (Fig. 18), published an extensive report on Lavoisier collection of instruments:[57]

> [Lavoisier's] chemical laboratory and his *cabinet de physique* have been piously conserved by Mme Lavoisier's family, and I am indebted to Etienne de Chazelles, their present fortunate owner, who has given me the pleasure, not to say the happiness, of 'allowing me to get to know the collection by making an inventory and handling one by one, understandably not without emotion, all the objects that remind us of the prolific work of by the immortal founder of Chemistry [...]. Let us now open the cabinet. *The large amount of apparatus accumulated there shows the respectful care with which the family of Lavoisier have felt it a duty to save even occasionally commonplace objects from destruction.*[58] (Fig. 19)

Truchot described in some detail the mercury gasometer, the precision balances, the most important thermometers and barometers, and a few other instruments, thus providing the first scientific assessment of the collection and making it very clear that an incredibly large and important part of Lavoisier's instruments was still in the hand of the heirs. In the subsequent decades, Étienne Bérard de Chazelles (1836–1923), the son of Léon, still tried to keep the collection together, both at the château de la Canière and at his Parisian residence.

57 Truchot (1879).

58 "Son laboratoire [Lavoisier's] de Chimie, son cabinet de Physique ont été pieusement conservés par la famille de Mme Lavoisier, et je dois à M. Étienne de Chazelles, qui en est actuellement l'heureux possesseur, le plaisir, je devrais dire le bonheur, d'avoir pu en prendre connaissance, en dresser l'inventaire et toucher un à un, non sans émotion facile à comprendre, tous ces objets qui rappellent les travaux féconds de l'immortel fondateur de la Chimie. [...] Ouvrons maintenant le cabinet. La grande quantité d'appareils qui y sont accumulés témoigne du soin respectueux avec lequel la famille de Lavoisier a cru devoir préserver de la destruction des objets quelquefois vulgaires". Ibid., pp. 291–292, italics my own.

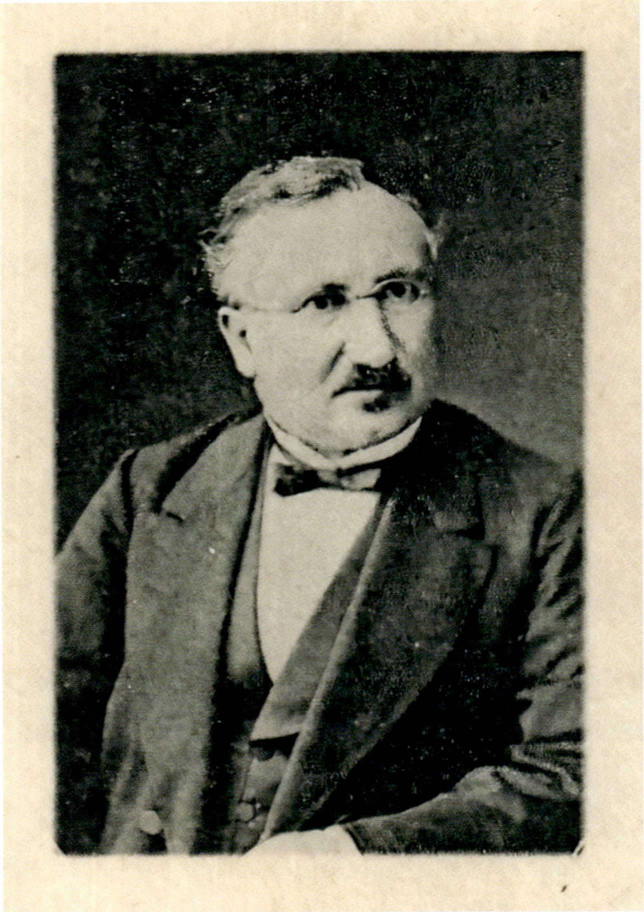

FIGURE 18 Photographic portrait of Pierre Truchot in 1900
PRIVATE COLLECTION

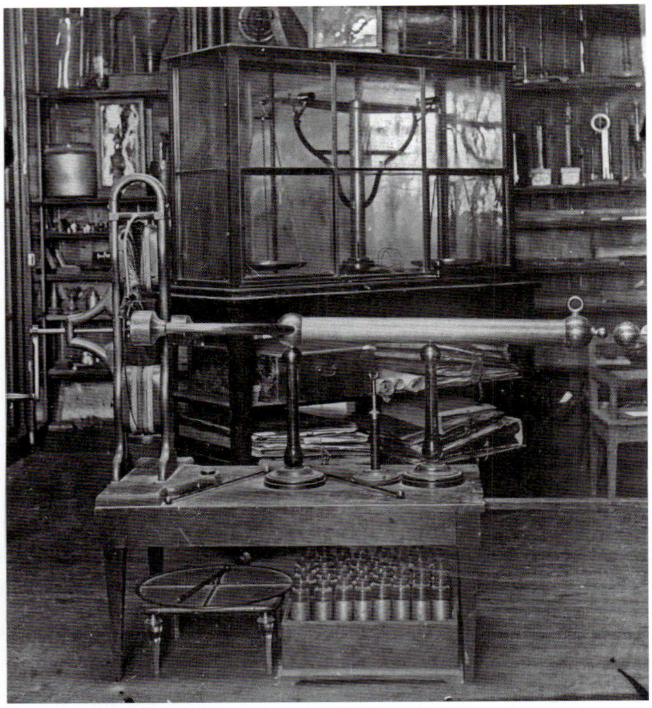

FIGURE 19 Fortin's electrical machine and other instruments stored in the château de la Canière at the end of the nineteenth century
COURTESY OF THE ARCHIVES DE L'ACADÉMIE DES SCIENCES – PARIS

After Etienne de Chazelles' death his possessions were inherited by his three children Pierre Bérard de Chazelles (1878–1934), Catherine Bérard de Chazelles (1881–1971), and Magdeleine Bérard de Chazelles (1884–1944). As a result, the famous portrait of Lavoisier and his wife was sold to John Rockefeller in 1925. It was probably in this period that "much of the 'stock apparatus', such as flasks, retorts and so on, was sold".[59]

A part of the collection, consisting of books, manuscripts, and instruments went to Pierre de Chazelles, who moved it to the château of La Champfortière in the department of Sarthe. On 1st October 1935, a few years after his death, his widow, Nelly de Vanssay, opened a museum exhibiting "les reliques de Lavoisier restées dans la famille" in the castle's annexe:

> these priceless and moving mementos include a substantial archive of papers of the illustrious founder of modern chemistry, his entire library, and the largest gathering of his laboratory instruments, most notably the scales.[60]

It is difficult to imagine that this small and short-lived private museum would have been particularly successful. Other instruments and some of the manuscripts were still in the castle of La Canière, where Henry Guerlac saw them in August 1939.[61] What happened to these instruments is not fully known, though they were probably moved to La Champfortière before La Canière was seized by the German occupying forces sometime around 1943. The dangers of the war led to another move, and in 1944 Nelly Vanssay de Chazelles took Lavoisier's instruments

59 Letter from Douglas McKie to Pierre Dupont, 23 July 1951. MS McKie (1947–1952). McKie acquired information on the recent history of Lavoisier's collection in several interviews with Nelly de Vanssay, the widow of Pierre de Chazelles.

60 "Ces inestimables et émouvants souvenirs comprennent un lot important de papiers de l'illustre fondateur de la chimie moderne, toute sa bibliothèque, le plus grand nombre enfin de ses instruments de laboratoire et notamment les balances". Printed leaflet of the inauguration sent to the Académie des sciences on 3 December 1935. MS Musée Lavoisier (1935).

61 Guerlac (1979), p. 96. Unfortunately, there is no description of the collection of instruments among Guerlac's papers: MS Guerlac-Perrin (1939–1987).

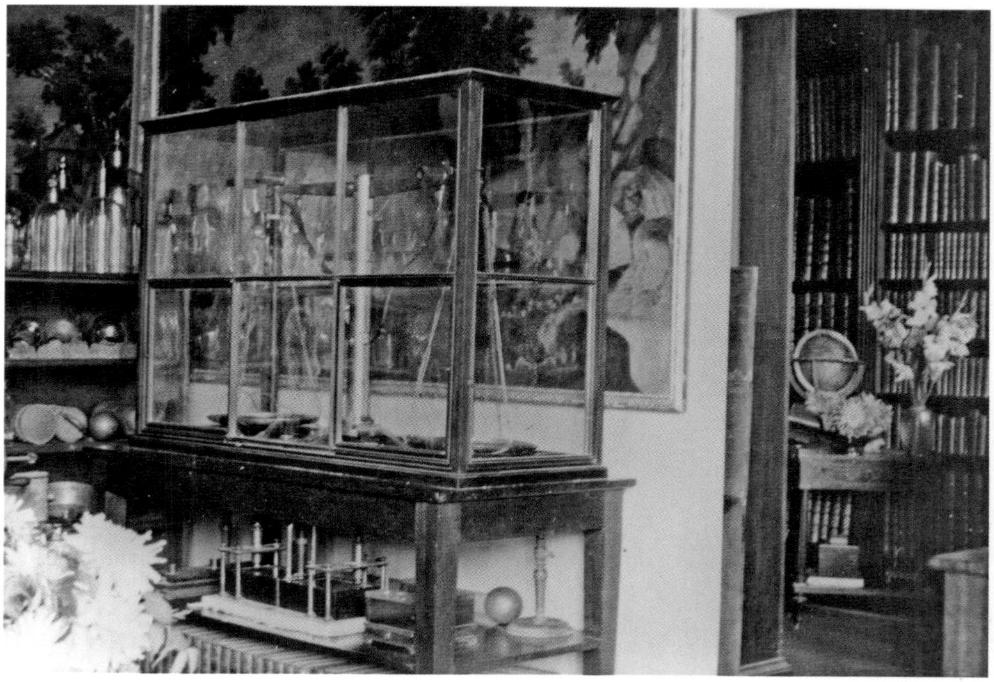

FIGURE 20
Lavoisier's collection of instruments and book at Champaissant (September 1951)
COURTESY OF THE HAGLEY MUSEUM & LIBRARY

and library to a small residence at the Champaissant, near Le Mans (Fig. 20). Although the house was too small for such a vast collection, she managed to keep it safe and in a relatively good condition.

In Paris in the 1920s and 1930s, no special attention was given to Lavoisier's collection. The great Paris Universal exhibition of 1937, which featured a large number of science and technology displays, projected a vision of the future. By contrast, in the same year Gabriel Bertrand of the Académie des sciences reconstructed a working replica of Lavoisier's grand apparatus for the analysis of organic substances, which was exhibited among the "reconstitutions historiques" at the newly opened Palais de la découverte. The De Chazelles' small private musée of 1935 and Bertrand's replica were part of the preparations for the forthcoming celebrations of the bicentenary of Lavoisier's birth in 1943. The dramatic events following the outbreak of World War II and the German invasion of France in the spring of 1940 did not prevent this original project; on the contrary, the Lavoisier celebrations became an emblem of the resistance to the German occupation. Lavoisier's chemical revolution trumped Georg Ernst Stahl's phlogiston theory, and this was an indisputable historical evidence.

The exhibition Lavoisier, organised by Gabriel Bertrand and the Académie des sciences, was held at the Palais de la découverte in Paris between November 1943 and January 1944.[62] In these extremely difficult circumstances the organisers' tireless efforts resulted in a truly spectacular event which exhibited for the first time a substantial number of Lavoisier's instruments (more than 150 altogether), minerals (78), books, manuscripts, memorabilia, and other items illustrating his life and scientific work (Fig. 21).[63] Many of the objects were lent by Nelly Vanssay de Chazelles, and they came from the small museum created in 1935. An interesting attempt was made to reconstruct Lavoisier's working atelier with its original objects, just as it had been portrayed by David (Fig. 22). Several portraits, many from private collections, were also exhibited. In addition to its unsurpassed scope, the most interesting and innovative aspect of this exhibition was the re-enactment of many of Lavoisier's experiments with the original instruments, for instance those on the combustion of phosphorous and on the combustion of oil (Fig. 23). It was because of this exhibition that Maurice Daumas, a graduate in literature, became interested in instruments; a few years later he would be a pioneer in the study of eighteenth-century scientific instruments.[64]

This magnificent exhibition celebrating the chemical revolution as a phenomenon of French history and a result of Lavoisier's discoveries was a resounding success,

62 Lavoisier (1943).

63 Ch. Maurain, vice-président of the Académie des sciences, remarked: "C'était une entreprise difficile, surtout dans les circonstances actuelles, que de présenter un tableau convenable d'une œuvre aussi vaste et aussi variée." Lavoisier (1943), pp. 16–17.

64 In 1941 Daumas published a biography (Daumas, 1941), which was exhibited in 1943.

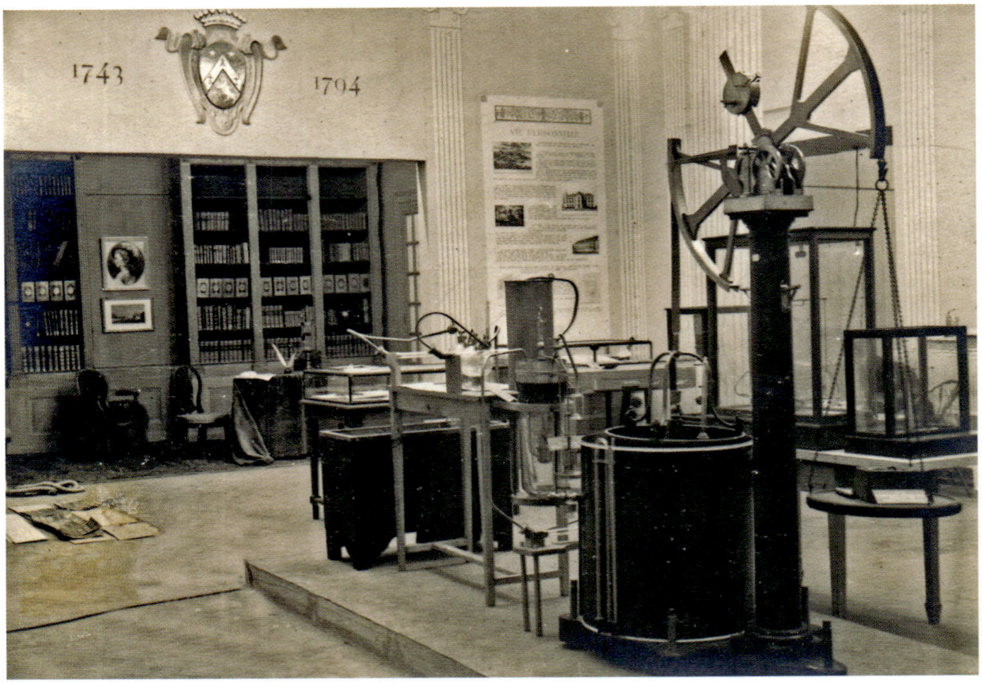

FIGURE 21
A view of the main hall of the Lavoisier's exhibition at the Palais de la découverte (1943)
PRIVATE COLLECTION

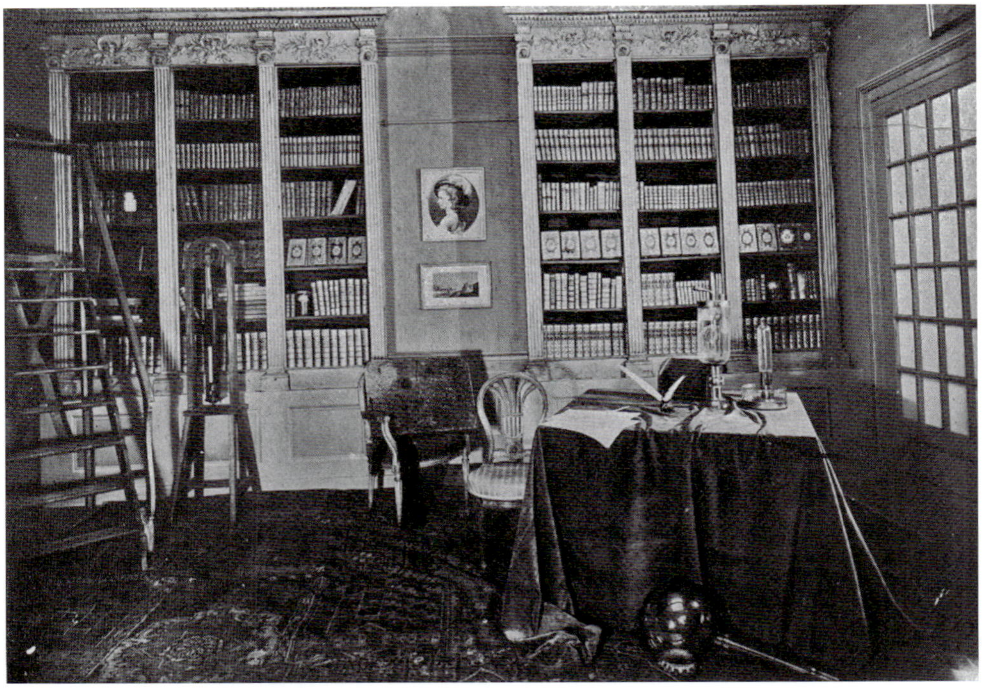

FIGURE 22
A reconstruction of David's portrait at Lavoisier's exhibition at the Palais de la découverte
FROM LAVOISIER (1943)

and as expected, reactions in the German press were not particularly enthusiastic.[65] This long-lasting cultural conflict between French and German chemistry may even have been the main reason for the Nazis' destruction of the Lavoisier statue that had been erected at La Madeleine in 1900.

The aims of the 1943 exhibition went beyond the vindication of Lavoisier's role in the history of chemistry and the indirect cultural war against the German occupation: with her generous loans Nelly de Vanssay also wished to make the collection so widely known that she could sell it for a good price after the war had ended. Indeed, between 1947 and 1956, she sold the memorabilia, instruments,

65 A long review of the exhibition was published in the German paper *Pariser Zeitung*; it vindicated the importance of Stahl's chemistry. A copy of the article is preserved in the Archives de l'Académie des sciences – Paris. Dossier Biographique Lavoisier.

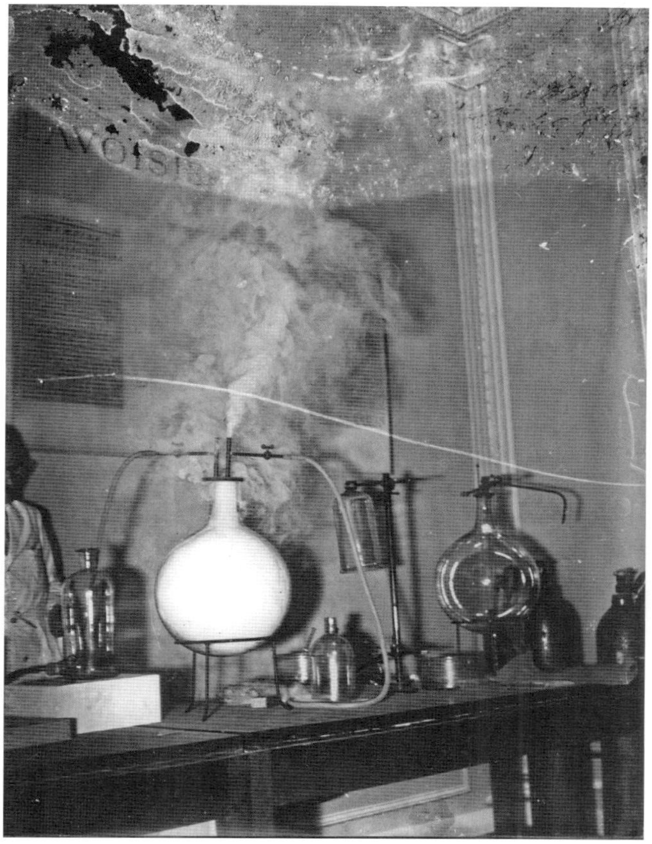

FIGURE 23 A reconstruction of Lavoisier's experiment on the combustion of phosphorous with his original apparatus at the Palais de la découverte
COURTESY OF THE ARCHIVES DE L'ACADÉMIE DES SCIENCES – PARIS

books, and manuscripts in her possession and any efforts from different patrons, institutions, and scholars to prevent its dispersal sadly failed.

In 1947 Pierre Samuel Dupont – the president of the chemical corporation Dupont that had been founded towards the end of the eighteenth century by one of Lavoisier's laboratory assistants – asked the British historian of science Douglas McKie to find the Lavoisier collection and make preparations for its deposit in a French public institution. In order to facilitate a transfer of the instruments that was satisfactory for everyone concerned, the idea of establishing a foundation ("Fonds Lavoisier") or a museum was put forward. Madame de Chazelles would have been a member of the board of trustees.

When the acquisition and the cataloguing were concluded, the next problem was the identification of a suitable home for the collection. Since both McKie and Dupont still hoped to persuade Madame de Chazelles to sell the manuscripts and books that remained in her possession, they tried to find a location for the creation of a "Museum Lavoisier". The French Minister of Education suggested,

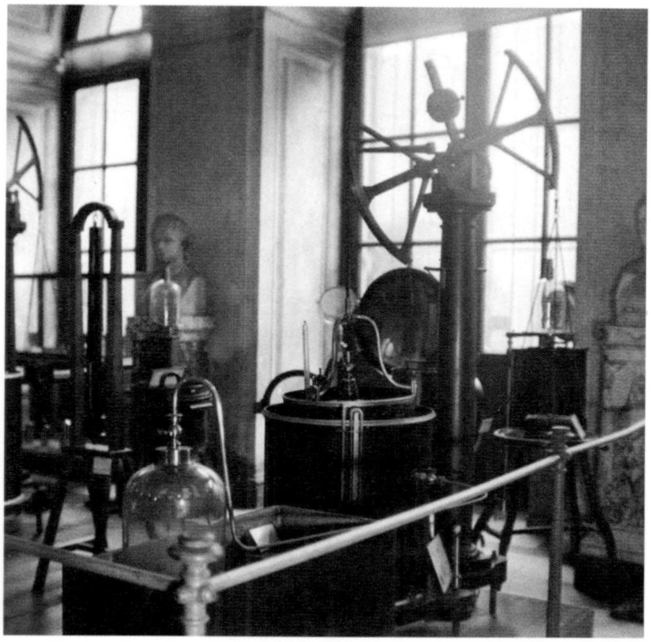

FIGURE 24 The exhibition of Lavoisier's instruments at the Musée des arts et métiers in 1954
COURTESY OF MAM. © MAM

among others, the beautiful eighteenth-century building Folie Saint-James, which was in Neuilly-sur-Seine near Paris.[66] Another suggestion, considered "quite unsuitable" by McKie, was the Maison de la chimie in Paris.[67] In the end, a less satisfactory solution was settled upon:

> [t]hinking over the question of location of the collection, there was an earlier idea of the Musée des Arts et Métiers – it is a good place, but very full. There is already some of Lavoisier's apparatus there and it might be a good thing in one way to have it all in one place, but I had hoped, when I heard of the Folie Saint-James, that eventually the material from the Arts et Métiers might be transferred there with your collection. There would be more room there and a finer setting – it would be a special place for a special collection. At the Arts et Métiers, however, there is a very good man, Maurice Daumas, whom I met and whose work I know, and he has made a special study of Lavoisier. My sole objection to it is that (1) it is not a place for a special collection and (2) it is so full that is very unlikely that the whole

66 This was the residence of the financier Claude Baudard de Saint-James (1738–1787).

67 Letter from McKie to Dupont, 7 April 1952: MS McKie (1947–1952).

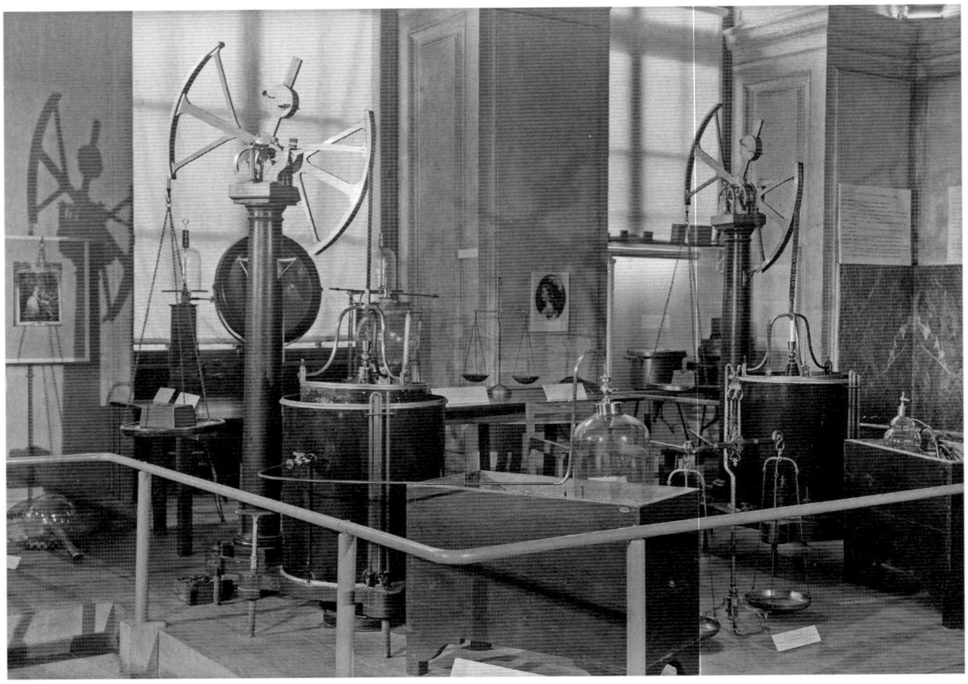

FIGURE 25
The exhibition in the Salle de l'écho of Lavoisier's instruments at the Musée des arts et métiers
COURTESY OF MAM. © MAM/ PHOTO ALAIN DOYÈRE

> collection could be mounted as a permanent exhibition, which, I feel, is most desirable.[68]

McKie's fears turned out to be well founded. In fact, after the Musée des arts et métiers took possession of Dupont's donation in the summer of 1952, the display of the whole collection changed radically thanks to Daumas' efforts. Lavoisier's collection was placed closer to the entrance and occupied a large hall – by far the largest devoted to an individual scientist (Figs. 24–25). When I had the opportunity to visit the collection in the mid-1980s its arrangement seemed to be very much like that originally conceived by Daumas in the early 1950s. Dupont had hoped to be able to purchase the entire collection (including manuscripts, books, and memorabilia), but the heirs, aware of their rising value, placed the transaction into the hands of antiquarian dealers in both Paris and London. An important auction organized by Drouot (Paris) took place in 1956, and most of the manuscripts and books were purchased by Denis Duveen, the son of the famous British art dealer. Duveen eventually sold the Lavoisier collection in 1962, for an astronomic sum (still kept confidential by the University) to Cornell University, and between 1963 and 1968 a couple of small exhibitions were held in Ithaca and New York.[69] Nevertheless, this splendid collection is still largely unexplored.

When this arrangement took place the history of science was becoming an independent discipline, but Lavoisier's collections, after miraculously surviving, were dispersed in the following institutions:

Institution	Items
Archives of l'Académie des sciences – Paris	ca. 5,000 manuscripts
Muséum Lecoq – Clermont-Ferrand	ca. 4,000 minerals and fossils
Kroch Library (Cornell University) – Ithaca, NY	ca. 1,100 items, mostly books from Lavoisier's library, numerous manuscripts, some memorabilia, and a couple of instruments
Musée des arts et métiers – Paris	ca. 600 items, mostly instruments, and some natural specimens and furniture
Bibliothèque municipale et interuniversitaire de Clermont-Ferrand	ca. 250 manuscripts
Bibliothèque de l'institut – Paris	89 books and manuscripts
Bibliothèque interuniversitaire de Bordeaux	64 books from Lavoisier's library

68 Ibid., p. 2.
69 Guerlac (1963).

(*cont.*)

Institution	Items
Lavoisier's heirs (Paris and Jozerand)	books, Madame Lavoisier's sepias and self-portrait; Lavoisier's furniture and an unknown number of manuscript documents; Jacques Paulze's manuscripts (?)
Hagley Library and Museum, Harvard University, Muséum national d'histoire naturelle	a few instruments and manuscripts, donated or sold by Lavoisier's heirs between 1856 and 1956

The collection which had once lived in Lavoisier's laboratory and which, in its entirety, traced the first great revolution in chemistry, is now dispersed. The instruments, without the manuscripts recording their use, rapidly became nothing more than mysterious and silent objects as well as rhetorical symbols of an obscure form of experimentation. They were left in the storage of the Musée des arts et métiers and progressively decayed. The manuscripts which classified collections of natural history, fossils, and minerals became devoid of meaning, and the minerals and the library inevitably lost their original importance.

It is clear that the ultimate fate of Lavoisier's collection was not just a dispersion of documents and objects, but also the result of a museological classification that ignored both the cultural function of scientific heritage and the material context of science. In many respects, historians have had to cope with the negative effects of this classification, and have often been forced to study experiments without access to the instruments, natural history collections, and specimens. Consequently, the history of science has become a text-based discipline relying on the sole and principal authority of the text. The strength and influence of this approach has largely conditioned the cultural appraisal of scientific heritage until today, and a tremendous effort is still necessary to overcome a tradition which is so firmly established. The approach of the historian of science seems to be the opposite of that applied by the archaeologist, who reconstructs any historical context from the objects, which are also a guide to the reading of textual evidence. The historian of science, by contrast, interprets the meaning of the objects from their descriptions. The separation of the textual and material sources of the history of science is particularly obvious in the new arrangement of Lavoisier's collection of instruments at the Musée des arts et métiers, which is in many ways less successful than Daumas' exhibition of the 1950s. Instruments have been disassembled, thus losing any historical meaning; they have been arranged arbitrarily; and they are not explained in ways that would help visitors understand their functions (Fig. 26).

The present catalogue both aims to reconstruct the material context of Lavoisier's experimental approach to chemistry and envisages a new and necessary project for a redesign of the hall at the Musée des arts et métiers: Lavoisier's instruments should be displayed in such a way that their historical meaning can be finally understood by the public. It is hoped that the realisation of this task would release the collection from all the stratifications of politics, ideology, nationalism, and hagiography that accompanied its troubled history, and emphasise the central importance of the historical and cultural contexts of its origins.

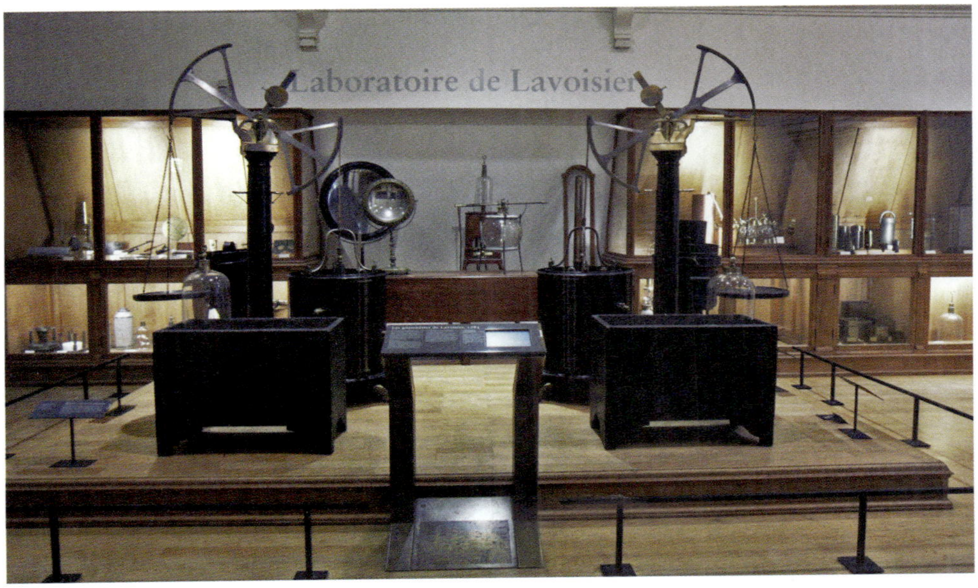

FIGURE 26
Lavoisier's hall at the Musée des arts et métiers in 2016

Inventory of Lavoisier's Residence and Laboratory on the Boulevard de la Madeleine (1796)

The following document is a partial transcription of the inventory prepared by the Parisian notary Pierre Charles Gondouin in 1796 in response to some of Lavoisier's relatives' claims for a share of Lavoisier's inheritance.[1] In order to solve this legal dispute, which ultimately confirmed that Madame Lavoisier was the sole heir, Gondouin produced a copy of a number of documents attesting the legal position of Marie Anne Lavoisier,[2] as well as the most comprehensive inventories of Lavoisier's Parisian residence made by the Revolutionary authorities between January and November 1794, and a detailed list of documents assessing the property of his financial and real estates.[3] The Archives Nationales in Paris preserve the originals of some of these inventories, and those which are of particular relevance to the present volume, do not appear to have survived in any other copies. The Archives Nationales also own a partial copy of the present inventory, but the hand is even less legible than that of the present copy, and some lacunae render it less complete.[4]

This documents is exceptional in several respects. Indeed, it offers an extraordinarily detailed survey of Lavoisier's laboratory and estate on the Boulevard de la Madeleine, which was his residence from the summer of 1792 to early Spring 1794.[5] The house seems to have been less spacious than that at the Arsenal, and Lavoisier was forced to move parts of his collection into the cellar (glassware) and to the second floor (minerals and fossils). However, the arrangement of the chemical laboratory and physical cabinet faithfully followed that of the Arsenal, and a careful reading of the lists adds such detail to the description of the laboratory that we may imagine it as if from a photograph.[6] The inventories of Lavoisier's collection of instruments prepared by Quinquet, Leblanc, Berthollet, Fortin, Charles, and Lenoir provide a unique overview of the laboratory: due to the size of the collection it took several experts' combined effort to come up with a total valuation figure.[7]

Quinquet's inventory of pharmaceutical items is of particular value, as it informs us not only of the large quantity of instruments which were confiscated, but also of the cost of several chemicals such as muriatic acid, nitric acid, chloridric acid (*acide marin*), *acide de manganese*, nitre, turpentine, mercury, platinum, clove, vinegar, and other substances, as well as drugs and metals. Most of the items listed by Quinquet never seem to have been returned to Marie Anne Lavoisier, and since their present location, if any, is unknown, it seems likely that they were dispersed among various Parisian institutions.

The inventories are also extremely useful because they allow a comparison between the values assigned to different kinds of commodities, thus providing more precise information about the relative cost of chemical apparatus.[8]

As it will be apparent by browsing the lists of the inventories, the collection which has survived at the MAM is only a small section of Lavoisier's laboratory, and most of the more or less ordinary chemical apparatus which was used by Lavoisier and his assistants on a daily basis can be found only in these precious documents.

Proces verbal de comparution.
L'an Quatrième de la Republique françoise le sept Prairial[9] trois heures de relevée.
Huiteme cote quatre.
1°. À la requête de Clement Augez Devillers Citoyen françois demeurant à Paris rue Croix des Petits Champs section de la Halle aux Bleds, habile à se porter heritier pour moitié dans la ligne maternelle d'Antoine Laurent Lavoisier membre de la cidevant Academie des Sciences, son cousin issu de germain.

Representé par François Lambert Breton Citoyen françois demeurant à Paris Place Vendome à ce present, comme fondé de sa procuration speciale à l'effet des presentes passé devant

1 In my transcription I have omitted only those parts already published elsewhere and those not directly connected with the inventory of the physical objects. I kept my interventions in the text to a minimum: the punctuation has been adjusted slightly, most of the unnecessary capital letters have been eliminated, and the spelling of any words which would otherwise be incomprehensible has been modernised. Gondouin's inventory is MS Gondouin (1796). I thank Francesca Antonelli, Patrice Bret and Christine Lehman for the revision of my transcription.

2 Such as, for instance, her certificate of marriage.

3 Gondouin did not, however, transcribe all 18 documents pertaining to Marie Anne Lavoisier in full.

4 MS Gondouin (1796b). Because of its poor condition, the manuscript cannot be consulted but it has microfilmed and a digital reproduction has been recently made available.

5 Marie Anne Lavoisier continued to live at this address until 1804.

6 A reading of these inventories ought to go hand in hand with those presented in Appendices 2 and 3.

7 See Chapter 4 in the present volume, pp. 73–88.

8 The commodities include furniture, artworks, books, utensils, carriages, different kinds of wine, clothes, towels, blankets, silverware, jewellery, two guns, a rifle and timber.

9 26 May 1795.

 DOI:10.1163/9789004511217_008

Thion l'un des notaires soussignés qui en à la minute et son confrere le vingt six Floreal dernier enregistré par Camusat le premier Prairial.

2°. À celle de Emnanuel Marie Augez Monthredon, Ancien Receveur des droits d'enregistrement a Varenne, demeurant aud. lieu Department de la Meuse.

Representé par Jean Bertez employé à l'annulation des assignats demeurant à Paris rue St Thomas du Louvre N° 270 à ce present comme fondé de la procuration specialle aussi à l'effet des presentes, passée devant le même notaire qui en à la minute et son collegue [1v] le vingt neuf Thermidor de l'an Trois[10] enregistrée par Penot le Trois fructidor suivant.

3°. À celle de François Romand, ancien Payeur General de l'Armée de l'Ouest et dame Julie Augez sa femme de lui autorisée par la procuration ci apres datée.

Representés par Charles François Nicolas Ancel C.en françois demeurant à Paris rue des Moineaux Section de la Butte des Moulins à ce present comme fondé de leur procuration commune speciale aussi à l'effet des presentes passée devant le même notaire qui en à la minute et son confrere. Le unze Fructidor de l'an Trois,[11] enregistrée par Penot le treize.

4°. En la presence de Philibert Riviere homme de loi demeurant à Paris rue de Gaillon Section Lepelletier N° 843, Commissaire du bureau du domaine nationale du departement de la Seine stipulant en cette qualité les droits de Cesar Alexandre Elie Augez Blaru habitant de St. Domingue Commandant de la Gendarmerie nationale, resident aud. lieu depuis l'année mil sept cent soixante quatorze et dont on n'a pu se procurer les pouvoirs et le certificat de residence à cause de la difficulté de la Communication maritime ainsi que l'ont declaré toutes les autres parties. [2r] Nommé pour representer les absents à l'inventaire et autres operations de la succession dud. C. Lavoisier, par arrêté dud. Bureau du six Germinal dernier, dont une copie signée Guillotin et Letourneur representée par led. Riviere, lui à été à l'instant rendue à sa requisition lesd. Augez Monthredon, et Augez Blaru et la Citoyenne Romand habiles à se porter heritiers conjointement pour l'autre moitié dans la ligne maternelle dud. feu Lavoisier leur cousin par representation de Charles Alexandre Augez dit Lavoye leur père.

5°. Et encore ledit Citoyen Riviere ensad. qualité appellé pour l'absence des heritiers paternels si aucuns il y a les Citoyens Gondouin & Thion no.res à Paris soussignés se sont transportés accompagné de Pierre Vallet ancien huissier priseur, demeurant à Paris rue Grenier st. Lazare Section de la Reunion N° 698, en une maison située à Paris, sur le Boulevard de la Madelaine N° 243 section de la place Vendome, dans l'appartement que la Citoyenne Marie Anne Pierrette Paulze veuve dud. Lavoisier y occupe, à l'effet de proceder à l'inventaire et prisée des meubles argenterie & deniers comptants, et à la [2v] description des titres papiers et renseignements de la succession dud. Lavoisier décédé à Paris le premier prairial de l'an deux.[12]

A l'instant est comparue lad. Citoyenne V.e Lavoisier
Laquelle à dit que les biens de la succession de son mary consistent uniquement dans des propres de communauté legaux ou Conventionnels attendu que tous les meubles et les conquets de la communauté qui à été entre elle et lui, appartiennent à elle seule tant de son chefs comme commune qu'en vertu de la convention formant l'art. 2 de son contrat de mariage qu'elle va representer, et dont elle demande l'inventorié prealable, qu'en ce qui concerne les propres dont est uniquement composée la succession de son mary ils sont deja constaté par l'etat et inventaire sommaire qui en a été fait lors de la remise que le C. Lalleman à faite aux Archives du Bureau du Domaine National du Departement de la Seine, le huit brumaire de l'an Trois,[13] des titres et papiers de lad. succession qu'il avait dans ses mains lesquels titres et papiers ainsi que led. inventaire, lad. v.e Lavoisier offre de representer à l'instant pour etre ou recollés ou de nouveau decrits et inventoriés, lesd. papiers lui ayant été rendus en vertu de l'arreté du bureau du domaine nationale et ont tous les comparants, [3r] signé avec lesd. No.res la minute des presentes demeurée aud. C. Gondouin.

Surquoi lesdits fondés de procurations ont observé que l'intention ou l'intérêt des presomptifs heritiers etant de jouir du benefice d'inventaire, ce qui impose la necessité de faire inventaire dans les formes ordinaires, ils requierent que cet inventaire soit fait.

Et lad. C.ne v.e Lavoisier à observé de sa part qu'il y à trois heritiers de la ligne paternelle qui sont trois soeurs du nom de Lavoisier demeurantes à Villers Cotterets, qu'elle est instruite que ces trois heritieres doivent envoyer incessament leurs procurations pour les operations de la succession et qu'il serait à propos d'attendre l'arrivée de ses pouvoirs.

A quoi led. Cit. Riviere en sa d. qualité à repondu qu'etant requis & commis pour representer les absents, les devoirs de sa mission ne lui permettent point de surseoir, sauf aux absents à excercer leurs droits lors qu'elles en justifieront, en consequence il requiert qu'il soit procédé de suite aud. inventaire.

Et lad. C.ne Lavoisier à dit que son empressement à faire tout ce qui peut convenir aux interets des heritiers, lui fait adopter sans peine, l'acomplissement d'une formalité qu'elle croit en [3v] soit inutile, et fatiguante pour elle, ne pouvant que lui rappeller de funestes souvenirs, mais elle observe que cet inventaire de tous les meubles effets et objets de la sucession de son mari à été deja fait par les Commissaires nationaux du departement de la Seine, et par ceux des differentes Commissions, ou

10 17 August 1795.

11 29 August 1795.

12 20 May 1794. In fact, Lavoisier was executed on the 8th of May.

13 29 October 1794.

agences, lorsqu'après l'apposition et au fure et à mesure de la levée des scellés apposés sur lesd. effets par le commissaire du departement, et croisés ensuitte par le Commissaire de l'agence du domaine national, les differents commissaires se sont mis en possession desd. meubles et effets mobiliers dont la majeure partie à été enlevée et transportée dans differens Bureaux et Comités, d'où ils lui ont été restitués et remis lorsque par un arrêté du domaine national, elle à été réintegrée dans ses droits de propriétés resultant de son contrat de mariage, en consequence elle à l'instant representé avec une expedition de son contrat de mariage dix huit pieces, dont les quatre premieres sont les expeditions d'apposition et levée de scellés, et les autres sont des inventaires des differents effets delad. succession et une notte de quelques declarations d'actifs et de quelques effet mobiliers faits par quelques personnes [4r] du Bureau du domaine nationale observant seulement qu'il y à quelques uns de ces inventaires qui ne con tiennent point de prisée, et que dans d'autres qui en contiennent il y à quelques objets qui n'ont point été prisés.

Laquelle representation elle fait pour les heritiers après en avoir pris communication, en fasse l'usage convenable soit en annexant lesd. inventaires à ces presentes dont ils doivent faire partie soit en les constatant d'une autre maniere et sauf à eux à faire faire en sa presence par led. C. Vallet la prisée des objets dont ces inventaires ne contiennent point l'estimation et à lad. C.ne Lavoisier signé ainsi que toutes les autres parties en pareil endroit la minute des presentes.

Et à l'instant le Citoyen Riviere aud. nom après avoir pris communication de tous les inventaires qui ont été faits par le bureau du domaine national et des divers etats de mobiliers remis partiellement par diverses autorités qui en etaient depositaires, et après avoir verifié que tout le mobilier decrits dans lesd. inventaires et etats est le seul qui puisse etre considéré comme existant après le décés dud. feu Lavoisier et que la plus grande portion d'y celui y a été [4v] prisée, et enfin attendu qu'il est important d'accelerer les presentes operations et d'eviter frais inutiles à requis que tous lesd. inventaires, etats et autres pieces generalement quelconques que representera la Citoyenne veuve Lavoisier soient annexées à ces presentes dans l'ordre qui paroitra le plus convenable, declarant s'en rapporter au contenu desd. pieces sauf au C. Vallet expert priseur, ou à tous autres experts appellés à cet effet à donner une prisée à la portion du mobilier qui aurait été decris esd. pieces sans cette prisée et a signé en pareil endroit la minute des presentes.

A leur egard, les fondés de pouvoirs declarent qu'ils adoptent egalement ces inventaires et etats pour eviter un nouveau detail du mobilier requerans seulement le recollement des objets non estimés et que la prisée en soit faite paticulierement à l'effet de quoi ils accedent au requisitoire cidessus fait par le Commissaire national et ont signée en pareil endroit la minute des presentes.

En consequence les notaires soussignés ont donné acte aux comparants, des dires [sic] requisitions et consentemens cidessus, pour procéder aud. inventaire, tous les comparants se sont ajourné au neuf du present mois quatre [5r] heures de relevée et ont le Commissaire national et les autres parties signé avec led. Vallet et les notaires en pareil endroit la minute de presentes demeurée au C. Gondouin l'un des notaires soussignés.

Thion Gondouin

Et led. jour neuf Prairial de l'an Quatre[14] quatre heures de relevée jour et heure, indiqués par le procès verbal des autres parts pour etre procédé à l'inventaire y mentionné.

A la requête de lad. Marie Anne Pierrette Paulze veuve dud. Antoine Laurent Lavoisier, cidevant nommée et domiciliée à ce presente à cause de la communauté de biens qui à subsistée entre elle et led. feu son mary, et ayant seule droit aux biens de lad. communauté suivant l'art deux de leur Contrat de mariage, et creanciere de la succession de son mary.

Plus à la requête de Clement Augez Devillers denommé aud. procès verbal habile à se dire et porter heritier pour moité dans la ligne maternelle dud. deffunt Lavoisier son cousin sur lequel led. Augez avoit le germain.

Representé par led. François Lambert Breton [5v] à ce present, son mandataire, suivant sa procuration enoncée aud. procès verbal.

Plus à la requête dud. Emmanuel Maria Augez Monthredon denommé aud. procès verbal, representé par le d. Jean Bertez, à ce present son mandataire, suivant la procuration enoncée aud. procès verbal.

Comme aussi à celle dud. François Romand et dame Julie Augez son epouse, aussi cidevant denommé qualifiés.

Representés par led. Charles François Nicolas Ancel, à ce present leur mandataire, suivant leur procuration aussi cidevant enoncée.

Comme aussi en la presence dud. Philibert Riviere en sad. qualité de Commissaire du bureau du domaine National, ce present stipulant les droits de Cesar Alexandre Elie Augez Blaru, habitant de St. Domingue Commandant de la Gendarmerie nationale.

Lesd. Augez Monthredon, Augez Blaru et lad. C.ne Romand, habiles à se porter heritiers conjointement pour l'autre moitié de la ligne maternelle dud. feu Lavoisier leur Cousin issu de germain par representation de Charles Alexandre Augez dit Lavoye leur père.

Et encore led. Riviere ensad. qualité stipulant les [6r] droits 1°. D'antoinette Françoise Lavoisier V.e de Nicolas Charles Antoine Parisis 2°. De Marie Charlotte Lavoisier, femme de Jean Nicolas Papillon.

14 28 May 1796.

3°. Et de Marie Anne Lavoisier femme de Jean Lentz. Toutes trois habiles à se dire et porter heritieres chacunes pour un tiers dans la ligne patenelle dud. deffunt Lavoisier, dont elles etaient cousines ayant le germain sur lui, demeurantes lesd. C.nes Parisis, Papillon, et Lentz, ainsi que leurs maris, à Villers Cotterets ou ils ont tous leurs domicile ainsi que l'a declaré la C.ne v.e Lavoisier comparante.

Led. C. Riviere nommé pour representer les absents au present inventaire par arrêté du bureau du domaine national du departement de la Seine, datté et enoncé au procès verbal de l'autre part.

À la conservation des droits et interets des parties et autres qu'il appartiendra, il va etre par lesd. notaires soussignés, procedé à l'inventaire des meubles et effets mobiliers qui n'ont point été prisés et estimés par les etats et inventaires mentionnés aud. procès verbal comme aussi à la description des papiers conformement audit procès verbal, le tout dependant de la [6v] communauté d'entre led. Lavoisier et sa veuve, et la succession dud. Lavoisier décédé à Paris le premier prairial de l'an deux et qui se trouvent dans l'appartement, que lad. Lavoisier occupe dependant d'une maison située à Paris boulevard de la Madelaine section de la place Vendome, sur la representation qui sera faite par la Citoyenne Lavoisier, des papiers et de ceux des meubles effets mobiliers et argenterie, compris auxd. inventaires et etats decrits aud. procès verbal, qui n'y ont point été estimés et prisés après serment par elle fait [d]es mains des notaires soussignés, de representer tout ce qui en existe à sa reconnoissance, lesquels objets, elle affirme etre avec ceux prisés et estimés auxd. etats et inventaires, les seuls dont l'existence lui soit connue affirmant n'en avoir caché, détourné, vu ni sçu qu'il en ait été detourné aucun.

Les objets qui seront sujets à prisée seront estimés à valeur fixe, en egard au cours du temps, ainsi qu'il suit, par led. Pierre Vallet, ancien huissier priseur, denommé qualifié et domicilié aud. procès verbal, à ce present.

Et ont toutes les parties comparantes signé en pareil endroit la minute des presentes.

Thion Gondouin
Suivent:
[7r] Le contrat de mariage, etats et inventaires representés et mentionnés aud. procès verbal des autres parts dont la Citoyenne Lavoisier fait de nouveau la representation pour etre decrit au present inventaire.

Premierement l'expedition en papier du contrat de mariage dud. feu Citoyen Lavoisier et sa veuve passé pardevant Duclos Dufresnoy et son confrere notaire à Paris le quatre decembre mil sept cent soixante unze en marge delaquelle est une mention qui à été insinuée à Paris au bureau près le Tribunal du premier arrondissement le dix neuf Ventose de l'an Trois,[15] par Penot, suivant lequel contrat les lors futurs ont stipulé communauté de biens entr'eux suivant la coutume de Paris sous la modification entre autres, ci après expliquée il a été dit que dans le cas que la future epouse survivroit led. futur epoux et Jean Antoine Lavoisier son père (ce qui est arrivé) la totalité des biens meubles et immeubles de la communauté appartiendroient, à lad. lors future epouse dans le cas ou il n'y auroit point d'enfant vivant issu dud. mariage, les dettes anterieures aud. mariage ont été stipulées payable par l'auteur d'y ocelles et sur ses biens personels, led. Lavoisier a fait la declaration des biens qui lui appartenoient [7v] enoncés aud. contrat; son pere lui a fait la donation y mentionnée la Citoyenne Lavoisier y à été dottée par son pere d'une somme mobiliaire, led. Lavoisier à doué sa femme de quatre mille livres de rente de douaire prefix dans le cas ou par le décès dud. Lavoisier il n'y auroit ouverture aud. douaire que dans six années du jour dud. contrat, dont a la requisition des parties il n'a été fait de plus amples descriptions lad. expedition inventoriée par led. Gondouin pour piece unique, vu cy V.N.

Item dix huit pieces qui sont celles representées par la Citoyenne Lavoisier aud. procès verbal et communiquées auxd. parties.

La premiere est une expedition delivrée par les C.ens Guillotin et Rennesson membres du bureau du domaine nationale du departement de Paris du procès verbal d'apposition de scellés chez le C. Lavoisier et recherche de papiers, fait par les commissaires du departement, les huit Pluviose[16] quatre et cinq Germinal de l'an Deux.[17]

La deuxieme est expedition d'un procès verbal delivrée par lesd. Rennesson et Guillotin esdittes qualités le vingt quatre Ventose de l'an Trois,[18] de levée de scellés et mise en possession du mobilier et objets sous les scellés en faveur delad. v.e [8r] Lavoisier, en datte du unze Ventose de l'an Trois.[19]

La troisieme est l'expedition delivrée par le C. Némel membre dud. Bureau du domaine national le vingt quatre Ventose de l'an Trois,[20] du procès verbal d'apposition de scellé de l'agence du domaine national, dans la maison dud. C. Lavoisier, contenant levée desd. scellés dud. Departement en datte du treize Fructidor de l'an Deux.[21]

La quatrieme est l'expedition delivrée par les mêmes membres du Bureau du domaine national le vingt quatre Ventose de l'an Trois,[22] du procès verbal de visitte faite par

15 9 March 1795.
16 25 January 1794.
17 24–25 March 1794.
18 22 February 1795.
19 1 March 1795.
20 14 March 1795.
21 30 August 1794.
22 14 March 1795.

le commissaire de l'Agence nationale dans la maison dud. Lavoisier, au sujet des domages qui peuvent etre arrivés, occasionnés par la commotion du feu de Grenelle en date du Quinze fructidor de l'an deux.[23]

La cinquieme est une expedition delivrée par les mêmes Commissaires le vingt quatre Ventose de l'an Trois[24] de l'inventaire et prisée, fait le unze Prairial de l'an Deux,[25] par l'un des Commissaires du departement des objets de pharmacie qui se trouvoient dans la maison dud. feu Lavoisier suivant lequel, ils ont été prisés et estimés la somme de cinq mille deux cent Quatre vingt quatorze livres dix neuf sols cy 5294 l – 19s.

La sixieme est une expedition delivrée par les mêmes Commissaires led. jour vingt quatre Ventose de l'an Trois[26] de l'inventaire de deux voitures [8v] suivant lequel elles ont été prisée la somme de quinze cent livres led. inventaire en datte du vingt deux Prairial an Deux.[27] cy 1500 l.

La septieme est une copie delivrée par le C.en Audry secretaire general de la Commission Temporaire des arts de l'inventaire sans prisée fait le vingt six Prairial de l'an Deux[28] par les membres de la Commission Temporaire des arts Commissaires pour la section de chimie, des objets de chimie, en ustencils, vazes et choses en dependantes.

La huitieme est une autre copie conforme delivrée par led. Audry secretaire General, de la notice des vazes et objets de chimie inventoriés par Leblanc le vingt Brumaire de l'an Trois,[29] dans la maison dud. deffunt Lavoisier conformement à l'arrêté du Comité d'instruction Publique, en datte du quinze du même mois,[30] lequel inventaire à été fait sans prisée.

La neuvieme est un etat delivré par les membres de l'agence de mines, des vazes et ustenciles de chimie proventant du laboratoire dud. deffunt, remis à l'Agence des mines en consequence des ordres des comités reunis du Salut public, instruction publique et Travaux publics led. etat datté du vingt neuf Thermidor de l'an Trois[31] ne contenant aucune prisée ni estimation.

La dixieme est un autre etat delivré le huit [9r] Fructidor de l'An Trois[32] par les membres de l'agence des mines, de quelques objets provenant du même laboratoire, qui avoient été remis à l'agence des mines en consequence des ordres des mêmes Comités.

La onz.e est une expedition delivrée par lesd. Rennesson et Guillotin le vingt quatre Ventose de l'an Trois,[33] de l'inventaire fait le unze Messidor de l'an Deux[34] par un Commissaire du departement de differens meubles de cabinet qui avoient été destinés pour le cabinet d'instruction publique, suivant lequel lesd. meubles ont été estimés la somme de deux mille quatre cent livres cy 2400 l.

La douzieme est une autre expedition delivrée par lesd. Commissaires membres du bureau du domaine national le vingt quatre Ventose de l'an Trois,[35] d'un inventaire fait par un commissaire du departement, de meubles et portions de bibliotheque pour etre portés au Bureau des comités de salut public, suivant lequel les objets qui y sont decrits ont été estimés la somme de deux mille sept cent quatre vingt livres led. inventaire en datte du quinze Messidor an Deux.[36] cy 2780 l.

La treizieme est une copie de livrée par led. Aubry secretaire de la Commission Temporaire des arts, d'un inventaire fait par Charles des instruments de phisique et de chimie du Cabinet [9v] dud. deffunt Lavoisier, lequel inventaire ne contient aucune prisée et estimation.

La quatorzieme est une copie delivrée par les mêmes membres du bureau du domaine national ledit jour vingt quatre Ventose de l'an Trois,[37] l'inventaire fait par le Commissaire du departement du Bureau du domaine national, le vingt quatre Vendemiaire de l'an Trois[38] et jours suivants, du mobilier etant alors dans la maison dudit feu Lavoisier, en presence du Citoyen Lalleman fondé de la procuration delad. veuve Lavoisier et conseil de l'union des creanciers de la succession dudit Lavoisier, et aussi en presence de Denis notaire à Paris, et de Trameaux maçon commissaire de l'union desd. creanciers, suivant lequel inventaire, les meubles et effets mobiliers ont été estimés quatorze mille cent quatre vingt cinq livres, cy 14185 l.

Un necessaire et vingt un demi brillants jaunes ont été estimés huit cent quatre vingt dix livres, cy 890 l.

Il à été constaté qu'il s'etoit trouvé en numeraire la somme de seize cent vingt neuf livres dix sols, cy 1629 l. – 10 s.
Et en assignats la somme de trois mille trois cent soixante trois livres cinq sols cy 3363 l. – 5 s.
Les objets decrits aud. inventaire et qui n'ont [10r] point été prisés sont 1° page six, les objets d'histoire naturelle.

2° page sept les statues, modeles en platre, peintures et objets semblables.

3° pages huit et neuf, et il a été declaré à ce sujet par les gardiens des scellés que les lits avoient été inventoriés par le Comité d'instruction publique auquel ils avoient été destinés.

23 1 September 1794.
24 14 March 1795.
25 30 May 1794.
26 14 March 1795.
27 10 June 1794.
28 14 June 1794.
29 10 November 1794.
30 15 November 1794.
31 16 August 1795.
32 25 August 1795.
33 24 March 1795.
34 27 June 1794.
35 24 March 1795.
36 27 June 1794.
37 24 March 1795.
38 15 October 1794.

4° pages vingt huit & vingt neuf, l'argenterie dont il a été seulement fait la description sans prisées ni pesées, et cinquante deux pieces medailles, anciennes monnoyes et monnoye etrangere jointe à l'argenterie.

La quinz.[e] est une copie delivrée par Lecamus de l'inventaire, des cartes et plans geographiques trouvés dans la maison dud. deffunt Lavoisier, mis en reserve par la Commission Temporaire des arts et rendus à la veuve par la Commission des Travaux publics, sauf ceux qui sont nottés à la marge, comme manquants, led. inventaire en datte du dix neuf Thermidor de l'an Trois,[39] fait en presence d'un commissaire du bureau du domaine national lequel inventaire ne contient aucune prisée ni estimation des objets y contenus.

La seiz.[e] est une copie delivrée par les membres de l'agence des mines de l'etat general des livres dud. deffunt Lavoisier, et qui avoient été mis [10v] à la disposition de l'agence et qui ont été restitués a la C.[ne] Lavoisier lequel etat contient simplement le detail sans estimation ni prisée dans l'ordre des matieres.

La dixseptieme, est l'original d'un etat et inventaire sommaire des titres et papiers appartenant à la succession dud. Lavoisier, remis au departement de Paris par le C. Lalleman, au pied duquel est la reconnaissance que le C. Berion chef des archives du Bureau du domaine nationale avoit donné audit Lalleman du depot fait desd. pieces auxd. Archives.

Et la dixhuitieme est une notte des declarations faites au Bureau du St Esprit au proffit de la succession dud. Lavoisier formant actif dans sa succession, sçavoir 1° par carré d'une so.[e] de six cent livres.

2° Par Lafond Ladebas de quarante cinq actions de la Caisse d'Escompte, portant cinq mille vingt six livres de rente.

3° par le C. Lebel de cinq à six tapis qui lui avoient été confiés.

4° Par Jumel d'une somme de quatre mille neuf cent quatre vingt seize livres.

5° Par Fortin de quarante trois, et une statue d'argent, pesant vingt neuf marcs deux onces deux gros.

6° Et par Dumoutiers differentes pieces d'etoffes lesd. pieces cottés et paraphées par premiere et derniere et inventoriée l'une comme l'autre – DEUX

Des pieces composant la cotte deux cidessus les [11r] parties esd. noms et qualités ont requise l'annexe à ces presentes des cinq, six, sept, huit, neuf, dix, unze, douze, treize, quatorze, quinze et seizieme pieces qui sont lesd. etats et inventaires des objets mobiliers desd. communautés et succession tant des objets prisés et estimés que de ceux qui ne l'ont point été, pour former un seul et même corps d'inventaire avec le present,

En consequence duquel requisitoire ces pieces au nombre de douze sont demeurées ci jointes après que sur icelles il a été fait mention de leur annexe par les notaires soussignés.

Il resulte de l'examen fait à l'instant par toutes les parties de ses etats et inventaires que les seuls objets dont la prisée et estimation n'a pas été faite sont 1° les objets de chimie, compris dans l'inventaire inventoriée sous la septieme piece,

2° des vazes et autres objets de chimie, compris dans l'inventaire sous la huitieme piece, etant observé que les objets, relatés dans les neuf et dix.[e] pieces, font parties de ceux decrits dans la septieme et huitieme pieces,

3°. Les objets de phisique detaillés dans l'inventaire formant la douzieme piece.

4°. Les objets d'histoire naturelle, les statues et peintures, argenterie et monnoye etrangeres, decrits dans l'inventaire formant la quatorz.[e] piece

5°. Les cartes et plans geographiques, decrits [11v] en l'inventaire formant la quinzieme piece.

6°. Et enfin les livres compris en l'etat formant la seizieme piece.

Desquels objets seulement les parties requierent le recollement et la prisée.

Il à été vacqué à tout ceque dessus depuis lad. heure de quatre jusques à celle de neuf sonnée par double vacation, à la requisition des parties.

Ce fait la piece unique de la cotte premiere les quatre premieres et deux dernieres de la cotte deux sont restés en la garde et possession delad. C.[ne] Lavoisier, du consentement des autres parties la vacation pour la continuation du present inventaire, à été ajournée du consentement de toutes les parties au primidi ujnze du present mois huit heures du matin et ont signé en pareil endroit la minute des presentes.

SUIT LA TENEUR DESD. ANNEXES

5° piece de la cotte 2

L'an Deuxieme de la Republique françoise une et indivisible le unze Prairial[40] neuf heures du matin, en vertu d'un Arrêté du departement de Paris, en datte du vingt deux Germinal[41] dernier portant que par le C. Gerard l'un de ses Commissaires pour la vente du mobilier national, il sera fait remise à la decharge du departement au Citoyen Quinquet de tous [12r] les objets de pharmacie, instruments pharmaceutiques et autres ustenciles et objets y relatifs qui pourront être par lui requis pour le service de l'infirmerie et hospice des malades de la Conciergerie etablie au cidevant Eveché tant dans la maison des cidevant Soeurs Grises de St Lazarre rue du Faubourg du Nord, que dans tous autres etablissemens nationaux, autorise en consequence le C. Gerard à se transporter dans lesd. lieux, et partout ou besoin sera avec led. C. Quinquet à lever les sçellés qui y pourront se trouver apposés sur les pharmacies etablies dans lesd. lieux, et a faire au C. Quinquet et sous leur decharge toute

39 6 August 1795.

40 16 September 1794.

41 11 April 1794.

remise necessaire des objets qu'il pourra avoir besoin et requerir pour le service et etablissement cidevant mentionné, à la charge toutes fois par led. C. Quinquet de faire acquitter les frais auxquels pourront donner lieu lesd. operations par le ministre de l'interieur sur les fonds qui ont pu être mis à sa disposition pour led. etablissement, en vertu des arrêtés du Comité de salut public des quatre et cinq Ventose[42] dernier.

Sur l'invitation des Citoyens Quinquet et Ray moi Claude Gerard Commissaire dud. departement de Paris, y demeurant rue du Four Section de Mutius Scevola n° 312 me suis rendu en la maison et demeure du nommé Lavoisier cidevant fermier General condamné à la peine de mort, sise Boulevard [12v] de la Madelaine Section des Picques n° 243, accompagné des citoyens Claude François Berard, rue des Fossez Germain, Section des Gardes Françoises n° 228, et Jacques Abraham Cellerin, demeurant aussi à Paris rue des Vieux Augustins, Section de Guillaume Tell, tous deux commissaires de la municipalité ainsi qu'ils m'ont justifié de leurs pouvoirs ou etant, nous avons trouvés les citoyens Antoine Arnou Quinquet demeurant à Paris, Marché au Poirés, section des Marchés, n° 335 et Ray aussi demeurant à Paris, maison du ci-devant eveché, actuellement hospice du Tribunal revolutionnaire, tous deux munis de pouvoirs de la commission Temporaire des Arts adjoint au Comité d'Instruction publique, et du C. Hermanne Commissaire de l'administration Civile de Police et Tribunaux, à l'effet de pourvoir à l'etablissement de l'hospice cy dessus designé lesquels en vertu de leurs pouvoirs, nous ont requis de leur faire remise et delivrance des objets etant en lad. maison qui leur sont necessaires pour l'etablissem.[t] dont ils sont chargés.

Et encore les citoyens Marc Louis Godet et Nicolas Caty tous deux gardiens des scellés apposés en lad. maison [et ouverture des portes necessaire],[43] ainsi qu'il nous l'ont tous justifié par pouvoir [13r] en datte du huit Pluviose dernier[44] signé Liot et Scellé du cachet du departement auxquels deux gardiens nous avons exibé nos pouvoirs et requis de nous faire la representation des scellés apposés en lad. maison, et ouverture des portes necessaires pour nous mettre à même de remplir la mission dont nous sommes chargés, ce qu'ils ont fait aussitot et nous avons fait remise auxd. C.[ens] Quinquet et Ray ainsi qu'il suit, scavoir:

Dans l'antichambre
Une pendule en marqueterie de Josué Pannier à Paris sur son pied, garniture et ornement de cuivre prisé la somme de cent cinquante livres cy 150 l.

Dans une salle à manger à droite un flacon de verre blanc rempli d'acide muriatique prisé vingt deux livres, cy 22 l.

Un vieux panier dans lequel est une dame jeane en verre noire prisé cinq livres, cy 5 l.

Un petit sceau de cuivre pesant sept livres et demi prisé quatorze livres, cy 14 l.

Sept [missing word][45] dans leurs etuis de fer blanc prisés vingt une livres, cy 21 l.

Un flacon de verre blanc rempli d'amalgame. Six autres petits remplis de differentes drogues le tout prisé sept livres, 7 l.

une cucurbite de verre blanc quatre petites [13v] cloches et une […] aussi de verre blanc, une bouteille de gomme elastique le tout prisés dix livres cy, 10 l.

Un appareil de jarre elastique prisé soixante douze livres, cy 72.

Un petit reservoir en plomb monté dans une chaise de bois peint sur ses pieds et robinet de cuivre[46] prisé la somme de cent livres, cy 100 l.

Dans un antichambre eclairé de deux croisées sur le jardin

Deux tables à dessus de marbre rapporté prisé cent livres, cy 100 l.

Un barometre d'observation prisé soixante livres, cy 60 l.

Une lampe à Quinquet prisée dix huit livres, cy 18.

Dans le cabinet de phisique
Une paire de balance de cuivre, fleau et potence de fer poli cinq poids de huit marcs en cuivre prisé cent vingt livres, cy 120 l.

Un flacon contenant du vinaigre prisé trois livres, cy 3 l.

Une petite paire de balance avec sa pottence en cuivre jaune fleau en fer prisé cinq livres, cy 5 l.

Un petit flacon dans lequel est de l'acide nitrique environ trois onces, deux vases de verre dont un à pied prisé cinq livres, cy 5 l.

Un petit vaze et son pillon de porcelaine blanche [14r] prisé douze livres, cy 12 l.

Un trebuchet ou balance prisé vingt cinq sols 1l 5s.

Une romaine[47] dehalain [Delalain?][48] prisée soixante douze livres, 72 l.

Un higrometre de Ruché [Richer][49] prisé cent livres, cy 100 l.

Un eudiometre de Fontana prisé cinquante livres, 50 l.

Un Ballon de verre blanc monté sur son chassis en fer le tout prisé douze livres, cy 12 l.

Deux tabourets en bois d'acajou prisés dix livres, cy 10 l.

Une machine electrique dont le plateau à vingt quatre pouces un conducteur en cuivre deverni de quatre pieds de

42 22–23 February 1794.
43 This passage was cancelled.
44 27 January 1794.

45 Hydrometers?
46 A pneumatic trough?
47 Probably a steelyard.
48 This might refer to the steelyard MAM inv. no. 19891-0000-, for which we have not been able to identify the maker. See pp. 194–195.
49 This seems to refer to the hair hygrometer; MAM inv. no. 20106-0000-. See p. 299.

long y compris les bras un carillon electrique et sa chaine,[50] un electrometre en yvoire, un second electrometre à boulle de cuivre tenant à la machine, un exitateur en cuivre, un pistolet de fer blanc verni, une bouteille de leyde, un carreau à signe Sage, une isoloire, une pointe en cuivre, un fouet, un tube de verre avec boule et crochet en cuivre jaune, un amalgame une jeu electrique, deux tableaux magiques et un tabouret electrique, le tout prisé ensemble quatre cent livres, cy 400 l.

Un petit tuyaux en cuivre pesant deux livres prisé quatre livres, cy 4 l.

Dessendus dans une salle à manger à coté de la cuisine

Dix bouteilles de verre vert à large ouverture prisé vingt livres, cy 20 l.

Premier laboratoire

Nous nous sommes occupés à trayer les objets propres **[14v]** auxd. C.ᵉⁿˢ Quinquet et Ray pour pouvoir les decrire et qui nous à occupés jusqu'à quatre heures après midi que nous nous sommes retirés pour prendre notre repas, et avons laissé le tout à la garde des citoyens gardiens des autres parts denommés qu'ils reconnoissent et ont signé après lecture, ainsi signés Godet, Caty, Hermane, Berrard, Quinquet et Gerard.

Et le même jour unze Prairial[51] cinq heures après midi, nous chargés de pouvoirs et commissaires cidessus et des autres parts denommés et domiciliés en vertu de nos pouvoirs respectifs et de notre ajournement nous sommes remis en la maison du nommé Lavoisier ou nous avons trouvés les C. Godet et Caty gardiens des scellés apposés en Icelle lesquels nous ont fait ouverture de la porte des laboratoires ou nous sommes entrés, et avons parachevez le traitement des objets propres à l'etablissement de l'hospice du Tribunal revolutionnaire dont sont chargés les Citoyens Ray et Quinquet, après quoi nous les avons decrits prisés et detaillés ainsi qu'il suit, scavoir:

Premier laboratoire

Quatorze tamis de differentes grandeurs prisés sept livres, cy 7 l.

Une fontaine de grais [grès] de deux voies d'eau dans sa chemise d'ozier prisée dix huit livres, cy 18 l.

Un canon de fusil, huit spatule de fer, un **[15r]** pilon idem un mortier de fonte, une pelle à feu, un petit forgeon trois petits cerceaux de fer, deux cornues de tole, un sendrier [*sic*], huit petites barres de fourneau, deux mortiers de granit avec leur pillon en bois, un petit mortier et son pillon de verre, une matelle de porfire, cinq petits creusets d'Allemagne de differentes grandeurs, deux pillons de verre et une boite de bois blancs renfermant de platines alliées et poussées au feu à cinq gobins [*sic*] le tout prisé cinquante livres, cy 50 l.

Un sceau et un poelon en cuivre rouge pesant unze livres prisés vingt deux livres, cy 22 l.

Soixante quatre pieces de verre, flacons, vazes, cornues &. le tout prisé ensemble cinquante livres, cy 50 l.

Sept bouteilles de quatre pintes de verre vert prisées la somme de sept livres, cy 7 l.

Cent quarante six bouteilles de verre blanc à col renversé de differentes grandeurs prisés cent quarante six livres, cy 146 l.

Trente cinq flacons de diverses grandeurs pris la somme de cinquante livres, cy 50 l.

Quatre vazes de verre dont un a pied prisé huit livres, cy 8 l.

Dix sept bouteilles de verre ordinaire prisées la so.ᵉ de huit livres dix sols, cy 8 l. 10 s.

[15v] Deux bains de sable, plus un en tolle de diverses grandeurs prisés sept livres dix sols, cy 7 l. 10 s.

Second laboratoire

Un fourneau de reverberre en terre cuite prisé seul la somme de douze livres, cy 12 l.

Un alombic [*sic*] en cuivre avec son bassin en etain un bassin en cuivre pesant cent quarante livres et prisé la somme de deux cent quatre vingt livres, cy 280 l.

Dix grilles de fourneaux deux trepieds, deux soufflets et differentes petites barres de fer ne meritant description prisée vingt livres, cy 20 l.

Sept bouteilles de verre noir dans deux desquelles il y à du gaudron, et cinq autres sans etiquettes contenant des liqueurs nous ayant paru de peu de valeur prisés le tout ensemble quarante livres, cy 40 l.

Item vingt quatre flacons de differentes grandeurs depuis quatre pintes jusques à trois chopines prisés cent quatre vingt livres, cy 180 l.

Dans quatre desquels est contenu de l'acide nitreux evalués a douze livres à raison de quarante sols la livre font la so.ᵉ de vingt quatre livres, cy 24 l.

Dans dix autres s'est trouvé acide marin evalué au poids de cinquante livres pesant à raison de quarante sols la Livre font cent livres, cy 100 l.

[16r] Dans deux autres s'est trouvé environ quatre livres d'acide nitreux fumant à quatre livres la livre fait seize livres, cy 16 l.

Dans deux autres s'est trouvé, ainsi que dans les six derniers differentes liqueurs sans etiquetes et inconnus, auxquels nous n'avons pu donner aucune estimation.

Après quoi se sont trouvés trois armoires desquels nous n'avons pu nous procurer les clefs et pour en faire l'ouverture avons envoyé chercher le C.ᵉ Luquet serrurier demeurant marché d'Aguesseau, Section des Piques, N° 270 qui à l'instant s'est transporté, en cette maison, et nous en à fait l'ouverture et pour ce lui avons payé la somme de une livre dix sols en presence de tous les Citoyens denommés au present.

50 It is not clear from the description if this is Fortin's electrical machine.

51 30 May 1794.

Ce qui nous à conduit jusques à l'heure de neuf du soir que nous nous sommes retirés après nous être ajournés à demain neuf heures du matin et avons laissé letout à la garde des C.^{ens} Godet et Caty gardiens delad. maison, ainsi qu'ils l'ont reconnu et ont signé avec nous après lecture, ainsi signés Godet Caty, Hermane, Quinquet, Berrard & Gerard.

Et le douze Prairial même année[52] neuf heures du matin nous Quinquet et Ray chargés de pouvoirs des autres parts enoncés, Gerard Commissaire du departement et Berrard Commissaire de la municipalité des **[16v]** autres parts domiciliés, en vertu de nos pouvoirs et de notre ajournement nous sommes reunis en la maison du nommé Lavoisier des autres parts enoncés ou etant nous avons trouvés les Citoyens Caty et Godet gardiens de lad. maison avec lesquels nous nous sommes transportés dans le second laboratoire, ou nous avons continué l'inventaire et prisée des objets de pharmacie propres à l'etablissement de l'hospice dont sont chargés les Citoyens Ray et Quinquet ainsi qu'il suit scavoir:

Une cuve de marbre sur une table en bois de chêne prisée quarante livres, cy 40 l.

Dans la cuve de marbre il s'est trouvé du mercure que nous avons mis dans neuf bouteilles de verre ordinaire pour pouvoir en constater le poids ce qui a produit la quantité de cent quatre vingt dix huit livres, prisés à huit livres la livre fait la somme de mille cinq cent quatre vingt quatre livres,[53] cy 1584 l.

Quarante sept vaisseaux de verre blanc de differentes grandeurs prisés vingt trois livres dix sols, cy 23 l. 10 s.

Cent trente flacons de differentes grandeurs, estimés cent vingt huit livres quatorze sols, cy 128 l. 14 s.

~~Dans trois desquels flacons il y a de l'acide nitreux mélé de mercure prisé neuf livres, cy 9 l.~~[54]

~~Quinze bouteilles contentant differentes gommes prisées compris les bocaux cinquante livres, cy 50 l.~~[55]

[17r] Dans trois desquels flacons il y a de l'acide nitreux mélé de mercure prisé neuf livres, cy 9 l.

Quinze bouteilles contenant differentes gommes prisées compris les bocaux cinquante livres cy 50 l.

Sept livres pesant de cloux de giroffle dans differentes bouteilles prisés quatre vingt quatre livres, cy 84 l.

Cinquante petites bouteilles et boccaux de differentes grandeurs prisés vingt cinq livres, cy 25 l.

Quarante tant boccaux que flacons, contenant diverses dissolutions et petrifications estimés ensemble vingt cinq livres, cy 25 l.

Un morceau de nitre pesant. Trois livres estimé dix livres dix sols la livre fait la somme de sept livres dix sols, cy 7l. 10 s.

Minium dix livres estimé vingt sols la livre fait la somme de dix livres, cy 10 l.

Trois pintes de vinaigre concentrés par la gelée estimés trois livres la peinte [*sic*] fait neuf livres, cy 9 l.

Seize livres de Goudron estimés huit livres, cy 8 l.

Quatre pintes de residu de distillations de huile de therebentine estimé quatre livres, cy 4 l.

Une livre d'acide de manganese estimé trois livres, cy 3 l.

Quatre onces de sapolite estimés à quatre livres, 4 l.

Seize flacons des differentes grandeurs estimés à vingt quatre livres, cy 24 l.

[17v] Vingt quatre bouteilles et boccaux de differentes grandeurs contenant diverses drogues sans etiquetes dont deux contenant du vitriol bleu estimé douze livres, cy 12 l.

Vingt cinq livres de mercure sous forme de differentes chaux estimés cinquante livres, 50 l.

Cinquante flacons de differentes grandeurs contenant des liqueurs non etiquetées estimés soixante livres, cy 60 l.

52 31 May 1794.

53 This enormous quantity of mercury (nearly 40 kg) was returned to Marie Lavoisier in October 1795 by Étienne Geoffroy Saint-Hilaire with the following letter: "Citoyenne,

Un plan d'attribution arreté par les comités de la Convention avoit disposé du cabinet de notre respectable et trop malheureux maitre Lavoisier en faveur de trois établissements nationaux le Muséum d'histoire naturelle, les écoles des travaux publics, et l'agence des mines. Je fut chargés par mes collegues les professeurs du Muséum de recevoir en leur nom la partie qui leur étoit accordé. Le souvenir des vertus et des rares et utiles talents du grand homme que nous venions de perdre, étoit présent à mon esprit; je consentis cependant à me charger de ces soins, non sans une répugnance qui m'a beaucoup couté, mais à cause de l'obeissance que je devois à mes collegues réunis et aux lois impérieuses du régime sous lequel avons vecu.

Je me rendis deux fois chez vous, Madame, d'abord pour convenir avec les commissaires des deux autres établissements de ce que nous avions à faire, la seconde fois, j'emportai quatre bouteilles de mercure pesant à peu près 80 l. [39.1 kg] autant que je puis m'en rappeler, et deux flacons ou étoit contenu du cinabre.

J'imaginai que ma quittance vous seroit remise lors de la restitution de vos biens et que vous reclameriez de nous ce que nous avions enlevés.

Mais votre silence me fait craindre que ma quittance se soit égarée et alors je m'empresse de vous accuser les quantites pour lesquelles j'ai donné un récépissé. J'ai fait mettre à part et même chez moi tout ce mercure, depuis que j'ai su que la nation ordonnoit de rendre. Vous pouvez donc, Madame, le faire prendre quand il vous plaira, mais je vous previens que, quoique ce mercure vous appartienne évidemment, nous ne pouvons vous le remettre sans que nous soyons autorisés par la Commission d'instruction publique.

Agréez Madame, l'hommage de mon respect.

Geoffroy

Au Muséum d'hist. Nat.

Rue de Saint Victor."

29 Vendemiaire an 4 [21 October 1795]." MS Geoffroy (1795).

54 This entry was cancelled, but an identical one was added at the beginning of the following page.

55 This entry was cancelled, but an identical one was added at the beginning of the following page.

Le montant de l'estimation des objets choisis jusques à ce moment se monte à la somme de quatre mille sept cent soixante neuf livres dix neuf sols et attendu qu'il est quatre heures après midi, nous nous sommes retirés pour prendre notre repas et avons laissé le tout à la garde des Citoyens gardiens des autres parts nommés, tels qu'ils le reconnoissent et ont signé avec nous après lecture ainsi signé Caty, Godet, Quinquet, Berrard & Gerard.

Et le même jour douze Prairial[56] cinq heures après midi nous chargés de pouvoirs et commissaires denommés dans la vacation de ce matin, nous sommes reunis en la maison susdesignée, ou nous avons trouvés les Citoyens Caty et Godet Gardiens delad. maison avec lesquels nous sommes descendus dans une piece souterainne dite magasin de verrerie, où conjointement avec le c. Quinquet nous avons procédé au treillement de differentes bouteilles, boccaux, et autres verrerie propres à l'etablissement **[18r]** dont il est chargé, les avons inventoriés et prisée ainsi qu'il suit scavoir.

Un fourneau de forge de terre cuitte prisé dix livres, cy 10 l.

Un idem de reverbere prisé dix livres, cy 10 l.

Un idem de fusion prisé dix livres, cy 10 l.

Sept [...], huit aludels, onze cornus de grais assortis prisés le tout ensemble dix sept livres, cy 17 l.

Soixante pieces de verrerie blanc tant cornus, entonnoirs, alambiques, et autres pieces. Quatre vingt dix livres, cy 90 l.

Vingt quatre verres à pattes, bouteilles et flacons prisés dix livres, cy 10 l.

Cinquante six bouteilles de verre vert de differentes grandeurs prisés cent quarante livres, cy 140 l.

Trois petites chaudieres de fonte, un petit etouffoir de tolle et cinq petits poeles le tout prisé vingt livres, cy 20 l.

Deux pots de fayance prisés quatre livres, 4 l.

Cent vingt quatre bouteilles de pinte remplis d'eau de mer[57] prisés soixante deux livres, cy 62 l.

Un panier d'ozier dans lequel sont cent petites bouteilles de verre prisé cinq livres, cy 5 l.

Un allonge de verre blanc prisé six livres, cy 6 l.

Unze cloches de verre blanc prisé trente cinq livres, cy 35 l.

Vingt deux bouts de tuyaux de poele prisés **[18v]** unze livres, cy 11 l.

Un preservatif à sept carrés en bois dont cinq doubles en plomb, une mauvaise caisse de bois blanc rempli de foin, et deux paniers d'ozier prisés quinze livres, cy 15 l.

Quatre ballons de verre blanc prisés quatre vingt livres, cy 80 l.

Qui sont tous les objets que le C. Quinquet à trouvés propres à l'etablissement dont il est chargé, dont la prisée se trouve monter à la somme de cinq mille deux cent quatre vingt quatorze livres dix neuf sols, cy 5294 l. 9 s.

Detous lesquels objets le C. Quinquet des à present et en vertu de ses pouvoirs nous donne bonne et valable decharge pour le departement de Paris pour lesd. effets meubles et verreries etre par lui transportés à l'hospice du Tribunal revolutionnaire etabli au cidevant eveché, et nous à requis de lui delivrer expedition du present ce que nous luis avons promis pour demain neuf heures du matin.

En consequence autorisons au nom du departem.^{t} les c.^{ens} Caty et Godet gardiens des scellés meubles et effets delad. maison à laisser librement sortir tout cequi est designé au present fait et arrété en lad. maison led. jour douze Prairial même année[58] à neuf heures du soir et ont signé après lecture Caty, Godet, Berrard, Quinquet, & Gerard, et au dessous est ecrit, enregistré à Paris **[19r]** le dix huit prairial de l'an deux[59] par debit de vingt six livres dix sols signé Barthélemy. Certifié pour copie conforme à l'original, delivrée par nous membre du Bureau du domaine national du departement de Paris ce vingt quatre Ventose l'an Trois[60] de la Republique françoise une et indivisible signé Guillotin et Rennesson.

6.e piece de la cotte 2

L'An deuxieme dela Republique françoise une et indivisible, le duodi vingt deux Prairial[61] onze heures du matin en execution de la loi du cingt cinq mars mil sept cent quatre vingt douze, en consequence des pouvoirs à nous donnés par le Departement de Paris, nous Antoine Pierre Lamoye commissaire du departement de Paris, assisté des Citoyens Julien Berton et Louis François Docaigne, tous deux commissaires de la municipalité, et accompagné du Citoyen Jean Charles Trouville, l'un des administrateurs des voitures de la Republique, nous sommes transportés en une maison sise à Paris boulevard de la Madelaine N° 243, cidevant occupée par Lavoisier mort par la Loi, ou, sous une remise nous avons trouvé une diligence à l'angloise à ressorts en C avec fond gros bleu, le haut d'une moulure doré doublée en draps gris blanc, les galons bleu serain et ponceau le train vermillon **[19v]** les ferrures couleur d'acier prisée et estimée mille livres, ci 1000 l.

Un vis avis à l'anglaise ressorts en fond bleu foncé avec petit alentour blanc les corps dorés, garni en velours d'Utrecq à etoile bleu et blanc, les galons de même, le train gris les serrures en moulures dorée prisée et estimée cinq cent livres, cy 500 l.

Lesquelles voitures nous avons presentement remis et livrées aud. C.^{en} Trouville ainsi qu'il le reconnoit et s'en charge au nom delad. administration au moyen dequoi le C. Godet gardien

56 31 May 1794.

57 This probably refers to *eaux mère*, i.e. the result of the cristallisation of a compound submitted to the operation of dissolution.

58 31 May 1794.

59 6 June 1794.

60 14 March 1795.

61 10 June 1794.

etabli des scellés apposés en lad. maison se trouve bien et valablement dechargé.

Fait à Paris les jour et an que dessus en presence desd. C.[ens] Berson et Docaigne qui ont signé avec nous ainsi signé Damoye, Docaigne, Berson, Trouville, et Godet et au dessous est ecrit enregistré à Paris le vingt trois Prairial de l'an Second.[62]

f°2 et caze 3e debet sept livres dix sols signé Rennesson et Guillotin.

Certifié pour copie conforme à l'original delivrée par nous membres du bureau du domaine national du departement de Paris ce vingt quatre Ventose l'an Trois.[e63] de la Republique f.[se] une et indivisible signé Rennesson et Guillotin.

7.e piece de la Cotte 2
Le vingt six Prairial l'an Second[64] de la Republique françoise, une et indivisible, six heures du soir, en la maison de Lavoisier condamné, boulevard [20r] de la Madelaine, nous soussignés membres de la Commission Temporaire des Arts, Commissaires pour la section de chymie, avons inventoriés les objets suivants:

1° Dans une piece à gauche, au bas de l'escalier qui descend à la cuisine, scavoir dix petits tamis, mauvais la plupart, les un en crin les autres en soie.

2° Dans la caze des premieres rangées de tablettes, vingt huit pieces de fayance et terre cuite, tant petits pots que cuvettes, especes de plats creux et une assiete, trente deux pieces de verres, la plupart en petits recipients pour l'appareil pneumatochimique au mercure. Quinze petites colonnes plusieurs desquelles sont torses, en cristal et en verre blanc; trois pillons, cinq petites capsules une petite molette de verre dix sept autres petites capsules de verre profondes; quatorze autres vazes de verre en petits sceaux dans l'un desquels se trouvent environ trois à quatre livres de salpêtre; vingt sept vazes tant en petit matras qu'en verre à pattes, petites cornues boccaux, un petit chapitaux, et un peu de sulphate de fer dans un des boccaux, un creuset de grès, un petit mortier de marbre [20v] et un de fayance.

Sur la planche du haut de ces tablettes, douze vazes de verre blanc consistant en cinq grands recipients, une cucurbite et d'autres petits recipients.

La seconde rangée de tablettes, ou la caze suivante environ deux cent vazes de differentes grandeurs, tant bouteilles que sceaux, bocaux flacons, dont quelques unes bouchés à l'emery; contenant pour laplupart differents produits de chimie, cinq à six creusets de terre, quatre petites boetes de bois, quelques torchons, des sacs de peaux contenant de la corne en morceaux et en vapeurs, quelques tubes de verre et une echelle de thermometre en cristal.

Sur les tablettes de la caze suivante, soixante et quelque pieces, tant boccaux que bouteilles, tubes cornues, bouteilles servant aux operations sur le gas; cruches de grez & la plupart de ces vazes contenant des produits, ensuivant le même coté, le long du mur, une fontaine de grez de moyenne grandeur, habillée d'ozier et en bon etat.

À droite de la porte dans la même pièce une armoire ou espece de buffet surmonté de tablettes d'abord sur la table couvrant cette même [21r] armoire, environ cent quinze vazes grands la plupart en flacons dont quelques uns sont bouchés à l'emery et presque tous ces vazes contenant differents produits.

Par suite le premier messidor[65] nous avons continué le susd. inventaire dans la premiere partie de la derniere armoire, dont nous avons parlé, à droite de la porte, six grandes cloches ou recipients en verre blanc dont une à felure à son fond recouverte d'un enduit et une autre une virolle en cuivre, dans l'autre pan de l'armoire vers la porte, six grandes cloches ou recipients egallement de verre et une cucurbitte.

Dans la premiere rangée des cazes dont cette armoire est surmontée, six boccaux de differentes formes dont deux contiennent des produits, un petit flacon et un sceau contenant dix areomètres de differentes grandeurs et dont quelques uns sont cassés. Sur la seconde rangée de tablettes au même endroit, environ trente six flacons laplupart bouchés à l'emery, et environ vingt trois bocaux, letout contenant differens produits les uns solides, les autres fluides.

La suivante ou troisième rangée de tablettes, environ quatre vingt flacons bouchés a l'émery, environ [21v] quatre vingt seize boccaux ou autre vazes de verre letout contenant aussi differents produits, deux mortiers de bronze, un pilon, deux petites boetes de bois, dont une contient deux petits poids de fonte.

Sur le carreau de la même piece, environ quatre vingt bouteilles, dont quelques unes très grandes des moyennes et bouteilles ordinaires; environ une douzaine de flacons dont quelques uns bouchés à l'emery; differents autres vazes tels que cornues de verre, cucurbites, matras &[a]. un assez grand mortier de fonte et son pilon, deux mortiers de marbre, un petit de terre, un seau de cuivre et huit casserolles de même metal quelques grands boccaux de fayance, un modele de vaisseau, quelques aludelles, [...] et cornues en terre, quelques parties d'un fourneau à reverberre, trois grand tamis, sept especes de bassines en tole garnies de leurs mains ou poignées. Sur la cheminée un bocal moitié plein d'une substance saline et cinq boetes en bois environ vingt poids en fonte, la plupart petits.

Dans la piece qui suit immediatement celle que nous venons de parcourir une pareille armoire en forme de buffet; d'abord aux tablettes qui la surmontent, sur la premiere [22r] caze de ces tablettes, environ soixante dix huit flacons, dont plus de moitié bouché à l'emery, quarante cinq bocaux, letout

62 11 June 1794.
63 14 March 1795.
64 14 June 1794.
65 19 June 1794.

contenant des produits, deux vazes composés chacun de demi pieces assemblées.

Sur les tablettes de la caze suivante, trois flacons assez grands contenant de [l'oxide] de mercure par l'acide nitrique; l'un en contient tres peu, et les deux autres nous ont paru en contenir environ quarante ou cinquante livres. Soixante autres flacons environ et quinze fioles à medecine letout contenant divers produits, un bocal et un flacon contenant ensemble huit livres de platine; cinquante cinq bocaux contenant differents produits. Trois entonnoires de verre sur les tablettes de la 3e caze, environ vingt sept flacons et vingt neuf bocaux contenant differents produits; deux grands entonnoirs de verre, un moyen ballon, quatorze tubes en terre à porcelaine, un morceau de nitre de cinq à six livres à peu près.

L'armoire surmontée des susd. tablettes composée de trois parties ou pans. Dans la première soixante dix boccaux environ de differentes grandeurs et contenant des produits un pot de grez et une boete de bois, dans la seconde [22v] partie quarante deux flacons environ, vuides; quatre entonnoirs de verre, quatre pilons de verre, deux allonges, deux petits ballons et trois petites cornues ces dernieres en terre. Dans la troisieme partie de cette même armoire, huit recipients de differentes grandeurs, deux entonnoirs, deux flacons letout vuide.

Dans une caze enfoncée dans le mur à la suite de la susd. armoire; sur les tablettes dont elle est garnie cinquante cinq flacons environ, dont une grande partie contient des produits; à peu près quatre vingt huit boccaux, contenant aussi des produits; un grand matras, et un ballon vuide, cinq cornues de verre luttées, quatre cornues de terre, et environ une vingtaine de creusets. Une armoire suivante et tenante à deux faces de mur; d'abord sur la table qui la recouvre environ deux cent cinquante cinq flacons de toutes grandeurs laplupart bouchés à l'èmery et contenant differents produits; environ quarante neuf boccaux contenant aussi differents produits; deux grands ballons, un très grand matras, un cylindre de deux pieds et demi d'hauteur, sept petites capsules, quelques autres vazes de figure [23r] irreguliere, letout de verre, et sept grandes bouteilles de verre contenant differents produits.

Dans la premiere partie de cette armoire, interieurement, deux flacons contenant des produits, une grande cucurbitte et un petit sceau. Dans la deux.[e] partie, deux matras, deux ballons, un recipient, un alambic et dix entonnoirs de differentes grandeurs, le tout de verre. Dans la troisieme partie de la même armoire, vingt quatre verres à pattes, onze petits flacons vuides sans bouchons; dans la quatrieme partie rien; dans la cinquieme, un tres grand verre a patte de verre commun, trois alambics, deux ballons, trois recipients et trois cornues, letout de verre; dans la sixieme partie, dix bouteilles ordinaires vuides.

Sur les tablettes établies au dessus dela ditte armoire, premiere et petite caze en commençant par le coté de la caze enfoncée dans le mur, vingt deux flacons petits, la plupart contenant divers produits et douze boccaux. Sur la caze des rangées suivantes environ cinquante flacons dont plusieurs sont petits, quelques uns bouchés à l'émery, quatre bouteilles et quarante huit bocaux, le tout de verre et contenant divers produits. [23v] La troisieme caze c.a.d. celle qui suit, dix sept flacons, dont une petite partie bouchés à l'emery, deux fioles, une bouteille, et environ dix sept boccaux, le tout de verre, et une partie contenant des produits. Sur les tablettes de la quatrieme caze environ quarante trois flacons, sept fiolles, seize boccaux, letout de verre et la plupart contenant des produits. Sur les tablettes de la cinquieme et derniere caze, environ vingt flacons, cinq fiolles et huit boccaux contenant aussi des produits.

Objets se trouvant sur le carreau de lad. piece.

D'abord, quatre tables en bois, sur l'une desquelles se trouve une cuve de marbre pour l'appareil au mercure; neuf bouteilles de pinte remplies de mercure coulant, excepté l'une d'elle ou il s'en trouve à peu près les trois quarts de la bouteille. Sur deux autres de ces tables, se trouve une seconde cuve de marbre, quelques boccaux, cornues flacons &.[a] une grande cuve montée en bois et revetue au dedant de tolle en fer blanc, deux bassines de cuivre, dont l'une grande et l'autre moyenne, un alambic de cuivre de moyenne grandeur deux fourneaux à reverbere; deux etouffoirs, un grand et un [24r] petit deux mauvais soufflets à mains, cinq grandes bouteilles de verre commun, quatre grands pots de fayance; un panier rempli de petit matras à col très long, suspendus a des cloux le long du mur; quelques parties de fourneaux et cheminées telles que grilles rondes, trepieds &[a]. Dans la meme piece se trouve encore un mauvais poele en fonte surmonté d'une capsule en tole pour le bain de sable, un grand couvercle en cuivre perforé pour servir aux evaporations qui se font au bain marie.

Sur un support traversant cette même piece, quatre ballons de moyenne grandeur et deux grands recipients.

Dans une piece dont la porte ou entrée se trouve dans une espece d'antichambre commun entre la cuisine et les pieces cidessus inventoriées, se trouve une quantité considerable de verrerie et autres matieres en vazes de toutes especes pour l'usage de la phisique et de la chymie, dont l'inventaire en detail ne pouvait se faire sans un tres grand danger à cause de leur entassement ou amoncelement confus, et qui, eu egard aupeu d'etendue de la piece comparativement aux quantités des [24v] objets, ne pouvoient etre deplacés sans accident.

Dans la piece que nous avons designée comme antichambre commun à la cuisine et aux pieces contenant les objets ci-dessus inventoriés, correspondent les portes desd. dernieres pieces, sur lesquelles le Commissaire du departement (Le Citoyen Binet) a aprés notre inventaire fait, apposé le scellé et dressé procès verbal que nous avons signé.

Dans cette même piece, dite antichambre se trouve une presse en bois propre aux usages domestiques et à la pharmacie.

De tout ce que dessus avons dressé le present, le premier Messidor l'an Deuxieme[66] de la Republique Francoise une et indivisible, Signé Leblanc & Berthollet plus bas est ecrit pour copie conforme Signé Oudry secretaire General.

8 p. de la Cotte 2
Notice des vazes & objets de chimie inventories par Leblanc, le dix neuf Brumaire[67] dans la maison du Citoyen Lavoisier conformement a l'arrêté **[25r]** du Comité d'instruction publique en datte du Quinze du même mois:[68]

Especes	**matieres**	**quantités**
Une serie de cornues des premieres grandeurs	verre blanc	25
Idem dans les differentes grandeurs qui suivent jusques au plus petits	idem	150 environs
Idem de differentes grandeurs luttées	idem	17 environs
Idem de differentes terres et grez, et de matieres et de differentes grandeurs luttées et non luttées	terre	118 environs
Petit creusets, pot a boeur [beurre] de Bretagne	idem	80 environs
Creusets ordinaires de differentes grandeurs	terre de differentes sortes	400 et plus
Couvercles de creusets	idem	20 environs
Cucurbittes, aludelles, et chapiteaux	en terre	60 pieces environs
Entonnoirs de toutes grandeurs	verre blanc	42 a peu pres
Ballons de toutes grandeurs et a differentes tubulures	idem	22 a peu pres
Matras de toutes grandeurs, la plus grande partie a long col	idem	100 et plus
Recipients, plusieurs tres grands, trois garni en cuivre, deux avec leurs robinets, trois felés	idem	44
Allonges pour l'appareil de Wolf, cilindres	idem	33

[25v]

Vazes cylindriques dont deux apattes	idem	3
Oeufs pour la phisique	cristal	3
Cucurbittes	verre	14
Chapitaux	idem	28
Alambics	idem	2
Pelicans	idem	4
Eolipile ou burettes à huille essentielle	idem	1
Flacons bouché a l'émery	cristal	22
Flacons tres grands et à double tubulure	idem	9
Recipients à double fond de deux grandeurs differentes	idem	8
Flacons à deux et trois gouleaux	verre blanc	9
Entonnoirs à plusieurs ventres	idem	3
Flacons à gouleaux renversés depuis la capacité de six à huit pintes jusques a deux onces	cristal	900 environs
Flacons la plupart à large ouverture	id.	23

Ces premiers objets qui sont etablis par series, dans une piece au rez de chaussée, seront prisés demain primidi; le même jour l'inventaire delaissé sera continué, il existe encore beaucoup de flacons en verre blanc et d'autres objets **[26r]** dans le magasin à verrerie. Il en existe aussi dans le laboratoire une grande quantité, la plupart en crystal, contenant des produits, plusieurs autres objets y sont encore en assez grand nombre, et ne seront bien connus qu'après les avoir etablis par ordre, et otés de l'etat d'amoncelement dans lequel il se trouve. Un objet qui me paroit meriter une attention tres particuliere, est une assez grande quantité de mercure. Neuf bouteilles ordinaires contiennent de cette substance, dans l'etat coulant, huit pleines jusques au col, et la neuvième au dela de la moitié, deux flacons en contiennent une assez grande quantité dans l'etat de mercure rouge par l'acide nitrique. Peut être seroit il prudent de mettre au plutot ces derniers objets à couvert de tous accident. Je ferai en sorte de donner de suite l'etat de ce travail à mesure qu'il se continuera mais il seroit bon que le Museum d'histoire naturelle, l'agence des mines et la Commission des Travaux publics, se consertassent des à present pour cette premiere partie; Le local qu'elle occupe devient necessaire pour une partie **[26v]** des objets qui restent à inventorier le vingt Brumaire l'an Trois[69] de la Republique une et indivisible signé Leblanc, pour copie confome Signé Oudry Secretaire General.

66 19 June 1794.
67 9 November 1794.
68 3 November 1794.
69 10 November 1794.

9e p. dela Cotte 2
Etat des vazes et ustencils de chimie provenant du laboratoire du Citoyen Lavoisier qui ont été remis à l'agence des mines en consequence des ordres des Comités reunis de salut public, Instruction publique et des Travaux publics, du vingt neuf Frimaire an Trois.[70]

Dix grandes bouteilles dites gouleau renversé vingt deux moyennes. Idem, quarante sept petites. Idem, cinq grandes cloches de verre dont une avec robinet de cuivre, treize antonnoirs de diverses grandeurs, quatorze cornues de verre de diverses grandeurs, quatre cornues de verre lutées, cinq capsules de verre, unzes ballons, quatorze matras, six fioles à medecine dix allonges, trois cucurbittes et six chapitaux, treize cornues de grez et porcelaine de diverses grandeurs, trois aludelles avec un chapitaux en terre, quatre vingt sept creusets tant moyen que petit, cinq grande creusets, un pelican, un mortier de granit sans pillon, trois cucurbittes de grez.

Le present etat certifié conforme à celui dressé lors des remise au laboratoire de l'agence des [27r] mines des objets y mentionnés a Paris le vingt neuf Thermidor l'an Troisieme[71] de a Republique Les membres de l'agence des mines. Signé Lelievre & Ipp Gillet./.

10 piece de la Cotte 2
Etat de quelques objets provenant du laboratoire du Citoyen Lavoisier qui ont été remis à l'agence des mines, en consequence de l'ordre des comités reunis de Salut public, Instruction publique et des Travaux publics du vingt neuf Frimaire an Trois,[72]

Scavoir:

Trois areometres dans leur etui de fer blanc[73] le present etat certifié conforme à celui dressé lors de la remise au laboratoire de l'agence des mines des objets y mentionnés.

Paris le huit Fructidor de l'an Trois de la Republique,[74] les membres de l'agence des mines, signé Lelievre, & Ip. N. Gillet./.

11 Piece de la cotte 2
Ce jourd'huy onze Messidor l'an Second[75] de la Republique françoise une et indivisible à six heures du matin.

Nous Jean Baptiste Alligre Citoyen françois Commissaire du departement de Paris aux inventaires et ventes des biens des emigrés dem.[t] [27v] rue Bayeulle [Bailleuil] N° 183 section des Gardes françoise en vertu des pouvoirs qui nous sont delegués par le C. Julien Dubois representant du peuple delegué par Le Comité de Salut public de la Convention nationale pour l'execution de son arrêté du deux floreal, à l'effet de fournir à la commission executive de l'instruction publique etablie au petit Luxembourg, tous Les meubles et effets quelconques qui lui sont necessaires pour les bureaux et logemens des Commissaires et adjoints; de nous transporter en consequence dans toutes les maisons nationales ou ils pourroient se trouver, visé egalement le susd. ordre par l'agent national aud. departement, en execution duquel et sur la nouvelle declaration faite par le Citoyen Payan, Commissaire delad. Commission et Fourcade adjoint, par l'organe du C. Chaix chef du Bureau delad. Comptabilité de lad. Commission, qu'il manque encore beaucoup de Bureaux tant à Cylindre que plats, linge chaises en maroquin et autres faux lits de maitre comptes, bibliotheques, tables et autres objets qui lui seront indiqués par led. Chaix, qui nous à accompagné pour en faire choix, nous sommes transportés accompagnés des Citoyens Antoine Thuillier et François Leonard Menet, Commissaires municipaux, ainsi qu'ils l'ont justifié par l'exibition de leurs commissions [28r] duement signées et enregistrées et led. C. Chaix et notre secretaire Greffier, Boulevard dit de la Madelaine, maison N° 243, qu'occupoit le nommé Lavoisier cidevant Fermier Général, puni de mort, à l'effet d'en distraire pour l'usage de lad. Commission, tous les objets qui seront propres à icelle; ou etant avons trouvés les Citoyens Marc Louis Godet et Nicolas Caty, gardiens des scellés du departement, auxquels nous avons fait part du susd. ordre, et les ayant requis de nous representer les scellés confiés à leur garde, ils nous ont conduits dans un antichambre donnant sur la cour et nous ont representé un scellé apposé sur une porte qui communique dans une autre piece. Après l'avoir trouvé sain et entier, tel qu'il avoit été mis par Binay, Commissaire a la suite dela Commission Temporaire des arts nous l'avons levé, et nous etant introduits dans la susd. piece nous avons trouvés deux scellées, l'un sur une porte à droite et l'autre sur une porte à gauche, lesquels ayant été reconnus sains et entiers, tels qu'ils avoient été apposés par led. Binay, nous avons levé successivement ainsi qu'un autre scellé du même Binay, que nous avons aussi trouvé sain et entier, dans la chambre à gauche sur une porte de communication à une troisieme chambre; et après avoir examiné, les meubles qui [28v] garnissent les susd. quatre pieces led. C. Chaix chargé par lad. Commission de l'instruction publique a fait choix pour icelles des meubles et effets ci après detaillés, inventoriés et prisés. Scavoir:

Dans la troisieme chambre à gauche un grand secretaire à cylindre[76] en bois d'acajou plaqué avec ses tiroirs, garni de balustrades et bordures en cuivre doré, avec les serrures et clef, presque neuf, de cinq pieds huit pouces de longueur sur deux pieds dix pouces de largeur, prisé a [mille] douze cent livres, cy 1200 l.

---

70 19 December 1794.
71 17 August 1795.
72 19 December 1794.
73 This description fits MAM inventory no. 19960-0000-. See p. 322.
74 25 August 1795.
75 29 June 1794.
76 This description corresponds to the roll top desk, MAM inventory no. 20576-0000-. See catalogue p. 390.

Item un porte papier en acajou plaqué de quatre pieds six pouces de haut et trois pieds trois pouces de largeur prisé à cent livres, cy 100 l.

Plaqué de trois pieds six pouces de hauteur et deux pieds quatre pouces de largeur avec tiroirs prisé à cent vingt livres, cy 120 l.

Item une table pliante plaqué de trois pieds an acajou.

Item un pupitre à ecrire debout en acajou plein de quatre pieds quatre pouces de longueur sur quatre pieds de largeur prisés à cent cinquante livres, cy 150 l.

Item une autre table ronde pliante en acajou, de deux pieds de diametre, sur un seul pied, prisée à cinquante livres, cy 50 l.

Item une autre table ronde en acajou, dessus en cuir noir de deux pieds de diametre avec un tiroir, et deux tablettes à coulisse avec un double couvert garni aussi en drap vert et dessus de marbre gris prisée à cent livres, cy 120 l.

Item une table de bois de rose plaqué avec un dessus de maroquin noir et son tiroir, trois pieds de **[29r]** longueur sur deux pieds de largeur prisée à quatre vingt livres cy 80 l.

Item douze chaises peintes en gris couvertes de maroquin vert, et rembourées en crin prisées à cent cinquante livres, cy 150 l.

Item une chaise de bois peint, couverte en cuir rouge usé, et rembourée en crin prisée à dix livres, cy 10 l.

Item une autre table ronde à deux pliants en bois peints façon d'acajou de trois pieds neuf pouces de diametre prisée à quarante livres, cy 40 l.

Item deux lampes à quinquet sur deux chandeliers de cuivre doré prisées soixante livres, cy 60 l.

Item six chandeliers de cuivre doré de differentes grandeurs prisés à quatre vingt livres, 80 l.

Item deux servantes à dessus de marbre blanc et en acajou de deux pieds quarré, garnies de leurs tablettes et de deux sceaux en fer blanc et d'un tiroir sur quatre pieds à roulettes, prisées à deux cent livres, cy 200 l.

Item une petite chiffonniere en bois de placage à dessus de marbre blanc veiné avec quatre tiroirs une tablette et ses quatre pieds à roulettes prisée à soixante livres, cy 60 l.

Tous lesquels effets cidessus detaillés, inventoriés et prisés, nous avons fait descendre dans la cour à l'aide de six journaliers mandés exprès, pour **[29v]** etre chargés sur des voitures et les lampes portées à bras par les susd. journaliers au Petit Luxembourg, et remis au C. Payan, pour l'usage delad. Commission, qui signera le present avec les Citoyens Julien Dubois representant du peuple et Chaix chef du bureau de la Comptabilité de l'instruction publique ici present, pour notre decharge, au moyen de qui avons donné et donnons bonne et valable decharge auxd. C.ens Godet et Caty, gardiens desd. scellés et avons de suite fait fermer les croisées et portes par nous ouvertes et reapposés les scellés sur icelle, ainsi que sur la porte d'entrée, avec le sceau du departement N° 33 que nous avons de nouveau confiés à la garde et surveillance des susd. C.ens Godet et Caty, qui ont promis de les representer de même toutes les fois qu'ils en seront requis et avons payé auxd. six journaliers et deux charretiers pour la descente et transport des susd. meubles et effets la somme de soixante douze livres qu'ils ont requis, et que nous leur avons payée en presence des susd. commissaires municipaux et du C. Chaix, ainsi que des gardiens de tout quoi nous avons fait redigé et clos le present procés verbal à la redaction duquel et au transport des susd. meubles et effets il a été vacqué depuis lad. heure du six du matin jusques à celles de six de rélévée sans interruption et ont signé et paraphé avec **[30r]** nous lesd. Commissaires municipaux, le C. Julien Dubois, Representant du Peuple, le Citoyen Payan Commissaire de l'instruction publique, le C. Chaix chef du bureau de la Comptabilité de cette Commission, et lesd. C.ens Godet et Caty, gardiens, lesd. jour et An que dessus dont acte ainsi signés, Caty, Godet, Paul Chaix, Thuillier, Menet & Alligre et en marge est ecrit: Enregistré à Paris le douze Messidor de l'an Second[77] f.° 7 N. N°. cazes, en debet de la somme de seize livres signé Macoan.

Certiffié pour copie conforme à l'original, delivré par nous membres du bureau du domaine national du departement de Paris, ce vingt quatre ventose l'an Troisieme[78] de la Republique francoise une et indivisible Signé Rennesson & Guillotin.

Ce jourd'huy quintidi quinze Messidor l'an Second[79] de la Republique francoise une et indivisible sept heures du matin, nous Jean Marie François Bezauts demeurant à Paris rue de la Michodiere N° 7 section Lepelletier, Commissaire du departement de Paris aux inventaires et ventes du mobilier des emigrés, en vertu d'un arrêté du comité de salut public de la Convention nationale du sept du present mois, qui autorise le C. Leconte a faire choisir dans les maisons nationales designées dans la copie collationnée d'un extrait du registre des arrêtés du Comité de salut publique, à nous adressé pour copie conforme à l'original, signée Bouseau Secretaire **[30v]** adjoint dud. departement de Paris, et notamment dans la maison du nommé Lavoisier et en vertu d'une Lettre du Citoyen Jacquier agent national du district près le departement de Paris, en datte du neuf du present mois à nous addressée en original le même jour, qui nous autorise, confomement au susd. arrêté du Comité de Salut public, de nous transporter dans plusieurs maisons et notamment dans celle dud. Lavoisier à l'effet de reconnoitre et lever les scellés du departement et de mettre led. C. Leconte à même de choisir des meubles et portions de bibliothèque qui peuvent etre necessaires au nouvaau local que le comité de Salut public doit occuper dans le palais national d'en faire inventaire et la juste prisée et de les remettre à la disposition dud. C. Leconte sur bonne et

77 30 June 1794.

78 14 March 1795.

79 3 July 1794.

valable decharge, de nous faire assister de commissaires et à dresser du tout proces verbal.

Pour mettre lesd. arrêtés et lettre à excution nous nous sommes transportés accompagné du C. Petit notre Secretaire Greffier, et assistés des C.ens Grenier & Legras Commissaire de la municipalité de Paris pour etre presente auxd. inventaires et ventes ainsi qu'ils nous l'ont justifié par la representation qu'il nous ont faite des pouvoirs qui leurs ont été donnés par le corps municipal de la Commune de Paris [31r] en datte pour led. C. Grenier des trente un aôut et trois septembre mil sept cent quatre vingt treize enregistré à la Commission administrative des biens nationaux division des emigrés, Le dix sept dud. mois de septembre; et pour led. C. Legras en datte du vingt six Brumaire[80] dernier, enregistré à la même Commission le octodi de la troisieme decade dud. mois de Brumaire, que nous leur avons à l'instant rendus ainsi qu'ils le reconnoissent sur le boulevard dit de la Madelaine N° 243 section des Piques, dans la maison cidevant habitée par Le nommé Lavoisier tombé sous le glaive de la loi, ou etant nous nous sommes adressés à un citoyen touvé dans la loge duportier delad. maison, que nous avons invité de nous indiquer si le gardien des scellés apposés en icelle y etoit present et nous ayant repondu qu'il étoit lui même un des deux gardiens et que son collegue etoit absent pour l'instant, il nous à dit se nommer Nicolas Caty, dem.t. ordinairement rue neuve du Luxembourg N° 153 section des Piques, et avoir été etabli gardien des scellés apposés sur le mobilier dud. feu Lavoisier, ainsi qu'il nous l'a justifié par la representation qu'il nous a faite de l'attestation qui lui en a été donnée par le C. Liot, notre collegue qui à fait lad. apposition de scellés le huit Pluviose dernier que nous lui avons rendue ainsi qu'il le reconnoit, et après lui avoir donné connoissance de notre mission et [31v] communication de nosd. pouvoirs, nous l'avons requis de nous representer lesd. scellés sains et entiers pour mettre led. C. Vacquier à portée de faire son choix dans le mobilier etant sous iceux, et y ayant consenti il nous à requis de lui donner acte de soud. Consentement et lesd. C.ens Vacquier, Grenier et Legras de leur comparution ce que nous leur avons accordé et ont signé avec nous après lecture faite ainsi signés Legras, Caty, Godet, Grenier, L.F. Vacquer & Bezauts.

Après quoi led. C. Caty, gardien nous à conduit dans une piece servant de salle à manger, ayant vue par deux croisées sur le jardin de lad. maison dans la quelle communiquent deux portes sur lesquelles etoient apposés les scellés et après les avoir reconnus sains et entiers et comme tels levés et otés ceux apposés sur la porte à gauche en entrant dans lad. salle à manger, et en avoir fait ouverture avec la clef representée par led. C. Caty nous sommes entrés dans une piece servant de cabinet, ayant vue par deux croisées sur led. jardin et deux autres sur lesusd. Boulevard, où etant nous avons egalement reconnus sains et entiers et comme tels levés et otés les scellés apposés sur une porte communiquant dans led. cabinet et après en avoir fait ouverture avec la clef representée par led. C. Caty nous sommes entrés dans un autre cabinet ayant vue par deux croisées sur le susd. Boulevard.

Dans lequel cabinet led. C. Vacquer a fait choix [32r] des objets ci après designés.

1° d'un corps de bas de bibliotheque en bois d'acajou, de trois pieds huit pouces de hauteurs, treize pieds et demi de longueur, et vingt pouces de largeur, ouvrant à six portes garnies de leurs serrures et d'une seule clef, et dans son milieu ouvrant encore à deux portes en dedans garnies d'une serrure et sa clef prisé avec son serre papier de pareil bois mille livres, cy 1000 l.

Duquel corps de bas de bibliotheque nous avons retiré tout ce qu'il contenoit que nous avons deposé savoir les cartons contenant des papiers et differens autres petits ustencils sur le parquet dans le cabinet, et quelques sacs et bourses dans lesquels s'est trouvé du numeraire nous les avons renfermés dans un carton que nous avous ficelé et sur lequel nous avons apposés les scellés du departement et quelques paquets contenants, suivant les etiquettes etant sur iceux des papiers monnoyé, nous les avons egalement renfermés dans deux cartons que nous avons ficelés et sur lesquels nous avons de même apposés les scellés du departement; lesquels trois cartons sont restés dans le susd. cabinet ainsi que le reconnoit led. C. Caty, gardien et Marc Louis Godet second gardien survenu à l'instant.

2° Six fauteuils de bois d'acajou à forme quarré, quant aux sieges, et rondes quant aux dossiers, à jour et fond sanglé garnis de crin, couverts de [32v] velours cramoisi a fleurs et dessus de même couleurs prisés quatre cent quatre vingt livres cy 480 l.

Qui sont les seuls objets choisis par led. C.en Vacquer dans led. cabinet qu'il en à fait sortir à l'instant.

Et après etre sortis d'icelui et en avoir fermé la porte avec la clef restée entre les mains desd. C.ens gardiens nous avons egallement reapposés les scellés sur icelles au cachet du departement N° 77 et après etre sortis dud. premier cabinet et en avoir fermé la porte restées entra les mains desd. C.ens gardiens, nous avons egallement reapposés les scellés sur icelle.

Ce fait et après avoir vacqué à ceque dessus depuis lad. heure de sept jusques à celle de trois heures sonnées nous avons renvoyez la vacation pour la continuation desd. operations à cejourd'huy quatre heures de relevée avec invitation auxd. C.ens Grenier, Legras et Vacquer de vouloir bien s'y trouver cequils ont promis faire: et avons laissé les scellés cidessus reapposés à la charge et garde desd. C.ens Caty et Godet qui s'en sont volontairement chargés pour les representer à qui de droit lorsquils en seront legalement requis.

Et ont lesd. C.ens Vacquer, Legras, Grenier, Caty, et Godet signés avec nous après lecture faite, ainsi signé Grenier, Legras, Caty, Godet, L.F. Vacquer et Bezauts.

80 16 November 1794.

Et lesd. jour et an que dessus quatre heures de rélévée, nous commissaires susd. et soussignés accompagné et assisté comme dessus nous sommes transportés [33r] en la maison susdesignée ou etant nous avons requis les C.ens Caty et Godet, gardiens de nous representer à l'instant sains et entiers les autres scellés apposés dans lad. maison pour nous mettre a portée de continuer les susd. operations et y ayant consentis il nous ont requis de leur donner acte de leur dit consentement, et lesd. C.ens Grenier, Legras & Vacquer de leur comparution ce que nous leur avons accordé et ont signé avec nous après lecture faite, ainsi signé Legras, Grenier, Caty, Godet, L.F. Vacquer et Bezauts.

Après quoi il nous ont conduits dans la susd. piece servant de salle à manger, au devant d'une porte à droite y communiquante sur laquelle etoient apposés les scellés du departement, et après les avoir reconnus sains et entiers et comme tels levés et otés et avoir fait ouverture delad. porte avec la clef representée par lesd. C.ens gardiens et nous sommes entrés dans une piece servant de sallon ayant vue par deux croisées sur Le jardin delad. maison, dans lequel led. C. Vacquer a fait choix des objets ci après decrits:

3° Quatre corps de bas de bibliotheques en bois d'acajou de chacune sept pieds et demi de long de seize pouces et demi de large de trois pieds quatre pouces de haut garnis de chacun deux tablettes en bois de sapin, l'epaisseur desquelles est plaqué en bois d'acajou prisés huit cent livres, cy 800 l.

[33v] Nous avons ensuite reconnus sains et entiers et comme tels levés et otés les scellés apposés sur une porte communiquante dans lesusd. sallon et après en avoir fait ouverture avec la clef representée par lesd. C.ens gardiens, nous sommes entrés dans une chambre à coucher ayant vue par deux croisées sur le jardin delad. maison dans laquelle s'est trouvé un secretaire que lesd. C.ens gardiens nous ont dit renfermer les papiers de la v.e dud. Lavoisier, et etre en etat d'arestation et que la clef dud. secretaire etoit entre les mains d'un des membres dud. comité revolutionnaire de la Section des Piques qui avoit precedemment apposés les scellés sur icelui qui ont levés depuis, n'y ayant trouvé que ceux du departement vu laquelle declaration nous avons ecrit sur le champ aux Citoyens membres composants led. comité revolutionnaire pour les inviter de nommer un ou plusieurs d'entre eux à l'effet d'apporter la clef dud. secretaire et de vouloir bien etre present à l'ouverture dicelui et lesd. membres y ayant adhéré, ils en ont chargés le C. Jean Baptiste Georges demeurant rue Caumartin N° 751 membre dud. Comité revolutionnaire, lequel nous à representé la clef dud. secretaire avec laquelle nous en avons fait l'ouverture en sa presence après avoir reconnus sains et entiers et comme tels levés et otés les scellés du departement apposés sur icelui nous avons egalement vuidé les papiers et autres objets qui se sont trouvés dans led. secretaire en la [34r] presence dud. C. Georges et nous les avons renfermés dans les deux tiroirs d'une petite commode de bois plaqué etant dans lad. chambre, et après les avoir fermés avec la clef restée entra les mains desd. C.ens gardiens, nous avons apposés les scellés sur iceux au cachet du departement.

Et au moyen de la remise que led. C. Georges nous à faite dela clef dud. secretaire, led. Comité Revolutionnaire en demeure bien et valablement dechargé et ledit C.en Vacquer ayant fait choix dud. secretaire nous l'avons fait sortir delad. chambre, lequel va etre designé et prisé.

4° Un secretaire à cylindre de bois d'acajou orné de plaque, anneaux et bordures de bronze doré avec galleries aussi de bronze doré, ayant un pied cassé et racommodé, de trois pieds quatre pouces de haut, de quatre pieds neuf pouces et demi de long et deux pieds quatre pouces de large, prisé trois cent cinquante livres, cy 350 l.

Nous avons ensuite reconnus sains et entiers et comme tels levés et otés, les scellés apposés sur une porte communiquante dans une chambre à coucher et après en avoir fait ouverture avec la clef representée par lesd. Citoyens gardiens, nous sommes entrés dans un cabinet ayant vue sur led. Jardin, dans lequel led. C. Vacquer à fait choix de ce qui suit:

5° Un bureau de bois plaqué garnis de trois tiroirs trouvés ouverts et vuides et sans clef, couverts de maroquin noir, garni de plaque, annaux et bordures de cuivre de cinq pieds quatre pouces de long sur deux pieds [34v] onze pouces de large et deux pieds six pouces de haut, prisé cent cinquante livres, cy 150 l.

Et après etre sortis dud. cabinet et en avoir fermé la porte entre les mains desd. C.ens gardiens, nous avons reapposés les scellés sur icelle, au cachet du departement.

Et après etre sortis dud. sallon et avoir fermé la porte qui communique dans lad. salle à manger avec la clef restée entre les mains des d. Citoyens Gardiens, nous avons reapposés les scellés sur icelle au cachet du departement.

Qui sont tous les meubles et effets dont le Citoyen Vacquer a fait choix lesquels nous lui avons remis et livrés à l'instant ainsi qu'il le reconnoit et s'en charge.

Au moyen dequoi lesd. C.ens Caty et Godet en demeurent bien et valablement dechargés, mais bien des scellés par nous cidessus reapposés ainsi que de ceux apposés precedemment dans lad. maison et de tout le mobiler etant en evidence, lesquels y ont volontairement consentis pour les representer à qui de droit lorsquils en seront legalement requis et ont lesd. C.ens Georges membre dudit comité Revolutionnaire de la Section des Piques, Grenier et Legras Commissaire de la municipalité, Vacquer inspecteur des fournitures et du mobilier de la Convention nationale, Caty & Godet, Signés avec nous après lecture faite et après avoir vacqué à ceque dessus depuis lad. heure de quatre jusques [35r] à celle de neuf sonnés, ainsi signé Georges, Grenier, Caty Godet, Legras, L.F. Vacquer et Bezauts et au dessous est ecrit enregistré à Paris le dix sept Messidor l'an Second[81] f.° 27 N° caze 7 reçu quatorze livres dix sols Signé Macoan

Certifié pour copie conforme à L'original delivrée par nous membres du Bureau du domaine national du departement de Paris ce vingt quatre Ventose l'an Troisieme[82] de la Republique françoise une et indivisible signé Guillotin & Rennesson.

13 piece de la Cotte 2
Phisique
Inventaire des instruments de phisique et de chimie du cabinet de Lavoisier cidev.[t] Fermier General et de L'Academie des Sciences Boulevard de la madelaine.[83]

Grand gazometre avec ses accessoirs et dependances par Meigné le jeune.

N.[a] Pour les details et l'intelligence de cet instrument capital et precieux, il faut consulter les elemens de chymie de Lavoisier, ainsi que pour les quatre suivans.

Grand appareil pour la combustion des huiles, appareil pour la fermentation, tres compliqué, appareil pour la combustion de l'esprit de vin, deux calorimetres, avec lampes en fer blanc, grande machine peumatique à deux corps de pompe par fortin avec barometre d'epreuve [35v] et une platine de lampe, deux tres grands ballons à peser l'air; grande balance de laboratoire, cinq recipients à douille et trois a boutons, vingt thermometres dont six à esprit de vin, barometres de Mossy monté en acajou, mais en mauvais etat, l'air rentré dans le tube, lampe à air ou gaz inflammable, autre lampe à air mais incomplette, appareil de Moth[84] pour impregner l'eau d'acide carbonique incomplet, trois cuvettes de fer blanc verni contenant en tous une vingte.[e] de flacons, machine electrique de vingt quatre pouces avec son conducteur et sa table, autre machine electrique de vingt deux pouces avec son conducteur posé sur une table sans pieds, batterie de soixante quatre petits boccaux dans une boete avec leurs fils de communication, grande jarre electrique, autre jarre plus petite, petite Bouteille de Leyde, tabouret en isoloir, trois tableaux magiques, pistolet de Volta en fer blanc, deux carillons à trois timbres, autre carillon a deux timbres, electrometre de Saussure, deux autres electrometres, petite presse electrique, appareil pour briser le verre, un tube etincelant, petit extitateur de cuivre, vaze pour l'inflammation de l'esprit de vin, petit isoloir avec virole terminée en fourche, eudiometre de Volta, condensateur de Volta [36r] à plateau de marbre diametre 22 pouces, miroir concave de vingt deux pouces tournant sur son pied, autre plus petit miroir concave de onze pouces tres gatés, prisme de verre verdatre monté sur un pied, grand prisme de verre de France non monté, prisme creux pour les liquides monté sur un pied, prisme quadrilatere creux avec cloison diagonale, miroir cilindrique avec ses cartons, miroir prismatique, machine hydrostatique ditte de Pascal, syphon à jet d'eau dans le vuide, vis d'archimede pour l'eau, autre vis pour une balle de plomb, marmitte de Papin, autre tres belle marmitte de Papin, c'est la plus considerable qu'on connaisse, grand pezon à ressort d'Hanin marquant cent soixante cinq livres, balance particuliere de Scanegati, avec ses poids dans une boete, petite balance de Scanegati à cadran poids de marcs savoir deux de seize marcs un de douze marcs, un de huit marcs et une boete de quatre marcs à poids rentrant et six petits poids de quatre onces et une once, petite balance avec son support, mauvaise petite balance dans son etui, joli trebuchet de la chine dans son étui avec quarante quatre poids de figures et grosseurs différentes, vingt quatre areometres en verre avec leurs etuis de fer blanc, treize areometres en [36v] metal cuivre et fer blanc grands et petits, une quantité de syphons et tubes communicaux en verre, dix bassines en fer blanc pour le transport des gaz, deux especes de refrigerens en fer blanc peint en noir, grande cuve hydro pneumatique garnie en plomb, autre grande cuve en chêne garnie en plomb pour le gazometre, autre cuve pour les mêmes experiences pour le gazometre du N° 69, table à souffler le verre, table à travailler et pollir les verres, ancien gazometre de Fortin, deux grandes armoires à quatre venteaux garnis de verre, grand recipient pour la combustion du gaz hydrogene, un respiratoire anglois en etain, autre respiratoire en fer blanc verni, mouvement appellé le briquet dans le vuide, boussole d'arpenteur, regle surmontée de pinnules, pinnules d'arpenteur, rapporteur à deux demi cercles et deux regles mobiles, deux eprouvettes à pandre à ressort, une equerre, petite equerre en quart de cercles par Baradelle, astrolabe, sextant de Bird sans lunettes, petit cadran à boussolle en argent par Lemaire, autre mauvais petit cadran dans une boëte d'yvoire, compas à verge de bois, hygrometre à plusieurs cheveux par Richer, petit miroir multipliant, tournant pour les allouetes, thermoscope particulier fait avec un tube [37r] sur une regle en cuivre graduée et plongeant dans un long recipient, pieces pour des experiences pneumatiques au mercure, belle regle anglaise divisée dans un etui en acajou, recreations magnetiques telle que le petit signe ingenieux, l'oracle, la boete aux chiffres, le petit peintre, jolie boussole de declinaisons montée sur un carreau de pierre polie, faite en allemagne, autre boussolle de declinaisons montée en cuivre rouge portée sur un plan de marbre blanc, par Fortin, il y manque l'aiguille aimantée et sa suspension. Barreaux aimantés dans une boete de chêne, gros ballon à tige garnie de viroles, robinet et tube contenant une eprouvette à mercure, boete d'acajou contenant une petite pharmacie, ballon de verre bleu d'un pied

81 5 July 1794.

82 14 March 1795.

83 The original inventory, kept at the Archives nationales – Paris, is transcribed in Appendix 3. In fact, a few differences emerge between the two documents, some of which (the prices) are of some importance. Since the original document given in appendix 3 is the more complete of the two, I have not added explanatory notes here.

84 John Mervin Nooth.

de diametre, deux petits gueridons d'acajou, petit modele d'appareil distillatoire, grande table en acajou ployante longueur huit pieds, et largeur quatre pieds, et table ovale d'ardoise de trois pieds huit pouces, sur deux pieds huit pouces, deux autres tables en acajou à etages longueur cinq pieds largeur vingt huit pouces, mortier et pillon de fayance blanche, douze mauvaises petites balances huit ensemble, deux petits creusets et une petite cuillère en **[37v]** platine, petite lampe en cuivre argenté, autre petite lampe pour eolipyles, deux tres petits soufflets montés sur une planchette d'acajou, pompe ou seringue en cuivre, un sceau en cuivre, regle de bois et de glace avec de long tubes de metal pour des experiences thermometriques.

Un barometre à deux tubes par Meignier, autre barometre à deux tubes par Meignier, deux baroscopes à liquide à quatre tubes immergés dans la même cuvette, autre baroscope en syphon monté sur une planche graduée en cuivre, quatre planches de debris de barometre, barometre a syphon portant un thermometre fait en Allemagne, petite machine electrique à plateau d'environ un pied et renfermée dans un grand recipient pour des demonstrations d'electricité dans le vuide, pendule de Rivas, boete peinte en verd propre à etre accrochée à la muraille, pendule à seconde excentriques dans sa boete de bois noir, grande balance dont le fleau à trois pieds de long avec une paire de grands bassins en cuivre, une paire de plateau de verre de cinq pouces de diametre, un simple plateau en verre, cette balance est dans une cage de bois d'acajou, une balance a fleau de dix huit pouces de long propre à peser **[38r]** au plus deux marcs avec une paire de bassins en verre dans une cage en bois d'acajou, autre balance pour les petites pesées de dix pouces de fleau avec une paire de bassins en verre sous une cage de verre.

N.ª Ces trois balances sont deposés a la commission des poids et mesures.

A Paris le vingt Brumaire de l'an Trois[85] de La Republique francoise une et indivisible, Signé Charles, Fortin, Lenoir.

N.ª depuis l'inventaire terminé il a été remis au depot par le C. Fortin.

Deux Loupes de verre de quinze pouces sans monture, une lunette acromatique à tuyau en cuivre, non terminée, cette lunette avoit été commencée par Rebours, elle est sans tuyaux oculaire. Signé Charles pour copie conforme Signé Oudry secretaires.

14ᵉ piece de la cotte 2

Cejourd'huy vingt quatre Vendemiaire l'an Troisieme[86] de la Republique françoise une et indivisible neuf heures du matin; moi Jean Nicolas Cristophe Bonnevie commissaire provisoire du Bureau du domaine national du departement de Paris; En vertu des pouvoirs qui m'ont été donnés par led. Bureau suivant son Arrêté du huit de ce mois, à l'effet 1° de reconnoitre et lever les scellés apposés chez Lavoisier, **[38v]** cidevant Fermier General condamné Boulevard de la madelaine N° 243 section des Piques, 2° de faire inventaire par distinction des meubles et effets qui appartenoient aud. Lavoisier, et de ceux appartenant à sa veuve, 3° de mettre lad. veuve Lavoisier en jouissance de tous lesd. meubles et effets à son usage et qui se trouveroient dans son logement particulier, je me suis transporté enlad. maison accompagné des Citoyens Germain Marigaut et Philippe Louanitier, tous deux commissaires civils de la Section des Piques que j'ai requis pour m'assister dans mes operations, où etant j'ai trouvé les Citoyens Nicolas Caty et Marc Louis Godet tous deux gardiens desd. scellés ainsi qu'ils me l'ont justifié; je leur ai fait part de ma mission.

J'ai aussi trouvé le C. Pierre Lalleman demeurant à Paris rue des Quatre Fils Section de l'Homme armé fondé de la procuration general de Marie Anne Pierrette Paulze veuve dud. Antoine Laurent Lavoisier, passée devant Gibé qui en à la minute et son confrere notaires à Paris le neuf Prairial dernier, enregistrée le treize, lequel a dit qu'ils compare pour etre present aud. inventaire, tant en qualité de procureur de lad. veuve Lavoisier que comme conseil de l'union des creanciers dud. Lavoisier; nommé alad. charge par deliberation desd. creanciers, et qu'il requiert que la vacation pour proceder aud. inventaire soit remise **[39r]** au vingt sept du present afin qu'il puisse prevenir les commissaires de l'union desd. creanciers de comparoitre et assister aud. inventaire conformement a la loi concernant l'administration et la vente des biens des emigrés, article huit section quatre et article dix huit du paragraphe deux.

En consequence j'ai remis la vacation au vingt sept du present mois sept heures du matin, et à led. C. Lalleman Signé avec lesd. Marigaut et Louaintier Commissaires civils, lesd. Catyet Godet.

Gurdieute et moi. Ainsi signé Lalleman Marigaut, Louaintier, Caty, Godet et Bonnevie.

Et led. jour vingt sept Vendemiaire, aud. an sept heures du matin je me suis transporté avec lesd. C.ᵉⁿˢ Marigaut et Louaintier, commissaires civils, en la demeure dud. deffunt Lavoisier, ou j' ai trouvé le C. Lalleman denommé et qualifié de l'autre part sont comparus les Citoyens Charles Nicolas Denis notaire public aux departement et residence de Paris, demeurant rue de Grenelle section du Bonnet rouge, et Pierre Tramaux, entrepreneur de batimens demeurant rue de Lalicorne section de la Cité, tous deux commissaires de l'union des creanciers dud. Lavoisier, lesquels ont dit qu'ils comparent pour etre present aud. inventaire en leurd. qualités conformement à loi et ont signé, ainsi signé Tramaux et Denis.

En consequence jai procédé au present inventaire en presence desd. C. Lalleman, Denis et Tramaux **[39v]** et desd.

85 10 November 1794.
86 15 October 1794.

Commissaires civils, ainsi qu'il suit sur la representation qui me sera faite desd. scellés par lesd. Caty et Godet gardiens d'iceux.

Sur ma requisition est comparu le C. Ely Eugene Levigneur, negociant demeurant à Paris rue Bayeulle N° 236 section des Gardes francoises, commissaire nommé par la Commission de Commerce pour le choix des meubles et effets precieux dans les maisons des emigrés et condamnés, lequel m'a requis de lever les scellés, et faire avec lui la visite du mobilier dud. Lavoisier pour qu'il puisse faire le choix des objets qui pourroient etre mis en requisition pour lad. Commission du Commerce et a signé ainsi signé Levigneur.

Est aussi comparu le C. Jacques Guillaume, membre de la Commission de revision des comptes des cidevant Fermiers Generaux à l'effet de faire a fur et à mesure de la levée desd. scellés, la recherche des papiers dud. Lavoisier qui pourroient etre relatifs auxd. comptes et a signé ainsi signé Guillaume.

Pour faciliter et accelerer la recherche dudit Citoyen Levigneur j'ai levé les scellés apposés sur le logement personnel dud. Lavoisier savoir:

Dans un antichambre au rez de chaussée, j'ai reconnu sains et entiers, levés et otés les scellés apposés sur la porte d'une chambre à droite ayant vue sur le jardin. J'ai aussi reconnu sains et entiers, levés et otés les scellés apposés sur la porte d'une autre chambre a gauche **[40r]** ayant aussi vue sur le jardin.

Dans lad. chambre j'ai aussi reconnu sains et entiers levés et otés les scellés apposés sur une porte d'une autre chambre ayant vue sur le Boulevard; led. C. Levigneur à trouvé dans lad. chambre une pendule mouvement de Furet[87] montée sur quatre colonnes et cassolettes, sur le haut, de marbre blanc ainsi que le soc et un amour entre les colonnes, letout ornés de guirlandes, rubans galleries decoupées à jour, perles et frizes en bronze doré dor moulu et omate; laquelle pendule il a marquée pour etre laissée a la disposition de la commission du Commerce et l'ont estimée de valeur de six cent livres, cy 600 l.

Led. C. Levigneur n'ayant pas trouvé autre chose dans lesd. chambres, les portes ont été refermées et j'ai sur icelles reapposé les scellés comme ils etoient avant la levée.

Et attendu qu'il n'y a pas d'autre endroit dans lad. maison qui ait été occupé par led. Lavoisier particulierement ainsi que nous nous en sommes assurés par les renseignements que nous avons pris; Led. C. Levigneur a Signé à cet endroit et s'est retiré ainsi signé Levigneur.

À l'instant les C.ens Denis et Tramaux et nous, ont dit que les d. C.en et C.ne Lavoisier ayant été communs en biens, tous leurs meubles et effets mobiliers, leur appartenoient sans distinction **[40v]** qu'ils requierent en consequence qu'il soit procédé au present inventaire en distinguant seulement les habits et linge à l'usage dudit Lavoisier ceux à l'usage de sa veuve, et les meubles et effets qui etoient à L'usage de tous deux sauf à lad. veuve Lavoisier à faire après le present inventaire telles reclamations que bon lui semblera et ont signé ainsi signé Tramaux et Denis.

Du consentement dud. C. Lalleman aud. nom j'ai procédé auxd. inventaire et prisée dans led. ordre, ainsi qu'il suit:

Dans une chambre à coucher, au rez de chaussée ayant vue sur le jardin.

Objets communs.

Premierement une couchette en bois peint sur roulettes, à equerres avec deux dossiers couverts de toille de cotton, un fond sanglé, un sommier de crin et deux matelats de laine couverts de futaine, un lit et un traversin de coutil rempli de plumes, une couverture piquée, deux couvrepieds, dont un petit d'etreteau, une courte pointe et les rideaux et pentes de l'alcove aussi en toille de coton garnis de franges prisés cinq cent livres, cy 500 l.

Item une commode de bois de placage garnie en cuivre avec dessus de marbre prisée soixante livres, cy 60 l.

Une autre petite commode à deux tiroirs en bois d'acajou avec ornemens de cuivre doré et dessus de **[41r]** marbre blanc prisée deux cent livres, cy 200 l.

Item une petite chiffonniere en bois d'acajou avec une table de marbre blanc, une table ronde sur un seul pied, aussi en bois d'acajou, deux petites chiffonnieres en bois de placage, un petit ecran en bois de noyer prisée quatre vingt livres, cy 80 l.

Item deux petits flambeaux de cuivre doré à jour pris a seize livres, cy 16 l.

Item deux pincettes et une pelle de fer prisés dix livres, cy 10 l.

Item six fauteuils couverts en tapisserie prisés soixante livres, cy 60 l.

Item trois autre fauteuils couverts d'etoffes de soie, prisés trente livres, cy 30 l.

Item trois chaisses foncées d'ozier, prisées douze livres, cy 12 l.

Item deux vazes de porcelaine en mauvais etat, une statue en platre prisés seize livres, cy 16 l.

J'ai vaqué à ceque dessus depuis lad. heure de sept du matin jusques à celle presente de quatre de relevée sonnées par triple vacation; le present inventaire sera du consentement de tous les sus nommés comparans continué demain sept heures du matin, et ont signé avec les Commissaires civils lesd. Caty et Godet et moi ainsi Signé Trameaux, Guillaume, Marigault, Denis, Louaintier, Caty, Godet, Lalleman, et Bonnevie. **[41v]** Et le vingt huit dud. mois de Vendemiaire l'an Troisieme[88] de la Republique francoise sept heures du matin, je me suis transporté à la demeure dud. Lavoisier, accompagné desd. Citoyens Commissaires Civils ou j'ai continué de proceder aud. inventaire en presence desd. C.ens Lalleman, Denis et Tramaux et nous et dud. C.en Guillaume membre de la Commission pour la revision des comptes des cidevant Fermiers Generaux; ainsi qu'il suit:

87 The king's clock maker.

88 October 19, 1794.

Dans lad. chambre à coucher au rez de chaussée ayant vue sur le jardin.

Jai reconnu sains et entiers levés et otés les scellés apposés sur la commode de bois de placage, inventoriée sous l'article deux cidessus il ne s'y est trouvé que des papiers qui ont été examinés par led. C. Guillaume et moi et attendu qu'il n'y à rien de relatif a la Ferme Generale, j'ai remis lesd. papiers dans lad. commode et j'ai sur icelle reapposé les scellés comme ils etoient pour rester jusqu'au transport que je ferai faire desd. papiers, conformement à mon pouvoir.

J'ai aussi reconnus sains et entiers levés et otés les scellés apposé sur une porte à gauche dela cheminée delad. chambre, communiquant à une autre chambre à coté ayant aussi vue sur le jardin.

Dans [42r] lad. chambre servant de bibliothèque.

Meubles communs.

Item une bergere, sept fauteuils et trois chaises couverts de velours cramoisi, une chiffonniere en bois de chêne à plusieurs tiroirs, prisés quatre vingt livres, cy 80 l.

Et attendu qu'il ne se trouve plus que des livres dans lad. chambre, les croisées et la porte ont été refermées et j'ai reapposé les scellés sur la ditte porte comme ils etoient avant la levée.

J'ai remis à faire la description et la prisée de lad. bibliotheque lors de la comparution du Commissaire de la comnission Temporaire des arts que j'ai requis; observant que les gardiens m'ont declaré qu'il a été fait par lad. Commission Temporaire des Arts un inventaire desd. bibliotheques et autres objets concernant les arts et sciences.

Dans un petit cabinet à coté de la chambre à coucher, servant de garde robbe.

Item une petite armoire en bibliotheque en bois noirci, avec ornemens de cuivre, un grand fauteuil, deux coussins et un tabouret de velours cramoisi, une petite table de bois de placage à un tiroir et dessus de marbre, prisés quatre vingt livres, cy 80 l.

Habits à l'usage de la C.^{ne} Lavoisier.

Item un pierrot et son jupon de taffetas noir, [42v] une robbe dite chemise en taffetas gris, un pierrot et son jupon de Sagati, une robe et un tablier de toille d'orange, un paquet de morceaux de toille et et etoffes, prisés comme tres vieux soixante livres, cy 60 l.

Item trois vieux manchons prisés.

Meubles communs.

Item une garniture de lit en toille de Jouy, prisée cent vingt livres, cy 120 l.

Item un miroir de deux morceaux de glace l'un de quarante un pouces et l'autre de vingt trois pouces le tout de haut sur vingt neuf pouces de large prisé.

Dans un autre cabinet à coté ayant vue sur la cour.

Meubles communs.

La face d'un petit corps de bibliotheque en bois d'acajou à deux portes, un petit corps de tiroirs en bois de placage prisés deux cent livres, cy 200 l.

Dans le corps de tiroirs s'est trouvé une collection de coquillage et autres objets d'histoire naturelle, j'ai remis à en faire la designation et la prisée ainsi que de quelques gravures et modeles en platre, lorsque le C. Commissaire de la Commission Temporaire des arts sera ici present.[89]

Item un petit matelat d'otomane, deux petits oreillers remplis de plumes et couverts de bazin rayé, prisé quarante cinq livres, cy 45 l.

Item quatre vieux fauteuils couverts d'etoffe de [43r] soie prisés quarante livres, cy 40 l.

Item deux chenets et une tenaille en fer garnis de cuivre prisés douze livres, cy 12 l.

Dans un autre petit cabinet a coté ayant aussi vue sur la cour.

Meubles en communs.

Item une pelle et une tenaille en fer garnies de cuivre, deux chandeliers de cuivre en forme de colonnes quarrées, une petite table de bois noirci à un tiroir prisée seize livres, cy 16 l.

Item une table en bois d'acajou à deux tiroirs servant a resserer des ustenciles de peintures et des couleurs prisée.

Item une glace sur la cheminée de trente neuf pouces de haut sur vingt neuf de large prisée.

Item six chaises couvertes de maroquin verd, prisée quatre vingt dix livres, cy 90 l.

Item un petit coffre en bois de marqueterie dans lequel sont plusieurs pots de medicaments prisés douze livres, cy 12 l.

Les autres objets qui se trouvent dans led. cabinet etant des statues, modeles en platre, peintures et autres semblables objets; j'ai remis à en faire la description et la prisée lorsque le commissaire de la Commission Temporaire des arts que j'ai requis sera ici present.[90]

Dans une salle a coté ayant aussi vue sur la cour et servant d'antichambre

[43v] Une petite table de bois noirci, une petite toillette de bois de noyer, une echelle double en bois de chêne, quatre chevalets pour peinture, deux jalons avec plaques de cuivre, un pied de niveau d'eau, prisé a vingt livres, 20 l.

89 In fact, this natural history collection seems not to have been inventoried. The reference to a collection of shells is quite surprising, unless it refers to the ca. 1,000 'coquilles fossiles' which were part of the mineralogical collection now at the Muséum Lecoq in Clermont Ferrand. However, as it will become clear from further references in the inventory, the mineralogical collection was placed in a different room, and since it is unlikely that Lavoisier kept items collected in 1767 during his mineralogical travels with Guettard separate, it is reasonable to assume that this was a different collection.

90 A detailed inventory of these items was apparently not prepared.

Item une petite table plyante couverte d'un drap vert prisée livres, cy.

Item deux fauteuils et quatre chaises couverts de velours cramoisi prisés quatre vingt livres, 80 l.

Item cinq fauteuils couverts de tapisseries, deux autres vieux fauteuils couverts d'etoffe de soie prisés quatre vingt dix livres, cy 90 l.

Item un lustre de verre de Boheme monté sur cuivre prisé douze livres, cy 12 l.

Item quatre morceaux de peau servant de marche pieds prisés dix livres, cy 10 l.

Dans une autre antichambre à coté, meublée en communs

Item un coffre en forme de banquette couvert de velours cramoisi, trois autres banquettes couvertes de pareil velours, quatre fauteuils et trois chaises foncés de paille, prisés vingt livres, cy 20 l.

Item trois petits sceaux et deux cuvettes de cuivre argenté, prisés huit livres, cy 8 l.

Item deux douzaines d'assiettes depareillées, trente autres pieces de fayance et verreries en differens ustencils de menage, prisés quinze **[44r]** livres, cy 15 l.

Item une petite pendule antique portant le nom de Josué Pannier[91] à Paris dans sa boete de bois avec ornemens de cuivre prisée quatre vingt seize livres cy 96 l.

J'ai vacqué à ce que dessus depuis lad. heure de sept du matin jusques à celle presente de une de relevée sonnée par double vacation.

Le present inventaire sera du consentement de tous les comparants susnommés à duodi prochain sept heures du matin, et ont signé avec moi ainsi signé Tramaux, Guillaume, Lalleman, Louaintier, Marigaut, Denis, Caty, Godet et Bonnevie.

Et led. jour deux Brumaire an Troisieme[92] de la Republique francoise une et indivisible sept heures du matin je, Commissaire du bureau du domaine nationale du departement de Paris, me suis transporté en la demeure dud. Lavoisier, accompagné desd. C.^{ens} Marigaut et Louaintier, Commissaires civils; où etant j'ai continué de proceder aud. inventaire en presence desd. C.^{ens} Lalleman, Denis, Tramaux et Guillaume ainsi qu'il suit:

Dans une salle au rez de chaussée ayant vue sur le jardin j'ai reconnus sains et entiers levés et otés les scellés apposés sur la porte d'une autre salle ayant aussi vue sur le jardin, il s'est trouvé dans lad. salle.

Meubles en commun.

Une pelle et une pincette de fer un petit **[44v]** ecran de bois de noyer en mauvais etat, un petit coffre fort en bois de rose avec garnitures et ormemens de cuivre, quatre chaises couvertes de velours cramoisi, un vieu fusil, cent vingt deux livres de poids de fonte prisés quarante livres, cy 40 l.

Il s'est trouvé dans lad. salle plusieurs paquets de livres destinés pour le Comité d instruction publique et d'autres sans indication de destination; les C.^{ens} Caty et Godet, gardiens, m'ont declaré qu'il avoit été fait un inventaire estimatif de tous les livres de la bibliotheque dud. Lavoisier.[93]

Ensuite les volets des fenetres et la porte delad. salle ont été fermés et j'ai reapposés les scellés du Bureau sur lad. porte comme ils etoient avant la levée.

Dans lad. salle au rez de chaussée ayant vue sur le jardin

Meubles en commun.

Item deux lits de sangle, six petits matelats de bourre et laine couverts de toille à carreaux, trois traversins de coutils remplis de plumes, deux couvertures de laine blanche, prisés cent quarante livres, cy 140 l.

Item un canapé, deux petits oreillers, six fauteuils et deux coussins, le tout couvert de satin fond cramoisi, prisés cent vingt livres, cy 120 l.

Item un petit ecran à deux feuilles, en bois de **[45r]** noyer, une petite epinette en mauvais etat prisés dix livres, cy 10 l.

Il ne s'est plus trouvé dans lad. salle que deux tables dessus de marbre de rapports que lesd. Caty et Godet m'ont déclaré avoir été mis en requisition pour le Garde meuble national, et cinq tableaux peints sur toille dont un portrait; j'ai remis à faire la designation et la prisés desd. objets lors de la comparution du Commissaire de la Commission Temporaire des arts, que j'ai requis.[94]

J'ai reconnu sains et entiers levés et otés les scellés apposés sur une porte à gauche de celle d'entrée delad. salle, et ouverture faite de cette porte avec la clef étant en la possession du gardien je suis entré dans une autre salle ayant aussi vue sur Le jardin, dans laquelle se sont trouvés les objets ci après designés.

Meubles en commun.

Item deux ottomanes et quatre fauteuils couverts de tapisserie, deux bergeres et six fauteuils couverts de differentes etoffes de soie prisés comme vieux, deux cent quarante livres, cy 240 l.

Item une tenaille de cheminée, une echelle de bibliotheque,[95] un petit devidoir, prisés quinze livres, cy 15 l.

91 Lavoisier apparently owned two clocks made by Panier.

92 23 October 1794.

93 Detailed references to these inventories may be found in Beretta (1995).

94 No detailed inventory of Lavoisier's paintings seems to have survived.

95 This description fits the surviving wooden ladder preserved at Cornell University's Kroch Library, Lavoisier manuscript collection, Division of Rare and Manuscript Collections, 4712. The ladder was exhibited and photographed (Fig. 1) in Lavoisier (1943).

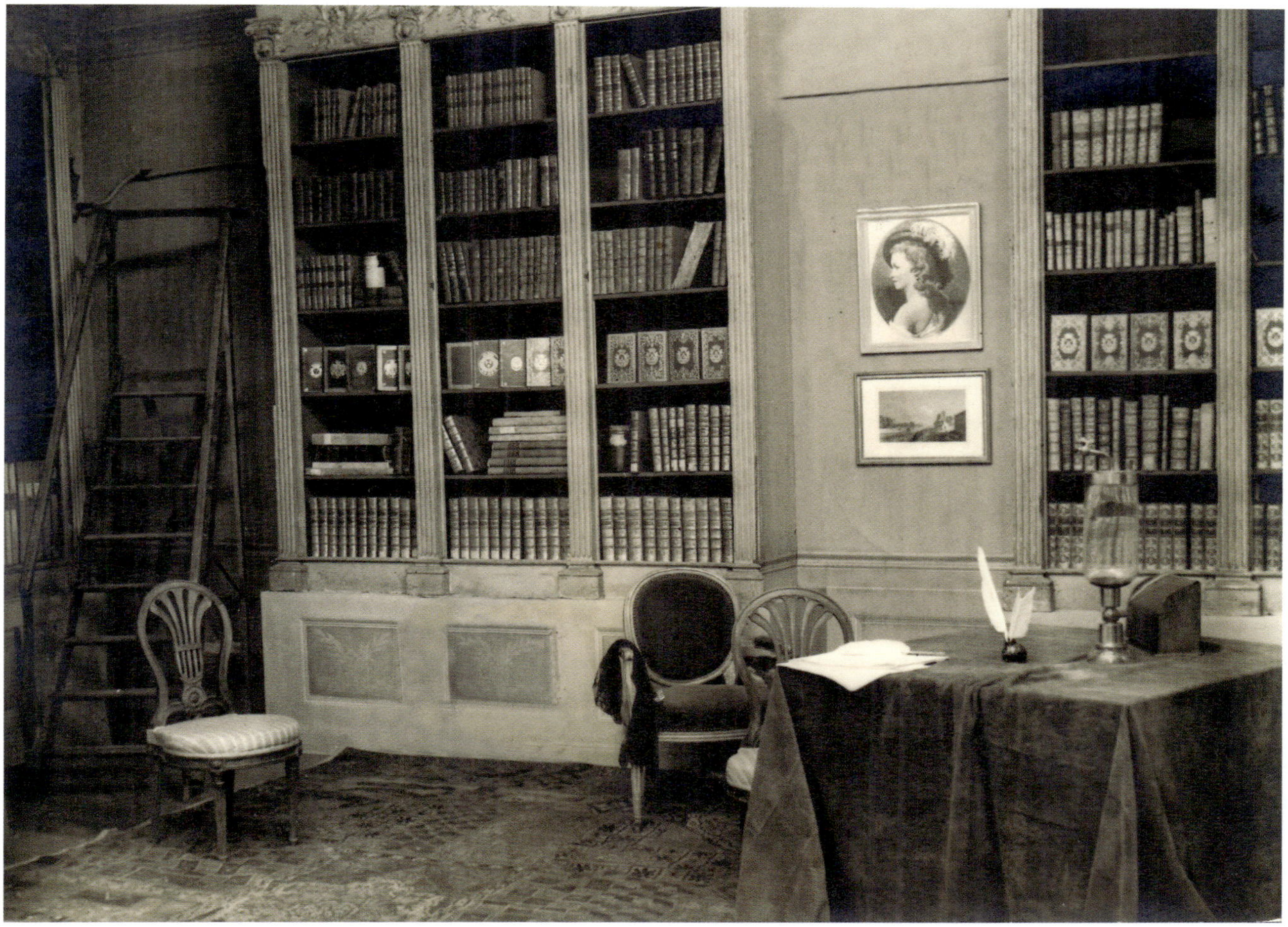

FIGURE 1 Lavoisier's cabinet in the 1943 exhibition. Lavoisier's library ladder is visible on the left
PRIVATE COLLECTION

J'ai reconnu sains et entiers levés et otés les scellés [45v] apposés sur les portes du bas d'une armoire servant de bibliotheque, il s'y est trouvé les objets ci après inventoriés et prisés.

Item une plaque d'argent gravée de differens sujets de devotion, laquelle plaque j'ai laissé dans led. bas d'armoire jusques à la fin du present inventaire que je la remettrai à la monnoye conformement à mon pouvoir.

Item une canne de jet à pomme d'or, deux couteaux à manches plaqués d'argent, quatre bourses brodée en or le tout prisés soixante livres, cy 60 l.

J'ai aussi reconnu sains et entiers levés et otés les scellés apposés sur un autre bas de bibliotheque entre les croisées du jardin. Il s'y est trouvé les objets ci après inventoriés et prisés.

Item un coffre fort garni de fer ouvert et vuide, un petit coffre couvert de chagrin dans lequel sont six flacons de cristal, prisés seize livres, cy 16 l.

J'ai refermé la porte de la premiere desd. deux armoires et j'ai sur icelle reapposé le scellé comme il etoit avant la levée.

Dans lad. salle sont encorre les objets ci après communs aud. Lavoisier et à sa veuve.

Item huit rideaux de croisée en toille de cotton blanche, quatre petits rideaux de mousseline, une lampe dite Quinquet à quatre [46r] faces avec un plateau de verre prisés ensemble cent quarante livres cy, 140 l.

Il s'est aussi trouvé une petite boete en bois de noyer fermée à clef que l'on ma dit etre un necessaire de campagne, attendu que les gardiens n'ont pas lad. clef et qu'il n'y a personne dans la maison qui l'ait, ainsi qu'ils me l'ont declaré, j'ai fait mander le c. Cheuzet serrurier demeurant marché Lepelletier, qui a fait l'ouverture delad. boete at auquel j'ai payé quinze sols pour son salaire.

Les objets renfermés dans lad. boete etant partie d'argent, d'acier, de cristal et de porcelaine, je l'ai fait refermer et j'ai apposé dessus les scellés du bureau aux bouts d'une bande de papier, pour lesd. objets inventoriés et prisés a la fin du present, avec ceux de même nature qui pourroient encore se trouver sous les scellés.

Plus il s'est trouvé dans lad. salle, une suite de bibliotheque qui à aussi été invèntoriée et prisée ainsi que lesd. gardien le declarent.

Plus neuf liasses de cartons attachés avec des rubans de fil et scellé du sceau du bureau et de celui du departement, les etiquetes desd. cartons indiquent que les papiers qui y sont renfermés, concerne la chimie, la phisique, l'agriculture, la mineralogie et une terre de la salle en vendemois.

En consequence lesd. cartons ont été laissés dans le même etat ou je les ai trouvés, sauf **[46v]** à faire faire leur transport aux differens lieux de leur destination après la comparution du Commissaire de la Commission des arts que j'ai requis.

J'ai reconnus sains et entiers levés et otés les scellés apposés sur la portes d'une autre salle donnant dans celle ou je suis et ayant vue sur le Boulevard.
Le objets qui y sont renfermés vont etre designés et prisés comme il suit:

Meubles en commun.
Item une pelle, une pincette et forts chenets en fer, une ottomane et huit fauteuils couverts des tapisseries, un petit bureau de bois d'acajou, deux petites tables longues en differents bois de couleur, un sceau de taule peint, environ quatre rames de papier à lettre, prisés trois cent livres. cy 300 l.

Item un petit pendule en cartel portant le nom de Musson[96] à Paris, prisée cent trente livres, cy 130 l.

Item deux petites paires de pistolets, un moulle à balles et un petit clef pour les charger, un ecran a huit feuilles avec fond de taffetas verd, un autre simple avec fond de velours cramoisi, prisés soixante livres, cy 60 l.

Item un très petit necessaire de toilette couvert **[47r]** de maroquin rouge, garni de ses ustencils en bois et ivoir &.c que j'ai prisé vingt livres, cy 20 l.

Item deux flacons de gros verre, un autre de verre blanc de differentes grandeurs dans lesquels sont environ trois livres de tabac dont une livre d'Espagne prisés avec differens petits ustencils en verrerie fayance et ferraille trente livres, cy 30 l.

Et attendu qu'il ne s'est plus trouvé dans lad. salle que des cartons contenant des papiers des assignats et du numéraire, et quelques bijoux que j'ai mis dans un carton pour etre inventoriés et prisés avec les autres bijoux qui pourront encore se trouver, les volets des fenetres et la porte de lad. salle ont été refermés et j'ai reapposé les scellés du Bureau sur lad. porte, comme ils etoient avant la levée.

J'ai vacqué à l'arrangement des objets cidessus decrits, à leur inventoriée et prisée et aux autres operations indiquées depuis lad. heure de sept du matin jusqu'a celle presente de quatre de relevée sonnée par triple vacation, le present inventaire sera continué du consentement de tous les comparans susnommés, primidi onze du present mois, et ont signé avec moi ainsi signé Guillaume, Travaux, Lalleman, Louaintier, Marigaut, Denis, Caty, Godet & Bonnevie.

Et led. pour onze Brumaire sept heures **[47v]** du matin moi Commissaire du Bureau du domaine nationale du departement de Paris, me suis transporté en la maison Lavoisier accompagné dudit Citoyen Louaintier commissaire civil et dud. Nicolas L'Huillier aussi commisaire civil de la Section des Piques que j'ai requis pour m'assister dans mes operations, attendu l'absence dud. C. Marigaut; ou etant jai continué de proceder aud. inventaire, en presence desd. C.ens Lalleman, Denis, Tremaux & Guillaume.

Dans un petit cabinet ayant son entrée sur le perron de l'escalier à gauche, et vue sur le Boulevard.

Meubles en commun.
Item un petit lit de fer en forme de baldaquin, garni en cuivre, un fond sanglé, un sommier de crin, deux matelats de laine couverts de toille à carreaux, un lit et un traversin de coutil rempli de plumes, un autre traversin de crin couvert en toille grise, une couverture de toille de cotton piquée, la courtepointe, les dossiers du lit, la garniture du baldaquin, ses rideaux, et la teinture delad. chambre en toille d'orange fond blanc à petite fleurs, prisés sept cent livres, cy 700 l.

Item deux vieux fauteuils couverts en tapisserie, un fauteuil couvert de maroquin rouge, six chaises d'ozier, deux petits sieges garnis de leur sceaux de fayence, prisés trente livres, cy 30 l.

[48r] Item une petite commode en bois de placage a dessus de marbre et deux tiroirs ouverts et vuides, une petite table de nuit a tables de marbre prisés vingt quatre livres, cy 24 l.

Item une vieille table de toillette en bois de placage prisée douze livres, cy 12 l.

Item un secretaire en bois de placage prisé trente livres cy 30 l.

Et attendu que led. secretaire est fermé et que les gardiens m'ont dit n'avoir jamais eu la clef, et ne pas scavoir où elle peut etre, j'ai mandé le C. Heuzet serrurier demeurant marché Lepelletier lequel a fait l'ouverture de la partie basse dud. secretaire et du tiroir du haut, ou il ne soit rien trouvé, à l'egard du corps dud. secretaire led. Heuzet n'a pas pu l'ouvrir, moi je me suis assuré autant qu'il est possible en soulevant led. secretaire, et en lui donnant du mouvement, que la partie qui est fermée est egalement vuide, en consequence il n'a pas été fait de nouvelles tentatives pour l'ouvrir afin de ne pas l'endomager considerablement j'ai payé aud. C. Heuzet quinze sols pour son salaire.

A l'usage du C.en Lavoisier
Item trente neuf chemises de toille blanche garnies de mousseline à usage d'homme prisées deux cent quarante livres, cy 240 l.

96 It is not clear whether the clock was made by Pierre Masson or, more likely, his son Louis.

Item quinze mouchoirs de toille blanche, quatre cravattes de mousseline, quatre paires [48v] de bas de cotton blanc une autre de couleur, deux paires de bas de soye noire, prisés vingt huit livres, cy 28 l.

Item deux robbes de chambre et deux vestes de piqué de marseille, deux culottes de Bazin, trois gillets de futaine, trois calleçons de toille, quatorze paires de chaussons aussi en toille, un bonnet de cotton, six pieces d'estomach prisés vingt cinq livres, cy 25 l.

Item trois robbes de chambre, trois vestes et une culotte de Bazin, une robe de chambre et une veste d'indienne fond sablé, un habit de drap brun, un autre habit de tricot couleur puce; quatre vieilles paires de souliers prisés quarante huit livres, cy 48 l.

Dans une petite chambre à l'entresol ayant vue sur le Boulevard et à laquelle on communique par un petit escalier derobé.

Meubles en commun
Item une pelle, une pincette, une tenaille a feu le tout en fer garni de cuivre doré, deux petits chandeliers aussi en cuivre doré, prisés seize livres, cy 16 l.

Item une petite table en bois d'acajou, un vieux secretaire de bois de noyer, deux vieux fauteuils couverts de velours cramoisi, prisés cent livres, cy 100 l.

A l'usage de la C.ne Lavoisier
Item cinquante chemises de toille blanche, non [49r] garnies à usage de femme prisées quatre cent livres, cy 400 l.

Item six bonnets ronds de mousseline et linon garnis de dentelle, sept vieux fichus de Linon et ourgandie, un fichu de linon en forme de chemise, quatre autres vieux fichus de linon, prisés vingt quatre livres, cy 24 l.

Item six jupons de bazin et futaine, six camisoles de nuit, deux robbes de chambre de mousseline, quatre peignoirs, quatre tabliers de toillette, quatre corsets de bazin, cinq paires de poches de bazin, douze paires de bas de cotton blanc, dix mouchoirs rouge des Indes, dix huit mouchoirs blancs, dix fichus de mousseline, dix autres vieux fichus servant de serre tête, trois pièces d'estomach, trente cinq serviettes de garde robbe, six paires de bas de soye blanche quatre vieilles paires de souliers à usage de femme, prisés deux cent quatre vingt livres, cy 280 l.

Item une robbe et son jupon de poulx soie noire, une autre robbe et son jupon de satin noir, une autre robbe de maz de St maure noir, une robbe d'etoffe de soie bleue camelée, une robbe de velours violet, deux pierrots de bazin anglais, quatre autres pierrots blancs dont un de mousseline brodée, une autre pierrot de mousseline brodée, un autre pierrot de cirsacat rayé, une robbe en fourreau de mousseline brodée, un pierrot de linon brodée, une robbe de [49v] taffetas rayé, une robbe et un jupon de crepe blanc un pierrot de gaze, une robbe d'etoffe de soie bleue, un mantelet de taffetas rose garni de dentelle de maline brodée, un autre mantelet de mousseline brodée blanc, un tablier d'ourgandie et un de linon baptiste, deux jupons de toille blanche prisés ensemble cinq cent cinquante livres, cy 550 l.

À l'instant led. C. Guillaume membre de la commission de revision des comptes des cidevant Fermiers Generaux, a dit qu'il alloit se retirer attendu que sa presence ici est inutile parceque la mission de la Commission se bornoit à l'examen des comptes et administrations de la cidevant Ferme Generale, que d'ailleurs le principal travail de la Commission est fait et que les papiers concernant la cidevant Ferme Generale qui pourroient se trouver parmi ceux dud. Lavoisier sont absolument inutile a lad. commission et en effet Led. C. Guillaume s'est retiré après avoir signé en cet endroit, ainsi signé Guillaume.

J'ai reconnu sains et entiers levés et otés les scellés dud. bureau apposés sur la porte d'une autre petite chambre à l'entresol ayant aussi vue sur le Boulevard, ouverture faite de lad. porte avec la clef representée par les gardiens, j'ai pareillement reconnu sains et entiers levés et otés les scellés du Bureau apposés sur les volets de la croisée.

[50r] Les objets trouvés en lad. chambre ont été inventoriés comme il suit:

Meubles en commun
Item une couchette de bois peint en gris, un fond sanglé, un sommier de crin, deux matelats de laine couverts de futaine, un lit et deux traversins de coutil remplis de plumes, un baldaquin à la turque avec dossier et bonne graces de damas cramoisi, les rideaux en serge cramoisie prisés deux cent quarante livres cy 240 l.

Item une petite pendule dite cartel dans sa boete de cuivre en couleur prisées quatre vingt dix livres, cy 90 l.

Item un grand chiffonnier en bois de noyer, une petite commode en bois de placage à dessus de marbre et deux tiroirs ouverts et vuides, une petite table de nuit en bois de noyer, un bidet à dossier garni de son sceau de fayance, prisés quatre vingt livres, cy 80 l.

Item quatre fauteuils couverts de velours cramoisi prisés quarante livres, cy 40 l.

A L'usage dud. Lavoisier
Item deux habits et une veste de drap noir, un habit veste et culotte d'étoffe de soie noire, un habit veste et culotte d'etoffe de soie camelée et rayée jaune et violets, un autre habit et une culotte de croisée de soie bleue, un autre habit de camelot de soie rayée, un habit de taffetas flambé, [50v] un habit et une veste de drap ecarlatte, un surtout de drap bleu, une veste et culotte de drap blanc, trois vieilles vestes d'etoffes de soie or et argent, une culotte de camelot vert et une autre de velours camelé et rayé, prisés à cent quatre vingt livres, cy 180 l.

Et ne s'etant plus rien trouvé à inventorier dans lad. chambre les volets de la croisée et de la porte d'entrée ont été refermés et j'ai sur iceux reapposé les scellés du Bureau comme ils etoient avant.

Dans une petite chambre au second etage servant de garde meuble

Meubles en commun
Item un vieux bois de lit, un fond sanglé, deux vieux coffres, quatre vieilles tables couvertes de tapis verd, une table de bois de sapin sur son pied en bois de chêne, deux vieux paravants, un ecran en tapisserie dans sa monture en bois doré, plusieurs boetes vuides et differents bouts de bois, prisés quarante livres, cy 40 l.

Item une garniture de lit en damas cramoisi prisé cent dix livres, cy 110 l.

Item une autre garniture de lit en damas cramoisi prisé avec huit parties de rideaux et plusieurs morceaux de pareille etoffe, deux cent cinquante livres, cy 250 l.

Item huit parties de rideaux de taffetas cramoisi et blanc, prisés quarante huit livres, 48 l.

Item deux garnitures de lit en indienne prisés [**51r**] deux cent livres, cy 200 l.

Item quatre pieces de draperie en damas de trois couleurs avec glands et cordons prisés soixante livres, cy 60 l.

Item une garniture de lit de toille de cotton prisé cent livres, cy 100 l.

Item trois vieilles couvertures de laine, une vieille courtepointe, un morceaux de grosse toille, une piece de toille d'ecorse d'arbres, prisés soixante livres, cy 60 l.

Item deux roulleaux de moquette propre à faire des tapis, un roulleau de toille verte, prisée cent livres, cy 100 l.

J'ai vacqué à ceque dessus depuis lad. heure de sept du matin jusques à celle de quatre de relevée sonnée par triple vacation.

Le present inventaire sera continué quintidi prochain sept heures du matin et ont les Citoyens ici presents signé avec moi ainsi Signé, Louaintier, Lhuissier, Caty, Godet, Lalleman, Denis, Tramaux et Bonnevie.

Et led. jour quinze Brumaire sept heures du matin moi commissaire du bureau du domaine national me suis transporté en la maison dud. Lavoisier, accompagné desd. Citoyens Louanitier et Commissaires civils, ou etant, j'ai continué de proceder aud. inventaire en presence desd. C.^{ens} Lalleman, Denis et Tramaux seulement, attendu la declaration faite par led. C. Guillaume dans le cours des vacations de la derniere journée.

[**51v**] Dans une petite chambre au second etage à coté de celle servant de garde meuble et cidessus indiquée.

Meubles en Commun
Item un bois de lit à colonnes, un fond sanglé, un sommier de crin couvert de toille à carreau, deux matelats de laine couverts de futaine, un traversin de coutil rempli de plumes, prisés cent vingt livres, cy 120 l.
Item cinq fauteuils couverts de velours cramoisi, un bureau de bois noirci à trois tiroirs ouverts et vuides une petite table ployante couverte d'un vieux drap vert, un corp de tablettes en bois de sapin, prisés soixante quatre livres, cy 64 l.
Item une paire de chenets à double branches, une pelle et une pincette de fer, deux petits chandeliers de cuivre argenté prisés douze livres, cy 12 l.
Dans une autre chambre a coté.

Meubles en commun
Item trois corps d'armoire à portes pleines par bas et portes vitrées par haut, un autre corps d'armoire en bois sapin avec portes pleines peint en gris, plusieurs planches tablettes prisées cent livres cy 100 l.

Item trois fauteuils et quatre vieilles chaises couvertes de velours et tapisserie, un fauteuil de [**52r**] canne, une grande table de bois de sapin sur deux treteaux de bois de chêne, un petit corps de tiroirs en bois de placage, une petite table et plusieurs boetes de bois blanc prisés vingt quatre livres, cy 24 l.

Il se trouve encore dans lad. chambre beaucoup d'objets de mineralogie et d'histoire naturelle qui seront inventoriés lorsque les commissaires de la Commission Temporaire des arts (que j'ai requis plusieurs fois) seront presents.[97]

Dans le coridor
Item un corps d'armoire en bois de sapin à cinq portes à deux battants, prisée soixante livres, cy 60 l.

Dans lad. armoire

Item deux cent cinquante volumes de livres depareillés et qui sont sujets de devotion et autres de nulle valeur, lesquels n'ont point été inventories avec la bibliotheque,[98] ainsi que lesd. Godet et Caty me l'ont declaré lesd. livres prisés cinquante livres, cy 50 .

Item il s'est aussi trouvé dans lad. armoire cinquante deux cartons dans lesquels sont des fleurs et herbages secs[99] et quelques feuilles de papier blanc j'ai prisé lesd. cartons vingt livres, cy 20 l.

Dans une chambre vis à vis led. corps d'armoire et à gauche du coridor.

97 The École centrale des travaux publics that had been entrusted, among other things, with Lavoisier's large collections of minerals and fossils (some 4,000 items) until Marie Anne Lavoisier managed to retrieve them in late 1795; see Tron (1996). I was unable to find any inventory of the collection from the Revolutionary period. Interestingly, probably due to a lack of space, the mineralogical collection which was also used in the chemical laboratory at the Arsenal was moved to a room on second floor.

98 Interestingly the religious books, probably inherited from Lavoisier's father – a jansenist – were separated from the main library and put into a cabinet on the second floor.

99 This could be a *herbarium vivum*.

Item une couchette à bas pilliers sur roulettes [52v] un fond sanglé, trois vieux matelats de laine et bourre, couverts de differentes toilles, un lit de coutil rempli de plumes, un traversin, une vieille couverture de laine, une vieille courte pointe piquée, prisés soixante livres, cy 60 l.

Item une vieille armoire de differents bois, un buffet de bois peint en gris avec dessus de marbre, une petite table, prisés avec un vieux fauteuil seize livres, cy 16 l.

Dans une chambre à côté

Item un bois de lit à colonnes, une paillasse, deux matelats de mauvaise laine couverts de toille à carreau, un lit et un traversin de coutil rempli de plumes, une couverture de laine blanche, et un couvre pied d'indienne piqué, prisés quatre vingt livres, cy 80 l.

Item la housse et les rideaux du lit en siamoise flambée prisée vingt quatre livres, cy 24 l.

Item une petite table de bois noirci, trois chaises foncées de paille prisés cinq livres, cy 5 l.

Il s'est encore trouvé dans lad. chambre une armoire de bois de noyer, un petit buffet de bois de chêne, un coffre et une cassette, un peu de linge marqué L.M. et quelques habits, qui appartiennent au C. Louis Antoine Masselot[100] qui couchoit dans lad. chambre et qui est actuellement avec la C.ne Lavoisier, le tout ainsi que la C.ne Françoise Pelletier V.e Foulon,[101] cuisinière de lad. C.ne Lavoisier, et ici presente l'a declaré.

[53r]

Dans une autre chambre à coté

Item une couchette à bas pilliers sur roulettes, cinq vieux matelats, un lit et un traversin de coutil rempli de plumes, deux couvertures de laine, une courtepointe et deux parties de rideaux en toille indienne fond bleu, une tringle de fer courbée, prisés cent vingt livres, cy 120 l.

Item une vieille armoire en bois de noyer, un vieux buffet de bois peint en gris, une table de nuit en bois de noyer, une autre petite table, un fauteuil foncé de paille, prisés vingt quatre livres, cy 24 l.

Il s'est ancore trouvé dans lad. chambre quelques habits à usage de femme et du linge aussi à usage de femme et marqué L.G. que lad. femme Foulon m'a declaré lui appartenir.

Dans une autre chambre à côté

Item une couchette à deux dossiers, un fond sanglé, un sommier de crin, deux matelats de laine couverts de toille à carreaux, un lit et un traversin de coutil remplis de plumes, une couverture de laine, un couvre pied d'indienne piquée, prisés cent livres, cy 100 l.

Item une armoire en bois de noyer, un petit buffet aussi en bois de noyer, une petite table de pareil bois, deux fauteuils foncés de paille, quatre bouts de planche tablettes, prisés vingt quatre livres 24 l.

Item la tenture de lad. chambre en siamoise rayée bleu et blanc prisée douze livres, cy 12 l.

[53v]

Dans une autre chambre à côté

Item deux armoires de bois de chêne, une autre de bois de noyer, un petit corps de tablettes, une petite table de bois blanc, prisées quatre vingt dix livres, cy 90 l.

Dans lesd. armoires

Item cinq paires de drap de toille blanche prisés cent vingt livres, cy 120 l.

Item trente sept autres paires de draps de toille commune prisés cinq cent cinquante livres, cy 550 l.

Item six rideaux de croisées en toille de coton, prisés trente livres, cy 30 l.

Item quatorze tayes d'oreillers, prisés dix livres, cy 10 l.

Item quinze nappes de toille unie de differentes grandeurs prisées quatre vingt dix livres, cy 90 l.

Item dix huit nappes ouvrées de differentes grandeurs prisées cent vingt livres, cy 120 l.

Item douze douzaines de serviettes ouvrées prisées cent quatre vingt livres, cy 180 l.

Item quinze douzaines de serviettes unies prisées cent quatre vingt livres, cy 180 l.

Item trente six essuie main prisés vingt sept livres, cy 27 l.

Item dix huit autres essuie mains de grosse toille prisés douze livres, cy 12 l.

Item trente deux tabliers de domestique prisés [54r] vingt quatre livres, cy 24 l.

Item dix tabliers plissés prisés avec un paquet de vieux linge ne meritant description quinze livres, cy 15 l.

Item deux nappes et quatre douzaines de serviettes damassée prisées cent huit livres, cy 108 l.

Item une couverture de toille de coton blanche prisée quinze livres, cy 15 l.

Dans une autre chambre à côté

100 Masselot was Lavoisier's most important servant (*principal domestique*) for twenty years (1772–1792), until he was dismissed due to Lavoisier's temporary financial difficulties after his removal from the Régie des poudres in September 1791 and the subsequent order for him to leave the Arsenal; see LC, vol. 7, p. 154. Masselot was rehired in 1796.

101 Marie François Foulon, née Peltier. In addition to Masselot and Foulon, the Lavoisiers had another servant Jean-Baptiste Girardot. On this see the *Inventaire du domicile de Lavoisier* (dated 27 January 1794), published in LC, vol. 7, pp. 420–422. In the tax declaration for 1791 (prepared on 12 April 1792) Lavoisier stated that he had in his service at the Arsenal "une femme de chambre, une cuisinière, un cocher et trois laquais" (Ibid, p. 74).

Item une courchette à deux dossiers, une paillasse, deux matelats de laine couverts de toille à carreaux un traversin et un oreiller de coutil rempli de plumes, une courtepointe et les garnitures des dossiers en indienne prisés cinquante livres, cy 50 l.
Item une table en bois de sapin avec son pied et un tiroir de bois de chêne, une autre petite table, deux petites chaises foncées de paille, une vielle corbeille couverte et doublée de taffetas prisés cinq livres, 5 l.

Dans un petit cabinet à côté
Il s'est trouvé quelques tableaux ebauchés, j'ai remis à en faire la designation et prisées lorsque le Commissaire de la Commission Temporaire des arts sera present.

Dans le grenier
Item sept paires de chenets de fer dont deux garnies de cuivre, trois poeles percées, une à frire, environ cinq cent livres de ferraille ne meritant description prisés cent vingt livres, 120 l.

Item une cheminée à la turque en mauvais etat, un petit poele de fonte prisés six livres, cy 6 l.

Item deux petits buffets un petit corps [54v] d'armoire en bois de chene, deux vieilles tables dont une couverte de drap vert, une petite chaise d'aisance, six vazes de fayance, prisés vingt livres, cy 20 l.

Item dix huit tablettes de bois de sapin, plusieurs faces d'armoires, une tenture de chambre en papier cramoisi collé sur toille prisés cinquante livres, cy 50 l.

Item une beignoire à cylindre, en cuivre rouge, prisée quatre vingt livres, cy 80 l.

Item plusieurs morceaux de boiseries, vingt vieilles chaisses depaillées, un grand nombre de differens ustencils meubles demembrés et morceaux de bois trainans qu'il est impossible de ranger dans led. grenier et que j'ai estimé valoir soixante livres, cy 60 l.

Jai vacqué à ce que dessus depuis lad. heure de sept du matin jusques à celle presente de quatre de relevée sonnée par triple vacation le present inventaire sera continué demain sept heures du matin, et ont les Citoyens presens signé avec moi ainsi signé L'Huillier, Louaintier, Caty, Godet, Lalleman,Tramaux et Denis.

Et led. jour seize Brumaire sept heures du matin moi Commissaire du Bureau du domaine nationale du departement de Paris, me suis transporté en la maison dud. Lavoisier accompagné desd. Citoyens Louaintier et L'Huillier commissaires [55r] civils, ou etant j'ai continué de proceder aud. inventaire en presence desd. C.ens Lalleman, Denis et Tramaux ainsi qu'il suit:

Sur le pallier de l'escalier qui conduit à la cuisine

Item une table ovale et deux allezes en bois d'acajou, le pied en bois d'acajou et en bois de noyer prisée avec deux autres allezes en bois de chene soixante livres, cy 60 l.

Dans la Cuisine et dans plusieurs petites pieces en dependantes

Item trois pelles et trois pincettes, une paire de gros chenets, deux landrieres, un tourne broche avec sa chaine et son poids de fonte, trois broches et une pelle à glacer, le tout de fer, une lechefrite, un couperet, quatre hachoirs, une romaine en fer et un pezon avec poids et crochets de fer, deux chandeliers de fer, trois poeles aussi de fer, prisés quarante huit livres, cy 48 l.

Item deux grands et trois petits flambeaux de cuivre jaune, quatorze casseroles à queue, dix couvercles, une casserole ovale, deux poissonnieres, deux poeles d'office, un poelon d'office, deux petites marmites, une bassinoire, un moule à aspic, deux tourtieres, deux moyens chaudrons, une passoire en cylindre avec ses pieds en fer, un poelon, prisés ensemble cent quatre vingt livres 180 l.

[55v] Item quatre ecumoires, une cuillere à pot, une cuillere à coquemart, quatre cuilleres a degraisser, une cuillere percée prisés six livres, cy 6 l.

Item une cuisiniere et deux chandeliers en forme de lampe de fer blanc prisés dix livres, cy 10 l.

Item six bouillottes, un bassin, une cuvette, une boulle à ris, le tout d'etain, une petite fontaine de fer battu, une theyere de cuivre bronzé, prisés huit livres, cy 8 l.

Item grand buffet de bois peint, un petit buffet de bois d'hetre, une petite table en bois de noyer, un coffre servant de saloir, une petite table de bois blanc sur deux treteaux, prisés seize livres, cy 16 l.

Item un petit pressoir en bois d'orme et de noyer prisé avec six vieilles chaises foncées de paille quarante livres, cy 40 l.

Item un fleau de fer, quatre plateaux de cuivre avec leurs cordages, deux paires de chenets de fer avec leurs garnitures de cuivre doré prisés deux cent livres cy 200 l.

Item deux fontaines de grez, un mortier de marbre blanc, un pilon de bois, un baquet en tormelerie prisés trente livres, cy 30 l.

Item cinq marmites de terre, douze plats et trois douzaines d'assiettes de fayance commune, et vingt pieces de fayance et verrerie et differents ustencils [56r] de menage, prisés dix livres cy 10 l.

Lad. f.e Foulon cuisiniere de la C.ne Lavoisier, m'a déclaré que deux tables de cuisine, un bloc de bois, le pied du mortier, deux buffets et un petit corps d armoire de differens bois appartiennent au proprietaire de la maison, et ne dependent point de la succession dud. Lavoisier.[102]

Dans une petite piece au bas de l'escalier de la cuisine servant d'office

Item vingt un plateaux de cuivre argenté avec leur glaces dont une de cassée, prisés cinquante livres, cy 50 l.

Item vingt petits plats et assiettes de porcelaine, dix tasses à caffé et leurs soucoupes, six autres tasses à caffé et leurs

102 The house belonged to the banker Lecoulteux de La Noraye; see LC, vol. 7, 273–275.

soucoupes, deux petits pots à crême, deux soupieres et leurs plats le tout de porcelaine en mauvais etat et partie cassée prisés vingt quatre livres, cy 24 l.

Item deux grands sceaux, dix autres petits sceaux de fayance, dix huit tasses et leurs soucoupes de terre blanche, deux douzaines de verre à pied, douze autres pieces de fayance et verrerie en differens ustencils de menage prixés vingt livres, cy 20 l.

Item une paire de flambeaux de trois branches, une paire de flambeaux à double branches, une paire de flambeaux simples, le tout de cuivre doré, un autre flambeaux de cuivre argenté à trois bobeches, plusieurs bobeches [56v] de chandeliers de cuivre argenté, deux rechaux et moyen sceau, aussi en cuivre argenté, prisés avec une petite fontaine de cuivre peinte en gris ainsi que sa cuvette, soixante livres, cy 60 l.

Item cinq verres à lampes sur leurs pieds de bois prisés cinq livres, cy 5 l.

Item une etuve en bois de chene avec grilles de fer, un corps d'armoire à quatre portes en bois de sapin, un autre corps d'armoire et un buffet en bois de chêne prisés trente livres, cy 30 l.

Il s'est encore trouvé dans led. office de l'argenterie que j'ai transporté dans la salle au rez de chaussée ayant vue sur le Boulevard pour etre inventoriés avec d'autre argenterie qui y est, à cet effet j'ai reconnu sains et entiers levés et otés les scellés apposés sur les portes communiquant a lad. chambre et j'ai reapposé lesd. scellés comme ils etoient.

Dans la cave

Il s'est trouvé peu de vin que j'ai inventorié et prisé comme il suit, attendu que j'ai inutilement appelé un commissaire degustateur.

Item deux cent bouteilles de vin de bourgogne rouge prisées avec les flacons cent cinquante livres, cy 150 l.

Item neuf cent bouteilles de vin des Jurançon rouge et blanc, de Champagne rouge et blanc, de grave de Bordeaux de Palma rouge, du Cap blanc de Château [57r] Grillet, de Malvoisi, de Grenache, ded.[t] André de l'Hermitage et autres prisés ensemble dix huit cent livres, cy 1800 l.

Item quinze cent bouteilles de gros verre vuides, prisés trois cent livres, cy 300 l.

Dans une autre cave servant de bucher

Item six voies de bois à bruler, un tas de souches que j'ai evalué à deux voies et quelques morceaux de bois trainant, prisés cent trente livres, cy 130 l.

Sous la remise

Plusieurs parties de tables et de boiserie que j'ai estimé de valeur de quinze livres, cy 15 l.

De retour dans la salle au rez de chauesée ayant vue sur le jardin.

J'ai reconnu sains et entiers levés et otés les scellés du bureau apposés sur la porte d'une autre chambre à gauche ayant aussi vue sur le jardin et sur le Boulevard.

J'ai pareillement reconnus sains et entiers levés et otés les scellés apposés sur la porte d'une autre chambre ayant vue sur le Boulevard.

Etant entré dans lad. chambre à l'effet d'inventorier l'argenterie, les bijoux, le numeraire et les assignats qui sont dans lad. chambre, j'ai trouvé deux cartons sur lesquels est ecrit, monnoye en papier, à en juger par les etiquettes qui sont sur les paquets, un autre carton [57v] sur lequel est ecrit, numeraire et un autre carton garni de maroquin rouge sur lequel est une notte qui indique qu'il renferme des effets trouvés dans un grand secretaire à cilindre en bois d'acajou enlevé par le C. Alligre commissaire du departement pour l'usage de la comptabilité de l'instruction publique etablie au petit Luxembourg, sur lequels quatre cartons sont apposés les scellés du departement; j'ai aussi trouvé un petit coffre de bois de noyer dans lequel sont des assignats que j'ai laissé sans les compter pour les joindre s'il y a lieu à ceux qui doivent etre dans les cartons cidessus indiqués.

Jai vacqué à tout ce que dessus depuis lad. heure de sept du matin jusques à celle presente de une de relevée par double vacation, le present inventaire sera continué le dix sept du present mois avec un commissaire du departement que je requerrerai pour reconnoitre et lever les scellés apposés sur lesd. cartons, les volets & portes desd. chambres ont été refermés et j'ai reapposé les scellés du Bureau comme ils etoient et ont les Citoyens presens Signé avec moi ainsi signé L'Huillier, Louanitier, Denis, Tramaux, Lalleman, Godet, Caty et Bonnevie.

Et led. jour dix sept Brumaire neuf heures du matin, moi Commissaire du bureau du domaine national du departement de Paris me suis transporté accompagné desd. C.[ens] Louaintier et L'Huillier commissaires [58r] civils en la maison où demeuroit led. Lavoisier où etant j'ai continué de proceder aux operations cidessus commencées, en presence desd. C.[ens] Lalleman, Denis et Tramaux, ainsi qu'il suit:

Dans la salle au rez de chaussée ayant vue sur le jardin j'ai reconnu sains et entiers levés et otés les scellés du bureau apposés sur la porte d'une autre chambre à gauche ayant aussi vue sur le jardin et sur le Boulevard.

Jai pareillement reconnu sains et entiers levés et otés les scellés apposés sur la porte d'une autre chambre ayant vue sur le Boulevard.

Dans lad. chambre

Est comparu le C.[en] Pierre Claude Rauselan, Commissaire du departement, lequel a procédé a la reconnoissance et levée des scellés apposés sur les cartons cidessus designés, ainsi qu'il suit:

Les scellés apposés sur le carton garni de maroquin rouge ont été par led. C. Rauselan reconnus sains et entiers levés et otés dans led. Carton il s'est trouvé 1° un paquet enveloppé d'une feuille de papier sur laquelle est ecrit, remis à M^r. De la Chaume le huit juin mil sept cent quatre vingt douze par M^r. Lavoisier comme appartenant à M^r. Auger Deblarn coheritier de son pere avec M^r. Romand, dans led. paquet se sont trouvés neuf cent livres en dix huit feuilles d'assignats de cinq livres et une **[58v]** lettre du C. Romand adressée au C.^en Lavoisier et indiquant l'envoy d'une somme de huit cent quatre vingt douze livres un sol six deniers. J'ai laissé led. paquet dans l'etat ou il etoit, et j'employe cet article pour lad. somme de neuf cent livres sans les reclamations dud. C. Romand 900 l.

2° Cent quarante livres en assignats de dix et quinze sols, 140 l. 15 s.

3° – Neuf cent quatre vingt six livres deux sols en numeraire, 986 l. 2 s.

Lesd. assignats et numeraire ont été laissés dans led. carton ainsi que quelques papiers, à l'egard d'une paire de boucle d'argent, d'une lunette d'approche, montée en argent, quatre petites cuilleres de vermeille dont une cassée, de dix sept jettons d'argent et de vingt un petits brillants jaune, je les ai mis avec l'argenterie et les bijoux pour etre inventoriés ci après.

J'ai sur led. carton reapposé les scellés du Bureau au bout d'un ruban de fil.

Led. C. Rausselon a egalement reconnu et levé les scellés apposés sur les autres cartons cidessus designés ainsi que sur un autre carton garni de maroquin rouge sur lequel est une note qui m'a fait presumer qu'il y avoit dedans quelques objets à inventorier.

J'ai aussitot remplacé les scellés du departement par ceux du Bureau que j'ai apposé sur lesd. cartons avant d'apposer les scellés du Bureau sur led. carton garni de maroquin rouge, j'ai inventorié ce qui s'y trouve ainsi qu'il suit:

[59r] Il s'est trouvé 1° en pièces d'argent de six livres la somme de cinquante quatre livres, cy 54 l.

2° Et quinze livres cinq sols en assignats, 15 l. 5 s.

Plus un paquet cachêté sur lequel est ecrit, M^r. de Lauraguais, 21-7-3.

Plus un autre paquet sur lequel est ecrit, M^r. Dechatlest 4-6-5.

En enfin quelques papiers et des billets de la cidevant Maison de secours.

J'ai laissé tous lesd. objets dans led carton sur lequel j'ai apposé les scellés du Bureau.

Jai vacqué à ce que dessus depuis lad. heure de neuf du matin jusques à celle presente de midi les volets et les portes desd. chambres ont étés refermés et jai reapposé sur lesd. portes les scellés du Bureau comme ils etoient. La vacation pour la continuation du present inventaire est remise au vingt deux du present mois et ont les Citoyens presens signé avec led. Rauselan et moi ainsi signé L'Huillier, Louaintier, Rauselan, Lalleman, Denis, Tramaux, Caty, Godet et Bonnevie.

Et led. jour vingt deux brumaire six heures de relevée, moi commissaire susd. me suis transporté en la maison ou demeuroit led. condamné Lavoisier, avec lesd. C.^ens L'Huillier et Louaintier, Commissaires civils, où etant j'ai continué de procéder aux operations cidessus commencées en presence desd. C.^ens Lalleman **[59v]** Denis et Tramaux ainsi quil suit:

Dans la salle au rez de chausée ayant vue sur le jardin

J'ai reconnu sains et entiers levés et otés les scellés apposés sur la porte d'une autre chambre ayant aussi vue sur le jardin.

J'ai pareillement reconnu sains et entiers levés et otés les scellés apposés sur la porte d'une autre chambre ayant vue sur le Boulevard.

Dans lad. chambre
Est comparu le C. Jean Joseph Barriere orfevre jouaillier, demeurant à Parie rue du Mail N° 11 Section de Guillaume Tell, lequel à procédé à l'estimation des bijoux dependants de la succession dud. Lavoisier et de la communauté de biens qui a existé entre lui et lad. C.^ne sa veuve ainsi qu il suit.

J'ai reconnu sains et entiers levés et otés les scellés apposés sur la boete de bois de noyer indiquée ensuite de l'article, quarante quatre cidessus, et qui s'est trouvée etre un necessaire composé de douze pieces en argent comme boettes, rechaux &^a et le surplus des pieces dud. necessaire, garnie en argent le tout estimé par led. C. Barriere, avec quatre petites cuilleres à caffé aussi d'argent à la somme de huit cent livres, cy 800 l.

Item vingt un demi brillants jaunes estimés quatre vingt dix livres, cy 90 l.

À l'egard du surplus des bijoux et argenterie **[60r]** attendu la difficulté de transporter des poids et balances et que d'ailleurs la valeur en sera constatée par le procès verbal du transport qui en sera fait à la monnoye, il n'en sera fait ici qu'une simple description.

Item trente cinq cuilleres, trente trois fourchettes, une cuillere a soupe, douze cuilleres à ragout, une jatte, neuf cuilleres à caffé, une cuillere à sucre, une pince à sucre, sept lardoires, une caffetiere d'argent montée en fer, une autre petite caffetiere en marabout, quatre petites cuillieres de vermeil, dix sept jettons d'argent, une grande paire de boucles à entourage d'argent, une grande lunette d'approche garnie en argent, un cachet en or, cassé,vingt six gros boutons et dix sept petits en mauvais argent, douze couteaux de table à manche d'argent dans leur boete couverte en cuir.

Tous lesquels objets j'ai joint à la plaque d'argent et aux deux couteaux inventoriés sous les articles 41 et 42 cidessus, pour cet effet j'ai reconnu sains et entiers levés et otés les scellés apposés sur un bas d'armoire dans la salle ayant vue sur le jardin et etant auprès de celle où je suis, lesquels scellés j'ai aussitot reapposés avec le cachet du Bureau.

Et n'ayant plus rien à inventorier du fait dud. Citoyen Barriere il a signé en cet endroit et s'est **[60v]** retiré ainsi signé Barriere.

Ensuite j'ai continué de proceder comme il suit:

Dans lad. seconde chambre ayant vue sur le Boulevard

J'ai reconnu sains et entiers levés et otés les scellés apposés sur le carton sur lequel est ecrit numeraire.

Dans led. carton il s'est trouvé

1° Un sac contenant deux cent cinquante deux livres en pieces de six et trois livres, l'etiquete dud. sac est ainsi conçue, deux cent cinquante deux livres en ecus de trois livres et six livres de chez Mr. Dupommeau le quatorze août mil sept cent quatre vingt onze, lad. somme sera comprise au present inventaire sauf la reclamation de qui'il appartiendra, cy 252 l.

2° et la somme de cinq cent quatre vingt neuf livres dix sols en pieces d'or d'argent et monnoye ayant cours, cy 589 l. 10 s.

Plus il s'est encore trouvé dans led. carton cinquante deux pieces qui sont medailles anciennes monnoyes & monnoyes etrangeres lesquelles j'ai joint à l'argenterie, à l'effet de quoi j'ai levé les scellés apposés sur led. bas d'armoire et je les ai de suite reapposés.

J'ai remis le numeraire dans led. carton sur lequel j'ai reapposé les scellés du bureau.

Ensuite jai reconnus sains et entiers levés et otés les scellés apposés sur les deux autres cartons **[61r]** sur lesquels est ecrit, monnoye en papier à en juger par les etiquettes qui sont sur les paquets

J'ai en effet trouvé dans lesd. cartons plusieurs paquets cachetés avec le sceau de la section de l'Arsenal, sur chacun desquels sont des etiquettes qui m'ont fait presumer qu'il n'y à dans lesd. paquets que des billets patriotiques de lad. section de l'Arsenal qui ne peuvent servir que pour le compte qu'il peut y avoir à faire entre la Section de l'Arsenal et la succession dud. Lavoisier. En consequence j'ai laissé lesd. paquets dans l'etat où je les ai trouvés et j'ai reapposé les scellés sur lesd. cartons pour le transport en etre fait aux archives du Bureau avec les autres papiers dud. Lavoisier et en faire tel usage qu'il appartiendra.

Item dans un petit coffre de bois de noyer il s'est trouvé en assignats de cinq livres et au dessous, la somme de trois mille deux cent huit livres, cy 3208 l.

J'ai apposé les scellés du Bureau sur led. coffre et j'ai mis une etiquet indiquant la somme qui y est renfermée.

À l'instant led. C. Lalleman aud. nom à dit que sans prejudicier aux reclamations que la Citoyenne Lavoisier pourra faire d'aucuns **[61v]** des objets cidessus inventoriés, il demande des à present qu'il soit fait remise à lad. C.ne Lavoisier 1°du necessaire inventorié article cent quarante quatre Cidessus 2° et d'un grand tableau qui represente son portrait et celui dud. deffunt son mary,[103] led. tableau est dans la chambre où j'opere actuellement.

Et attendu que tous les meubles et effets etant dans le logement desd. C.en et C.ne Lavoisier sont inventoriés excepté les objets de mineralogie et de peinture pour lesquelles j'ai inutilement requis des Commissaires de la Commission Temporaire des arts qu'à l'egard des objets de chimie, le Citoyen Leblanc, membre de la ditte Commission, en fait actuellement l'inventaire en vertu de pouvoirs particuliers, avec le C.en Binay mon collegue, j'ai clos et arrêté mon present proces verbal dans l'etat où il se trouve sauf à y ajouter ce qu'il appartiendra.

Indication des papiers à inventorier.

Le transport des papiers dud. Lavoisier aux archives du Bureau ne peut se faire qu'après l'enlevement que la Commission Temporaire des arts doit faire faire, de ceux concernant les arts et sciences, autrement on mettroit beaucoup de desordre dans lesd. papiers.

Les portes et les fenetres et volets desd. salles ont été refermés et j'ai reapposé les **[62r]** scellés du Bureau comme ils etoient jetés ou laissés en la garde desd. C.ens Caty et Godet auxquels j'ai enjoint d'etre exacts à leur poste et de ne laisser sortir aucuns meubles ni paquets appartenant à la succession dud. Lavoisier que sur la representation d'un ordre suffisant.

J'ai vacqué jusques à neuf heures de relevée sonnées, et ont ent lesd. C.ens Caty et Godet, Lalleman, Denis et Tramaux signés avec lesd. commissaires civils et moi ainsi signés L'Huillier, Louaintier, Caty, Godet, Lalleman, Denis, Tramaux et Bonnevie, et au dessus est ecrit enregistré a Paris le vingt huit Brumaire de la Troisieme année.[104]

F° 133 N° caze 7 Debet cent deux livres dix sols signé Macoan.

Certiffié pour copie conforme à l'original delivrée par nous membres du Bureau du domaine national du departement de Paris ce vingt quatre Ventose l'an Trois[105] de la Republique francoise une et indivisible signé Guillotin et Tennesson.

15e piece de la cotte 2

Inventaire des cartes et plans geographiques trouvés dans la maison de Lavoisier, condamné Boulevard de la Madelaine mis en reserve par la Commission Temporaire des Arts, et rendus à la **[62v]** Citoyenne Lavoisier par la Commission des Travaux publics conformement à l arrêté du comité d'instruction publique en date du dix Prairial l'an Troisieme[106] de la Republique francoise.

Collection de la carte de France de Cassini collée sur toille, à charmerel dans trente boetes in quarto contenant cent vingt huit feuilles seulement, et en mauvais etat mais sur lesquelles on à indiqué la nature des terres par des caracteres mineralogiques.[107]

Un porte feuille grand in folio, contenant vingt trois cartes manuscrites d'une carte de france reduite d'après la carte de

103 David's portrait of Monsieur and Madame Lavoisier.

104 18 November 1794.

105 19 March 1795.

106 29 May 1795.

107 A set of Cassini's map not entirely corresponding to this description is now preserved at the Cornell University Library; see Beretta (1995), p. 134.

Cassini et que l'on croit avoir été destinée à former une carte mineralogique.[108]

Un autre porte feuille semblable, contenant quelques plans de paris, plusieurs exemplaires d'une carte de la France par Zanoni[109] in 4° et des epreuves d'une autre carte de la france projettée pour l'usage des fermes.

Un autre porte feuille conenant trente cartes communes et en mauvais etat, des diverses parties de l'Europe et de la France.

Atlas formé de quinze cartes de Delisle un volume in f° en parchemin verd.

Plan de Paris par Jaillot, relié en veau format in quarto.[110]

[63r] Geographie ancienne de Danville pour l'histoire de Rollin 1er volume in quarto.

Carte mineralogique de la France par Dupain Triel, in quarto en feuilles.

Tableau de la navigation interieure de la France et carte generale des fleuves et rivieres de France par Dupain Triel, sur toille et une sur gorge.

Cinq feuilles de la carte de France en vingt feuilles par Capitaine, sur lesquelles on a rapporté des caracteres mineralogiques mais en mauvais etat.[111]

Carte de la France en quatre feuilles sur toille par Capitaine.

Deux feuilles du cours du Rhin, par Le Rouge, sur lesquelles on à appliqué dee caracteres mineralogiques.

France par Bonne une feuille sur toille.

Plan de Paris par Roussel neuf feuilles.[112]

Bretagne, par Lafosse, sur toille.

Orleannois par Juillot, sur toille.

Planispheres celeste par C. Degy en deux feuilles.

Provinces unies par Lafosse une feuille.

Environs de Paris premiere feuille de la carte de France.

Cours du Rhin du fort Louis à Mayance.

Plan du canal de Douay à Arras manuscrit.

Plan du fief de Longpont, manuscrit.

[63v] Tableau methodique des mineraux, manuscrit sur toille.

La generalité de Paris par Denos un volume in quarto.

Un model de vaisseau avec tous ses grements mais gaté par l'humidité et en mauvais etat.

Le present inventaire fait en presence du C. Binay commissaire du Bureau du domaine national, qui en a inseré copie dans son procès verbal, et après l'estimation faite par le C. Vignon, ces objets ont été remis au C. Nadran pour etre transportés au depot national des cartes et plans, Maison d'Harcourt Rue de l'Université, ce six vendemiaire l'an Trois[113] de la Republique francoise une et indivisible.

Signé Buache

N.ta Le model du vaisseau porté au N° 26 a et remis au depot de la Commission de la marine et des colonies, Rue de Varennes Maison Dorset, vis à vis La Rue de Bourgogne signé Buache.

Pour Copie Conforme signé Oudry Secretaire G.al.

Recu de la Commission des Travaux publics d'après l'arrêté du Comité d instruction publique en datte du dix Prairial l'an Trois.e[114] de la Republique.

Les objets mentionnés au present etat, excepté ceux portés aux N.os 10 & 22 envoyés aux differentes commissions nommées dans la note en marge cidessus et ceux des N.os 25 & 24 de la recherche desquels **[64r]** on s'occupe.

À Paris ce vingt neuf Thermidor l'an Trois.e[115] de la République françoise.

Approuvé pour etre rendu conformement à l'arrêté du Comité d'instruction Public en datte du dix prairial pour copié conforme signé Lecamus.

Pour copie conforme la Comission signé Lecamus.

16e piece de la cotte 2
Etat général des livres du Citoyen Lavoisier qui avoient été mis à la disposition de l'agence des mines et qui ont été restitués le vingt six Thermidor an Trois[116] à la Citoyenne Lavoisier, en execution de l'arrêté du Comité d'instruction publique en datte du dix Prairial, an Trois[117] de la Republique.[118]

[72v]
Onze Prairial[119] huit heures du matin en consequence de l'indication à ce jour lieu et heures les notaires soussignes se sont transportés en la demeure de lad. C.ne Lavoisier avec

108 These are the maps which would be published in the *Atlas minéralogique de la France*, now preserved at the Cornell University Library, Lavoisier QE 381.F8 G93+. The maps, hand-coloured, were partially annotated by Lavoisier.

109 Giovanni Antonio Rizzi Zannoni.

110 Bernard Antoine Jaillot, *Plan de la ville de Paris et de ses faubourgs / dédié au Roi par … B. Jaillot … 1748 … corrigé et augmenté* (Paris, 1778).

111 This and subsequent maps by Louis Capitaine are listed in the catalogue of the BNF at the link https://data.bnf.fr/15302409/louis_capitaine/.

112 *Paris, ses fauxbourgs et ses environs où se trouve le détail des villages, châteaux, grands chemins pavez et autres, des hauteurs, bois, vignes, terres et prez, levez géométriquement / par le Sr Roussel* (Paris, 1730–1739).

113 September 27, 1794.

114 29 May 1795.

115 17 August 1795.

116 13 August 1795.

117 29 May 1795.

118 Fols 64r to 72r contain a list of scientific books from Lavoisier's library (here omitted) on mineralogy, metallurgy, chemistry, physics, natural history, mathematics, geography, and travel, economy, and periodicals, which I identified and whose fate I reconstructed in detail in Beretta (1995). The quantitative and qualitative evalutation of this collection from fols 74v–75r has been transcribed.

119 30 May 1795.

led. C. Vallet à l'effet de continuer le present inventaire même requestes, presences et demeures que cidessus ainsi quil suit.

Prisée des objets compris aux inventaires annexés qui avaient été decrits auxd. inventaires sans etre prisés:

Premierement tous les objets de chimie detaillés dans l'inventaire du vingt six Prairial l'an Second[120] formant la septieme piece de la cotte deux du present inventaire ont été à l'instant recollés et s'etant trouvé en nature par l'evenement dudit recollement ont été prisés par led. Vallet d'après L'avis du C. Charles à ce present lequel a mis sur led. inventaire. La valeur de chacque objet article par article et sommé enfin dud. inventaire qui a été signé par lui à la somme de quatre mille quatre cent cinquante trois livres cy 4453 l.

Item les vazes et objets de chimie compris dans l'inventaire du dix neuf Brumaire[121] formant la huitieme piece de la Cotte deux du present inventaire ont été à l'instant recolés et s étant trouvés, en nature par l'evenement dud. recollement ont été prisés par led. C. Vallet de l'avis dud. C. Charles lequel à mis sur led. inventaire la valeur de chacque objet article par article et somme enfin d'icelui qui a été par lui **[73r]** signé à la somme de troix mille six cent soixante dix sept livres, cy 3677 l.

Item tous les instruments de phisique et de chimie detaillés en l'inventaire formant la troisieme piece de la cotte deux du present inventaire, lesquelles après recollement faits sur led. inventaire s étant trouvés en nature ont été par led. Vallet d'après l'avis dud. C. Charles lequel a mis sur led. inventaire la valeur de chacque objet article par article et sommés en fin d'icelui qui a été par lui signé a la somme de vingt mille trois cent vingt sept livres, cy 20327 l.

Item une collection incomplette de morceaux d'histoire naturelle en coquillage, mineraux, mine metallique, vegetaux, petrification, madrepore et autres. Le tout en desordre et en mauvais etat mentionné en l'inventaire du vingt quatre vendemiaire an trois faisant la quatorzieme pièce de la cotte deux prisées huit cent livres, cy 800 l.

Item seize modelles en platre mentionnes aussi au même inventaire prisés vingt quatre livres, cy 24 l.

Item un tableau peint sur toille par Coipel[122] representant l'amour precepteur[123] un autre **[73v]** autre tableau aussi peint sur toille representant la mort des enfans de Sedecius, un autre tableau aussi peint sur toille representant le depart de Coriolan pour l'exil.[124] Ces deux derniers peint par Geraudais,[125] tous trois dans leurs bordures de bois dorés mentionnés aussi au même inventaire prisés ensemble six cent livres cy 600 l.

Item trente cinq cuilleres trente trois fourchettes à bouche, une cuillere à soupe, douze cuilleres à ragout, une cuillere a sucre et six attelais, le tout d'argent poinçon de Paris pesant ensemble trente marcs cinq onces six gros, fixé à juste valeur et sans crue comme vaisselle platte à raison de cinquante livres treize sols six deniers. Le marc de laquelle argenterie il est aussi fait mention dans led. inventaire revenant lad. quantité aud. prix à la somme de quinze cent cinquante cinq livres huit sols six deniers, cy 1555 l. 8 s. 6 d.

Item neuf cuilleres à caffé, quatre autres cuilleres à caffé de vermeille, le tout d'argent d'Allemagne pesant cinq onces deux gros, fixé a juste valeur et sans crue à raison de trente neuf livres dix sols neuf deniers, de laquelle argenterie il est pareillement fait mention dans led. inventaire revenant lad. quantité aud. prix à la somme de vingt cinq livres dix huit sols dix **[74r]** deniers, cy 25 l. 18 s. 10 d.

Item deux caffetieres et un compotier d'argent poinçon de Paris pesant ensemble quatre marcs quatre onces un gros, fixé à juste valeur et sans crue comme vaisselle montée à raison de cinquante livres deux sols quatre deniers le marc revenant lad. quantité aud. prix à la somme de deux cent vingt six livres six sols deux deniers, 226 l. 6 s. 2 d.

Item dix sept jettons d'argent pesant un marc deux onces cinq gros fixé a juste valeur et sans crue à raison de cinquante livres dix sept sols six deniers le marc revenant lad. quantité aud. prix à la somme de soixante sept livres six sols, cy 67 l. 6 s.

Item une paire de boucle de souliers à tours d'argent chappe de fer,[126] une lunette d'approche de peau de chien de mer garnie en argent, un cachet d'or brisé vingt six gros et dix sept petits boutons montés en argent et douze couteaux à manchè d'argent mastiqués prisés le tout ensemble cent quarante quatre livres, cy 144 l.

Item un benitier d'argent prisé cent livres, cy 100 l.

Item vingt quatre medailles d'argent, quinze petites medailles et monnoyes etrangeres en or et treize pieces de monnoye en cuivre, prisés le tout ensemble trois cent quatorze livres, cy 314 l. **[74v]**

120 14 June 1794.

121 9 November 1794.

122 The French artist Charles-Antoine Coypel (1694–1752) son of Antoine (1661–1722).

123 *L'Amour Precepteur*; 1740, private collection. Private communication by David Pullins (The Metropolitan Museum).

124 Anne-Louis Girodet de Roucy-Trioson's *concours de Rome* submissions, serving as pendants, described as 'le depart de Corioluas [*sic*] pour l'Exil' (*Les Adieux de Coriolan à sa famille*, 1786; The National Gallery, Washington) and 'la mort des Enfans de Sedeias [*sic*]' (*Nabuchodonos.or fait tuer les enfants de Sédécias en presence de leur père*, 1787; musée Tessé, Le Mans. Private communication by David Pullins (The Metropolitan Museum).

125 It is in fact Girodet de Roucy-Trioson.

126 This description seems to fit that of the shoe buckles preserved in the Kroch Library of the Cornell University. Lavoisier manuscript collection, 4712, Box 29.

Item une bassine et sa spatule d'argent pesant vingt neuf marcs deux onces deux gros fixé à juste valeur et sans crue comme vaisselle plate à raison de cinquante livres treize sols six deniers, le marc revenant lad. quantité aud. prix à la somme de quatorze cent soixante treize livres deux sols six deniers, 1473 l. 2 s. 6 d.

Item les cartes et plans geographiques detailles en l'inventaire du dix praireal formant la quinzieme piece de la cotte deux du present inventaire après avoir été recollés et s'étant trouvés en nature ont été prisés par led. Vallet d'après l'avis qu'il à pris du C. Santus libraire la somme de deux cent livres, cy 200 l.

Item recollement fait des livres de mineralogie et métallurgie detaillés en l'etat formant la seizieme pièce de la cotte deux du present inventaire au nombre de cent huit volumes de differents formats ont été aussi de l'avis dud. C. Santus prisés par led. Vallet la somme de de deux cent dix livres, cy 210 l.

Item recollement pareillement fait des livres de chimie detaillés au même etat au nombre de trente quatre volumes de differens formats d'après l'avis dud. C. Santus ont été prisés par led. Vallet la somme de trente livres, cy 30 l.

Item recollement pareillement fait des livres de phisique detaillés au même etat au nombre de cent [75r] dix huit volumes de differens formats s'étant trouvés en nature ont été du même avis prisés par led. Vallet la somme de cent livres, cy 100 l.

Item recollement pareillement fait des livres d'histoire naturelle detailles au même etat au nombre de trente sept volumes aussi de differents formats s'etant trouvés en nature ont été du même avis prisés la somme de quatre vingt seize livres, cy 96 l.

Item recollement pareillement fait des livres de mathematiques detaillés au même etat au nombre de neuf volumes s'etant trouvés en nature ont été du même avis prisés la somme de dix huit livres, cy 18 l.

Item recollement pareillement fait des livres de geographie et voyages au nombre de quarante cinq volumes detaillés au même etat ont été du même avis prisés quarante huit livres, cy 48 l.

Item et enfin recollement fait des livres d'economie et collection, au nombre de deux cent cinq volumes designés dans le même etat s'etant trouvés en nature ont été du même avis prisés la somme de deux cent livres, cy 200 l.

Continuation de l'inventaire pour les objets non inventoriés.

Nous a été ensuite representé par lad v.[e] Lavoisier pour etre compris aud. inventaire les objets ci après detaillés.

Item sept aulnes de drap de Silezie prisés trente cinq livres, cy 35 l. [75v]

Item quatre aulnes de drap vert dragon prisés quatre vingt seize livres, cy 96 l.

Item une aulne de drap mélé prisée vingt quatre livres, cy 24 l.

Item quatre aulnes de velours de cotton bleue reteint prisées vingt quatre livres, cy 24 l.

Item huit aulnes de velours de cotton noir reteint prisées trente deux livres, cy 32 l.

Item trois aulnes de drap bleue commun prisées quarante cinq livres, cy 45 l.

Item huit aulnes en deux coupons de siamoise rayée prisées seize livres, cy 16 l.

Item six pieces de Nanquin prisées trente livres, cy 30 l.

Item un manteau de cocher de gros drap bleue soixante douze livres, cy 72 l.

Item cinq tapis de pied dont un grand en bon etat et les quatre autres petits et tres vieux prisés soixante douze livres, cy 72 l.

Il a été vacqué à tout ce que dessus depuis lad. heure de huit jusques à celle de deux sonnées par double vacations à la requisition des parties ce fait les objets mobiliers cidessus decrits recollés et inventoriés sont restés du consentement des autres parties en la garde et possession de lad. C.[ne] Lavoisier et la vacation du present inventaire à été remise à cejourd huy quatre heures de relevée et ont signé en pareille endroit la minute des presentes et le ded. Jour [76r] onze Prairial de l'an Quatre,[127] quatre heures de relevée, Il va etre par les notaires soussignés, es mêmes requêtes et presence que cidvant, en consequence de l'ajournement pris par la precedente vacation, procédé à la continuation du present inventaire de la maniere suivante:

Les parties ont requis lesd. notaires de faire sur l'etat et inventaire de papiers formant la dix sept.[e] pièce de la cotte deux cidessus, la verification des liasses qui y sont ennoncées et que lad. v.[e] Lavoisier à l'instant representées afin de voir si tous les papiers compris aud. etat se trouvent dans lesd. liasses representées et de pouvoir ensuite inventorier ces papiers suivant l'ordre naturel qu'il convient de leur donner et qui n'a pas été observé lors de la confection de cet etat, lesd. papiers ayant été simplement enliassés avec divers melanges qui en rendroient le depouillement tres difficile, en consequence de laquelle requisition, lesd. notaires ont procedé a la verification du tout par l'evennement de laquelle il s'est trouvé que toutes ses liasses et pieces portées audit etat existent à l'exception de l'expedition en parchemin du partage de la succession de la C.[ne] Lalaure, et celle de l'acte du vingt six aoûst mil sept cent quatre vingt fait entre son mary et le C. Lalaure concernant son desistem.[t] de l'usufruit des biens de la succession de sa femme.[128]

127 30 May 1796.

128 The inventory continues up to fol. 92v, with a list of documents concerning the properties and estates belonging to Lavoisier with no direct relevance to the physical and scientific objects preserved in the residence.

APPENDIX 2

Inventory of Lavoisier's Laboratory by Nicolas Leblanc (1794)

On 6 February 1794, a decree issued by the Convention Nationale ordered the members of the Commission temporaire des arts and the Commission d'instruction publique to "inventorier et faire réunir dans des dêpots convenables, tous les objets de sciences et arts provenant, soit des maisons ci-devant religieuses, soit des émigrés, soit des conspirateurs, soit de la liste civile".[1] The commission charged with this daunting task was composed of several of Lavoisier's former colleagues at the Académie des sciences and some of his most important instrument makers. Among its 32 members we find Claude Louis Berthollet, Ferdinand Berthoud, Gaspard Monge, Louis Marie Richard, Jacques Alexandre César Charles, Félix Vicq d'Azyr, René Louiche Desfontaines, Etienne Lenoir, Nicolas Fortin, Jean Henri Hassenfratz, and Nicolas Leblanc; all of them, with the exception of Leblanc, collaborated with Lavoisier on various scientific and technological projects. It took some time before the commission could focus on Lavoisier's collection, but it seems certain that the work began relatively early in the process. As shown in the inventory published in Appendix 1, in early 1794 the apothecary Quinquet was charged by the commission with compiling an inventory of the pharmaceutical items that were to be confiscated at Lavoisier's residence on the Boulevard de la Madeleine.

After this preliminary investigation of Lavoisier's chemical laboratory made on 14 June 1794, Claude Louis Berthollet and Nicolas Leblanc began to compile a more comprehensive list of instruments.[2] However, the enormous quantity of instruments, tools, vessels, and chemical substances made a more detailed inquiry and inventory necessary. Accordingly, on 5 November 1794, the Commission d'instruction publique asks Leblanc to draw up "l'inventaire détaillé des objets qui se trouvent dans le laboratoire de chimie de Lavoisier".[3] During the first visit to Lavoisier's residence, on 9 November, Leblanc counted 2,145 retorts, cucurbits, tanks, stills, and pelicans, as well as 3,129 jars and 549 flasks containing large quantities of different chemical substances. During the following visit, Leblanc listed 112 glasses, more than 820 vessels, and 600 small vases containing unidentified substances of which 150 bore labels; 400 glass tubes and alembics of different dimensions, several aerometers, and some 300 bottles. Leblanc's inventory of about 7,000 objects did not take into account all those precision instruments that he was unable to identify but whose importance he recognised.[4]

What follows is the transcription of Leblanc's inventory dated 5 November 1794, published by Auguste Anastasi at the end of his biography of Leblanc.[5]

The interest of these documents lies both in the evaluation of the chemical apparatus and the additional information provided by Leblanc on the chemicals he found in Lavoisier's laboratory. In addition to those listed in the documents presented in Appendix 1 there are varying quantities of the following substances: sulphuric acid, oxygenated muriatic acid, antimony, sulphate of potash, sal ammoniac, resins, manganese, charcoal, enamels (*émaux*), crystals, amalgams, metals such as lead and bismuth, phosphorous, soda, aluminium sulphate, arsenic, iron, and sulphates.

INVENTAIRE DU LABORATOIRE DE LAVOISIER

Le 19 Brumaire, l'an III[6] de la République française une et indivisible.

Chargé par l'arrêté du Comité d'instruction publique en date du 15, article IV, de faire l'inventaire détaillé des objets qui se trouvent dans le laboratoire de chimie de Lavoisier, boulevard de la Madeleine, j'ai, ce même jour dix-neuf, accompagné des citoyens Binay, commissaire du bureau de l'agence du domaine national de Paris; Louaintier et Guillemard, commissaires civils, de la section des Piques; Caty et Gaudet, tous deux gardiens aux scellés dans ladite maison du condamné Lavoisier, procédé à ce même inventaire en retirant des caveaux où ces mêmes objets se trouvaient en plus grande partie amoncelés et confusément mêlés; j'ai, dis-je, fait transporter ces objets dans une pièce au rez-de-chaussée, pour y être établis par séries suivant l'ordre distinctif de ces mêmes objets; d'où il résulte

1° Cornues de toutes grandeurs en verre blanc, creusets de différentes grandeurs et espèces, couvercles, cucurbites, aludelles, chapiteaux, entonnoirs, ballons de toutes grandeurs et à différentes tubulures, matras, récipients, allonges, vases cylindriques, alambics et pélicans, éolipiles, flacons de toutes grandeurs bouchés à l'émeri, flacons à un ou plusieurs goulots, grande et petite capacité.

1 Anastasi (1884), p. 33.
2 MS Berthollet, Leblanc (1794).
3 Anastasi (1884), p. 221.
4 For these see Appendix 3.
5 Anastasi (1884), pp. 221–230.
6 9 November 1794.

 | DOI:10.1163/9789004511217_009

En tout 2145 pièces environ, estimées 1227 livres, Par suite, le 21 brumaire,[7] accompagné des citoyens Dinay, commissaire de l'agence du domaine national de Paris Louaintier et L'Huillier, commissaires civils de la section des Piques; Caty et Gaudet, gardiens des scellés, tous présents;

Objets transportés dans une pièce voisine au rez-de-chaussée, bocaux de différentes grandeurs, bouteilles rondes, petits mortiers, pilons, bocaux de forme à goulots renversés, flacons à deux goulots, plateaux, couvercles et capsules, cylindres à deux couvertures, pélicans, ballons à plusieurs tubulures, vases en forme de récipient pour des expériences au mercure, petits évaporatoires et seaux, creusets de petits volumes, cônes, lanternes, pièces en manchons, allonges courbées;

En tout 3129 pièces environ, estimées 848 livres 1 sol.

Le 22,[8] accompagné des mêmes commissaires et gardiens que la veille, opérant par en bas dans les pièces dites laboratoires, d'abord sur la suite des grandes tablettes qui se trouvent à hauteur en commençant par la droite en entrant.

Vases contenant des liqueurs alcalines en différents étals, flacons dont onze bouchés à l'émeri et un petit ballon, le tout contenant de l'acide sulfurique.

Flacons contenant de l'acide aciteux [*sic*], quelques parties de potasse, etc.

Vases contenant du nitre cristallisé;

Flacons contenant différents sulfates et sels de peu de valeur;

Vases contenant du nitre cristallisé;

Flacons bouchés à l'émeri contenant différents produits;

Vases de différentes grandeurs contenant différents produits; suite d'une analyse sur la girofle de différents pays;

Vases contenant de l'antimoine diaphonique, sulfate de potasse, sel ammoniac, nitrate de chaux; Bocaux de moyenne grandeur, contenant un assortiment de gommes, résines, telles que laque, galipot, sandaraque, copal, karabé, etc.;

48 flacons bouchés à l'émeri, depuis la capacité d'environ 10 pintes jusqu'à demi-selier, contenant l'acide muriatique oxygéné et ordinaire, l'acide nitrique aussi dans différents états.

Vases de gros verre, flacons contenant des produits de sucre fermenté, de l'huile essentielle de térébenthine et vinaigre concentré par la gelée. Sur les tablettes de la petite armoire enfoncée dans le mur de la seconde pièce, différents vases contenant différents produits avec leurs étiquettes; Une partie de verrerie dont quelques pièces en cristal;

En tout 549 pièces, pour 838 livres 2 sols.

Cette partie d'inventaire et prisée ayant été close, nous, commissaires et artistes priseurs soussignés, certifions que les objets sus-énoncés se montent au nombre d'environ quatre mille neuf cent quarante pièces, et les prisées se montent à environ deux mille neuf cent trois livres.
Signé Leblanc, Melan, L'Huillier, Louaintier,
Binay, Godet, Caty, Fourcy.

Observations[9]
Il reste encore une grande partie d'objets à reconnaître et à décrire, qui feront une seconde partie d'inventaire; mais il parait nécessaire de prévenir qu'il se trouve parmi ces objets des séries de substances, telles que les charbons, les fontes, les cristaux blancs et coloriés, les émaux, les amalgames, etc., qui peuvent encore mériter quelque attention. On les a rassemblés autant qu'il a été possible de le faire. Il existe en outre une quantité assez considérable d'objets sans étiquette et qui généralement ne pourraient être reconnus qu'au moyen d'un examen particulier, qui ne peut se faire dans ce travail d'inventaire qui nous occupe. Il est possible que, parmi ces derniers objets, il se trouve des choses importantes pour l'art ou bien sous tout autre rapport. La suite de cet inventaire sera terminée d'ici à très peu de jours.

Signé LEBLANC.

Le 27 et le 29,[10] accompagné des commissaires de l'agence et de la section, suite et seconde partie de l'inventaire détaillé des objets du laboratoire de Lavoisier.

Dans la pièce à droite du second antichambre, environ cent soixante-dix livres de mercure coulant dans neuf bouteilles ordinaires et environ soixante livres d'oxyde rouge de mercure par l'acide nitrique en douze ou quatorze vases différents, le tout estimé deux mille trente livres, bouteilles de pintes, grands bocaux, fioles à médecine, boite contenant des tubes et petits matras. 112 verres & expériences. 620 courtines. Clarquets et petits bocaux. Un tonneau rempli de manganèse (150 liv.) deux tonneaux remplis de charbons de terre dans deux états différents; dans l'antichambre qui précède le magasin, une presse; dans la première pièce du laboratoire, 4 fourneaux à réverbère assortis, un fourneau de Macquer, une partie en fer comprenant des chaudières, fourneaux à atanor, étouffoirs, cornes en tôle, trépieds, chevrettes, moulins à café, spatules, grands et petits mortiers de fonte avec leurs pilons, grils, une forte cornue en fonte en deux pièces (ci 90 liv.). Partie de cuivre comprenant un grand alambic, des bassines, casseroles, seaux-évaporatoires à bain-marie, un mortier de bronze, deux grilles d'étain, etc. (ci 430 liv.). Différents métaux, tels que plomb, bismuth, zinc et débris d'expériences, une petite fontaine en grès, différents petits supports en bois de gaïac, dans une boite de bois, des

7 11 November 1794.
8 22 November 1794.
9 In Leblanc's hand.
10 17–18 November 1794.

bouteilles de gomme élastique, dans une autre boitte pareille 3 petites planches de cuivre gravées à l'usage d'un ouvrage de chimie.

Dans la pièce du fond, une cuve à expériences hydropneumatiques dont le fond est en cuivre verni et monté sur un châssis de bois, cette cuve ayant 4 pieds de long, 2 pieds de large et 14 pouces de profondeur, garnie de deux tablettes et supports en cuivre étamé (ci 200 liv.). Deux cuvettes en marbre qui paraissent avoir été appropriées aux expériences au mercure.

Plusieurs morceaux de glace brute, parmi lesquels il s'en trouve une provenant de celle qui a servi à former la lentille connue sous le nom de lentille de Trudaine (ci 8 liv.).[11]

Sur les tablettes d'une espèce d'armoire enfoncée dans le mur, bocal contenant de la platine (2 livres environ, ci 400 liv.).

Un flacon enveloppé d'une boite en fer-blanc, beau phosphore (1 livre environ, ci 50 liv.); un grand bocal contenant une nouvelle partie de nitre (8 livres, ci 9 liv.).

Capsule de verre contenant de la soude purifiée (3 livres à peu près, ci 3 liv.). Dans un petit bocal, environ 4 livres d'antimoine cru.

Dans un grand bocal, 6 livres environ de sulfate d'alumine (ci 10 liv.). Une grande bouteille contenant environ 6 livres d'huile de térébenthine (ci 9 liv.). Une balance composée de son fléau en fer, les cordes et quatre plateaux en cuivre (ci 95 liv.).

Total des prisées de ce jour 3717 liv. 16 sols.

Le 29[12] du même mois, étant accompagné des commissaires du bureau de l'agence du domaine national de Paris et civils de la section des Piques.

Dans la première des deux pièces du laboratoire, 120 vases de différentes sortes, contenant des matières et produits, comme particulièrement une série de fonte (ci 50 liv.).

A gauche et derrière la porte d'entrée, 350 vases en verre blanc contenant des objets non désignés (ci 50 l.). A droite et au-dessus de la tablette précédente, plus de 350 vases différents, contenant différents objets et quelques aréomètres (ci 40 liv.). Dans la pièce qui suit, sur la tablette à hauteur d'appui, quelques petits vases contenant de petites parties de poudre à canon sur la rangée au-dessus, 300 vases, contenant des objets non désignés, c'est-à-dire sans étiquettes.

Sur la troisième rangée, 150 vases contenant des parties d'amalgame, alliages, sulfates, parties d'arsenic, oxydes de plomb, mine de fer, essais de cristaux blancs et coloriés (ci 22 liv. 10 sols).

Le long du mur suivant, dans la même pièce, sur la tablette à hauteur d'appui, 300 vases, flacons en verre blanc bouchés à l'émeri, contenant des objets non désignés.

Trois paniers contenant de petits matras et bocaux au nombre de 400 environ, grands et petits tubes en verre blanc (111 [*livres*?]).

Dans différents endroits dans les deux pièces du laboratoire, bouteilles depuis la capacité d'environ 10 pintes jusqu'à 2 pintes, verre noir, 100 environ, 11 grands pots de faïence, cuvettes; dans un panier au rez-de-chaussée, 150 petites cornues et petits matras en verre blanc.

Total des prix auxquels ont été estimés les matières ou substances et les verres quatre mille trois cent trente-quatre livres seize sols (ci 4334 liv. 10 sols).

Nous soussignés, l'un marchand faïencier et de verrerie pour la physique et la chimie, demeurant rue Quincampoix, maison de Beaufort, l'autre apothicaire chimiste, demeurant rue Coquillière, certifions avoir été requis par le citoyen Leblanc et avoir fait l'estimation et prisée des objets énoncés sur les procès verbaux revêtus de nos signatures, dans le laboratoire et magasins de Lavoisier, maison de Gouteux-Lamorage, Boulevard de la Madeleine, le tout en présence des commissaires désignés auxdits procès-verbaux.
A Paris, le premier Frimaire, l'an Troisième[13] de la République française une et indivisible.

Signé Fourcy et Melan.

Toutes les parties de vases neufs, principalement la verrerie et les creusets, sont dans les deux pièces du rez-de-chaussée une partie est restée en bas dans le magasin.

Les parties comprises dans les deux pièces du laboratoire sont en général les substances et les produits avec différents instruments.

J'ai désigné sous le nom d'objets connus ceux qui, il l'œil simple ou par leur étiquette, pouvaient en effet être facilement reconnus, et je les ai séparés des autres avec un arrangement par séries et à peu près.

J'ai désigné sous le nom d'inconnus ou de peu d'importance ceux qui manquaient d'étiquette, ou qui pouvaient exiger une plus longue attention ou même des opérations pour pouvoir les reconnaître; chaque rangée dans le laboratoire porte des inscriptions indicatives.

Je crois devoir faire remarquer que la partie des objets appelés inconnus ou non désignés est très étendue, qu'elle comprend très certainement des parties intéressantes qu'il serait bon d'examiner attentivement. Il serait possible que quelques-unes d'elles appartinssent à des opérations importantes et qui peuvent se

11 It is extremely interesting that Lavoisier owned a piece of Trudaine's large burning lens which was made by Charpentier in 1775 and used by the chemistry section of the Académie des Sciences between 1775 and 1776. On this, see the catalogue MAM inventory no. 19884-0000-, see pp. 405–406.

12 19 November 1794.

13 21 November 1794.

trouver dans les différents ouvrages de Lavoisier, qui, comme on le sait, traitait la chimie et beaucoup d'autres parties de science et d'art, avec des connaissances et des succès peu communs.

J'ai prévenu la commission temporaire des arts et le comité d'instruction publique qu'il existait beaucoup d'objets d'histoire naturelle,[14] dont une grande partie est encore emballée, et que, dans la même pièce où se trouvent ces mêmes objets, se trouvent encore beaucoup de papiers et de cartons, dont les étiquettes indiquent des titres, des mémoires, notes, et qui pourraient peut-être présenter aussi des objets de la plus grande importance.

Signé Leblanc.

FIN

14 It is not clear what these objects might be. As pointed out in the inventories transcribed in Appendix 1, the commissioners found one collection of shells and one of minerals and fossils, but none of them refer to their conservation.

Inventory of Lavoisier's Precision and Chemical Instruments (1794)

As we have shown in Appendix 2, Leblanc's inventory did not take into account any precision instruments that he was unable to identify. Significantly, this part of the laboratory was not inventoried by a chemist but by two physicists, Jacques A. Charles and Jean Charles Pierre Lenoir, with the assistance of the instrument maker Nicolas Fortin, who had constructed sophisticated instruments for Lavoisier since the late 1770s.

What follows is a complete transcription of the inventory of Lavoisier's instruments compiled by Charles, Lenoir, and Fortin in November 1794.[1]

I have tried to match all the instruments with those presented in our catalogue, but in many cases the commissioners' descriptions are so succinct that I have only been able to propose educated guesses. This is the case, for instance, for some of the balances, thermometers, barometers, and hydrometers.

Apart from these difficulties, regrettably many important instruments, such as Fortin's first gasometer, have not survived, but are only documented in the exceedingly concise entries in the following list.

Inventaire des Instrumens de Physique et de Chymie du Cabinet de Lavoisier cy devt Fermier Générale et de l'Académie des Sciences, Boulevard de la Madeleine.

8000 l.[2] 1. Grand gazomètre avec accessoires et dépendances,[3] exécuté par Meigniér [*sic*] le J.[e4]
N.[a] Pour les détails et l'intelligence de cet instrument capital et précieux, il faut consulter les elemens de chymie de Lavoisier,[5] ainsi que pour les 4 suivans.
1600 l. 2. Grand appareil pour la combustion des huyles.[6]
3. Appareil pour la fermentation, *très compliqué*.[7]
4. Appareil pour la combustion de l'esprit de vin.[8]
600 l. 5. Deux calorimetres; avec lampe en fer blanc.[9]
800 l. 6. Grande machine pneumatique à deux corps de pompe de Fortin avec baromètre d'epreuve et une platine de rechange.[10]
7. Deux tres grands ballons à peser l'air.[11]
8. Grande balance de laboratoire.[12]
9. Cinq récipiens à douille et trois à bouton.
10. Vingt thermometres dont six à esprit de vin.
11. Barometre de Mossy monté en acajou, mais en mauvais état, l'air rentré dans le tube.[13]
12. Lampe à air ou gaz inflammable.[14]
13. Autre lampe à air, *mais incomplette*.
14. Appareil de Moth[15] pour imprégner l'eau d'acide carbonique, *incomplet*.[16]
15. Trois cuvettes de fer blanc verni contenant en tout une 20.e de flacons.
16. Machine électrique de 24 pouces avec son conducteur et sa table.[17]
17. Autre machine electrique de 22 pouces avec son conducteur posé sur une tablette sans pieds.
18. Batterie de 64 petits bocaux dans une boëte avec leurs fils de communication.[18]
19. Grande jarre électrique.[19]
20. Autre jarre plus petite.
21. Petit bouteille de Leyde.
22. Tabouret ou isoloir.

1 MS Charles, Lenoir, Fortin (1794); a transcription without notes was published in Beretta (2005), pp. 330–334.
2 The sum of 8,000 livres refers to the value estimated by the commissioners. Also, the values of instruments nos. 2, 5, 6, 121, 122, and 123 were estimated. The numbering of the instruments is that of the commissioners.
3 MAM inv. nos. 07547-0001-001-, 07547-0002-001-; see catalogue pp. 340–346.
4 Pierre Bernard Mégnié also known as Mégnié le jeune.
5 The reference is to Lavoisier (1789).
6 This refers to MAM inv. n. 07549-0000-; see catalogue pp. 351–353.
7 MAM inv. nos. 07551-0000-; see catalogue p. 351. Since Nicolas Fortin was one of the commissioners charged with this inventory, and at the same time the maker of both apparatuses for vinous fermentation, the statement that the machine was considered to be very complicated is quite surprising. Admittedly Lavoisier's experiments on vinous fermentation were quite elaborate.
8 MAM inv. nos. 07550-000-, 19998-0000-; see catalogue p. 347.
9 MAM inv. nos. 07520-0000-, 07547-0004-001-; see catalogue pp. 229–232.
10 This is probably the pneumatic pump in MAM inv. nos. 07517-0001, 07517-0002; catalogue pp. 221–225. Quite surprisingly only one pneumatic pump made by Fortin was described.
11 This is likely to be the kind of balloon, of Lavoisier's invention, that was described by David below his portrait of the Lavoisiers with the hydrometer.
12 It is not clear which balance is described here. Fortin's "grande balance" is described below at no. 121. For a similar object see MAM inv. no. 19986-0000-, catalogue p. 189.
13 This barometer never made it to the collection of MAM.
14 This is probably the electric lighter in MAM inv. nos. 19999-0000-, 19972-0000-, 20120-0000-; see catalogue p. 264.
15 In fact, this is John Mervin Nooth; see Nooth (1775).
16 Lavoisier probably owned Magellan's perfected version of the apparatus; see Magellan (1777).
17 This probably Fortin's electrical machine; MAM inv. nos. 20110-0001-, 20110-0002-, 20110-0003-, 20110-0004- and 20122-0000-. See catalogue p. 249–252.
18 MAM inv. nos. 20114-0000-, 20203-0000-. See catalogue p. 256.
19 This could be the large cylindrical Leyden jar; MAM inv. n. 20115-0000-. See catalogue p. 255.

 | DOI:10.1163/9789004511217_010

23. Trois tableaux magiques.[20]
24. Pistolet de Volta en cuivre.
25. Pistolet de Volta en fer blanc.[21]
26. Deux carillons à trois timbres.
27. Autre carillon à 2 timbres.
28. Electrometre de Saussure.[22]
29. Deux autres electrometres.
30. Petite presse electrique.
31. Petit appareil pour briser le verre.[23]
32. Un tube étincelant.[24]
33. Petit excitateur de cuivre.
34. Vase [pour] l'inflamm. de l'esp. de vin.
35. Petit isoloir avec virole terminée en fourche.[25]
36. Eudiometre de Volta.[26]
37. Condensateur de Volta à plateau de marbre d'env.n 22 p.e.[27]
38. Miroir concave de 22 pouces tournant sur son pied.[28]
39. Autre plus petit miroir concave de 11 pouces très gâté.[29]
40. Prisme de verre verdatre monté sur pied.[30]
41. Grand prisme de verre de France non monté.[31]
42. Prisme creux pour les liquides monté sur son pied.[32]
43. Prisme quadrilatere creux avec cloison diagonale.
44. Miroir cilindrique avec ses cartons.[33]
45. Miroir prismatique.
46. Machine hydrostatique ditte de Pascal.[34]
47. Syphon à jet d'eau dans le vuide.
48. Vis d'Archimedes pour l'eau.[35]
49. Autre vis pour une balle de plomb.[36]
50. Marmite de Papin.

20 These are probably MAM inv. nos. 20128-0001-, 20128-0002-, 20128-0003-; See catalogue p. 257.
21 This probably Volta's electrical gun MAM inv. no. 20123-0000-. See catalogue p. 337.
22 This is probably MAM inv. no. 20117-0000-. See catalogue p. 263.
23 This is probably MAM inv. no. 20048-0000-. See catalogue p. 390.
24 This is probably MAM inv. no. 20132-0000-. See catalogue p. 261.
25 This is probably MAM inv. no. 20119-0000-. See catalogue p. 258.
26 This is probably MAM inv. no. 20004-0000-. See catalogue p. 154.
27 This probably refers to MAM inv. nos. 20121-0000-, 20206-0001-, and 20134-0000-. See catalogue p. 252.
28 This refers to MAM inv. no. 20080-0000-. See catalogue p. 329.
29 This probably refers to MAM inv. no. 20079-0002-. See catalogue p. 329.
30 MAM inv. no. 20089-0000- or MAM inv. nos. 20043-0000- and 20090-0000-. See catalogue p. 241.
31 This is probably MAM inv. no. 20088-0000-. See catalogue p. 241.
32 This is probably MAM inv. nos. 20090-0000- and 20043-0000-. See catalogue p. 242.
33 This probably refers to the semi-cylindrical anamorphic mirror in MAM inv. nos. 20092-0000-, 20226-0000-. See catalogue p. 241.
34 This refers to Pascal's apparatus for the hydrostatic paradox, MAM inv. nos. 20097-0000-, 20002-0000-. See catalogue p. 214.
35 MAM inv. no. 20100-0000-. See catalogue p. 215.
36 This is probably MAM inv. no. 20099-0000-. See catalogue p. 215.

51. Autre très belle marmite de Papin. C'est la plus considérable que l'on connoisse.[37]
52. Grand pezon à ressort d'Hanin marquant 165 livres.[38]
53. Balance particulière de Scanegati, avec ses poids dans une boëte.
54. Petit balance de Scanegati à cadran.[39]
55. Poids de marcs savoir deux de 16 marcs, une de douze marcs, un de 8 marcs, un de 4 marcs et une boëte de 4 marcs à poids rentrans et six petits poids de quatre onces et de 1 once.[40]
56. Petite balance avec son support.
57. Mauvaise petite balance dans son étui.
58. Joli trebuchet de la chine dans son étui avec 44 poids de figures et grosseurs différentes.[41]
59. 24 aréometres en verre avec leurs étuis de fer blanc.
60. 13 aréometres en métal cuivre et fer blanc grands et petit.
61. Une quantité de syphons et tubes pour le transport des gaz.
62. Dix bassins en fer blanc pour le transport des gaz.
63. Deux espèces de refrigérens en fer blanc peint en noir.
64. Grande cuve hydropneumatique garnie en plomb.
65. Autre grande cuve en chêne garnie en plomb p.r le gazom.[42]
66. Autre cuve pour les mêmes expériences p.r le gazometre du N° 69.
67. Table à souffler le verre.
68. Table à travailler et polir les verres.
69. Ancien gazomètre de Fortin.[43]
70. Deux grandes armoires à 4 ventaux garnis de verre.
71. Grand recipient pour la combustion du gaz hydrogène.
72. Un respiratoire anglois en etain.
73. Autre respiratoire en fer blanc verni.
74. Mouvement appellé le briquet dans le vuide.
75. Boussolle d'arpenteur.[44]
76. Regle surmonté de pinnules.[45]
77. Pinnules d'arpenteur.[46]
78. Rapporteur à 2 demi cercles et deux regles mobiles.[47]
79. Deux éprouvettes à poudre à ressort.[48]
80. Une equerre.

37 This is probably MAM inv. no. 20098-0000-. See catalogue p. 308.
38 This is related to MAM inv. no. 19896-0000-. See catalogue p. 194.
39 MAM inv. no. 19894–0000-. See catalogue p. 193.
40 This is probably MAM inv. nos. 19898-0000- and 19897-0000-. See catalogue p. 199.
41 This is probably MAM inv. no. 19893-0000-. See catalogue p. 195.
42 This is probably MAM inv. no. 07547-0001-002-. See catalogue pp. 346–347.
43 This entry probably refers to the gasometers made in 1783.
44 This is probably MAM inv. no. 20217-0000-. See catalogue p. 205.
45 This is probably MAM inv. no. 20138-0000-. See catalogue p. 204.
46 This is probably MAM inv. no. 20140-0000-. See catalogue p. 205.
47 This is probably MAM inv. no. 20135-0000-. See catalogue p. 208.
48 This is probably MAM inv. no. 19895-0000-. The other seems to be missing. See catalogue p. 306.

81. Petite equerre en quart de cercle par Baradelle.[49]
82. Astrolabe.[50]
83. Sextant de Bird sans lunettes.[51]
84. Petit cadran à Boussolle en argent par Le Maire.[52]
85. Autre mauvaise petit cadran dans une boete d'yvoire.
86. Compas à verge de bois.
87. Hygrometre à plusieurs cheveux par Richer.[53]
88. Petit miroir multipliant tournant pour les allouettes.
89. Thermoscope particulier fait avec un tube sur une regle en cuivre graduée et plongeant dans un long recipient.[54]
90. Piece pour des expériences pneumatiques au mercure.
91. Belle regle angloise divisée dans un étui en acajou.
92. Recreations magnétiques telles que le petit Cigne ingénieux, L'oracle,[55] la boëte aux chiffres, le petit peintre.
93. Jolie boussolle de declinaison montée sur un carreau de pierre polie, faite en Allemagne.[56]
94. Autre boussolle de declinaison montée en cuivre rouge portée sur un plan de marbre blanc, par Fortin. Il y manque l'aiguille aimentée et sa suspension.[57]
95. Barreaux aimantés dans une boëte de chêne.[58]
96. Gros ballon à tige garnie de virole, robinet et tube contenant une éprouvette à mercure.
97. Boëte d'acajou contenant une petite pharmacie.
98. Ballon de verre bleu d'un pied de diamètre.
99. Deux petits guéridons d'acajou.[59]
100. Petit modèle d'appareil distillatoire.[60]
101. Grande table en acajou ployante long.r 8 pieds et largeur 4 pieds; et table ovale d'ardoise de 3 p. 8 pou. sur 2 p. 8 po.[61]
102. Deux autres tables en acajou à 3 etages, longueur 5 pieds largeur 8 pouces.[62]
103. Mortier et pilon de fayance blanche.[63]
104. Douze mauvaises petites balances liées ensemble.
105. Deux petits creusets et une petite cuillière en platine.[64]
106. Petite lampe en cuivre argenté.
107. Autre petite lampe pour les colipyles.
108. Deux très petits soufflets montés sur une planchette d'acajou.[65]
109. Pompe ou seringue en cuivre.[66]
110. Un seau en cuivre.
111. Regle de bois et de glace avec de longs tubes de metal pour des experiences thermométriques.
112. Un barometre à deux tubes par Meigniez.[67]
113. Autre barometre à deux tubes par Meignez.[68]
114. Deux baroscopes à liquide à 4 tubes immergés dans le même cuvette.
115. Autre baroscope en syphon monté sur une planche graduée en cuivre.
116. Quatre planches de debris de barometre.
117. Barometre à syphon portant un thermometre, fait en Allemagne.[69]
118. Petite machine electrique à plateau d'environ un pied et renfermée dans un grand récipient pour des demonstrations d'électricité dans le vuide.
119. Pendule de Rivas boëte peinte en verd propre a etre accroché à la muraille.
120. Pendule à secondes excentriques dans une boëte de bois noir.
121. Grande balance dont le fléau a trois pieds de long avec une paire de grands bassins en cuivre, une paire de plateaux de verre de 5 pouces de diametre, un simple plateau de verre; cette balance est dans une cage en bois d'acajou, 2200 l.[70]
122. Une balance à fleau de 18 pouces de long propre à peser au plus deux marcs, avec une paire de bassins en ver; dans une cage en bois plaqué, 800 l.[71]
123. Autre balance pour les petites pesées de dix pouces de flanc avec une paire de bassins en verre sans une cage de verre, 500 l. N.a Ces trois balances sont deposés à la Commission des poids et mesures.

49 MAM inv. nos. 20137-0000-, 20149-0000-. See catalogue p. 206.
50 MAM inv. no. 20142-0000-. See catalogue pp. 200–203.
51 John Bird.
52 This was exhibited in Lavoisier (1943). This item was part of the collection of M. Lammot du Pont Copeland in 1952, but its present location its unknown.
53 This is probably MAM inv. no. 20106-0000-. See catalogue p. 299.
54 This is probably MAM inv. no. 20005-0000-. See catalogue p. 228.
55 This is probably the game *Oracles merveilleux*, now kept at the Kroch Library, Cornell University, Ithaca NY, Lavoisier 4712 Box 29. See catalogue p. 394.
56 MAM inv. no. 20136-0000-. See catalogue p. 210.
57 This is probably MAM inv. no. 20105-0000-. See catalogue p. 248.
58 This is probably MAM inv. no. 20049-0000-. See catalogue p. 247.
59 This is probably MAM inv. no. 20047-0000-. See catalogue p. 317.
60 This one of the three models of a distilling apparatus in MAM inv. nos. 20171-0000-, 20176-0000-, 20177-0000-. See catalogue p. 318.
61 This is the large laboratory table in MAM inv. no. 20574-0000-. See catalogue p. 301.
62 These are probably MAM inv. nos. 20575-0001- and 20575-0002-, See catalogue p. 390.
63 These are probably MAM inv. nos. 20169-0000-, 20170-0000-. See catalogue p. 310.
64 One of these "crucibles" (in fact, a salt cellar with spoon) is now at the Kroch Library, Cornell University, Ithaca NY, Lavoisier 4712 Box 29. See catalogue p. 377.
65 This is probably MAM inv. no. 20174-0000-. See catalogue p. 260.
66 This is probably the pump which was used to empty the balloon and pipes in the 1785 experiments on the synthesis of water.
67 MAM inv. no. 07658-0000-. See catalogue p. 290.
68 MAM inv. no. 08761-0000-. See catalogue p. 288.
69 MAM inv. no. 19952-0000-. See catalogue p. 293.
70 MAM inv. no. 19887-0000. See catalogue p. 190.
71 This is probably MAM inv. no. 19885-0000-. See catalogue pp. 192.

A Paris le vingt Brumaire de l'an Trois[72] de la Republique francoise une et indivisible.
Charles Fortin Lenoir
N.a Depuis l'inventaire terminé, il a été remis au Depot par le Cit. Fortin
1. Deux loupes de verre de 15 pouces sans monture.
2. Une lunette acromatique à tuyau en cuivre, non terminée. Cette lunette avoit été commencé par Rebours.[73] Elle est sans tuyau oculaire.

72 10 November 1794.
73 Le Rebours.

Charles.
Je soussignée reconnais avoir reçu du Citoyen Charles les objets détaillés dans l'inventaire ci dessus
A Paris ce quatre Fructidor an Trois.[74]
Paulze veuve Lavoisier

74 21 August 1795.

APPENDIX 4

Select Inventory of Marie Anne Lavoisier's Residence in the Rue d'Anjou (1836)

The following document is a partial transcription of Marie Anne Lavoisier's *Inventaire après décès* and concerns Lavoisier's collection of instruments.[1] This was apparently kept in the library and another room close to it. According to the plan of the residence in the Rue d'Anjou (see Figs. 6–7 in Chapter 5) there may have been additional room devoted to the *cabinet de physique*, also located close to the library. The descriptions of Lavoisier's instruments are quite brief but nonetheless interesting in many respects. The valuation for some of the apparatus is surprisingly low: Méigné gasometers were valued at only 60 francs, 40 francs less than Fortin's electrical machine. The most expensive instrument was Fortin's balance at 300 francs – a price which probably accounted for the reputation gained by Fortin in the early decades of the nineteenth century. The list also includes a calorimeter made by or for Rumford. Marie Anne Lavoisier and Rumford divorced in 1809, and, as the catalogue shows, she kept a few of his instruments. Interestingly, Rumford's *Inventaire après décès* includes a "calorimètre de feu M.r de Lavoisier monté sur une table en bois".[2] As the divorce was amicable, this exchange of instruments must have been deliberate. Unfortunately, the current location of Rumford's collection is unknown.

The list here transcribed also includes chemical glassware and an unspecified number of Lavoisier's chemicals, shells minerals, fossils, and volcanic rocks.[3]

Inventaire après le décès de Mme la Ct^{esse} de Rumford

[3r] *Dans une pièce au fond du second corridor éclairée sur le jardin servant de petit office*

[…]

Un mortier en marbre avec son pilon et son billot prisés dix francs ci: 10

Un fourneau en tôle et faïence avec ses tuyaux prisé quinze francs ci: 15

[…]

[41r] *Dans une pièce au premier étage éclairée sur une terrasse au-dessus de la galerie des fleurs*: […]

[41v] *Dans un cabinet noir faisant suite à droite*: […]

Un lot de tubes et divers instruments de physique et de chimie en un lot de caoutchouc, prisé le tout six francs ci 6 […]

[43v] *Dans une pièce à droite de la chambre à coucher précedemment decrite, servant de garde meuble*: […]

[43r] Un carton contenant différents instruments de *musique* mathematiques et autres objets ne méritant description prisés quatre francs ci […]

[50v] […]

Dans la commode de la chambre à coucher de Mme de Rumford: […]

[52r] Une boite d'instruments de mathematiques en argent, prisée quinze francs ci 15

Cinq lunettes avec un memento en acier, prisés deux francs ci 2

Une boussole en argent prisée vingt francs ci 20

Une boussole en ivoire prisée un franc ci 1 […]

[58r] *Dans les tiroirs du Bureau du petit salon ci-devant decrit*: […]

[58v] Deux petits bassins avec une petite cuiller le tout en platine, propre à des operations de chimie, pesant cinquante grammes et prisé à raison de cinquante centimes le gramme vingt cinq francs ci 25[4] […]

[61r] *Dans une pièce au premier étage du corps de batiment habité par Mr et Mme Masselot*:

Dans un cabinet à côté: […]

[61v] Un lot de vieux bois, fer et fonte, ne méritant autrement description prisé quinze francs ci 15

Deux vieux fauteuils, des alambics et un lot d'instruments en cuivre destinées à la chimie, le tout prisé cinquante francs ci 50 […]

[66v] *Dans les différentes salles et casiers de la Bibliothèque*:

Un appareil pour les gazs composé de plusieurs ballons à robinets, prisé la somme de cinquante francs ci 50[5]

Une cuve en bois doublé en plomb, egalement destiné aux opérations chimiques sur les gazs, prisée dix francs ci 10

Un petit appareil à gaz hydrogène en plomb et fer blanc, prisé la somme de cinq francs ci 50

Deux grands gazomètres de Lavoisier avec tous leurs accessoires, prisés ensemble la somme de soixante francs ci 60[6]

Une machine électrique de vingt deux pouces prisée soixante francs ci 60

1 MS Paulze-Lavoisier (1836). The transcription is by Francesca Antonelli, whom I thank for providing me with this document. The whole inventory is published in Antonelli (2021).

2 MS Thompson (1814), item no. 264.

3 The description given here is more analytical than that given in Appendix 1.

4 One of these "petits bassins" (in fact, a salt cellar with spoon) is now at the Kroch Library, Cornell University, Ithaca NY, Lavoisier 4712 Box 29. See catalogue p. 377.

5 This refers to MAM inventory nos. 07549-; see catalogue pp. 351–353.

6 MAM inv. nos. 07547-0001-001-, 07547-0002-001-; see catalogue pp. 340–346.

 | DOI:10.1163/9789004511217_011

Une autre machine electrique de vingt quatre pouces, sur sa table, prisé cent francs ci 100[7]

Un second appareil pour les gazs composé de plusieurs flacons à robinets, prisé la somme de vingt francs ci 20[8]

Une grande balance en cuivre de Fortin sous sa case de verre, prisée la somme de trois cent francs ci 300[9]

Une autre balance plus petite, egalement [**67r**] sous case de verre, prisée cent francs ci 100[10]

Une balance à essais prisée la somme de soixante francs ci 60

Un lot assez considérable de cloches et flacons en verre, dont plusieurs à robinet, le tout prisé trente francs ci 30

Deux prismes de verre, prisés ensemble cinq francs ci 5

Un lot considérable de thermomètres, aerometres, et gazomètres, prisé ensemble quinze francs ci 15

Un briquet à rouage pour la machine pneumatique prisé la somme de quarante francs ci 40

Plusieurs flotteurs en cuivre prisé ensemble dix francs ci 10

Plusieurs vases en cuivre, pour les expériences sur la chaleur, prisé ensemble six francs ci 6

Une petite meule prisée six francs ci 6[11]

Plusieurs supports en bois et cuivre prisés ensemble quatre francs ci 4

Une batterie électrique de soixante petits bocaux, prisée six francs ci 6[12]

Une boussole à lunette prisée la somme de vingt francs ci 20

Deux microscopes, dont un Dellebare et l'autre microscope solaire, prisés ensemble quarante francs ci 40[13]

Une petite électricité contenue dans un [**67v**] ballon de verre, prisée dix francs ci 10

Une machine pneumatique de Fortin et quelques accessoires, le tout prisé deux cents francs ci 200[14]

Un grand appareil à gaz garni de robinets prisé cinq francs ci 5

Un appareil hydrostatthique de Pascal prisé dix francs ci 10[15]

Une aiguille de déclinaison prisée cents francs ci 100

Un anneau astronomique prisé la somme de cinq francs ci 5[16]

Deux eprouvettes de Regnié [Régnier] prisées ensemble dix francs ci 10[17]

Une petite aiguille d'inclinaison prisée huit francs ci 8

Une boite renfermant plusieurs thermometres prisée six francs ci 6

Un briquet à gaz hydrogène prisé la somme des quinze francs ci 15

Un lot d'appareils d'electricité prisé six francs ci 6

Un grand condensateur en marbre et cuivre prisé dix francs ci 10[18]

Trois calorimètres prisés ensemble soixante francs ci 60

Un grand miroir de deux pieds prisé cent francs ci 100[19]

Deux bouteilles de quinze pouces chacune montées sur leur pied prisées ensemble la somme de cent francs ci 100

Un miroir concave en metal, ayant [**68r**] un pied de diamètre, prisé la somme de cinquante francs ci 50

Un lot considerable de verreries, cloches et flacons ballons le tout prisé dix francs ci 10

Une casserole et une marmite en cuivre prisées ensemble quinze francs

Une balance ordinaire prisée quinze francs ci 15

Deux calorimètres en cuivre prisés ensemble dix francs ci 10[20]

Un lot considérable de vases en cuivre et fer blanc ayant servi aux expériences de M^{r}. de Rumford sur la chaleur prisé dix francs ci 10

Plusieurs bassines et marmites en cuivre prisée douze francs ci 12

Un hygromètre de Saussure prisé vingt francs ci 20[21]

Un calorimètre de Rumford et son thermomètre, prisés ensemble vingt francs ci 20

Deux miroirs en metal et leurs cartons, le tout prisé quinze francs ci 15[22]

Dans une pièce au-dessous de la Bibliothèque, a côté des serres: [...]

Un lot très considerable de cornues [**68v**], ballons et matras, le tout prisé la somme de vingt francs ci 20

Un lot aussi très considerable de bocaux et flacons dont une grande partie renferment differentes substances, le tout prisé vingt cinq francs ci 25[...]

[**69r**] *Dans un médailler ci-devant décrit est trouvé dans le petit salon de Mme de Rumford*: [...]

Dans un médailler aussi ci devant decrit est trouvé dans la chambre à coté de Mr. Masselot: [...]

Un lot nombreux de coquillage de tout genre, soigneusement étiquetés, numérotés et classés par familles et espèces, le tout prisé la somme de cent francs ci 100

Dans un autre médailler aussi précédemment décrit et trouvé dans la même pièce:

7 This probably Fortin's electrical machine; MAM inv. nos. 20110-0001-, 20110-0002-, 20110-0003-, 20110-0004- and 20122-0000-. See catalogue pp. 249–252.

8 MAM inv. nos. 07551-0001-, 7551-0002-; see catalogue p. 351.

9 MAM, inv. no. 19887-0000-. See catalogue pp. 190.

10 This is probably MAM inv. no. 19885-0000-. See catalogue p. 192.

11 This is probably MAM inv. no. 20056-0000-. See catalogue p. 396.

12 MAM inv. nos. 20114-0000-, 20203-0000-. See catalogue p. 256.

13 MAM inv. nos. 20085-0001-, 20085-0002-, 20085-0003-. See catalogue p. 246.

14 This is probably the pneumatic pump in MAM inv. nos. 07517-0001-, 07517-0002-; catalogue pp. 221–225.

15 MAM inv. nos. 20097-0000-, 20002-0000-. See catalogue p. 214.

16 This is probably the astrolabe MAM inv. no. 20142-0000-. See catalogue pp. 200–203.

17 MAM inv. no. 20234-0000- (scraper). See catalogue p. 397. MAM inv. no. 19895-0000- (gunpowder tester). See catalogue p. 306.

18 This could be MAM inv. nos. 20134-0001-, 20134-0002-. See catalogue p. 253.

19 This could be MAM inv. no. 20080-0000-. See catalogue p. 329.

20 MAM inv. nos. 07520-0000-, 07547-0004-001-; see catalogue pp. 229–232.

21 This is probably MAM inv. no. 20106-0000-. See catalogue p. 299.

22 MAM inv. nos. 20091-0000-, 20092-0000-. See catalogue p. 241.

Une collection d'echantillons de marbre de peu valeur, prisé avec une plante marine parfaitement conservée et se trouvant [**69v**] dans le tiroir supérieur au dit casier, la somme de douze francs ci 12

Dans une chambre précedemment décrite se trouvant à côté de l'appartement de Mr. Masselot: [...]

Un lot très considérable de bocaux et de cartons renfermant differentes plantes médicales très anciennes et des matières chimiques, le tout à cause de sa vetusté prisé seulement la somme de quatre-vingt francs ci 80 [...]

Mineraux trouvés dans des caisses en boit sous les remises où sont les voitures et dans la bibliothèque. [...]

Un lot très considerable consistant principalement en mines de plomb, de fer et d'argent de différentes natures et couleurs, prisé le tout à cause de son ancienneté seulement à la somme de cent cinq francs ci 150

Un autre lot très considerable consistant principalement en cristal de roche, bois petrifiés, pierres et extraites des différentes sols de la France, des montagnes étrangères et production des différentes bouches de Volcan, le tout prisé la somme de cent franc ci 100

Tous les dits mineraux classés numerotés et soigneusement étiquetés de la main de Mr Lavoisier.

Un autre lot très considerable de boit petrifié et pierres de toute nature [**70r**] amoncelés sans ordre sans etiquettes le tout prisé cinquante francs ci 50 [...]

Inventory of Lavoisier's Instruments Acquired by the Conservatoire des arts et métiers in 1864

The following document records the acquisition of Lavoisier's instruments by the Conservatoire des arts et métiers. The collection had been given to the Académie by the heirs of his widow in June 1836,[1] where they were placed on deposit for more than 25 years. There are no known records of a public exhibition of Lavoisier's collection at the Institut and this explains why, until recently, historians have thought that it was donated to Conservatoire des arts et métiers as late as 1866. The Académie's decision to dispose of the instruments dates to 1864,[2] and the instruments received their inventory numbers at the Conservatoire in 1866. This document was compiled at the Conservatoire before 1866. Although it is not dated, the list refers to instruments which were never inventoried, or if some of them were, they were not explicitly connected to Lavoisier when the inventory number was given. During our research in the warehouse of the MAM we were able to identify a few pieces of glassware which are documented in the present list but had not been assigned an inventory number. It is, therefore, possible that other instruments from the list were similarly not inventoried in 1866, and are still somewhere in the warehouse.

Another point of interest in this list lies in instruments whose description suggests an assemblage of pieces that where eventually separated. It is not clear whether these descriptions were made from accompanying manuscript notes that are now lost, or if they represent an interpretation by the members of the Académie des sciences in June 1864.[3] This is particularly the case for the first item described, the gasometer, which is presented together with one of the two calorimeters but without the two pneumatic troughs which are present elsewhere.

1 Appareils de Lavoisier que possède le Conservatoire des Arts et métiers donnés par l'Académie des Sciences

1° Un appareil pour la détermination de la chaleur dégagée dans la combustion de l'hydrogène et la formation de l'eau.[4]

Cet appareil comprend

2 grands gazometres construits par Méigné le jeune et un calorimètre. Le calorimètre est celui qui a servi dans les expériences sur la respiration des animaux et dans la détermination des chaleurs spécifiques des gaz.[5]

2° Un calorimètre de mêmes dimensions.[6]

3° Appareil pour l'étude des fermentations tel qu'il est dessiné dans son Traité élémentaire de chimie.[7]

4° Appareil analogue au précédent et comprenant 1 ballon à long col, 4 flacons laveurs et 2 tubes desséchant.[8]

5° Deux cuves à eau destinées à recueillir les gaz dégagés dans les deux appareils précédents.[9]

6° Appareil ayant servi pour l'analyse des produits de la combustion des huiles et de l'éther, comprenant: la lampe, le vase dans lequel d'effectue la combustion, un serpentin destiné à recueillir la vapeur d'eau, 9 ballons servant de flacons laveurs, de 4 tubes desséchants et deux autres flacons laveurs. (La lampe à éther n'a pas été retrouvée).[10]

7° Le grand ballon muni de sa garniture dans lequel Lavoisier effectuait la recomposition de l'eau.[11]

8° Des pièces détachés de l'appareil de Priestley destiné à oxider les métaux dans l'oxygène à l'aide du verre ardent et un support paraissant être celui du verre ardent.[12]

---

1 Letter from Dupin to L. De Chazelles: "A Monsieur le Comte de Chazels [*sic*] //

Paris 27 juin 1836 // Monsieur //Je me suis empressé de communiquer à l'Académie l'offre que vous lui faites d'instruments avec lesquels l'illustre Lavoisier a constaté plusieurs de ses grandes découvertes.

Je m'empresse de vous informer que l'Académie accepte votre offre faite en votre nom et en celui de Madame de Chazels [*sic*], avec une reconnaissance vivement sentie. Je m'estime heureux d'avoir à vous exprimer ce sentiment. //Veuillez adresser les machines à M Cardot, agent de l'Académie, à l'Institut.

Le Président de l'Académie //Baron Charles Dupin." MS Dupin (1836). On 11 July Cardot took the instruments from the residence of the late Marie Anne Lavoisier in the Rue d'Anjou and brought them to the Academy.

2 Register (1864).

3 See chapter 5, p. 99.

4 MS *Appareils* (1864–1866).

5 MAM inv. nos. 07547-0001-001-, 07547-0002-001-, and the calorimeter MAM inv. no. 07547-0004-001-. See catalogue pp. 340–346 and p. 232.

6 MAM inv. no. 07547-0004-001-. See catalogue p. 232.

7 MAM inv. nos. 07551-0001-, 7551-0002-. See catalogue pp. 350–351.

8 MAM inv. nos. 07550-0000-. See catalogue p. 350.

9 It seems that the reference here is to the two pneumatic troughs that went with the gasometers: MAM inv. nos. 07547-0001-002- and 07547-0002-002-. See catalogue p. 344.

10 MAM inv. nos. 07549-. See catalogue pp. 351–353. This apparatus was apparently more complete in 1864 than it is today.

11 MAM inv. no. 07548-0000-. See catalogue p. 340.

12 A depiction of this piece of apparatus was published at the end of 1773 in Lavoisier (1774), plate 2, fig. 8. The apparatus was not inventoried in 1866, and its present location is not known.

 | DOI:10.1163/9789004511217_012

9° Support en bois et 2 tubes en cuivre relatifs aux expériences de combustion du soufre et du phosphore dans un ballon rempli d'oxygène.

10° Le ballon et la cloche ayant servi à la détermination de la densité des gaz.[13]

11ème Une balance de 3 pieds de longueur construite par Fortin et pouvant peser de 15 à 20 livres avec la précision d'un ½ grain. Diamètre des plateaux en cuivre rouge = 0,43 m.[14]

12° Une petite balance à main pouvant peser une livre à la précision du grain [elle est sensible à 0,02 g].[15]

13° 2 plateaux servant à manœuvrer les cloches sur la cuve à eau.

14° Diverses pièces de verrerie consistant en: 2 chapiteaux des cucurbites en verre.[16] Une cucurbite en verre vert.[17] 4 cornues de diverses grandeurs.[18] Un grand ballon à deux orifices fermés par des garnitures en bois. Un petit ballon à long col. 2 cloches s'adaptant sur le support des cuves à eau. 3 flacons. Un récipient en verre vert et à 3 tubulures.[19] Un matras d'essayeur. Une pièce en verre avec garniture en cuivre ayant la forme ci jointe

15° 14 supports en bois garnis des cordes.[20]

7 areomètres en cuivre[21]

Dont un déplace 4,92

2 autres déplacent 2,14

1 de forme prismatique 1,42

et 2 autres plus petits.

3 boites remplis de plomb fondu et pesant l'une 1 livre et l'autre 16 livres.

13 The only glass recipients with an inventory number compatible with the date in which this inventory was made are MAM, inv. nos. 07547-0002-004- and C-2017-0070-. See catalogue p. 353 and 357.

14 This description corresponds only in part with MAM inv. no. 07545-0000- (see catalogue p. 188). The origin of the attribution to Fortin is not known.

15 MAM inv. no. 07544-0000-. See catalogue p. 187.

16 Only one has been found: MAM inventory no. C-2017-0071-. See catalogue p. 352.

17 MAM inv. no. C-2017-0069-. See catalogue p. 351.

18 One of these may be MAM inv. no. 36645-0000-, that is an alembic corresponding to Lavoisier (1789), vol. 2, plate III, fig. 13. I would like to thank Denis Pruvrel for bringing this instrument to our attention. See p. 352.

19 MAM inv. no. C-2017-0068. See catalogue p. 387.

20 This entry seems to refer to the straw rings recently inventoried as MAM inventory nos. C-2017-0064-, C-2017-0072-, C-2017-0073-, C-2017-0074-, C-2017-0075-, C-2017-0076-. See catalogue p. 317.

21 This entry probably refers to the set of six constant volume hydrometers in MAM inv. nos. 07508-0001-, 07508-0002-, 07508-0003-, 07508-0004-, 07508-0005-, 07508-0006-. See catalogue pp. 319–321. The seventh hydrometer in prismatic form seems to have been lost.

APPENDIX 6

Biographical Dictionary of Lavoisier's Instrument Makers and Suppliers of Chemicals

The list presented below presents those of Lavoisier's instruments makers which he references in his published and unpublished works or in his correspondence, which are mentioned in the inventories of his laboratories made during the Revolution, and those evidenced in the collection of instruments which we have catalogued.[1] Since only a small part of Lavoisier's laboratory has survived, it is likely that a larger number of makers was involved in the making of Lavoisier's instruments than those listed here. I have also decided to include a few makers of instruments which were part of Marie Anne Lavoisier's collection, or possibly Benjamin Thompson, Count Rumford's. In addition to instrument makers I have also included the pharmacists who supplied Lavoisier with chemicals – the very nature of chemical laboratories and central role played by chemical substances, as well as the fact that several apothecaries (above all Baumé and Quinquet) were also known inventors and suppliers of chemical apparatus, justify their inclusion in the list. Last but not least, the list provides information about all of Lavoisier's known laboratory assistants even when, as in the case of Adet, their role in the construction of instruments is not documented.

For each maker and supplier (listed in alphabetical order) I have provided a short biography and a reference to the relevant instruments and chemicals owned by Lavoisier.

Adams, George Jr. (1750–1795). The son of the optician and instrument maker George Adams (1720–1773), George Jr. continued the family business in London and specialised in making mathematical instruments, electrical devices, and globes. He published several sale catalogues of instruments, the most comprehensive in 1791.[2] He also made a few chemical instruments, such as a perfected version of Volta's eudiometer.[3] For Lavoisier he made a Henley's quadrant electrometer (Fig. 1) in 1780.[4]

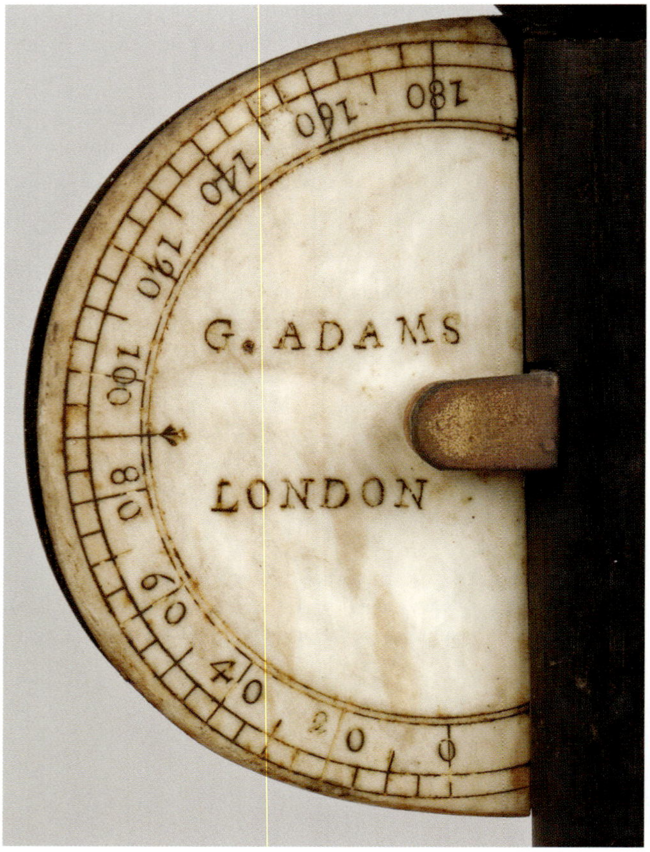

FIGURE 1 Adams' signature (inv. no. 20109-0000-)
© MAM/PHOTO FRANCK BOTTÉ

Adet, Pierre Auguste (1763–1834). Physician and, from the early 1780s, Lavoisier's laboratory assistant. In 1787 Lavoisier involved him in a project on the continuation of the *Description des arts et métiers*, and Adet was supposed to write the volumes devoted to the following arts: *brasseur, batteur et tireur d'or* (together with Berthollet), *bandagiste* (together with Gengembre), and *l'art des glaces au teint* (together with Gengembre).[5]

Alban, Léonard (1741–1803). Founder of the chemical manufactory Peeters et Alban, also known as *Javel*, in association with Jean Baptiste Peeters, in 1776. Lavoisier purchased 180 *livres* of vitriolic acid from him in November 1784.[6]

1 Unless otherwise stated, biographical information for Lavoisier's instrument makers was mostly drawn either from Daumas (1989), Marcelin (2004), Beaudouin, Brenni, Turner (2018) or from LC.
2 Adams (1797).
3 Museo Galileo, Florence, inventory no. 930/a.
4 MAM inv. no. 20109-0000-. See catalogue p. 262.
5 LC, vol. 5, pp. 279–285.
6 LC, vol. 4, pp. 47–48.

 | DOI:10.1163/9789004511217_013

Argand, François-Pierre-Aimé (1750–1803).[7] Swiss natural philosopher and inventor. After studying with Horace-Bénédict de Saussure Argand moved to Paris. In 1780 he patented a lamp with an increased intensity of light. The invention was appropriated by the Parisian pharmacist Antoine Quinquet (1745–1803), which resulted in an extended controversy over the true originator of the lamp.[8] Lavoisier owned several of these lamps, which in the inventories are named 'lampes Quinquet'.[9] It is likely that the lamps owned by Lavoisier were made by Quinquet (see below entry Quinquet).

Artaria, Carlo (fl. late eighteenth/early nineteenth centuries). Maker of meteorological instruments from Geneva and based in Mannheim, where he worked for the local Academy of Sciences. He collaborated with Benjamin Thompson, and Lavoisier's collection of instruments lists a thermometer (Fig. 2) and a barometer which might have belonged to Thompson.[10]

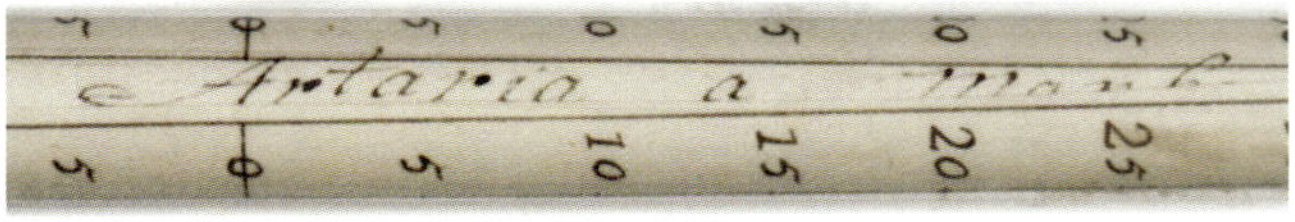

FIGURE 2 Artaria's signature (inv. no. 19917-0000-)
© MAM/PHOTO FRANCK BOTTÉ

Assier-Perica, Antoine (1730/31-pre-1806). Also known as Assier-Perricat. Parisian engineer and instrument maker, he became a skilled maker of meteorological instruments. During the 1770s he had a shop "Au Barometre Royal Rue St Antoine au coin de la Rue de Fourcy". Around 1775 Lavoisier purchased a thermometer "approuvé de l'Accademie [*sic*] Royale des Sciences" (Fig. 3).[11] He owned a similar one which is not dated.[12] Assier-Perica's son Auguste (born in 1767) continued the profession and distinguished himself as an equally skilful instrument maker.[13] According to the Danish astronomer Thomas Bugge (writing in 1798), Auguste was "a very well-known glass-blower" showing his skill "every evening in the Palais Royal".[14]

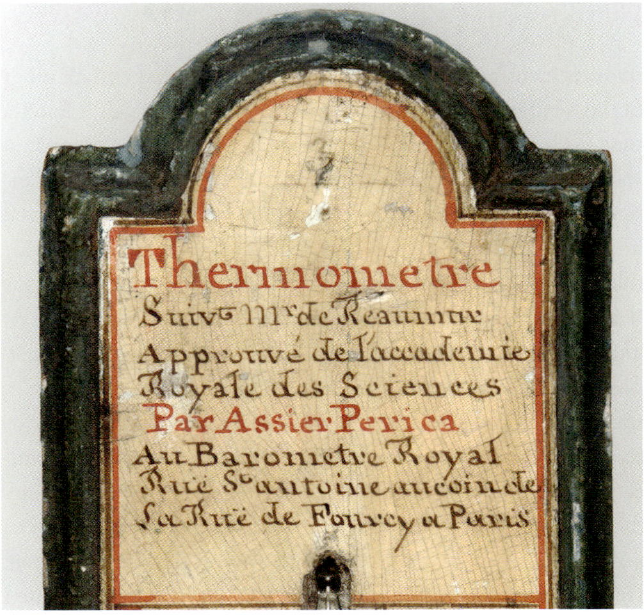

FIGURE 3 Assier-Perica's signature (inv. no. 19936-0000-)
© MAM/PHOTO FRANCK BOTTÉ

Baradelle, Jean Louis Jacques (1752–1794). The son of Nicolas Jacques and a maker of mathematical instruments with shops at the Au Quartier Anglais Marcelin (Paris). Lavoisier owned a surveyor level marked "Baradelle fils Paris 1770" for land surveying (Fig. 4) which was possibly designed to be used in Lavoisier's geological work.[15]

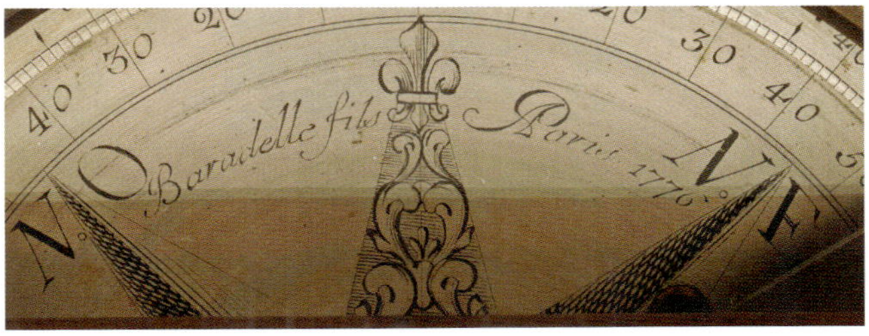

FIGURE 4 One of the two signatures by Baradelle on his surveyor level (inv. no. 20137-0000-)
© MAM/PHOTO FRANCK BOTTÉ

7 Wolfe (1999).
8 Quinquet was the author of the inventory of Lavoisier's pharmaceutical instruments in 1794. On Quinquet, see Dorveaux (1919).
9 LO, vol. 3, pp. 91–100 and Appendix 1, p. 129.
10 MAM inv. no. 19917-0000- (thermometer). See catalogue p. 277. MAM inv. no. 19952-0000- (barometer). See catalogue p. 293.
11 MAM inv. no. 19936-0000-. See catalogue p. 270.
12 MAM inv. no. 19937-0000-. See catalogue p. 268.
13 "M. Assier – Péricat, ingénieur pour les instrumens de physique en verre, demeurant rue des Prêtres St.-Germain-l'Auxerrois, n° 14, place de l'Ecole, près le Pont-Neuf, continue de fabriquer tous les appareils de physique et de chimie, tels que baromètres, thermomètres, hydromètres et aréomètres; le gravimètre de M. Guyton-Morveau; le thermoscope de M. de Rumfort; le tube de sûreté de M. Welther; l'aréomètre pour les salpêtres de M. Rifiault; les pèse-monnaies, les alcalitypes, berthollimètres, nouveaux thermomètres à air, baromètres à robinet, etc. M. Assier-Péricat, élève de son père, et successeur de Bétally, est connu par son application à donner à ses instrumens le degré de précision qui en fait le mérite, et qui lui a mérité la confiance de ceux qui en font usage". *Annales de chimie* 79 (1811): 224.
14 Crosland (1969), p. 172.
15 MAM inv. no. 20137-0000-. See catalogue p. 206.

Baradelle, Nicolas Alexandre, known as Baradelle l'ainé (fl. ca. 1740–1791). Engineer to the King and merchant maker of mathematical instruments, with a shop on the Quai de l'Horloge Paris. In 1787 Lavoisier purchased an electroscope from him.[16]

Baumé, Antoine (1728–1805). Parisian pharmacist and member of the Académie royale des sciences from 1773 onwards.[17] Baumé took on a shop in the Rue Saint-Denis, and in 1763 moved to a spacious site in the Rue Coquillière, where he created a pharmacy, drugstore, and manufactory of chemical products. In 1768 he gained public acclaim for his chemical hydrometer, a device that was still in use in the nineteenth century. By the early 1770s Baumé had become the most important supplier of chemicals and chemical equipment in Paris. Due to his successful business Baumé was familiar with the chemical trades present in the French capital, and in 1766 he wrote the following entries for the *Dictionnaire des arts et métiers*: "apothicaire, artificier, cabaretier, chaufournier, confiseur, dégraisseur, distillateur d'eaux fortes, épicier, faiencier, ferblantier, fournaliste, fumiste, orfèvre, parfumeur, platrier, potier d'étain, potier de terre, saunier, vernisseur, verrier".[18] In the entry on furnace makers he pointed out: "Il existe à Paris un très petite communauté de fournalistes: ils ne peuvent vendre leurs produits qu'aux personnes qui ont le droit de fondre les métaux et de distiller, ou avec la permission de la Monnaie, formalité qui n'est pas en générale pas observée".[19] From 1775 onwards Lavoisier purchased chemicals from Baumé on several occasions.[20] Lavoisier also owned at least six hydrometers of the Baumé type, but it is not known whether they were made by Baumé or by Cartier, who claimed to have created the same invention.[21] While Baumé never converted to the new chemistry he continued to collaborate with Lavoisier, and in 1788 they worked together at the Arsenal, where Baumé proposed a new procedure for refining saltpeter.[22]

Bavière, Jacques. Trader in Basel based at *l'enseigne aux trois pots rouge*. He provided Lavoisier with a barometer on 31 July 1767.[23]

Beringer, David (1756–1821).[24] A German dial maker born either in Nuremberg or in Dieppe. He was active in Augsburg around 1776 and in the neighbourhood of Nuremberg in 1798. He became known for building cubic wooden sundials. Lavoisier owned a sundial; its current location is unknown.[25]

Bernière, Claude (d. 1783). *Contrôleur général des ponts et chaussées*. In 1773 Bernière constructed an alcohol-filled glass lens of large diameter (2.6 m) for Trudaine de Montigny at Saint-Gobain. He also designed the elaborate mount constructed by Charpentier (see entry below). The lens was used by Lavoisier and Macquer on various occasions in 1775 and 1776.[26] The cost of the lens experiment is recorded in LC, vol. 3, pp. 649–650. The Revolutionary inventory transcribed in Appendix 2 reports that Lavoisier's laboratory still included one lens from the original apparatus.[27]

Berthoud, Ferdinand (1727–1807). A famous clock maker, *mécanicien de la Marine*, and instrument maker of the Académie des sciences, Berthoud was elected a member of the first class at the Institut National on 13 December 1795. Lavoisier, who was familiar with Berthoud's work in clock making, owned a set of eleven magnets made by his workshop, which he acquired in 1790.[28]

Bird, John (1709–1776). Mathematical instrument maker known for the high precision of his astronomical instruments.[29] In 1745 Bird opened his own business on the Strand in London. In March 1771 Borda gave Lavoisier a quadrant made by Bird, which the former compared with a quadrant lent to him by Cassini.[30] It is not clear if Bird's instrument described in the Revolutionary inventories of Lavoisier's estate is the same one that was used in 1771.[31]

16 MAM inv. no. 20117-0000-. See catalogue p. 263.
17 Delunel (1805); Dorveaux (1918); Davy (1955).
18 Macquer Ph. (1766).
19 Cited in Davy (1955), p. 94.
20 LC, vol. 2, p. 533; LC, vol. 3, pp. 558–559; ibid., p. 575.
21 MAM inv. no. 19956-0001-0006- (Baumé hydrometers). See catalogue p. 323. Cartier (1783); Baumé (1797), pp. 401–405.
22 Davy (1955), p. 129.
23 MS Lavoisier (1767–1788), vol. 3, fol. 79v Grimaux (1888), p. 20.
24 Chandler (1969).
25 The instrument arrived at the MAM in 1952, and its inventory no. was 20144-0000- (McKie 280). See catalogue p. 393.
26 LC, vol. 22, pp. 377–384; Smeaton (1987).
27 Appendix 2, p. 151.
28 MAM inv. nos. 20053-0001-, -0002-, -0003-, -0004-, -0005-, -0006-, -0007-, -0008-, -0009, -0010-, -0011-. See catalogue p. 248.
29 Hellman (1932).
30 "M. Borda m'avait en outre confié un quart de cercle de ... pouces de rayon divisé par Bird avec lequel je pouvais évaluer les angles à 10 ou 12 secondes près. J'avais un autre instrument qui m'avait été prêté par M. de Cassini, mais dont je ne me suis pas servi parce qu'il ne donnait pas une précision aussi grande. J'ai su depuis que le quart du cercle de Bird avait quelques erreurs dans sa division; mais je ne crois pas qu'elles aient pu influer beaucoup sur la justesse de mes résultats au moyen de ligne. J'avais déterminé avec l'instrument des Ponts et chaussées des points de niveau à l'horizon sur lesquels je ramenais à chaque observation le zéro du quart de cercle". Lavoisier (1792), p. 209.
31 See Appendix 3, p. 155.

Bleuler, John (ca. 1757–1829). London maker of optical and mathematical instruments. He opened his shop around 1790. Lavoisier owned a pocket thermometer by Bleuler.[32] (Fig. 5)

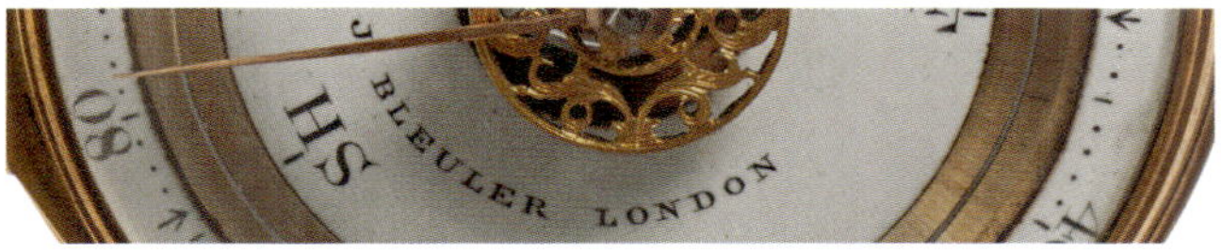

FIGURE 5 Bleuer's signature (inv. no. 20577-0000-)
© MAM/PHOTO FRANCK BOTTÉ

Borda, Jean Charles (1733–1799). A mathematician and military engineer, Borda entered the Académie royales des sciences in 1764. He collaborated with Lavoisier and Laplace on their experiments on the dilatation of metals in 1781–1782, as well as in May 1793 (with the assistance of Lenoir). He vindicated the authorship of an instrument made by Lennel .[33]

Brancas, Louis-Léon-Félicité de, 3rd Duke de Lauraguais (1733–1824). Following his career in the military Brancas joined the Académie royale des sciences in 1758 as *adjoint mécanicien*. In the early 1760s he developed a keen interested in applied chemistry, and in 1764 he founded a porcelain manufactory in collaboration with Jean Darcet which used a new recipe of his own invention, based on the recently discovered Kaolin, but was short-lived and closed in 1768. Brancas was also credited with having one of the best equipped chemical laboratories in Paris. Lavoisier owned one crucible from the Brancas-Lauraguais manufactory, which appears to be the only one that survives today.[34] (Fig. 6)

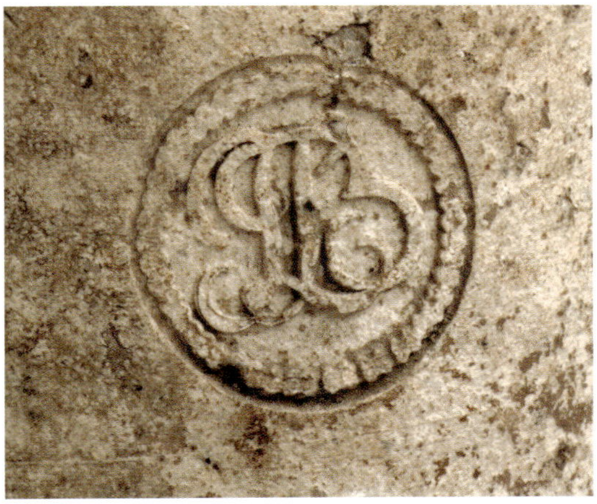

FIGURE 6 Brancas' stamp (inv. no. 20040-0000-)
© MAM/PHOTO FRANCK BOTTÉ

Brander, Georg Friedrich (1713–1783). A German mathematician and physicist, Brander also became a skilled and well-known maker of precision instruments.[35] In ca. 1781 Lavoisier purchased a surveyor's sextant from him.[36] (Fig. 7)

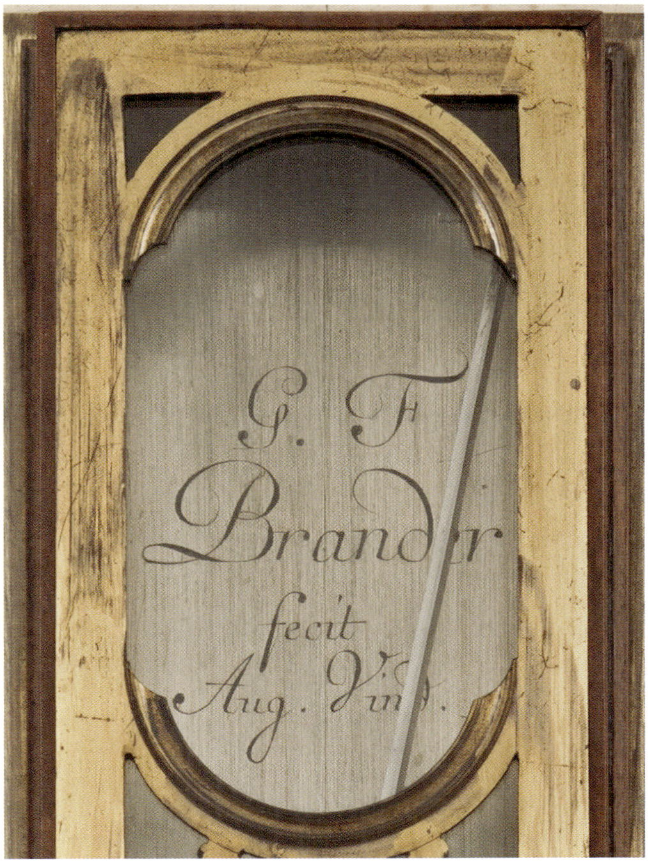

FIGURE 7 Brander's signature (inv. no. 20136-0000-)
© MAM/PHOTO FRANCK BOTTÉ

Bucquet, Jean-Baptiste-Michel (1746–1780). A physician by education, Bucquet became a professor of chemistry at the École de Médecine in Paris in 1776. He collaborated extensively with Lavoisier. On 26 October 1776 he delivered two small glass bells, four small curved retorts, four *livres* of mercury and (via Baumé) two *onces* of precipitated *per se* to Lavoisier.[37]

Cadet de Gassicourt, Louis Claude (1731–1799). Pharmacist and Rouelle's pupil. In the early 1760s he opened a shop as *marchand apothicaire* in the Rue Saint-Honoré, at its corner with the Rue de Grenelle-Saint-Honoré (today Rue Jean-Jacques-Rousseau).

32 MAM inv. no. 20577-0000-. See catalogue p. 280.
33 Daumas (1955), pp. 155–156.
34 MAM inv. no. 20040-0000-; Chavagnac (1906), pp. 399–405. See catalogue p. 306.
35 Brachner (1983).
36 MAM inv. no. 20136-0000-. See catalogue p. 210.
37 LC, vol. 3, p. 580.

He supplied Lavoisier with five *livres* of ether and 3 *onces* of oil of rosemary in February 1787.[38]

Canivet, Jacques (1720–1773). Canivet was the nephew of Claude Langlois, opened a shop of mathematical instruments on the Place du Marché Neuf in 1743, and moved to the Quai de l'Horloge on the Île de la Cité in 1747, where he opened his shop with the famous signboard *A la sphère*. In 1762 Canivet sold Lavoisier a brass alidade for a plane table, the first instrument recorded as having been purchased by Lavoisier.[39] (Fig. 8) Canivet also made a copper recipient for Lavoisier's experiments on the specific gravities of fluids (1768).[40]

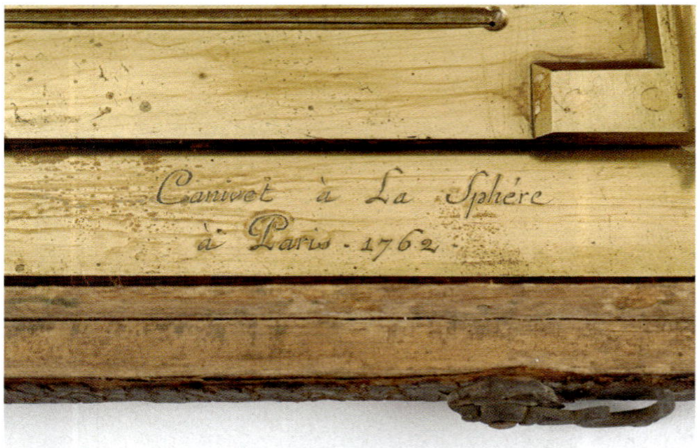

FIGURE 8 Canivet's signature and address (inv. no. 20138-0000-)
© MAM/PHOTO FRANCK BOTTÉ

Cappy, Toussaint (fl. 1760s–1775). Cappy was a maker of thermometers and barometers, calling himself in advertisements "fabricateur d'instrumens de physique", and had a shop in the Rue et Place Royale in Paris.[41] In 1768 his son was associated with Mossy, Cappy's nephew, and continued in the same shop.[42] They got a patent from the Académie des sciences in May 1775. In the mid-1760s Lavoisier purchased, probably from Cappy senior, two mercury thermometers and two thermometers with wine spirit, which he used in 1767 during his scientific travels to Alsace and Lorraine with Guettard.[43] Cappy also graduated Lavoisier's hydrometers (made between 1766 and 1768).[44] The MAM holds three thermometers and one manometer.[45] (Figs. 9a and 9b)

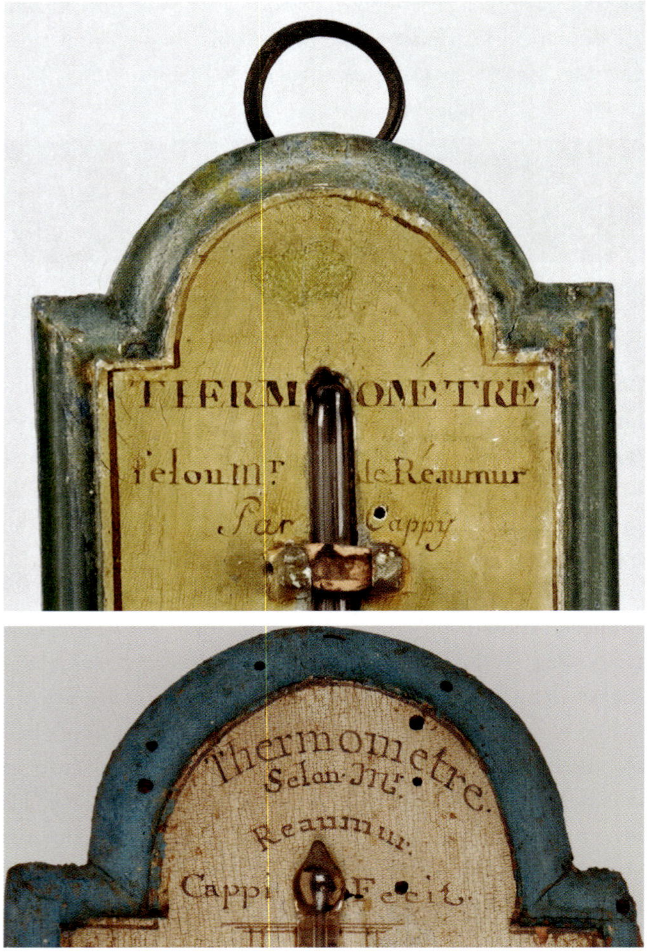

FIGURES 9A AND 9B Cappys's signatures (inv. nos. 19925-0000-, 19934-0000-)
© MAM/PHOTO FRANCK BOTTÉ

Carochez, Noël Simon (ca. 1745–1813/14). Parisian maker of optical and mathematical instruments.[46] He had a shop on the Pont Notre-Dame near Saint-Denys-de-la-Charte. Lavoisier owned a simple magnifier with two lenses that came from Carochez.[47] (Fig. 10)

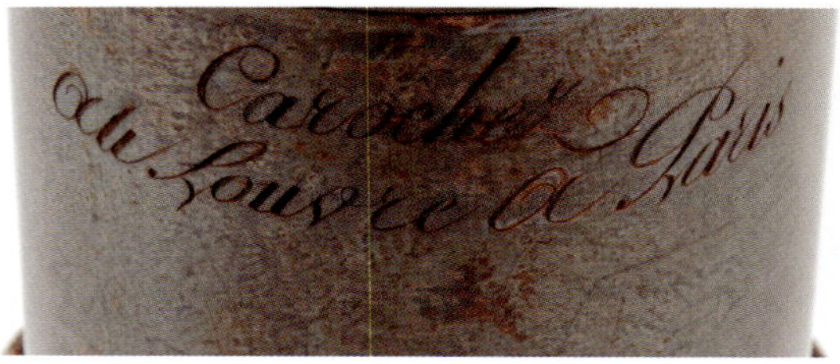

FIGURE 10 Carochez's signature and address (inv. no. 20087-0000-)
© MAM/PHOTO FRANCK BOTTÉ

38 LC, vol. 5, p. 10.
39 MAM inv. no. 20138-0000-. See catalogue p. 204.
40 LO, vol. 3, p. 438.
41 *Journal de politique et littérature* 13/2 (1775):513–514.
42 Avis (1775).
43 LC, vol. 1, pp. 44, 65, and 77.
44 LO, vol. 3, p. 446; see also ibid., pp. 708–709.
45 MAM inv. nos. 19925-0000- (made in 1768); 19926-0000- (ca. 1770); 19951-0000- (ca. 1775). See catalogue pp. 266, 271 and 225.

46 Chapin (1971).
47 MAM inv. no. 20087-0000-. See catalogue p. 399.

Cartier (fl. mid-eighteenth century). Instrument maker probably residing in Paris, who is exclusively known for the construction of hydrometers. In 1768 Lavoisier and Brisson reported unfavourably on Cartier's metal hydrometer before the Académie royale des sciences.[48] In August 1771 Cartier proposed a new glass hydrometer to the Académie, which was an instant success and was adopted by Lavoisier at the Régie des poudres in 1776. However, the invention was contested by Baumé, who felt that his invention of 1768 had been plagiarised (see entry Baumé).

Casbois, Nicolas (1728–1795). Also known as Dom Casboix. Member of the Académie in Metz and prior of the Abbay of Saint Symphorien. Casbois made comparable hygrometers.[49] He also introduced some important innovations to the making of barometers, and Lavoisier adopted his method in the construction of a barometer with a flat surface.[50]

Cassini, César-François (1714–1784). Astronomer, cartographer, and director of the Observatoire de Paris.[51] In 1770 Lavoisier made observations and carried out tests with Cassini's instrument for measuring latitude.[52] About a year later (on 28 March 1771) he measured the height of Mont Valérien with another instrument devised by Cassini.[53]

Caumont, Jean (1736–1800). Famous French cabinet maker. His furniture in the Louis XVI period and Transition styles were elegant, extremely well made, and included a remarkable variety of materials. Lavoisier owned a roll top desk made by Caumont in 1774.[54]

Chabrol de Murol (active fl. late eighteenth/ and beginning of the early nineteenth centuries). Astronomer and instrument maker. Lavoisier (?) owned a Torricellian barometer signed and marked "Paris, 1768". (Fig. 11) A mark added in 1821 suggests that the barometer might have belonged to Marie Anne Lavoisier.[55]

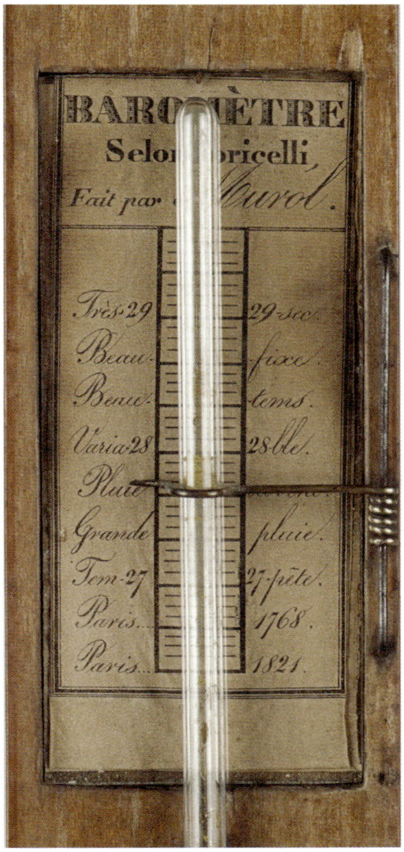

FIGURE 11
Chabrol de Murol's signature (inv. no. 19953-0000-) © MAM/PHOTO FRANCK BOTTÉ

Charpentier, François-Philippe (1734–1819). The son of a bookbinder, Charpentier began to work as an engraver, and in his spare time created several inventions, the most successful of which was a perfected version of the Papin digester. Sometime before 1774 the king appointed him *mécanicien du roi* and granted him an apartment close to the Jardin de l'Infante at the Louvre. Lavoisier purchased a Papin digester from Charpentier for 780 *livres* in February 1776.[56] The digester was used also in the 1783 experiments on the decomposition of water.[57]

Chemin, Nicolas (d. 1780). *Adjusteur de la Monnaie* and balance maker. He opened a shop at No. 4 Rue de la Ferronerie, which was still active in 1819.[58] Lavoisier purchased two small portable balances which he used during his experiments. One balance was used for the experiments on the presumed transmutation of water into earth, and in 1773 Lavoisier declared it to be extremely accurate.[59] Lavoisier used Chemin's balance

48 LO, vol. 4, pp. 17–20.
49 *Observations sur la physique* 29 (1786): 349–352.
50 MS Lavoisier (1772–1788), vol. 6, fols. 98–105; Lavoisier (1783a).
51 Wolf (1902).
52 LO, vol. 4, pp. 55–67.
53 LO, vol. 5, pp. 205–214.
54 MAM inv. no. 20576-0000-. See catalogue p. 390.
55 MAM inv. no. 19953-0000-. See catalogue p. 295.

56 LC, vol. 3, pp. 544–549, and pp. 559–560. The instrument made by Charpentier appears to be MAM inv. no. 20098-0000-. See catalogue p. 308.
57 LO, vol. 2, pp. 334–359.
58 Rapport (1819a).
59 MAM inv. no. 07544-0000-. See catalogue p. 187. LO, vol. 2, p. 17.

again in the 1774 experiments on the calcination of tin in closed vessels.[60] He also owned a set of weights by Chemin.[61]

Chevalier, Charles (1804–1859). Optician and instrument maker with a shop at the Palais Royal in Paris. The Lavoisier collection at the MAM includes a camera 'clara', patented by Chevalier.[62] (Fig. 12) As the only known Parisian by the name of Charles Chevalier was active in the first half of the nineteenth century, the instrument was probably owned by Marie Anne Lavoisier or by her heirs.

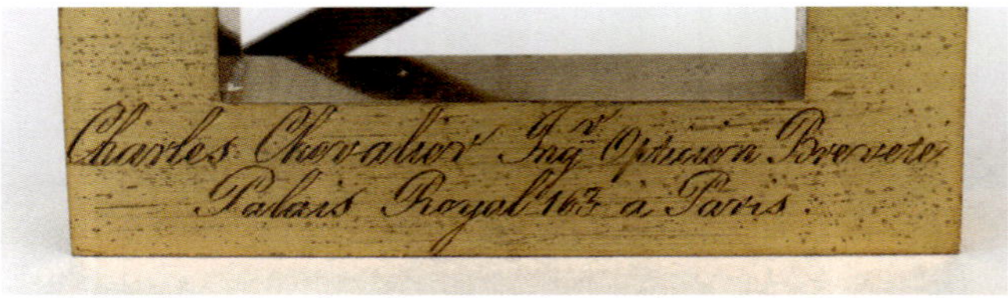

FIGURE 12 Chevalier's signature and address (inv. no. 20108-0000-) © MAM/PHOTO FRANCK BOTTÉ

Clair, Pierre (1804–1870). Maker, inventor and entrepreneur. Around 1840, he made a reconstruction of the apparatus used by Laplace and Lavoisier in 1781 to demonstrate the linear dilatation of metals.[63] This model does not belong to the original Lavoisier collection, but was included in the collection at a very early stage by the MAM curators.

Crichton, James (fl. 1785–1835). Famous instrument maker based in Glasgow. Lavoisier owned four mercury thermometers by Chrichton.[64] (Fig. 13)

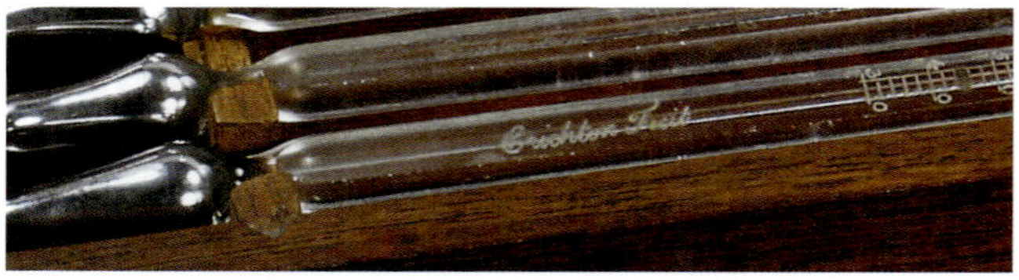

FIGURE 13 Chrichton's signature (inv. no. 19912-0001-) © MAM/PHOTO FRANCK BOTTÉ

Delalain (fl. Eighteenth century). No further information about this maker has survived other than Lavoisier owned a Roman balance.[65]

Delamarche, Charles François (1740–1817). Famous French cartographer, known as a maker of globes and geographical and astronomical instruments. Lavoisier owned a Copernican armillary sphere that appears to have been made by Delamarche.[66]

Dellebarre, Louis-François (1726–1805). Dellebarre was born in Paris and worked in Holland for many years. He was best known for his optical instruments and, in particular, for a new type of microscope which he presented to the Académie des sciences in 1777. In spite of their commercial success, Dellebarre's microscopes had poor optical performance. Lavoisier bought one such microscope in 1765 for his experiments on gypsum.[67] He bought a new microscope from Dellebarre in December 1778.[68]

Deluc, Jean-André (1727–1817). Swiss meteorologist and naturalist based in England.[69] In 1780 Lavoisier used a De Luc thermometer in a study on the action of heat on gases.[70]

Deparcieux, Antoine (1703–1768). French mathematician and member of the Académie des sciences from 1758 onwards. He invented several machines and devices. Lavoisier knew and used Deparcieux's hydrometer.[71]

De Wendel, François-Ignace (1741–1795). Founder and director of the Fonderie royale of Le Creusot. On 14 March 1789 he sent Lavoisier a box containing several specimens of hematite.[72]

Dinon (fl. Eighteenth century). Lavoisier owned a barometer and thermometer inscribed "Barometre Thermometre de Reaumur par Dinon", purchased sometime during the 1760s.[73]

Dumotiez, Pierre François (1743–1817) and **Louis Joseph (1757–1815).** Parisian instrument makers with a shop in No. 2 Rue du Jardinet. According to the Danish astronomer Thomas Bugge's account of 1798, "Dumotiez [was] very skilful in making all kind of physical instruments, not only according to the usual French pattern, as they [were] sketched by Nollet, Sigaud de la Fond, and Brisson, but also according to other drawings submitted. [...] His prices [were] very low".[74] The Dumotiez' also prepared

60 LO, vol. 2, pp. 108–109.
61 MAM inv. no. 19901-0000-. See catalogue p. 197.
62 MAM inv. no. 20108-0000-. See catalogue p. 408.
63 MAM inv. no. 03069-0000-. See catalogue p. 239.
64 MAM inv. no. 19912-0001-0004-, see p. 273.
65 Appendix 1, p. 121.

66 MAM inv. no. 20228-0000-. See catalogue p. 203.
67 The microscope, now lost, is cited in LO, vol. 3, p. 141.
68 LC, vol. 3, pp. 631–632; MAM inv. nos. 20085-0001-, 20085-0002-, 20085-0003-. See catalogue p. 246.
69 Hübner (2010).
70 LO, vol. 2, p. 222.
71 In Lavoisier (1767), pp. 154–155, Lavoisier compares Deparcieux' instrument with his own, new, portable hydrometer (devised around 1767).
72 LC, vol. 6, pp. 28–31.
73 MAM inv. no. 19947-0000-. See catalogue p. 296.
74 Crosland (1969), p. 171.

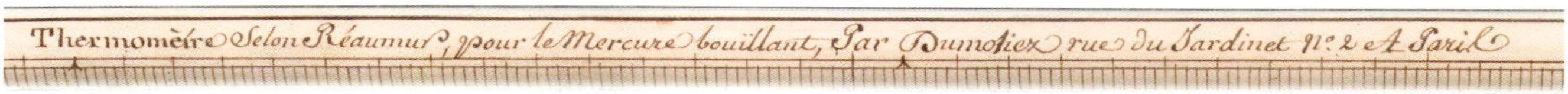

FIGURE 14 Dumotiez's signature and address (inv. no. 19945-0000-) © MAM/PHOTO FRANCK BOTTÉ

firework shows using hydrogen at the Lycée Répulicain.[75] One of the two brothers,[76] the maker of Guyton de Morveau's disinfecting apparatus, was called "Ingénieur en Instrumens de physique" in 1804.[77] In 1801 Luigi Valentino Brugnatelli reported that Dumotiez, "eccellente macchinista", constructed several pneumatic apparatuses for the study of gases, and that they were exceedingly expensive.[78] Lavoisier owned two thermometers (Réaumur's scale), one of which was lost in 1952 during the MAM collection's move.[79] (Fig. 14)

Du Pont de Nemours, Éleuthère Irénée (1771–1834) began to work in Lavoisier's laboratory of the Arsenal in 1787 when he was admitted as a pupil to the newly established École des poudres, an institution that had been strongly recommended by Lavoisier since he had joined the Régie.[80] During his apprenticeship, Du Pont also took courses at the Collège Royal (in physics with Lefèvre-Gineau, in mineralogy with Daubenton) and at the Louvre with Charles. He eventually continued his studies in practical chemistry at the refinery in Paris and at the gunpowder manufactory in Essonnes. Even though we may assume that Du Pont worked as Lavoisier's assistant, it is not known if he had any active role in the conception of new instruments.

Fahrenheit, Daniel Gabriel (1686–1736). Dutch-German-Polish instrument maker, glass-blower, and natural philosopher. Lavoisier used a glass Fahrenheit hydrometer in 1777.[81]

Fallot, Jérôme (1747-?). This former *garçon de laboratoire* of Macquer was employed in Lavoisier's laboratory in around 1784, and was put in charge of the preparation of experiments.[82] During this period he was also called as "garde des experiences à l'Academie royale des sciences", and in 1791 he was appointed to "Inspecteur des poudres et salpetres".[83]

Fois dit La Rose, Louis (fl. Eighteenth century). Maker of mathematical instruments based in Clermont Ferrand and active towards the end of the eighteenth century. Lavoisier owned a surveyor cross by this maker.[84] (Fig. 15)

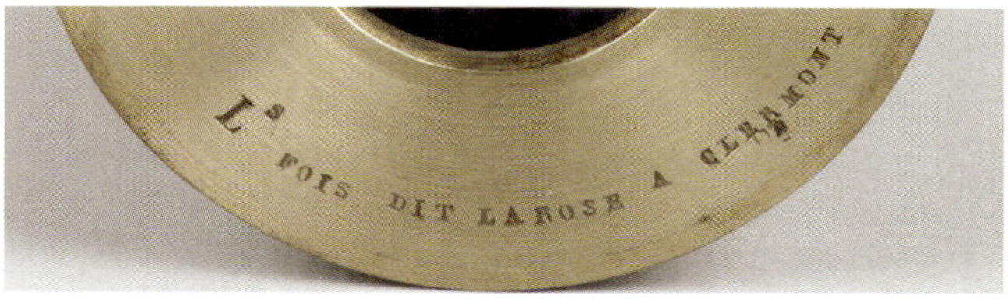

FIGURE 15 Fois' signature and address (inv. no. 20217-0000-) © MAM/PHOTO FRANCK BOTTÉ

Fontana, Felice (1730–1805). Italian natural philosopher and director of the Florentine Reale Museo di fisica e storia naturale (opened in 1775).[85] Fontana made several chemical instruments, the most important of which was the eudiometer (1775). He met Lavoisier during a long stay in Paris (January 1776–July 1778). Lavoisier owned one of Fontana's eudiometers. Fontana had also found that the red-hot charcoal extinguished in mercury, held in a glass tube immersed in a bath, would cause a great quantity of air to be absorbed. Subsequently, many other European naturalists started to use this method (which was also enthusiastically adopted by Joseph Priestley) to experiment with red-hot charcoal and various types of air; and in 1780 Fontana himself discovered that, if one extinguished red-hot charcoal in a glass bell full of water, inflammable air (hydrogen) was liberated. The device invented by Fontana was illustrated in David's double portrait of the Lavoisiers.[86] (Figs. 16 and 17)

75 Ibid., pp. 163–164.
76 The name was not specified. It is likely that by this time only one of the two brothers was still active.
77 *Bulletin de la Société d'encouragement pour l'industrie nationale* (Paris: Madame Huzard, 1804), p. 131.
78 Brugnatelli (1997), pp. 192, 194, and 206.
79 MAM inv. no. 19945-0000-. See catalogue p. 280.
80 Bret (1994), pp. 231–250; Bret (1995); see also Dujarric de la Rivière (1954).
81 Lavoisier (1777a), p. 259.
82 LC, vol. 4, p. 33. Bottée, Riffault (1811), p. 332.
83 Lehman (2019), p. 251.
84 MAM inv. no. 20217-0000-. See catalogue p. 205.
85 Knoefel (1984).
86 Beretta (2001).

Fortin, Nicolas (1750–1831). Instrument maker with a shop on the Place de la Sorbonne in Paris.[87] He had only two assistants: Jean Louis Cheron and Louis Marie Frédérique Pinson.[88] According to Thomas Bugge's account of 1798, Fortin was a "very good craftsman" but required "better payment than Dumotiez, because his workshop [was] not so well staffed and equipped".[89] Fortin became one of the most important suppliers for Lavoisier. In a letter to Achard dated 6th December 1789 Lavoisier declared that Fortin was the "ouvrier" he employed "le plus habituellement".[90] Fortin's earliest invoice for Lavoisier dates from 1st March 1784 and concerns his famous air pump.[91] Lavoisier owned two of these, one made between 1779 and 1782, and the other in 1792.[92] (Figs. 18a and 18b) From an invoice sent

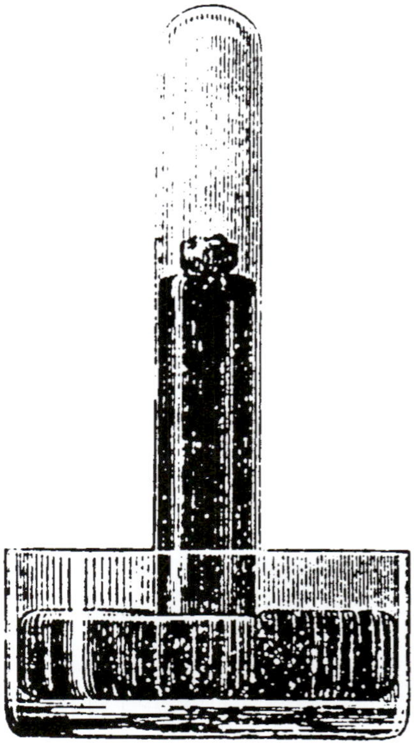

FIGURE 16
Felice Fontana's device for absorbing gases (1777) in a reproduction made in 1900 by the Italian historian of chemistry Icilio Guareschi

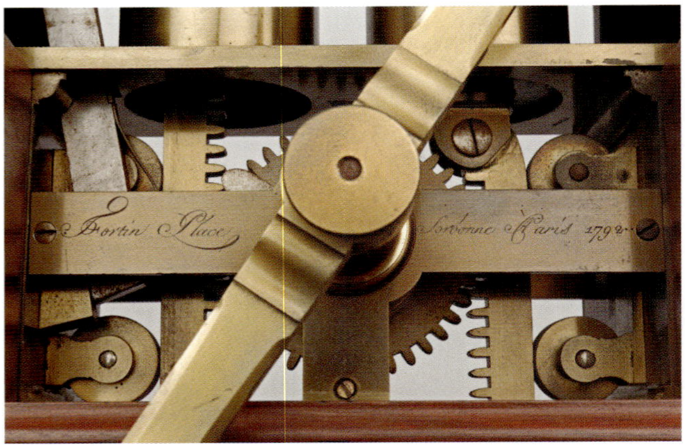

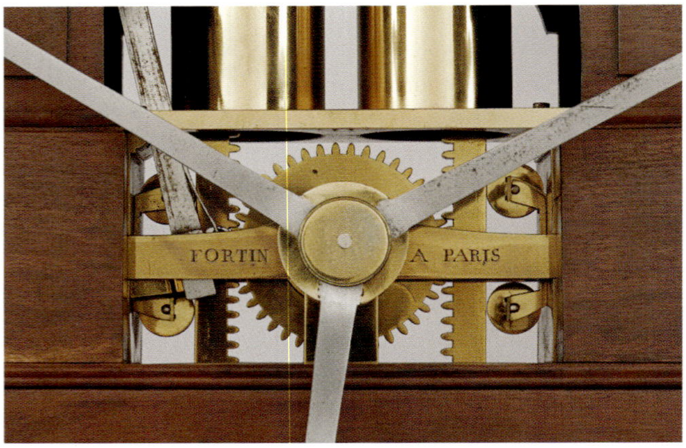

FIGURES 18A AND 18B Fortin's signatures on the two air pumps he made for Lavoisier (inv. nos 07517-0001-, 19904-0000-)

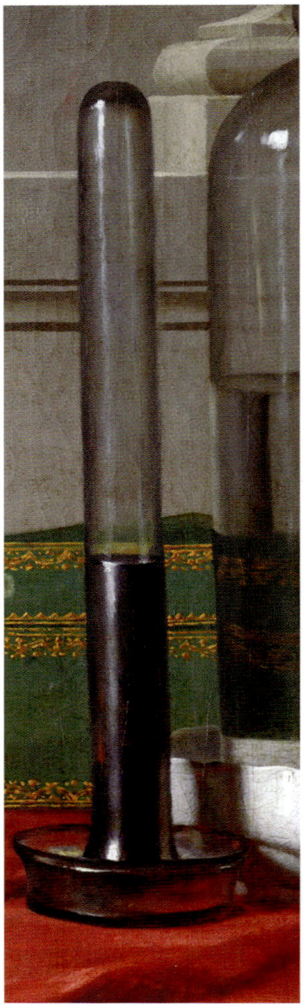

FIGURE 17
Detail of David's portrait of the Lavoisier's illustrating Fontana's device

87 On Fortin, see Turner (1989a).
88 LC, vol. 7, p. 391.
89 Crosland (1969), p. 171.
90 LC, vol. 6, p. 90.
91 LC, vol. 4, p. 11.
92 The MAM owns the one made in 1792 (inventory no. 07517-). See catalogue pp. 221–225.

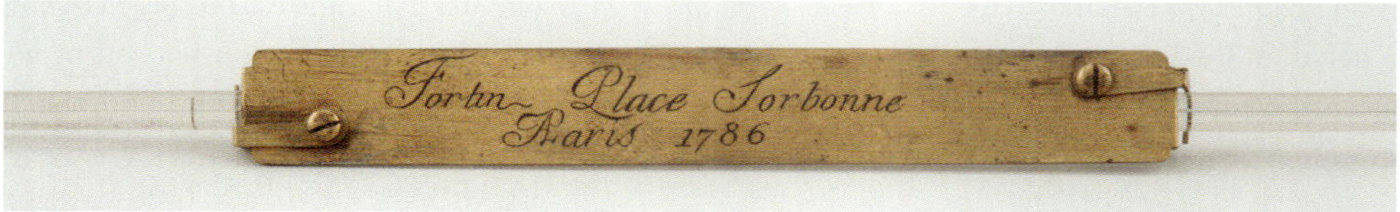

FIGURE 19 Fortin's signature and address on one of the thermometers he made for Lavoisier in 1786 (inv. no. 19908-0001-)
© MAM/PHOTO FRANCK BOTTÉ

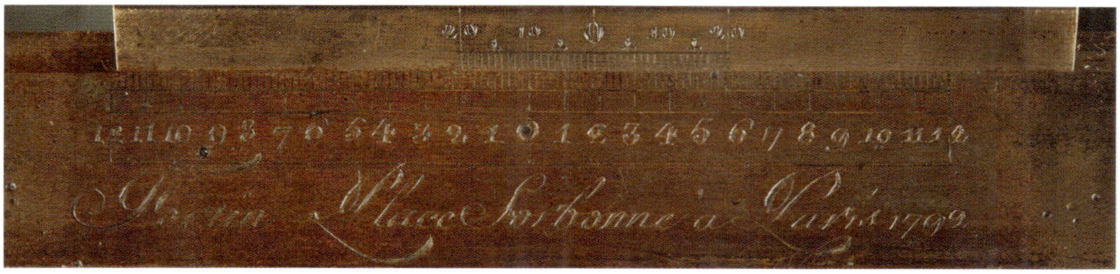

FIGURE 20 Fortin's signature and address (inv. no. 20105-0000-)
© MAM/PHOTO FRANCK BOTTÉ

on 20 July 1785 it is clear that Fortin provided Lavoisier with most of the equipment used in the experiments on water.[93] The total cost of the equipment amounted to 483 *livres*. In April 1786 Fortin provided Lavoisier with an electrical machine (172 *livres*), two pipes, and an apparatus for collecting nitrous air.[94] In June 1788 he sent another invoice for apparatus used in experiments on the synthesis of water at the Collège royal on 27 May and 7 June.[95] In August 1788 Fortin finished his large-scale apparatus for experiments on oil combustion[96] (1109 *livres*), a great balance (600 *livres*),[97] some apparatus for ethers (520 *livres*) and another for the combustion of carbons (110 *livres*).[98] Between 1788 and August 1789 he supplied Lavoisier with two apparatuses for wine fermentation.[99] On 9 May 1790 he sent an invoice for 480 *livres* for a few pieces of equipment and the readjustment of existing apparatus.[100] On 1 December 1792 he sent an invoice for a total of 300 *livres* for the construction of several machines that are not further specified.[101] In addition to several thermometers, (Fig. 19) Lavoisier also owned a declinometer, (Fig. 20) a compass and a gasometer by Fortin.[102]

Fourché, C. (fl. 1777–1811). Balancier and *ajusteur de la Monnaie*. He took over Chemin's shop in No. 4 Rue de la Ferronerie in 1780. In 1788 Fourché made decimal subdivisions of the *livre poids de marc* for Lavoisier and named this series of weights "pile cylindrique pyramidale de Lavoisier".[103]

Fraser, William (fl. 1777–1824). A well-known maker of mathematical instruments with a shop at 3 New Bond St., London. He also made instruments for Henry Cavendish. Lavoisier owned a pedometer with compass by Fraser, which he acquired around 1785.[104] (Fig. 21)

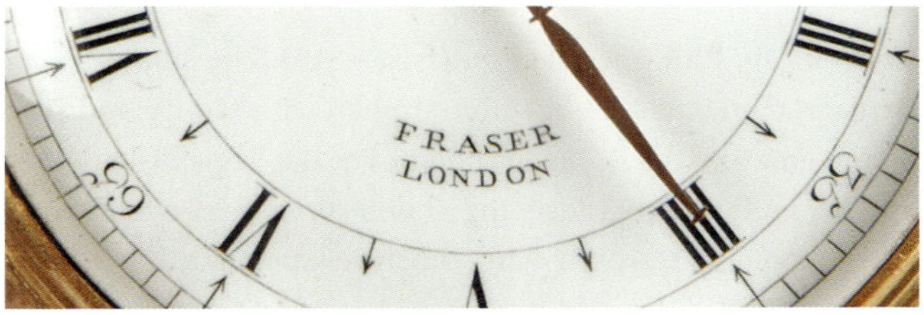

FIGURE 21 Fraser's signature (inv. no. DW0619)
HARVARD UNIVERSITY, COLLECTION OF HISTORICAL SCIENTIFIC INSTRUMENTS

Furet, Jean-Baptiste-André (fl. Eighteenth century). *Horloger du Roi* with a shop situated in the Rue Saint-Honoré. Furet became famous the late 1770s and early 1780s both for the fine decoration of his clocks and for their precise mechanisms. Lavoisier owned a pendulum clock made by Furet, which was

93 LC, vol. 4, pp. 137–140.
94 The electrical machine was used in March 1786; Berthelot (1890), p. 304. LC, vol. 4, pp. 212–213 (pipes and apparatus).
95 LC, vol. 5, pp. 176–177.
96 MAM inv. no. 07549-0000-. See catalogue pp. 351–353.
97 MAM inv. no. 19887-0000-. See catalogue p. 190.
98 LC, vol. 5, pp. 209–210.
99 LC, vol. 6, p. 64 and MAM inv. nos. 07550-0000- and 07551-0000-. See catalogue pp. 350–351.
100 LC, vol. 6, pp. 141–142.
101 LC, vol. 7, p. 169.
102 For the compass see Appendix 3, p. 155; for the "Ancien gazometre de Fortin", see Appendix 3, p. 154.

103 MAM inv. no. 19900-0000-. LO, vol. 1, p. 249. See catalogue p. 197.
104 Harvard University, Collection of Historical Scientific Instruments DW0619. See catalogue p. 210.

valued at the considerable sum of 600 *livres* in 1794 thanks to its rich decoration.[105] In early 1793, on behalf of the Bureau de consultation des arts et métiers, Lavoisier rejected Furet's proposal for a new method of printing paper money (*assignats*).[106]

Gaibert, see Goubert

Gallonde, Louis Charles (1715–1771). Parisian clock and instrument maker with a shop "aux Galeries du Louvre". Lavoisier owned a spirit thermometer made by Gallonde in 1768–1769.[107] (Fig. 22)

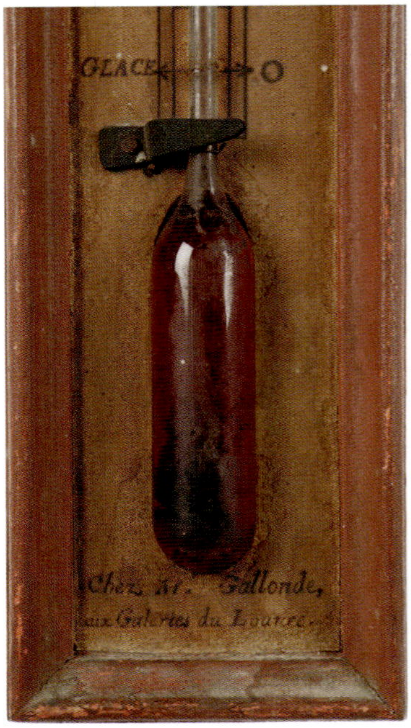

FIGURE 22 Gallonde's signature and address (inv. no. 19935-0000-) © MAM/PHOTO FRANCK BOTTÉ

Gengembre, Philippe Joachim (1764–1838). The son of a *concierge* at the Louvre, Gengembre studied with Monge and Sage before entering the École des poudres at the Régie des poudres and, as early as 1785, teaching chemical courses which combined updated research with practical chemistry.[108] In 1803 Gengembre became "inspecteur général" at La Monnaie. As a young chemist he collaborated with Lavoisier on several experiments, and often helped him to assemble apparatus.[109]

Goubert, Guillaume Charles (1747–1824). Also known as Gaibert. Engineer and maker of meteorological instruments of the Société royale de médecine and of the Académie de Dijon. He had shops in Paris in the Rue Dauphine, called l'Anjou, and in the Rue St André des Arts. In 1781 he published a famous *Description et usage des baromètres, thermomètres et autres instrumens météorologiques*, which Lavoisier had in his library.[110] Lavoisier purchased a thermometer from him which was made in 1774, and a hygrometer from around 1780.[111] (Fig. 23)

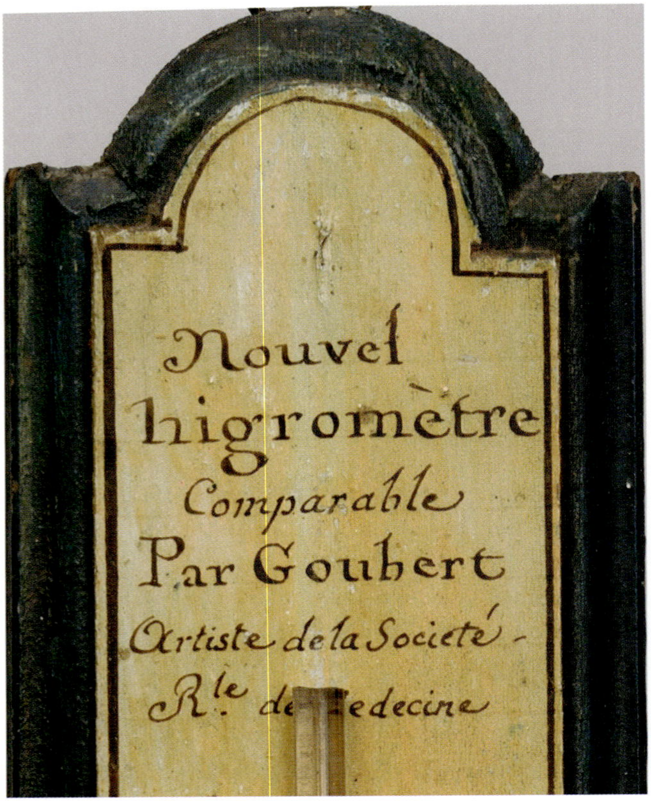

FIGURE 23 Goubert's signature (inv. no. 19932-0000-) © MAM/PHOTO FRANCK BOTTÉ

Grosset, Laurent (fl. 1675–1724). A balance maker active in Lyon towards the end of the seventeenth century, Grosset made portable folding scales which became quite popular. Lavoisier owned one of them.[112]

Hall, James (1761–1832). Scottish geologist. Hall studied chemistry under the guidance of Joseph Black. He worked in Lavoisier's laboratory during his stay in Paris in 1791, and there contributed to the perfecting of the furnace made by Séguin for the fusion of platinum.[113]

Hanin (father and son, fl. 1760s–1790s). Hanin the elder was a locksmith who also made balances.[114] Buffon (in a likely

105 See Appendix 1, p. 134.
106 LO, vol. 4, pp. 703–704.
107 MAM inv. no. 19935-0000-. See catalogue p. 267.
108 On Gengembre, see Bret (1994).
109 LC, vol. 4, p. 6; LC, vol. 6, p. 153.
110 Beretta (1995).
111 MAM inv. nos. 19938-0000-, 19932-0000-. See catalogue p. 268 and p. 298.
112 McKie 327. See catalogue p. 195. On Grosset, see Turner A.(2018), p. 288.
113 Chaldecott (1968), p. 34.
114 Hanin (1768).

reference to him) considered him as an "habile artiste" in making balances.[115] Lavoisier owned a large spring balance by Hanin, but it is not clear whether it was made by father or son.[116] Hanin the younger was also a well-known balance maker.[117]

Hassenfratz, Jean-Henri (1755–1827). Hassenfratz first trained as a carpenter under the guidance of Nicolas Fourneau, whose technical education fostered Hassenfratz' interest in solving technical problems.[118] After further training with Monge he became a mining inspector, and began to work regularly in Lavoisier's laboratory at the Arsenal in the early 1780s. While at the Arsenal, Hassenfratz received regular financial support from Lavoisier. The French chemist often supported artisans who found themselves in financial difficulties.[119]

Jacob, Georges (1739–1814) was one of the most famous cabinetmakers of his time. He became a master cabinetmaker in 1765. An innovator, he was among the earliest cabinetmakers in France to use mahogany. He made a large number of carved, gilded, and painted pieces of furniture for the French royal châteaux. Lavoisier owned two console tables by Jacob.[120]

Janety, Marc Etienne (ca. 1750–ca. 1823). Royal goldsmith of Louis XVI until 1792 and instrument maker. He lived in the Rue de Colombier (today Rue Jacob) and sold part of his estate to Fourcroy and Vauquelin so that they could open their workshop for the production of chemicals there.[121] He became famous for his experiments on platinum and the method, based on the use of arsenic, of rendering it malleable. In 1790 Lavoisier, via Séguin, purchased some artefacts from Janety which served in his experiments on the fusion of platinum.[122] (Fig. 24) He showed two of them before the Académie des sciences, and the platinum salt cellar with spoon now preserved at the Kroch library seems to be one of them.[123]

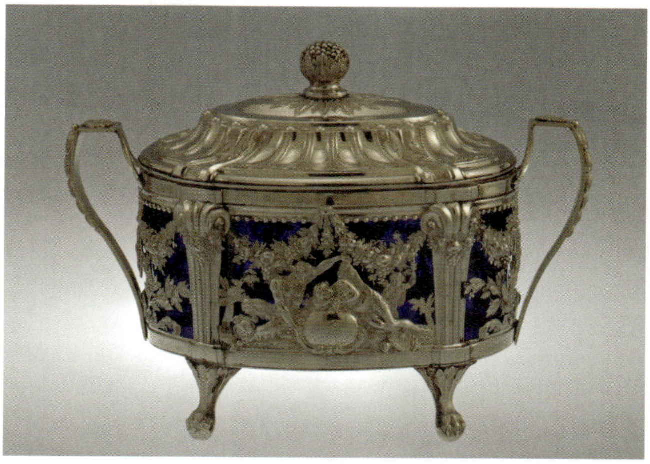

FIGURE 24 Platinum and blue glass sugar bowl made by Janety in 1786
COURTESY THE METROPOLITAN MUSEUM – NEW YORK

Knight, Gowin (1713–1772). British natural philosopher, inventor, and scholar, known for his discovery of a process for forming strongly magnetised steel, for which he was awarded the Royal Society's Copley Medal in 1747. Lavoisier owned two long, rectangular magnetic bars made according to Knight's principle.[124]

Langlois, Claude (1703–1756). An engineer at the Académie des sciences, Langlois opened a shop at the Galleries du Louvre in 1736. He specialised in mathematical instruments. Lavoisier owned a graphometer by Langlois.[125] (Fig. 25)

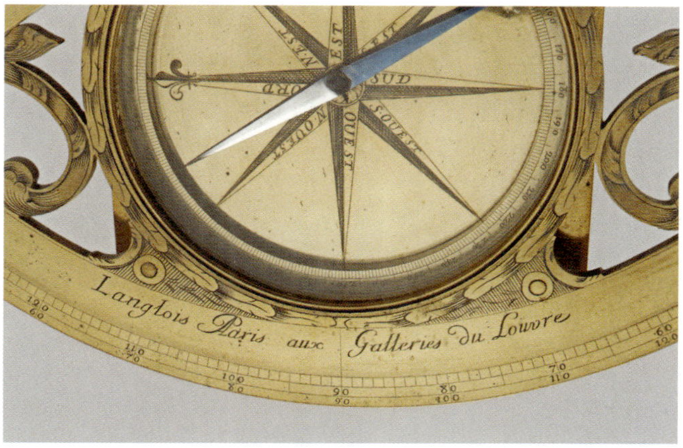

FIGURE 25 Langlois' signature and address (inv. no. 20135-0000-)
© MAM/PHOTO FRANCK BOTTÉ

115 "[L]a troisième espèce [de balance], qu'on appelle peson ou balance à ressort, n'a pas besoin de poids, et donne la pesanteur des masses par un index numéroté; enfin la quatrième espèce est celle où l'on emploie un seul poids attaché à un fil ou à une chaîne qu'on suppose parfaitement flexible, et dont les différens angles indiquent les différentes pesanteurs des masses. Cette dernière sorte de balance ne peut être d'un usage commun, par la difficulté du calcul et même par celle de la mesure des angles; mais la troisième sorte, dans laquelle il ne faut point de poids, est la plus commode de toutes pour peser de grosses masses. Le sieur Hanin, habile artiste en ce genre, m'en a fait une avec laquelle on peut peser trois milliers à la fois, et aussi juste que l'on pèse cinq cents livres avec une autre balance". Buffon (1817–1818), vol. 5, p. 432.
116 See Appendix 3, p. 154.
117 Hanin (1791).
118 Grison (1996).
119 Ibid., pp. 84–85.
120 MAM inv. nos. 20575-00001-, 20575-00002-. See catalogue p. 390.
121 Kersaint (1966), p. 282.
122 Chaldecott (1968), p. 33; Lavoisier (1790).
123 Kroch Library, Cornell University, Ithaca NY, Lavoisier 4712 Box 29. See catalogue pp. 382–383.
124 MAM inv. no. 20049-0000-. See catalogue p. 247.
125 MAM inv. no. 20135-0000-. See catalogue p. 208.

Launoy, Jean Baptiste (fl. Eighteenth century). Well-known Parisian merchant and importer of chemical products.[126] Lavoisier purchased 24 *livres* and 14 *onces* of platinum for the sum of 680 *livres* and 10 *sols* from him, some of it on behalf of the Académie des sciences.[127]

Lefebvre (dates unknown). Unidentified Parisian instrument maker. Lavoisier owned a brass standard measure of volume by him.[128]

Le Gaux (Legaux), Pierre (1748–1827). "M. Le Gaux [from Metz], avocat en Parlement, plein de talent pour l'observation. & génie pour la perfection des instrumens".[129] He was known for his thermometers, and Lavoisier's correspondent Louis Cotte often praised the accuracy of his meteorological instruments. Lavoisier owned a storm glass by Le Gaux.[130]

Le Maire, Pierre (fl. 1730–1760). French maker of astronomical instruments based in Paris "au nouveau quartier anglais". Lavoisier owned a silver "petit cadran à boussolle".[131]

Lennel, Louis-Pierre Florimond (1743–1781) -). Engineer to the king and maker of mathematical instruments. Lennel owned a shop on the Quai des Grands Augustins in Paris. Lavoisier owned a precision ruler by Lennel, which he used for experiments measuring thermal expansion together with Laplace in 1782.[132] Borda eventually established himself as the originator of this instrument.[133]

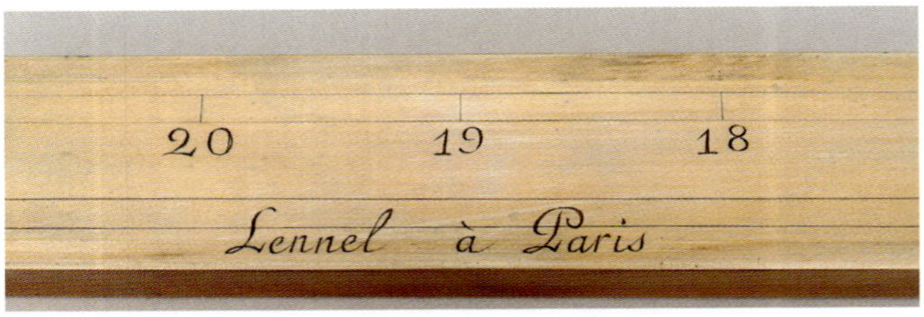

FIGURE 26 Lennel's signature (inv. no. 19907-0000-) © MAM/PHOTO FRANCK BOTTÉ

Lenoir, Etienne (1740–1830). Famous Parisian maker of mathematical and astronomical instruments.[134] He had a shop in the Rue de la Pelleterie, and from 1787 onwards a large workshop in the Rue Basse des Ursins. It seems that Lavoisier used this instrument maker's services only in connection with the activities of the Comité des poids et mesures, for experiments on the dilatation of platinum he conducted together with Borda.[135] Lenoir made a comparator for these experiments.[136]

Le Rebours, Noël Jean (1761–1840) Parisian optician and astronomical instruments maker active between the end of the eighteenth century and the first two decades of the nineteenth. Le Rebours was highly appreciated by the astronomers of the Académie des sciences, particularly for the quality of his achromatic lenses. Lavoisier had one unfinished achromatic lens by Le Rebours.[137]

Magellan, Jean Hyacinthe de (1722–1790). This wandering Portuguese savant settled in London at the beginning of the 1760s after travelling through Italy and France, and soon gained a reputation as a scientist, instrument maker, inventor, and, above all, trader of instruments and apparatus. In 1770 he was elected a fellow of the Society for the Encouragement of Arts, Manufactures, and Commerce, and in 1774 of the Royal Society; in 1780 he was also a co-founder of the Coffee House Philosophical Society, an informal society dedicated to discussing new discoveries in pneumatic chemistry. Magellan was a supplier of English chemical and physical instruments for Trudaine de Montigny and, in the 1770s, Lavoisier. Lavoisier probably owned Magellan's perfected apparatus for making mineral water, which had been invented by Nooth in 1775 (see entry Nooth *ad vocem*).[138]

Magny, Alexis (1712–ca. 1777). Engineer, clock and instrument maker based in Paris. Lavoisier owned a barometer by Magny which he used in his meteorological campaign of 1778.[139]

126 In 1788 Felice Fontana purchased minerals from Jean Baptiste Launoy: MS Fontana (1788), c. 399.
127 LC, vol. 4, pp. 64–65.
128 MAM inv. nos. 07542-0028-001- and 07542-0028-002-. See catalogue p. 187.
129 *Histoire de la Société Royale de Médecine* (1779) [but published in 1782], p. 177.
130 MAM inv. no. 20030-0000-. See catalogue p. 299.
131 Appendix 3, p. 155. This item was part of the collection of M. Lammot du Pont Copeland in 1952, but its present location its unknown. See catalogue p. 393. On Le Maire, see Morrison-Low, Schechner, Brenni (2017), pp. 148–156.
132 MAM inv. no. 19907-0000-. See catalogue p. 184.
133 Daumas (1955), p. 155.
134 Turner (1989).
135 "Les expériences pour la dilatation relative du cuivre et du platine se firent l'année suivante, du 24 mai au 5 juin, dans le jardin de la maison que M. Lavoisier occupoit alors sur le boulevard de la Nouvelle-Madeleine, et les bornes qu'on y avoit solidement établies pour cet objet ont subsisté tant qu'on a cru que leur conservation pourroit être utile. On verra tous les détails de ces diverses expériences dans deux mémoires de Borda". Delambre (1806–1810), vol. 1, p. 21.
136 Daumas (1955), p. 156.
137 See Appendix 3, p. 156.
138 Magellan (1777); see Appendix 3, p. 153.
139 LC, vol. 3, p. 626.

Mégnié, Pierre Bernard (1751–1807). Also known as Mégnié le Jeune. "Ingénieur en instruments mathématiques". He owned a shop at the Cour de commerce, Foire St. Germain in Paris. He was extensively used by Lavoisier, whom he called his patron ("protecteur") and benefactor ("bienfaiteur").[140] The first recorded instrument he made for Lavoisier was a barometer used in Lavoisier's meteorological campaign of 1778;[141] (Fig. 27) a pair of gasometers followed in 1787 (at the cost of 7554 *livres*).[142] (Fig. 28) Lavoisier also owned a mercury thermometer and a glass recipient used for the experiments on the composition of water (1786).[143] (Fig. 29) During the winter of 1781–1782 he contributed to the construction of the Ramsden optical pyrometer, which Lavoisier and Laplace used in their experiments at the Jardin de l'Arsenal.[144]

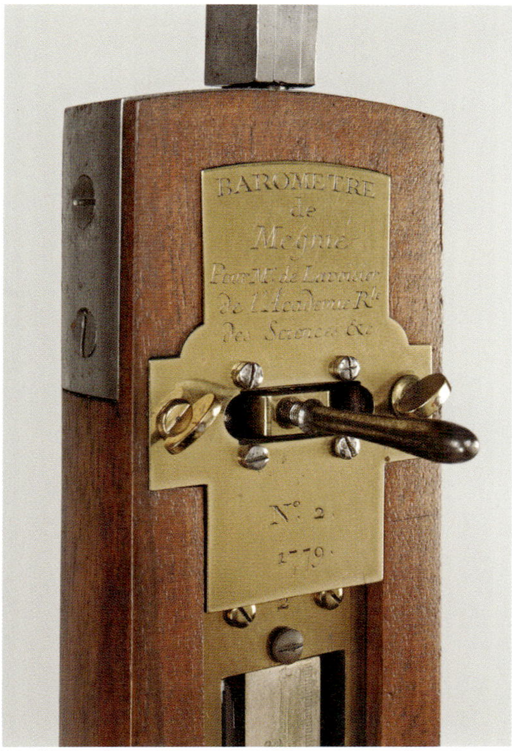

FIGURE 27 Mégnié's signature and address (inv. no. 19949-0000-)
© MAM/PHOTO FRANCK BOTTÉ

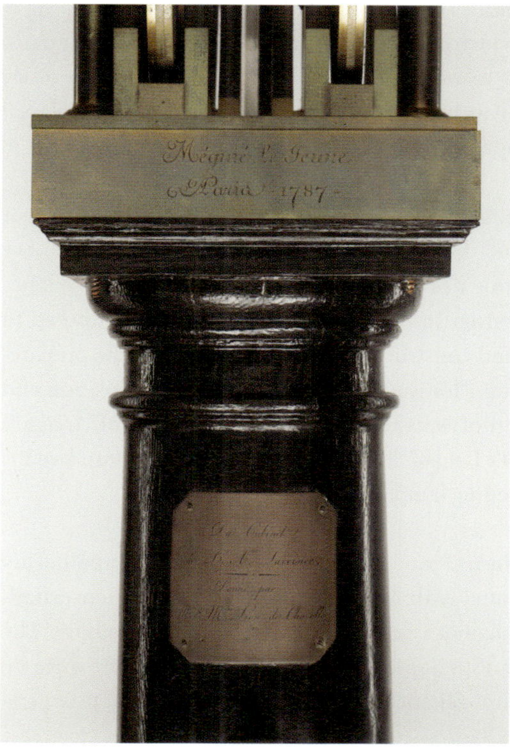

FIGURE 28 Mégnié's signature (inv. no. 07547-0001-001)
© MAM/PHOTO FRANCK BOTTÉ

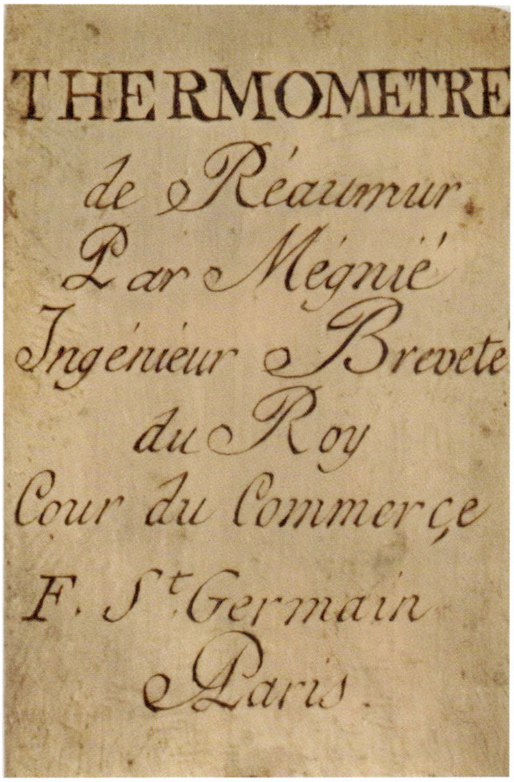

FIGURE 29 Mégnié's signature and address (inv. no. 19914-0000-)
© MAM/PHOTO FRANCK BOTTÉ

140 LC, vol. 4, pp. 66–67.
141 MAM inv. no. 07658-0000-. This was followed one year later by two more barometers (inventory nos. 08761-0000- and 19949-0000-); see LC, vol. 3, pp. 623–624. See catalogue pp. 288–290.
142 LC, vol. 5, pp. 52–55; MAM inv. no 07547-0001-001 and 002.
143 MAM inv. no. 19914-0000- (thermometer). Musée national d'histoire naturelle Paris, inventory no. OA 289 (glass recipient). See catalogue p. 348.
144 LO, vol. 2, pp. 739–764.

Mégnié the elder (l'ainé) (fl. Eighteenth century). Nothing is known of this instrument maker who, as suggested by Daumas, was probably related to Pierre Bernard Mégnié.[145] Lavoisier owned two balances by this maker, with only the assay balance in the collection of the MAM surviving.[146]

Meusnier de la Place, Jean-Baptiste (1754–1793). Engineer to the king and, from 1776 onwards, a corresponding member of the Académie des sciences in Paris.[147] He began to collaborate with Lavoisier in the summer of 1783 and contributed to the concept and construction of the gasometer,[148] the preparation of apparatus for the experiments on the composition and decomposition of water,[149] and the improvement of Argand's and Quinquet's lamps.[150] He also helped Lavoisier improve his oxygen enamelling lamp.

Mitouard, Pierre-François (1733–1786). Parisian apothecary who collaborated with Macquer and Lavoisier in their experimental campaign of 1772 on the combustion of diamonds. The same year, on 10th September, he sold Lavoisier "une once de beau phosphore venant d'Allemagne, pour 45 livres, prix de la facture".[151]

Mossy (fl. 1775–1821). Nephew of Cappy, was associated with Cappy fils in 1775 and began working in the shop in the Rue et Place Royale. As an instrument maker Mossy specialised in thermometers and barometers. Around 1780 he opened a shop on the Quai Pelletier-à-la-Croix-d'Or, and became a maker of instruments of the Académie des sciences. In the *Instruction sur l'établissement des nitrières et sur la fabrication du saltpetre* Lavoisier refers to a new type of hydrometer made in series for the Arsenal by "Moussy, constructeur d'instruments de physique de l'Académie des sciences".[152] Lavoisier also bought five thermometers (the last made in 1789)[153] (Fig. 30) and one barometer from him.[154]

Mudge, John (1721–1793). This English physician based in Plymouth was also a skilled instrument maker. He won the

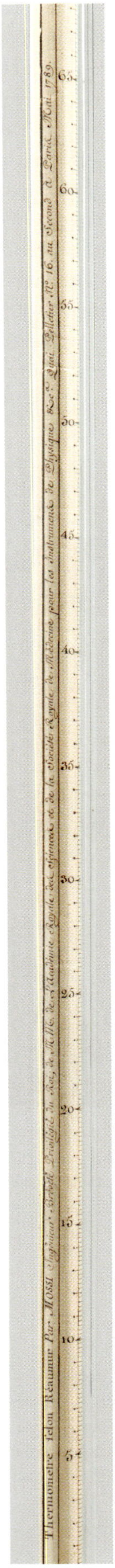

FIGURE 30 Mossy's signature and address (inv. no. 19929-0000-)
© MAM/PHOTO FRANCK BOTTÉ

145 Daumas (1955), p. 135.
146 MAM inv. no. 19885-0000-. See catalogue p. 192.
147 Belin (1992).
148 LC, vol. 4, pp. 87–88.
149 LC, vol. 4, pp. 299–303.
150 LO, vol. 3, pp. 91–100.
151 MS Lavoisier (1768–1788), vol. 14, fol. 38.
152 Lavoisier (1777). LO, vol. 5, pp. 428–429; MAM inv. no. 19956-0000-. See catalogue p. 323.
153 MAM inv. nos. 19926-0000-, 19911-0000-, Observatoire de Paris, inventory no. 141, MAM 19910-0000- and 19929-0000-. See catalogue pp. 271, 274, 276 and 279.
154 MAM inv. no 19948-0000-. See catalogue p. 194.

Copley Medal in 1778 for an essay on reflecting telescopes, and one year later he invented a useful inhaler which was in Lavoisier's instrument collection, but was lost in 1952 during the collection's move to the MAM.[155]

Musson, Pierre and Louis (fl. Eighteenth century) Pierre Musson, the father of Louis, became *Maître horloger* in 1746 and opened a shop on the Place Dauphine. Louis became *Maître horloger* in 1770, opened a shop in the Rue Neuve St Merry, became a successful clockmaker, but went bankrupt in 1783. Lavoisier owned a pendulum clock of Musson make, which is described in the inventory prepared by Gondouin in 1796 and was valued at 130 *livres*.[156]

Nairne, Edward (1726–1806). Nairne was an optician and scientific instrument maker based in London. Elected a member of the Royal Society in 1776, Nairne patented a new type of electrical machine in 1782.[157] Lavoisier purchased, via Magellan, an air pump by him in 1776.[158] His library also included a pamphlet by Nairne.[159]

Naudier or **Nodier, Pierre (fl. second half of the Eighteenth century).** Tinsmith employed by Lavoisier from 1774.[160] Naudier had an important role in the making of the calorimeter[161] and was involved in the construction of the apparatus used by Lavoisier and Meusnier de La Place for their experiments on the decomposition of water (April 1784).[162] He was also involved in the activities of the Commission des poids et mesures.[163]

Noël, Nicolas (1712–1783). Benedictine priest, known as 'Dom Noël', with a productive scientific instrument workshop at the abbey of St. Germain-des-Prés, Paris. Louis XV entrusted Noël with the care of his collection of physical instruments at La Muette. Lavoisier owned a clockwork-driven flint lock apparatus for producing sparks in vacuum.[164]

Nooth, John Mervin (1737–1828). English physician and army officer who, by recommendation of Benjamin Franklin and William Hunter, was elected a member of the Royal Society in 1774 for his improvement of the electrical machine. His most important and popular invention was the apparatus "for impregnating water with fixed air", which Lavoisier also owned.[165]

Panier, Josué (Pannier). Parisian clock maker who lived in the first half of the eighteenth century and had shops on Quai Neuf (1713), on Quai Pelletier (1716 and 1729), and in the Rue de la Monnaie (1726). Lavoisier owned two domestic pendulum clocks.[166]

Passemant, Claude Paris (1703–1763) and **Claude-Siméon (1702–1769).** Famous Parisian makers of optical, mathematical, astronomical, and glass instruments. Lavoisier owned a thermometer *à l'esprit de vin* made by them which he used during his travel in Alsace in the summer of 1767.[167]

Périer, Jacques-Constantin (1742–1818) and his brother **Auguste-Charles Périer.** Entrepreneurs active in Paris.[168] In the 1750s they attended Nollet's lectures at the Collège de Navarre for seven years. In 1778 they founded the successful Compagnie des eaux de Paris, and in 1781 they were the first to import Watt's steam engine to France. They gained the patronage of the Duc d'Orléans for whom they prepared a series of models of chemical manufactories and of a chemical laboratory in 1783.[169] In their foundry at Chaillot they had eight different workshops, and in one of these they built large reverberatory furnaces. Between 1792 and 1793 they produced five cannons per day. In 1798 the Danish astronomer Thomas Bugge reported that the two brothers "built between Porte aux Pierres on the Seine and the Grande Rue de Chaillot, an excellent gun-factory, in which iron cannons and other objects are cast; they also built fine, large workshops, where they make all kinds of machines. [...] Everything is cast so well in Perier's foundry that it would be hard to find anything better produced in the best English foundries".[170] In 1807 they employed some 300 workers.[171] Jacques Constantin was the first artisan to become a member of the Académie des sciences, as *adjoint mécanicien surnuméraire*. He was appointed on 19 January 1784. In April of the same year, Lavoisier purchased a pipe of red copper from him which would be used in the experiment on the decomposition of water.[172]

Pluvinet, Jean-Baptiste-Charles (fils) (1754–1814). The pharmacist and druggist Pluvinet had a shop in the Rue des Lombards,

155 See Appendix 3, p. 154.
156 See Appendix 1, p. 138.
157 Bertucci (2001).
158 LC, vol. 3, pp. 568–571. The pneumatic pump was used by Lavoisier in chemical experiments on gases conducted together with Trudaine at Montigny in October 1776 (LC, vol. 3, p. 578).
159 Nairne (1777); Beretta (1995).
160 LC, vol. 2, p. 458.
161 MAM inv. no. 07520-0000-. See catalogue pp. 229–232.
162 LC, vol. 4, p. 18, pp. 26–29, and 33.
163 LO, vol. 6, p. 693.
164 MAM inv. no. 20141-0000-. See catalogue p. 227.
165 Nooth (1775). See Appendix 3, p. 153. It is, however, likely that Lavoisier owned a modified version of this apparatus made by Magellan (see entry).
166 See the confiscation inventory, Appendix 1, p. 136.
167 MS Lavoisier (1767–1788), vol. 2, p. 17.
168 Payen (1969).
169 Payen (1969), pp. 51–51.
170 Crosland (1969), p. 141.
171 Payen (1969), p. 204.
172 LC, vol. 4, pp. 17 and 151.

the Parisian street with the highest density of apothecary shops. In May 1788 Lavoisier purchased from him 517.5 *livres* of impure mercury for the considerable amount of 1335 *livres*.[173] In the following years Pluvinet continued to supply Lavoisier with chemicals.[174] Being the cousin of André Siméon Olivier Dupin, Lavoisier's great accuser during the Revolution, he made efforts to save the chemist from the guillotine. According to Cadet de Gassicourt's testimony, written several years after Lavoisier's death, "Pluvinet avait beaucoup connu Lavoisier; il lui fournissait, depuis plusieurs années, les drogues et les réactifs de son laboratoire. Enfin, il prenait un grand intérêt à ce savant".[175]

Polichot. A maker of weights who cannot be otherwise traced.[176]

Quinquet, Antoine (1745–1803). Pharmacist. In 1779, after being an assistant to Baumé for nine years, Quinquet opened his own shop in the "Rue du Marché aux Poirées, au coin de la porte de l'ancienne Halle au Blé". Interested in natural philosophy he invented a new lamp in 1783, which was used to light large public spaces effectively. The lamp's authorship was contended by Argand, who was probably its actual inventor (see entry Argand). In 1784 Quinquet's lamp was used at the Comédie Française for the première of Baumarchais' *Le mariage de Figaro*. Lavoisier used it and reported favourably on its efficiency.[177] Lavoisier owned at least two of these lamps.[178] A contemporary witness claimed that Lavoisier was the inventor of the lamp who passed on the secret of the invention his protégé to Quinquet.[179] In 1784 Quinquet collaborated with Louis Caullet de Veaumorel in the construction and sale of mathematical instruments. They offered, among other things, a cheaper version of Nairne's electrical machine for which they had the exclusive *privilège*.[180] This was soon copied by Francesco Bianchi, who had a shop at No 55 Rue Saint-Honoré, where he delivered courses in natural philosophy. During the confiscation of Lavoisier's belongings in 1794, Quinquet was in charge of compiling the catalogue of the pharmaceutical collection.[181]

Ramsden, Jesse (1735–1800). Famous instrument maker based in London.[182] Lavoisier purchased a barometer from him in 1778, in late 1781 (?) a pyrometer and in 1786 an electrical machine.[183]

Régnier, Edme (1751–1825). Engineer and inventor based in Paris. Régnier is famous for his spring-dynamometer, but he also perfected devices used to assay various kinds of gunpowder. Lavoisier owned a small scraper signed 'Regnier' and a gunpowder tester by him.[184] (Fig. 31)

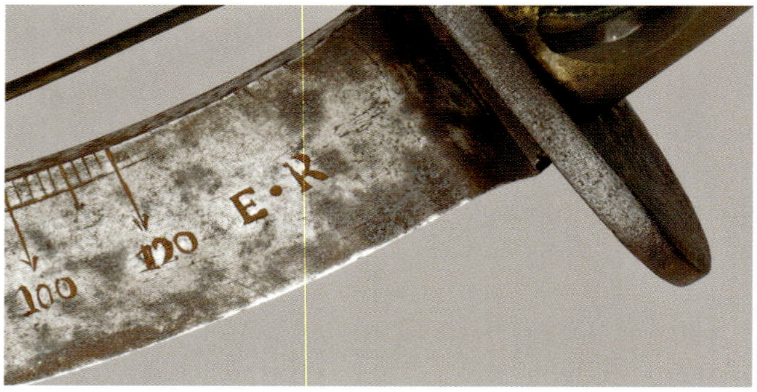

FIGURE 31 Régnier's stamp with his initials (inv. no. 19895-0000-) © MAM/PHOTO FRANCK BOTTÉ

Richer, Jean François (1743–ca. 1820). Parisian instrument maker specialising in meteorological instruments. Richer was also a specialist in the art of dividing instruments.[185] Lavoisier owned a de Saussure's hygrometer (1788) and a hair hygrometer (Richer's invention).[186] (Fig. 32) The collection also included a part from an unidentified instrument which Richer probably made for Rumford.[187]

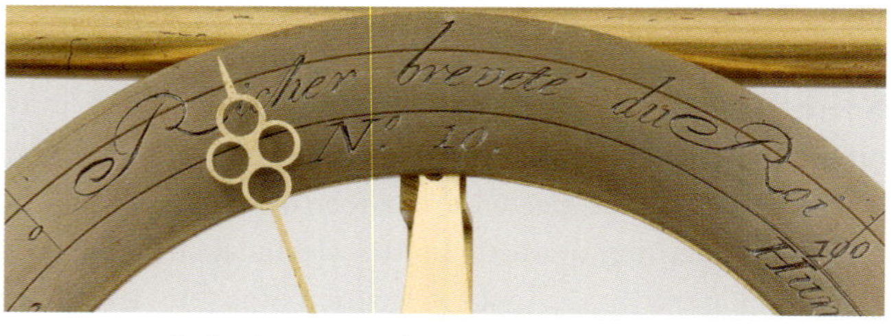

FIGURE 32 Richer's signature (inv. no. 19895-0000-) © MAM/PHOTO FRANCK BOTTÉ

173 LC, vol. 5, pp. 171–172. Pluvinet eventually also sent Lavoisier some sample of clove (LC, vol. 5, pp. 173–174).
174 LC, vol. 7, p. 437.
175 Cited in Kersaint (1966), p. 71.
176 MAM inv. no. 19898-0000-. See catalogue p. 199.
177 Lavoisier (1781c).
178 Appendix 1, p. 129.
179 After having assisted with Beaumarchais' *Mariage de Figaro* at the Comédie française in May 1784, the Baroness d'Oberkirch remarked: "La salle était éclairée par une nouvelle invention due à M. Quinquet, qui avait fort bien réussi, et à laquelle il a donné son nom. Cette lumière douce, vive, exempte de fumée, est d'ailleurs peu dispendieuse: elle est généralement adoptée aujourd'hui. On assure que M. Quinquet doit le secret de cette découverte à M. Lavoisier, fermier générale et grand chimiste. Il en a fait cadeau à son protégé pour l'enrichir, et, en effet, ce dernier est maintenant tout à fait à son aise. M. de Lavoisier dépense une partie de sa fortune en expériences scientifiques; il est gendre de M. Paulze, président de la Ferme générale et l'un des hommes les plus estimé de la finance". Oberkirch (1854), vol. 2, p. 36.
180 The machine cost 192 *livres*. On the *privilège*, see Dorveaux (1919), p. 44.
181 See Appendix 1, p. 115.
182 McConnell (2007).
183 LC, vol. 3, p. 629 (barometer). Lavoisier (1783a) (pyrometer). MAM inv. no. 20110-0000- (electrical machine). See catalogue pp. 249–252.
184 MAM inv. no. 20234-0000- (scraper). See catalogue p. 397. MAM inv. no. 19895-0000- (gunpowder tester). See catalogue p. 306.
185 Richer applied his skills to the divisions on Lavoisier's barometer and thermometer made by Mossy.
186 MAM inv. no. 20106-0000- (hygrometer). See catalogue p. 299. Appendix 3, p. 155.
187 McKie 123.

Rivas (fl. mid-eighteenth century). Clockmaker, probably Parisian. Rivas is best known for his attempt to perfect the mechanism of pendulum clock, on which the Académie des sciences reported favourably in 1749.[188] Lavoisier owned a pendulum clock.[189]

Rothenberger, Christoph Jobst. Master instrument maker in Nuremberg from 1787 onwards. Lavoisier owned a set of weights by him.[190]

Rotton, P.W. English cutler active at the end of the eighteenth and beginning of the nineteenth centuries. Lavoisier owned a candle snuffer.[191] (Fig. 33)

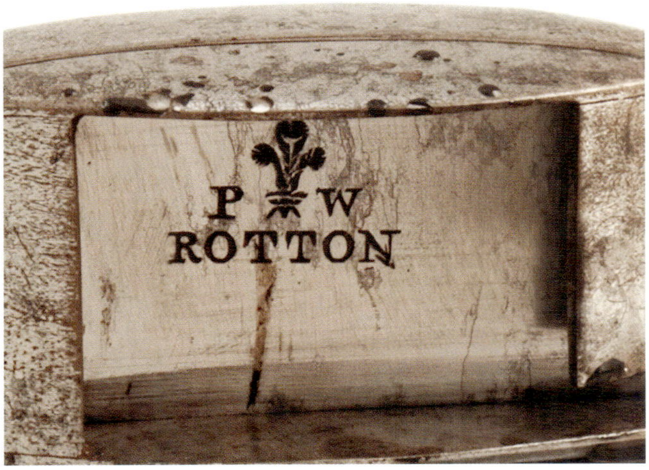

FIGURE 33 Rotton's signature and stamp (inv. no. 20211-0000-) © MAM/PHOTO FRANCK BOTTÉ

Scanegatti, Flavien Marie (d. 1793).[192] Also known as Scanégatty or Scanegaty. (Fig. 34) Instrument maker of Italian origin, probably from Milan, who was based in Rouen, where he delivered courses in experimental physics. He was particularly well-known as a maker of meteorological instruments. In 1773 he presented himself as a *Méchanicien célèbre et membre de l'Académie des Sciences de Rouen*. Of this "excellent souffleur"[193] Lavoisier owned two balances.[194] Only one, a spring balance made and signed in 1789, is preserved at the MAM.[195] (Fig. 35)

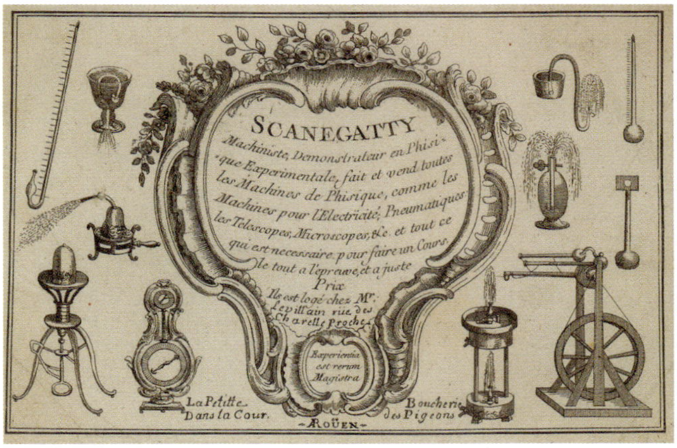

FIGURE 34 Trade Card of Flavien Marie Scanegatti ca. 1775. Waddesdon (National Trust) Bequest of James de Rothschild, 1957. Acc no: 3686.1.65.123 PHOTO: WADDESDON IMAGE LIBRARY, UNIVERSITY OF CENTRAL ENGLAND DIGITAL SERVICES

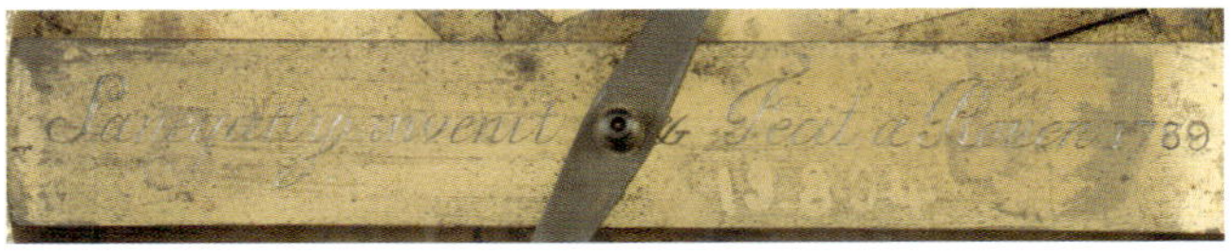

FIGURE 35 Scanegatti's signature (inv. no. 19894-0000-) © MAM/PHOTO FRANCK BOTTÉ

Séguin, Armand (1767–1835). Introduced by Fourcroy to Lavoisier in the mid-1780s, Séguin initially worked in the laboratory of the Arsenal as a *garcon de laboratoire*.[196] However, he soon became one of Lavoisier's favourite assistants, and from 1789 he was involved in several crucial experiments. On 20 September 1789 he was in charge of assembling Fortin's apparatus for combustion.[197] In November 1790 he assembled the apparatus used in the experiments on human respiration;[198] he also had a prominent role in making the scales invented by Santorio Santorio for weighing the loss of fluids during transpiration more accurate.[199] One year later he followed the experiments on the fusion of platinum, and built and, together with Meusnier and James Hall, made a furnace attaining high temperature which was eventually highly appreciated by the Commision des poids et mesures.[200] He also made a new type of eudiometer in 1792.[201] Before 1795 Séguin became a hugely successful entrepreneur by developing a new method for preparing leather in a few days; the leather "formerly underwent a

188 Delaunay (2017), vol. 2, p. 228.
189 See Appendix 3 p. 155.
190 MAM inv. no. 19897-0000-. See catalogue p. 198.
191 MAM inv. no. 20211-0000-. See catalogue p. 391.
192 See the comprehensive biography by Morvan Becker (2010), pp. 779–789; see also Thébaud-Sorger (2018), oo. 89–90.
193 LO, vol. 3, p. 394.
194 Appendix 3.
195 MAM inv. no. 19894-0000-. See catalogue p. 193.
196 On Séguin, see Mercier (1976). On the collaboration between Lavoisier and Séguin, see Beretta, (2001a).
197 LC, vol. 6, pp. 70–71.
198 LC, vol. 6, pp. 192–193.
199 Beretta (2012).
200 On Meusnier and James Hall's role see Chaldecott (1968), p. 33. For the furnace, see LC, vol. 6, pp. 327–328. On the appreciation, see Chaldecott (1968), pp. 50–51.
201 Séguin (1814).

preparation of two years".[202] In 1798 he proposed a new type of gasometer which met the approval of the Institut.[203]

Subrin. Mineralogist and friend of Hassenfratz, who introduced the latter to Lavoisier in April 1786. Subrin eventually became for a short period one of Lavoisier's laboratory assistants.[204]

Tor[…], J.H. Swiss maker of barometers from Lausanne. Lavoisier owned a syphon barometer for domestic use by this maker.[205]

Volta, Alessandro (1745–1827). Professor of natural philosophy at the University of Pavia. Volta devised several electrical instruments that were used by Lavoisier.[206] In March and April 1782 Lavoisier and Laplace used Volta's condenser (electrometer) in their experiments on electricity and the vaporisation of fluids.[207] Lavoisier also owned Volta's electrical guns (one in copper and in iron), one eudiometer, and one hydrogen lamp.[208]

Wedgwood, Josiah (1730–1795). English potter and instrument maker. On 18 August 1791 Wedgwood offered Lavoisier a specimen of a retort made of a "most refractory" clay;[209] and on 12 September Lavoisier ordered 48 retorts of different sizes from him.[210] (Fig. 36a and 36b) In the same year Lavoisier also purchased two of Wedgwood's pyrometers.[211]

Welter, Jean Joseph (1763–1852). 'Velter' is mentioned in Lavoisier's *Registres de laboratoire* in connection with the experiments of March 1788.[212] It is likely that he was Jean Joseph Welter,[213] the industrial chemist who, in the late eighteenth century, produced large quantities of gunpowder at Meudon.[214] He eventually opened a chemical factory near Valenciennes, and was also involved in the successful improvement of textile manufacture.[215] Welter was further a maker of chemical apparatus. At the beginning of the nineteenth century he became Berthollet's, Laplace's, and Guy Lussac's occasional collaborator.[216]

202 Crosland (1969), p. 150.
203 Séguin (1798) and Duchesne (1801), vol. 3, p. 139.
204 LC, vol. 5, p. 286. Subrin was called back by his family in Lyon in August 1787 (personal communication by Patrice Bret).
205 MAM inv. no. 19954-0000-. See catalogue p. 292.
206 Bellodi *et al.* (2002).
207 Lavoisier (1781d).
208 Appendix 3 p. 154.
209 LC, vol. 6, p. 314.
210 LC, vol. 6, p. 323.
211 Chaldecott (1968), p. 44.
212 "On s'est rassemblé au laboratoire: MM. Landriani. Assenfratz [*sic*]. Velter. Lavoisier." Berthelot (1890), p. 307.
213 LC, vol. 5, pp. 141 and 220.
214 Bottée, Riffault (1811), p. 333.
215 Chaptal (1819), vol. 2, p. 9.
216 Crosland (1967), pp. 238–239.

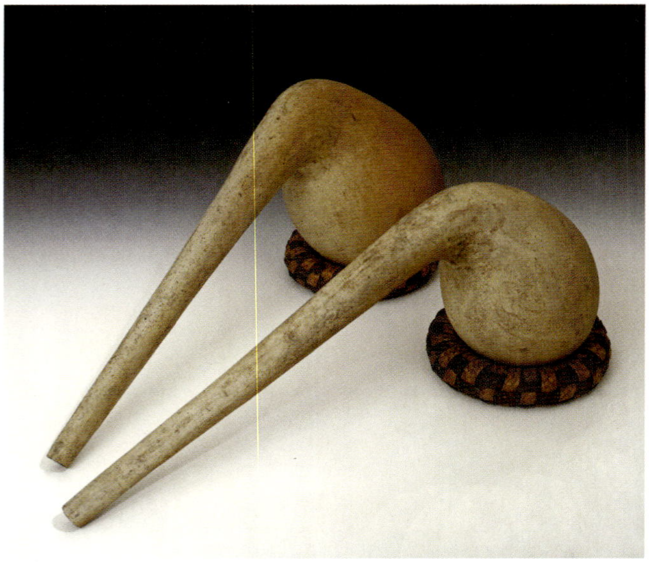

FIGURE 36A Wedgwood's clay retorts from the collection of Felice Fontana
COURTESY MUSEO GALILEO – FLORENCE

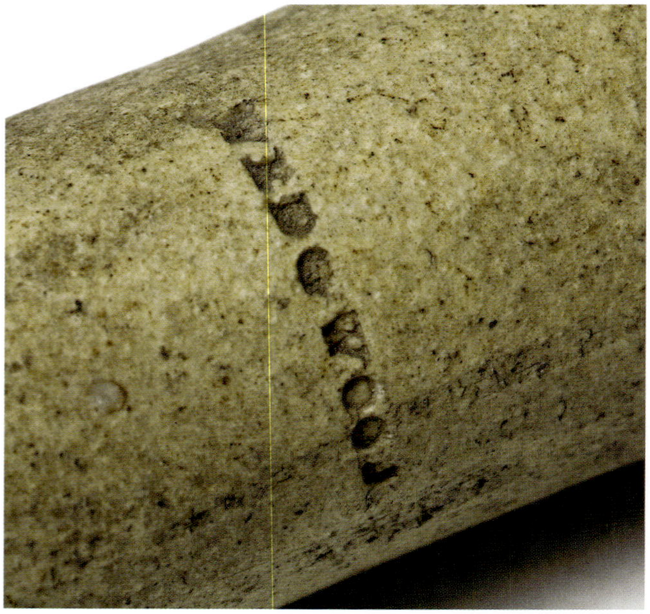

FIGURE 36B Wedgwood's stamped signature on one of his retort
COURTESY MUSEO GALILEO – FLORENCE

Wilkes John (fl. late eighteenth/early nineteenth centuries). English professional pen cutter and author of *The art of making pens scientifically*. Lavoisier owned a quill pen cutter bearing the inscription "Wilkes".[217]

217 MAM inv. no. 20071-0000-. See catalogue p. 398.

PART 2

Descriptive Catalogue of the Collection of Lavoisier's Instruments in the Musée des Arts et Métiers

(Marco Beretta, Paolo Brenni)

Introduction to the Catalogue

The following catalogue includes the analytical description of Lavoisier's instruments, chemicals, minerals and miscellaneous artefacts accessioned by the Musée des Arts et Métiers (hereafter MAM) in 1864 and 1952. We have also included the instruments withdrawn from the donation by Lavoisier's heirs in 1952, those that have been lost and the few instruments we have been able to locate outside the Musée. In addition, we have described the instruments and artefacts that belonged to Rumford and Madame Marie Anne Lavoisier, as well as some items from the de Chazelles family, as they originally belonged to the collection.

For ease of reading, we have decided, following other examples,[1] to divide the catalogue by discipline into the following sections: metrology, mathematics and related subjects, experimental physics, meteorology, chemistry, miscellaneous instruments, and artefacts and fragments. Although this division is consistent with eighteenth-century disciplinary boundaries, Lavoisier changed the realm of chemical instrumentation by adapting apparatus and methods from other disciplines – primarily experimental physics – for chemical experimentation. This approach, together with compelling evidence when available, led us to allocate individual instruments to specific sections, and these choices will be explained both in the short introduction to each section and in the individual item descriptions.

The individual entries consist of the following parts:

Inventory number: some inventory numbers do not follow the logical sequence. Over time, some instruments were damaged, fell apart or were disassembled, so that curators occasionally assigned different inventory numbers to different parts of the same instrument. Our reassembly of these instruments has retained the original inventory numbers.

Name of the instrument: given in bold.

Dating of the instrument: a precise dating is possible only when an instrument bears a maker's manufacturing date, or when surviving letters, invoices and other types of documents mention and clearly identify a specific instrument. In most cases, the dating is only approximate, since instruments were often manufactured according to the same design for several decades.

MAM accession date: the MAM accessioned Lavoisier's most important instruments in 1864, but the largest number entered its collections with the donation of 1952, and a few further additions were made on other occasions.

Name of the maker(s): the name is recorded whenever the instrument is signed. For unsigned instruments we occasionally suggest makers from the evidence given in extant manuscript or printed documents.

Dimensions: dimensions are generally reported in centimetres, from the longest to the shortest part. When possible the weight of the instrument has been given. For circular, cylindrical and spherical instruments only the radius and width (or height) are indicated.

Materials: the most important materials are listed. In certain cases, when it was not possible to be specific, a more generic term is used (for example: metal).

Descriptive note: this includes the physical description of the item as well as a short explanation of its functions and use, followed by its historical context.

Bibliography: when possible, the bibliography includes the most important manuscript sources ("MS"/"MSS") as well as printed primary and secondary sources.

Most of the items are accompanied by at least one illustration.

All entries in the following catalogue are by Marco Beretta and Paolo Brenni, except for the description of the astrolabe (written by Anthony Turner) and that of the missing instruments (written by Douglas McKie in 1952).

1 Such as Turner, Levere (1974).

 | DOI:10.1163/9789004511217_014

Metrology

This section includes several items which were used by the Commission of Weights and Measures, created by the Académie royale des sciences in 1790 for the decimal reform. It also features Lavoisier's precision balances, which he is known to have used for chemical experiments. Explicit references to experiments using these balances are rare – in fact, owing to their sensitivity, the balances were kept not in Lavoisier's chemical laboratory but in his *cabinet de physique*. We have therefore included them in this metrology section. The balances are also known to have been used by the Commission of Weights and Measures in its work.

Measures of Length

20073-0000- (Fig. 1)
Folding rule
Second half of 18th c.
MAM accession date: 1952
15.3 × 0.4 × 0.3; 5 g
Ivory, brass

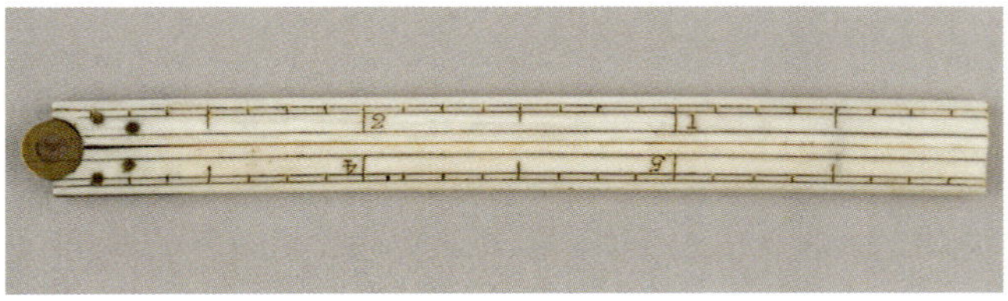

FIGURE 1 (inv. no. 20073-0000-)
© MAM/PHOTO FRANCK BOTTÉ

Ivory rule to measure half an English foot comprising two sections hinged together with a brass joint. The scale (on both sides) is divided into eighths of an inch. The scale is marked with numbers at every inch from 1 to 5.

MS Pommier (2007), no. 305

20074-0000- (Fig. 2)
Folding rule
ca. 1780
MAM accession date: 1952
32.5 × 1 × 0.3 (open); mass 5 g
Ivory, brass

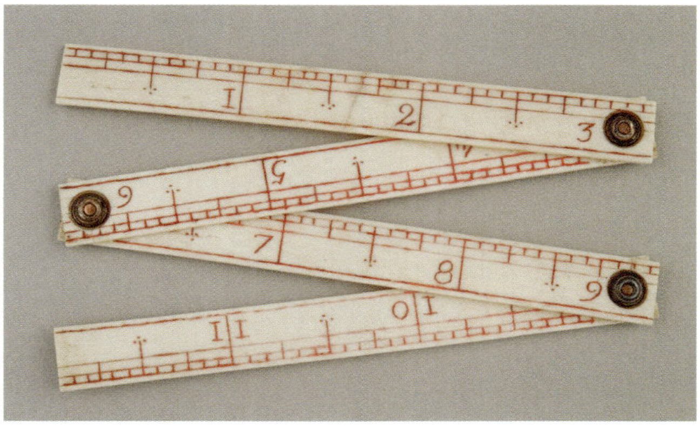

FIGURE 2 (inv. no. 20074-0000-)
© MAM/PHOTO FRANCK BOTTÉ

Folding ivory rule to measure one Paris foot comprising four sections which are hinged with brass rivets. The length of its scale corresponds to one Paris foot (ca. 32.48 cm). It is divided into inches and lines (1 inch = 12 lines), and the scale is marked with numbers at every inch (ca. 2.77 cm).

MS Pommier (2007), no. 6

20075-0000- (Fig. 3)
Folding rule
ca. 1780
MAM accession date: 1952
59.6 × 1.1 × 0.3 (open); 20 g
Ivory, brass

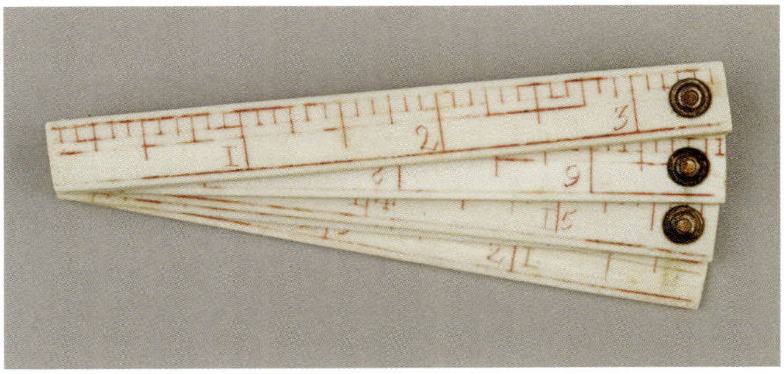

FIGURE 3 (inv. no. 20075-0000-)
© MAM/PHOTO FRANCK BOTTÉ

Folding ivory rule comprising seven sections hinged with brass rivets. One side bears a scale of 22 Paris inches in length, divided into inches and lines (1 inch = 12 lines). The

 | DOI:10.1163/9789004511217_015

scale is marked with numbers at every inch (ca. 2.77 cm). The other side shows six subdivisions marked with the capitals "T", "Q", "D/T", "D/Q", "D" and "S". These likely correspond to different fractions.

19907-0000- (Figs. 4a and 4b)
Precision rule
ca. 1782
MAM accession date: 1952
Louis-Pierre-Florimond Lennel
Ruler: 111.5 × 2.9 × 2.7; 2.33 kg
Box: 114.5 × 5 × 5; 810 g
Brass, steel wood

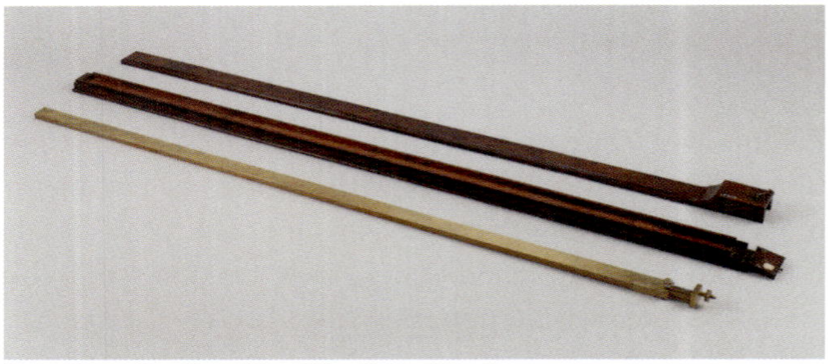

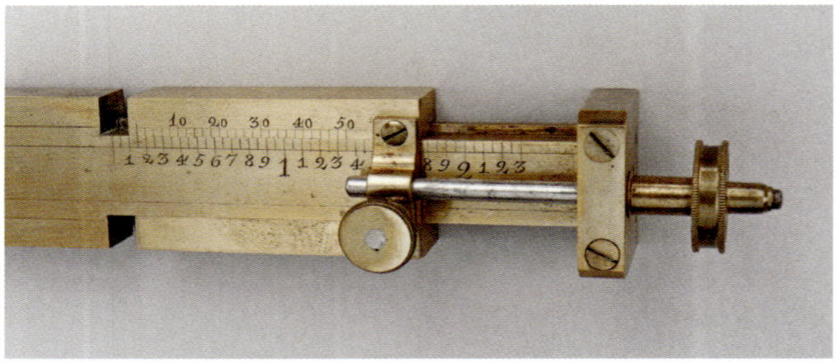

FIGURES 4A AND 4B (inv. no. 19907-0000-)
© MAM/PHOTO FRANCK BOTTÉ

Long brass bar of rectangular section divided (from left to right) into Paris inches from 37.5 to 0. The scale is marked with numbers at every inch (except 0). After 0 the bar has a smaller section with a sliding nonius. This section follows a decimal scale from 0 to 2.3 inches, which is marked at every 0.05 of an inch, and numbered at every tenth of an inch.[1] The nonius has a scale with 25 divisions allowing a reading precision of 0.002 inches. On its top is a small jaw with a screw that allows the locking of a short horizontal steel axle. The latter penetrates a wheel at the right end of the bar. The bore of the latter and a portion of the axle are both threaded. When the axle is locked, the rotation of the wheel produces a translation of the nonius. The ruler bears the inscription "Lennel à Paris". The instrument is preserved in a wooden box with a sliding cover.

It is not clear how this ruler was used. According to Truchot and Daumas, this device was related to the investigation of the dilatation of metals by Lavoisier and Laplace in 1782.[2]

Truchot (1879), pp. 26–27; Daumas (1955), pp. 154–156; Daumas (1989), p. 262

Measures of Volume

07542-0001- to 07542-00029-003- (Fig. 5)
Series of standard measures of volume
Third quarter of 18th c.
MAM accession date: 1864–1866
Brass, wood

Series of parallelepiped vessels made of brass used as (non-metric) standard measures of liquid volume. They are placed on modern wooden bases. This set was used by Lavoisier and his colleagues of the Académie royale des sciences probably between 1790 and 1793, during the active period of the Commission of Weights and Measures.

Note on the French liquid volume units:

1 cubic foot = 1728 cubic inches = 24.28 litres
1 pinte (pint) = 1/36 cubic foot = 48 cubic inches = ca. 0.952 litres

07542-0001-
Vessel: 50 × 22 × 22.5; 10.59 kg

The opening of the vessel, which has a capacity of 1024 cubic Paris inches, is closed with a plug. It bears the inscription "1024 P°".

07542-002-
Vessel: 40.3 × 32.8 × 17; 11.86 kg

The opening of the vessel, which has a capacity of 864 cubic Paris inches (1/2 cubic foot), is closed with a plug. It bears the inscription "1/2 PI· C // DE 864 PO' C·".

07542-0003-
Vessel: 32.83 × 28.5 × 17; 7.95 kg

1 This scale is quite unusual: normally Paris inches are subdivided by lines 1/12 of an inch apart.

2 Truchot's examination of the ruler was apparently quite superficial, as he wrongly described the scale and its nonius.

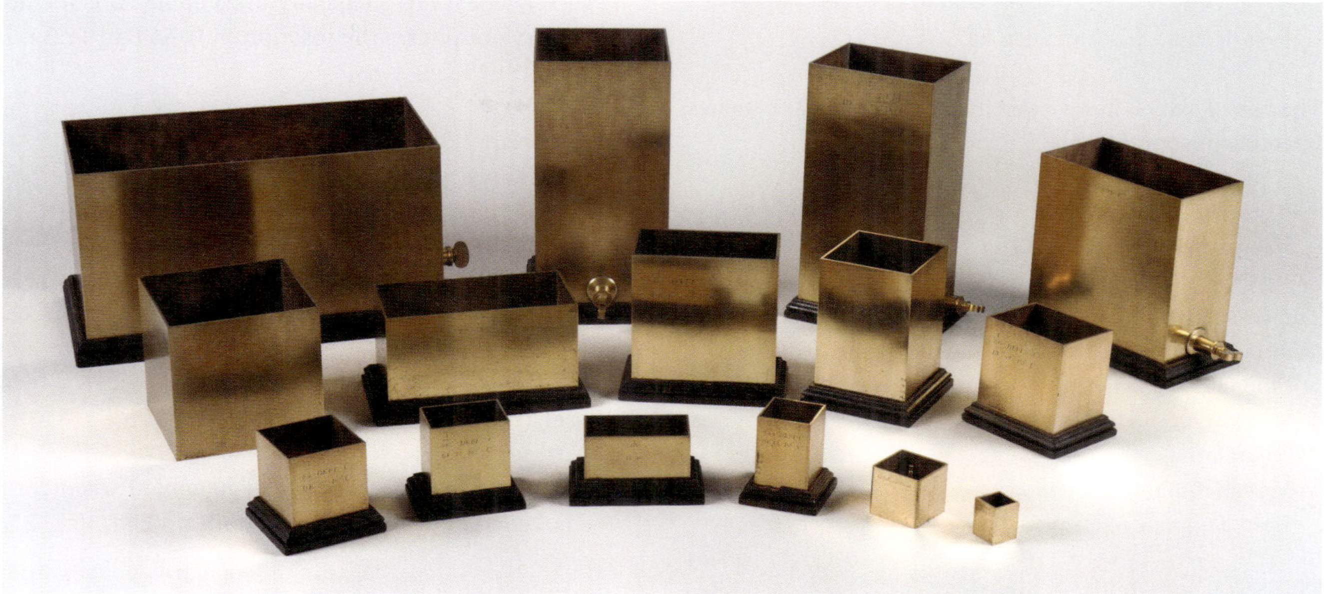

FIGURE 5 (inv. nos. 07542-0001- to 07542-00029-003-)
© MAM/PHOTO FRANCK BOTTÉ

The opening of the vessel, which has a capacity of 576 cubic Paris inches (1/3 cubic foot), is closed with a plug. It bears the inscription "1/3 DEPI· C · // DE 576 · PO' · C·".

07542-0004-
Vessel: 28 × 22.2 × 22; 5.86 kg

The opening of the vessel, which has a capacity of 512 cubic Paris inches, is closed with a plug. It bears the inscription "512 P°".

07542-0005-
Vessel: 23 × 17 × 33; 6.14 kg

The opening of the vessel, which has a capacity of 432 cubic Paris inches (1/4 cubic foot), is closed with a plug. It bears the inscription "1/4 DEPI · C · // DE 432 · PO · C ·".

07542-0006-
Vessel: 32.9 × 29.5 × 17.8; 9.74 kg

The vessel, with a capacity of 384 cubic Paris inches, shows an opening that is closed with a plug, and two opposing handles. It bears the inscription "SEPTIER DE 384 · PO'· C ·".[3]

07542-0007-
Vessel: 32.8 × 22.7 × 11.5; 4.67 kg

The opening of the vessel, which has a capacity of 288 cubic Paris inches (1/6 cubic foot), is closed with a plug. It bears the inscription "1/6 DEPI · C · // DE 288 · PO · C ·".

07542-0008-
Vessel: 28.5 × 22 × 11.5; 4.37 kg

The opening of the vessel, which has a capacity of 256 cubic Paris inches, is closed with a plug. It bears the inscription "256 P°.".

07542-0009-
Vessel: 16.8 × 16.8 × 16.5; 3.16 kg

The vessel, which has a capacity of 216 cubic Paris inches (1/8 cubic foot), bears the inscription "1/8 DEPI · C · // DE 216 · PO'· C ·".

07542-0010-
Vessel: 22 × 16.8 × 11.4; 3.23 kg

The vessel, which has a capacity of 192 cubic Paris inches (1/9 cubic foot), bears the inscription "1/9 DEPI · C · // DE 192 · PO'· C ·".

07542-0011-
Vessel:16.7 × 16.5 × 11.4; 2.17 kg

The vessel, which has a capacity of 144 cubic Paris inches, bears the inscription "1/12 DEPI · C · // DE 144 · PO'· C ·".

3 The *septier* or *setier* was a measure of volume.

07542-0012-
Vessel: 22.2 × 11.4 × 11.2; 2.11 kg

The vessel, which has a capacity of 128 cubic Paris inches, bears the inscription "128 P°".

07542-0013-
Vessel: 17 × 16.1 × 8.9; 2.44 kg

The vessel, which has a capacity of 108 cubic Paris inches (1/16 cubic foot), bears the inscription "1/16 DEPI · C · // DE 108 · PO'· C ·".

07542-0014-
Vessel: 16 × 11.5 × 11.5; 2.04 kg
Base: 14.5 × 14 × 2.5; 315 g

The vessel, which has a capacity of 96 cubic Paris inches (1/18 cubic foot), bears the inscription "1/18 DEPI · C · // DE 96 · PO'· C ·".

07542-0015-
Vessel: 16.5 × 11.4 × 8.6; 1.46 kg

The vessel, which has a capacity of 72 cubic Paris inches (1/24 cubic foot), bears the inscription "1/24 DEPI · C · // DE 72 · PO'· C ·".

07542-0016-
Vessel: 16.6 × 8.8 × 8.8; 1.72 kg

The vessel, which has a capacity of 54 cubic Paris inches (1/32 cubic foot), bears the inscription "1/32 DEPI · C · // DE 54 · PO· C ·".

07542-0017-
Vessel: 11.5 × 11.2 × 8.7; 1.72 kg

The vessel, which has a capacity of 48 cubic Paris inches (1/36 cubic foot, 1 pint), bears the inscription "1/36 DEPI · C · // DE 48 · PO'· C ·".

07542-0018-
Vessel: 11.4 × 11 × 6; 820 g

The vessel, which has a capacity of 32 cubic Paris inches (1/54 cubic foot), bears the inscription "1/54e // 32 P.o".

07542-0019-
Vessel: 8.7 × 8.5 × 8; 610 g

The vessel, which has a capacity of 27 cubic Paris inches (1/64 cubic foot), bears the inscription "1/64 DEPI · C · // DE 27 · PO'· C ·".

07542-0020-
Vessel: 8.6 × 8.4 × 5.8; 495 g

The vessel, which has a capacity of 18 cubic Paris inches (1/96 cubic foot), bears the inscription "1/96 DEPI · C · // DE 18 · PO'· C ·".

07542-0021-
Vessel: 11.3 × 5.8 × 5.2; 415 g

The vessel, which has a capacity of 16 cubic Paris inches (1/108 cubic foot), bears the inscription "1/108e // 16 P°".

07542-0022-
Vessel: 8.3 × 5.9 × 5.9; 370 g

The vessel, which has a capacity of 12 cubic Paris inches (1/144 cubic foot), bears the inscription "1/144 DEPI · C · // DE 12 · PO'· C ·".

07542-0023-
Vessel: 8.6 × 8.4 × 3.2; 400 g

The vessel, which has a capacity of 9 cubic Paris inches (1/192 cubic foot), bears the inscription "1/192 DEPI · C · // DE 9 · PO'· C ·".

07542-0024-
Vessel: 8.4 × 6 × 3.2; 310 g

The vessel, which has a capacity of 6 cubic Paris inches (1/288 cubic foot), bears the inscription "1/288 DEPI · C · // DE 6 · PO'· C ·".

07542-0025-
Vessel: 8.4 × 3.2 × 3.2; 235 g

The vessel, which has a capacity of 3 cubic Paris inches (1/567 cubic foot), bears the inscription "1/576 DEPI · C · // DE 3 · PO'· C ·".

07542-0026-
Vessel: 5.7 × 3.1 × 3.1; 120 g

The vessel, which has a capacity of 2 cubic Paris inches (1/864 cubic foot), bears the inscription "1/864e //2 P.O".

07542-0027-
Vessel: 4.4 × 3.3 × 3.3; 145 g

The vessel, which has a capacity of 1.5 cubic Paris inches, bears the inscription "ˋROQUILLE // DE 1 PO ½ C↓".[4]

07542-0028-001-, 07542-0028-002-
Standard measure of volume with cover plate
Second half of 18th c.
MAM accession date: 1864–1866
Lefebvre
Vessel: 6 × 6 × 5.7; 350 g
Top plate: 5.7 × 5.4 × 5.4; 55 g
Brass

This standard measure of volume, without a wooden base, seems to be of a different manufacture than the previously listed measures. It is a cubic brass vessel accompanied by a square plate with a vertical holding bar that was used for compressing the content of the vessel (probably a powder). It bears the inscription "LeFebvre AParis."

07542-0029-001-, 07542-0029-002-, 07542-0029-003
Wooden bases
Second half of 18th c.
MAM accession date: 1864–1866
07542-0029-001-: 13.8 × 14.3 × 2.5; 295 g
07542-0029-002-: 11.3 × 11 × 2.5; 170 g
07542-0029-003-: 8.6 × 5.6 × 2.5; 70 g
Wood

These wooden bases originally supported two parallelepiped brass vessels (now missing) which were used as standard measures of volume.

20156-0000- (Fig. 6)
Standard measure of volume
Late 18th c.
MAM accession date: 1952
25.5 × 21 × 19.5; 5.31 kg
Wood, brass, fabric

Cubic brass container. Each wall shows a vertical engraved scale divided into Paris inches and lines. A removable L-shaped brass strip with two uprights is inserted in the box. Its use is not clear. The container is covered by a square glass plate and inserted into a wooden box with a hinged lid lined with green velvet. The top of the lid bears the manuscript inscription "12 17".

The cubic container was probably used as standard measure of volume.

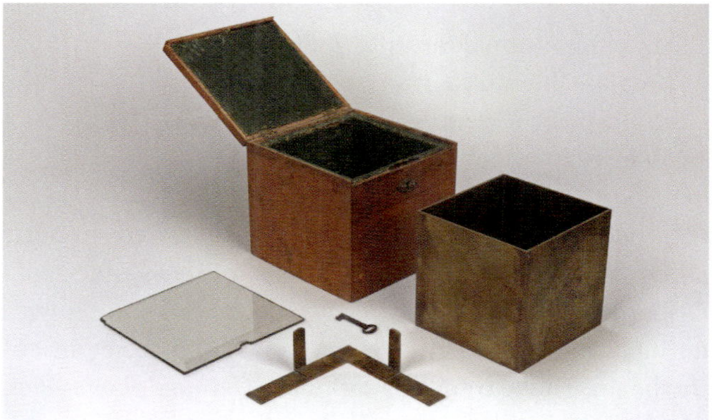

FIGURE 6 (inv. no. 20156-0000-)
© MAM/PHOTO FRANCK BOTTÉ

Measures of Weight

07544-0000- (Fig. 7)
Balance

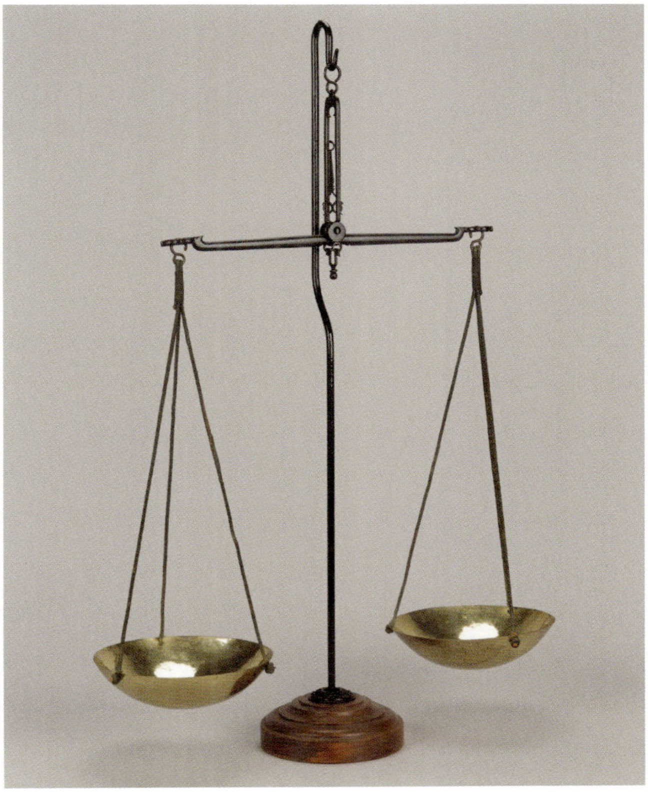

FIGURE 7 (inv. no. 07544-0000-)
© MAM/PHOTO FRANCK BOTTÉ

4 The *roquille* is a measure of volume for liquids corresponding to 1.5 cubic inches.

ca. 1770
MAM accession date: 1864–1866
Nicolas Chemin (d. 1780)
70 × 49 × 17; 1.06 kg
Steel, brass, wood, thread

A turned wooden base with a vertical iron stand supports an equal-arm balance. The steel beam is suspended in a ∩-shaped frame and has a vertical decorated pointer with a ring at the top. A light weight hangs from the centre of the beam. The extremities support two brass plates, each of which is attached to three threads. The balance is not signed, but it is attributed to the maker Chemin.

In 1773, Lavoisier assessed the accuracy of the balance as follows:

> At the same time, I had taken a very accurate scale, made by M. Chemin, *ajusteur de la Monnaie*. The scale was extremely sensitive and, even when loaded with five to six *livres*, it tipped at less than one grain. Despite all the care that had been taken in constructing the instrument, it was not totally flawless. When, after the two weights had been perfectly balanced, they were placed in another pan, there was always some difference to be found, but this could never result in an error, because by averaging the two weights found, one always obtained the same result, even several days apart. There were barely a few slight fractions of a grain of difference.[5]

Lavoisier (1773); Daumas (1950), p. 48; Daumas (1955), p. 133; Delacroix, Porte (1995), p. 298

07545-0000- (Fig. 8)
Balance
Second half of 18th c.

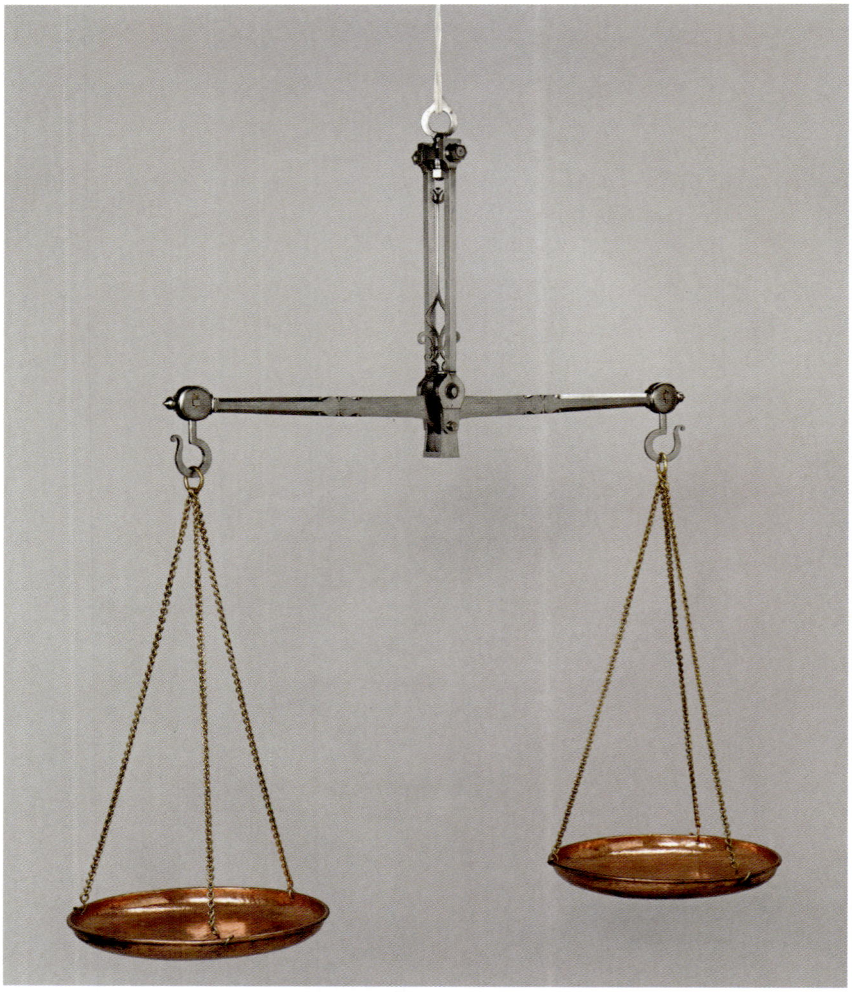

FIGURE 8
(inv. no. 07545-0000-)
© MAM/PHOTO FRANCK BOTTÉ

5 "Je m'étais muni, en même temps, d'une balance très-exacte, exécutée par M. Chemin, ajusteur de la Monnaie. Cette balance

MAM accession date: 1864–1866
106 × 65 × 10; 9.86 kg
Steel, brass, copper

Large equal-arm balance. The steel beam is fixed by a pivot in a rectangular frame composed of bolted steel bars. At its top, a tear-shaped ring allows the balance to be suspended. The vertical steel pointer is decorated at the bottom with a fleur-de-lis and at the top with another floral ornament. The copper plates are suspended by three brass chains.

19890-0000- (Fig. 9)
Balance
ca. 1775
MAM accession date: 1952
112 × 95 × 35.5; 7.27 kg
Steel, tin, brass, glass

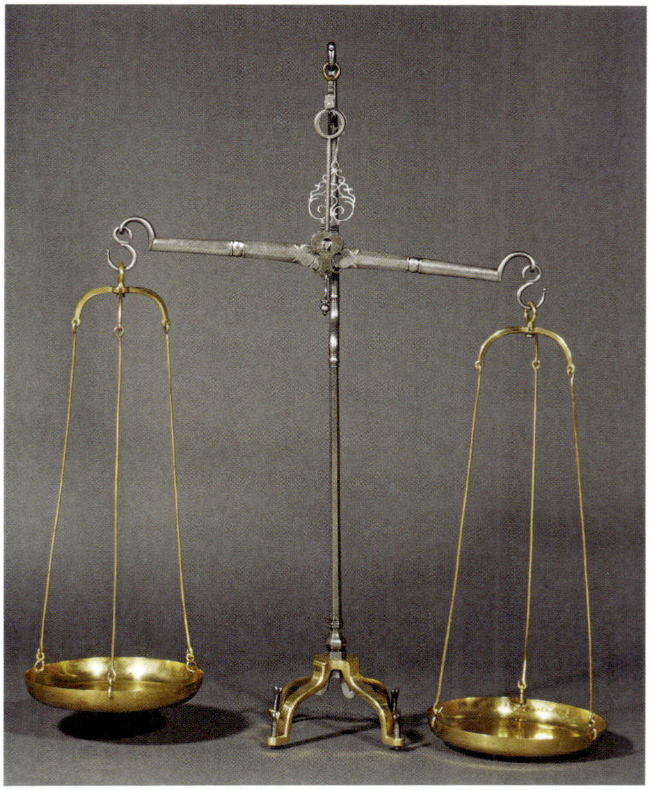

FIGURE 9 (inv. no. 19890-0000-)
© MAM/PHOTO PASCAL FALIGOT

Equal-arm balance suspended on a vertical iron upright mounted on a four-legged brass base. A levelling screw is inserted in each leg. The beam, whose fulcrum is suspended in a decorated frame, has two holes at its curved extremities. In each hole is inserted a hook with three tin bars chains supporting a brass plate. The decorated vertical pointer has a ring with a point. It is placed between the arms of a frame with two larger rings and a small vertical pin. When the balance beam is perfectly horizontal, the ring at the end of the pointer is concentric to the rings of the frame and its point is aligned with the pin.

Sigaud de la Fond (1775), vol. 1, pp. 181–183, plate XIII, fig. 1

19886-0000- (Fig. 10a, 10b)
Precision balance
ca. 1770–1780
MAM accession date: 1952
93.3 × 77.5 × 56; 32.1 kg
Brass, wood, glass, steel, fiber, plastic

Precision equal-arm balance mounted on a box of precious wood with decorated flower inlays and protected by a glass case. The box has two drawers with one iron key. The beam is mounted on a brass pillar and rests on two curved brass arms when the balance is not in use. Their vertical position can be adjusted with two micrometric screws. A long vertical pointer moves across a sector with a subdivided scale. An eyepiece with a magnifying lens fixed to the wooden top of the glass case allows a precise reading of the pointer's position on the scale. At the centre of the beam is a brass piece with four radial screws for adjusting the fulcrum's position. When a thread with a pearl is pulled, a lever releases a U-shaped brass holder, raising the fulcrum so that the balance is ready for use. The scale plates are not original but clear-plastic reconstructions. Each is suspended on a thin bar with two brackets and a ring. Two springs hold the plates in place.

Truchot assessed the precision of the balance in 1879.[6]

était extrêmement sensible, et, lors même qu'elle était chargée de cinq à six livres, elle trébuchait à moins d'un grain. Quelque soin qu'on eût pris dans l'exécution de cet instrument, il n'était pas absolument sans défaut: lorsque, après avoir établi un parfait équilibre entre les deux poids, on les changeait de bassin, il se trouvait toujours quelque différence, mais il ne pouvait jamais en résulter aucune erreur, parce qu'en prenant un milieu entre les deux pesanteurs trouvées on obtenait toujours, même après plusieurs jours de distance, le même résultat; à peine se trouvait-il quelques légères fractions de grain de différence". (LO, vol. 2, pp. 108–109).

6 The balance "rappelle nos bonnes balances de précision, et Lavoisier n'est pas allé trop loin en disant que, chargée de 600gr., elle trébuche sous l'effort de 0,005 gr. Elle est placée dans une cage vitrée, sur un meuble en marqueterie de bois de rose, sans pieds, mais avec des vis calantes. Le fléau, long de 0,50 m., est surmonté d'une aiguille de 0,20 m.; une lunette L fixe permet de suivre cette aiguille se mouvant sur un cadran. Les trois couteaux sont en acier, et les

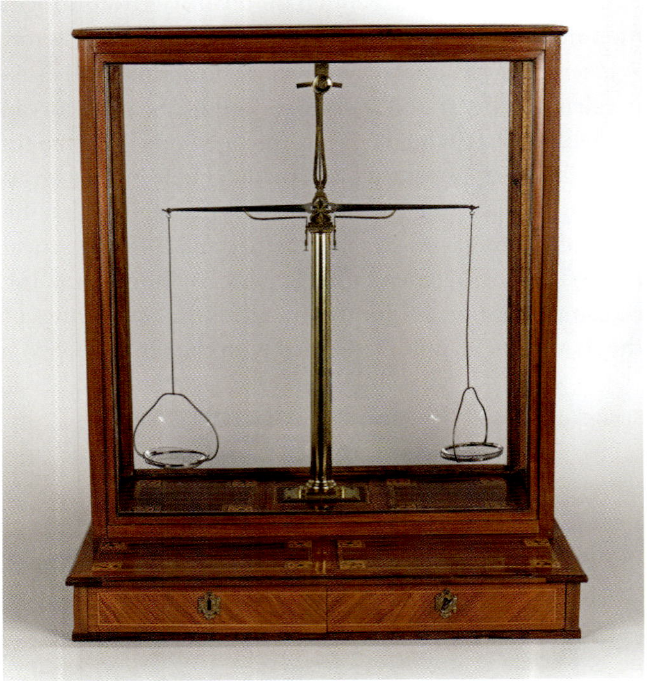

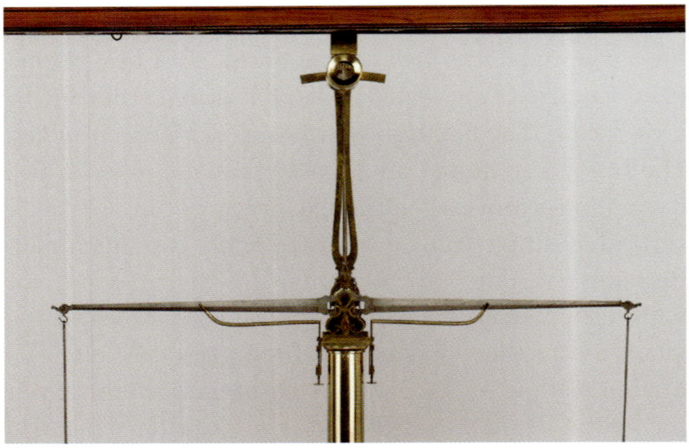

FIGURES 10A, 10B (inv. no. 19986-0000-)
© MAM/PHOTO FRANCK BOTTÉ

Truchot (1879), pp. 6–8; Daumas (1950), pp. 57–60

plateaux à étriers, suspendus par un seul fil d'acier, sont formés par des cuvettes de verre de 0,11 m. de diamètre. Le fléau, représenté à part, se soulève en tirant une cordelette qui fait mouvoir un levier articulé dans l'intérieur de la colonne. Pendant le repos, ce fléau est soutenu par des fourchettes placées sous chaque bras". Truchot (1879), p. 6.

19887-0000-, 19888-0000- (Figs. 11a, 11b, 11c, 11d, 11e)
Large precision balance
1788
MAM accession date: 1952
Nicolas Fortin
152 × 105 × 58; 77 kg
Brass, copper, steel, wood, glass

Large precision equal-arm balance[7] protected by a glass case with four opening windows. The beam is particularly long and its knife rests on a steel plate at the top of a brass column mounted on a tripod with three levelling screws. When not in use, the balance rests on two curved arms fixed to the column. At its bottom, a rotating lever with a wooden knob allows the beam to be lifted, making the balance operational. Each copper plate is suspended from the hooks at the end of the beam by means of three chains. (Fig. 11a) A plumb line is attached with thread at the top of the column and its brass bob is enclosed in a brass tube. The scale position is read by means of two magnifying eyepieces attached to the curved suspension arms. Thanks to the eyepieces, one can observe the position of two pointers (fixed to the beam) moving in front of two divided brass sectors. The scales bear 60 divisions (30-0-30). When the beam is perfectly horizontal, the pointer indicates 0. The balance is equipped with two additional plates made of glass (inv. no. 19888-0000-) (Fig. 11b). These plates rest on a ring of a thick brass wire with a ^-shaped suspension. A similar third glass plate with suspension (inv. no. 19889-0000-) (Fig. 11e) was also part of the balance set. This plate bears the engraved inscription "2 livres, 5 onces, 2 gros, 59 grains 50".

One of the curved arms bears the engraved inscription "Fortin AParis".

This balance, which cost the remarkable sum of 600 *livres*, was the largest laboratory balance used by Lavoisier. He described it in his *Traité élémentaire de chimie* (1789) as follows:

> As the determination of the weight of materials and products, before and after experiments, is the basis of all that can be done usefully and accurately in chemistry, no precision applied to the task can be excessive [...]. To obtain this degree of accuracy requires scales produced by a skilled craftsman and used with special precautions; above all, one must make sure never to use them in a laboratory, where they would inevitably rust and be damaged. They must be stored in a

7 The beam is nearly one metre long.

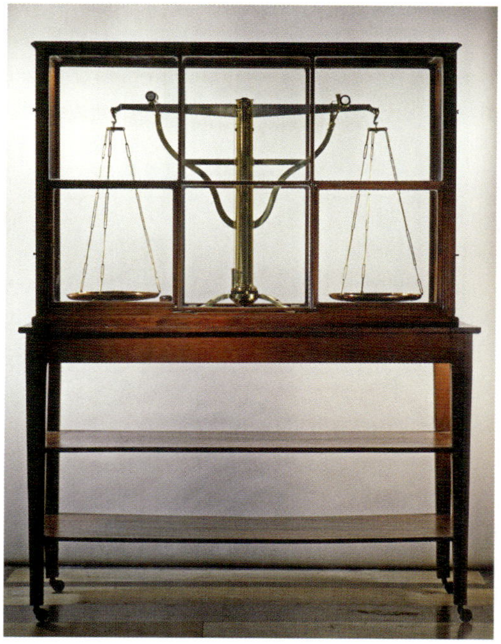

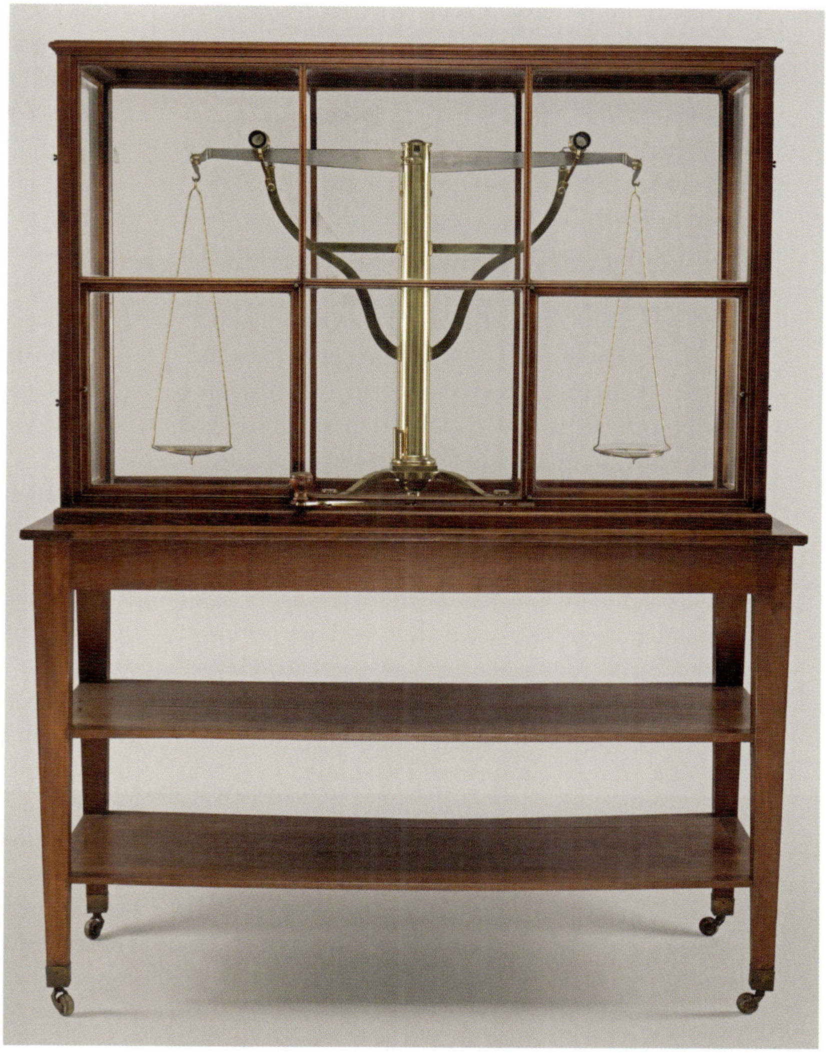

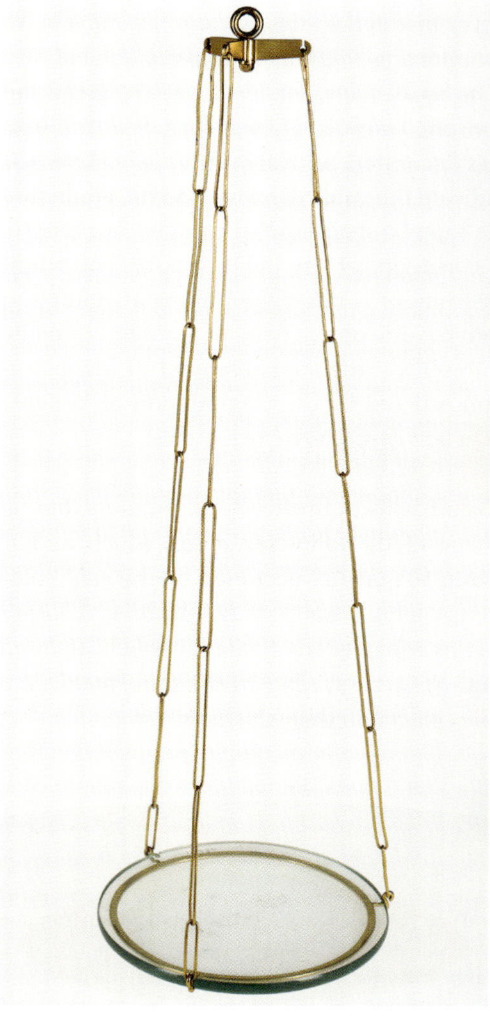

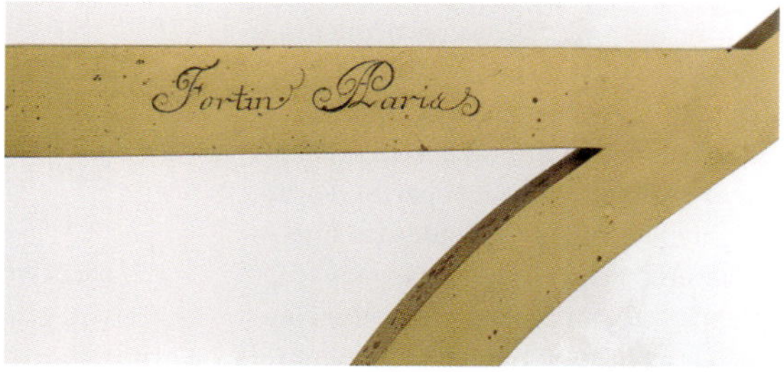

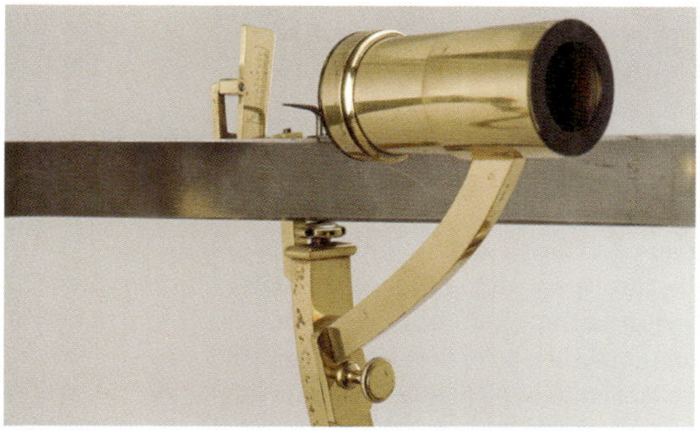

FIGURES 11A, 11B, 11C, 11D, 11E (inv. no. 19887-0000-)

separate room, where no acids are ever introduced. The scales I use have been built by M. Fortin; their arm is three feet long, and they combine all the safety and convenience features one could desire. I do not believe that there are any other scales – except for Ramsden's – that can compare for accuracy and precision. Besides this heavy scale, I have two others, which, like the first, are banned from the laboratory. One weighs up to 18 or 20 ounces, with an accuracy of one-tenth of a grain. The third weighs only up to one *gros*, and is very sensitive to one-512th of a grain.

I shall submit a specific note to the Academy describing the three scales, with details on the degree of accuracy obtained.[8]

Moreover, these instruments – one of which should be used solely for research experiments – do not obviate the need for less valuable ones for everyday laboratory work. There is always a need in the laboratory for a large scale with an iron arm painted black that can weigh whole earthenware vessels filled with liquid, and quantities of water of 40–50 pounds, to within half a *gros*; for a second scale that can weigh up to 8–10 pounds, to within 12–15 grains; and for a small hand-held scale, for weighing about one pound, to within one grain.[9]

Lavoisier (1789), vol. 2, pp. 333–334; Truchot (1879), pp. 293–296; Daumas (1950), pp. 57–60; Stock (1969); Levere (1992); Delacroix, Porte (1995); Jenemann (1997), pp. 40–41; Beretta (2003), p. 321 and pp. 333–334

19885-0000- (Figs. 12a and 12b)
Assay precision balance
Late 18th c.
MAM accession date: 1952
Mégnié the Elder (l'Aîné)
50.5 × 45.5 × 37; 8.4 kg
Wood, glass, brass, ivory, thread

Assay balance mounted on a rectangular box with a drawer and protected by a glass case.

A quadrangular brass column supports the knives of the steel beam, while a pair of decorated arms ensure that the latter remains horizontal when the balance is not in use. By pulling a thread with a small ivory sphere, one can move a lever inserted in the column. Its movement lifts the slot supporting the knives, making the balance operational. The small pans, made of watch-glasses, are suspended from the beam ends by means of a thin iron bar. The beam's decorated vertical pointer is broken: it is too short to indicate the value displayed on the graduated brass arc at the top of a steel support fixed to the column. The nominal sensitivity of the balance is 1/10 of a milligram. The instrument is not signed but is attributed to the maker Mégnié l'Aîné on the basis of a statement made by Lavoisier himself during an assessment of his instrument collection ordered by the Commission of Weights and Measures in late 1793. The reports states:

> Said Citizen Lavoisier further declared to us that he is the owner of the following items, which he has deposited in the facility occupied by the said commission: 1) A large scale [...] by Fortin; 2) Another scale built by Meynier the Elder, for weighing up to one pound; 3) An assay scale by the same maker, the three scales placed in their glass cases.[10]

8 Lavoisier never published this memoir, and this is the only known reference to his precision balances.

9 "La détermination du poids des matières et des produits, avant et après les expériences, étant la base de tout ce qu'on peut faire d'utile et d'exact en chimie, on ne saurait y apporter trop d'exactitude [...]. Il faut, pour arriver à ce degré de précision, des balances faites par un artiste habile et avec des précautions particulières; il faut surtout se faire une loi de ne jamais s'en servir dans un laboratoire, où elles seraient immanquablement rouillées et gâtées: elles doivent être conservées dans un cabinet séparé, où il n'entre jamais d'acides. Celles dont je me sers ont été construites par M. Fortin; leur fléau a trois pieds de long; et elles réunissent toutes les sûretés et les commodités qu'on peut désirer. Je ne crois pas que, à l'exception de celles de Ramsden, il en existe qui puissent leur être comparées pour la justesse et pour la précision. Indépendamment de cette forte balance, j'en ai deux autres, qui sont bannies, comme la première, du laboratoire; l'une pèse jusqu'à 18 ou 20 onces, à la précision du dixième de grain; la troisième ne pèse que jusqu'à un gros, et les 512es de grain y sont très-sensibles.

Je donnerai à l'Académie, dans un mémoire particulier, une description de ces trois balances, avec des détails sur le degré de précision qu'on en obtient.

Ces instruments, au surplus, dont on ne doit se servir que pour les expériences de recherche, ne dispensent pas d'en avoir d'autres moins précieux pour les ouvrages courants du laboratoire. On y a continuellement besoin d'une grosse balance à fléau de fer peint en noir, qui puisse peser des terrines entières pleines de liquide, et des quantités d'eau de 40 à 50 livres, à un demi-gros près; d'une seconde balance susceptible de peser jusqu'à 8 à 10 livres, à 12 ou 15 grains près; enfin d'une petite balance à la main, pesant environ une livre, à la précision du grain". Lavoisier (1789), vol. 2, pp. 333–334.

10 "Le dit citoyen Lavoisier nous a déclaré en outre qu'il est propriétaire des effets suivants qu'il a déposés dans le local occupé par ladite commission: 1° Une grande balance [...] par Fortin;

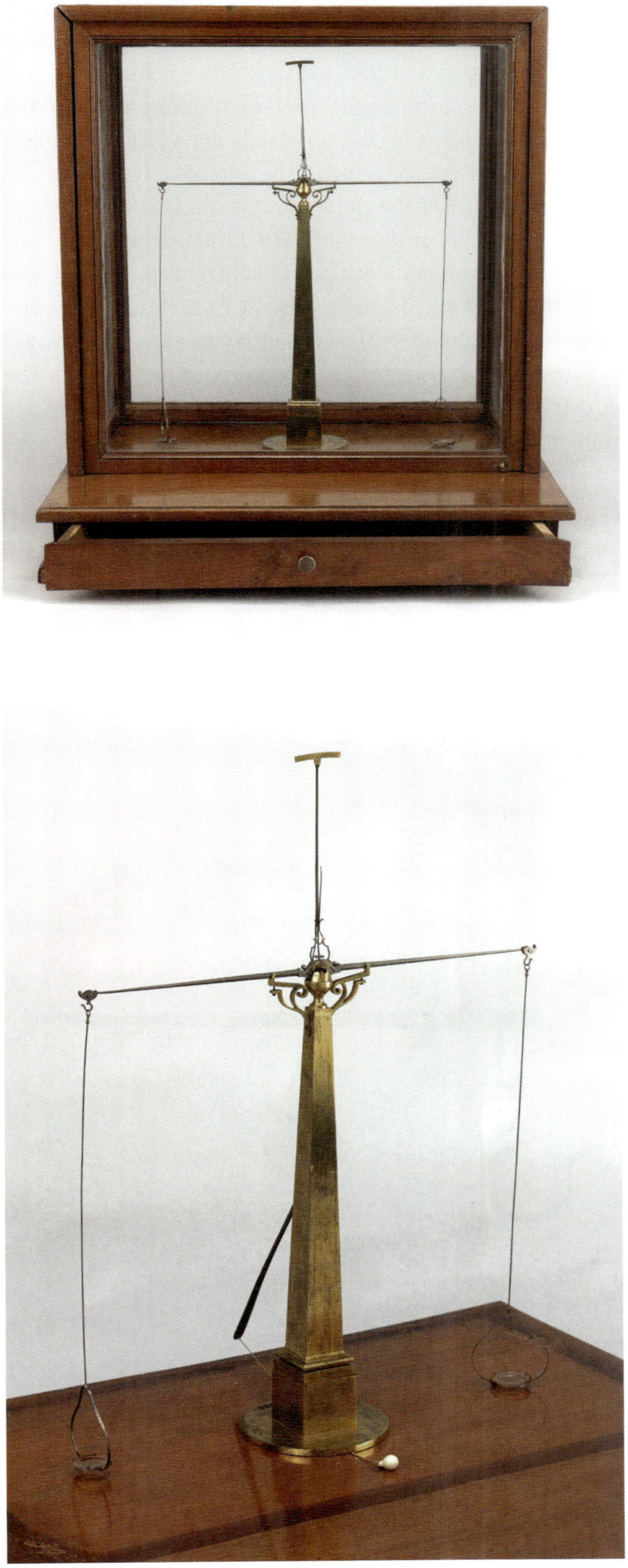

FIGURES 12A AND 12B (inv. no. 19985-0000-)
© MAM/PHOTO FRANCK BOTTÉ

Lavoisier used the assay balance (*balance d'essai*) throughout his career from 1768 to 1789. As proved by the statement cited above, this balance was used by the Commission of Weights and Measures in the early 1790s.

Truchot (1879); Daumas (1950), pp. 57–60; Daumas (1955), pp. 135–136; Beretta (2003), p. 334

19894-0000- (Fig. 13)
Spring scale
1789
MAM accession date: 1952
Flavien Marie Scanegatti (d. 1793)
19 × 15.2 × 2.5; 430 g
Brass, steel

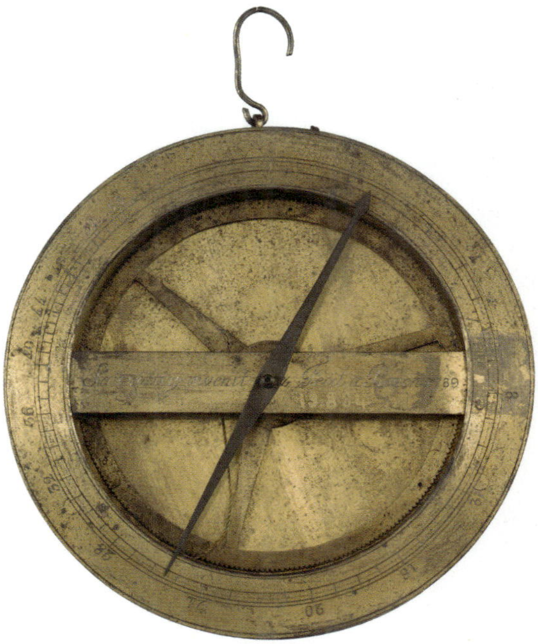

FIGURE 13 (inv. no. 19894-0000-)
© MAM/PHOTO FRANCK BOTTÉ

Brass circular spring scale with a suspension hook. A toothed wheel has two pulleys, one connected with a thread to a semi-circular spring, the other connected with a wire (now missing) for hanging the object to be weighed. A pointer fixed to the wheel indicates the weight on a non-linear circular scale graduated from 0 to 58 with

2° Une autre balance construite par Meynier l'aîné, pesant jusqu'à une livre; 3° Une balance d'essai du même auteur, lesdites trois balances dans leurs cages de verre". Beretta (2003), p. 334.

divisions every ¼ of a unit. A ratchet serves to block the wheel's position, displaying the maximum measured weight.

The front of the cross-bar bears the inscription "Sanegatty [*sic*] invenit et fecit a Rouen 1789" and on the back: "Lavoisier 54".

Beretta (2003), p. 332

19896-0000- (Fig. 14)
Hanin spring scale
Last quarter of 18th c.
MAM accession date: 1952
Hanin
5 × 56 × 36.5; 3.04 kg
Brass, iron

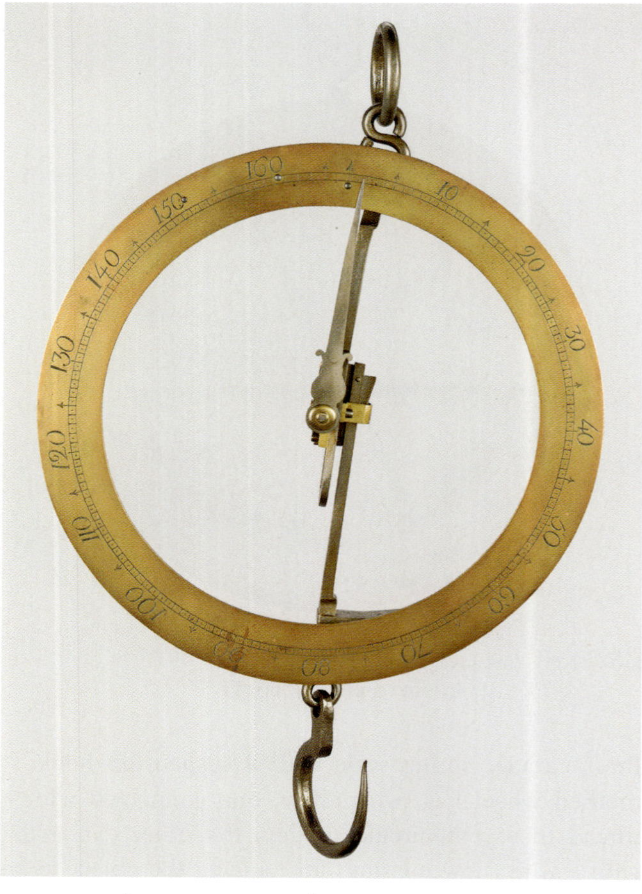

FIGURE 14 (inv. no. 19896-0000-)
© MAM/PHOTO FRANCK BOTTÉ

Large circular spring scale. On a circular crown made of brass there is a scale ranging from 0 to 165 (kg?) on which the figures are marked every ten units. A semi-circular spring with a hook for hanging the scale is attached to the back side of the crown. The spring has a vertical arm with a frame to which a pinion with a pointer is hinged. A second arm with a rack (engaged in the pinion) is fixed to the free end of the spring, which also bears a hook. A weight attached to the latter deforms the springs. The movement of the racks causes the pinion and its pointer to rotate.

This type of spring scale with a rack-and-pinion system was introduced by the French locksmith and balance maker Hanin in the 1760s. In 1791, Hanin's son was awarded 20 guineas by the Society for the Encouragement of Arts, Manufactures and Commerce in London for the construction of a similar scale with a dial reporting the units of weight of 14 European cities.

Hanin (1768), Hanin (1791), Hachette (1828), pp. 54–56, plate 2, fig. 5 and 6.

19891-0000- (Fig. 15)
Steelyard
Second half of 18th c.
MAM accession date: 1952
29.5 × 13.5 × 4; 395 g
Iron, brass, lead

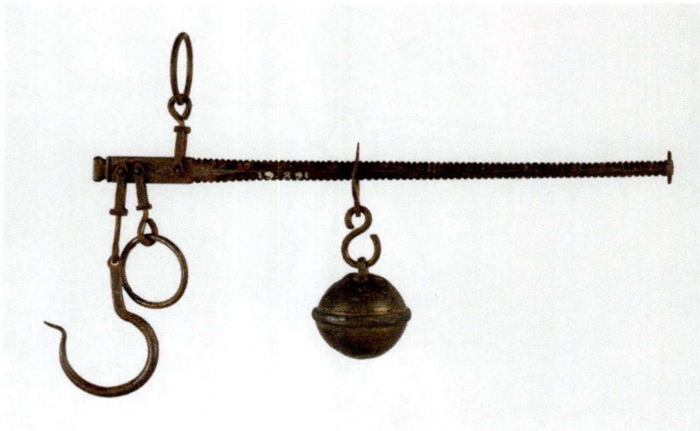

FIGURE 15 (inv. no. 19891-0000-)
© MAM/PHOTO FRANCK BOTTÉ

The steelyard is an unequal-arm balance. This model has two possible fulcrums materialized by a pair of suspension rings pivoting on the beam and pointing in opposite directions. The two parts of the beam flanking the rings form the arms. The shorter one has a pivoting bracket with a hook from which the object to be weighed is hung. The longer part has two rows of teeth on opposite sides corresponding to two scales (one numbered from 5 to 20, the other numbered 12, 16, 20). A spherical brass counterweight is suspended from a ring and its position along the scale determines the weight of the load. To use the second scale – and thus change the maximum weight to be measured – the steelyard is turned upside down. The hook swings around the end of the steelyard so that it hangs again below the arm. The ring acting as the second fulcrum now faces upward and can be used to suspend the steelyard. This type of steelyard is therefore called "reversible".

Owne (1922), pp. 101–113; Robens et al. (2014), pp. 169–173

19893-0000- (Fig. 16)
Chinese steelyard
Late 18th c.
MAM accession date: 1952
29.2 × 7 × 1.7; 135 g
Brass, ivory, wood, thread

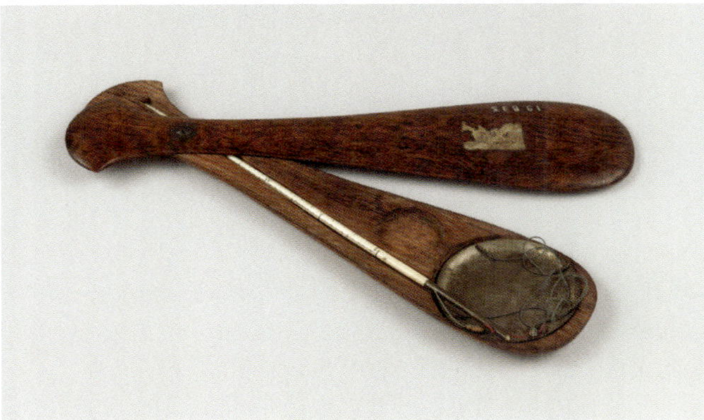

FIGURE 16 (inv. no. 19893-0000-)
© MAM/PHOTO FRANCK BOTTÉ

Chinese steelyard consisting of an ivory scale rod with a brass pan attached to it with four threads. The scale is marked with black dots. The weight, which originally slided along the rod, is missing. The steelyard is stored in a paddle-shaped wooden case, whose two halves are joined with a swivel. This type of steelyard was mainly used for weighing small pieces of precious metals, gems, and similar objects.

The case carries the barely readable inscription "Lavoisier // PBC". On the back of the box there is also a printed label, "Exposition Lavoisier // Palais de la découverte", indicating that the instrument was displayed at the Lavoisier exhibition held in Paris in 1943.

The steelyard was included in the 1794 inventory of Lavoisier's instruments made when his property was confiscated.

Lavoisier (1943); Beretta (2003), p. 332; Eric Robens et al. (2014), pp. 169–173

McKie 327
Balance-set for money-changing
Early 18th c.
Catalogued by Douglas McKie in 1952 but not included in the donation to the MAM
Laurent Grosset (active between 1675 and 1724)
Brass (?), wood
L. of beam 15.5; diam. of pans 4.5

Hand-held scale in a wooden case similar to no. 19892-0001-, 19892-0002-, with square-shaped weights. The painted inscription inside the case reads "LAVREN. GROSSET // RVE D.4 CHAPAVX. ALYON // A. LA. PLVME ROYALLE". A similar scale, also made by Grosset, is preserved at the Musée National de la Renaissance (Écouen), inv. no. ECL9032.

This item belonged to the collection of the late Pierre S. Dupont.

McKie 328
Hand-held coin scale
18th c. (?)
Catalogued by Douglas McKie in 1952 but not included in the donation to the MAM
Brass (?), wood
L. of beam 11.1; diam. of pans 4.9
Hand-held scale in a wooden case similar to no. 19892-0001-, 19892-0002-, with the same type of weights.

19892-0001-, 19892-0002- (Fig. 17)
Balance-set for money-changing
Late 18th c.
MAM accession date: 1952
Box: 18.5 × 6.5 × 2.2; 120 g
Brass, wood, steel, paper, thread

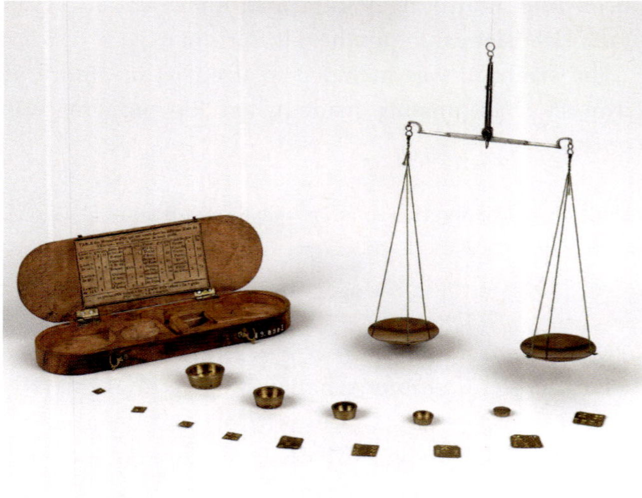

FIGURE 17 (inv. nos. 19892-0001-, 19892-0002-)
© MAM/PHOTO FRANCK BOTTÉ

Hand-held scale used for weighing coins. Small equal-arm scale (19892-0001-), comprising a steel beam with a vertical indicator and two brass pans. These are suspended from the arm with a ring and three threads. The scale is stored in a round-ended wooden box together with a pile of five cup-shaped brass weights (for a total of 8 grains = 1 ounce) (19892-0002-), and eight smaller square flat brass weights (from 24 to 1 *grains*).[11] The label on the inside lid reads "TABLE des Monnoies d'or qui ont cours dans les differens Etats de L'Europe, avec la désignation de leur poids".

The table indicates the weights of various coins of different countries and cities (France, England, Spain, Holland, Austria, Rome, Piedmont and Sicily, Parma, Prussia, Portugal, Genoa).

Musée National des Techniques (1990), pp. 46–47; Turner (2018), pp. 280–305

20232-0001-, 20232-0002- (Fig. 18)
Balance-set for money-changing
1786–1810
MAM accession date: 1952
Box and scale (20232-0001-): 11.7 × 5.2 × 2.2; 50 g
Weights (20232-0002-): 17.7 × 14.5 × 4: 10 g
Brass, wood, steel, paper, thread

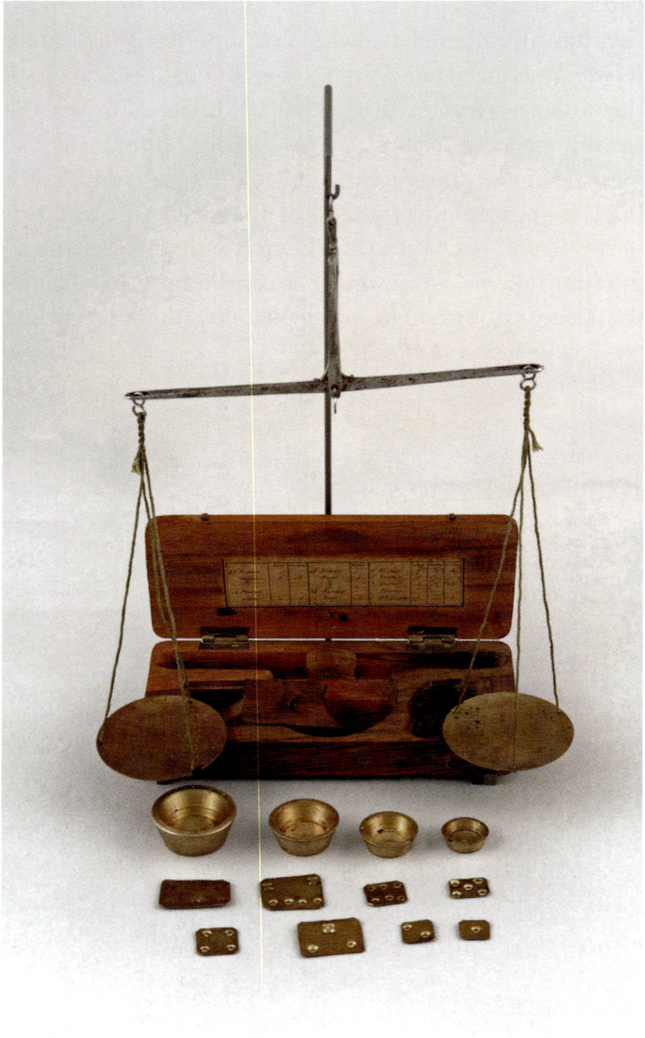

FIGURE 18 (inv. nos. 20232-0001-, 20232-0002-)
© MAM/PHOTO FRANCK BOTTÉ

Hand-held scale used for weighing coins. Small equal-arm scale comprising a steel beam with a vertical indicator and two brass pans. These are suspended from the arm with a ring and three threads. The scale is stored in a rectangular-ended wooden box together with a pile of four cup brass weights (from 4 *gros* to 1 *demi-gros*) and eight smaller square flat brass weights (from 24 to 1 *grains*).[12] The label on the inside lid indicates the weights of various French coins (francs) and of the Spanish piaster.

Musée National des Techniques (1990), pp. 46–47; Turner (2018), pp. 280–305

11 The *gros* was a traditional French weight equal to 3.824 g. The weight named *grain* is equal to about 53 mg. All the weights are inventoried under no. 19892-0002-.

12 All the weights are inventoried under no. 20232-0002-.

19902-0000- (Fig. 19)
Set of 36 Chinese weights
1700–1790
MAM accession date: 1952
Largest: 6 × 4.5 × 3; 375 g
Smallest: 1.2 × 0.7 × 0.5; 5 g
Bronze

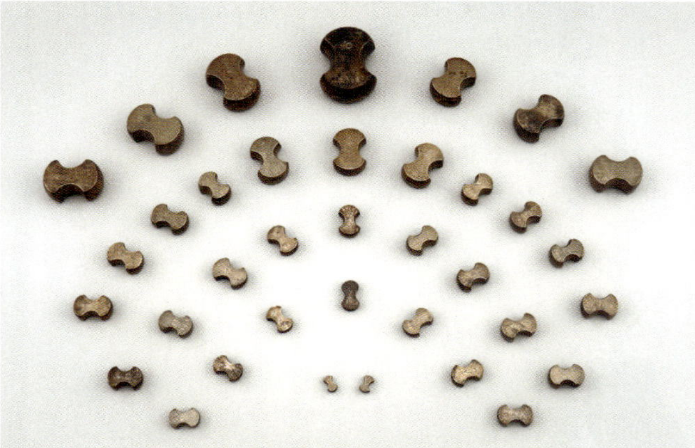

FIGURE 19 (inv. no. 19902-0000-)
© MAM/PHOTO FRANCK BOTTÉ

Series of 36 Chinese weights similar in shape to a violin body. They are marked with Chinese ideograms and Arabic numbers. There are 14 weights ranging from 10 to 1 *liăng*, 20 weights from 9 to 3 *qián* and 2 weights of 8 and 7 *fēn*. A *liăng* (ou *tael*) = 10 *qián* = 10 *fēn* (or condorin). It varied between 37.4 g and 37.8 g. These weights were originally contained in a bronze mortar (missing).[13]

Musée National des Techniques (1990), p. 127

19901-0000- (Fig. 20)
Set of 7 weights
1770–1780
MAM accession date: 1952
Nicolas Chemin (d. 1780)
Largest: 10.5 × 9.5 × 7.5; 2.3 kg
Smallest: 5.5 × 5 × 3.5; 480g
Iron

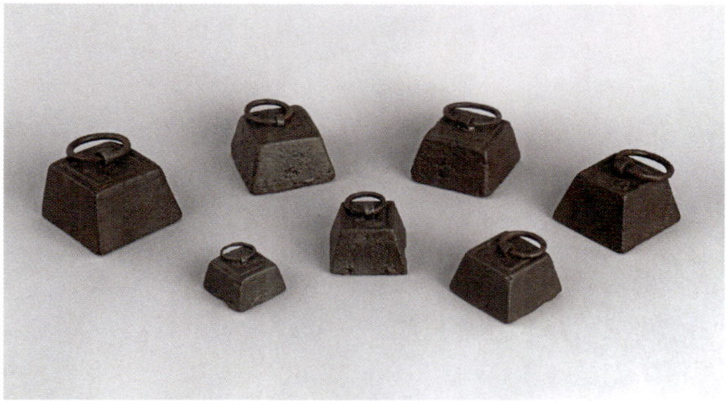

FIGURE 20 (inv. no. 19901-0000-)
© MAM/PHOTO FRANCK BOTTÉ

Series of seven iron weights: one of 6 pounds, three of 4 pounds, two of 2 pounds and one of 1 pound.[14] They are in the shape of truncated pyramids whose bottoms are loaded with a mass of lead. The leads are punched with one or more crowned Qs, denoting the French craftsman Nicolas Chemin, Lavoisier's favourite balance-maker in the early 1770s. All the weights are fitted with lift rings.

Musée National des Techniques (1990), p. 38

19900-0000- (Fig. 21)
Set of 8 cylindrical weights
1788
MAM accession date: 1952
C. Fourché (fl. 1777–1811)
20.5 × 9; 1.95 kg
Brass, wood, lead

Set of brass decimal weights in the shape of hollow cylinders. They are marked at the bottom: "1 LIVRE", "5/10", "3/10", "2/10", "1/10", "5/100", "3/100", "2/100". The weights are stacked inside one other on a circular wooden base loaded with a mass of lead. Any weight can be taken from the pile without moving the others. There are traces of the printed label indicating that the set was displayed at the "Exposition Lavoisier" in 1943. Lavoisier described the circumstances of the construction of these weights in *Traité élémentaire de chimie* (1789):

> I have always intended to obtain the division of the *livre poids de marc* into decimal fractions, and have only recently succeeded. M. Fourché, scale-maker, successor to M. Chemin, Rue de la Ferronnerie, has achieved this with great intelligence and exactitude,

13 The mortar has been missing since at least 1975.

14 One pound (*Livre*) is equal to ca. 489 g.

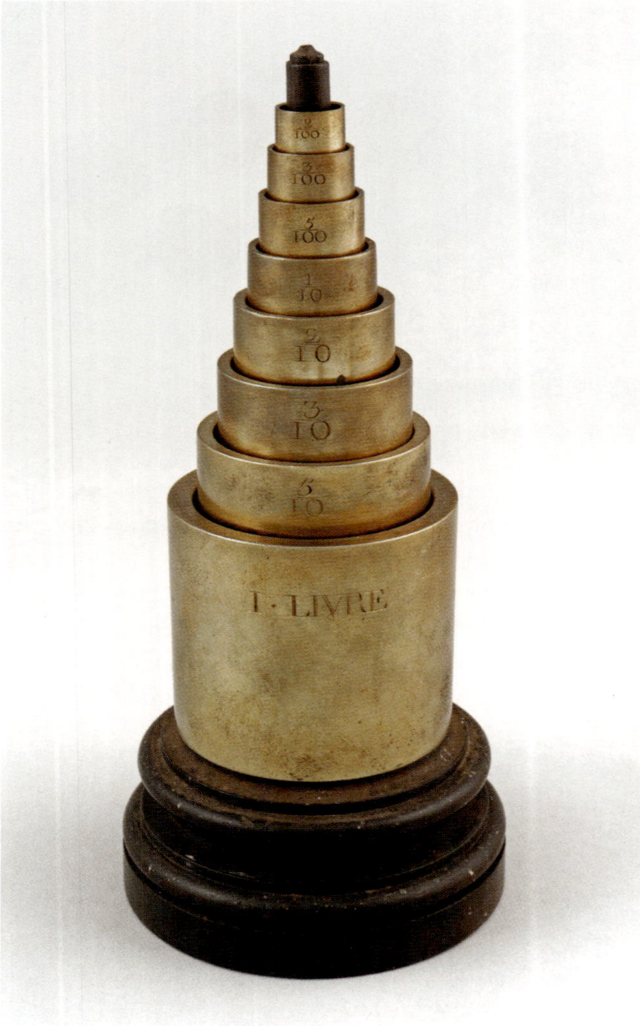

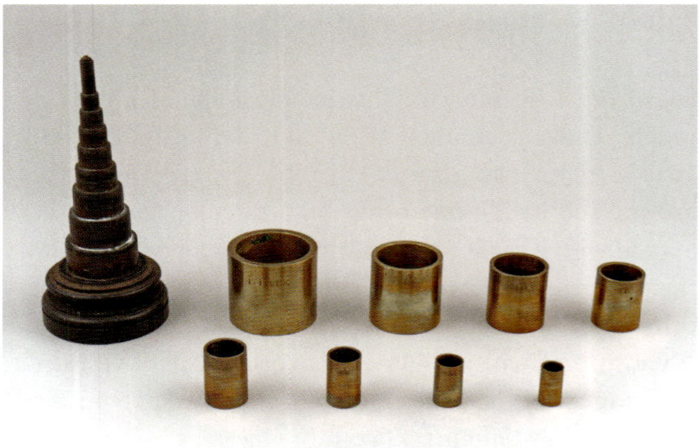

FIGURES 21A AND 21B (inv. no. 19900-0000-)
© MAM/PHOTO FRANCK BOTTÉ

> and I invite all those who conduct experiments to acquire similar divisions of the pound: even if they make limited use of the calculation of decimals, they will be surprised by the simplicity and ease that this division will offer in all their operations. In a special note for the Academy, I shall describe the precautions and care required for this division of the pound.[15]

Lavoisier (1789), vol. 2, pp. 329–330; Lavoisier (1943); *Musée National des Techniques* (1990), p. 49

19897-0000- (Fig. 22)
Pile of weights
ca. 1790
MAM accession date: 1952
Christoph Jobs Rothenberger
4.7 × 3.5 × 2.3; 125 g
Brass

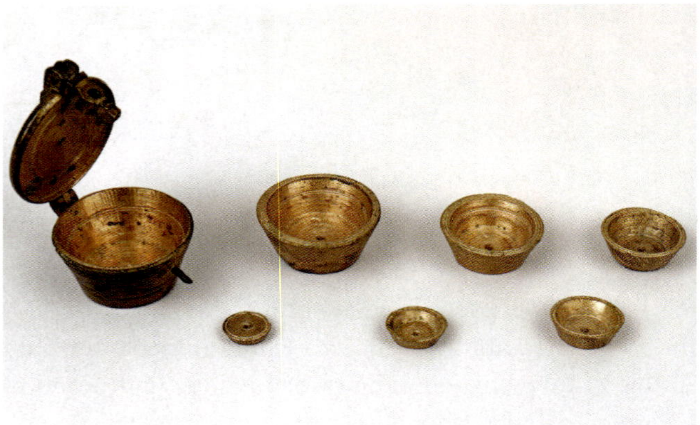

FIGURE 22 (inv. no. 19897-0000-)
© MAM/PHOTO FRANCK BOTTÉ

Set of seven small cup weights. The largest weight has a hinged lid and serves as container. The smallest is disk-shaped. Three are marked “8”, “4”, “2”, “2B”, “1”. The set’s total weight is ca. 122 g, or roughly 4 ounces. (1 ounce = ca. 30.6 g). The lid also features the punched figure of a rooster, denoting the maker Christoph Jobst Rothenberger of Nuremberg, who became master craftsman in 1787. The set was probably used by the Commission of Weights and

15 “J’ai toujours eu le projet de faire diviser la livre poids de marc en fractions décimales, et ce n’est que depuis peu que j’y suis parvenu. M. Fourché, balancier, successeur de M. Chemin, rue de la Ferronnerie, a rempli cet objet avec beaucoup d’intelligence et d’exactitude, et j’invite tous ceux qui s’occupent d’expériences à se procurer de semblables divisions de la livre: pour peu qu’ils aient d’usage du calcul des décimales, ils seront étonnés de la simplicité et de la facilité que cette division apportera dans toutes leurs opérations. Je détaillerai, dans un mémoire particulier destiné pour l’Académie, les précautions et les attentions que cette division de la livre exige”. Lavoisier (1789), vol. 2, pp. 329–330.

Measures, set up by the Académie royale des sciences in 1790.

Musée National des Techniques (1990), p. 27

19899-0000- (Fig. 23)
Set of 4 Spanish weights
Late 18th c.
MAM accession date: 1952
3.5 × 3.2; 110 g
Brass

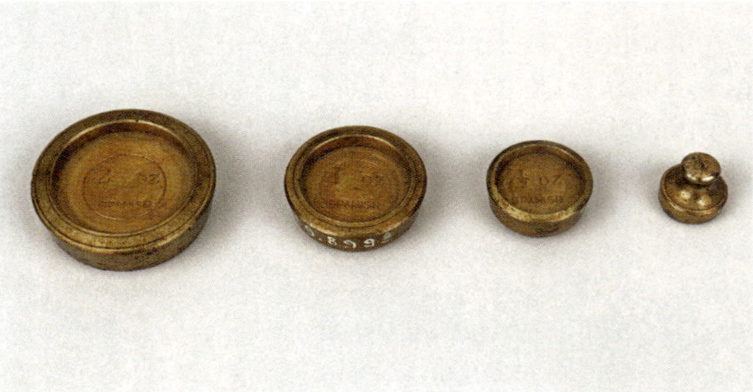

FIGURE 23 (inv. no. 19899-0000-)
© MAM/PHOTO FRANCK BOTTÉ

Set of 4 Spanish weights, fitting one inside the other in a small pile. They are marked "2 oz // Spanish", "1 oz // Spanish", "1/2 oz // Spanish", "1/4". One Spanish ounce (*onza*) is approximately equal to 28.8 g. This set was probably used by the Commission of Weights and Measures, set up by the Académie royale des sciences in 1790. Aimé Pommier, author of the inventory of the collection of weights at the MAM, dates this set to 1800–1820 without giving any specific reason.

Musée National des Techniques (1990), p. 119

19898-0000- (Fig. 24)
Set of 4 weights
1812–1815
MAM accession date: 1952
Polichot
4.8 × 2.6; 230 g
Brass

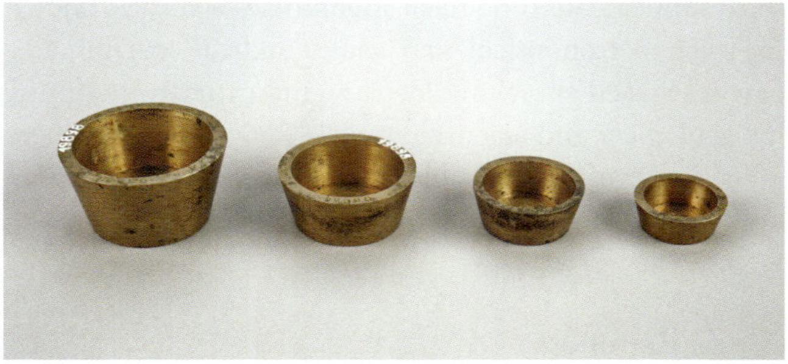

FIGURE 24 (inv. no. 19898-0000-)
© MAM/PHOTO FRANCK BOTTÉ

Set of four cup weights, fitting one inside the other in a small pile. The weights are marked in ounces and grams: "4 ONCES // 125 G", "2 ONCES // 62 G. 5", "1 ONCE // 31 G. 3", "4 GROS // 16 G". The weights are punched with the figure of an eagle and a crowned K, the latter indicating the French maker Polichot.

Musée National des Techniques (1990), p. 61

Astronomy, Surveying, Drawing and Mathematics

Lavoisier studied astronomy with La Caille while at the Collège Mazarin. His only work in the field dates from October 1763. Despite this limited involvement, Lavoisier often showed a keen interest in mathematical instruments, and several of the young craftsmen he recruited to make his most sophisticated chemical apparatuses were also known for their astronomical and mathematical instruments. Lavoisier used surveying instruments extensively during his geological and scientific travels (1763–1767) but there is evidence of his continued interest in a survey-based approach to field research until the end of his life.

Astronomy

20142-0000- (Figs. 1a, 1b, 1c)
Planispheric astrolabe
Second half of 16th c.
MAM accession date: 1952
28.7 × 22.8 × 4 (thickness); 1.28 kg.
Material: brass, paper

The astrolabe is not signed.

Obverse

The *rete* shows the positions of thirty-nine named stars by means of elegant, lightly curved pointers, one v-shaped notch (a negative pointer) and one pointer carrying three star indications. Each star is accompanied by its magnitude and its planetary symbols. The individual words in compound names are separated by dots, and dots are also placed at the beginning and end of each name. The following stars are shown, their names all punched in small capitals:

Caput serpens	2 *alpha Oph*	Caput Herculis	3 *alpha Her*
Ped[es] Ser[pens]		Corona S[eptrionalis]	2 *alpha Cr Bor*
Lanceator	1 *beta Lib*	S[inister] H[umerus] Bootis	3 *gamma Bo*
Dors[vm] vrs[a] maj	2 *alpha UMa*	Capvt dragonis	*gamma Dra*
Vvl[tvr] Cad[ens]	1 *alpha Lyr*	Spica virginis	1 *alpha Vm*
Cor leonis	1 *alpha Cor*	Lvcida Hidra	2 *alpha Hya*
Canis minoris	1 *alpha Cmi*	Canis maior	1 *alpha Ca Ma*
Cauda [leonis]	*beta Leo*	Dors[sum]	2 *delta Leo*
Cervix		Caput Ge[mini] An[tecedentis]	
Hircvs	1 *alpha Aur*	D[exter] L[watus] Persei	2 *alpha Per*
Cap[ut] Algol	2 *beta Per*	Vmb Andro[meda]	3 *beta And*
Pecta ca P	3	D[exter] hvmervs] Orionis	1 *alpha Ori*
S[inister] Pes Orionis	1 *beta Ori*	Occulvs Tauri	1 *alpha Tau*
Naris ceti	3 *alpha Cet*	Venter ceti	3 *gamma Ceti*
Cavda ceti	3 *beta Cet*	Crvs Pegasi	2 *beta Peg*
D[exter] ad Cephei	3 *alpha Cap*	Cavda cigni	2 *alpha cyg*
Aquila avt v[ultur] vo[lans]	2 *alpha Aqr*	Muscida Pegasi	3 *epsilon Peg*
H[umerus] Pegasi	2 *alpha Peg*	Crvs aquaria	3 *delta Aqr*
Cavda	3 *gamma Cap*	Lanx S libre[1]	2 *alpha lib*

The *limb* is cast in one piece with the front of the small (21 mm high), nearly circular *throne* and is riveted to the back plate, cast in one piece with the back of the *throne*. There is a delicate *shackle* carrying a swivel pin to which the suspension ring is attached. The inventory number is painted in white on the rim and stamped on the *mater*. Reading from the outer edge inwards, the *limb* carries:

1 the label 'Meridies' immediately below the *throne*
2 wind names: reading clockwise from beneath the centre of the throne, they are stamped in larger and smaller capitals. Here, the former are indicated by capitals and the latter by mixed upper- and lower-case letters. Points and stars are as on the instrument:
AVSTER VEL NOTVS
Avstro . Libo * Notvs
* Aphicvs * Libo

1 The use of an 'S' for 'sud' instead of an 'M' for *Meridonalis* confirms the French origin of this instrument.

 | DOI:10.1163/9789004511217_016

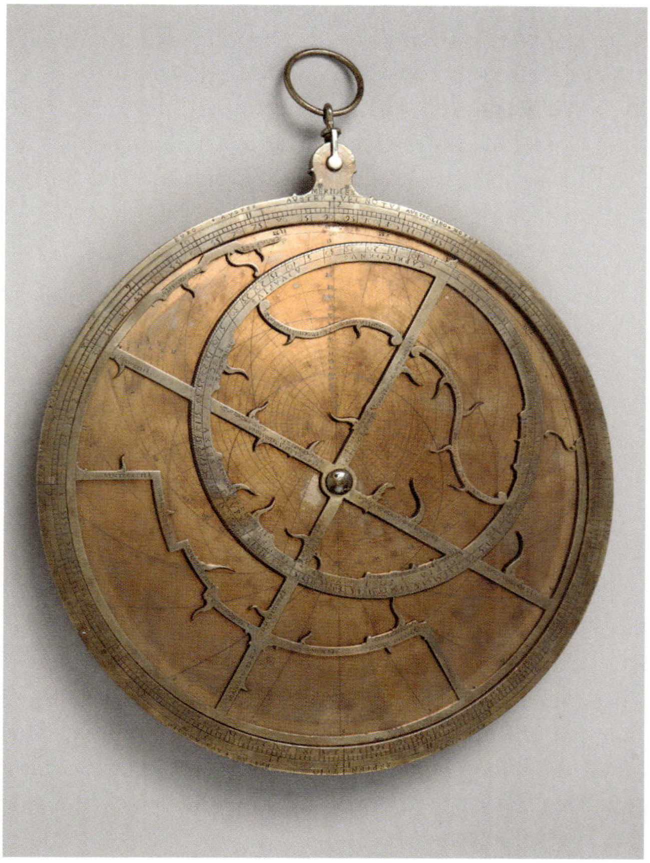

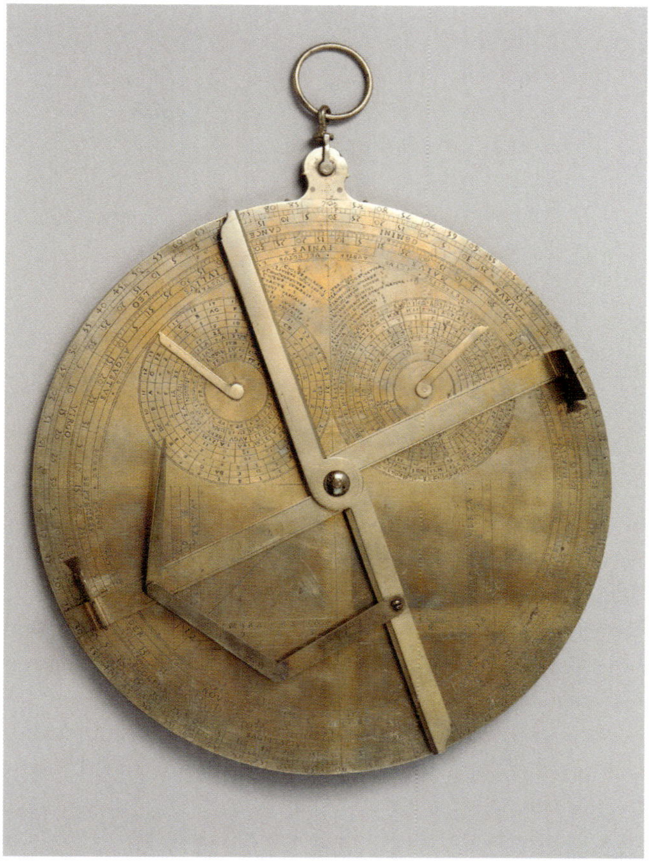

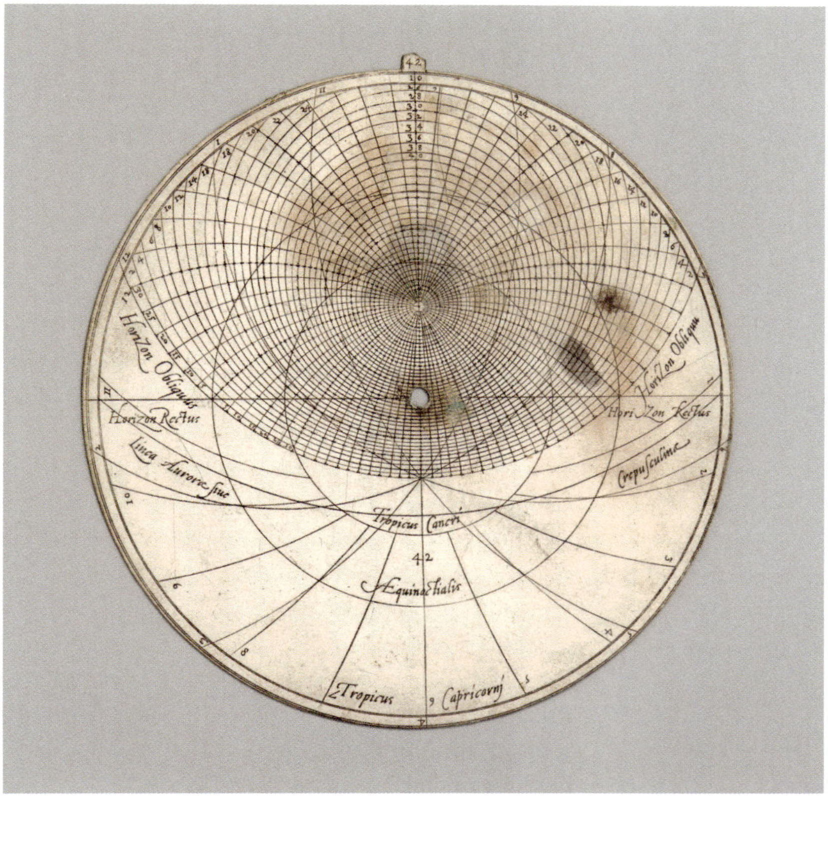

FIGURES 1A, 1B, 1C
(inv. no. 20142-0000-)
© MAM/PHOTO FRANCK BOTTÉ

FAVONIUS VEL ZEPHIRVS
Cho * rvs
Circ * ivs
SEPTENTRIO
Boreas * avt Aquilo
Vvltv * nvs
SVBSOLANVS
Evrvs *
Evro * avster

3 a double hour scale 1–12/1–12 reading to four minutes against the division scale shared with

4 a 360° scale reading from the east point (Subsolanus), and numbered every five degrees.

Plates

There are three plates of 208 mm diameter. On plates 1a,b and 2,a,b the almucantars are drawn for every three degrees and numbered above the pole, and the azimuths for every 10°. These diagrams are overlaid by dashed lines representing the houses of the heavens numbered 1M[ansionarius], 2M, and so on. Unequal hour lines numbered 1–12 are drawn beneath the horizon, and circles for the tropic of Cancer and for the Equator are traced.

1a is drawn for a latitude of 45°.

1b is drawn for a latitude of 47½°. It may be significant that this plate has remained so long under the *rete* without being moved that the *rete* pattern has offset onto it.

2a is drawn for a latitude of 48°

2b is drawn for a latitude of 54°

3 is a replacement plate of printed cardboard that is too small for the instrument (187 mm diameter). It derives from one of the editions of the printed paper astrolabe published by Philippe Danfrie.[2] It is drawn for latitude 42°, has the almucantars drawn for every 2° and the azimuths for every 5°; it carries the houses of the heavens, the direct and oblique horizons, a crepuscule line, and circles for the tropics of Cancer and Capricorn and for the Equator, all of which are named.

Reverse

Reading from the outer edge inwards, there are scales for:

1 degrees arranged in four quadrants of 90° with reversed readings, inwards and outwards from quadrant to quadrant.

2 & 3 a concentric zodiacal calendar with signs divided from 0 to 30 in 1° intervals, the months to single days (February 28 only). 0° Aries = 10.75 March.

4 Wind names reading outwards and clockwise from the centre line of the instrument below the throne. The points and stars are as marked on the instrument:

Avster * .
Austro libo * notvs
Aphicvs * . libo
Favovs ve * Aphicvs
Cho * rvs
Circ * ivs
Sept * entrio
Borvs * avt aquilo
Vvltvr * nvs
Sunso * lanvs
Evrvs *
Evro * Avser
* Velnotvs

The upper left quadrant contains a circular calendrical diagram with a rotatable index for which a key is provided immediately above in small capitals; each word is preceded by an asterisk, with a point between each word.

1 Ciclvs solaris
2 Littere D[o]m[ini]calis
3 Concvrentes
4 Ciclvs lunaris
5 Epacte
6 Termini pascalis
7 Menses
8 Regvlaris lvnaris
9 Regvlaris ferialis

The upper right quadrant contains a circular diagram, with a rotatable index, containing astrological information for which a key is provided immediately above in small capitals; each word is preceded by an asterisk, with a point between each word.

1 Termini
2 Mansion[um] of the Moon (represented by a symbol) G et natvra
3 Mansion[um] of the Moon (represented by a symbol) nvmervs
4 Triplicitatvm Domini
5 Faciervm domini
6 Signa Zodiaci et eorvm natvra
7 Domarvm domini
8 Exaltionvm domini
9 Gaudia planetarvm

The two lower quadrants contain a double shadow square to base twelve numbered by groups of three.

2 First edition 1578; first edition, re-issue adjusted for the Gregorian calendar, 1584; 2nd edition (by Jean Moreau), 1625. For Danfrie, see Turner (1989), pp. 22–49.

The instrument is accompanied by two alidades and two rules. One pair of these, although possibly dating from the late 16th or early 17th century, is clearly not original since each element is too short to reach the circumference of the instrument; the other pair is of the correct size, but since both the rule and alidade have lost their points, they are also unlikely to be original. An articulated arm on one of the rules is a later addition.

Commentary

Date

For the past fifty years at least, the instrument has been dated to the first half of the 16th century. This dating was made by Maurice Daumas on the basis of his correspondence with George Prin (1889–1959), telescope-maker and collector of early mathematical instruments. Prin pointed out that the positions of *Cor Leonis* and *Spica* were very consistent with the values given by Stöffler in his early issues of his *Elucidatio astrolabii* (numerous editions between 1512–1513 and 1619). Although the instrument's general style is not discordant with an early date, the list of wind names on the limb corresponds closely in both terminology and presentation with those found in a similar position on an incomplete single-plate French astrolabe dated 1543.[3] The value of 10.75 March for the Vernal equinox suggests a date somewhat later in the century as does the use of a flat-topped figure '8'. This is employed on an astrolabe of 1553 attributed to the clockmaker Jean Naze in Lyon[4] and on one by Louis Martinot at Sens in 1598. The latter instrument employs a calendar roundel similar to, but simpler than, the one found on the Lavoisier astrolabe.[5]

Location

Given the sequence of the existing plates, it is likely that the missing plate was drawn for 42° and 51°, which would yield a natural sequence of three-degree increments for the three plates. The sequence is broken by the specific latitude of 47° 30′, suggesting that this is the latitude of the location where the astrolabe was expected to be most frequently used. In France, this latitude passes south of Blois and north of Tours: a location in the Loire valley during the period of château construction there seems not unlikely.

Latitudes (modern) of selected royal or princely châteaux in the Loire valley:

Amboise	47° 25′ 47″	Chambord	47°36′58″
Plessis-lez-Tours	47° 22′ 57″	Angers	47° 28′ 12″
Tours	47° 23′ 49″	Blois	47° 35′ 8″
Beauregard	47° 32′ 13″	Brissac	47° 21′ 11″
Chaumont-sur-Loire	47° 28′ 45″	Cheverny	47° 28′ 1″
Clos Lucé	47° 24′ 36″	Baugé	47° 32′ 29″
Briare	47° 38′ 22″	Villesavin	47° 32′ 48″

Another – though perhaps less likely – locality for which the astrolabe could have been intended is Montbéliard (47° 30′ 37″).

Since Lavoisier had no antiquarian interest, it is not clear whether this astrolabe belonged to him or to his father. While at the Collège Mazarin, Lavoisier studied astronomy with La Caille and in October 1763 wrote his first scientific contribution on the Aurora Borealis (LO, vol. 4, pp. 1–7).

Daumas (1949); Michel (1976); d'Hollander (1999); Morrison (2007)

20228-0000- (Fig. 2)
Armillary sphere
Late 18th c.
MAM accession date: 1952
41 × 28 × 25; 425 g
Wood, paper, brass, iron

Copernican armillary sphere, made of wood, representing the solar system. The sphere is mounted on a turned wood pedestal. Five rotating vertical rings (*armillae*) represent Mercury, Venus, Mars, Jupiter and Saturn. A sixth, partly open ring marks the position of the Earth, represented by a small sphere rotating on an inclined axle inside a smaller ring. The Moon is a small disk held by a curved arm. The column supporting the Earth pivots on a bracket inserted in the vertical axle of the sphere. Thanks to a pair of pulleys and a connecting thread (missing), when the bracket rotated, the Earth revolved around the Sun and its axle remained parallel to itself during the entire revolution.

Two fixed orthogonal vertical rings represent the equinoctial and the solstitial colures. A horizontal ring materialises the celestial equator, and a band represents the ecliptic with the zodiacal signs.

The sphere is unsigned but very similar to those made by the French geographer, globe-maker and mapmaker

3 Instrument in a private collection. See Chayette-Cheval (2018), lot 22.

4 Royal Museums of Greenwich ASTO 571 (A40: CCA N° 429). See van Leempoel (2005), pp. 153–55.

5 Adler Planetarium and Astronomy Museum M31 (CCA N° 122). See Webster and Webster (1998), pp. 82–84.

Charles François Delamarche (1740–1817). An almost identical sphere is preserved at the National Maritime Museum in Greenwich (inventory ASTO 631).

Delamarche (1800); Dekker (1999), p. 173

Surveying Instruments

20138-0000- (Fig. 3)
Alidade for plane table
1762
MAM accession date: 1952
Jacques Canivet (1721–1773)
21.5 × 49.5 × 5.5; 370 g
Box 52.5 × 8.5 × 3.5; 860 g
Brass (instrument); leather, wood, iron (box)

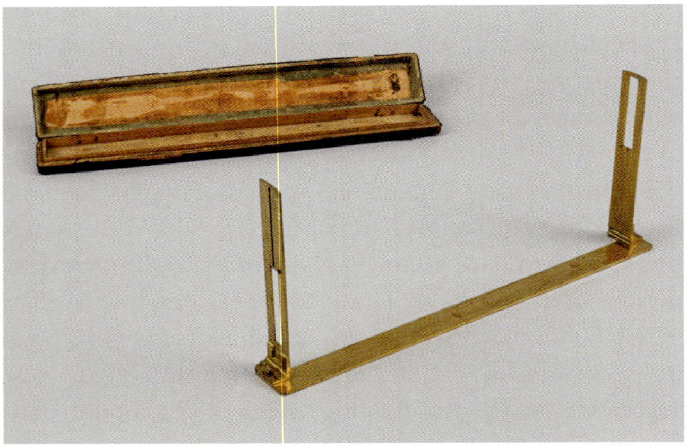

FIGURE 3 (inv. no. 20138-0000-)
© MAM/PHOTO FRANCK BOTTÉ

This alidade is used with a plane table for drawing charts and maps. It consists of a brass ruler with two folding sights at its ends. Each sight is composed of a brass rectangle with a vertical window and a narrow slit. A thin sighting wire (missing) was originally fixed vertically in the centre of the windows. The instrument bears the engraved signature: "Canivet à la Sphère // à Paris. 1762." The instrument is housed in a wooden box covered with decorated and embossed leather. A pair of iron hooks serve to close the box.

The surveyor would attach a drawing sheet to a horizontal plane table mounted on a tripod, and then place the alidade on it. Using the slits and the wires, he would point the alidade at the target and trace the corresponding line with a pencil on the paper.

Lavoisier probably used this instrument in October 1763 when he wrote from Villers-Cotterets an extremely

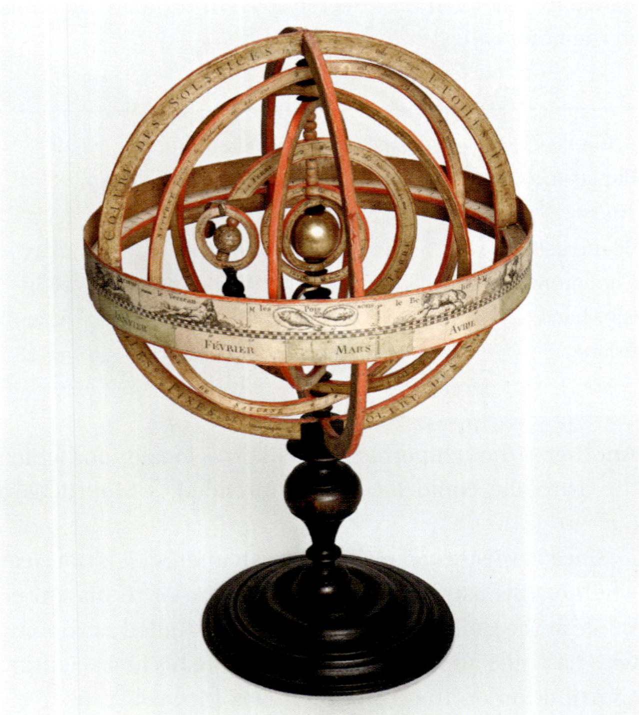

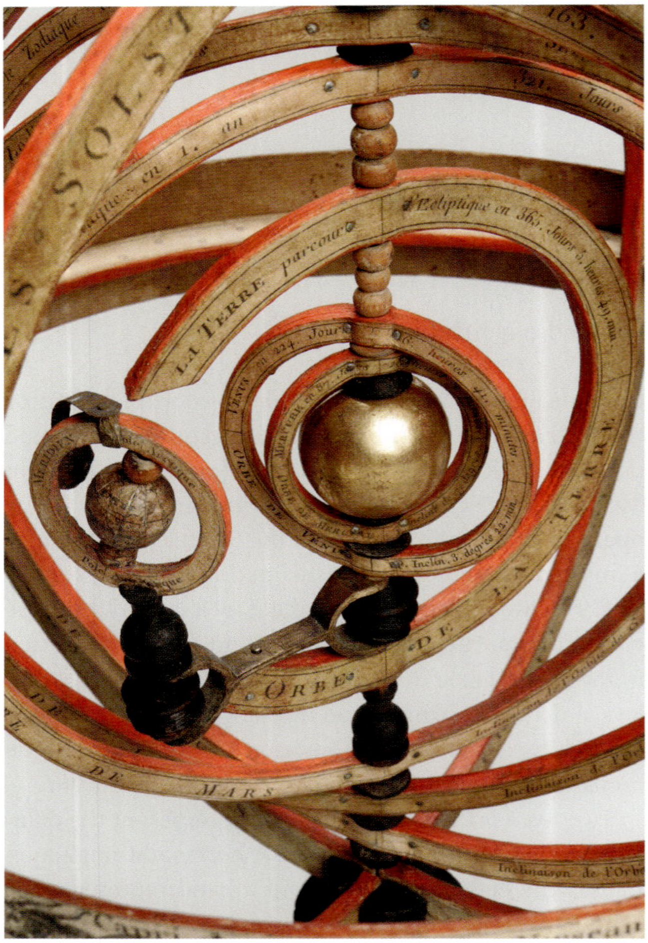

FIGURES 2A, 2B (inv. no. 20228-0000-)
© MAM/PHOTO FRANCK BOTTÉ

accurate report of an observable aurora borealis. On this occasion, he showed a remarkable familiarity with Flamsteed's star table and, with the help of the alidade, he located the streamers precisely with respect to the visible stars.

Lavoisier used Canivet's expertise on more than one occasion, as we may deduce from his reference to a copper vessel in a memoir of 1768 on the specific gravities of fluids. The text clearly indicates that he purchased the item many years before and that by 1768 he believed Canivet to be dead.

See also item 20140-0000-.

Marinoni (1751), pp. 5–35; Francœur (1835), pp. 7–8, plate I, fig. 10; LO, vol. 4, pp. 1–7; LO, vol. 3, p. 438

20140-0000- (Fig. 4)
Alidade for plane table
18th c.
MAM accession date: 1952
Alidade: 52.5 × 12 × 4.7; 590 g
Box: 55.5 × 16.5 × 8; 420 g
Brass, wood, leather

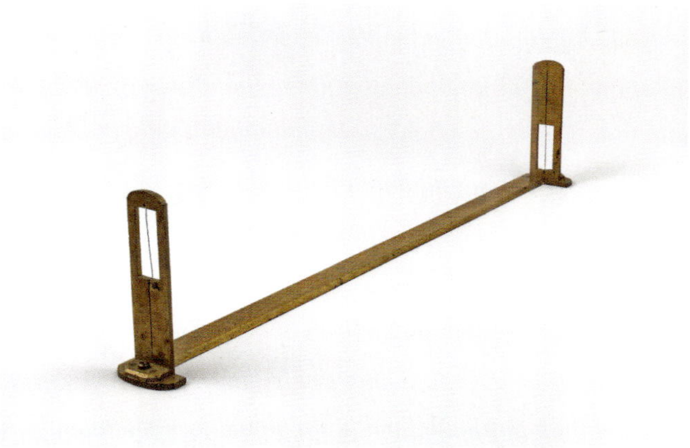

FIGURE 4 (inv. no. 20140-0000-)
© MAM/PHOTO FRANCK BOTTÉ

This alidade consists of a brass ruler with two vertical sights at its ends. Each sight is composed of a brass rectangle with a vertical window and a narrow slit. A thin sighting wire is fixed vertically in the centre of the windows. The instrument, which was used in the same manner as no. 20138-000-, is housed in a leather-coated wooden box (in poor condition).

Marinoni (1751), pp. 5–35; Francœur (1835), pp. 7–8, plate I, fig. 10

20217-0000- (Fig. 5)
Surveyor's cross
ca. 1790
MAM accession date: 1952
Louis Fois (dit Larose)
15.5 × 7.2; 640 g
Brass

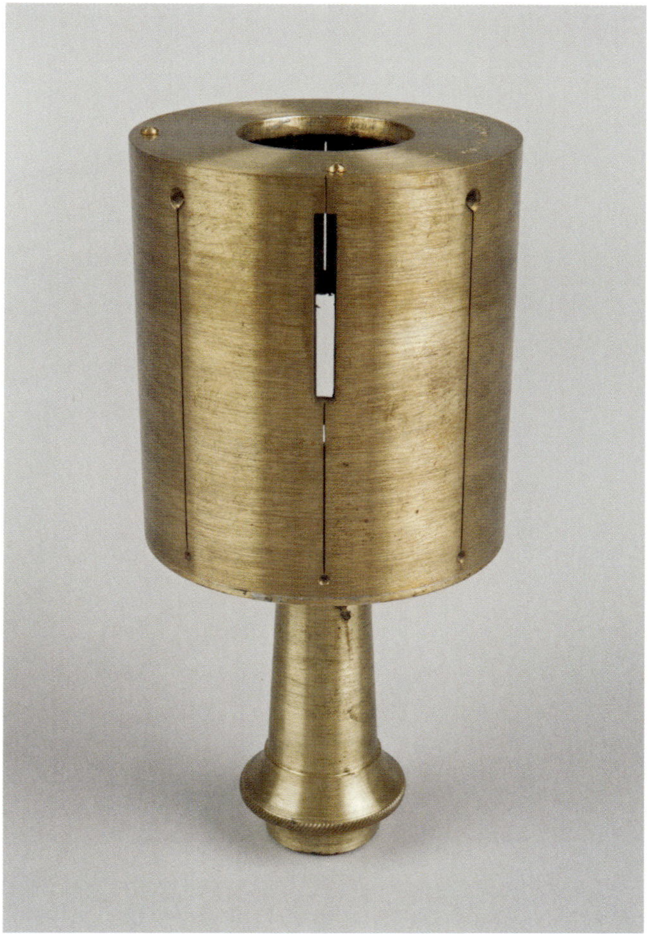

FIGURE 5 (inv. no. 20217-0000-)
© MAM/PHOTO FRANCK BOTTÉ

This surveyor's cross[6] is composed of a hollow brass cylinder fixed to a short tube to be inserted on a vertical stick or a tripod. The cylinder has eight vertical slits at 45°, of which four have a lower half larger than the upper half.

A surveyor's cross is a simple instrument used to establish straight lines and lay out lines at right angles or at multiples of 45° to one another.

6 This type of surveyor's cross (also called closed surveyor's cross) is more recent than the open type composed of four arms set at right angles to one another.

The top of the cylinder bears the punched inscription "L.s Fois dit la rose a Clermont". Other surveying instruments signed by Louis Fois in Clermont survive, but nothing is known about his career. His son (La Rose fils) was active in the 1820s after serving in Étienne Lenoir's workshop.

Marinoni (1775), pp. 10–39; Adams (1797), pp. 211–212, plate XIV, fig. 3; Francœur (1835), pp. 15–17, plate I, fig. 17–17 bis

20137-0000-, 20149-0000- (Figs. 6a, 6b, 6c)
Surveyor level
1776
MAM accession date: 1952
Jean Louis Jacques Baradelle (Baradelle fils) (1752–1794)
Level: 54 × 17 × 14; 3.79 kg
Box: 58 × 23 × 17; 4.26 kg
Brass, iron, wood, glass, agate

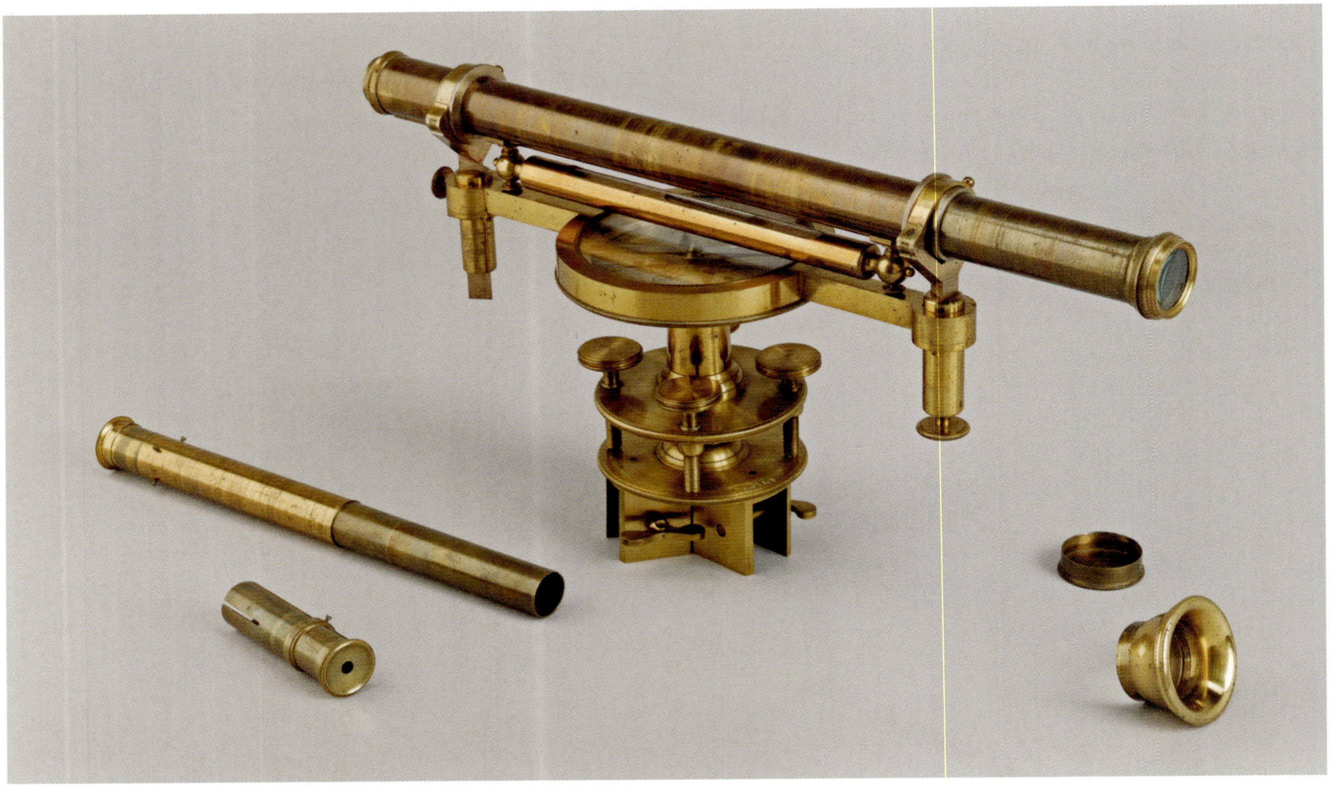

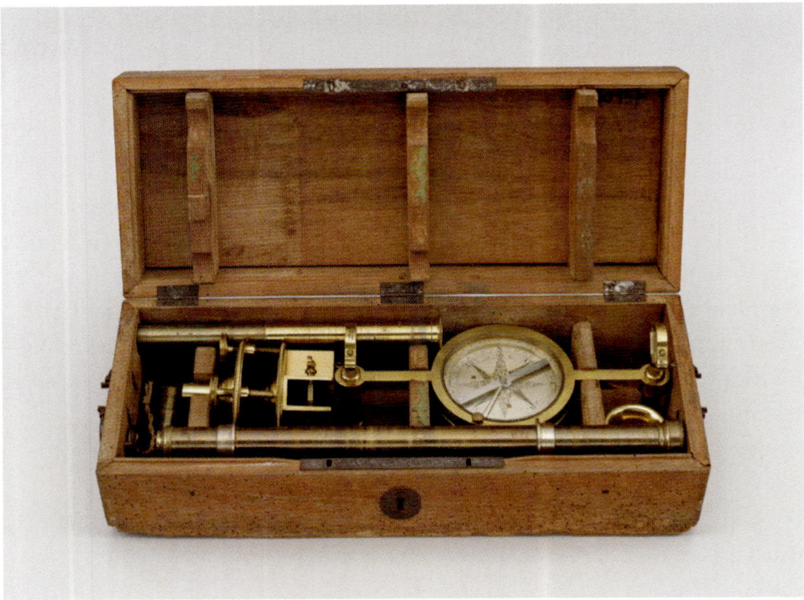

FIGURES 6A, 6B
(inv. nos. 20137-0000-, 20149-0000-)

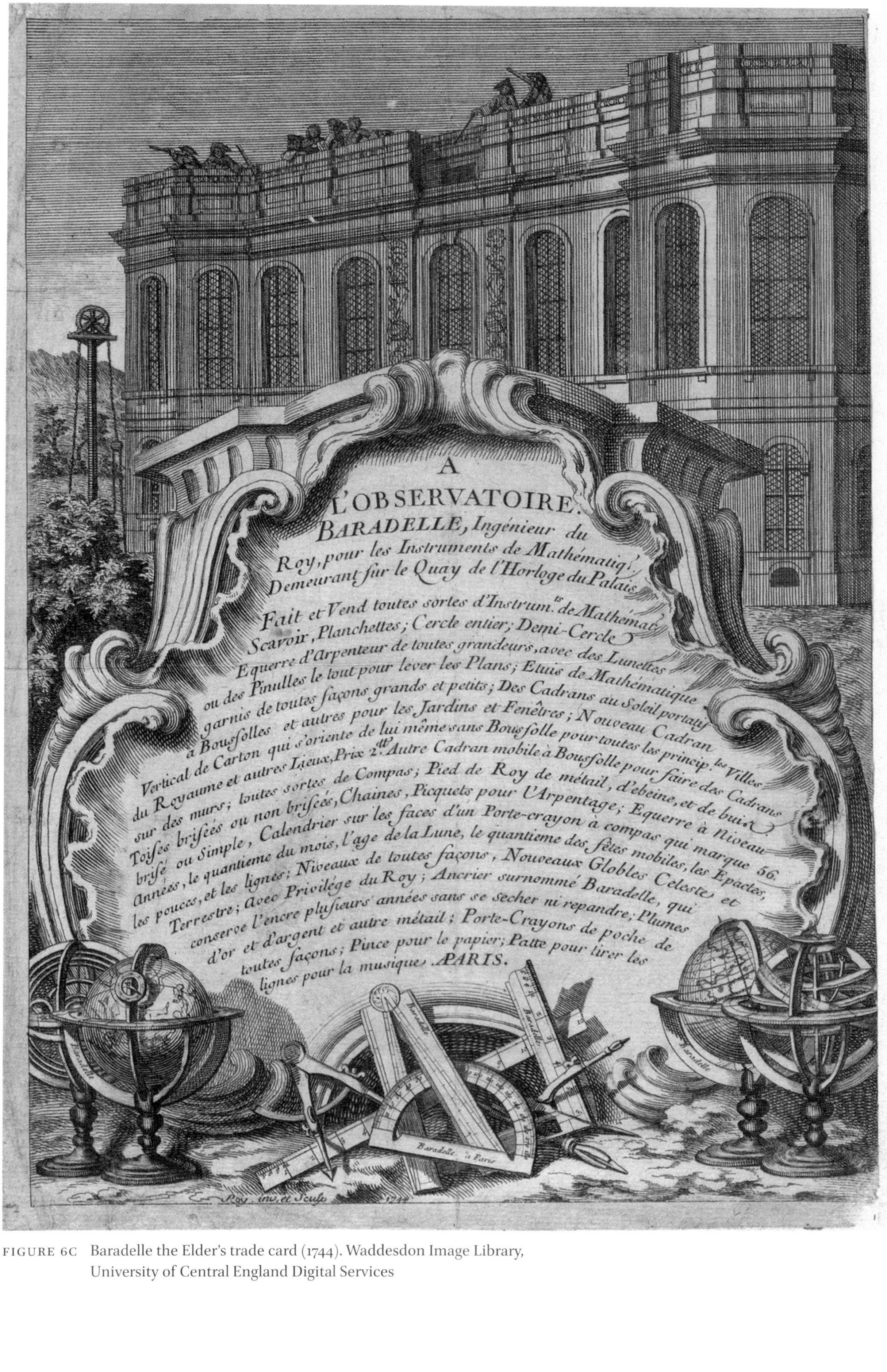

FIGURE 6C Baradelle the Elder's trade card (1744). Waddesdon Image Library, University of Central England Digital Services

Levelling is the process of finding the difference in level between two places. The instrument rotates on a vertical axis, because it is inserted by means of a conical pivot on the tube of a stand composed of two disks connected with four levelling screws.[7] Under the lower disk, three butterfly screws could be originally fastened on a wooden tripod (missing). The magnetic compass is inserted in a circular frame with two diametrical arms holding the vertical supports for the telescope. The compass dial displays a wind rose with the usual indications of the directions – "N, NO, O, SO, S, SE, E, NE" – and two concentric scales. The smaller one is divided into four quadrants (0°–90°); the external one is divided into two semicircles (0°–180°). The magnetic needle is a rectangular flat steel bar with two reference marks. A small agate cup rests on a vertical pin at its centre. A screw and a lever are used to block the needle when the instrument is not in use. The dial carries the inscription "Baradelle fils AParis 1776". The telescope rests on two Y-shaped supports with rotating blocking brackets. Their vertical arms are inserted into two tubes in the holes at the ends of the compass frame, and their exact heights can be set and fixed by means of a pair of screws. Because of the shape of the support, this type of level is often called a "Y" or "Wye" level.

The telescope has two different eyepieces with cross-wires and an achromatic objective. The latter is protected by a brass cap and a sunshade can be screwed on it. The tube is engraved with the inscription "Baradelle AParis". Under the telescope there is a bubble level. It is composed of a brass tube protecting a glass phial that contains a liquid (alcohol?) with an air bubble in it. The six lines engraved on the phial serve to centre the floating bubble. By turning the screw that holds the bubble level, one can adjust the parallelism between the level and the telescope.

The instrument is housed in a wooden box divided into different compartments (with traces of green velvet) and fitted with two iron handles and a locker screw.

This type of lever was introduced by the famous English instrument-maker Johnathan Sisson (1690–1747) in the early eighteenth century. It was later reproduced by many other makers and, with a few improvements, remained very popular throughout the 18th and 19th centuries.
It is unclear whether Lavoisier ever used the instrument. His geological surveys date from the late 1760s, and the memoirs on the height and strata of mountains that he presented at the Académie royale des sciences in 1792 relied on data he collected during the 1760s.

Gardiner (1737), pp. 108–125; Adams (1797), pp. 333–388, plate XVII, fig. 3; Bennett et al. (1982), pp. 12–13

20135-0000-, 20135-0001-, 20135-0002- (Fig. 7)
Graphometer
ca. 1750
MAM accession date: 1952
Claude Langlois (ca. 1700–ca. 1756)
30 × 23 × 12; 1.67 kg
Brass, steel, leather, wood

The graphometer was used by surveyors, geodesists and, more generally, mathematical practitioners to measure horizontal and vertical angles. It is composed of a semicircular limb with a double scale divided into degrees from 0° to 180° and from 180° to 0°. The limb is subtended by the diameter with two sights at its ends. Each sight consists of a rectangular plate with a rectangular window divided by a vertical metal blade and a narrow slit. A pivoting alidade with two other sights is attached to the middle of the limb diameter. At the ends of the alidade, two noniuses with 12 divisions each made it possible to read the angle with a precision of 5′. A magnetic compass is inserted in the instrument between curved and ornate arms. The compass has a circular scale divided into degrees from 0° to 360° and a wind rose with the inscriptions: NORD, N.EST, EST, S.EST, SUD, S.OUEST, OUEST, N.OUEST. The instrument is mounted on a ball-and-socket joint that could be inserted in a tripod. On the limb there is the signature "*Langlois AParis aux Galleries du Louvre*".

The instrument has a leather case[8] (in poor condition) with some decorations (fleur-de-lis).

Bion (1716), pp. 157–165, plate XIV; Lefèvre (1803), pp. 58–64, plate II, fig. 30, Francœur (1835), pp. 18–21, plate II, pp. 22–22 bis; Rocca, Launay (2018)

7 The level stand had been inventoried separately from the level under the inventory number 20149-0000-.

8 With the inventory no. 20135-0002-.

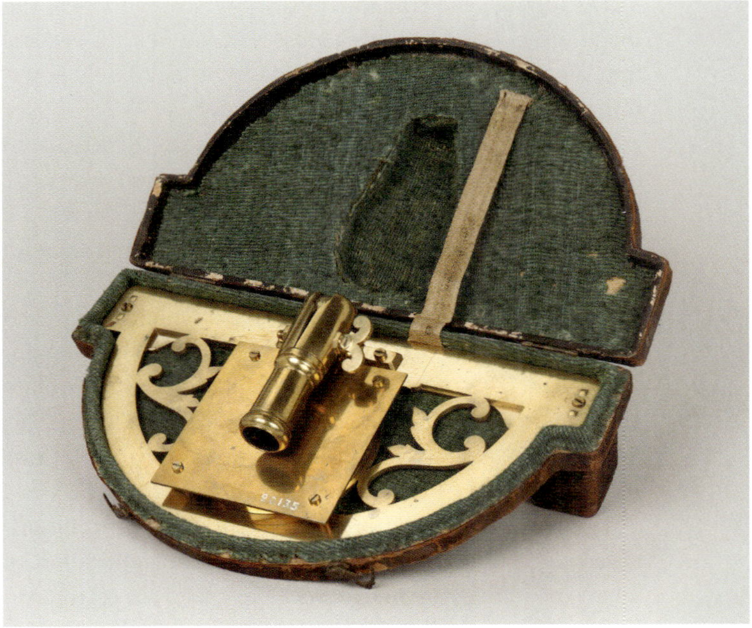

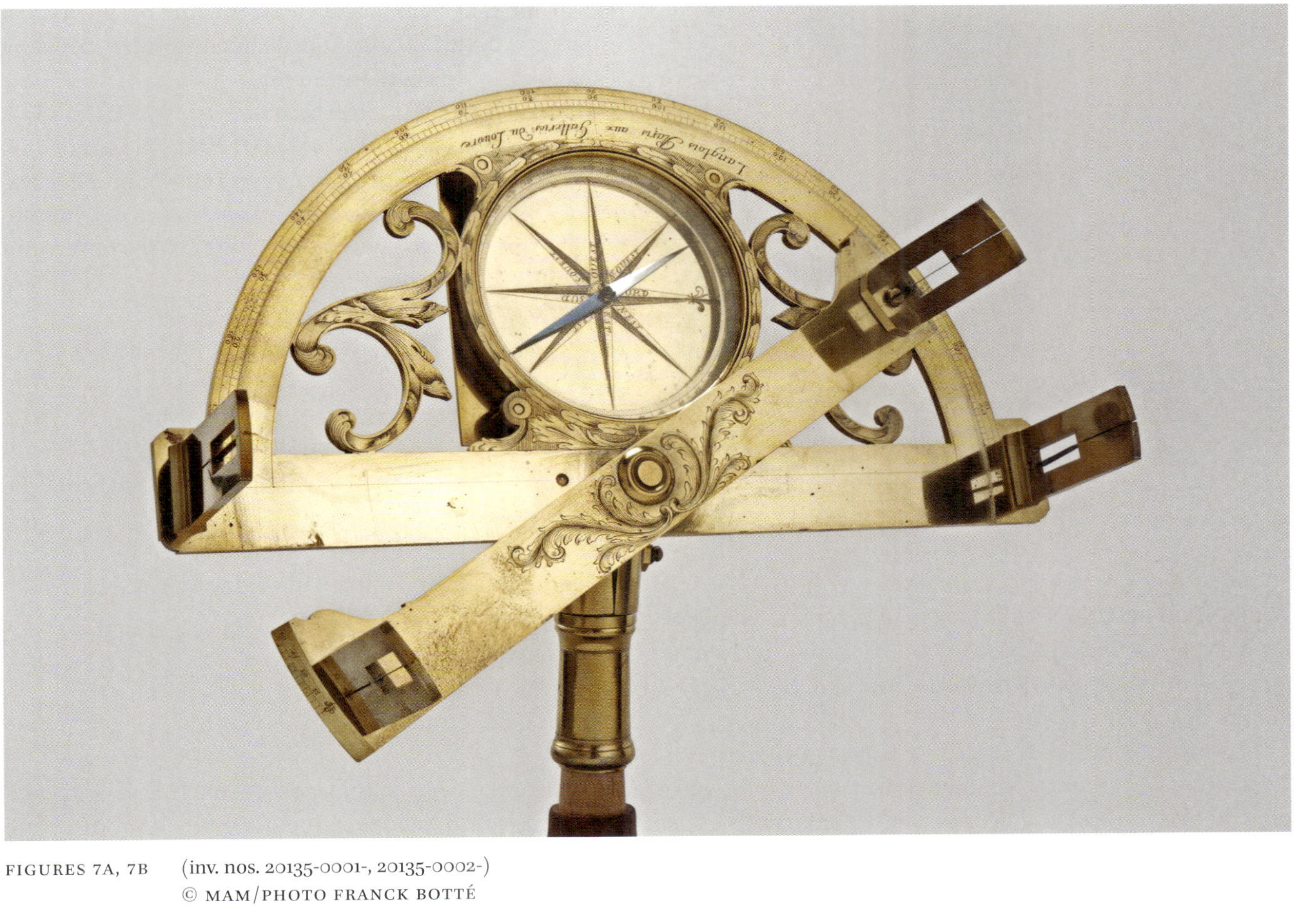

FIGURES 7A, 7B (inv. nos. 20135-0001-, 20135-0002-)
© MAM/PHOTO FRANCK BOTTÉ

20136-0000- (Fig. 8)
Declinometer
Georg Friedrich Brander (1713–1783)
1781
MAM accession date: 1952
34.5 × 21.5 × 9.5; 4.25 kg
Stone, wood, brass, steel, glass

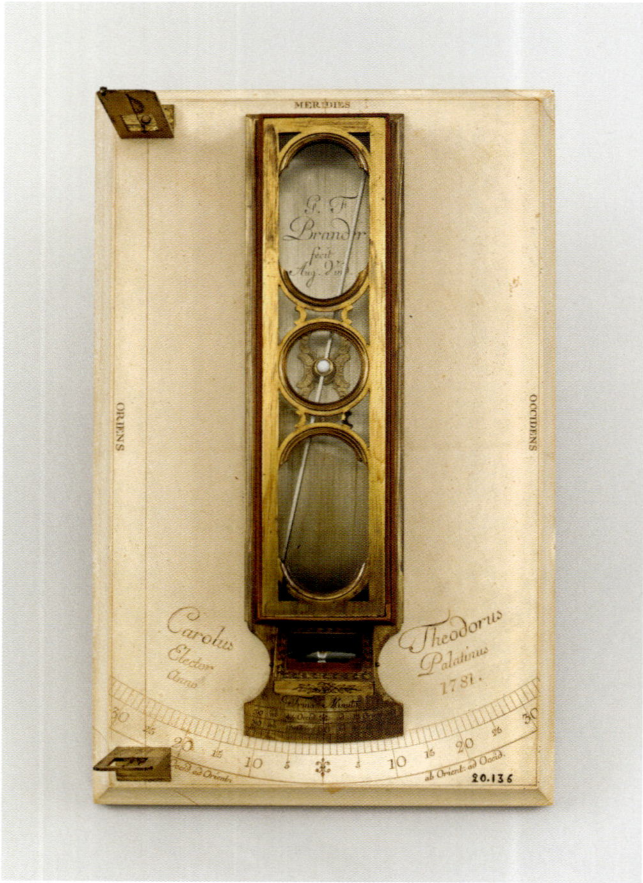

FIGURE 8 (inv. no. 20136-0000-)
© MAM/PHOTO FRANCK BOTTÉ

The declinometer is a special type of magnetic compass used to determine declination, which is the horizontal angle between the magnetic North Pole and the true geographic North.

The instrument has a sandstone base. On it there is a rotating wood and brass rectangular box with a glass window (upper face) enclosed in a decorated brass frame. The box contains a long magnetic needle with a central agate cup resting on a pivot. The base plate of the box carries the maker's signatur: "G. F Brander // fecit // Aug. Vind".

The box has a horizontal sector with the inscription "Trina Minuta". Under it there are a nonius with 3′ divisions and a double indication of the angle every 15′: "30 ab 45 Occid. ad 15 Orient. 30" and "30 ab 15 Orient. 60 ad 45 Occid. 30". A mirror inclined at 45° serves to view the magnetic needle through a glass window with a vertical line, allowing the needle's position to be accurately determined while observing the scale.[9]

The stone base is engraved with a scale divided into degrees 30°-0°-30°; the angles are indicated every 5°. The left side of the scale reads "ab Occid ad Orient.", the right side "ab Orient. ad Occid.". Using the nonius, one can determine the angles with a precision of 3′.

Two brass sights are fixed to the base, which is engraved with a line connecting them. Each sight has a square window divided by a vertical wire and a hole.[10] The plate also bears the inscriptions "ORIENS", "OCCIDENS", "MERIDIES" and "Carolus Theodorus // Elector Palatinus // Anno 1781".

To measure declination, the instrument has to be oriented along the direction of the astronomical meridian by means of the sights (and if necessary with an astronomical instrument). The box is then moved until the position of the needle coincides with the line on its vertical windows.[11] In this manner, the declination angle could be read with the scale and the nonius.

This instrument, like the barometer (19952-0000-), was made for the *Societas Meteorologica Palatina* founded by the Prince Elector Karl Theodor (1724–1799). Brander, the best German maker of his time, made several declinometers like this one, which were distributed to the observatories affiliated with the society.

Ephemerides (1783), pp. 78–80; Brander (1779); Brachner (1983), pp. 265–273

Harvard University – Collection of Historical Scientific Instruments (Figs. 9a, 9b)
DW0619
Pedometer with compass in gold watch case
ca. 1785
William Fraser (late 18th c.)
2.3 × 8 × 5.8
Glass, gold, enamel

The instrument has a gold pocket-watch case with two white enamelled faces and is equipped with a magnetic compass. The white enamelled front face has two dials.

9 This mirror, not present in Brander's original article, was added to the declinometers after 1779.
10 Originally a horizontal wire (missing) connected the tops of the sights.
11 The needle has to be parallel to the long sides of the box.

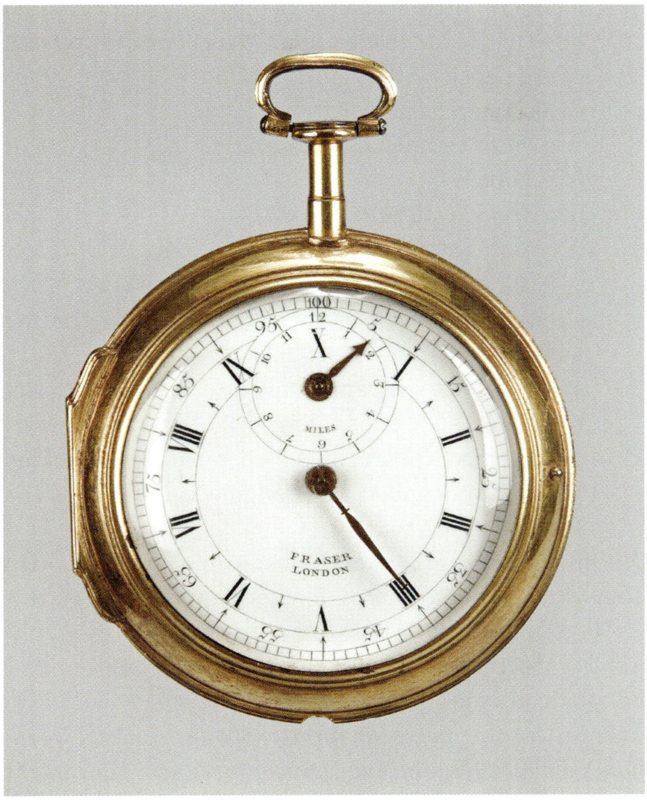

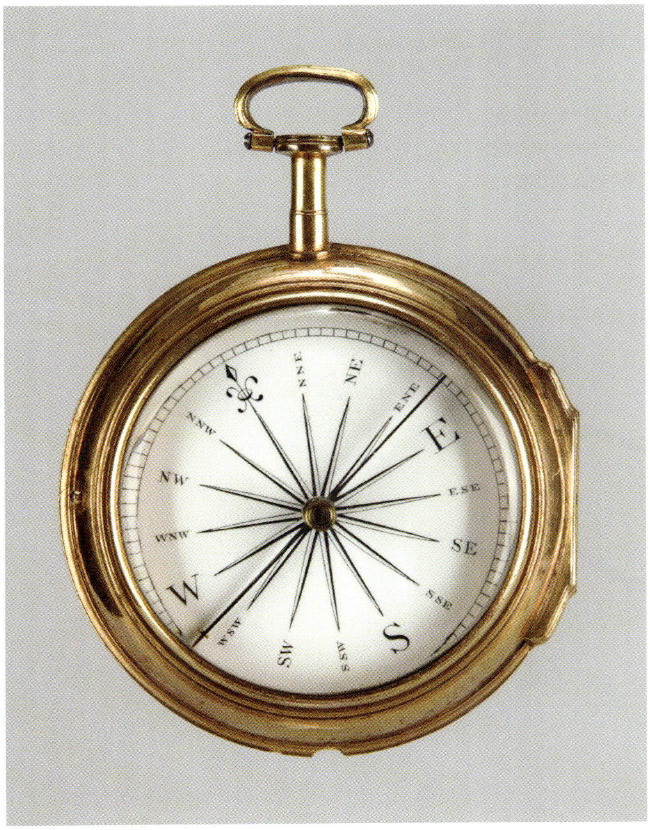

FIGURES 9A, 9B (inv. n. DW0619). Harvard University – Collection of Historical Scientific Instruments

The larger one measures up to a hundred steps and is marked with two scales. The inner scale has Roman numerals I through X (indicating the tens of steps). The outer scale has values ranging from 0 to 100. A gold hand with broken tip points to the scales. The smaller dial located at top centre has values ranging from 0 to 12 miles and bears the inscription "MILES". A gold hand with an arrow-shaped tip points to the scale. On the front face, there is also the inscription "FRASER/LONDON". The back compass face of the pedometer is printed with a sixteen-point compass rose. The tip of the rose pointing north finishes in a fleur-de-lis. The remaining cardinal and intercardinal directions are marked with letters: "E, ESE, SE, SSE, S, SSW, WSW, W, WWN, WN, NNW, NNE, NE, ENE". Around the circumference of the compass face is an unlabelled scale with 72 divisions (one for every five degrees). On top of the pedometer is a stem with an oval-shaped loop at the far end. A fine chain comes out of a hole in the bottom of the watch case and it acts on a ratchet that advances a toothed wheel that transmits motion to the hands on the dial face. The instrument was preserved in a small yellow (chamois?) pouch, perhaps not original.

The instrument was used to measure distance, in either steps or miles, and to determine one's direction of travel. The pedometer was put in the breeches pocket, after having been reset by pushing a small pin to the right of the front face. The small chain was attached by the waistband. Every pace caused the chain to act on the ratchet and to move the hand. This pedometer it was displayed at the Lavoisier exhibition in the Palais de la Découverte in 1943. It was acquired by in 1957 by David P. Wheatland[12] from E. Weil (London). It was not in the list of Lavoisier's item auctioned in Paris in March 1956 (*Souvenirs de Lavoisier* 1956). Presumably, it was sold separately by Madame de Chazelles to the Paris dealer Lucien Scheler, who was in direct contact with Lavoisier's heirs, and then to Weil.

Fraser (1780–1805); Lavoisier (1943); Wernimont (2019) pp. 123–124

12 The American scholar, author and collector David P. Wheatland (1898–1993) assembled the nucleus of objects that eventually formed the basis of the Collection of Historical Scientific Instruments.

Drawing and Geometrical Models

20139-0000- (Fig. 10)
Pantograph
Late 18th c.
MAM accession date: 1952
Box (closed): 68 × 13.5 × 9; 4.01 kg
Ebony, cardboard, brass, lead, wood, paper, silk thread, ivory

A pantograph is a drawing instrument used for copying, enlarging or reducing maps and drawings. It is composed of a rhomboid mechanical assembly with four linked arms. When the point fixed on one arm follows a line in a drawing, a pen inserted on another arm reproduces the same line. The reduction (or enlargement) ratio depends on the relative positions of the elements composing the apparatus (pen, point, arm hinges). Pantographs were used since the early seventeenth century, and this type of improved pantograph was described in 1743 by the French instrument-maker (ingénieur du Roi) Claude Langlois (1703–1760).

This instrument is composed of an articulated parallelogram with four ebony arms. Two arms are marked with scales indicating the reduction/enlargement ratio. The arms are connected with brass joints. There are several accessories:

a) an ebony ruler in its cardboard case. On it is the inscription “Regle”.
b) a rectangular lead weight with a hole. Used to maintain the pantograph fulcrum in a fixed position.
c) a pencil holder. On it is a small cup that could be filled with a few lead weights to regulate the pressure of the pencil on the drawing paper.
d) a tracer and its holder
e) two brass legs and a screw
f) four pencils
g) five wooden sticks
h) two brass rings
i) two sliders for the arms with ivory wheels. They allowed the arms to move smoothly on the drawing board.
j) two sliders for the pantograph arms with the inscriptions “E B D E”
k) a silk thread. It was connected to the pencil holder. When it was pulled, the pencil was lifted from the drawing paper.

The instrument and its accessories are housed in a wooden case with cardboard compartments.

The lid of one of the box compartments bears the handwritten inscription “Pinceaux Et Couleurs”; on another lid, one can read “..antes”. Some of the small pieces are wrapped in fragments of the newspaper *Nouvelles politiques* dated “Brumaire an 2” (November 1793). They might indicate the approximate date when the instrument was purchased by Lavoisier, but they might also refer to the confiscation of Lavoisier’s belongings.

Langlois (1744); *Encyclopédie* (1751–1772), vol. 20 (1763), section Dessin, plate III; Hambly (1988), pp. 130–133; Goebel et al. (2003), pp. 27–32

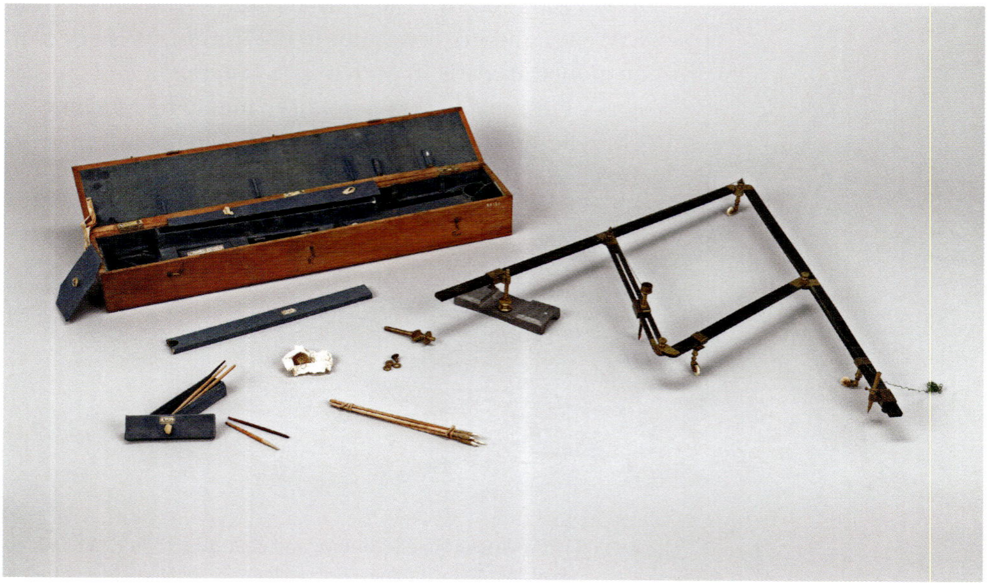

FIGURE 10
(inv. no. 20139-0000-)
© MAM/PHOTO FRANCK BOTTÉ

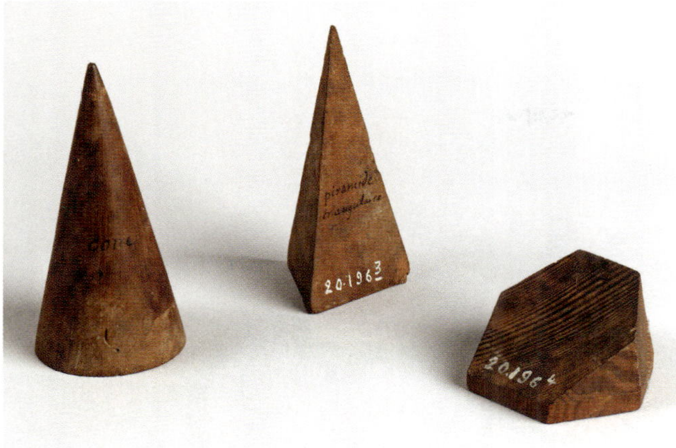

FIGURE 11 (inv. no. 20196-0002-, 20196-0003-, 20196-0004-)

20196-0002-
Model of cone
Second half of 18th c.
MAM accession date: 1952
10.6 × 5.2; 60 g
Wood

Wooden triangular cone used as a geometrical model. It bears a label with the inscription "cone" in Lavoisier's hand.

20196-0004-
Model of truncated prism
Second half of 18th c.
MAM accession date: 1952
6 × 5.7 × 4; 45 g
Wood

Wooden hexagonal truncated prism used as a geometrical model. The number "4" is inscribed on it in blue chalk.

20196-0003-
Model of pyramid
ca. 1760
MAM accession date: 1952
10.5 × 4.8 × 4.5; 30 g
Material: wood

Wooden triangular pyramid used as a geometrical model. It bears the inscription "piramide triangulaire", possibly in Lavoiser's hand.

Experimental Physics

This section of the collection includes various instruments related to experimental physics. It follows with few variants the classification presented by Jean Nollet in his *Leçons de physique expérimentale*. This choice is justified by the fact that Lavoisier attended Nollet's lectures in 1761 and eventually adopted the same way of presenting physical apparatuses. As the instruments all date from roughly the same period, we have preferred to follow a typological criterion rather than a chronological one in each class. In this section there are a few important instruments by Rumford that we have decided to keep together with Lavoisier's for reasons explained in the general introduction.

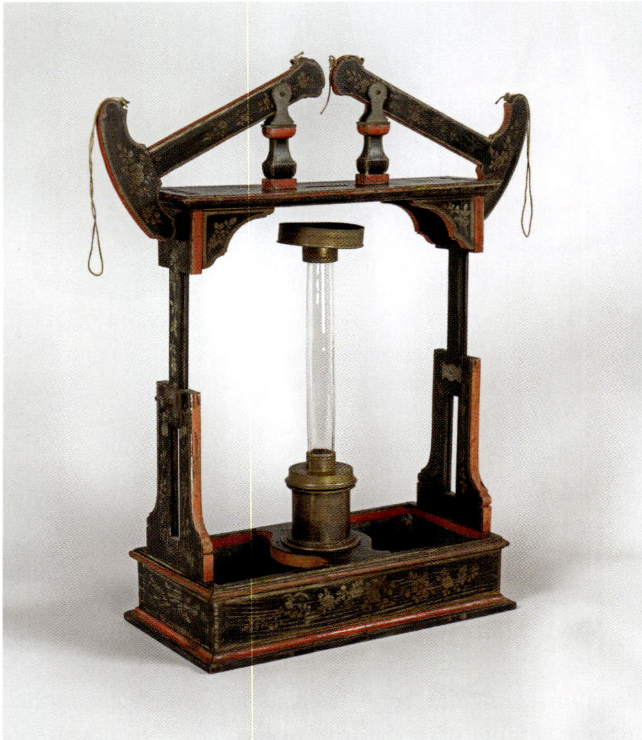

FIGURE 1A (inv. no. 20097-0000-)
©MAM/PHOTO FRANCK BOTTÉ

Hydrostatics and Hydraulics

20097-0000- , 20002-0000- (Fig. 1a, 1b)
Pascal's apparatus for the hydrostatic paradox
1760s
MAM accession date: 1952
80 × 68.5 × 31; 9.99 kg
(20002-0000-): 37 × 12; 2.01 kg
Wood, brass, lead, iron, thread, glass

The instrument known as Pascal's apparatus was used to demonstrate the "hydrostatic paradox": the pressure of an incompressible fluid at the bottom of a vessel varies with its depth and does not depend on the shape of the container.

A wooden box with two uprights supports a brass collar holding a large conical glass vessel. The box is lined with lead. Two wooden bars slide in the uprights and support a crosspiece. At the top of the crosspiece are two small columns, each supporting a brass bracket. Two levers are hinged to the brackets. Originally, two weights (missing) were attached to the outer side of the lever. Two strands of twine (missing) fixed on the opposite side served to keep a circular valve (missing) closed at the bottom of the conical vessel.[1] There is also a cylindrical glass tube (20002-0000-) (see Fig. 1b) with a brass collar at its base and an open brass cup, whose diameter is larger than that of the tube. The tube – and probably one or two more, now missing – could be fastened in place of the conical vessel.

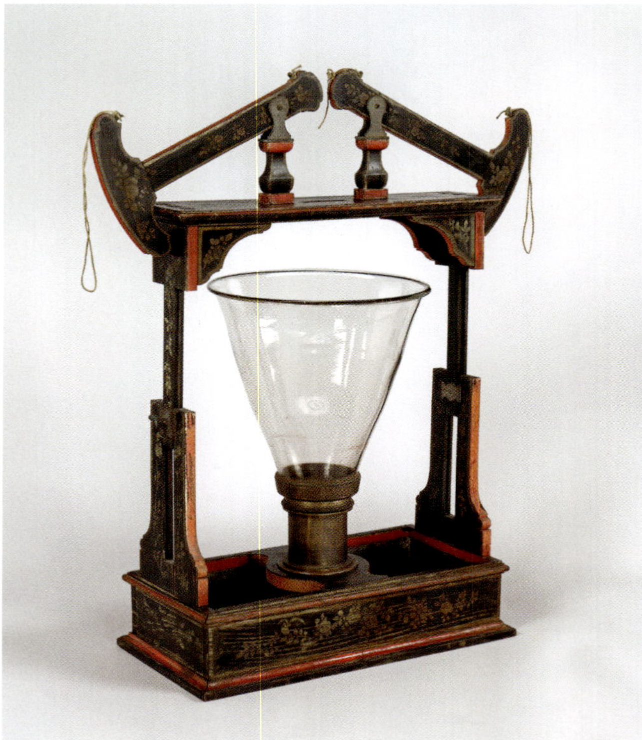

FIGURE 1B (inv. no. 20002-0000-)
©MAM/PHOTO FRANCK BOTTÉ

1 The tube was erroneously catalogued with a different inventory number (20002-0000-). A third tube, designed to show the effect of pressure acting laterally, is missing.

 | DOI:10.1163/9789004511217_017

One of the vessels was mounted on the apparatus and filled with water just before its weight caused the bottom valve to open. The same experiment was repeated with the other vessels. It was then possible to demonstrate that the valves opened when the water reached the same level in all the vessels. The pressure on the bottom thus did not depend on the shape of the vessel but only on the height of the liquid.

The wood surfaces painted in black and red and decorated with gold floral motifs was typical of the instruments described by Jean Nollet and Sigaud de La Fond.

Nollet (1743–1764), vol. 2, pp. 267–275 and plate III; Sigaud de La Fond (1784) vol. 2, pp. 288–295 and plate XIX, fig.1; Pyenson and Gauvin (2002), pp. 179 and 185

20100-0000- (Fig. 2)
Archimedes' screw
1760s
MAM accession date: 1952
47.5 × 33 × 32.5; 2.1 kg
Wood, iron, tin, lead

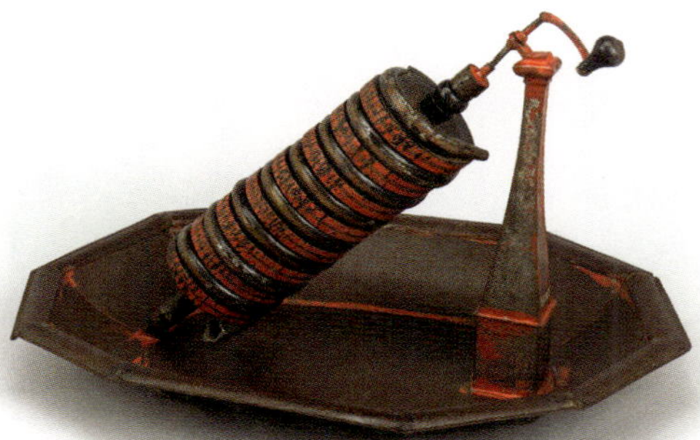

FIGURE 2 (inv. no. 20100-0000-)
©MAM/PHOTO FRANCK BOTTÉ

The Archimedes' screw is a type of water-raising machine generally used for irrigation. It consists of an inclined screw (a helical surface surrounding a central cylindrical shaft) inside a hollow pipe, whose bottom end is dipped in water. As the screw turns, its end scoops up some water, which is then pushed up the tube by the rotation of the helical surface until it pours from the top.

This model was made of a wooden cylinder with a lead pipe wound in a helical groove. The top of the cylinder shaft has a handle and is hinged at the top of a square pillar fixed on a polygonal metal tray. The bottom of the shaft pivots in the tray. The model is painted in black and red.

Sigaud de La Fond (1784), vol. 1, pp. 2018–219 and plate XVI

20099-0000- (Fig. 3)
Archimedes' screw
1760s
MAM accession date: 1952
41 × 15.5 × 23; 1.15 kg
Wood, iron

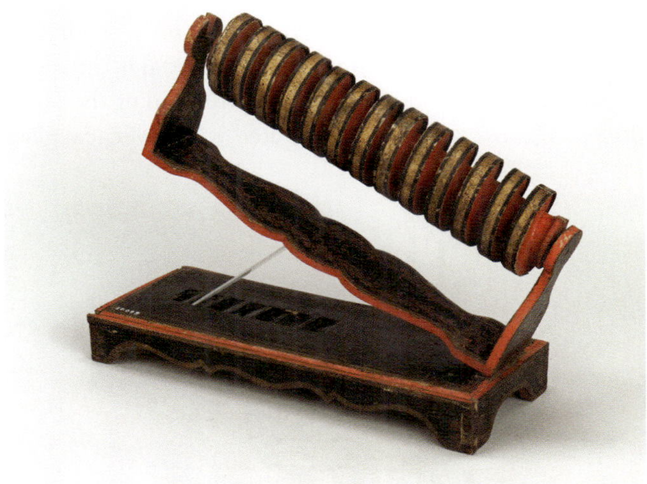

FIGURE 3 (inv. no. 20099-0000-)
©MAM/PHOTO FRANCK BOTTÉ

This model simply shows the working principle of the Archimedes' screw and is derived from inv. no. 20100-0000-. A wooden base holds a tilting wooden frame supporting the screw shaft, which consists of a cylinder with a helical groove. A small metal marble inserted at the bottom of the groove is lifted to the top by the rotation of the cylinder. The groove makes it possible to see the movement of the marble.

The model is painted in black, red and gold with floral ornaments typical of Nollet's and Sigaud's demonstration instruments. The hinged arm for setting the inclination of the screw is missing, and both ends of the screw are broken. A similar model, attributed to Nollet, has been acquired by the Stewart Museum.

Lavoisier likely bought a set of Nollet's models after attending Nollet's course in experimental physics in 1761. Throughout the 1760s, Lavoisier took Nollet's experimental physics as an extremely useful model for his approach to science.

Sigaud de La Fond (1784), vol. 1, pp. 2018–219 and plate XVI; Nollet (1743–1764), vol. 3, pp. 134–135, plate VIII; Pyenson, Gauvin (2002), p. 149 and plate IX; Beretta (1994)

20101-0000- (Fig. 4)
Siphon (incomplete)
1760s
MAM accession date: 1952
82 × 51 × 26; 2.57 kg
Brass, wood, tin

Apparatus demonstrating the action of a hydraulic siphon. A wooden base supports a square flat tin vessel and a decorated wooden upright. At the top of the latter there is a brass cup originally sealed with a vertical glass vessel (missing). A long brass pipe connects the bottom of the cup with the flat vessel; a shorter bent pipe runs from a nozzle at the centre of the cup to just above a small shelf inserted in the upright of the apparatus. A glass beaker (missing) originally rested on the shelf and the shorter pipe was inserted in it.

The beaker had to be filled with water and some water had to be introduced in the longer pipe. This made it possible to start the siphon: the water was sucked from the beaker, sprouted from the nozzle and then flowed down the long pipe into the vessel at the base of the instrument. The decorations of the wooden parts are typical of Nollet's and Sigaud de la Fond's instruments.

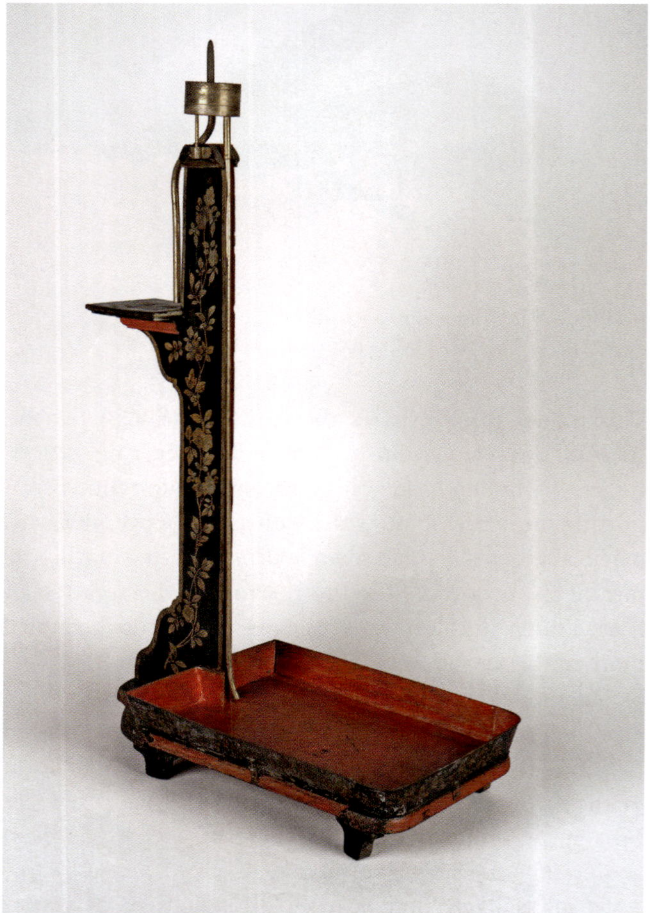

FIGURE 4 (inv. no. 20101-0000-)
©MAM/PHOTO FRANCK BOTTÉ

Nollet (1743–1764), vol. 2, pp. 332–334 and plate VII, fig. 34

20148-0000- (Fig. 5)
Basin
Second half of 18th c.
MAM accession date: 1952
21.5 × 10.7; 375 g
Tin, brass

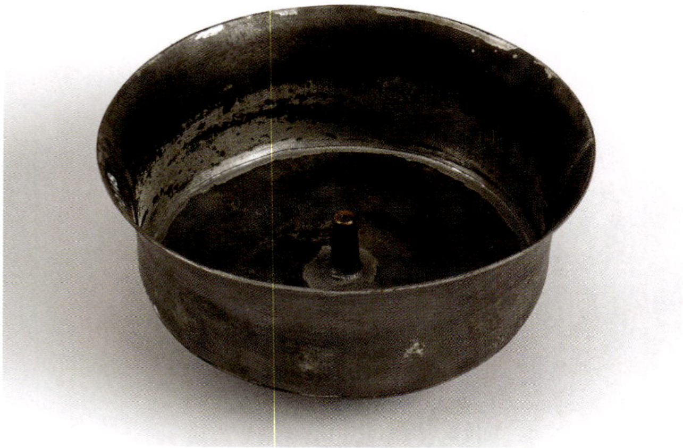

FIGURE 5 (inv. no. 20148-0000-)
©MAM/PHOTO FRANCK BOTTÉ

Tin basin with a central brass nozzle and a short tube. It is a fragment of an unidentified instrument (Hero's fountain)?

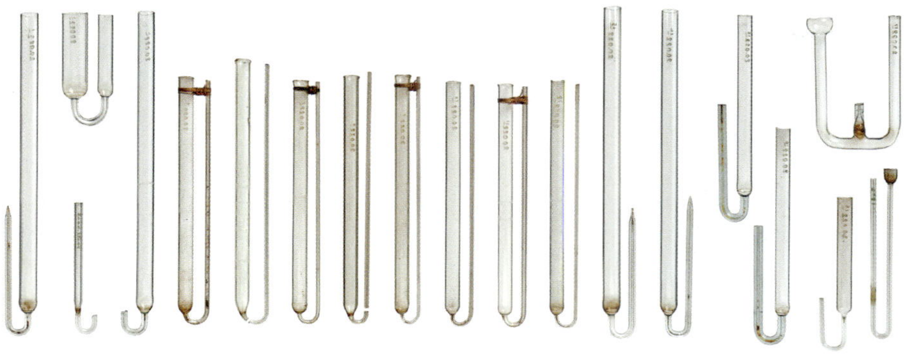

FIGURE 6 (inv. nos. 20022-0001-, 20022-0002-, 20022-0003-, 20022-0004-, 20022-0005-, 20022-0006-, 20022-0007-, 20022-0008-, 20022-0009-, 20022-0010-, 20022-0011-, 20022-0021-, 20022-0013-, 20022-0014-, 20022-0015-, 20022-0016-, 20022-0017-, 20022-0018-, 20022-0019-)
©MAM/PHOTO FRANCK BOTTÉ

20022-0001-, 20022-0002-, 20022-0003-, 20022-0004-, 20022-0005-, 20022-0006-, 20022-0007-, 20022-0008-, 20022-0009-, 20022-0010-, 20022-0011-, 20022-0021-, 20022-0013-, 20022-0014-, 20022-0015-, 20022-0016-, 20022-0017-, 20022-0018-, 20022-0019- (Fig. 6)
Set of nineteen U-shaped tubes
Third quarter of 18th c.
MAM accession date: 1952
Glass

Series of U-shaped tubes of different dimensions
20022-0001-: 29.5 × 3 × 2; 40 g
The pointed capillary arm is shorter than the other
20022-0002-: 10.5 × 4.5 × 2.3; 25 g
The two arms of different diameter are joined by a capillary
20022-0003-: 12 × 2.2 × 0.5; 10 g
The shorter capillary arm is broken. Paper label with "330"
20022-0004-: 29.5 × 3 × 1.5; 40 g
Similar to 0003-. Engraved inscription "4" at the top of the tube
20022-0005-: 23 × 3 × 1.3; 25 g
With a capillary arm consolidated by a piece of cork and a thread. Engraved inscription "1" at the top of the tube
20022-0006-: 23 × 1.3; 30 g
With a shorter capillary arm containing another capillary tube. Engraved inscription "4" at the top of the tube
20022-0007-: 22.5 × 3 × 1.3; 20 g
Almost identical to 0005. Engraved inscription "n° 2/2" at the top of the tube
20022-0008-: 22.5 × 2.5 × 1.5; 40 g
With a shorter capillary arm (broken in two pieces). Engraved inscription "7" at the top of the tube
20022-0009-: 23 × 2.3 × 1.3; 35 g
Almost identical to 005. Paper label with "5 3"
20022-0010-: 22.6 × 2.6 × 1.3: 20 g
Almost identical to 005. Paper label with: "329–6". Engraved inscription "3/3" at the top of the tube
20022-0011-: 22.5 × 2.7 × 1.5; 25g
Almost identical to 005. Paper label with: "329". Engraved inscription "8" at the top of the tube
20022-0012-: 23 × 2.5 × 1.5; 20 g
Almost identical to 005 but without cork and thread. Paper label with "329–8". On the glass: "6"
20022-0013-: 30.5 × 3 × 1.7; 40g
With a shorter, pointed capillary arm. Engraved inscription "3" at the top of the tube
20022-0014-: 30 × 3 × 1.5; 45 g
Similar to 0013. Paper label with: "331". Engraved inscription "2" at the top of the tube
20022–0015: 19 × 3 × 1.5; 30g
Similar to 0014-: Paper label with "345". Engraved inscription "10" at the top of the tube
20022-0016-: 20 × 3.2 × 1.3; 30 g
Similar to 0015-
20022-0017-: 12.2 × 8.5 × 2.5; 25 g
One of the vertical arms ends in a small funnel. There is a broken capillary tube on the horizontal portion of the tubes connecting the vertical arms.
20022-0018-: 14.5 × 3 × 1.5; 10 g
With a shorter capillary arm. Paper label with "344"
20022-0019-: 16 × 2.5 × 1.2; 15 g

Two identical vertical capillary arms. One of them ends in a small funnel.

The use of these tube is not clear. Some were probably part of more complex instruments. The tubes 20022-0005-, -0007-, -0009-, -0011- and -0012- are identical to the one normally used for showing the behaviours of liquids in capillary tubes.[2]

Nollet (1743–1764), vol. 2, pp. 405–407 and plate II, fig. 14; Sigaud de La Fond (1775), vol. 1, pp. 287–291 and plate XXII

Acoustics

20178-0000- (Fig. 7)
Speaking trumpet
1780–1790
MAM accession date: 1952
70 × 12.5 × 8.5; 315 g
Tin

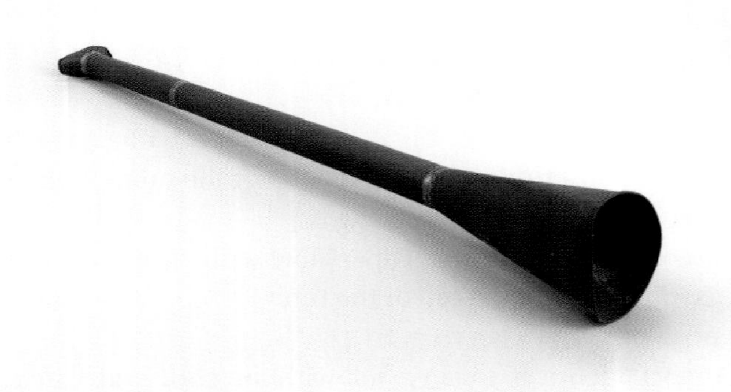

FIGURE 7 (inv. no. 20178-0000-)
©MAM/PHOTO FRANCK BOTTÉ

This speaking trumpet (*porte-voix*) is a device for amplifying a person's voice or other sounds and for directing it in a given direction. This megaphone is composed by a mouth piece connected to a slightly conical tube terminating with a funnel.

Lavoisier made several experiments in large spaces, such as those devoted to the dilatation of metals at the garden of the Arsenal, and this device might have effectively been used to give instructions on how to dispose the instruments used.

2 For example, when a U-tube with a capillary arm is filled of water, the level of the liquid is higher in the capillary arm than in the other. The narrower the bore of the capillary tube, the higher the water rises. With mercury the effect is reversed. Capillarity is the result of surface, or interfacial, forces.

s'Gravesande (1742), p. 650 and plate LXXIV, fig.3; Sigaud de La Fond (1775), vol. 2, pp. 149–151 and plate XIII; Chladni (1809), pp. 285–290; De Clercq (1997), pp. 84–85

Pneumatics

19904-0000- (Figs. 8a, 8b)
Air pump
1779–1782
MAM accession date: 1952
Nicolas Fortin (1750–1831)
140 × 76 × 50; 26 kg
Wood, brass, iron, glass, leather

Two-barrelled pneumatic pump mounted on a table and protected by a wooden case. One of the table legs has a brass levelling screw. The two vertical barrels and the horizontal board connecting them at the top are cast in a single piece. The pistons are moved up and down in their barrel by a rack-and-pinion system. The system's axle has a triple crank handle. The board on top of the barrels is covered by a sliding plate with a hole in its centre and two side valves, each enclosed in a cylinder. A fixed plate is mounted on the sliding plate. The butterfly stopcock connected to the pump plate is screwed into the fixed plate. The pump plate consists of a brass disk covered with a circular glass plate. The sliding plate has a bracket with a horizontal rack that engages in a pinion hinged on the top board. The rack at the top of an inclined bar also engages in this pinion, while its bottom end is activated by a second concentric wheel (mounted on the axle of the crank handle) with a ratchet.

The pump has a mercury manometer composed of a barometric tube (now broken) inserted in a wooden frame. A brass scale inserted in the frame is divided into lines from 21 to 29 Paris inches and a sliding nonius allows a reading precision of 1/10 of a line. Another scale ranging from 5 to 20 inches is engraved in one of the wooden bars of the frame. The top of the tube is closed by an iron stopcock connected to the base of the pump plate. The letters "F" and "O" on the cock indicate closed (*fermé*) and open (*ouvert*). The lower end of the tube terminates in a cylindrical beaker held in a frame. Its vertical position can be modified by a screw in order to calibrate the manometer. A pointed bar serves to determine the mercury level.

A brass element at the bottom of the mechanism carries the inscription "FORTIN A PARIS".

This peculiar air pump works as follows. When the piston in one of the barrels (say A) is lowered by turning the crank, the sliding plate at the top of the pump is positioned

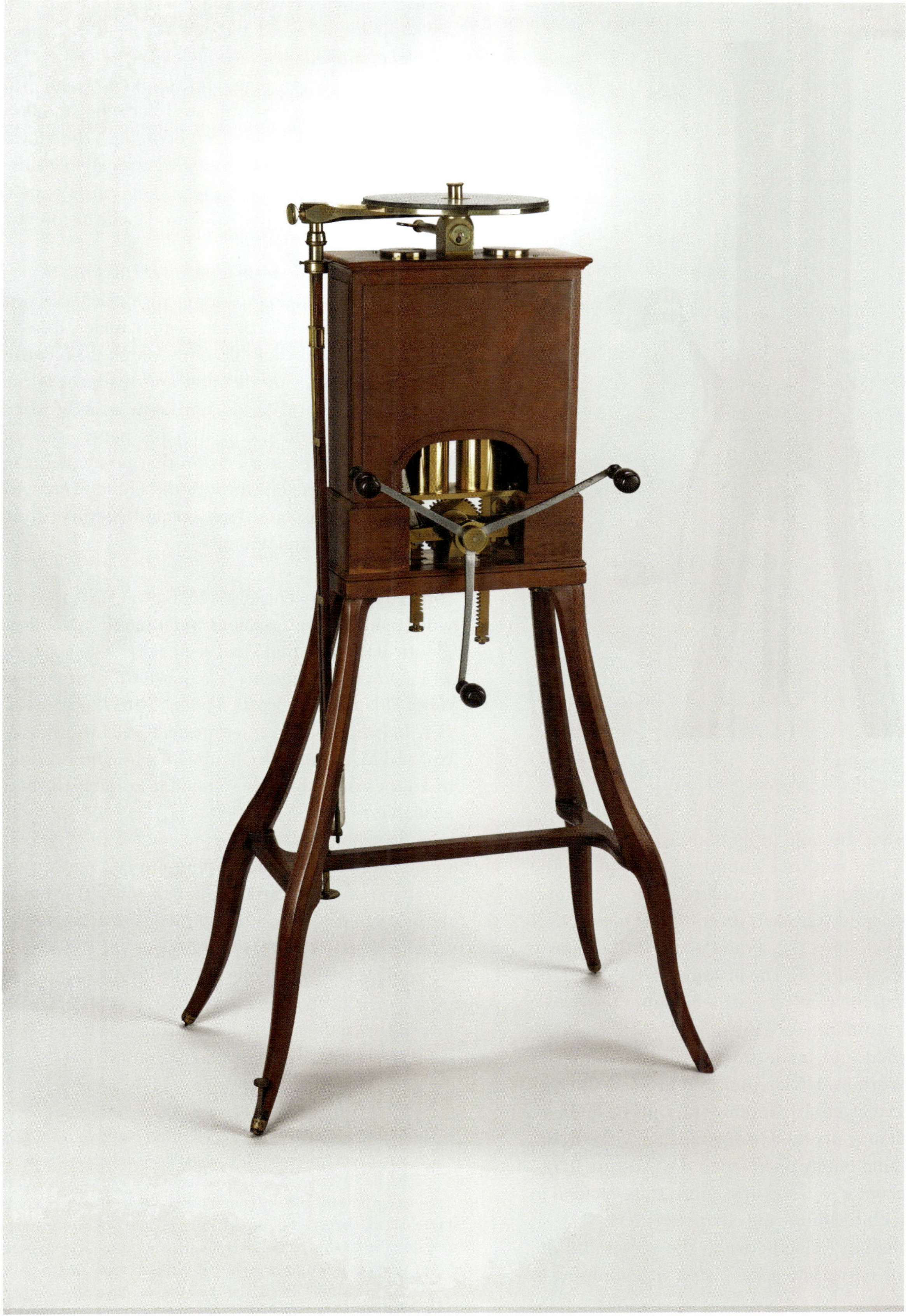

FIGURE 8A (inv. no. 19904-0000-)
©MAM/PHOTO FRANCK BOTTÉ

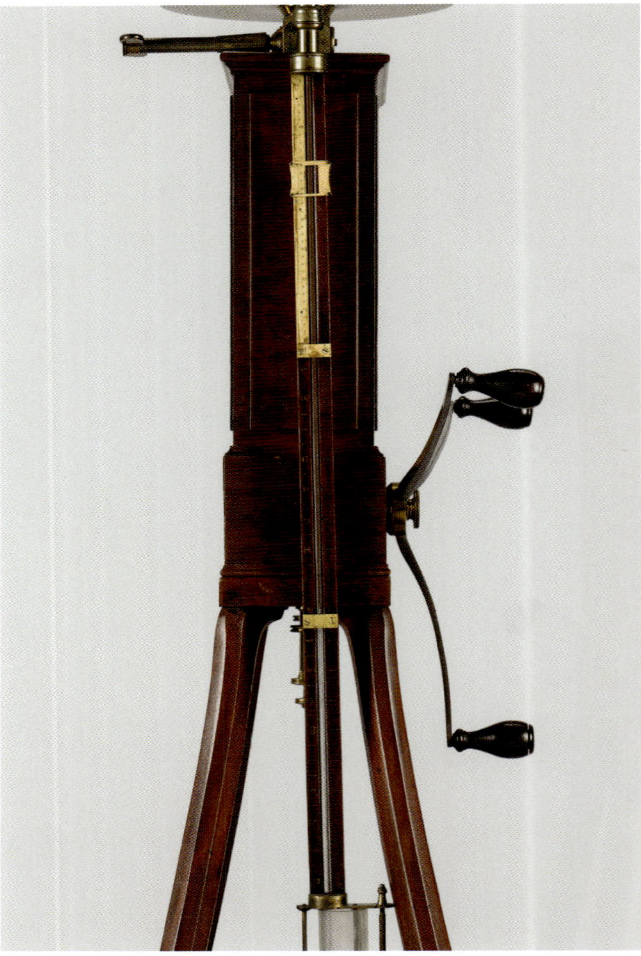

FIGURE 8B (inv. no. 19904-0000-)
©MAM/PHOTO FRANCK BOTTÉ

in such a way that the hole in its middle coincides with the opening of the stopcock supporting the plate: the receiver on the plate is thus evacuated. When the rotation of the triple crank handle is reversed, the mechanism of the inclined bar shifts the sliding plate so that a valve moves to the top of barrel A. The piston then rises, and the air extracted from the receiver is expelled from the barrel. Barrel B and its piston produce the opposite action.

Fortin designed and made this type of air pump in order to avoid certain defects that undermined the performance of several other contemporary pump models. One such problem concerned the exhaustion valve in the barrel, which could barely open when the pressure in the vessel to be evacuated was too low. Fortin thus decided to make a pump with the following characteristics:

a) no intermediary valve between the vessel and the body of the barrel when the piston was pumping in the air
b) no connection between the barrel and the vessel when the piston was expelling the pumped air
c) perfect contact between the top of the piston and the sliding plate when the piston was at the upper end of its stroke.

Fortin's air pump had an ingenious, original design, but its mechanism was more complicated than the standard pump, and the surfaces of the fixed board and the movable plates on it had to be machined with utmost precision. For these reasons Fortin's pump was likely more expensive than the more common models and its diffusion was therefore limited.

Fortin probably made the first pump of this kind in 1779 when he submitted it to the Académie royale des sciences. On that occasion, Lavoisier, together with Laplace, Brisson and Le Roy, reported favourably. On 1 March 1784, Fortin sent to Lavoisier a bill for cleaning and readjusting the pump (LC, vol. 4, p. 11). We do not know exactly when Lavoisier purchased the pump, and his papers do not record its use. However, it is certain that he owned it in November 1782 because on the 27th the French geologist Barthélemy Faujas de Saint-Fond visited Fortin's workshop at the Sorbonne and wrote:

> After supper we went to see M. Fortin. He is the one who made such excellent pneumatic machines, identical to those that I had seen at M. de Lavoisier's, which lower the mercury to a quarter-line in the test tube. This machine costs 25 *louis* with the accessories. It is one of the most perfect machines possible, and M. le Duc [de Chaulnes], who knows about instruments, assured me that it is absolutely the best available.[3]

Fortin made an almost identical pump in 1792 (see 07517-0001-, 07517-0002-). A pump – it is not clear whether it was the first or second – is also mentioned in the *Inventaire des Instrumens de Physique et de Chymie du Cabinet de Lavoisier* compiled by Charles, Fortin and Lenoir (see Appendix 3, p. 153). Its cost was estimated at 800 *livres*, a considerable price if compared, for instance, with the two calorimeters, estimated at only 600.

3 "Après le diné nous sommes allés voir M. Fortin c'est lui qui fait de si excellentes machines pneumatiques, pareilles à celle que j'avais vu chès [*sic*] M. de Lavoisier et qui font baisser le mercure à un quart de ligne à l'éprouvette celle-ci coutent avec l'assortiment 25 Louis c'est une des machines les plus parfaites qui puisse exister et monsieur le duc [De Chaulnes] qui se connait en instrument m'a assuré que c'est tout ce qu'on pouvoit faire de mieux". Comparato 2018, vol. 2, p. 195. In his PhD thesis, Comparato has transcribed Faujas' travel diary in Paris of 1782, which contains several interesting notes on Parisian scientific activities, instruments makers and natural collections.

In 1777 used an air pump during an experimental campaign on the vaporization of several liquids (water, ether and alcohol). These experiments demonstrated that the liquids could all be transformed, at different temperatures, into elastic fluids.

Lavoisier (1779); Biot (1816), vol. 1, pp. 128–140 and plate 12, fig. 45; Gehler (1831), vol. 6, pp. 572–573, plate IX, fig. 84; Daumas (1950), pp. 57–58; Lavoisier, Laplace (1982), p. xv; Beretta (2003); Comparato (2018), vol. 2, p. 198

07517-0001-, 07517-0002- (Fig. 9)
Air pump
1792
MAM accession date: 1864–1866
Nicolas Fortin (1750–1831)
184 × 76 × 47; 32 kg
Wood, brass, iron, glass, leather

This two barrelled pneumatic pump is almost identical to the one with the inventory no. 19904 and works in exactly

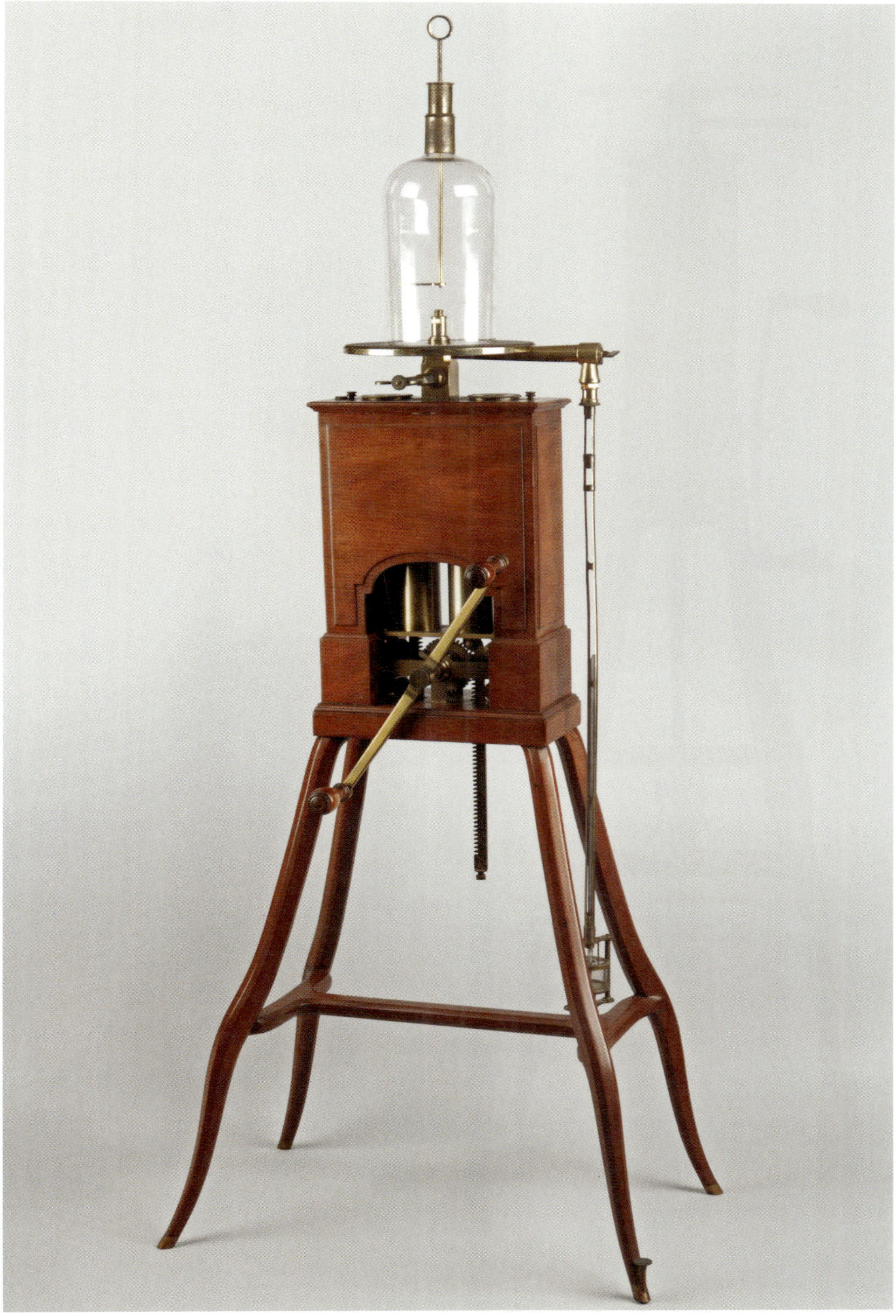

FIGURE 9A
inv. no. 07517-0001-
©MAM/PHOTO FRANCK BOTTÉ

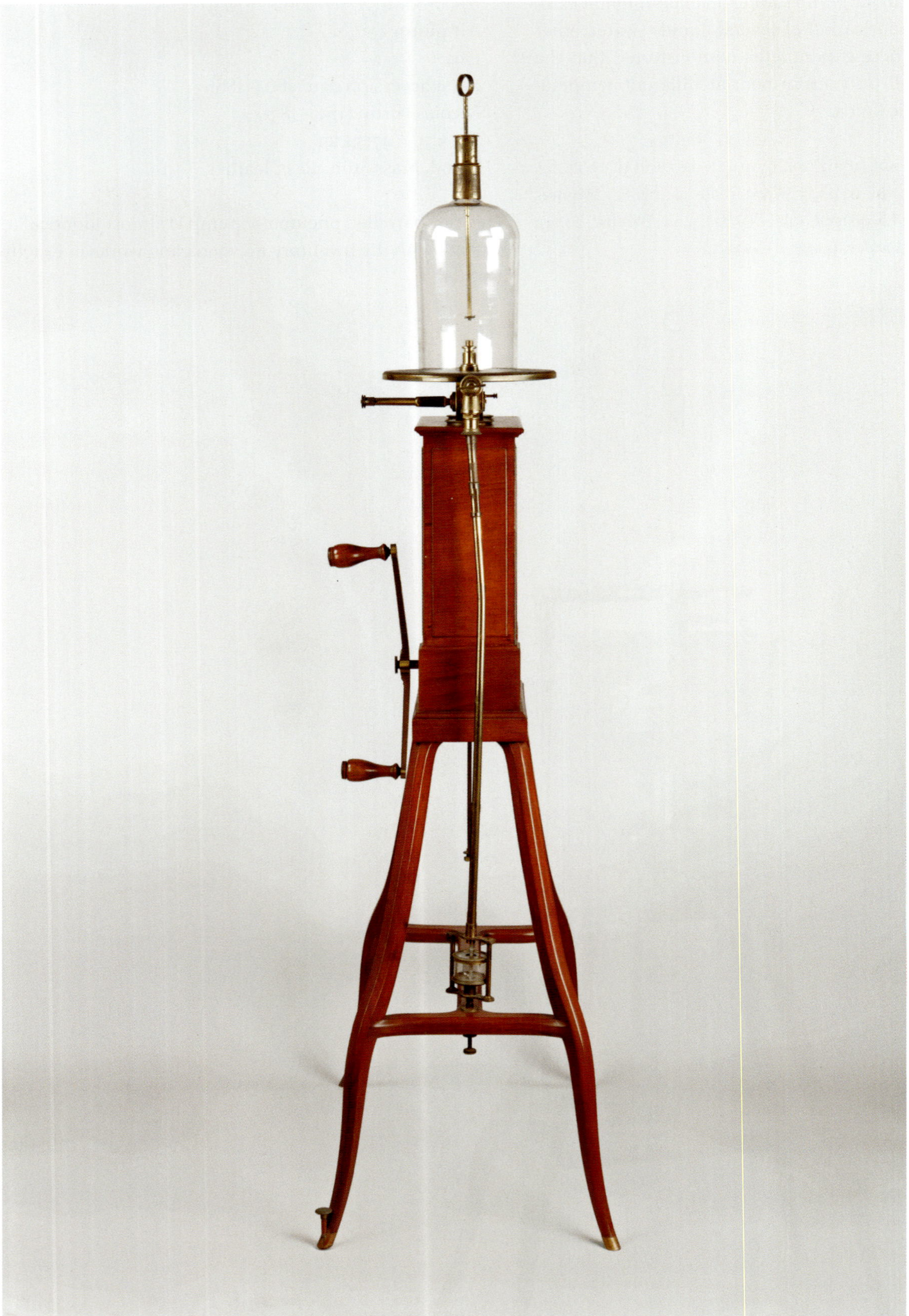

FIGURE 9B (inv. no. 07517-0001-)
©MAM/PHOTO FRANCK BOTTÉ

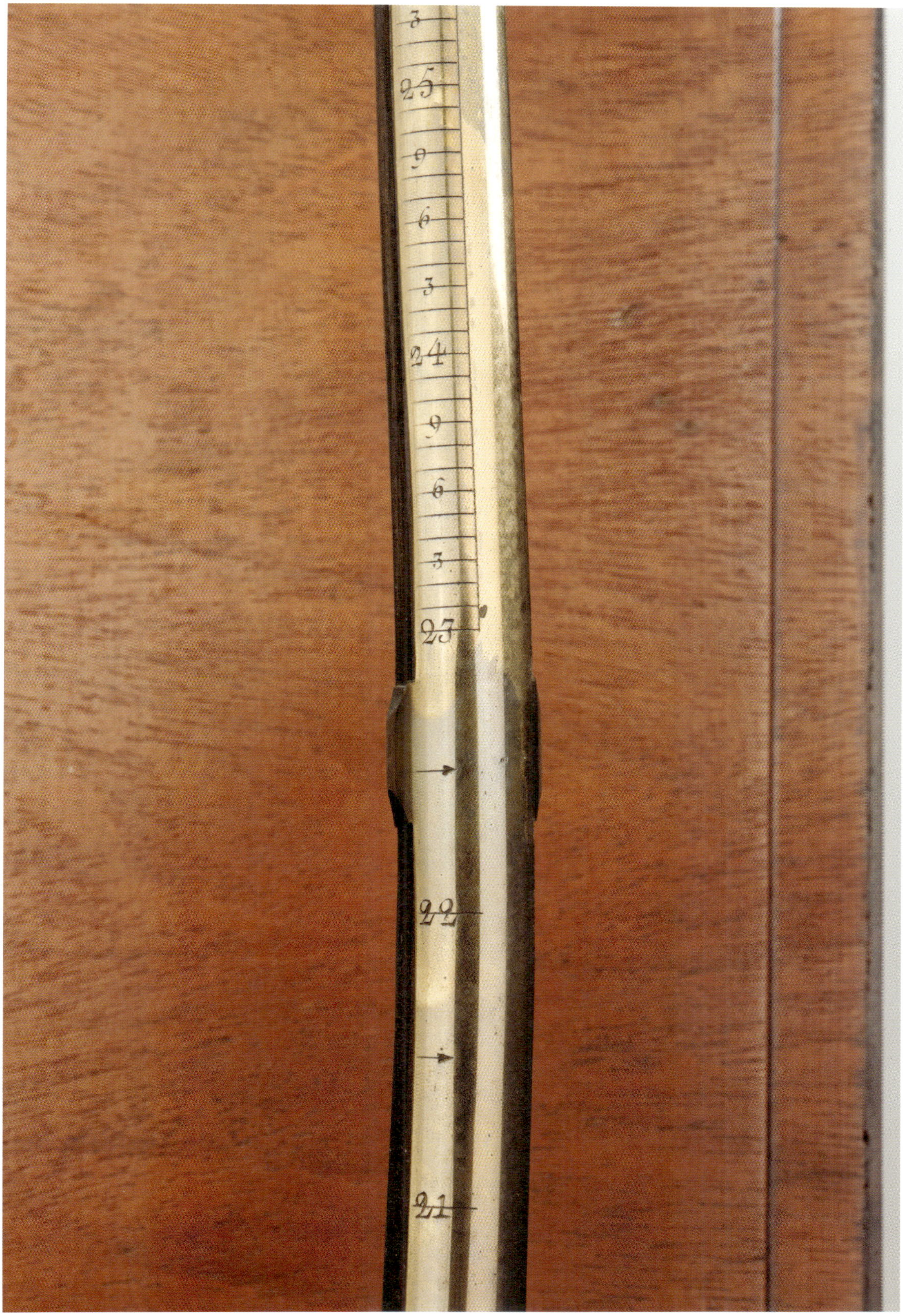

FIGURE 9C (inv. no. 07517-0002-)
©MAM/PHOTO FRANCK BOTTÉ

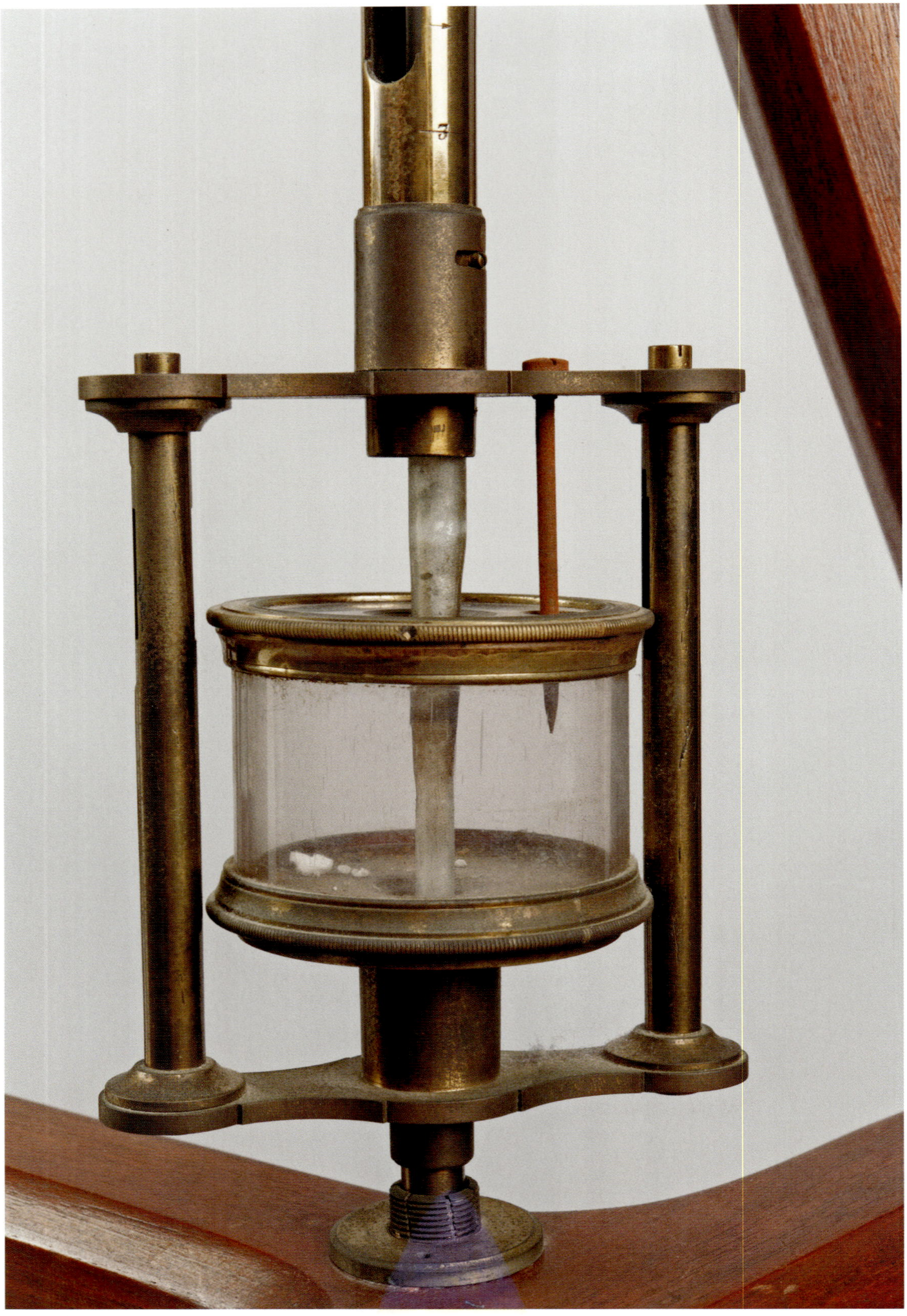

FIGURE 9D (inv. no. 07517-0002-)
©MAM/PHOTO FRANCK BOTTÉ

the same way. A few details are different: there is a double crank hand and the barometric tube (broken) of the manometer,[4] which is connected with a stopcock to the base of the pump plate, is protected by a brass pipe with windows.

A brass element carries the inscription "Fortin Place Sorbonne A Paris 1792".

Lavoisier (1779); Biot (1816), vol. 1, pp. 128–140 and plate 12 fig. 45; Gehler (1831), vol. 6, pp. 572–573, plate IX, fig. 84; Daumas (1950), pp. 57–58; Beretta (2003)

19951-0000- (Fig. 10)
Mercury manometer
Late 18th c.
MAM accession date: 1952
Cappy, Mossy (fl. ca. 1770–1810)
18.5 × 5 × 1.7; 180 g
Silvered brass, glass

Mercury manometer[5] consisting of a U-shaped tube sealed at one end and open at the other. The tube is mounted on a silvered brass plate by means of three brass hooks. There are two scales ranging from 0 to 2 inches and 3 lines divided every ½ line. Figures are marked every 3 lines (0, 3, 6, 9, 12, ...). The scale on the left of the tube runs from the 0 reference line to the top; the scale on the right runs from the 0 line to the bottom of the tube.

The U-tube was originally filled with mercury until it reached the 0 lines on both arms. The tube was then tilted so that the mercury could fill the closed arm completely.

The manometer was placed under a bell-jar on the plate of a pneumatic pump. When the pressure in the jar decreased, the level of the mercury in the closed arm of the manometer fell, while the one in the open arm rose. For example, a mercury level of 6 meant that the pressure in the jar corresponded to the pressure exerted by a column of mercury of 6 lines (approximately 6 × 2.256 = 13.56 mm/Hg). Developed by Robert Boyle in the late 1650s, this type of manometer remained in widespread use until the late 19th c.

Engraved inscription "Cappy et Mossy".

Biot (1816), vol. 1, pp. 132–133, and plate 2, fig. 43

4 The fragments of the glass tube are preserved under no. 07517-0002-.
5 This type of manometer was often called "éprouvette".

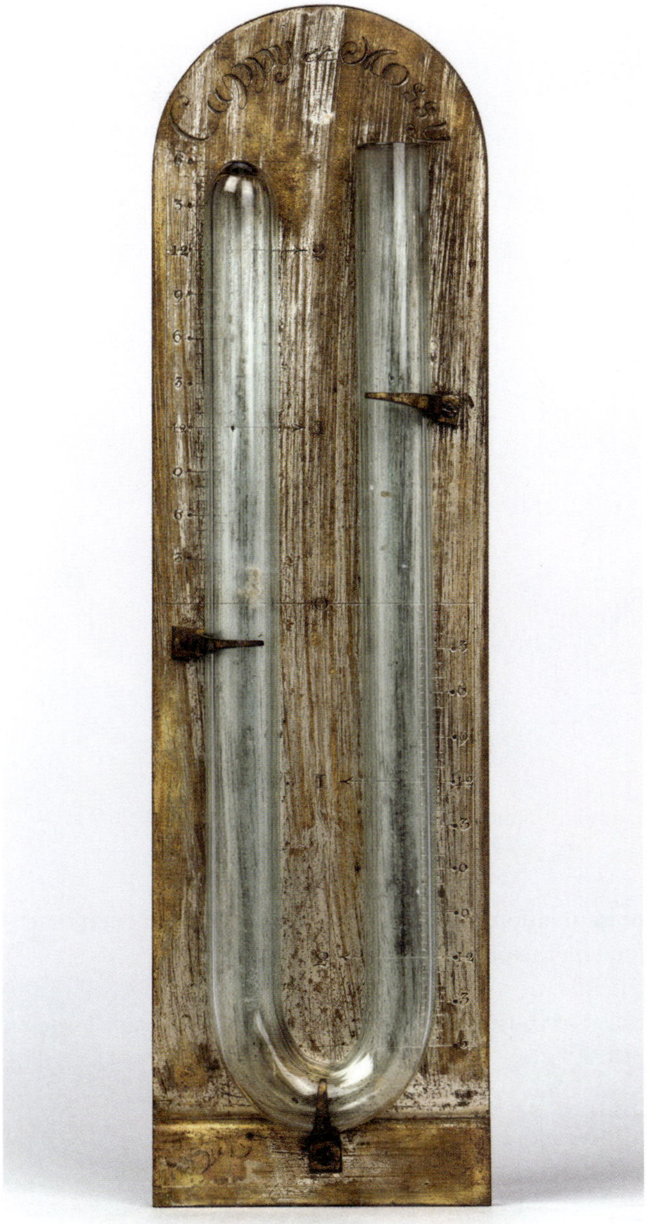

FIGURE 10 (inv. no. 19951-0000-)
©MAM/PHOTO FRANCK BOTTÉ

19965-0000- (Fig. 11)
Glass receiver
Second half of 18th c.
MAM accession date: 1952
48 × 15.5; 1.45 kg

Cylindrical glass receiver (bell) closed at the top by a brass collar with a stuffing box. A brass rod with a ring at the top passes thought the stuffing box. The lower end of the rod carries an orthogonal forked arm. The rod was used for

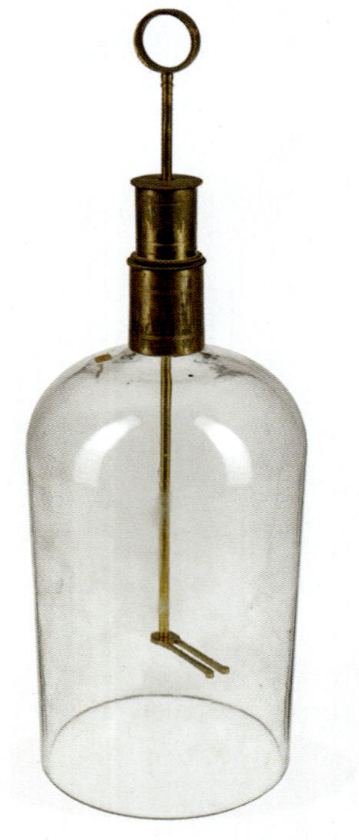

FIGURE 11 (inv. no. 19965-0000-)
©MAM/PHOTO FRANCK BOTTÉ

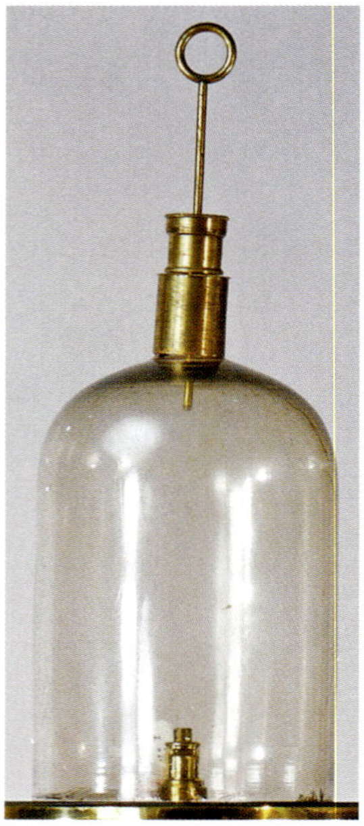

FIGURE 12
(inv. no. C-2013-0043-)
©MAM/PHOTO SYLVAIN PELLY

manipulations inside the receiver after it had been evacuated on the plate of a pneumatic pump.

Nollet (1743–1764), vol. 3, lesson X, plate III, fig.32 and lesson XI, plate I, fig. 8

C-2013-0043- (Fig. 12)
Bell jar
Second half of 18th c.
MAM accession date: 1952
53 × 19.5; 2.15 kg
Glass, brass, sealing wax

Glass bell-jar closed at the top by a brass collar with a stuffing box. A sliding rod with a ring at the top penetrates into it. The bell jar was evacuated with a pump. By manipulating the rod, it was possible to perform an experiment in the vacuum.

Nollet (1743–1764), vol. 3, lesson X, plate III, fig.32 and lesson XI, plate I, fig. 8

19969-0000- (Fig. 13)
Glass receiver
Second half of 18th c.
MAM accession date: 1952
34 × 18, 1.85 kg
Glass, brass

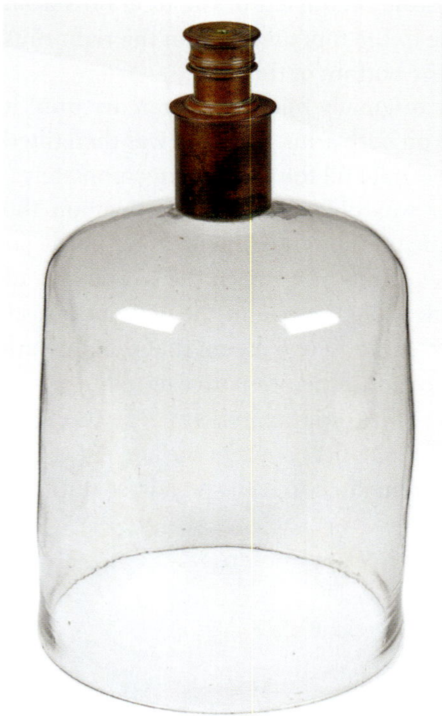

FIGURE 13 (inv. no. 19969-0000-)
©MAM/PHOTO FRANCK BOTTÉ

Cylindrical glass receiver (bell) closed at the top by a brass collar with a stuffing box. Originally a brass bar (missing) was inserted into the bell and used for manipulations inside it after the bell had been evacuated on the plate of a pneumatic pump.

Nollet (1743–1764), vol. 3, lesson X, plate III, fig.32 and lesson XI, plate I, fig. 8

19976-0000- (Fig. 14)
Glass receiver
Second half of 18th c.
MAM accession date: 1952
23.5 × 17.5; 5.19 kg
Glass, wax

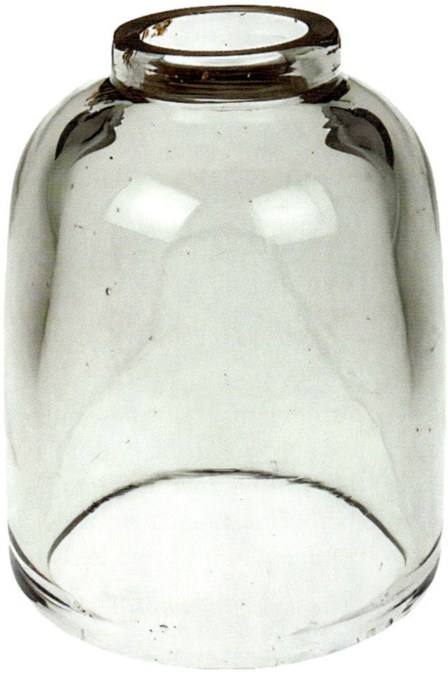

FIGURE 14 (inv. no. 19976-0000-)
©MAM/PHOTO FRANCK BOTTÉ

Receiver (bell) made of thick glass with a short open neck. The bottom rim shows traces of wax.[6] Because of the thickness of the glass, this receiver was used with a compression pump.

6 Wax was used to seal the receiver to a pump plate.

20205-0000- (Fig. 15)
Brass cap with rod
Late 18th c.
MAM accession date: 1952
32.5 × 10; 1 kg
Brass

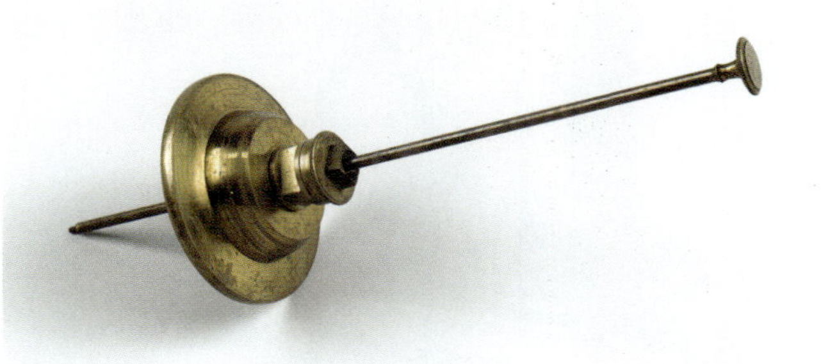

FIGURE 15 (inv. no. 20205-0000-)
©MAM/PHOTO FRANCK BOTTÉ

Massive brass cap. A sliding brass rod with a button at the top passes through a stuffing box screwed on the cap. The bottom end of the rod is threaded, indicating that part of the instrument is likely missing. This cap probably closed the neck of an evacuated bell jar and the rod allowed the user to perform manipulations in it.

De Clercq (1997), p. 80

20141-0000- (Fig. 16)
Clockwork-driven flint-lock apparatus for producing sparks in vacuum
1760–1780
MAM accession date: 1952
Nicolas Noël, known as Dom Noël (1712–1783)
22.8 × 19.5 × 17; 8.44 kg
Brass, steel, iron gilded wood

The apparatus produces sparks thanks to a clockwork-driven flint-lock. A spring-driven mechanism is enclosed in a decorated brass box mounted on a gilt wooden base. The mechanism is loaded with a brass key. A rotating arm pivots in an ornate frame fixed to the front of the box. Two curved steel sectors are attached to the ends of the arm. A brass fork is connected with the clockwork and serves to stop and start it. A decorated bent movable arm is fitted with a curved lever and a clamp that can hold a piece of flintstone. The front of the box bears the inscription “Dom Noel P[aris] fecit”.

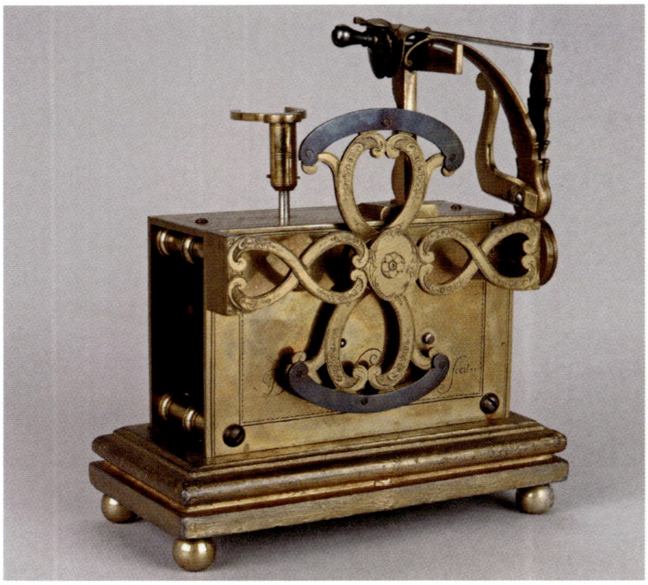

FIGURE 16 (inv. no. 20141-0000-)
©MAM/PHOTO FRANCK BOTTÉ

The apparatus was placed on the plate of a vacuum pump and covered by a special glass bell (missing). The bell had two bars with handles passing through a stuffing box in the bell's cap. The handles served to manipulate both the fork actuating the mechanism and the arm holding the flint. When the bell was filled with air, the rotating steel sectors hitting the flintstone produced an abundance of vivid sparks. When the pressure in the bell decreased, the sparks became steadily fainter and eventually disappeared. An almost identical apparatus is preserved (with its glass bell) in the collection of the Museum of Pavia University. It was used by the Italian physicist Carlo Battista Barletti (1735–1800).[7]

Sigaud de La Fond (1784), vol. 2, pp. 224–228 and plate XVII; Barletti (1794)

20005-0000- (Fig. 17)
Unidentified apparatus
Late 18th c.
MAM accession date: 1952
76 × 22 × 12; 2.68 kg
Glass, brass, tin

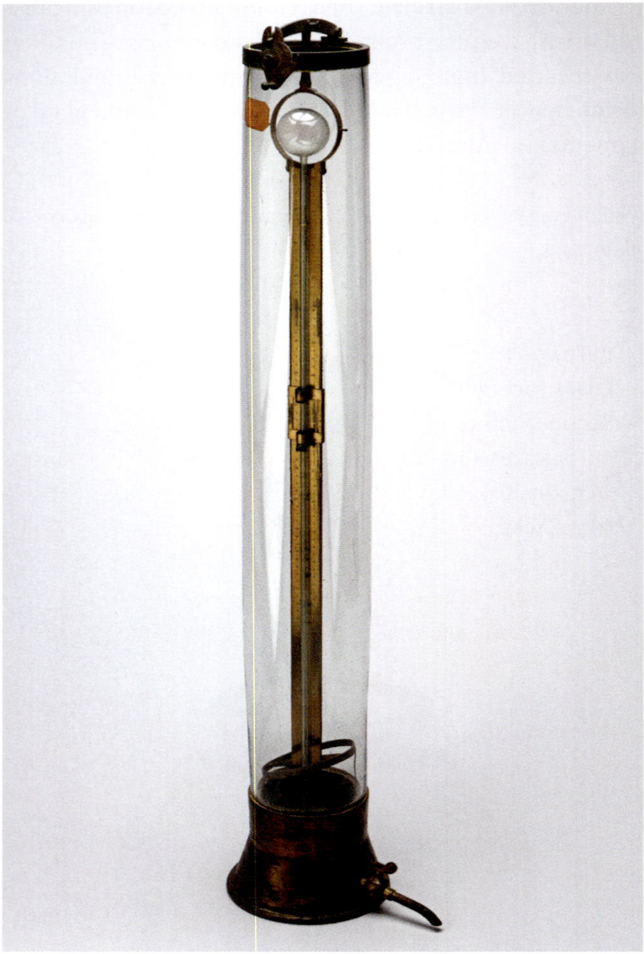

FIGURE 17 (inv. no. 20005-0000-)
© MAM/PHOTO FRANCK BOTTÉ

Cylindrical glass vessel with a brass base and a stopcock. A brass ring-shaped support at the bottom of the vessel holds a vertical tube with a bulb at the top and a small open cup at the bottom. A brass scale fixed to the tube is divided into half-lines from 1 to 20 Paris inches, with numerical markings every three lines. A vernier provides a reading precision of 1/20th of a line. A brass ring at the top of the scale is connected to a rod and a lever. The lever is hinged to the collar at the top of the glass vessel, and a rotating sector serves to block it firmly. An old paper label carries the inscription "19".

7 Barletti's experiments with this apparatus in Pavia on 20 June 1794 were attended by his colleagues Volta, Mascheroni, Spallanzani, Fontana and Brusati.

19996-0000-, 19906-0000- (Fig. 18)
Glass balloon with lateral pipe and stand
Late 18th c.
MAM accession date: 1952
61 × 42 × 21; 6.6 kg
Brass, glass, leather

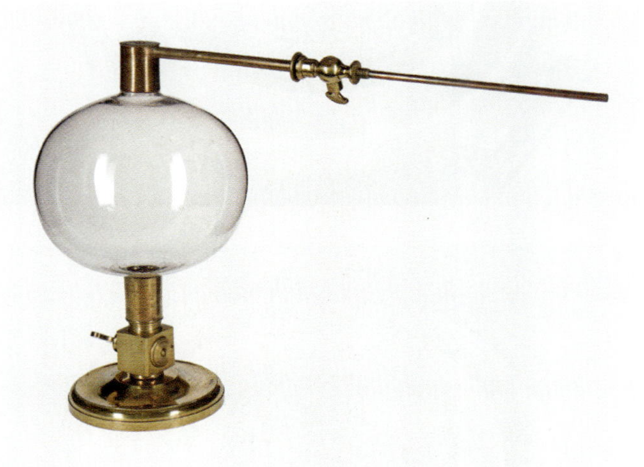

FIGURE 18 (inv. nos. 19996-0000-, 19906-0000-)
© MAM/PHOTO FRANCK BOTTÉ

Glass balloon with two opposite necks enclosed in a brass collar. The bottom balloon is screwed onto a stopcock mounted on a heavy round brass base.[8] The top balloon is connected with a pipe perpendicular to the collar. A second pipe is screwed onto to the first with a brass stopcock. The instrument was probably used on the plate of a pneumatic pump, but its function is not clear.

20104-0000- (Fig. 19)
Round base
1760s
MAM accession date: 1952
27.2 × 11; 1.53 kg
Wood, brass, leather

Round wooden base painted in black with floral gilt decorations. In its centre there is a short vertical brass screw in a collar and a remnant of a leather seal. In all certainty, this was the stand for a missing apparatus that, given the style, was probably made following the indications of Nollet or Sigaud de la Fond.

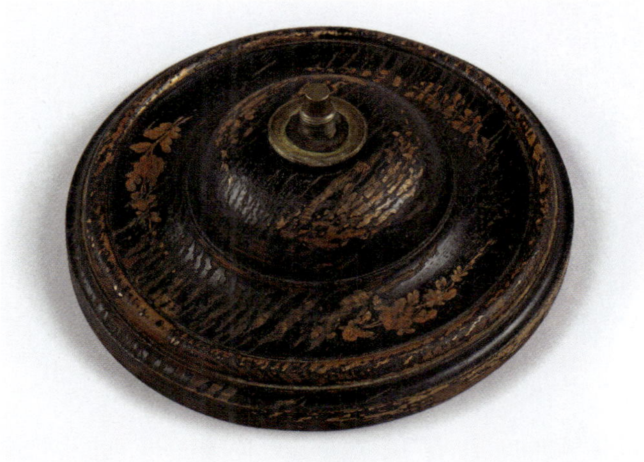

FIGURE 19 (inv. no. 20104-0000-)
© MAM/PHOTO FRANCK BOTTÉ

Thermology

07520-0000- (Figs. 20a, 20b, 20c, 20d)
Lavoisier's and Laplace's calorimeter
ca. 1782–1783
MAM accession date: 1864–1866
105 × 69 × 67; 21 kg
Zinc, iron, brass

Large calorimeter mounted on a tripod, basically composed of three coaxial metal painted vessels: outer (A), middle (B), and inner (C). Vessel C is a cylindrical basket closed by a lid consisting of a grid with a collar and handle. Four short arms maintain C in the centre of vessel B, whose conical bottom terminates with a brass stopcock. B, in turn, is inserted in vessel A. Its bottom terminates with a inclined tube and a second stopcock. A is closed by a round cover with a raised edge and a handle. Two other handles are attached to the wall of A for transporting the instrument. A flat circular tray is placed under the gasometer. The instrument is not signed but was probably made by the tinsmith Pierre Naudier.

This calorimeter was designed and used by Laplace and Lavoisier to measure the quantity of heat (*quantité de calorique*) ceded by a body when its temperature changed from T to 0° or the heat produced during a chemical reaction[9] or combustion.

To make the apparatus work, it was first necessary to fill the spaces between vessels A and B and between B and C

8 The base was catalogued separately from the rest of the apparatus with the number 19906-0000-. The instrument was placed on a straw ring (C-2017-0076-).

9 In this case, the reaction had to take place in a small glass balloon and the substances involved had to be cooled to 0° before the reaction.

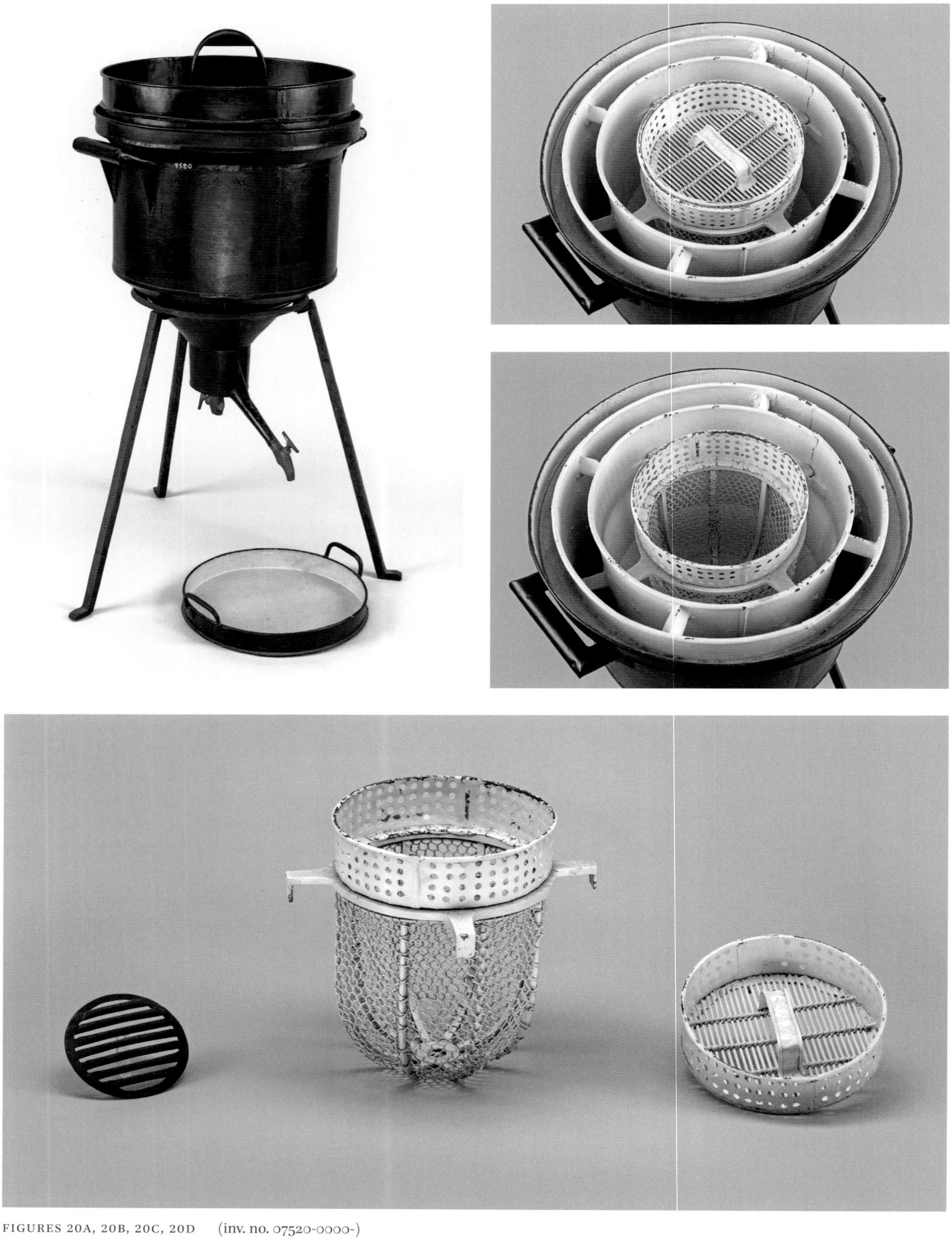

FIGURES 20A, 20B, 20C, 20D (inv. no. 07520-0000-)

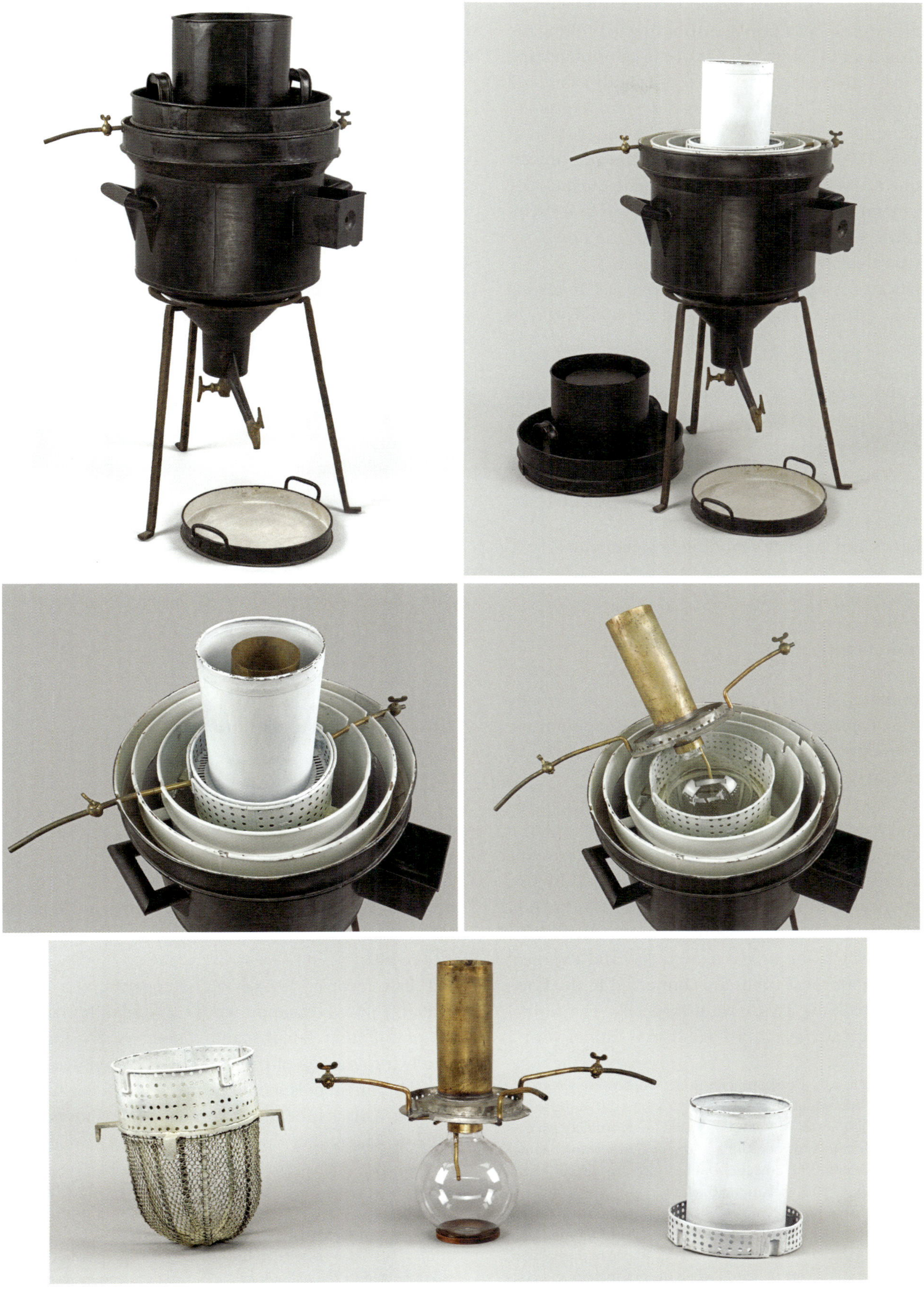

FIGURES 21A, 21B, 21C, 21D, 21E (inv. no. 07547-0004-001-)
© MAM/PHOTO FRANCK BOTTÉ

with crushed ice.[10] The ice between A and B was essential to prevent external heat from contributing to melting the ice between B and C. Next, the lids of the calorimeter and of C had to be covered with ice. Lastly, the body examined had to be placed in C[11] until it reached the temperature of 0°. The quantity of water produced by the liquefaction of ice in vessel B could be collected under the calorimeter and weighed. The weight was proportional to the quantity of heat generated by the cooling of the body or by a chemical reaction. Knowing T and the mass of the body, one could then calculate the body's specific heat.

In November 1794, the two calorimeters (see following entry) were estimated by Charles, Lenoir and Fortin at 600 *livres*.

Lavoisier and Laplace (1782); Daumas (1955), pp. 141–142; Lavoisier and Laplace (1982); Roberts (1991); Beretta (2003); Heering (2005)

07547-0004-001- (Figs. 21a, 21b, 21c, 21d, 21e)
Lavoisier's and Laplace's modified calorimeter
1783–4
MAM accession date: 1864–1866
117.7 × 76 × 74; 90 kg
Iron, zinc, brass

This calorimeter, similar to no. 07520-0000-, was partially modified by Laplace and Lavoisier in order to determine the heat developed by the synthesis of water produced by burning a current of hydrogen in oxygen.

Like no. 07520-0000-, it is composed of an external cylindrical vessel A mounted on a tripod containing a second coaxial vessel B, in which a perforated basket C is inserted. The bottoms of A and B are connected by two brass stopcocks.[12] In the basket there is a glass balloon (similar to no. 07548-0000-) whose neck is closed by a brass collar.[13] The latter is fixed to a tin disk and a vertical brass funnel. Four brass tubes are connected to the brass collar of the balloon. Two have a stopcock. The funnel and basket are enclosed in a second iron funnel with a perforated rim. A cylindrical cover with a wide edge and two handles closes vessel A. A pipe partially enclosed in a square box runs into the calorimetric vessel; a second pipe connects the box with the outside. A flat vessel with two handles under the calorimeter collects the water. There is also a broken balloon on a straw ring (no. 7547-0004-002), but it is not at all certain that the item forms part of the apparatus.

To make the calorimeter work, the spaces between A and B and between B and C were filled with crushed ice. Some ice was added on the top of the apparatus in the vertical funnel. The balloon (used exactly as no. 07548-0000-) was then evacuated by connecting one of its tubes to a pneumatic pump. Next, the balloon was filled with oxygen by means of another tube. Another pipe terminating with a nozzle in the centre of the balloon was used to introduce the hydrogen, which was ignited with a spark.[14]

The combustion of hydrogen produced a quantity of heat that, as in the other calorimeter (no. 07520-0000-), could be calculated by collecting and weighing the water produced by the liquefaction of ice in vessel B.

Lavoisier and Laplace (1782); Daumas (1955), pp. 141–142; Lavoisier and Laplace (1982); Roberts (1991); Beretta (2003); Heering (2005)

20154-0000-, 20153-0000- (Fig. 22)
Rumford combustion calorimeter
ca. 1812
MAM accession date: 1952
20154-0000-: 27.5 × 25 × 18: 1.29 kg
20153-0000-: 44 × 13.5 × 10.5: 1.25 kg
Copper, wood

This type of calorimeter, composed of two main elements (receivers), was presented by Rumford at the Institut de France in 1812.

The first receiver (20154-0000-) consists of a parallelepiped copper box supported by a wooden frame. Two tubes of different lengths and diameters are inserted vertically into the top of the box; one is closed by a cork. Under the box there is a hole connected to a flat serpentine made with a tube attached near the inside bottom of the box. The end of the serpentine protrudes laterally from the box.

10 In fact, one had to use melting ice at 0°. A cooler ice would falsify the measurements.

11 A special glass vessel (missing) was used for substances such as acid, which could corrode the apparatus. Furthermore, sometime to accelerate the process, which could take several hours, also C was filled with ice around the body.

12 One collects the water produced by the liquefaction of the ice used to refrigerate the apparatus; the other collects the water produced by the ice melted by the heat developed in the balloon.

13 The balloon is copy made in 1975 when the original 07547-0002-002- broke.

14 The electrode needed to produce the spark seems missing. It may have been inserted in the balloon through the shorter pipe connected to the balloon collar, but the actual arrangement is unclear.

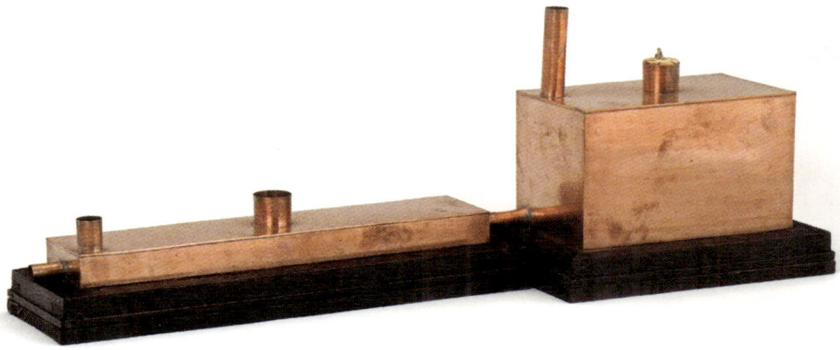

FIGURE 22 (inv. nos. 20154-0000-, 20153-0000-)
© MAM/PHOTO FRANCK BOTTÉ

The second receiver (20153-0000-) also consists of a parallelepiped copper box (longer and flatter than the first) with two short vertical tubes and supported by a wooden frame. This box, as well, contains a serpentine whose ends protrude from two opposite sides of the box.

Rumford used this instrument to measure the heat developed by the combustion of inflammable substances such as alcohol, ether, wood, wax, oil and charcoal, which were burned with various special lamps. These were placed so that the flame was under the mouth of the first serpentine. Before starting with the measurements, the receivers were filled with known quantities of water and a precision mercury thermometer was inserted into each of them through one of the vertical tubes. A given weight of the substance to be examined was then lit. After measuring the water temperature in the receivers at the end of the combustion (and knowing the initial temperatures), the apparatus determined the amount of heat generated.

It should be noted that Rumford used the two receivers only in the first measurements (combustion of wax). He added the secondary receiver to collect the heat that might still have been present in the combustion gases after they had passed through the serpentine of the primary receiver. But, having noticed that the temperature in the secondary receiver registered a slight increase only after an extremely long combustion, Rumford decided to get rid of it. All the other measurements were made with only one calorimetric box.[15]

Thompson (1813); Chevalier, Fau (1854), vol. II, plate 76, pp. 350–351; Pouillet (1856); Salleron (1864), p. 289

20150-0000- (Fig. 23)
Rumford's calorimetric vessel
ca. 1799
MAM accession date: 1952
49 × 23 × 22.5; 1.185 kg
Wood, brass

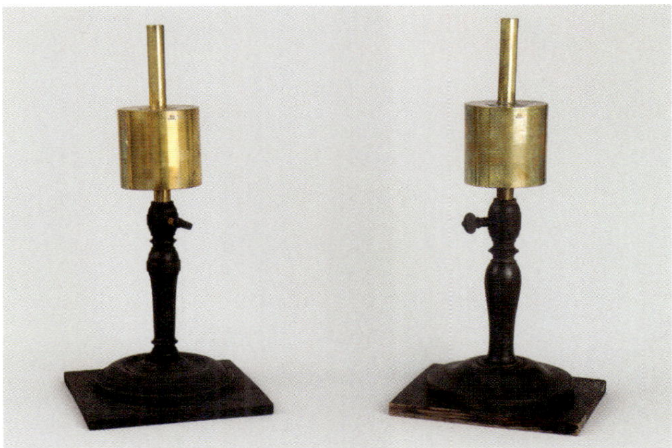

FIGURE 23 (inv. nos. 20150-0000-, 20151-0000-)
© MAM/PHOTO FRANCK BOTTÉ

15 Rumford did not publish a drawing of his calorimeter, but several nineteenth-century physics textbooks did. The known illustrations show only a single receiver, with one exception: the instrument maker Jules Salleron inserted an engraving of the apparatus with both receivers in his 1864 trade catalogue. By then, Rumford's calorimeter was only of historical and didactic interest.

Cylindrical brass vessel with a long neck and a wooden bar fixed vertically on its bottom. The bar is inserted telescopically in a column with a round base. A screw serves to set the vessel at the desired height. The vessel carries the punched inscription "I 10 LOD" and the sign "/".

This apparatus was presented by Rumford in 1799. He used four such instruments to determine the insulating properties of various materials such as cloth, fur and wool. The vessel would be filled with water and covered with the material to be tested, while a long mercury precision thermometer (missing) was inserted in the neck. The experimenter then determined how long the water took to cool by a given number of degrees. The slower the temperature decrease, the greater the insulating capacity of the tested material.

Thompson (1804)

20151-0000- (Fig. 23)
Rumford's calorimetric vessel
ca. 1799
MAM accession date: 1952
51 × 23.6 × 19.5; 1.16 kg
Wood, brass

While practically identical to no. 20150-0000-. The vessel carries the punched inscription "II 10 LOD" and the sign "//".

Thompson (1804)

20145-0000- (Fig. 24)
Cylindrical vessel
ca. 1799
MAM accession date: 1952
28 × 10.2; 522 g
Brass

Brass cylindrical vessel with a pipe. The vessel, filled with hot water, was probably used by Rumford as a heat source for his calorimetric experiments. The vessel bears the inscription "Aw 458.9 grm"

Thompson (1804)

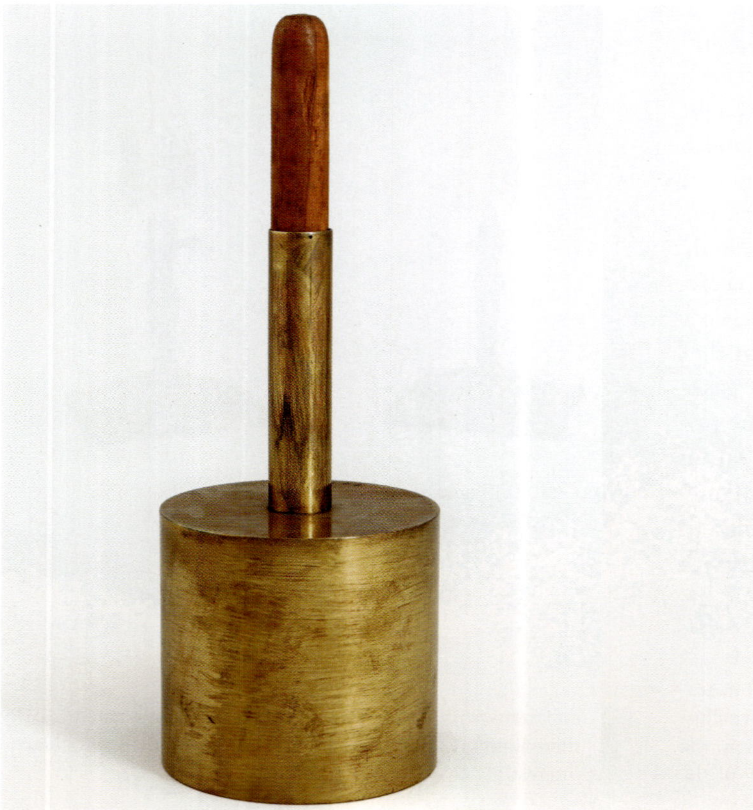

FIGURE 24
(inv. no. 20145-0000-)
© MAM/PHOTO FRANCK BOTTÉ

19942-0000-, 19943-0000- (Fig. 25)
Two of Rumford's radiant heat thermometers
MAM accession date: 1952
19942-0000-: 53.5 × 11; 215 g
19943-0000-: 53 × 11; 215 g
Glass, cardboard, wood, brass, pitch, cork

The two thermometers are almost identical. They are composed of a vertical glass tube with a small funnel at the top, closed by a cork. The bottom of the tube is inserted in a thermometric reservoir consisting of a double walled brass cone originally containing coloured spirit or linseed oil. The reservoir is enclosed in a wooden cylinder covered with paper and filled with eiderdown. A narrow wooden board is fixed to the tube. The boards carry some very faint inscriptions in an unknown hand:

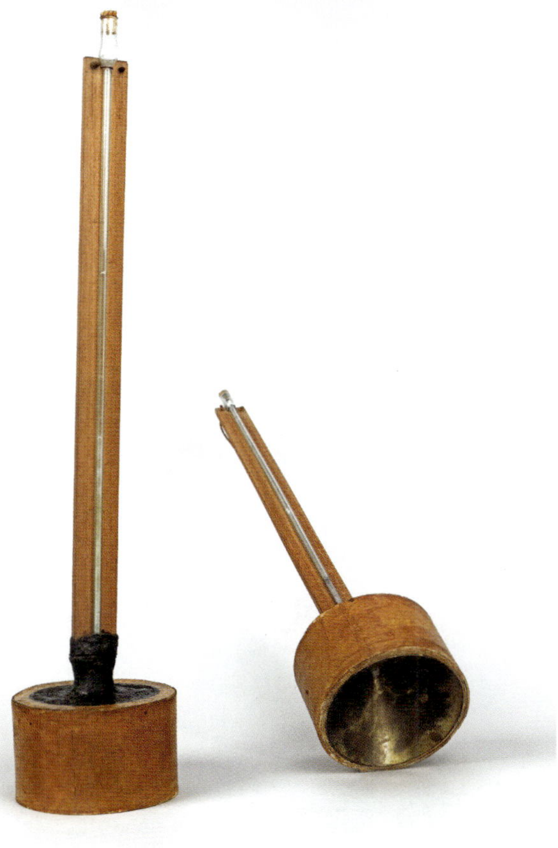

FIGURE 25 (inv. nos. 19942-0000-, 19943-0000-)
© MAM/PHOTO FRANCK BOTTÉ

19942-0000-: "32, 50, 56, 58 (with a figure 7 between these two last numbers), 60, 62, 64, 66, 67, 68" and a pencilled line near the top of the board.

19943-0000-: "50, [5]8, 68"

Rumford used these thermometers to studying radiant heat. The external cone, originally blackened, was directed towards a heat source, and the effects of the radiant heat could be measured by the change in height of the liquid column in the tube.

Thompson (1805)

19940-0000-, 19941-0000- (Figs. 26a, 26b)
Calorimetric thermometers
ca. 1780
MAM accession date: 1952
19940-0000-: 43.5 × 16.5 × 15.5; 585 g
19941-0000-: 44.5 × 16.2 × 15.8; 580 g
Glass, wood, paper, tin, cork

Two almost identical instruments. Both are composed of a wooden base supporting a cubic wooden box open at the top. The box was originally lined inside with tin foil, of which only few fragments remain.[16] A vertical support is fixed on a side of the box and on it there is a paper strip graduated from 50 to 63 with each division being divided into ten parts (the paper strip of 19941-0000- is damaged). A thermometric tube, fixed against the paper, terminates at the top with a small pear-shaped vessel with a roughly cut wooden stopper. The bottom part of the tube ends in a large spherical bulb with a concavity forming a sort of cup (the bulb of 19940-0000- is broken). The bulbs were originally filled with coloured alcohol.

The function of the two instruments is unclear. They were probably used as calorimetric thermometers. They could measure the heat produced by a chemical reaction taking place inside the cup of the bulb, if one accepted the notion that most of the heat was transferred to the alcohol. In this case, the instruments acted as a kind of rudimentary Favre and Silbermann calorimeter.[17]

16 The box may also have been filled with cotton, wool, fabric, straw or similar substances that, together with the tin foil, could have improved the thermal insulation of the calorimetric bulb.

17 A 20th-century handwritten label preserved with one of the instruments suggests this possibility. Pierre-Antoine Favre (1813–1880) and Jean-Thiébault Silbermann (1808–1865) proposed a calorimeter that was essentially a thermometer with a very large iron bulb filled with mercury. The thermochemical reactions to

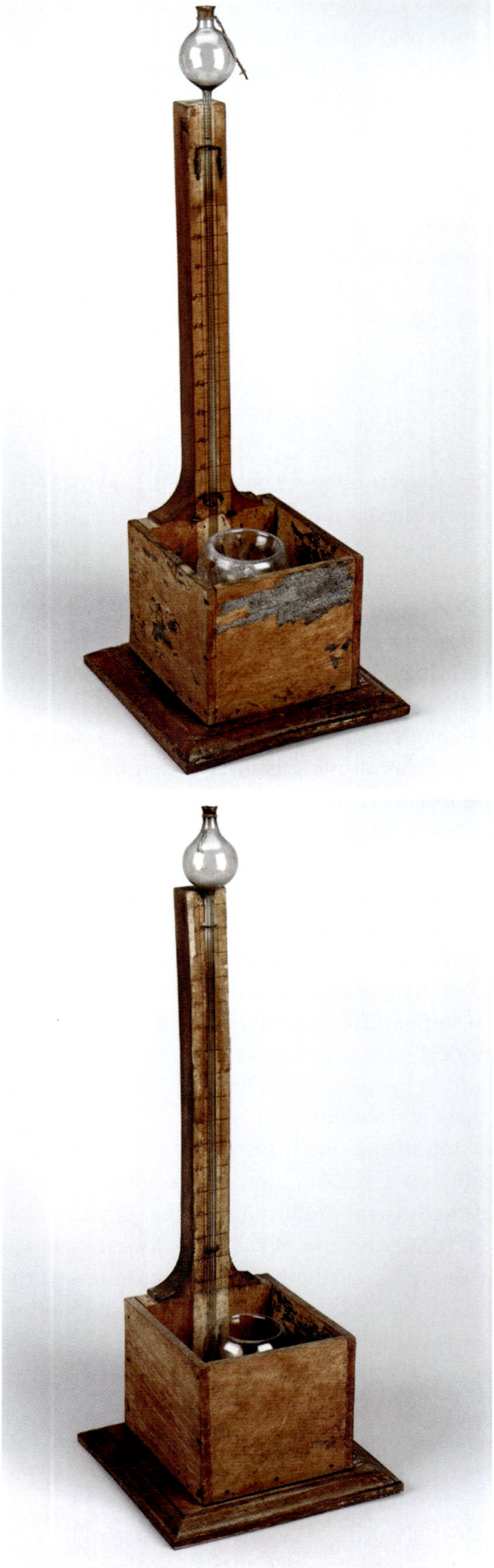

FIGURES 26A, 26B (inv. nos. 19940-0000-, 19941-0000-)
© MAM/PHOTO FRANCK BOTTÉ

20095-0000-, 20008-0002- (Fig. 27)
Rumford's thermoscope
Early 19th c.
MAM accession date: 1952
Support: 43.5 × 25.3 × 21.5; 1.48 kg
Tube: 38 × 16 × 4; 30 g
Wood, brass, iron, glass, fibre, paper

This thermoscope functions as a highly sensitive differential thermometer. A rectangular wooden base supports a T-shaped frame. On its horizontal bar there are two movable sliders whose position can be fixed with wooden screws. On them is inserted a U-shaped glass tube with a closed bulb at each end (no. 20008-0002-).[18] The tube originally contained a drop of spirit.

In a quadrangular column fixed on the base is inserted a vertical flag-shaped frame with a screen made of two

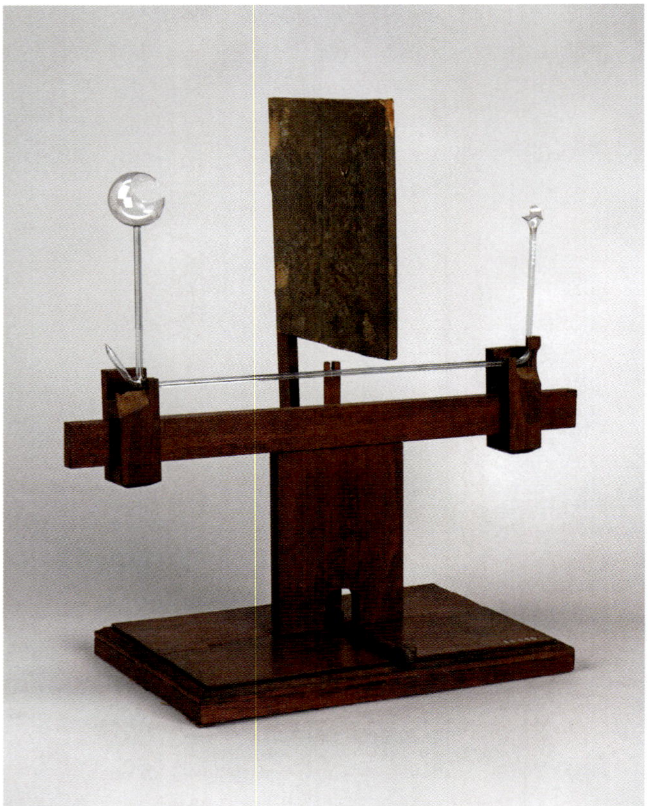

FIGURE 27 (inv. nos. 20095-0000-, 20008-0002-)
© MAM/PHOTO FRANCK BOTTÉ

be studied were produced directly in a cavity of the bulb. See, for example: Chevalier, Fau (1854), vol. 2, pp. 349–350, plate 76, fig. 956; Daguin (1861), vol. 2, pp. 367–368.

18 The tube was catalogued separately from its stand because the instrument was not assembled until the cataloguing work on the collection.

separate sheets of (damaged) metallized paper.[19] On the base a movable horizontal lever activates a vertical rod whose purpose is unclear.

The thermoscope made it possible to compare the relative intensities of two radiant-heat sources. These were aligned along the axis of the thermoscope, each in front of one of the two bulbs. If the bulbs were at the same temperature, the drop in the horizontal arm of the tube remained still.[20] A slight temperature increase in one of the bulbs expanded the air inside it, pushing the drop towards the other bulb.

The screen prevented one of the bulbs from the effects of the heat produced by the hot body placed near the other bulb.

In 1804, Rumford described his thermoscope, which he used in his studies on the nature of heat and its propagation. The instrument enabled him to compare the intensity of heat radiation[21] from different sources.

Thompson (1804); Chevalier, Fau (1854), vol. 1, pp. 120–121, plate 15, fig. 157; Chen (2005)

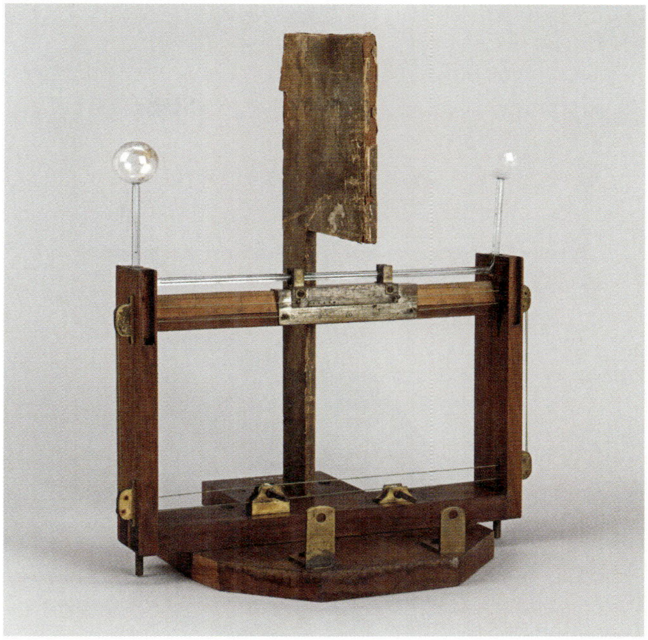

FIGURE 28 (inv. nos. 20094-0000-, 20008-0001-) © MAM/PHOTO FRANCK BOTTÉ

20094-0000-, 20008-0001- (Fig. 28)
Rumford thermoscope
Early 19th c.
MAM accession date: 1952
Support: 33 × 23 × 23; 1.48 kg
Tube: 32 × 12 × 3; 30 g
Wood, brass, glass, silvered brass, ivory, iron, paper

This thermoscope is similar to but more sophisticated than the previous one (no. 20095-0000). A vertical rectangular frame rests on an octagonal wooden base. A U-tube with two bulbs (no. 20008-0001-) is inserted on top of the frame.[22] One bulb is partly gilt and the tube still contains some liquid (probably spirit). The frame's horizontal bar is covered by a double paper scale divided into 1/10ths of units from 0 to 120. On it there is a silvered brass slider with a second scale divided into 1/10 of units with an indication every unit.[23] A second slider on top of the first has a nonius offering a precision of 1/100th of a unit and two holders made of brass strips. Both sliders can be moved separately with a system of strings passing on ten pulleys fixed on the frame and two small winches, whose handles are missing. A rectangular metallized paper screen is fixed to a vertical flag-shaped frame inserted in the base. This thermoscope works exactly like the previous one, and the graduated sliders were probably used to determine the displacement of the spirit drop, regarded as proportional to the difference in temperature between the heat sources.

Thompson (1804); Chevalier, Fau (1854), vol. 1, pp. 120–121, plate 15, fig. 157; Chen (2005)

20009-0000- (Fig. 29)
Tube of Rumford's thermoscope (broken)
Early 19th c.
MAM accession date: 1952
36 × 16 × 6; 50 g
Glass

19 Rumford used what he called "gilt paper". In fact this was a paper coated with a paint containing copper powder. The copper helped to provide a better screen for radiant heat.

20 Before the measurements and with the bulbs at the same temperature, the drop usually rested in the middle of the horizontal tube.

21 "Radiant heat" was the term used for what is known today as infrared radiation.

22 This tube, as well, was catalogued separately from its stand.

23 These units correspond to 2.3 cm and are therefore Paris inches, not English inches.

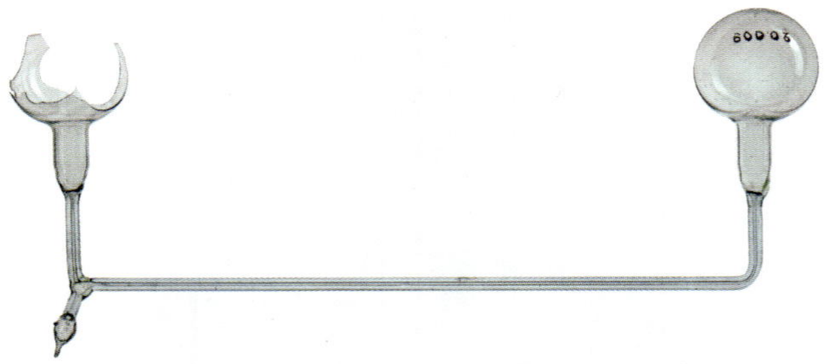

FIGURE 29 (inv. no. 20009-0000-)
© MAM/PHOTO FRANCK BOTTÉ

U-tube terminating with 2 bulbs, one of them broken. It is similar to the ones mounted on the thermoscopes 20095-0000- and 200094-0000-. This tube has no stand.

Thompson (1804)

20021-0001-, 20021-0002-, 20021-0003- (Fig. 30)
Three glass bulbs for showing the effect of the expansion of air
Second half of 18th c.
MAM accession date: 1952
20021-0001-: 23 × 7 × 5; 20 g
20021-0002-: 22 × 6.5 × 5.5; 15 g
20021-0003-: 18 × 5 × 5; 15 g
Glass

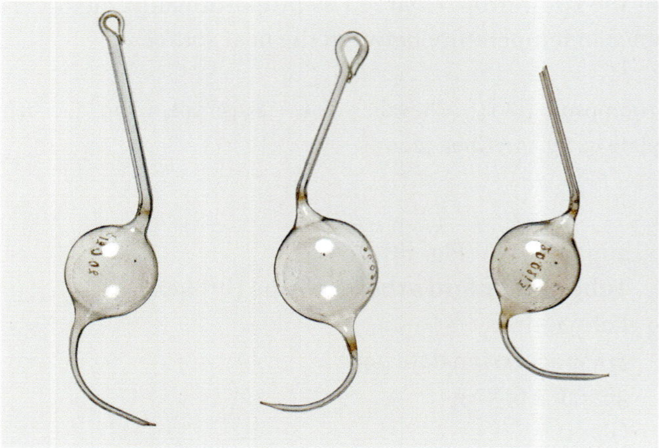

FIGURE 30 (inv. nos. 20021-0001-, 20021-0002-, 20021-0003-)
© MAM/PHOTO FRANCK BOTTÉ

Three almost identical glass bulbs. All have a curved, pointed capillary and a longer tube closed at its end.

This apparatus, introduced by Jean Nollet, was used to demonstrate in a spectacular manner the effect of the expansion of air. The bulb was first immersed in boiling water, causing the air contained in it to expand. The bulb was then taken out of the water and the opening of its capillary was dipped in alcohol. As the air cooled, its contraction caused some alcohol to be sucked in the bulb. Lastly, the bulb was immersed again in the boiling water. Because of the expansion of the air left in the bulb, the alcohol was pushed through the capillary, producing a jet of flame when it was ignited with a candle.[24]

Nollet (1743–1764), vol. 3, p. 356, plate V, fig. 26; Nollet (1770), vol. 3, pp. 20–21 and plate III, fig. 1

20096-0000- (Mckie 214) (Fig. 31)
Aeolipile
Late 18th c.
MAM accession date: 1952
22 × 10
Copper

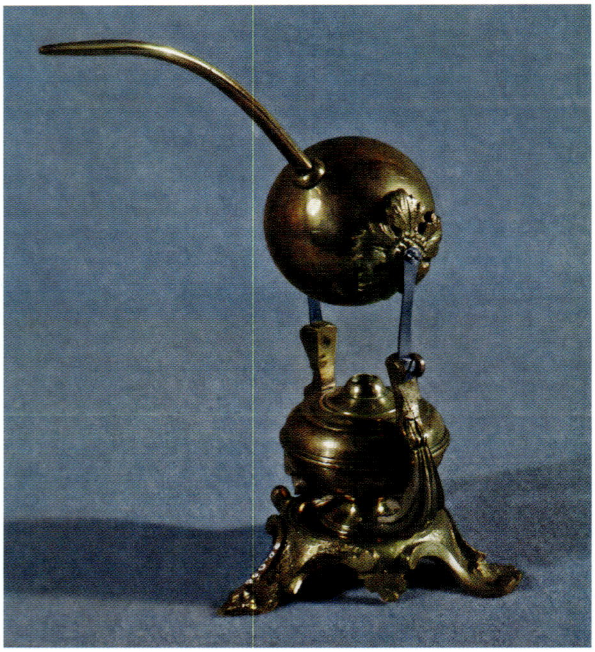

FIGURE 31 (inv. no. 20096-0000-)
© MAM/PHOTO 1960

24 Nollet proposed this experiment in his *Avis aux amateurs de physique*. In his earlier *Leçons de physique expérimentale*, he described the same experiment with a larger and more expensive metal sphere.

The aeolipile is an instrument, known since the antiquity, which was used for showing (and sometimes exploiting) the force of vapour.

A metallic spherical vessel with a long and curved tube ending in a nozzle, is supported by two iron uprights. These are fixed on a rococo decorated base holding a lamp.

The spherical vessel is gently heated and the air in it expands. The vessel is subsequently cooled while the end of the tube is immersed in water which partially fills the vessel. The vessel is then heated with the lamp until the water boils and a jet of vapour escapes from the nozzle. The vessel is finally turned upside-down, so that the pressure of vapour will violently expel the water from the nozzle creating a jet like in a fountain. Sometimes the experiment was repeated with wine spirit which was ignited. In this way it was possible to produce a jet of fire.

Two more common versions of this apparatus are the simple steam turbine and the aeolipile mounted on a small chariot. In the latter the action of the jet of vapour pushed the chariot in the opposite direction.

This item was stolen in February 1984.

Nollet (1743–1764), vol 4, leçon XII, pp. 88–89, planche II, figg. 8–9; Sigaud de la Fond (1775) vol. 1, pp. 310–313, planche 1, fig. 4; Daguin (1861), vol. II, pp. 334–335; Daumas (1960), p. 183; Turner, Levere (1973), pp. 273–274

20179-0000- (Fig. 32)
Copper sphere with glass pipe
Second half of 18th c.
MAM accession date: 1952
97 × 11.5; 530 g
Box: 86.5 × 4.5 × 5, 685 g
Copper, glass, wood, paper, thread, iron, leather

Hollow copper sphere joined to a straight brass pipe. The latter was connected to a glass tube with a broken leather joint partially wrapped with a string. On the glass tube there are two paper scales divided into Paris inches: one is marked at 32 and then from 40 to 160 (every 10 units), the other from 32 to 220. The tube is protected by a wooden box divided in two sections assembled with two iron screws. The use and function of this instrument are not known.

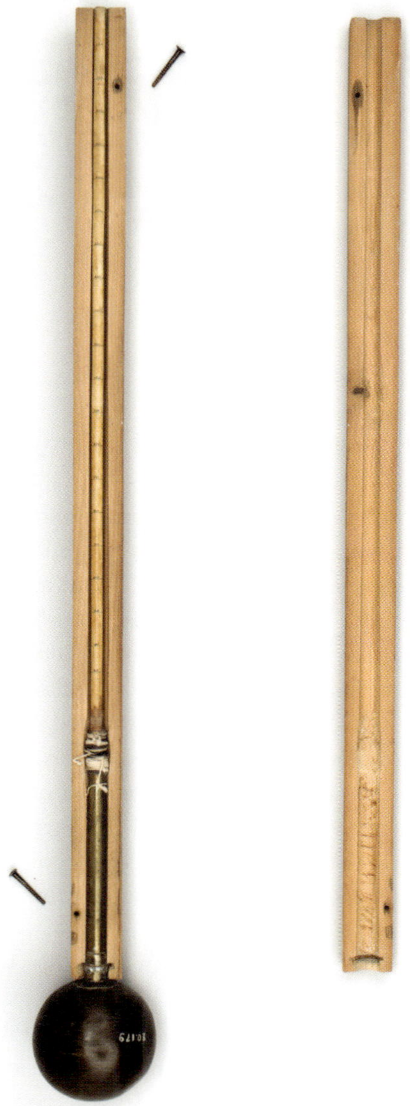

FIGURE 32 (inv. no. 20179-0000-)

03069-0000- (Fig. 33)
Model of Laplace's and Lavoisier's pyrometer
ca. 1840
MAM accession date: 1845
Pierre Clair (1804–1870)
53 × 61.5 × 45; 17.76 kg
Iron, steel, wood, brass, glass, zinc

Although this item does not belong to the Lavoisier collection, the curators have always included it with Lavoisier's instruments.

Reduced scale model reproducing Laplace's and Lavoisier's pyrometer (or dilatometer). A wooden base supports four rectangular pillars. Between them there is a rectangular frame with a brass vessel. Two pairs of steel

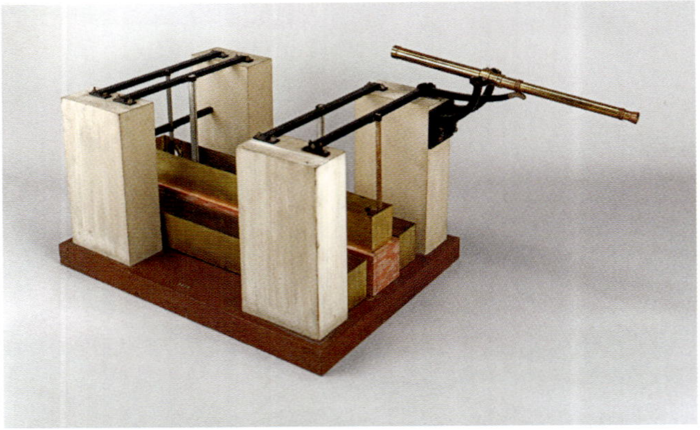

FIGURE 33 (inv. no. 03069-0000-)
© MAM/PHOTO FRANCK BOTTÉ

crosspieces resting on the pillars support the metal bar to be tested in the vessel. One end of the bar touches a third crosspiece; the other end is in contact with an arm connected to a rotating axle. This is hinged on the top of two pillars. A lever fixed to the axle touches a telescope that pivots on two supports fixed to one of the pillars.

The vessel was filled with water, first mixed with ice and then brought to different higher temperatures. The dilatation of the test bar moved the telescope, with which one could observe a divided scale placed 100 or 200 *toises*[25] away. Thanks to optical lever system, one could measure the slightest variations in the bar's length (a few hundreds of line).

The original instrument, presented at the Académie royale des sciences on 22 December 1781, was eventually used from January to September 1782. Lavoisier and Laplace performed experiments on the linear dilatation of bodies (made of various kinds of glass, copper, iron, steel, lead, tin, silver, gold and platinum) in the garden at the Arsenal. Although the results were never published during Lavoisier's lifetime, they were quite well known and Lalande reported on them in 1795. Lavoisier was about to publish them in 1793 when his *Mémoires de physique et de chimie* were in press at Dupont's printing house. Jean-Baptiste Biot, a frequent guest of Marie Anne Lavoisier's salon in rue d'Anjou, had the privilege of seeing the manuscript notes on the experiments and the drawing of the instrument. He provided a detailed description of both in the first volume of his *Traité de physique* (1816). The original instrument seemed to have been lost by then.

Lalande (1795), p. 185; Lavoisier (1805), vol. 1, pp. 246–280; Biot (1816), vol. 1, pp. 146–158; LO, vol. 2, pp. 739–764 and 776; LO, vol. 6, pp. 711–712; Ganot (1894), pp. 604–605

25 Until 1812, a French *toise* was equal to ca. 1.95 m.

Optics

20091-0000- (Fig. 34)
Polygonal anamorphic mirror
Second half of 18th c.
MAM accession date: 1952
27.5 × 11.3; 1.51 kg
Wood, speculum metal

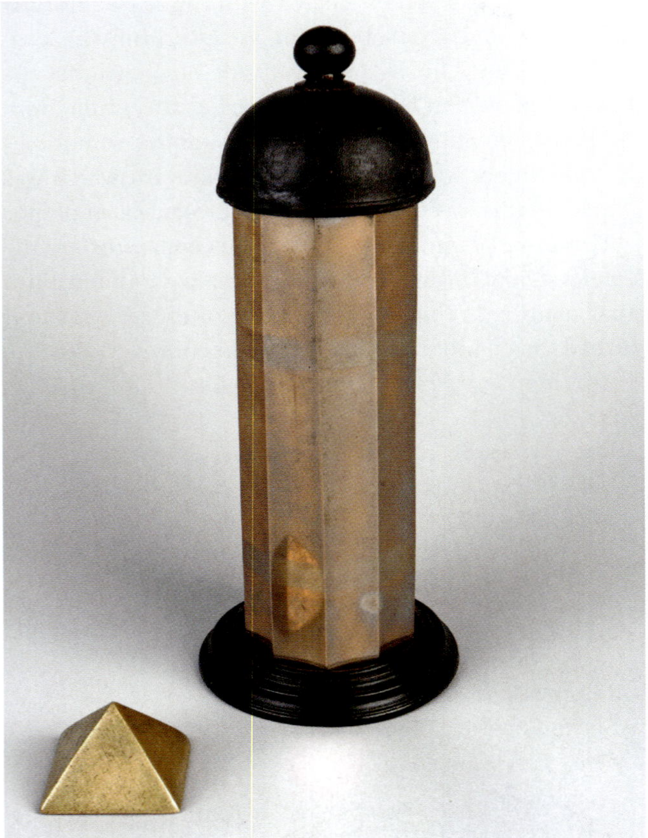

FIGURE 34 (inv. no. 20091-0000-)
© MAM/PHOTO FRANCK BOTTÉ

Five polished metal plates are mounted on a wooden pillar with a cap. The plates form a polygonal mirror, used to observe an anamorphic image (anamorphosis). This was a distorted drawing that could be properly interpreted only from its reflection in the mirror. The wood of the instrument is in poor condition.

Anamorphoses were very popular optical curiosities since the 17th century and were common in physics cabinets at least until the late 19th century.

Niceron (1638) pp. 83–99

20092-0000-, 20226-0000- (Fig. 35)
Semi-cylindrical anamorphic mirror
Second half of 18th c.
MAM accession date: 1952
28.5 × 9.6; 1.18 kg
Wood, speculum metal

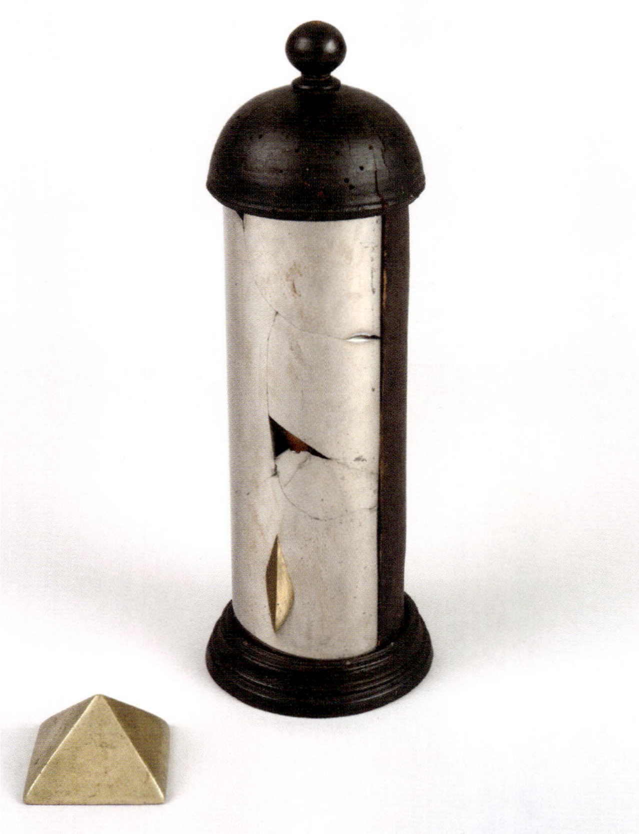

FIGURE 35 (inv. nos. 20092-0000-, 20226-0000-)
© MAM/PHOTO FRANCK BOTTÉ

On a wooden pillar with a cap there was a semi-cylindrical metal mirror. The mirror was broken into seven pieces (originally five, marked from 1 to 5) and was catalogued separately from its stand with the number 20226-0000-. This device is very similar to the one with the inventory number 20991-0000-.

Niceron (1638), pp. 83–99; De Clercq (1997), pp. 114–115

20088-0000- (Fig. 36)
Equilateral prism
Second half of 18th c.
MAM accession date: 1952
35.7 × 5 × 4.2; 950 g
Glass

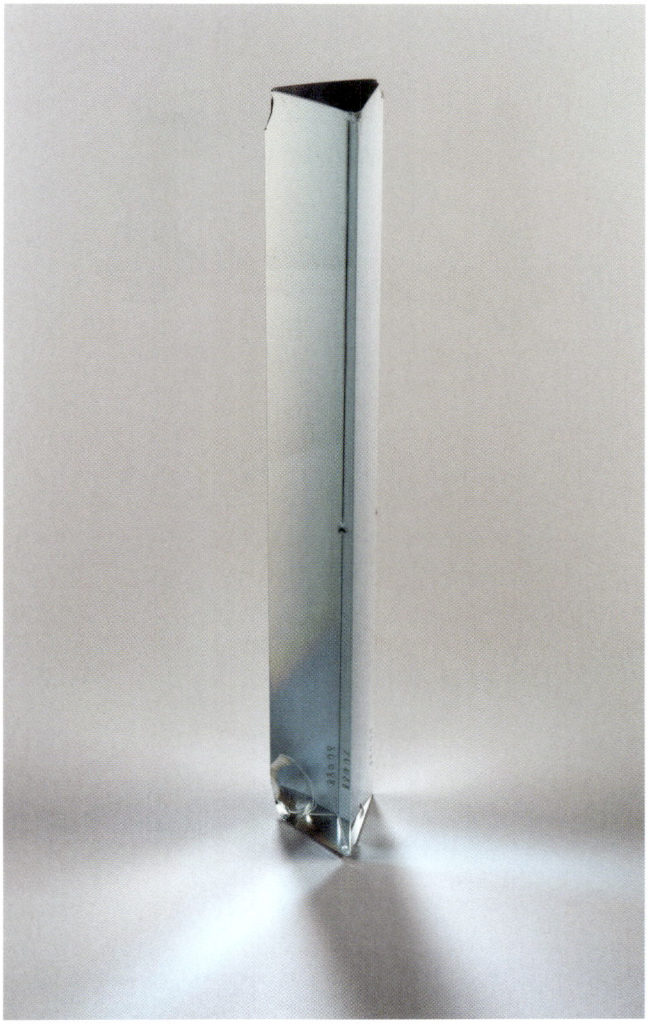

FIGURE 36 (inv. no. 20088-0000-)
© MAM/PHOTO FRANCK BOTTÉ

Near-equilateral prism made of slightly greenish glass with a few chips.

20089-0000- (Fig. 37)
Prism with stand
Second half of 18th c.
MAM accession date: 1952
33.5 × 23.5; diam. of the base: 13; 1.1 kg
Wood, brass, glass, lead

Horizontal equilateral glass prism with brass caps at its ends. The short axles of the caps pivot in a wooden frame. The frame is attached with a brass hinge to the top of a wooden column mounted on a circular base. The base, which is in poor condition, is loaded with lead weight.

This adjustable prism was used to decompose sunlight and produce its spectrum.

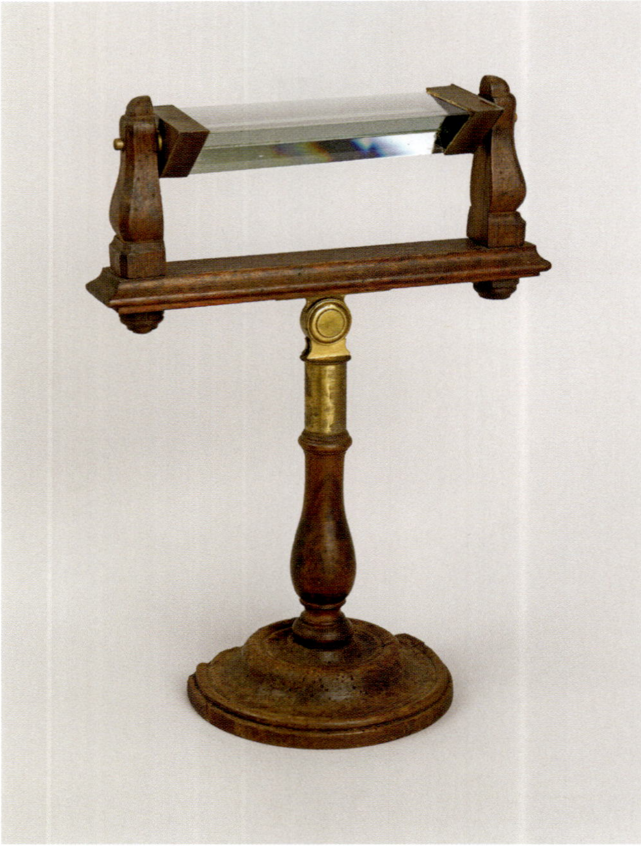

FIGURE 37 (inv. no. 20089-0000-)
© MAM/PHOTO FRANCK BOTTÉ

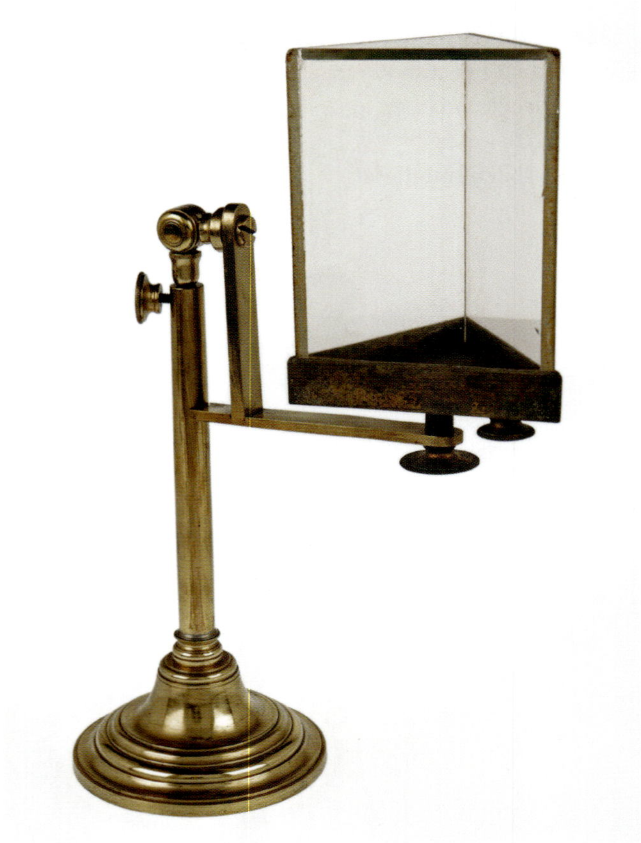

FIGURE 38 (inv. nos. 20043-0000-, 20090-0000-)
© MAM/PHOTO FRANCK BOTTÉ

Sigaud de La Fond (1784), vol. 2, p. 287, and plate XIX, fig. 3; De Clercq (1997), pp. 99–100 and 103; Turner, Levere (1973), pp. 283–284

20043-0000-, 20090-0000- (Fig. 38)
Liquid prism with stand
Late 18th c.
MAM accession date: 1952
29.5 × 21 × 10 cm; 2.02 kg
Lead, iron, brass, glass

A round brass base with a lead ballast supports a vertical tube. A round telescopic bar with an L-shaped arm attached to a rotating joint is inserted in the tube. The vertical position of the bar can be fixed with a screw. A hollow equilateral prism[26] made of glass plates has a brass base that pivots in the L-shaped arm. The prism base has a hole with a plug screwed into it.

26 The prism was inventoried separately from its stand with the number 20090-0000-.

The prism was filled with a refractive liquid and used to show the refraction and dispersion of light. When large pieces of good-quality optical glass were not available, liquid prisms offered an effective substitute.

Turner, Levere (1973); 282–283; De Clercq (1997), pp. 97

20093-0000- (Figs. 39a, 39b)
Rumford's photometer
Early 19th c.
MAM accession date: 1952
33 × 29 × 22; 3.12 kg
Wood, paper, iron, glass

This instrument, which measures the intensity of light (or the brightness of illuminated surfaces), is based on the principle that a brighter light would cast a deeper shadow. It is a simpler version of the photometer described by Benjamin Thompson (Rumford) in 1794.

A pentagonal wooden box painted in black is supported by a column mounted on a square base. The front face of the box has two lateral openings and a central one covered

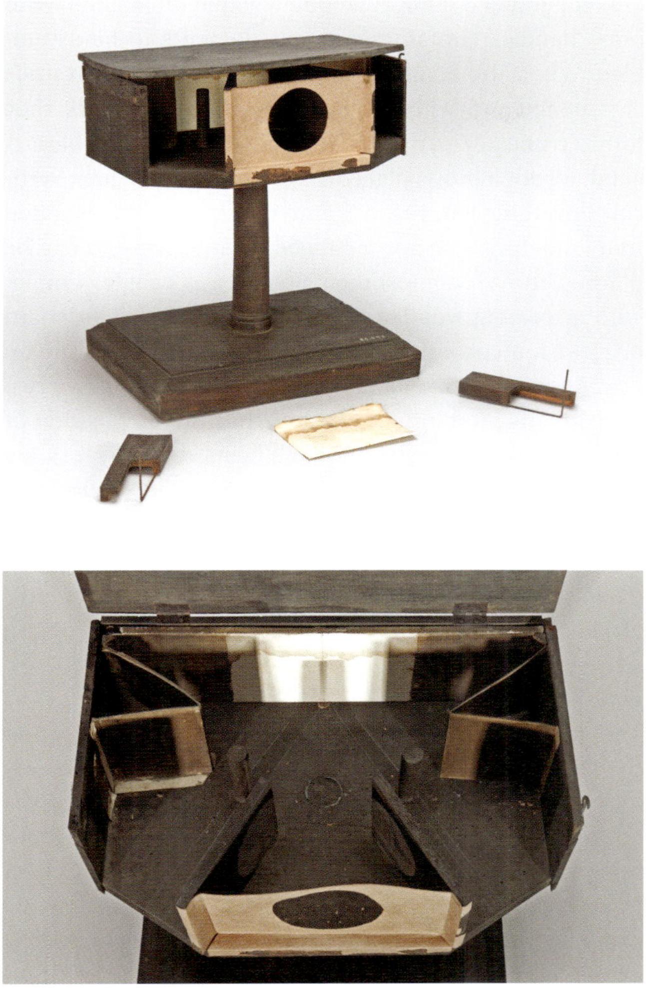

FIGURE 39 (inv. no. 20093-0000-)
© MAM/PHOTO FRANCK BOTTÉ

by a paper with a round hole. Inside the box, on the axis of each lateral opening there is a vertical cylindrical wooden rod and two folded pieces of cardboard act as a screen. The rear face has a glass window also covered with a paper sheet. Two wooden sliders with bent iron wires are joined to the photometer, but their use is unknown.

This photometer makes it possible to compare the relative intensities I_a and I_b of two light sources A and B. These are placed against the photometer's lateral openings. The cylindrical rods cast two shadows – one for each source – on the rear screen. By modifying the distances of A and B from the photometer (d_a and d_b), one could cast two shadows of the same depth. The difference between the two distances made it possible to determine the difference in intensity.

From the inverse-square law[27] of photometry, $Ia/Ib=(d_b/d_a)^2$. If I_a is taken as the reference source, the relative intensity of I_b will therefore be $I_b= I_a\ (d_a/d_b)^2$.
Rumford developed his photometer in order to find an economical and efficient lightning method.

Thompson (1794)

20235-0001- (deaccessioned) McKie 249 (Fig. 40)
Fragment of model of street lamp-holder
ca. 1765
MAM accession date: 1952 but the item was eventually donated to Pierre Dupont
18.2 × 10.9
Silver, iron

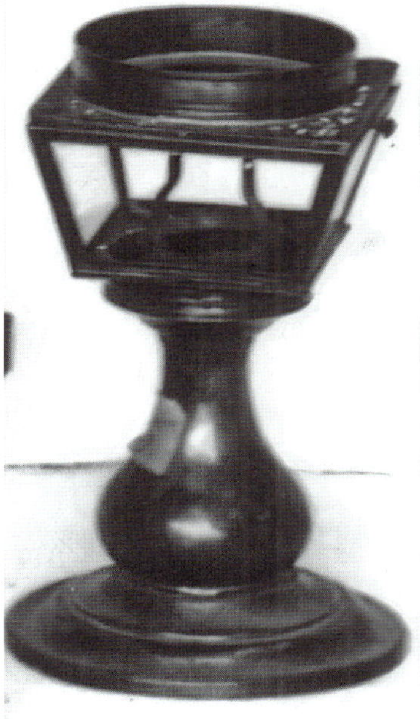

FIGURE 40
(inv. no. 20235-0001- or McKie 249)
PHOTO DOUGLAS MCKIE (1952). COURTESY HAGLEY MUSEUM

Silver model of street lamp-holder with a wrought-iron cage for the lamp. Top cover missing. The window frames are designed for glazing, and one face carries pieces of bevelled-edged glass. One face is hinged for opening and has a knob. The base and the top of the lamp-holder have a perforated design. Label on upright is inscribed "BC" followed by a number, possibly "8." This model was in the collection of Lammot du Pont Copeland.

27 In physics, the inverse-square law states that a given physical quantity or intensity is inversely proportional to the square of the distance from the source of that physical quantity.

20235-0002- (Fig. 41)
Fragment of model of street lamp-holder
ca. 1765
MAM accession date: 1952
5.3 × 4.2; 10 g
Iron

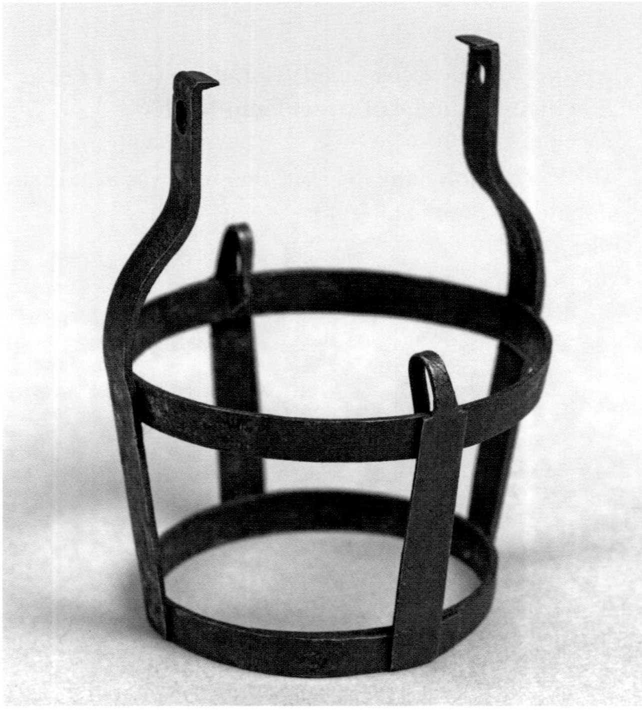

FIGURE 41 (inv. no. 20235-0002-)
© MAM/PHOTO FRANCK BOTTÉ

This basket-shaped iron part is probably a fragment of a model of street lamp-holder. Between 1764 and 1766, Lavoisier proposed several types of lamps to improve public lighting. He participated in the prize competition launched by the Académie Royale des Sciences de Paris, winning a gold medal for his memoir in 1766.

Lavoisier (1765); Bothereau (2018)

20103-0000-, 20050-0000- (Figs. 42a, 42b)
Camera obscura and optical diagonal machine (zograscope)
Late 18th c.
MAM accession date: 1952
95 × 68 × 49; 12 kg
Wood, brass, iron, glass, paper, cardboard, leather

The apparatus is composed of a wooden folding box with brass handles containing all the elements to be assembled. When the lid and two hinged lateral wooden walls are opened and fastened together with small hooks, they form a truncated pyramidal box with a rectangular base. A cubic wooden box is placed on top of it with a large vertical converging lens inserted into the front side. A wooden frame holding a mirror is hinged in the cube and can be placed vertically against the lens or at a 45° angle. A tablet with a rotating wooden ring and a telescopic cardboard tube with a converging lens close the top of the square box. A round mirror in a rectangular frame pivots on two uprights inserted into the tablet. The vertical position of the lens in the telescopic tube can be modified by rotating a screw (20050-0000-) thanks to a brass bar with an universal joint and a wooden handle. A rectangular frame is fixed horizontally on the cubic box by means of a pair of hooks. It originally supported a large piece of black cloth.

The apparatus can be used as a camera obscura or as a zogorascope.

a) Camera obscura

The square mirror was put in the vertical position and the vertical side of the base in front to the operator was removed. A sheet of drawing paper was then fixed on the bottom of the base. The image of a landscape, or a monument, or a person reflected by the round mirror (which was properly orientated) was focused on the drawing paper by the lens. The right focus could be achieved by acting on the rotating handle. The operator partially covered by the black cloth, which masked the parasite light, could observe and copy the image appearing on the paper. This type of camera obscura was very common in the 18th and early 19th centuries and it was widely used by artists and draughtsmen.

b) Zogorascope (or optical diagonal machine)

It is an optical device for enhancing the sense of depth perception from a flat picture. A printed coloured image representing a town, landmark, palace or historical event was placed on the bottom of the base. The square mirror was set at 45°. The image was thus reflected by the mirror through the horizontal lens, causing the picture to appear vivid and almost three-dimensional to the viewer in front of the lens. The optical diagonal machine – of which there were various designs – was an extremely popular optical amusement and an enormous number of engravings to be used with it (called optical views) were printed by several specialised workshops in London, Paris, Vienna, Augsburg and even in a small town such as Bassano in Italy.

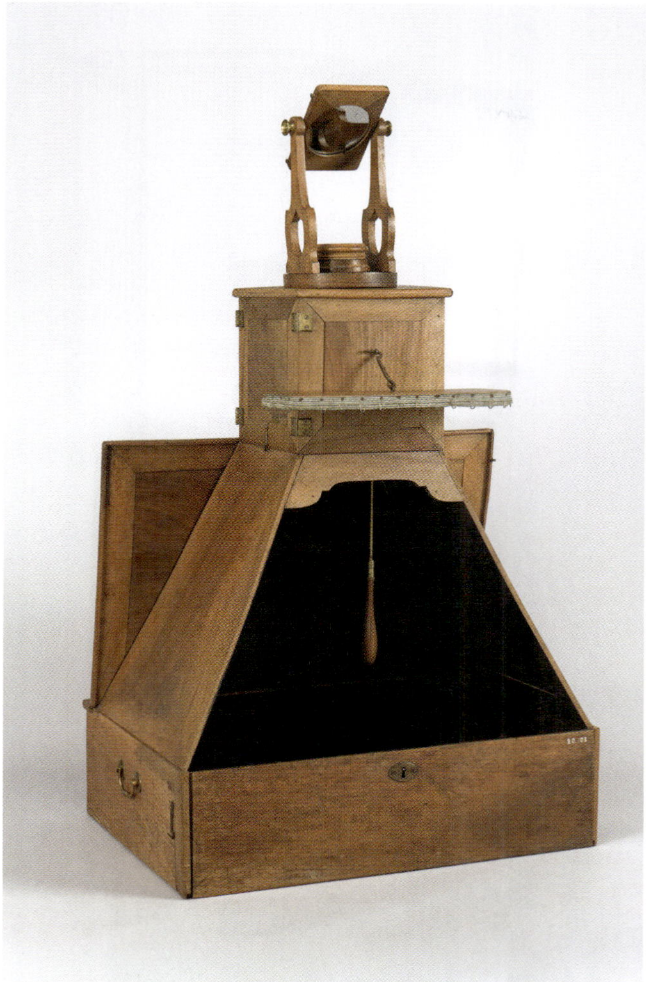

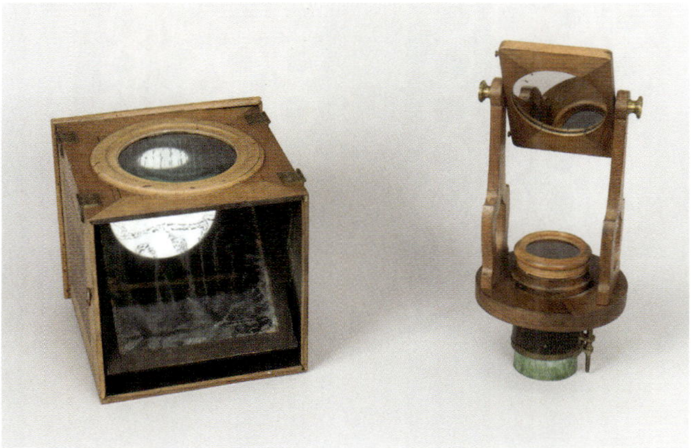

FIGURES 42A, 42B (inv. nos. 20103-0000-, 20050-0000-)
© MAM/PHOTO FRANCK BOTTÉ

Nollet (1743–1764), vol. 5, pp. 532–535, plate VI, fig. 6; Chaldecott (1953); Hammond (1981)

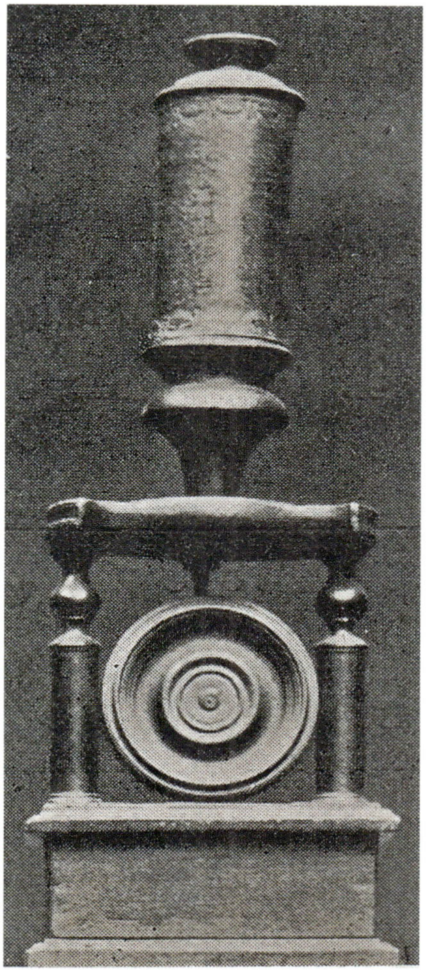

FIGURE 43 Lost microscope of the Muséum National d'histoire naturelle
PHOTO *MUSÉE CENTENNAL* (1900), P. 2

Muséum National d'Histoire Naturelle (Fig. 43)
Microscope
18th c. (?)

In 1918, the second issue of the *Bulletin du Muséum National d'Histoire Naturelle* (p. 1) announced the donation by Madame de La Bassetière of Lavoisier's microscope, displayed "parmi les souvenirs soigneusement gardés dans un pièce spéciale de la galerie de Zoologie". During the 1970s, before the item was either properly catalogued or photographed, it was stolen (communication from Véronique van de Ponseele). This microscope was perhaps the one exhibited at the Musée Centennal de la Chimie in 1900.

Musée Centennal (1900), p. 2

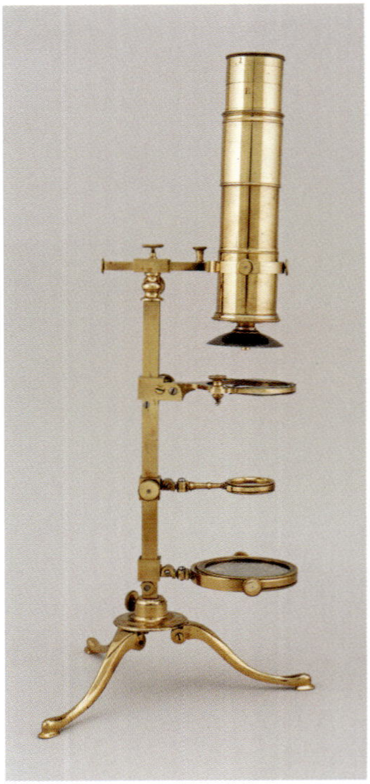
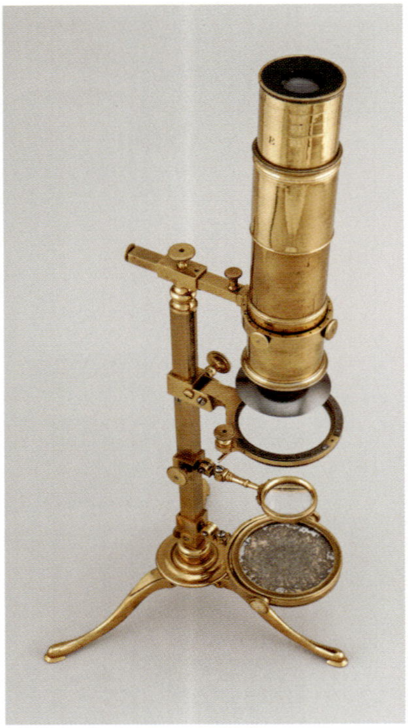
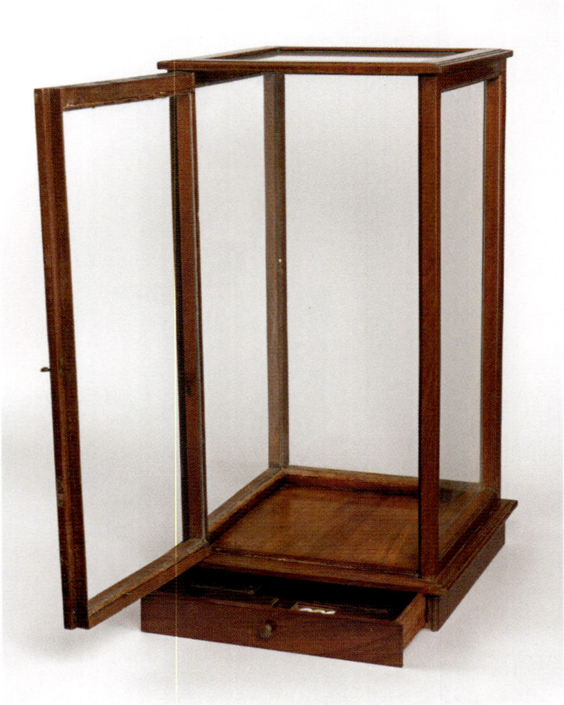
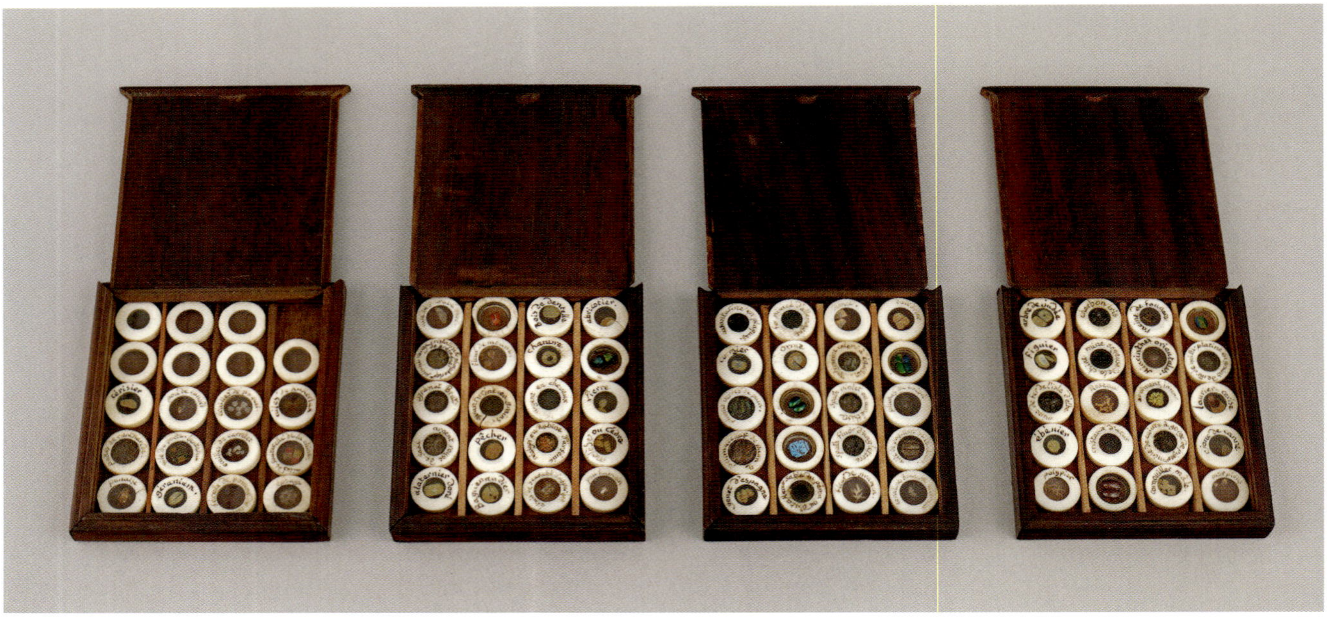

FIGURES 44A, 44B, 44C, 44D (inv. nos. 20085-0001-, 20085-0002-, 20085-0003-)
© MAM/PHOTO FRANCK BOTTÉ

20085-0001-, 20085-0002-, 20085-0003- (Figs. 44a, 44b, 44c, 44d)
Dellebarre's universal compound microscope with glass case and samples
ca. 1778
MAM accession date: 1952
Louis-François Dellebarre (1726–1805)
20085-0001-: 52 × 23.5 × 22.5; 1.92 kg
20085-0002-: (a single box) 9 × 8 × 1.2; 55 g
20085-0003-: 60.5 × 33 × 32.5; 6.25 kg
Brass, glass, silvered glass, wood, ivory

Compound microscope with a folding tripod hinged with a joint to a square pillar supporting the tube. On the lowest position of the pillar there is an illuminating glass mirror that can be oriented in a gimbal and is attached by a compass joint to a slider. A biconvex lens fixed to a second slider is used as a condenser to concentrate light on the specimen. The latter was placed on a circular stage hinged to a small box slider on the pillar. A rack-and-pinion mechanism provides vertical translation of the stage for focusing the specimen image. The tube is inserted in a ring (which can be opened) by means of a horizontal arm that slides in a rectangular box hinged at the top of the pillar. Thanks to the arm, the tube can be translated horizontally. The tube is made of sections marked (from top to bottom) "1", "3", "4", "E", "5", "D" and "A." The sections numbered from 1 to 5 form the eyepiece and each (except E) contains a converging lens.[28] Section A has an additional condensing lens, and under it are screwed a single-lens objective (marked "I") and a Lieberkühn reflector for illuminating opaque objects. The eyepiece tube marked "D" has a single lens.

The microscope is preserved in a rectangular glass case (inv. 20085-0003-) mounted on a base with a drawer containing seven flat wooden boxes of microscopic preparations (inv. 20085-0002-).[29] Each box stores approximately 20 mineral and organic specimens from different European countries mounted on ivory rings between two mica plates. The names of the specimens are handwritten on the ivory rings. One box contains only empty rings.

While unsigned, the microscope but was certainly made by Louis-François Dellebarre (1726–1805), one of the best French opticians of his time. He introduced a "universal microscope" in 1770 and Lavoisier purchased this instrument in 1778. The typical construction with a composite eyepiece (with up to 5 lenses) and a single low-power lens in the objective was based on the erroneous concept[30] that such a combination could correct optical aberrations. Even though it was not achromatic, this type of microscope won a substantial reputation in the late eighteenth century and was considered optically superior to those of some of the best English makers. However, its mechanical construction was defective, and – despite a strong magnification – 19th-century tests revealed that it had a poor resolving power.

Beginning in 1765 with his analysis on gypsum, Lavoisier used different types of microscopes for his research on the nature of minerals, his experiments on the combustion of diamonds and precious stones, his experiments on respiration and his work at the Commission of Weights and Measures. However, the microscope does not seem an instrument to which he paid his usual attention.

Dellebarre (1777); Marivetz, Goussier (1783), vol. 3, pp. 304–328; Andrieu (1937); LC, vol. 3, pp. 631–632; Clay and Court (1975), pp. 204–207; Daumas (1989), pp. 251–252; Turner (1989), pp. 202–204; Fournier (2003), p. 178

Magnetism and Electricity

20049-0000- (Fig. 45)
Magnetic bars
Second half of 18th c.
MAM accession date: 1952
31.5 × 7.5 × 2.5; 945 g
Steel, wood, brass

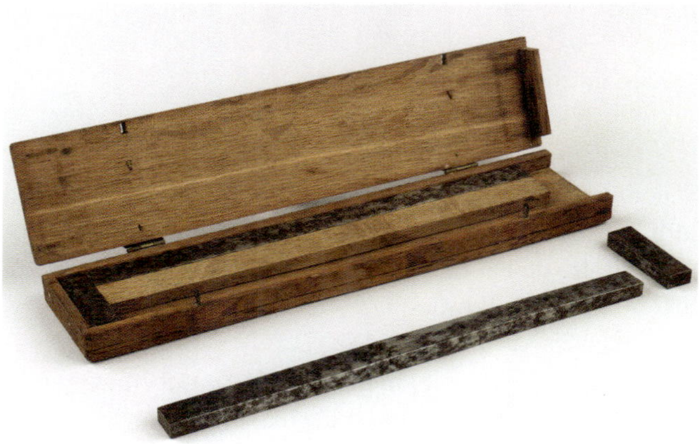

FIGURE 45 (inv. no. 20049-0000-)
© MAM/PHOTO FRANCK BOTTÉ

Two long rectangular magnetic bars with two shorter transversal bars are contained in a wooden box with a lid held by brass hinges. The poles are marked S and N although, owing to the rust covering the bars, only one S is clearly visible. The longest bars are separated by a rectangular piece of wood.

This rectangular arrangement, which preserves the bars' magnetism more effectively, was proposed by the English scientist and inventor Gowin Knight (1713–1772). Knight conducted extensive research on magnetisation. He was able to construct powerful magnets and significantly improved magnetic compasses.

Knight (1744 and 1746); Sigaud de La Fond (1784), vol. 2, pp. 421–423 and plate XXVIII

28 Section 2 with its converging lens is probably missing, as are the forceps for the specimen and perhaps other accessories that came with the instrument.
29 When the microscope was acquired, there were eight boxes.
30 Derived from Leonhard Euler's theoretical work on optical aberration.

20053-0001-, -0002-, -0003-, -0004-, -0005-, -0006-, -0007-, -0008-, -0009-, -0010-, -0011- (Fig. 46)
Set of 11 magnets
Late 18th c.
MAM accession date: 1952
Iron

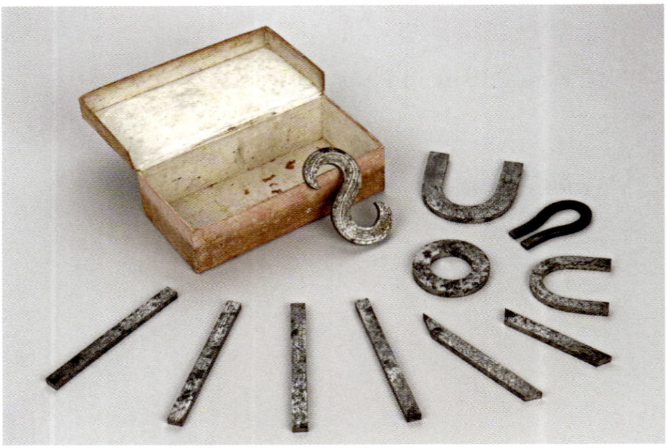

FIGURE 46 (inv. nos. 20053-0001-, 011-)
© MAM/PHOTO FRANCK BOTTÉ

Set of 11 iron magnets of different sizes and shapes, housed in a cardboard box with two labels bearing the inscriptions "Boite de F.d Bertoud"[31] and "1790". The magnets are inventoried as follows:

20035-0001-: U-shaped; 10.8 × 10.4 × 0.8 cm, 335 g
20035-0002-: elongated U-shaped; 9.5 × 5 × 0.6 cm; 105 g. Engraved "N"
20035-0003-: U-shaped; 8 × 7.8 × 0.7 cm; 145 g
20035-0004-: S-shaped; 12 × 6.5 × 0.8 cm; 230 g
20035-0005-: ring-shaped; diameter (ext.) 8 cm, thickness 1 cm; 245 g
20035-0006-: rectangular; 16.2 × 1.6 × 0.8 cm; 145 g. Engraved "N"
20035-0007-: rectangular; 16.2 × 1.5 × 0.8 cm; 145 g
20035-0008-: rectangular; 16.2 × 1.5 × 0.8 cm; 155 g
20035-0009-: rectangular; 16.2 × 1.4 × 0.8 cm; 140 g
20035-0010-: trapezoidal; 15 × 1.5 × 0.8 cm; 130 g
20035-0011-: trapezoidal; 11.7 × 1.5 × 0.8 cm; 90 g

The set was shown at the 1943 exhibition at the Palais de la Découverte.

Lavoisier (1943)

31 Probably Ferdinand Berthoud (1727–1807), the famous clock- and chronometer-maker who was Horloger Mécanicien du Roi et de la Marine.

20105-0000- (Fig. 47)
Coulomb declinometer
1792
MAM accession date: 1952
Nicolas Fortin (1750–1831)
58 × 32.5 × 27.5; 17.39 kg
Marble, steel, copper, wood, glass

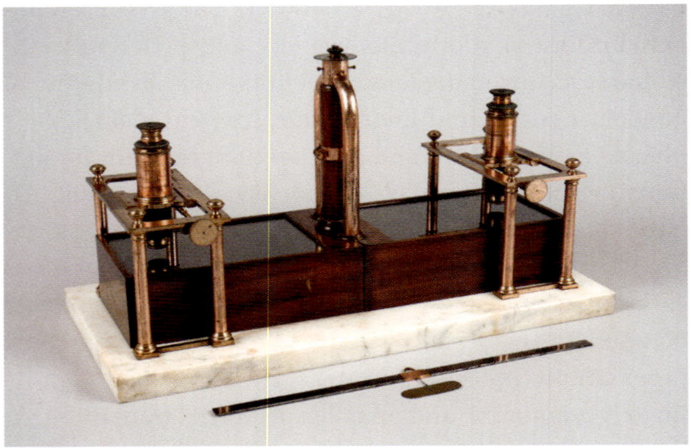

FIGURE 47 (inv. no. 20105-0000-)
© MAM/PHOTO FRANCK BOTTÉ

This instrument served to determine declination (also called magnetic variation), i.e., the angle between the magnetic meridian and the geographic meridian.

The declinometer is mounted on a marble slab. At its centre a ∩-shaped copper support carries a dial with a rotating knob, where originally a torsionless bundle of silk threads (missing) supporting the magnet was attached.[32] The latter consists of a long thin steel bar with a copper suspension ring at its centre and a light rectangular brass paddle attached under it. The paddle acted as hydrodynamic brake.[33] The suspension threads were protected by a wooden pipe and the magnetic needle is enclosed in a wooden box with two glass windows. Two identical compound travelling microscopes with a crosswire allowed the user to observe the longitudinal lines marked on the ends of the magnetic bar. Each microscope can be translated in a frame supported by four copper columns. The microscopes' positions can be adjusted by a pair of micrometric screws and can be read on two divided scales. These are 2 inches long and divided into 1/5ths of a line. Every line is marked (from 12 to 0 to 12). A nonius on the base

32 Depending on the type and weight of the magnetic needle, Coulomb used different materials for the suspension thread, such as silk, silver and steel.

33 The paddle, which was probably immersed in a small vessel (missing) filled with water, dampened the magnet's oscillations.

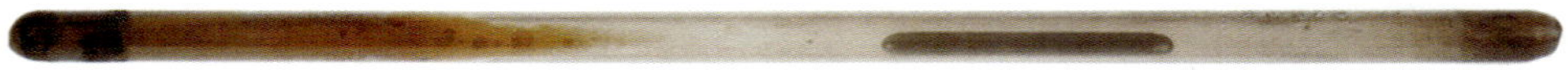

FIGURE 48
(inv. no. 20029-0000-)

of each microscope offers a reading precision of 1/100th of a line.[34]

There are two inscriptions "Fortin Place de la Sorbonne à Paris 1792" near the scales.

The declinometer was used to determine the magnetic meridian by measuring the angle between the direction of the magnet at rest[35] and a line representing the geographic meridian. The instrument could also be used to measure the variations in declination by periodically observing the magnet's position. Thirdly, the declinometer served to measure the intensity of the Earth's magnetic field. For this purpose, after the magnet had been displaced from its equilibrium position, it was necessary to determine the frequency of a given number of oscillations.

The French engineer and physicist Charles-Augustin de Coulomb proposed various magnetic instruments in the 1770s and 1780s. Thanks to his extensive studies on the torsion of wires and threads, he developed his famous electric torsion balance and preferred to use it instead of pivots for suspending the magnetic needles of his declinometers.[36]

Coulomb (1777); Coulomb (1785); Biot (1824), vol. 2, pp. 110–111; Good and Multhauf (1987) pp. 5–14; Basso Ricci et al. (1997), pp. 80–85

20029-0000- (Fig. 48)
"Mercurial phosphorus" tube
ca. 1750
MAM accession date: 1952
36.5 × 1.2; 85 g
Glass, tin, mercury

A closed glass tube whose ends are coated with tin foil contains some oxidised mercury.[37]

The tube, which was evacuated, was probably used to show the electroluminescence phenomenon called "mercurial phosphorus". When the tube is shaken the mercury sealed in it frictions the wall of the tube, generating electrical charges. These produce glowing discharges in the tube's rarefied atmosphere. The phenomenon was first observed in 1676 by the French astronomer and geodesist Jean Picard while conducting experiments on the vacuum space of a barometer in a dark room. It was subsequently studied by several natural philosophers such as Johann Bernoulli and Francis Hauksbee, who thought the glow was caused by electricity.

The tin foils were probably adde later and were used as external electrodes to connect the tube to an electrical machine.

Hauksbee (1705); Diebknecht (1716); Newton Harvey (1957), pp. 271–276

20110-0001-, 20110-0002-, 20110-0003-, 20110-0004-, 20122-0000-[38] (Figs. 49a, 49b)
Plate electrical machine
1786
MAM accession date: 1952
Nicolas Fortin (1750–1831)
113 × 64 × 156; plate diameter 66; 48 kg
Glass, brass, wood fabric, leather, horse hair

This is a typical late-18th-century plate electrostatic generator mounted on a table.[39] On it there is a wooden frame

34 This figure is purely theoretical because it is practically impossible to measure such a length with this system.

35 Coulomb suggested that this line could be materialized by a silver wire passing exactly under the focus of the microscopes (after the magnet had been removed).

36 Coulomb's thread-suspended compasses were adopted by the Paris Observatory in 1777. The type of declinometer described here was largely improved by the instrument maker Henry Prudence Gambey in the early nineteenth century and became quite popular.

37 The tube probably has micro-cracks, and the air that entered it oxidized the mercury.

38 The machine has been inventoried under different numbers: 20110-0001- electrical machine, 20110-0002- electrometer, 20110-0003- original broken disk, 20110-0004-fragments of the column. The number 20122-0000- designates a brass sphere that had been erroneously inventoried separately. In fact it forms part of the machine and has to be screwed onto the prime conductor.

39 In the nineteenth-century French and Italian literature, electrostatic plate machines are often called "Ramsden machines". But the famous English instrument maker Jesse Ramsden (1735–1800) was only one of many 18th-century makers who designed or made very similar plate machines. Lavoisier's electrical

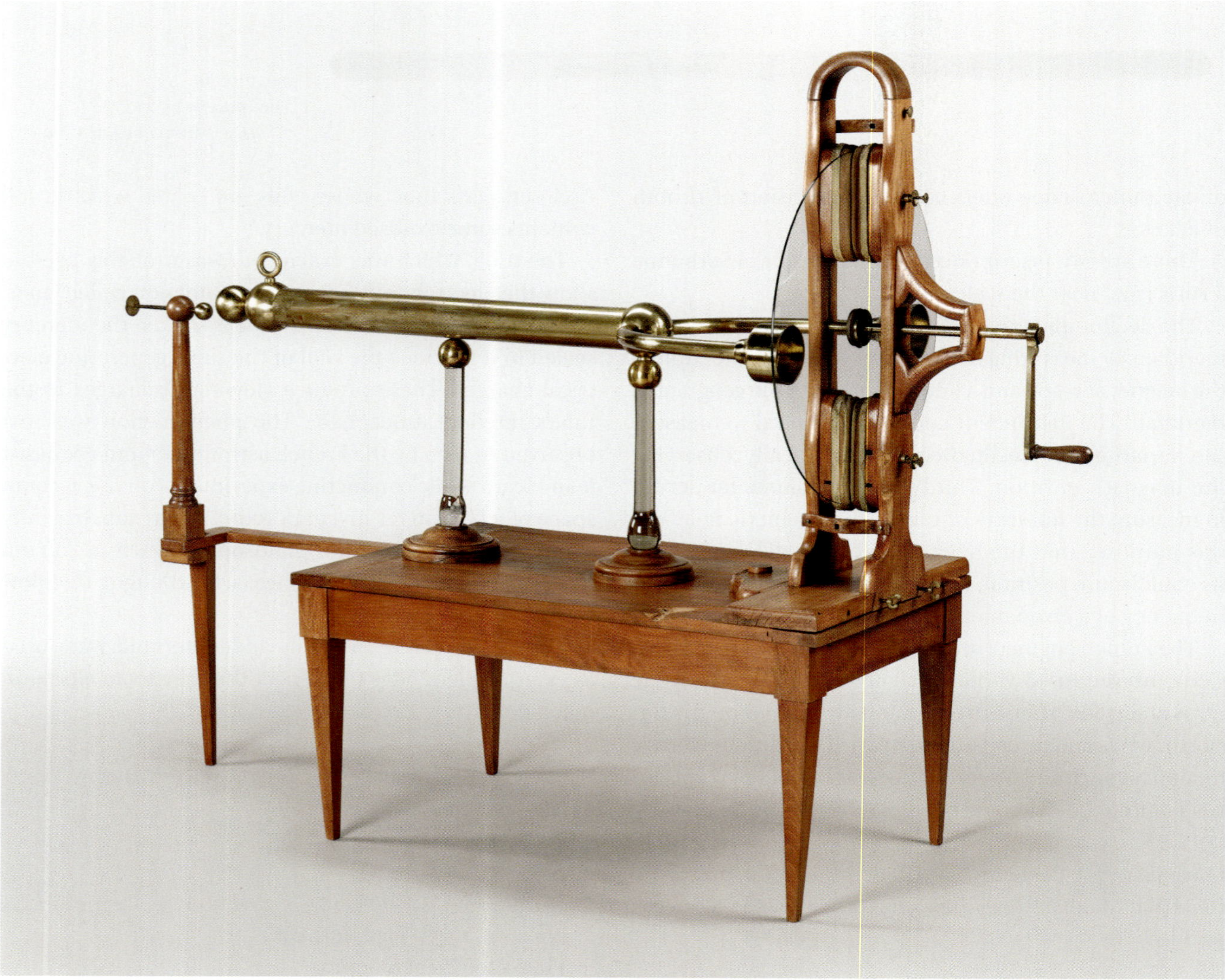

FIGURE 49A (inv. nos. 20110-0001-, 20110-0002-, 20110-0003-, 20110-0004-, 20122-0000-)
© MAM/PHOTO FRANCK BOTTÉ

composed of two uprights jointed at the top by an arc. The glass disk is blocked on a steel axle thanks to a pair of brass flanges. The axle is inserted in the uprights and can be turned with a crank. The frame bears two pairs of leather cushions (stuffed with horsehair) that clamp the plate. The cushion's pressure on the glass can be adjusted by a pair of screws acting on their wooden plates. Two glass pillars with turned wooden bases and brass spheres at the top support the cylindrical prime conductor. This has two bent brass pipes holding the collectors near the plate. Each collector consists of a brass cup with a pointed central spike. The conductor, topped by a ring, also carries a sphere of smaller diameter.[40]

The machine has a spark-gap Lane-type electrometer[41] mounted on a plank sliding under the table and supported by a wooden leg. On it is a column topped by a horizontal micrometric screw. One end of the screw holds a disk carrying a circular scale with 12 divisions; the other end holds a brass sphere. As the screw thread is ½ of a line, each division corresponds to 1/24th of a line.[42] Parallel to the screw is a scale with 40 divisions (lines).

machine is identical to the one described by Sigaud de La Fond (see bibliography).

40 Because the brass sphere was separate from the machine, it has its own inventory number: 20122-0000-.

41 Timothy Lane (1743–1807), an English chemist, became a fellow of the Royal Society. He presented his electrometer in 1766.

42 In fact, the possibility of measuring with a precision of 1/24th of a line with such a device is purely theoretical and not real.

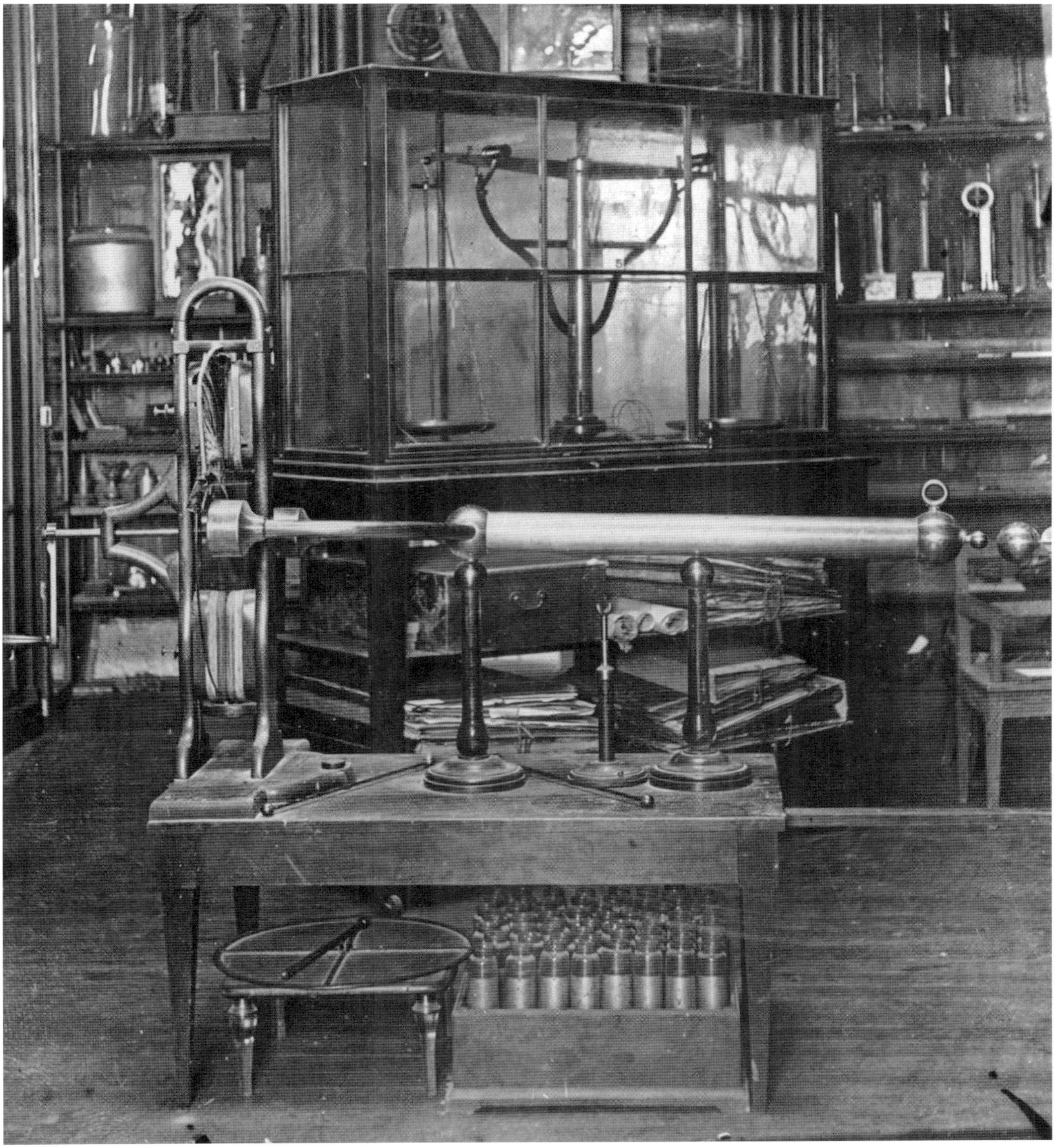

FIGURE 49B Fortin's electrical machine at the Château de la Canière around 1900
COURTESY ARCHIVES DE L'ACADÉMIE DES SCIENCES – PARIS

The electrical machine was used to generate high-voltage static electricity. The friction of the cushion on the rotating glass chargsed the plate with positive electricity. Thanks to the collector, these charges were neutralized and the prime conductor also became positive. The length of the spark between the electrometer's sphere and the conductor provided an estimate of the quantity of electricity produced by the machine and at the same time could limit it.

Lavoisier acquired the machine from Fortin in early spring 1786. In an entry of the 12th volume of his laboratory notebook, Lavoisier set forth the idea that electrical phenomena, like those explaining combustion, were caused by the decomposition of air and that electricity was a very slow combustion. Fortin's machine was used to make experiments in vacuum. The machine was also probably used in the series of experiments for the synthesis of water performed in March 1788 with the new pair of gasometers made by Mégnié. Its spark triggered the combination of hydrogen and oxygen. Fortin's machine was used during Lavoisier's and Séguin's experiments on human respiration (ca. 1790), as attested by one of the sepias made by Marie Anne Lavoisier illustrating them (see Chapter 3, fig. 19). The machine was photographed at the end of the nineteenth century, when still in the possession of Lavoisier's heirs at the Château de la Canière (Fig. 49b).

From the inventories made in 1794, it is clear that Lavoisier owned at least three electrical machines, of which Fortin's was the largest. A smaller one was used during the earliest experiments made with Meusnier de la Place on the composition and decomposition of water in spring 1784 and 1785 (see Chapter 2, fig. 20). There is no trace of this smaller model in later inventories.

The glass plate and a glass column were shattered during the Second World War and were subsequently replaced. The woodwork was also damaged by bullets fired from a British airplane engaged in attacking a German ammunition lorry. The original broken disk, a few fragments of the original glass column and some nails are preserved with the machine.

Ms. Lavoisier, Registre de laboratoire, vol. 12, fols. 39–77; Sigaud de La Fond (1784), vol. 4, pp. 310–322 and pp. 359–362 and plate XX; Daumas (1950); Daumas (1955), pp. 58–59; Hackmann (1978); LC, vol. 4, pp. 212–213; Beretta (2001), pp. 50–52

20121-0000-, 20206-0001- (Figs. 50a, 50b)
Electrophorus
1780s
MAM accession date: 1952
22 × 13.5; 665 g
Brass, glass, resin, wood

The electrophorus (or electrophore) is a simple apparatus for generating electrostatic charges by means of induction. It was described in 1775 by the Italian physicist Alessandro Volta but similar apparatuses were used at about the same time by other physicists such as Giovanni Francesco Cigna, Johann Carl Wilcke and Franz Ulrich Theodor Aepinus.

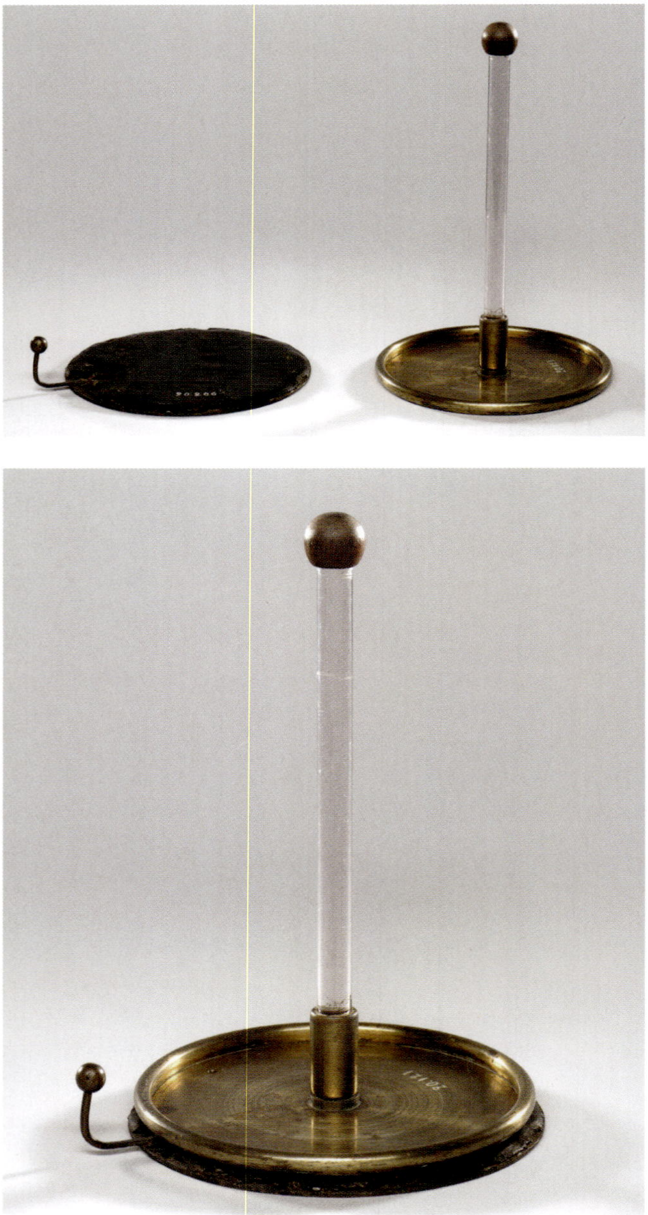

FIGURES 50A, 50B (inv. nos. 20121-0000-, 20206-0001-)
© MAM/PHOTO FRANCK BOTTÉ

The apparatus is composed of a round, resin-coated brass plate with a lateral electrode (dielectric cake)[43] and a brass disk with a glass handle terminating in a wooden knob. The cake[44] is electrically charged by rubbing it with

43 The cake was erroneously separated from the other element of the electrophorus (disk) and inventoried as no. 20206-0001-.

44 Many substances had been proposed for preparing the cake. Volta suggested various mixtures composed of wax, turpentine, rosin, pitch, tar and so on.

a fur or cloth. The disk is then placed over the cake and its electric charges are briefly grounded (for example by touching it with a finger). In this way, the positive charges in the disk are attracted to the side facing down toward the dielectric, charging it positively, while the negative charges are repelled to the side facing up, charging it negatively. The disk remains electrically neutral as a whole. Next, the upper side is briefly grounded (for example by touching it with a finger), draining off the negative charges: the disk is now positive. When it is lifted, the system's capacity drops and the potential difference between the two elements rises. With the charged disk, one can produce a spark or charge a Leyden jar (by repeatedly touching its internal armature with the disk).

The dielectric cake did not lose its charge in the process,[45] which could therefore be repeated many times. For this reason, Volta called his instrument a "perpetual electrophorus".

In March and April 1782, Volta was in Paris to present his instruments, including the electrophorus, at the Académie royale des sciences. During these visits, he also conducted experiments with Lavoisier and Laplace in the Arsenal garden. One experiment involved a condenser of electricity which, according to Volta, allowed the detection of even the smallest discharges of electricity from vaporisation and chemical effervescence. In his reconstruction of the first experiments with the condenser, Volta claimed that he tried, unsuccessfully, to test the instrument earlier in March when he communicated his method to the fellows of the Royal Society, but it was only one month later, in the company of Lavoisier and Laplace, that the experiment finally succeeded. The experiment was performed in the garden of Lavoisier's country residence on April 13. Lavoisier modified Volta's condenser by using a large marble plate on which the condenser disk was laid. An iron rod was attached to the disk, and the rod, in turn, was attached to a metal plate. By heating this plate and manipulating the condenser disk simultaneously, several signs of negative electricity became visible. A few days later, Lavoisier and Volta met again to perform experiments with the condenser on the effervescence produced by several gases, in particular fixed and nitrous airs. The detection of electricity during effervescence and vaporisation led Volta to conclude that electricity, like heat, was an essential and active agent of chemical operations. It was demonstrated that water vapour, for instance, contained a certain amount of positive electricity, leaving the negative one in the body. From this Volta inferred that (1) when bodies became vapours, they acquired a greater capacity to attract electricity and (2) when, after their condensation, they returned to the solid state, they lost their latent heat and electricity, which had served, as a surplus, to their change of state. By generalising this assumption, Volta explained the atmospheric discharge of electricity, such as lightning, as being due to the condensation of vapours.

Another type of electrophorus owned by Lavoisier (but no longer extant) is described by Barthélemy Faujas de Saint-Fond during his visit to the Arsenal on 28 December 1782:

> Monsieur de Lavoisier showed us the electrical experiments performed with Volta's conductor, a sort of electrophorus that makes the slightest degree of electricity highly perceptible. This electrophorus is composed of an upper plate of yellow copper, about twenty inches in diameter with a copper handle, but the lower plate is simply made of white marble. The upper plate is fitted with an insulated iron wire that runs all the way to a courtyard and reaches up to a height of some forty feet. The atmospheric electricity condenses on the plate, and when the plate is raised it produces sparks that are often very powerful; when there is little electricity, it manifests itself by means of M. Cavallo's electrometer.[46]

Volta (1918–29), vol. 3, pp. 87–108; Veau Delaunay (1809), pp. 71–75 and plate X, pp. 117–120; Biot (1016), vol. 1, pp. 564–566, plate VII, fig. 32; Beretta (2001a); Bellodi et al. (2002), pp. 40–51; Beretta (2003); Comparato (2010)

20134-0001-, 20134-0002 (Fig. 51)
Small table with marble disk for an electrophorus
Late 1782 (?)
MAM accession date: 1952
61 × 45.6; 13.01 kg
Tin, wood, brass, marble

45 In fact, the dispersion of the resin's charge of the resin in the atmosphere is very slow, and it was reported that some electrophoruses could maintain their charge for several months.

46 "Monsieur de Lavoisier nous a fait voir les expériences électriques faites avec le conducteur de Wolta [*sic*], c'est une manière d'électrophore qui rend le moindre degré d'électricité très sensible. Cet électrophore est composé d'un plateau supérieur en cuivre jaune d'environ vingt pouces de diamètre avec une manivelle en cuivre, mais le plateau inferieur est simplement en marbre blanc, l'on adapte au plateau supérieur un fil de fer isolé qui aboutit dans une cour et s'élève d'une quarantaine de pieds. L'électricité atmosphérique vient se condenser sur ce plateau et lorsqu'on l'élève il donne des étincelles souvent très fortes ou lorsqu'il y a peu d'électricité elle se manifeste par le moyen de l'électromètre de M. Cavallo". Comparato (2010).

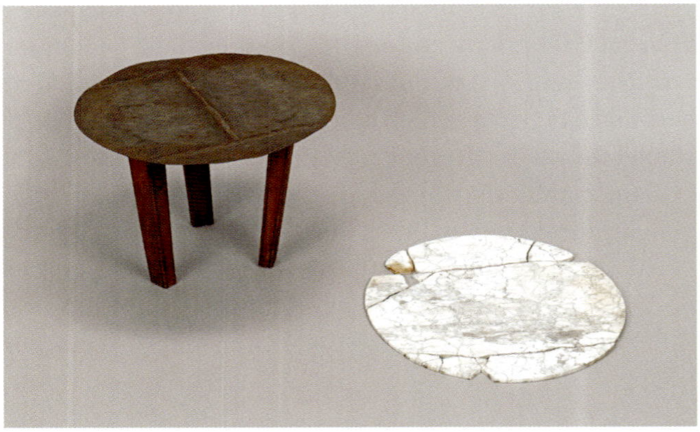

FIGURE 51 (inv. nos. 20134-0001-, 20134-0002-)
© MAM/PHOTO FRANCK BOTTÉ

Round tin plate[47] fixed to a brass ring and supported by three wooden legs. A small hook is fixed to the ring. The table supports a broken marble disk[48] that formed the dielectric "cake" of an electrophorus (see nos. 20121-0000- and 20206-0001-). The metal disk placed on it is missing. This apparatus could be part of the electrophorus mentioned in Faujas' journal, used by Lavoisier to conduct experiments on atmospheric electricity on 28 December 1782 (see full quotation in the preceding instrument description).

Comparato (2018), vol. 2, pp. 207–208

20220-0000-, 20133-0000-, 35367-0002- (Fig. 52)
Metal disk of an electrophorus
Late 18th c.
MAM accession date: 1952
50 × 25; 3.53 kg
Copper, brass, glass

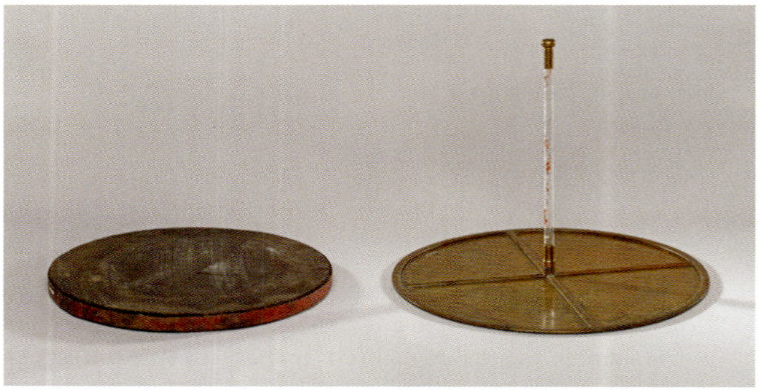

FIGURE 52 (inv. nos. 20220-0000-, 20133-0000-, 35367-0002-)
© MAM/PHOTO FRANCK BOTTÉ

Copper disk with a vertical handle made of glass screwed in its centre.[49] This is the mobile element of an electrophorus (see also no. 20121-0000-).

20116-0002- (Fig. 53)
Leyden jar
ca. 1780
MAM accession date: 1952
18.6 × 9; 215 g
Glass, tin, gold leaf, brass, cork, resin

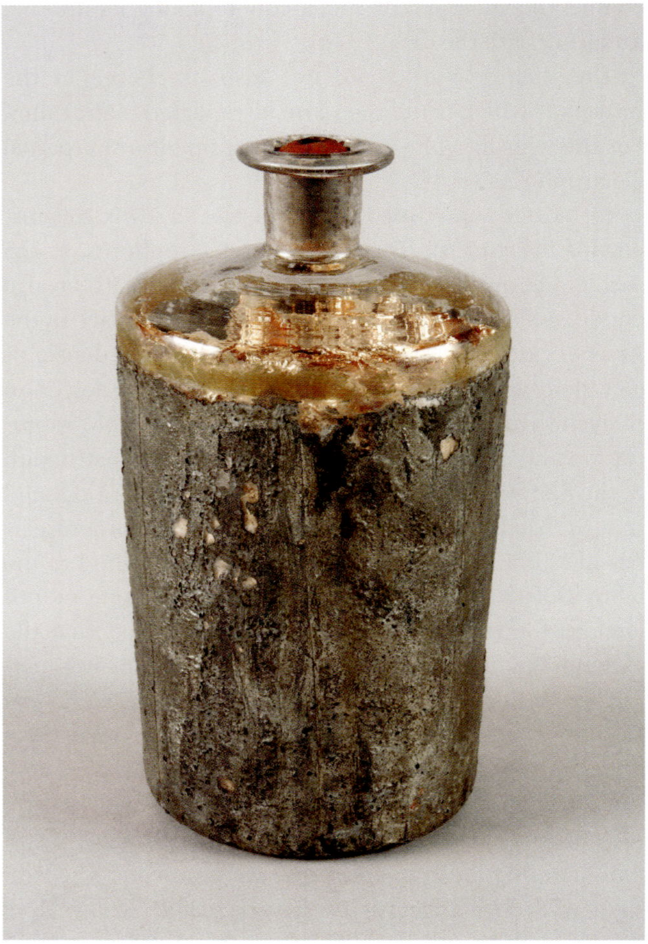

FIGURE 53 (inv. no. 20116-0002-)
© MAM/PHOTO FRANCK BOTTÉ

47 The plate is made of several pieces soldered together.
48 The cake was typically made with a mixture of resins, but marble plates were also used.
49 The handle was wrongly inventoried as a separate item under no. 20133-0000-.

Leyden jar with an external armature consisting of tin foil and an internal armature of tiny gold leaves. The thin neck is closed by a cork but the electrode that originally penetrated into it is missing.

20115-0000- (Fig. 54)
Leyden jar
Second half of 18th c.
MAM accession date: 1952
35.5 × 34; mass 3.9 kg
Glass, tin-foil, sealing wax

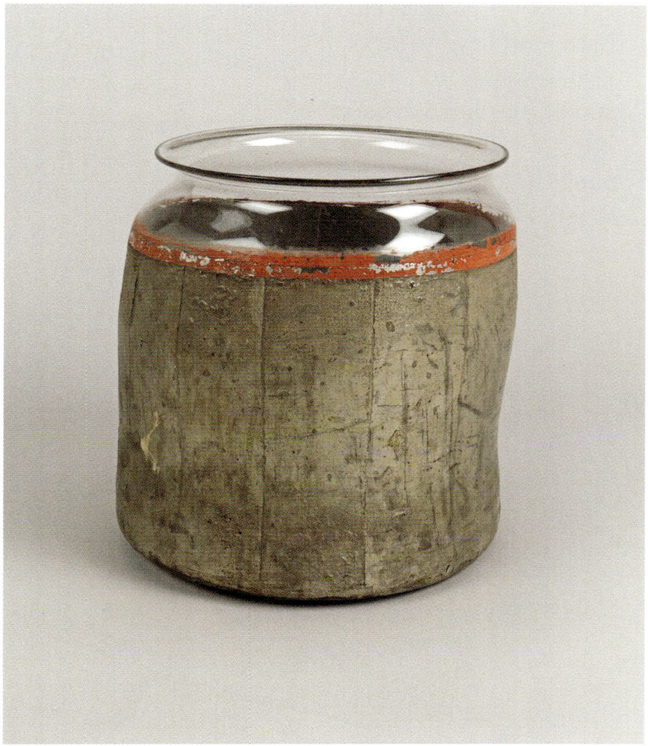

FIGURE 54 (inv. no. 20115-0000-)
© MAM/PHOTO FRANCK BOTTÉ

Large cylindrical Leyden jar with tin-foil armatures. The rim of the foil is coated with red sealing wax to reduce the dispersion of charges. There were probably more jars of this type in the collection.

20116-0003- (Fig. 55)
Leyden jar with bell
Late 18th c.
MAM accession date: 1952
24 × 11; 365 g
Glass, tin, gold leaf, brass, cork, resin

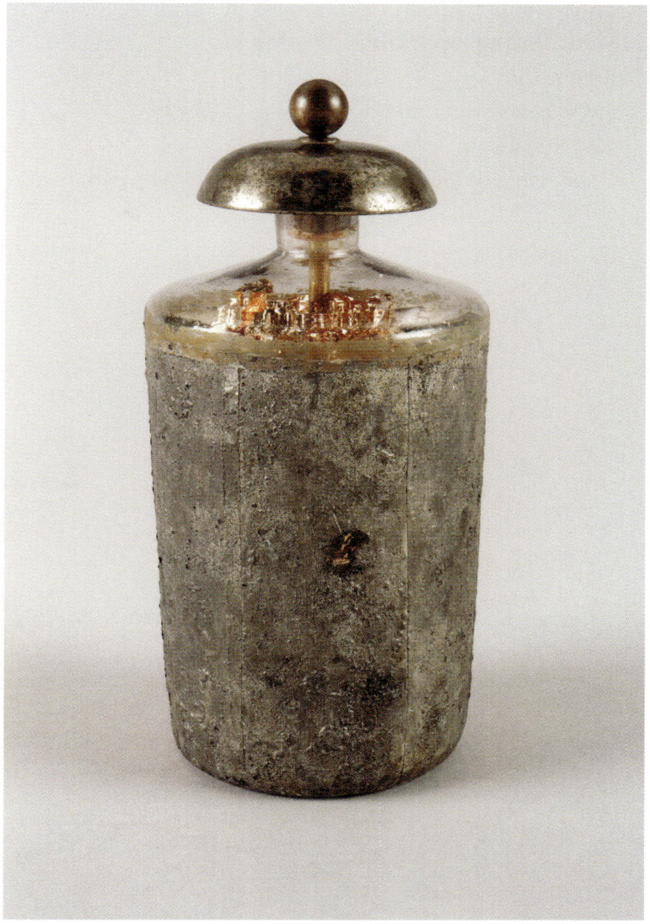

FIGURE 55 (inv. no. 20116-0003-)
© MAM/PHOTO FRANCK BOTTÉ

Leyden jar with an external armature made of tin foil and an internal armature consisting of gold-leaf fragments. The electrode penetrating the jar supports a metal bell and a small sphere. A small hook is fixed to the jar.

This jar is part of an incomplete instrument. Originally, a stand with a second bell and a pendulum with a small metal ball was connected to the jar's external armature.[50]

The pendulum was attracted by the bell in the internal armature. By striking it, the pendulum became electrically charged and was thus immediately repelled towards the second bell. By touching the latter, the pendulum was neutralised and was therefore attracted again by the charged bell. The pendulum's oscillation, which caused the bell to ring, continued until the jar was completely discharged.

Chevalier, Fau (1854), vol. 1, p. 223 and table 23, fig. 254

50 The entire apparatus was probably mounted on an ordinary wooden base.

20116-0001- (Fig. 56)
Scintillating (sparkling) Leyden jar
Late 18th c.
MAM accession date: 1952
24.5 × 8; 240 g
Glass, tin, gold leaf, brass, cork, metallic varnish, resin

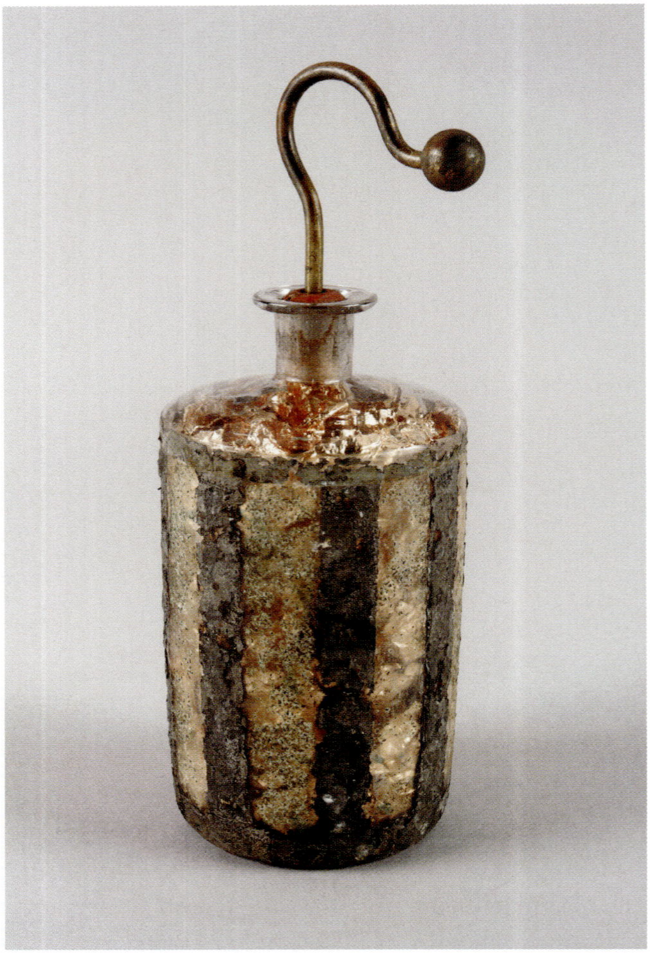

FIGURE 56 (inv. no. 20116-0001-)
© MAM/PHOTO FRANCK BOTTÉ

Special type of Leyden jar. The inner armature consists of pieces of gold leaves. A hooked electrode with a sphere is inserted in the bottle neck through a cork and touches the leaves. The external surface is covered by tin strips (damaged) forming eight vertical windows. The glass on these window sections is partially coated with a granulated varnish, probably containing a metal powder.[51] A small hook fixed at the bottom of the jar probably served to attach a small chain that touched the floor, grounding the external armature.

The jar was charged in the usual way by connecting its electrode with the prime conductor of an electrostatic machine.

When the bottle was discharged, it produced myriad tiny sparks that arched across the granulated metal varnish. The jar sparkled and appeared luminous in the dark.

Veau Delaunay (1809), pp. 60–61 and plate V, fig. 47–48

20114-0000- , 20203-0000- (Fig. 57)
Battery of Leyden jars
1770–1780
MAM accession date: 1952
Box: 56 × 56 × 17; 5.89 kg
Single bottle 27 × 5.6; mass 235 g
Wood, glass, tin foil, iron powder

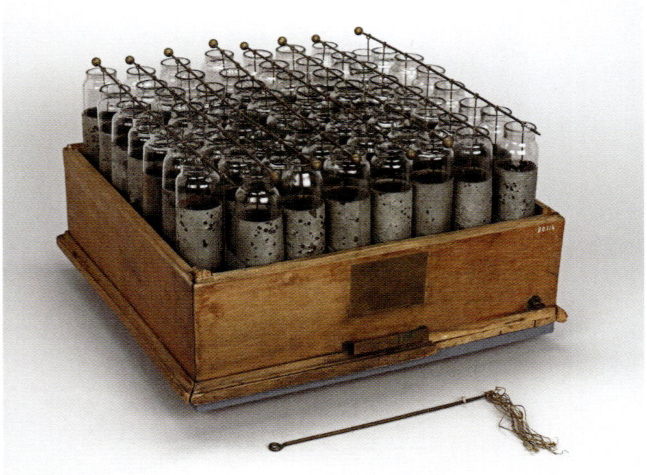

FIGURE 57 (inv. nos. 20114-0000- , 20203-0000-)
© MAM/PHOTO FRANCK BOTTÉ

Battery of 63 cylindrical Leyden jars (one missing) inserted in a wooden box. Their external armatures are made of tin foil, while the internal ones are made of metal powder (iron?) glued to the glass. On each row of eight jars there is a horizontal electrode with a small terminal sphere at each end (two spheres are missing).[52] Every electrode sup-

51 Most of the varnish is missing. This coating was often called "aventurine", but the term could denote several different substances: a variety of quartz characterised by bright inclusions of mica or other minerals; a type of glass with inclusion of brass powder; or a copal varnish mixed with brass powder (fake aventurine). The latter was often used as an armature for Leyden jars or plate condensers to produce spectacular luminous discharges. See *Dictionnaire du commerce et de l'industrie* (Bruxelles: Imprimerie de A. Cauvin, 1837), vol. 1, p. 197.

52 The bars and electrodes were separated from the battery of jars – probably when the instruments were moved in 1952 – and were catalogued under a different inventory number (20203-0000-).

ports eight shorter vertical bars inserted in the jars. Each bar bears a bundle of wire that ensures contact with the jar's internal armature. A ninth extra electrode (also with a bundle of wires) could be used with a single jar.

Priestley (1767), pp. 518–520 and plate 3

20128-0001-, 20128-0002-, 20128-0003- (Fig. 58)
Fulminating panes (Franklin panes)
ca. 1780
MAM accession date: 1952
All: 45 × 36 × 1.5; 0001 and 0002: 975 g; 0003: 220 g
Wood, tin foil, glass, brass

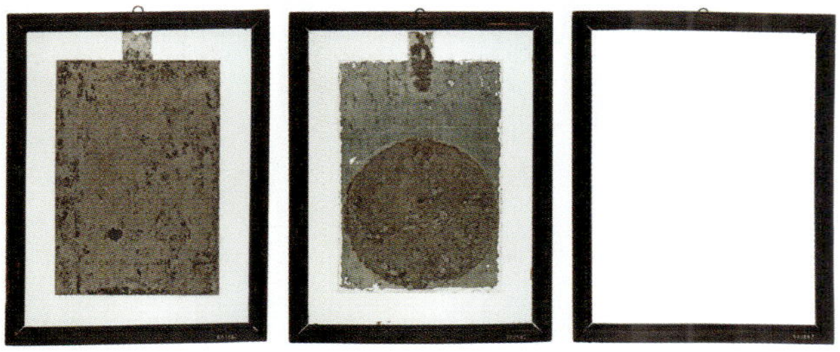

FIGURE 58 (inv. nos. 20128-0001-, 20128-0002-, 20128-0003-)
© MAM/PHOTO FRANCK BOTTÉ

This type of parallel plate condenser is often associated with the name of Benjamin Franklin. It consists of a glass plate inserted in a rectangular wooden frame with a suspension ring. The condenser armatures are made of tin foil applied to the surface of the glass pane. no. 20128-0001- has two rectangular armatures; no. 20128-0002- has a round armature and a rectangular one. The plate and the armatures of no. 20128-0003- are missing; only the wooden frame remains.

Veau Delaunay (1809), pp. 87–89 and plate VI, fig. 67

20112-0000- (Fig. 59)
Electrical discharger
Second half of 18th c.
MAM accession date: 1952
2.5 × 52 × 20; 330 g
Brass

Electrical "horn" discharger composed of two articulated curved brass bars with spheres at the top. Such dischargers were commonly used for discharging Leyden jars (by

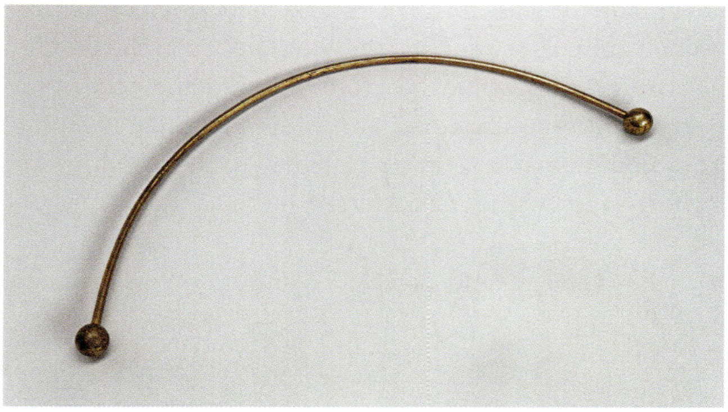

FIGURE 59 (inv. no. 20112-0000-)
© MAM/PHOTO FRANCK BOTTÉ

connecting the two armatures) or in electrotherapy for applying static electricity to a patient.

Sigaud de La Fond (1784), vol. 2, p. 336 and plate XXII; Veau Delaunay (1809), pp. 24–25 and plate III, figs. 11, 12 and 13

20125-0000- (Fig. 60)
Universal discharger
1780s
MAM accession date: 1952
26.5 × 7 × 13; 185 g
Brass, wood, ivory

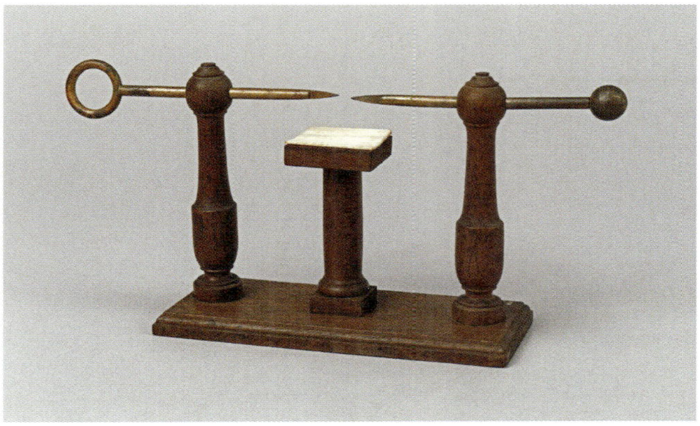

FIGURE 60 (inv. no. 20125-0000-)
© MAM/PHOTO FRANCK BOTTÉ

Two pointed brass electrodes are inserted horizontally (and can slide) at the top of a pair of turned wooden columns. These are fixed to a wooden base. Between them is a third shorter column supporting a small square table covered with an ivory plate. One electrode ends with a sphere, the other with a ring.

This instrument was used to perform various electrical experiments with electric sparks such as perforating a small plate of glass, firing a Volta's pistol or electrocuting a small animal. To produce the spark between the pointed electrodes, one would be grounded and the other connected to the prime conductor of an electrical machine. They could also be simply connected to the armatures of a charged electric jar or battery.

Sigaud de La Fond (1785), pp. 600–601, plate IX, fig. 11; Beretta (2001a)

20119-0000- (Fig. 61)
Insulating stand
Second half of 18th c.
MAM accession date: 1952
36 × 15.5; 1.72 kg
Wood, glass, brass, sealing wax

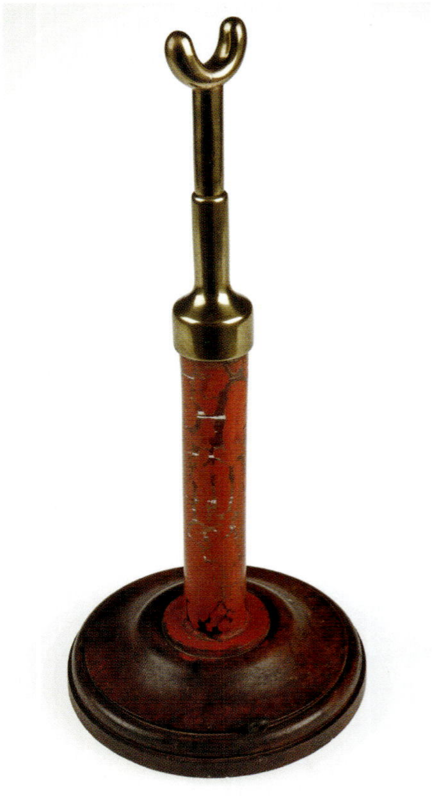

FIGURE 61 (inv. no. 20119-0000-)

A lead-weighted turned wooden base supports a glass column coated with red sealing wax. The column is topped by a brass cap with a vertical bar. A second bar terminating with a U-shaped support is screwed onto the first. This insulating stand was probably used for an electrical experiment.

20127-0000- (Fig. 62)
Insulating stool
Second half of 18th c.
MAM accession date: 1952
42.5 × 43 × 21.5 cm; 5 kg
Wood, brass, glass

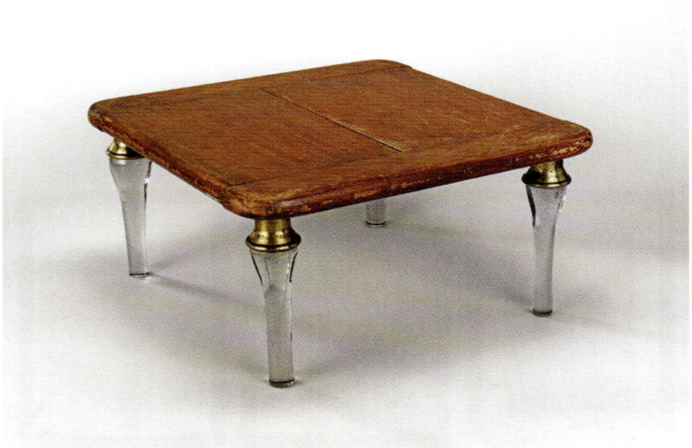

FIGURE 62 (inv. no. 20127-0000-)

This insulating stool has a wooden top and four glass legs. It was used in many electrostatic demonstrations. For example, an experimenter standing on the stool could be charged when he was connected with an electric machine.

Sigaud de La Fond (1784), vol. 2, pp. 312–313 plate XX; Veau Delaunay (1809), pp. 25–26 and plate III, fig. 14–15

20033-0000- (Fig. 63)
Glass rod
Second half of 18th c.
MAM accession date: 1952
Length 57.2, diam. 1.5; 130 g
Glass

Cylindrical glass bar probably used for electrostatic demonstrations. When rubbed with a silk cloth or a fur, it becomes positively charged.

20224-0000- (Fig. 64)
Brass sphere with stem
Late 18th c.
MAM accession date: 1952
13 × 5.5; 180 g
Brass

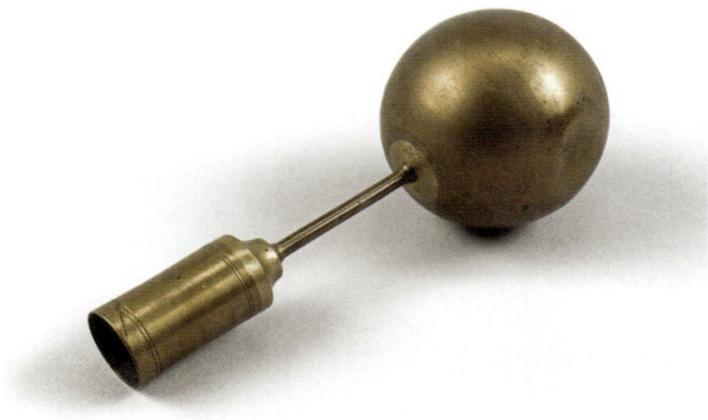

FIGURE 64 (inv. no. 20224-0000-)
© MAM/PHOTO FRANCK BOTTÉ

Hollow brass sphere with a small hole at the top. The sphere is joined to a short stem and a small hollow cylinder, into which a glass rod was probably inserted. The object was used as an insulated conductor for electrical experiments.

20202-0000- (Fig. 65)
Metallic bar with two hooks
Second half of 18th c.
MAM accession date: 1952
43 × 2 × 0.5; 50 g
Copper

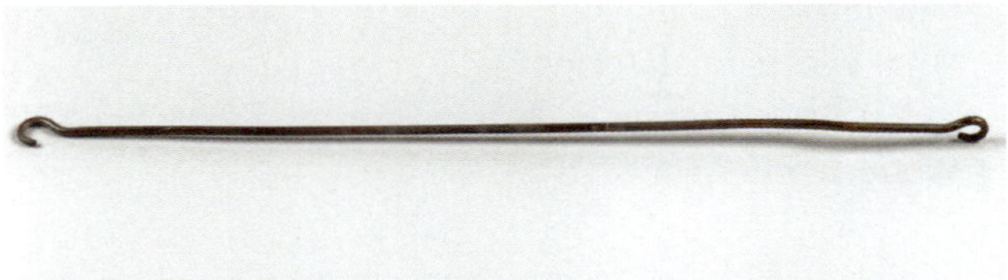

FIGURE 65 (inv. no. 20202-0000-)
© MAM/PHOTO FRANCK BOTTÉ

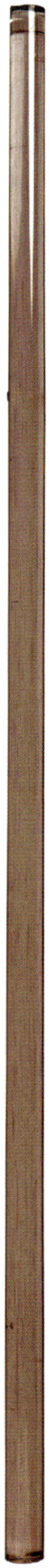

FIGURE 63
(INV. NO. 20033-0000-)
© MAM/PHOTO
FRANCK BOTTÉ

Long copper bar with two hooks at its ends. Probably used as conductor in electrical experiments, for example to connect the prime conductor of an electrostatic machine to a battery of Leyden jars.

Veau Delaunay (1809), pp. 22–23 and plate I, figs. 7–9

20111-0001-, 20111-0002- (Fig. 66)
Electrical conductors
Second half of 18th c.
MAM accession date: 1952
54.5 × 2.5 × 2.5; 300 g
Brass

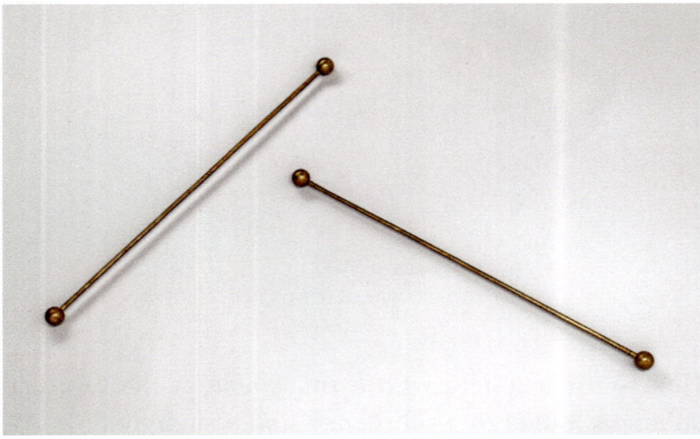

FIGURE 66 (inv. nos. 20111-0001-, 20111-0002-)
© MAM/PHOTO FRANCK BOTTÉ

Two identical brass conductors, each consisting of a thin bar with two small spheres at its ends. Used for electrical experiments.

Veau Delaunay (1809), pp. 22–23 and plate I, fig. 7–9

20174-0000- (Fig. 67)
Bellows for electroscopic powder
Late 18th c.
MAM accession date: 1952
32 × 9, 5 × 5.5; 145 g
Wood, leather, tin

Bellows (identical in design to the common fireplace tool) with a nozzle and a tin bi-conical container. A circular opening in the nozzle could be plugged with a cork. There is a small leather loop for hanging the tool. One of the handles is broken.

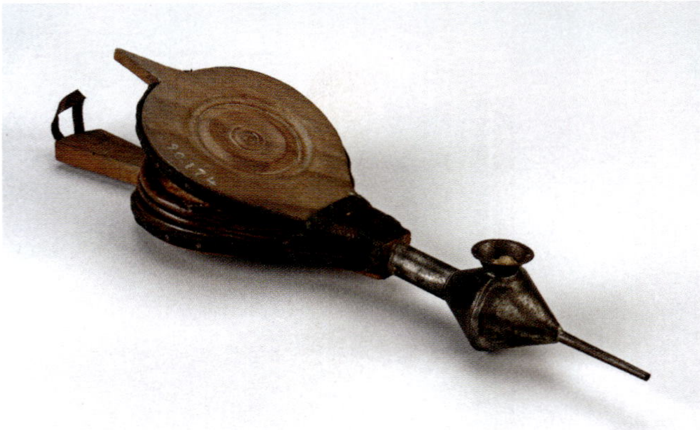

FIGURE 67 (inv. no. 20174-0000-)
© MAM/PHOTO FRANCK BOTTÉ

Probably used to dust electroscopic powder on a plate (made of resin, glass or other dielectric material) after it had been electrified with a machine or a Leyden jar. The powder, placed in the tin container, consisted of a mixture of orange minium (red lead, i.e., lead tetroxide) and yellow sulphur. Sprinkled by the nozzle, the minium tended to acquire a slightly positive charge, while the sulphur became negatively charged. The two powders thus accumulated on the areas of the plate with the opposite polarities, forming what were known as Lichtenberg's figures.[53] These were first observed by the German physicist and satirist Georg Christoph Lichtenberg.

Bennet (1789), p. 42

20118-0000- (Fig. 68)
Electric hail apparatus
1780s
MAM accession date: 1952
32 × 14.5; 1.23 kg
Brass, glass, pith

The bottom of a glass bell is closed by a brass base with a rising collar. The bell neck has a cap with a vertical sliding brass bar. This ends with a horizontal disk inside the bell, which contains several small pith balls.

When the electrode was connected with an electrostatic machine, the balls were alternately attracted and repelled by the disk. The instrument was used to show the

53 Through friction, minium becomes positively charged and sulphur negatively charged, producing the typical globular and fern-like patterns on the charged areas.

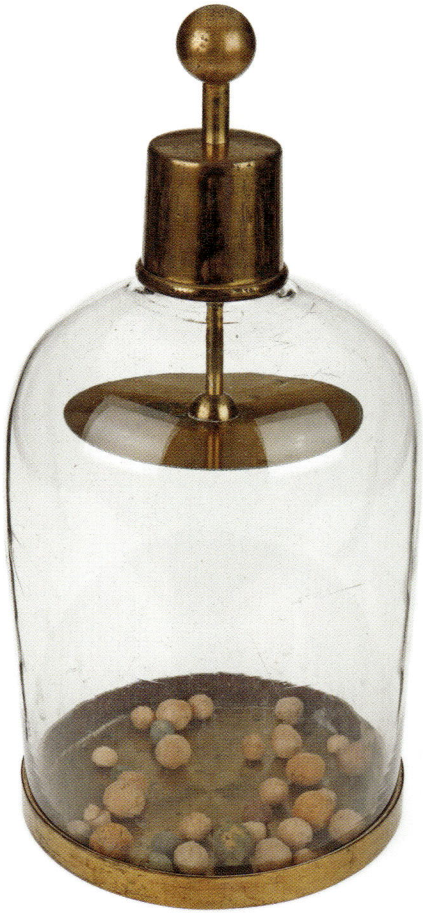

FIGURE 68 (inv. no. 20118-0000-)
© MAM/PHOTO FRANCK BOTTÉ

action of electrostatic force on light bodies and to illustrate the hypothesis of the Italian physicist Alessandro Volta, who supposed that hail had an electrical origin.

Veau Delaunay (1809), pp. 37–38, plate III, fig. 25, Chevalier, Fau (1854), vol. 1; pp. 236 and plate 25, fig. 281

20132-0000- (Fig. 69)
Sparking tube
Late 18th c.
MAM accession date: 1952
26.5 × 2.7; 180 g
Glass, brass, tin

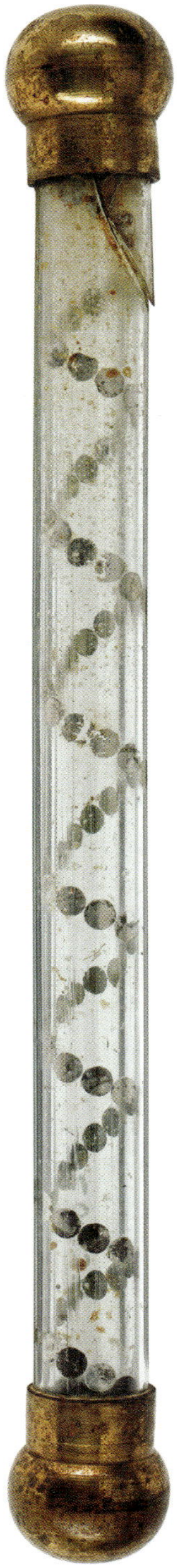

FIGURE 69
(inv. no. 20132-0000-)
© MAM/PHOTO FRANCK BOTTÉ

A series of small spangles of tinfoil forming a spiral are glued inside a glass tube closed by two brass caps.

When one cap is connected to the conductor of an electric machine (or of a Leyden jar) and the other is grounded, a series of small electric sparks arches over the intervals between the spangles and produces a luminous spiral.

Veau Delaunay (1809), pp. 55–56 and plate IV, fig. 42

20052-0000- (Fig. 70)
Press for electrical experiments
Second half of 18th c.
MAM accession date: 1952
12 × 9.5 × 6.5; 160 g
Wood

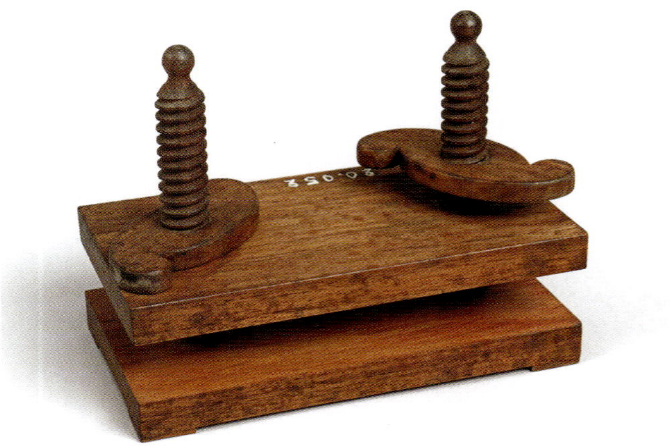

FIGURE 70 (inv. no. 20052-0000-)
© MAM/PHOTO FRANCK BOTTÉ

This instrument is used to show the melting of a very thin foil of gold (or of another metal) with an electrical discharge.

The wooden base has two vertical screws. A wooden plate with two holes is held by the screws and can be pressed against the base by means of a pair of shaped nuts.

A thin gold foil is inserted between two glass plates (or two pieces of cardboard) are clamped in the press. A powerful battery of Leyden jars is discharged through the metal foil, which is instantly vaporised by the electricity. The gold condenses on the glass, leaving a purple halo.

A variant of this experiment called "portrait of Franklin" was very popular in the nineteenth century. The gold foil would be evaporated through a cardboard stencil representing the face of Benjamin Franklin. The condensed metal would reproduce the portrait on a piece of white silk or on a sheet of paper.

Sigaud de La Fond (1784), vol. 2, pp. 384–306 and plate XXV, fig. 6; Veau Delaunay (1809), pp. 133–134 and plate IX, fig. 101

20109-0000- (Fig. 71)
Henley electrometer
George Adams Junior (1750–1795)
ca. 1780
MAM accession date: 1952
16 × 8 × 6.8; 60 g
Brass, ivory, wood

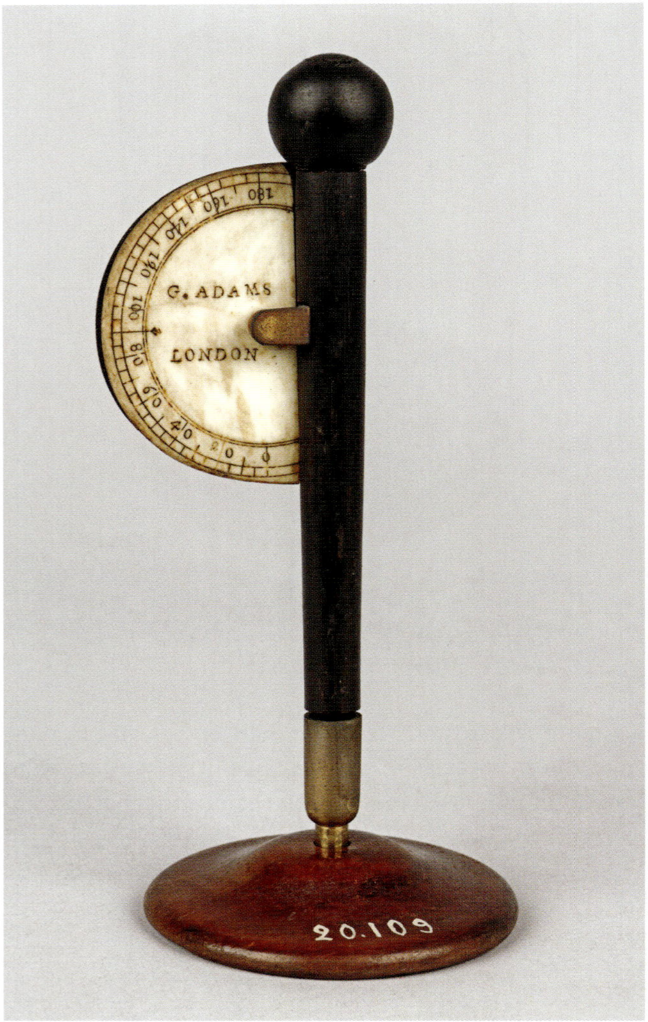

FIGURE 71 (inv. no. 20109-0000-)
© MAM/PHOTO FRANCK BOTTÉ

The electrometer invented in the early 1770s by the Englishman William Henley (?–1779), sometimes called Henley's quadrant, is in fact a single-pendulum electroscope.

The device consists of a round brass base supporting a wooden stem with an ivory (or bone) semicircular scale attached to it vertically. The scale ranges from 0° to 180° with divisions every 5°. A light pendulum made of straw and terminating with a pith ball (missing) was originally hinged in the centre of the quadrant. On it there is the inscription "G. ADAMS // LONDON".

When the base is electrified,[54] the pendulum moves away from the base and the angle of divergence from the vertical roughly indicates the amount of electric charge.

William Henley (1772); Adams (1785), pp. 42–42, plate II, fig. 6; Veau Delaunay (1809), pp. 67–68 and plate V, fig. 52

20031-0006- (Fig. 72)
Pocket electroscope
Late 18th c.
MAM accession date: 1952
11.5 × 0.8; 5 g
Glass, sealing wax, cork, fibre

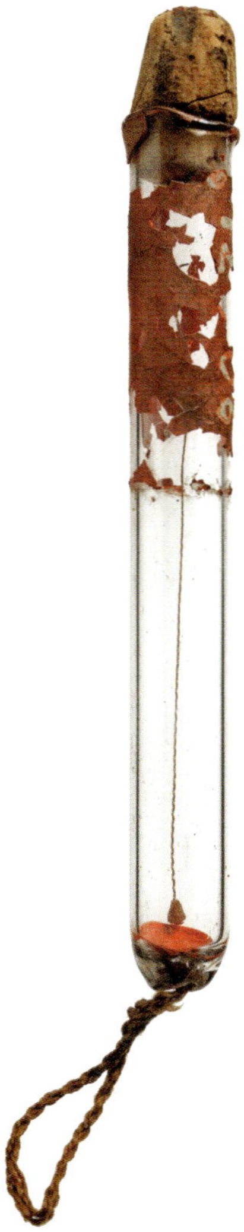

FIGURE 72 (inv. no. 20031-0006-) © MAM/PHOTO FRANCK BOTTÉ

54 These electrometers were often inserted into the prime conductor of an electrostatic machine.

A glass tube partially covered by shellac and closed by a cork contains a very light pendulum. A thread is fixed to one end of the tube.

The instrument was perhaps used as a simple electroscope. Near an electrically charged body, the pendulum charged by influence showed a deflection.

20117-0000-3 (Fig. 73)
De Saussure's pith-ball electroscope
1787
MAM accession date: 1952
Nicolas Alexandre Baradelle (Baradelle l'Aîné), (active ca. 1740–1791)
Electroscope 8.3 × 6.7 cm; 150 g; dome cover 8.5 × 7 cm; 70 g
Brass, glass, sealing wax, silk

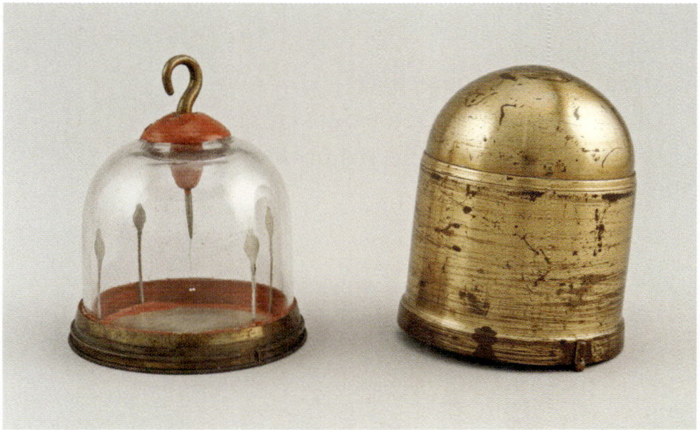

FIGURE 73 (inv. no. 20117-0000-) © MAM/PHOTO FRANCK BOTTÉ

Pith-ball electroscope on copper base in dome-shaped glass vessel. A brass dome protects the instrument when not in use. The underside of the base bears the inscription "Baradelle l'ainé a Paris N.° 1/ 1787".

Four metal strips are placed at equal distances along the inner edge of the glass dome, with corresponding scales marked on the copper base. The strips prevent the persistence of electric charges on the glass. The scales are divided into two symmetrical parts, reading 0 to 6 in both directions, with divisions every half-unit. The top of the glass vessel has a small hook connected to a small vertical bar bearing two thin metal threads that hold the pith balls.

The base is attached to the dome by means of sealing-wax.

The instrument is very similar to the one described by Horace Bénédict de Saussure and used for measuring atmospheric electricity. The pith balls diverged when electrified. For his measures, Saussure connected an open metal ring to the hook with a thin, flexible silver cord from which a small lead ball was suspended. He threw the ball in the air. Thanks to the ring, the ball on its cord would swing away from the electroscope, which remained charged. The ball's divergence indicated the quantity of atmospheric electricity. This type of electrometer was sometimes fitted with a long pointed vertical rod instead of a hook.

Lavoisier conducted experiments on electricity in spring 1786[55] but the known sources do not document any research in 1787–1788. However, it should be noted that many of his laboratory notes compiled after 1786 are lost.

De Saussure (1779–1796), vol. 2, pp. 194–199; Veau Delaunay (1809), pp. 69–70 and plate V, fig. 56; Hackmann (1978)

19999-0000-, 19972-0000-, 20120-0000- (Fig. 74)
Electric lighter
Late 18th c.
MAM accession date: 1952
Dimensions: 55 × 21.5 × 23; 4.4 kg
Brass, glass, wood, zinc, resins, shellac

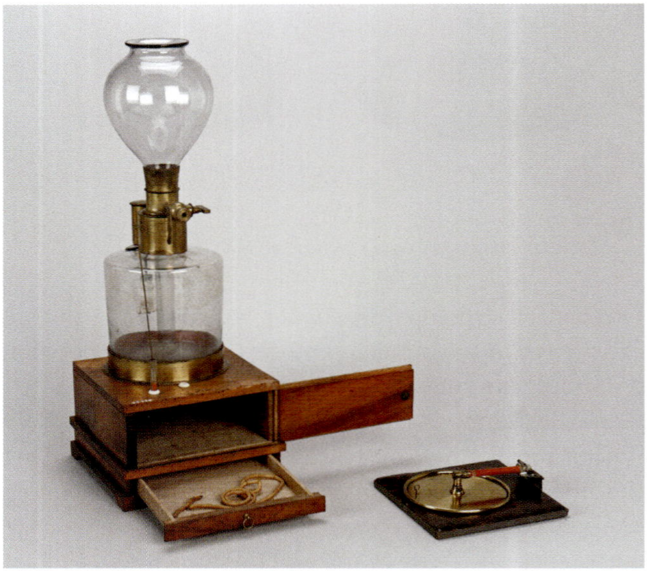

FIGURE 74 (inv. nos. 19999-0000-, 19972-0000-, 20120-0000)
© MAM/PHOTO FRANCK BOTTÉ

Electric hydrogen lighter. A cylindrical glass bottle (no. 19972-0000-) has a central neck with a brass collar and a stopcock with a small lever. A second, smaller lateral neck is closed by a brass lid and a zinc bar hangs from it. A pear-shaped funnel (no. 19999-0000-) is inserted into the collar of the central neck. The bottle is placed with a brass ring on top of a wooden box (no. 20120-0000-).[56] On the front is a compartment with a sliding cover; below it, a flat drawer containing an electrophorus. This is composed of a square flat metal box filled with resin cake and a brass disk on top of it. The disk is held by a glass bar (covered by sealing wax) hinged to one side of the box.

The funnel was filled with diluted sulphuric acid. When the stopcock was opened and the hydrogen contained in the bottle flowed from a nozzle (missing), the acid entered the bottle, partly covering the zinc bar and thus generating more hydrogen. A thread (missing) connected the stopcock lever to the disk. When the stopcock was opened, the disk would thus be lifted from the previously charged resin cake and would touch the bottom end of the vertical conductor inserted in the wooden box and connected to a spark gap (missing). The spark ignited the hydrogen and the flame was used to light a waxed wick, probably held by the spark gap.

Having discovered the "flammable air from the marshes" (methane) in 1776 and having observed that it could be ignited by an electric spark, Alessandro Volta proposed various lamps (or lighters) using methane or hydrogen. Other scientists and inventors also developed many different models. This lighter represents one of the most improved versions of the apparatus, which was widely used until the 1820s, when it began to be replaced by the Döbereiner lighter.[57]

Chevalier, Fau (1854), vol. 1; pp. 240–241 and plate 26, fig. 295; Brenni (2003)

55 MS Lavoisier (1772–1788), vol. 12, fols. 45–51.

56 The apparatus was disassembled and therefore the different pieces were catalogued separately because they were not recognized as belonging to a single instrument.

57 In 1823, the German chemist Johann Wolfgang Döbereiner (1780–1849) discovered that a jet of hydrogen on a platinum sponge catalysed a reaction with atmospheric oxygen. The reaction heated the catalyst and ignited the hydrogen, producing a gentle flame. Döbereiner's lighter was thus simpler because it did not require the presence of an electrophorus.

20126-0000- (Fig. 75)
Spark gap (part of a missing Volta's lighter)
Second half of 18th c.
MAM accession date: 1952
18.5 × 7.5 × 2; 85 g
Brass, glass, sealing wax

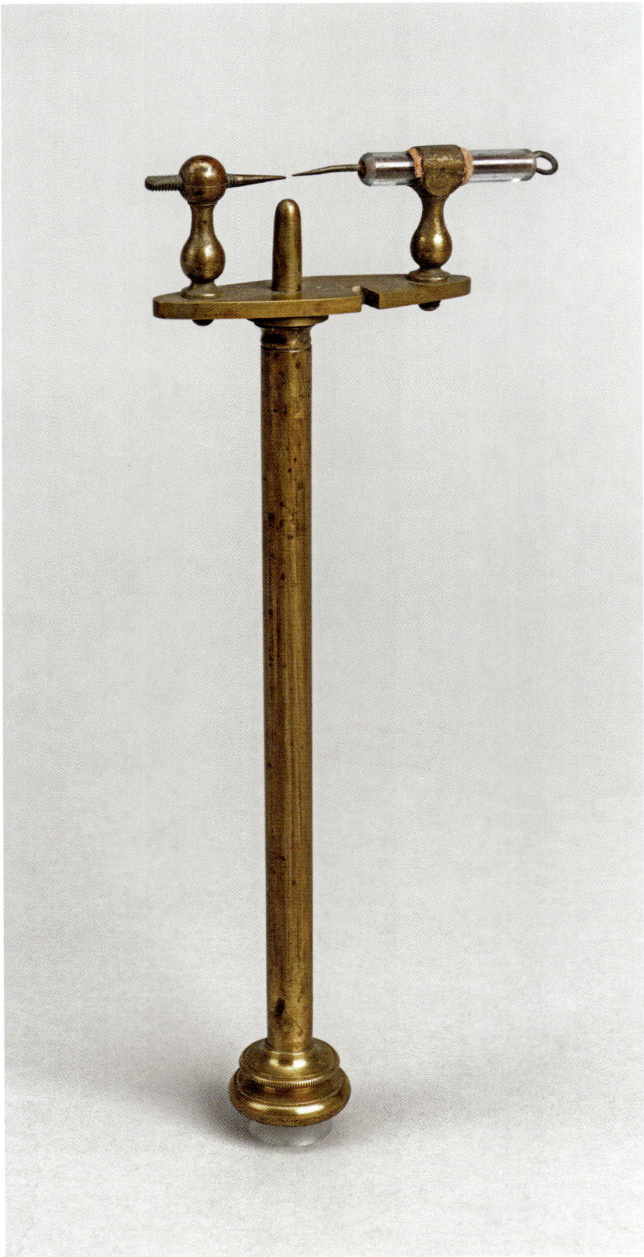

FIGURE 75 (inv. no. 20126-0000-)
© MAM/PHOTO FRANCK BOTTÉ

Brass plate mounted on a vertical rod with two small supports in which two pointed electrodes are inserted. One is insulated from the support by a glass pipe. This spark gap was part of a Volta's lighter that could not be identified. It was used to ignite the hydrogen generated by the lamp.

20124-0000- (Fig. 76)
Spark gap for Volta's lamp
Late 18th c.
MAM accession date: 1952
18 × 19 × 9; 130 g
Glass, brass

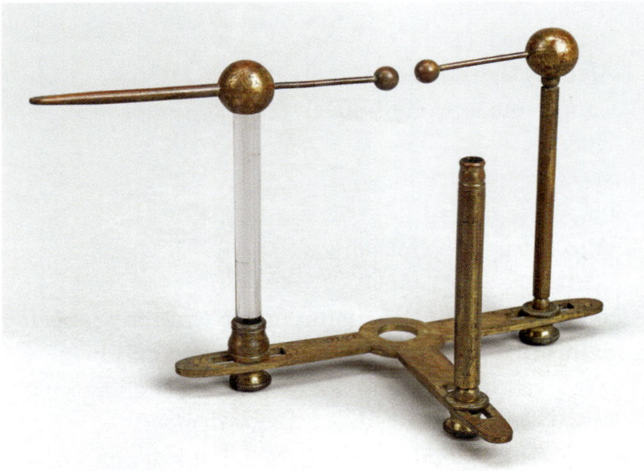

FIGURE 76 (inv. no. 20124-0000-)
© MAM/PHOTO FRANCK BOTTÉ

Spark gap of a missing Volta's lamp. It is composed of a T-shaped brass base with two columns, one made of brass, the other made of glass and therefore insulated. Each column supports a horizontal electrode with a sphere. The glass column can slide in a slot of the base. Between them, it was possible to produce the spark generated by the lamp's electrophorus. A brass pipe fixed to the base with a clamping screw held the candle to be lit by the hydrogen flame of the lamp. Lavoisier may have acquired this instrument in 1782, when Volta was in Paris, or soon after.

Beretta (2001a); Bellodi et al. (2002)

Meteorology

Lavoisier is known to have used thermometers and barometers in chemical and physical experiments as well. However, since it is almost impossible to determine which instruments he used for these purposes, we have classified all of them in the meteorological section.

Thermometers

19925-0000- (Fig. 1)
Alcohol thermometer
Cappy Toussaint (the elder)
1768
MAM accession date: 1952
69.7 × 7 × 3; 395 g
Wood, glass, alcohol, cork

Alcohol (red-coloured) thermometer with an elongated bulb inserted on a painted wooden frame and held by a pair of crosspieces. There are two scales. The first ranges from −5° to 125° and is divided into degrees with marks every 5°. The second ranges from 905 to 10125 and is also divided into degrees with marks every 5°.[1] There are several inscriptions written near and across the scales: "Glace" (0°), "Orangers" (6°), "Temperé" (10°), "Serre ch.^{de} Poëles" (15°), "C.^{bre} de malade" (17°), "Vers à soie" (19°), "Bains ordinaire" (26°), "Paris 1753" (30.5°), "Poulle Pondichery" (32.5°), "Senégal" (38.5°), "Eau bouillante" (110°) and "L'esprit de vin ce // Thermométre est de 1000 parties // condensé par le froid de la // Glace devient 1110 // dilaté par l'eau bouillante". The top of the frame carries the inscription "THERMOMÉTRE // selon Mr de Réaumur // Par Cappy"; the base is inscribed "Rüe & Place Royale Paris 1768". The back of the thermometer is marked with the number "36 109 ½".

Réaumur conducted extensive research in thermometry. After many experiments and tests, he chose an amount of a mixture of spirit and water such that at the freezing point of water the liquid had a volume of 1000 (value set at random) and at the temperature of boiling water a volume of 1080.[2] (Réaumur regarded this range of 80 units as

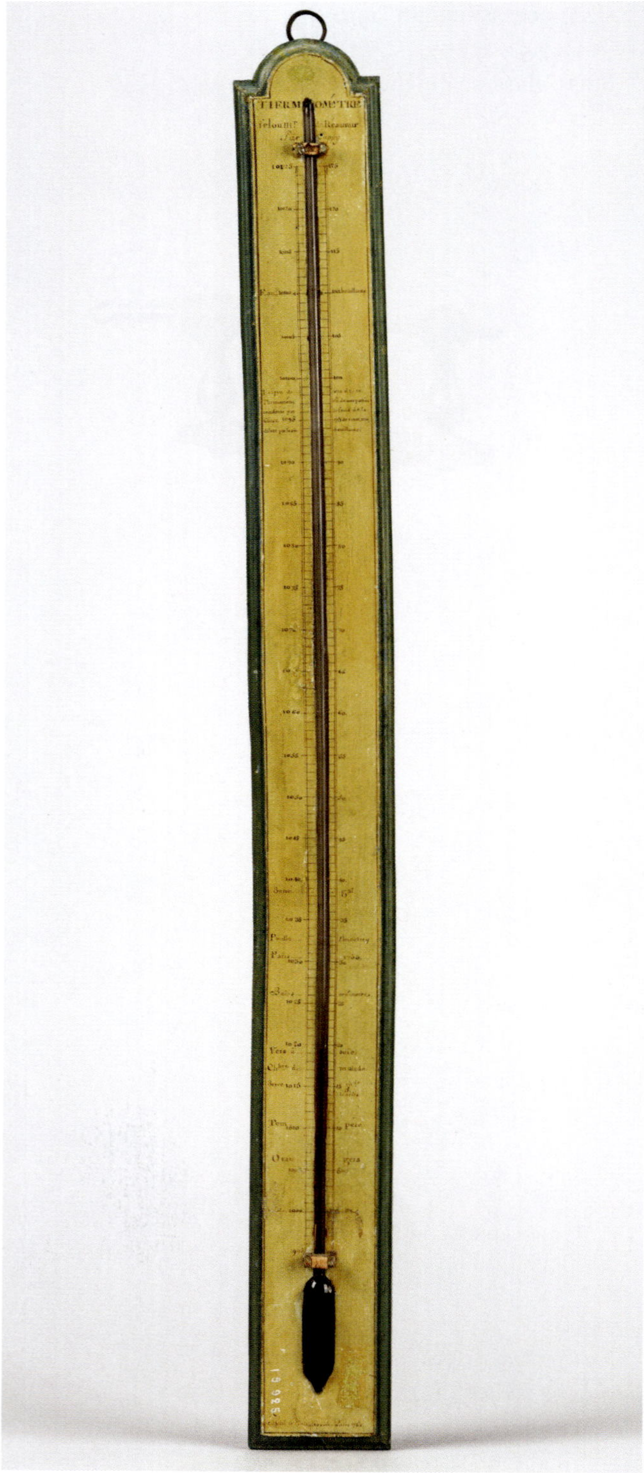

FIGURE 1 (inv. no. 19925-0000-)
© MAM/PHOTO FRANCK BOTTÉ

1 It is not clear why the marking 1095 is followed by 10100, 10105, 10110 ... instead of 1100, 1105, 1110 ... This may be an error by the maker.

2 In fact, 80° degrees was not the temperature of boiling water but that of the alcoholic mixture used.

 | DOI:10.1163/9789004511217_018

particularly convenient to divide into parts.) Accordingly, he marked two scales on his thermometer: one ranging from 1000 to 1080 and a parallel one from 0 to 80. The latter is commonly designated as the Réaumur scale.

In this thermometer in Lavoisier's collection, the composition of the thermometric liquid was different, so that the temperature of boiling water (100° centigrade) represented an increase of 110 parts on the freezing point of 1000. This explains why the point of the scales marked "Eau bouillante" corresponds to the marks 10100 (in reality 1110) and 110 on the scales.

Réaumur (1730); Réaumur (1731); Birembaut (1958); Middleton (1966), pp. 79–86; Holland and Stöhr (2013), 1.4.4 and 2.3.8

19935-0000- (Fig. 2)
Alcohol thermometer
1768–1769
MAM accession date: 1952
Louis Charles Gallonde (1715–1771)
51 × 7.5 × 2; 270 g
Wood, glass, paper, alcohol, brass

Thermometer with an elongated bulb, filled with coloured alcohol and mounted on a small wooden board with a reddish frame. A small metal indicator slides on the tube. The scale is divided into Réaumur degrees from 0° to 24° with markings every 5°.

The paper covering the board carries the inscriptions "Thermometre // Comparable à celui // de M. Réaumur. // N. 7", "chez Mr Gallonde // aux Galeries du Louvre" and, near 0°, "GLACE".

Lavoisier used Gallonde's thermometers early in his career. In 1767, he carried one during his mineralogical travels with Guettard, constantly measuring the temperature of the mineral waters he analyzed and comparing the results with those of two other thermometers.

In 1772, Lavoisier wrote a short memoir reporting his former collaboration with the astronomer Jean-Baptiste Chappe d'Auteroche (1722–1769). In 1769, the Académie Royale des Sciences sponsored Chappe's journey to California where he hoped to collect new and more accurate observations on the transit of Venus. On this occasion, Lavoisier asked his fellow academician to perform hydrometric tests with his new hydrometer, modeled on the one designed by Fahrenheit in 1724. Lavoisier also provided Chappe with two thermometers, one by Gallonde, the other by Cappy. Both were similar to a third thermometer by Gallonde that Lavoisier kept in Paris.

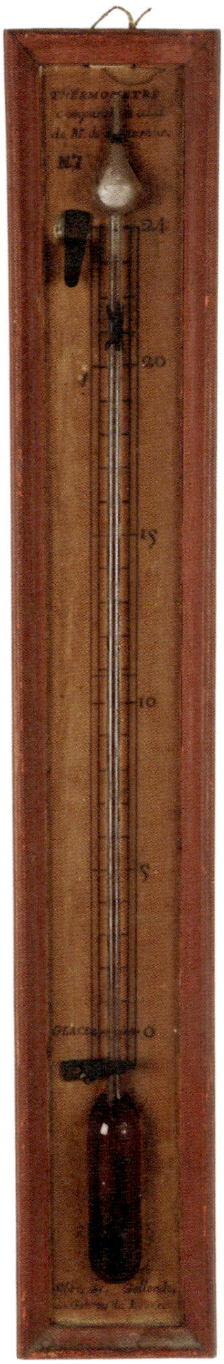

FIGURE 2
(inv. no. 19935-0000-)
© MAM/PHOTO FRANCK BOTTÉ

Ms. Lavoisier (1767–1768)

19934-0000- (Fig. 3)
Alcohol thermometer
ca. 1770
MAM accession date: 1952
Cappy Toussaint
51.5 × 8 × 1.5; mass 120 g
Wood, glass, alcohol

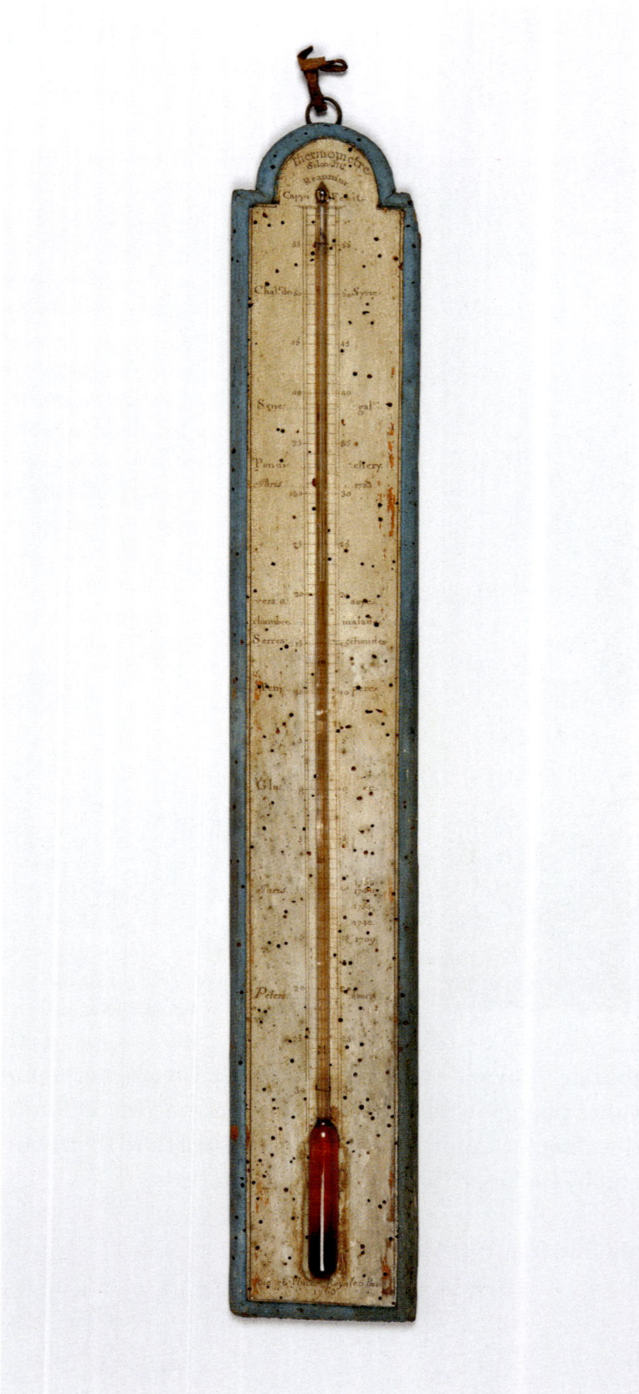

FIGURE 3 (inv. no. 19934-0000-)
© MAM/PHOTO FRANCK BOTTÉ

Alcohol thermometer mounted on a badly worm-eaten wooden tablet. The top carries the inscription "THERMOMETRE // Selon M.r Réaumur // Cappi Fecit". The scale is divided into degrees from −33° to 58°, with temperatures marked every 5 degrees.

Near the scale are the following indications: "Petersbourg" (−20.5°), "1709" (−14.5°), "1742" (−13.5°) "1754" (−12°), "1740" (−10.5°), "1763" (−10°), "Glace" (0°), "Temperé" (10°), "Serres chaudes", (5°) "chambre malade"(17°), "vers à soye" (19°), "Paris 1753" (30.5°), "Pondischéry" (32.5°), "Sénégal"(38.5°), "Chal.r de Syrie" (50°).

This thermometer was certainly made after 1763 and before 1776, when the coldest temperature was recorded in Paris.

19937-0000- (Figs. 4a, 4b)
Alcohol thermometer
ca. 1770
MAM accession date: 1952
Antoine Assier-Perica (1730/1–pre 1806)
41 × 9 × 2; mass 210 g
Glass, wood, alcohol, brass

Alcohol thermometer inserted in slot of a rectangular painted wooden tablet.[3] The bulb of the tube is a flattened spiral with 5 turns. Because of the slot, the thermometer and its bulb are visible on both side of the tablet and the scales (divided in degrees) from −17° to 51° are inscribed on the front and on the back of it.[4] On the front it is inscribed: "Thermometre // de Réaumur // par Assier Perica" and along the scale: "Paris 1709" (−15.5°), "1708" (−14.5°), "1707" (−13°), "1734" (−12°), "1740" (−11°), "GLACE" (0°), "Temperé" (10°), "Serre chaude", "Chambre malade" (17°), "Vers a soie" (19°), "Bains ord.^{re}" (26°), "Paris 1753" (30°), "Poul(15°) es" (32.5°), "Senégal" (38°), "Syrie" (50°). The inscriptions on the back: "Thermometre // de Réaumur // Approuvé de l'accademie // Royale des Sciences // Par Assier Perica" (top) and "Rue St antoine [au coin] // de La rue Four[cy]" (bottom).

Inscription on back scale: "Paris 1709" (−15.5°), "1708" (−14.5°), "1707" (−13°), "1734" (12°), "1740" (−11°), "Froid [undreadable date] " (10.5°), "GLACE" (0°), "Temperé" (10°), "Serre chaude" (15°), "Ch.^{bre} malade" (17°), "Vers a soie" (19°), "Bains ord.^{re}" (26°), "Paris 1753" (30°), "Poules qui couvent" (32.5°), "Senégal" (38°), "Syrie" (50°).

3 The paint is damaged and it is difficult to read the inscriptions.
4 The thermometer could be fixed in a window or perpendicularly to a wall so that it could be red on both sides.

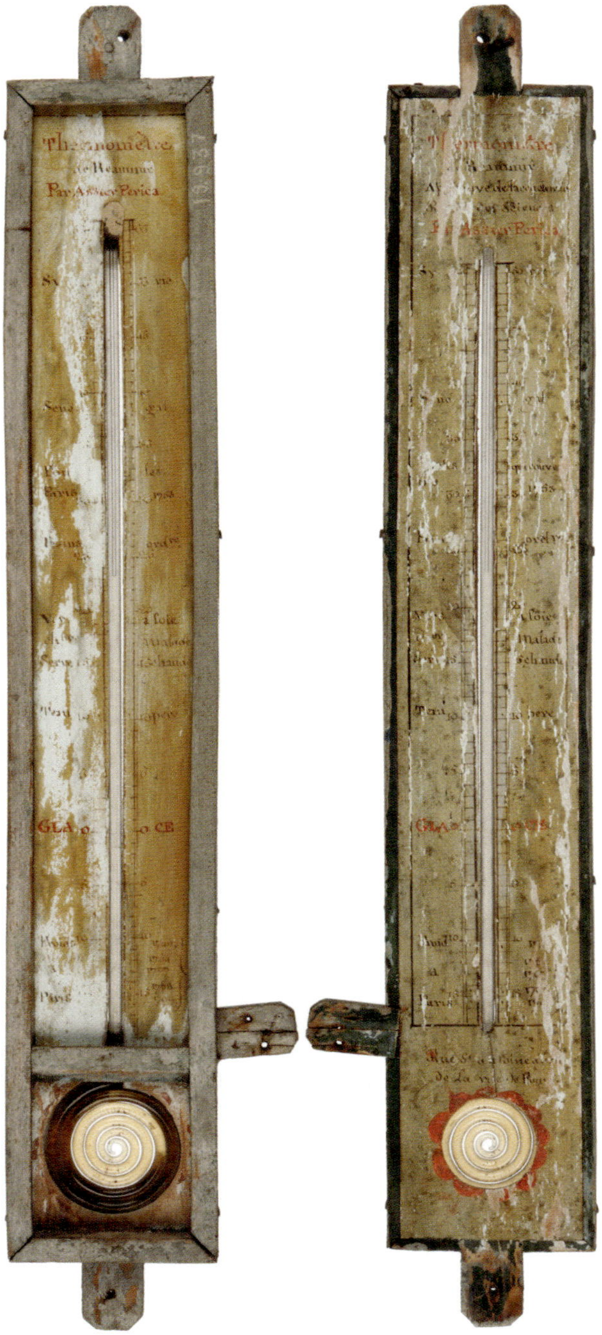

FIGURES 4A, 4B (inv. no. 19937-0000-)
© MAM/PHOTO FRANCK BOTTÉ

In January 1776, Lavoisier conducted several experiments and observations concerning the exceptional cold wave hitting Paris and the north of France. On this occasion he made systematic comparisons between the thermometers in his possession and was able to assess the accuracy of those made by Assier-Perica.

Lavoisier (1776)

19938-0000- (Fig. 5)
De Luc's mercury thermometer
1774
MAM accession date: 1952
Guillaume Charles Goubert (1747–1824)
45 × 6.5 × 2.5; 295 g
Wood, glass, mercury, sealing wax

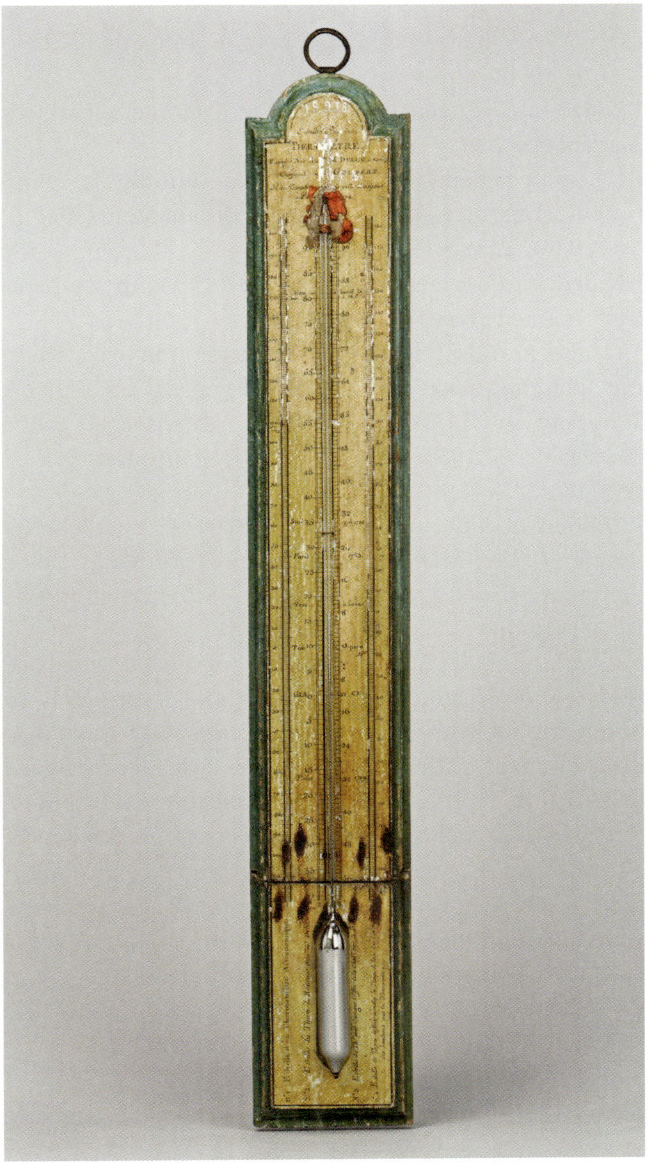

FIGURE 5 (inv. no. 19938-0000-)
© MAM/PHOTO FRANCK BOTTÉ

Mercury thermometer with an elongated bulb mounted on a painted wooden tablet with a suspension ring. The tablet, whose lower part can be folded, originally carried two other thermometers (missing). There are four scales marked on the tablet, of which nos. 2 and 3 are used with the central surviving thermometer; the other two applied to the missing thermometers.

a) "n° 1 Echelle d'un Thermométre Astronomique Suiv. m.r Deluc"

Divided into degrees from −130° to 207°, with temperatures marked every 10°

b) "N° 2 Echelle du Therm de Réaumur"

Divided into degrees from −41° to 90°, with temperatures marked every 10°

c) "N° 3 Echelle du Th. pour corriger l'effet de la Chal.r sur le Ba[romètre]"

Divided into degrees from −60° to 96°, with temperatures marked every 8°

d) "n° 4 Echelle du Therm. destiné a indiq. la Temp. de l'air dans la m[esure] // des hauteurs par le Baromètre."

Divided into degrees from −132° to 184°, with temperatures marked every 10°

There are also the following inscriptions related to the Réaumur scale: "Paris 1709" (−17°), "GLACE" (0°), "Temperé" (10°), "Vers à soie" (18°), "Paris 1753" (28°), "Senégal 1738" (38°), "Eau bouill.te // bar[ometre] à 28 P.ces" (80°).

At the top of the tablet there is the inscription "Echelles pour le // THERMOMÉTRE. // Purgé d'Air suiv.t M. Deluc de Genève // Construit par Goubert. // Rüe Dauphine [… is] celle l'anjous // à Paris 1774".

Jean-André de Luc (1727–1817) conducted extensive research on meteorology and largely contributed to improving its instruments. This thermometer was made in accordance with de Luc's design. Scale no. 1 served to measure temperatures for calculating atmospheric refraction, a value needed to correct astronomical observations – hence the name "thermomètre astronomique". Scale no. 3 was used to correct barometric readings, which were influenced by temperature. Scale no. 4 was used with a barometer to calculate altitude.

De Luc (1784), pp. 7–49

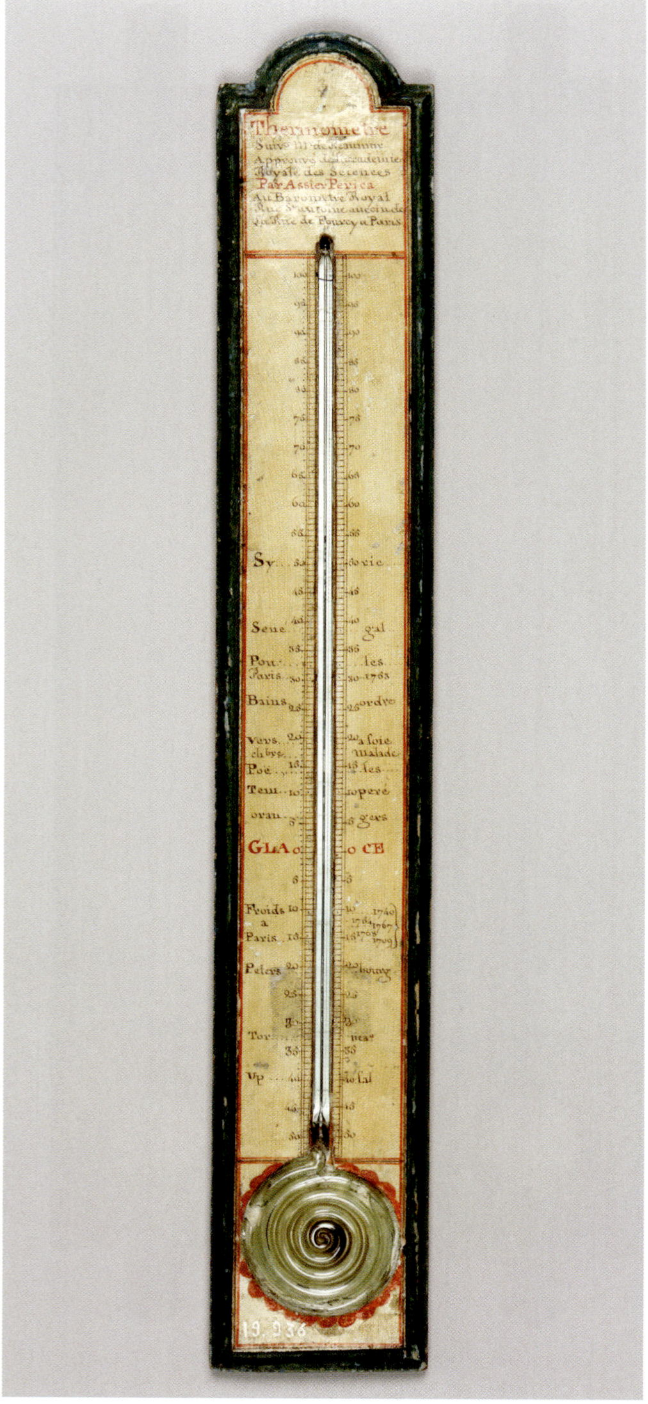

FIGURE 6 (inv. nos. 19936-0002-, 19936-0001-)
© MAM/PHOTO FRANCK BOTTÉ

19936-0002-, 19936-0001- (Fig. 6)
Alcohol thermometer
ca. 1775
MAM accession date: 1952
Antoine Assier-Perica (1730/1–pre 1806)
44 × 6.6 × 1.5; 195 g
Glass, wood, mercury

Alcohol thermometer inserted into the slot of a rectangular painted wooden tablet. The spiral bulb is broken and separated from the instrument (inventoried as no. 19936-0001-). The double scale is divided into degrees from −55° to 103°, marked every 5°. The top of the tablet bears the inscription "Thermometre // Suiv. M. de Reaumur // Approuvé de l'accademie // Royale des Sciences // par

Assier Pericat // Au Barometre Royal // Rue St antoine au coin // de la Rue de Fourcy a Paris". Inscriptions on the scale: "Upsal" (-40°), "Torneao" (-33°), "Petersbourg" (-21°), "Froid à Paris 1709" (-15.5°), "1768" (-14.5°), "1767" (13°), "1754" (12°), "1740" (10.5°)", "GLACE" (0°), "orangers" (6°), "Temperé" (10°), "Poële" (14°), "Ch.bre malade" (16.5°), "Vers a soie" (19°), "Bains ord.re" (26°), "Paris 1753" (30°), "Poules" (33°), "Senégal" (38.5°) and "Syrie" (50°). This thermometer was certainly made after 1767 and before 1776 (the lowest temperature recorded in Paris).

On the back of the tablet there is the following note by Lavoisier, written after 1776): "thermometre n° 7 // compare en 1776 avec // l'etalon de Mr de Reaumur // il est exact a la // glace fondant il donne // 9 1/2 aux caves de // l observatoire // 12. 2/3 quand l'etalon de // M de Reaumur donne // 11.91, 17 1/16 quand // l'etalon donne 16".
See note for item no. 19937-0000-.

Lavoisier (1776)

19939-0000- (Fig. 7)
Gilded frame for thermometer
1779
MAM accession date: 1952
38.5 × 13 × 4.5; 310 g
Wood, glass

This decorated frame of gilded wood originally contained a thermometer (missing). The upper part of the frame is ornamented with two cornucopia and an oval medallion containing the inscription "dedie // a Mr de // Lavozier // en. // 1779".

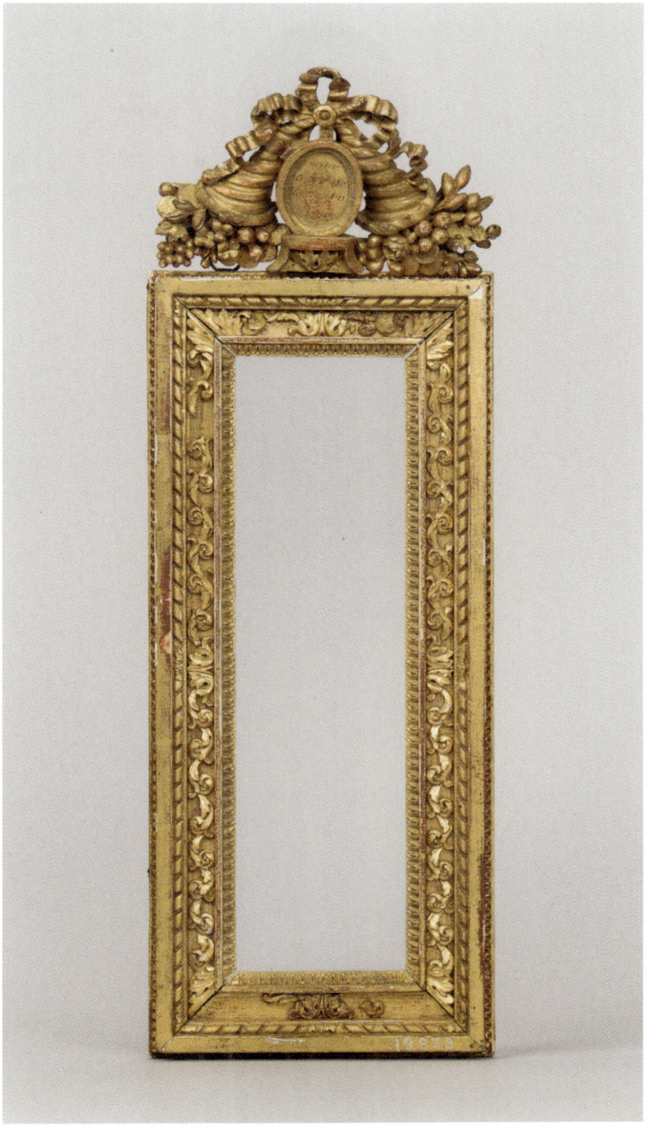

FIGURE 7 (inv. no. 19939-0000-)
© MAM/PHOTO FRANCK BOTTÉ

19926-0000- (Fig. 8)
Mercury thermometer
Cappy, Mossy (fl. 1770–ca. 1810)
ca. 1780
MAM accession date: 1952
28 × 4.3 × 1.5; 160 g
Glass, brass, traces of mercury

Mercury thermometer mounted on brass tablet with Réaumur and Fahrenheit scales. The bulb is broken and missing. The Réaumur scale ranges from -55° to 85°, the Fahrenheit scale from -80° to 212°. Both have divisions

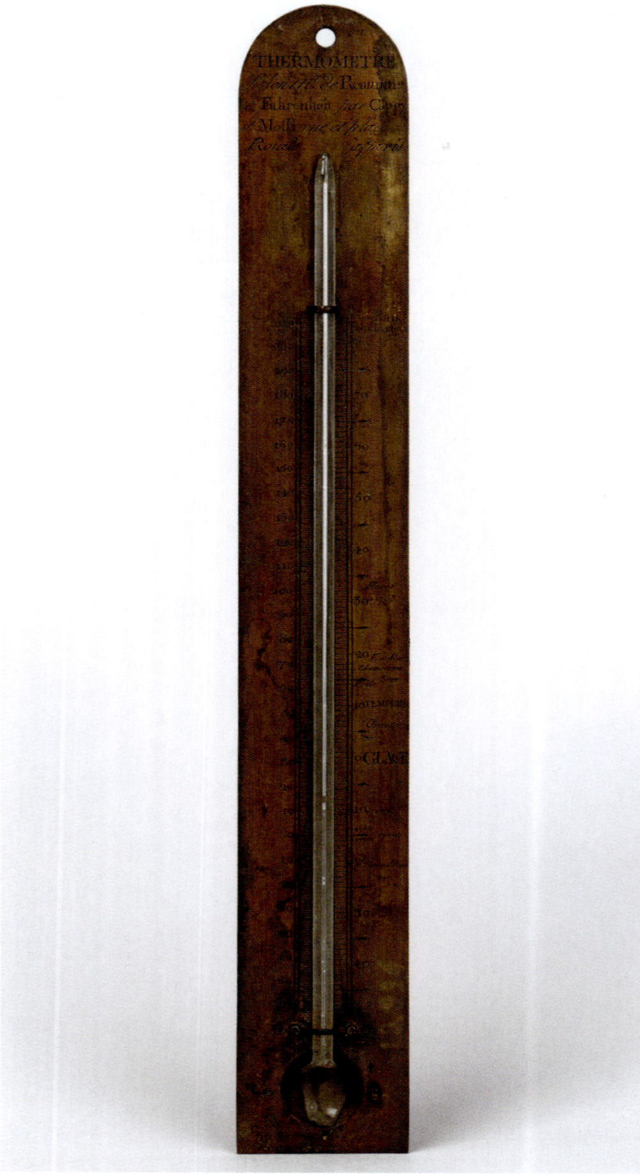

FIGURE 8 (inv. no. 19926-0000-)
© MAM/PHOTO FRANCK BOTTÉ

for every degree and the temperatures are indicated every 10°. The top of tablet carries the following engraved inscription: "THERMOMETRE // *Selon M.s de* REAUMUR // *et* FAHRENHEIT par CAPPY // *et* MOSSI *rue et place* // *Roiale à paris.*".

Inscriptions along the scales: "petersbourg" (–20.5° R), "1776" (–16° R), "1709" (–15° R), "1768" (–14° R), "1740" (–10.5° R), "GLACE" (0° R), "Orangerie" (5° R), "TEMPERÉ" (10° R), "Poële" (14° R), "Serre" (15° R), "Chambre m." (17° R), "V.e à Soie" (20° R), "Paris 1753" (30° R), "Eau bouillante" (80° R).

Despite the fact that Cappy and Mossy were highly reputed for their thermometers and barometers, very little is known about their activity. We know that Mossy (Cappy Senior's nephew) was first associated with his son (Cappy Junior) in 1775 and later became his successor.

At the beginning of his scientific career, during the 1760s and early 1770s, Lavoisier often used thermometers made by Cappy senior.

19931-0000- (Fig. 9)
Mercury thermometer
ca. 1780
MAM accession date: 1952
38 × 2.5 × 0.8; 145 g
Box: 50 × 4 × 2.5; 165 g
Glass, silvered brass, paper, wood, velvet

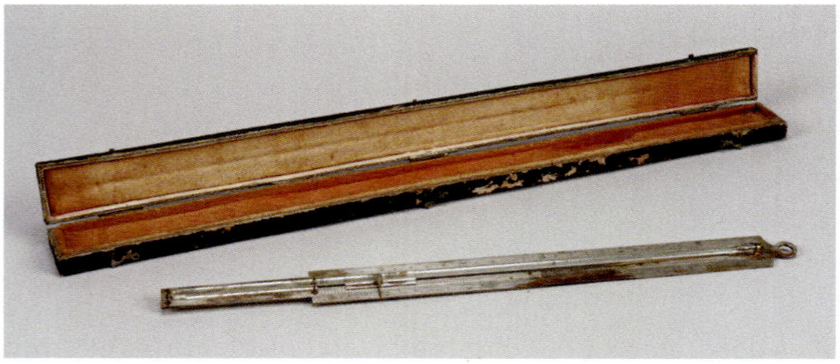

FIGURE 9 (inv. no. 19931-0000-)
© MAM/PHOTO FRANCK BOTTÉ

Mercury thermometer mounted on a silvered brass plate. The thermometer is broken and its bulb is missing. The scale ranges from –3.8° to 20° with divisions every fifth of a degree. The temperatures are marked every degree. A nonius sliding in a groove behind the thermometric tube allows a reading precision of 0.02°. A ring at the top is used for suspending the instrument. The thermometer is preserved in a wooden case, coated with black paper and lined with red velvet. The case originally contained another instrument.

19912-0000-, 19912-0001-, 19912-0002-, 19912-0003-, 19912-0004-, 19912-0005- (Fig. 10)
Set of four mercury thermometers in a box
ca. 1780
MAM accession date: 1952
James Crichton (active 1785–1835)
Case with thermometers: 23.8 × 3 × 6.2 cm; 270 g
19912-0001-: length 18.3 cm; 15 g
19912-0002-: length 18.6 cm; 15 g
19912-0003-: length 19.1 cm; 15 g
19912-0004-: length 22 cm; 25 g (not signed)
Glass, mercury, wood

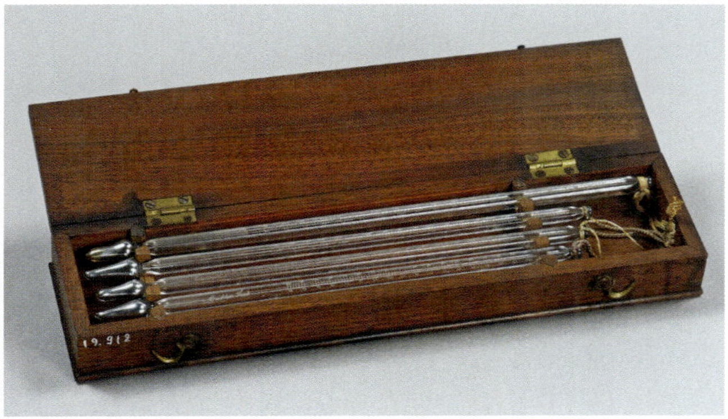

FIGURE 10 (inv. nos. 19912-0001-, -0004-)
© MAM/PHOTO FRANCK BOTTÉ

The set is contained in a mahogany case with a hinged lid (19912-0005-) and is protected by cotton wool. It comprises four mercury thermometers with pear-shaped bulbs. The Fahrenheit scales from 30° to 210° are engraved on the tubes and three are signed “Crichton fecit”. A suspension loop of thread (or ribbon) is attached to the knob at the top of each thermometer.

Displayed at the 1943 Palais de la Découverte exhibition marking the bicentenary of Lavoisier’s birth.

Lavoisier (1943); Middleton (1966), p. 135

19914-0000- (Fig. 11)
Mercury thermometer with stand
ca. 1780
MAM accession date: 1952
Pierre Bernard Mégnié the Younger (Mégnié le Jeune) (1758–1807)
33.6 × 13.2 × 6.7; 260 g
Brass, glass, mercury, iron alloy.

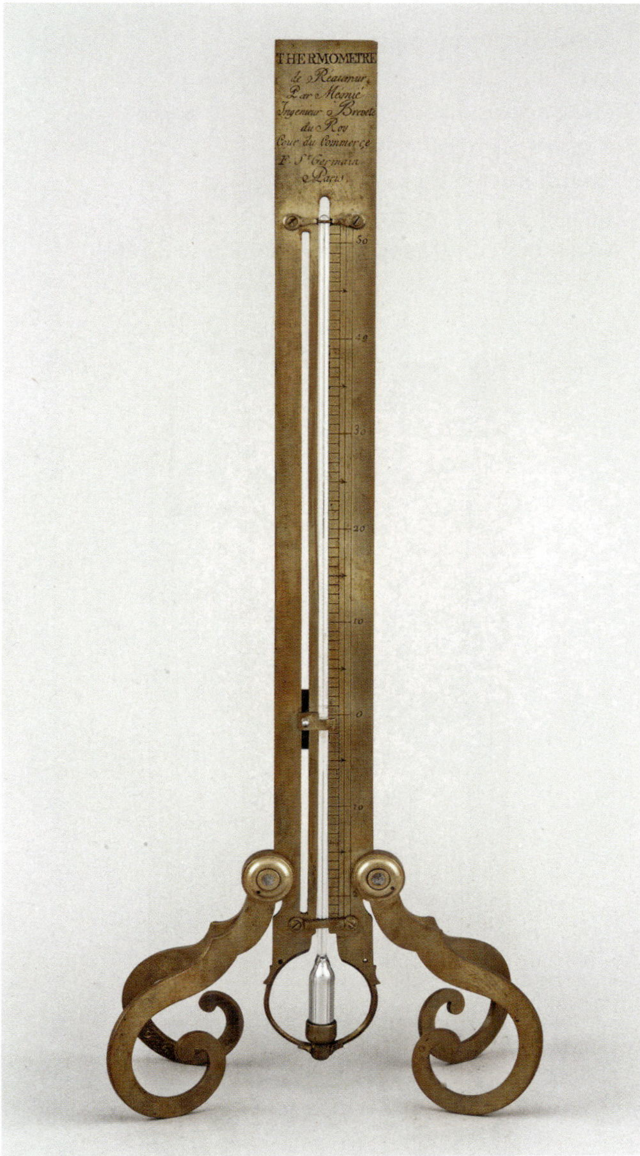

FIGURE 11 (inv. no. 19914-0000-)
© MAM/PHOTO FRANCK BOTTÉ

Mercury thermometer inserted in a brass frame with four ornate curved feet. On the frame there is a Réaumur scale from −23° to 52° divided into degrees with the temperatures marked every 10°. A small slider inserted in the frame acts as a nonius and allows readings with a precision of one-tenth of a degree.

The frame bears the inscription “THERMOMETRE // de Réaumur // Par Mégnié // Ingénieur Breveté // du Roy // Cour de Commerce // F. S.t Germain // Paris”.

Truchot (see reference below) rightly remarked that, because of the design, this was a domestic thermometer and not a laboratory instrument.

Truchot (1879); Lavoisier (1943)

19911-0000- (Fig. 12)
Mercury thermometer
1782
MAM accession date: 1952
Mossy (active ca. 1775–1821)[5]
Thermometer: 60 × 3 × 2.5; 280 g
Box: 63 × 4.7 × 3.5; 205 g
Glass, mercury, brass, wood, chamois leather

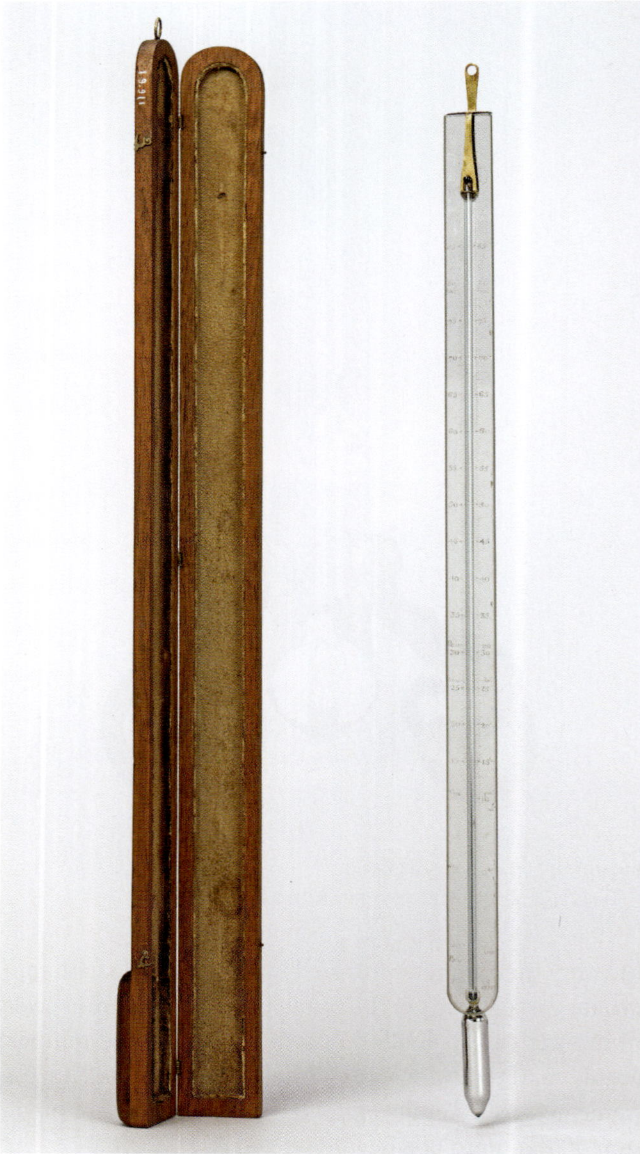

FIGURE 12 (inv. no. 19911-0000-)
© MAM/PHOTO FRANCK BOTTÉ

Mercury thermometer with an elongated bulb. It is inserted and fixed on a glass scale divided into degrees from −15° to 90°, with five markings for every degree. The temperatures are marked every 5° on both sides of the thermometric tube. At the top of the scale is a brass attachment for hanging the instrument. There are other indications along the scale, often marked with half the word on one side of the stem and the other half on the other side. At −15° is the marking "1776" (the year of a very cold winter in Paris); to the left of the fifteenth division above it, "Paris"; to the right of the twentieth division, "1740"; to the left of 0°, "Glace"; to the left of 10°, "Temperé;" to the left of 26°, "Bains ordin"; at 31°, "Paris" on the left and "1753" on the right. Below the 80° mark on the left, with an arrow pointing from 80°, are the words "Eau" and "bouill" (eau bouillante) on opposite sides.

The top of the scale bears the inscription "Par Mossy // Quay Pelletier à Paris // 1782".

The instrument is suspended from a hook in the wooden case lined with chamois leather. A second suspension ring is fixed at the top of the case.

Observatoire de Paris Inv. no. 141 (Figs. 13a, 13b)
Thermometer (broken) in metal frame
Mossy (active ca. 1775–1821), Jean François Richer (1743–ca. 1820)
1782 and 1785
76 × 30
Brass, glass

This mercury thermometer, called "Lavoisier's thermometer",[6] was installed in the cellar of the Paris Observatory in 1782.

The thermometer (now broken; tube missing) is inserted in a rectangular brass frame with three perpendicular protruding supports. The bulb is protected by a brass armature. The scale divided into tenths of degrees is engraved on a glass plate whose vertical position could be slightly modified with a screw and a key (missing?) when it was necessary to correct the readings. The frame supports were placed on the rim of a glass vase filled with sand so that the bulb was immersed in it. The following inscription was engraved on the thermometer: "par Mossy sous la conduite de Mrs. de l'Académie des sciences divisé par la machine de M. Richer, 1783".

This thermometer (as well as at least two others) was made with the greatest care under Lavoisier's direction for the Académie Royale des Sciences. It was used to determine whether the temperature in the Observatory cellar was truly constant. Its receipt was acknowledged

5 See also the thermometer 19226-0000-.

6 This designation appears to have been given by Arago.

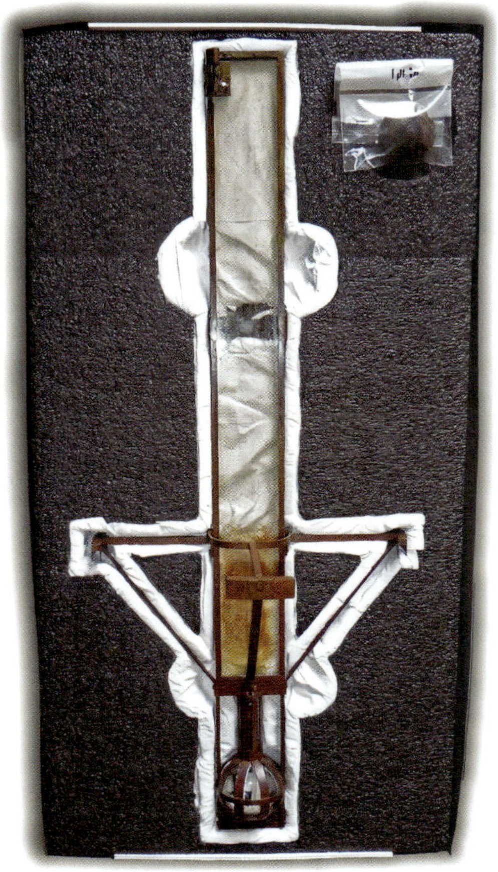

FIGURE 13A (inventory no. 141)
COURTESY OBSERVATOIRE DE PARIS

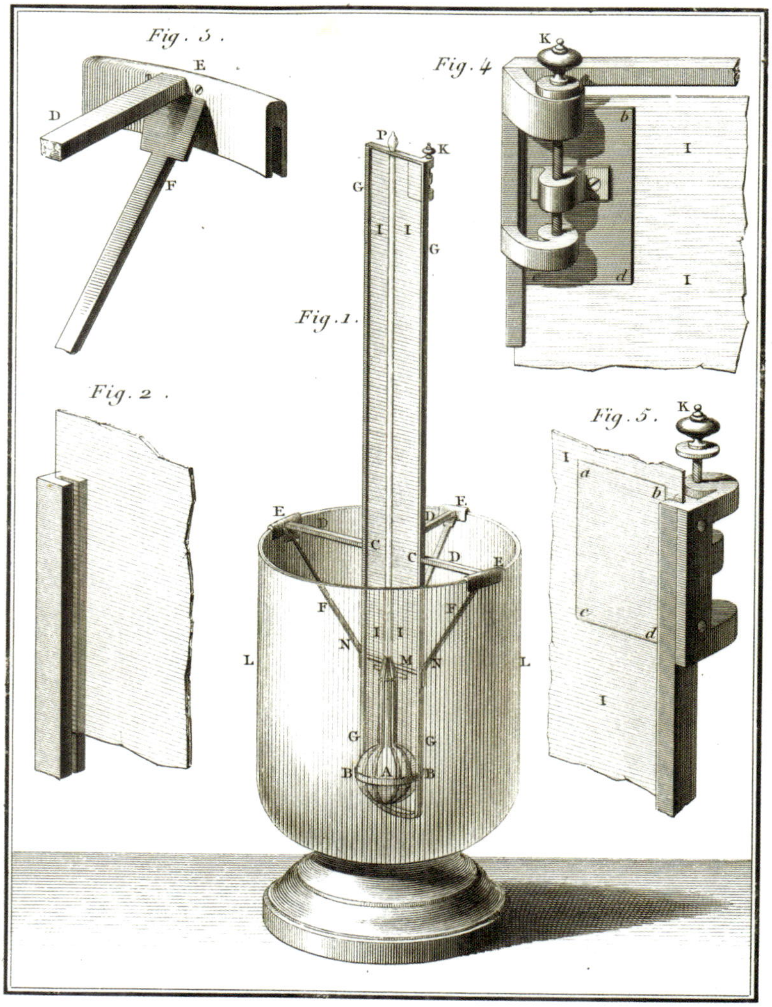

FIGURE 13B LO, vol. 3, plate 11

in mid-June 1782 by Cassini, who installed it in the cellar by the end of the month, Richer made a new scale at Lavoisier's request in March 1785. The thermometer vase was broken in 1792. The thermometer was then repaired and was still working in 1865. At the time, it continued to be compared with the thermometers made by Gay-Lussac, Dulong and Walferdin in the first half of the nineteenth century.

Lavoisier (1785); Arbey (1952)

19923-0000- (Fig. 14)
Mercury thermometer
1783
MAM accession date: 1952
Nicolas Fortin (1750–1831)
52.5 × 3 × 2.5; 275 g
Glass, brass, mercury

Mercury thermometer (the bulb is missing) fixed on a glass plate with a clamp and brass suspension. A Réaumur scale from −36° to 106° is engraved on the glass. Every degree is divided into five parts and the temperatures are marked every 5°. There is a series of inscriptions near different temperatures: "Petersbourg" (−20.5°), "1776" (−17°), "1709" (−15.5°), "GLACE" (0°), "Temperé" (10.5°), "Vers à soie" (19°), "Bains" (26°), "Paris 1753" (31.5°) and "Eau b.^{te}" (80°).

At the top of the glass scale is the inscription "thermometre // selon Réaumur // Par Fortin // Place Sorboñe // Paris 1783".

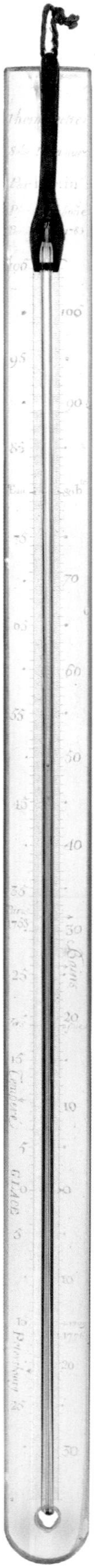

FIGURE 14
(inv. no. 19923-0000-)

19910-0000- (Fig. 15)
Mercury thermometer
1783
MAM accession date: 1952
Mossy (active ca. 1775–1821), Jean François Richer (1743–ca. 1820)
105.5 × 10 × 4 cm; 1.98 kg
Glass, brass, mercury

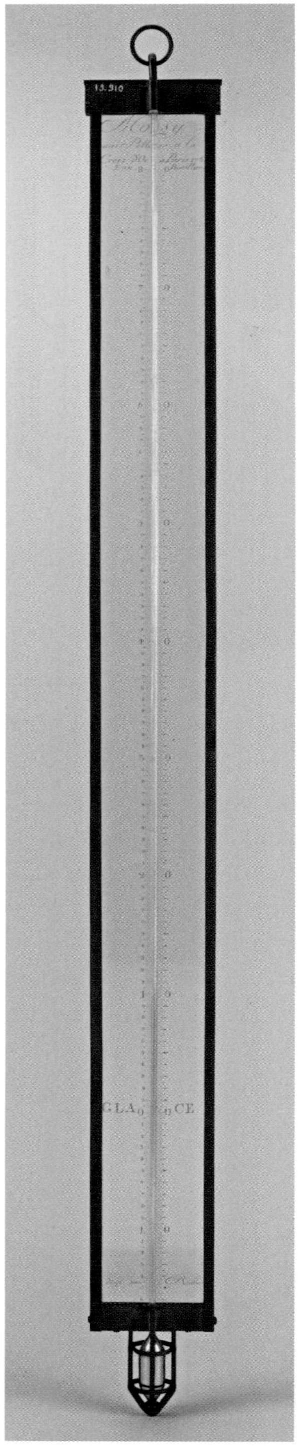

FIGURE 15
(inv. no. 19910-0000-)

Thermometric tube mounted on a glass plate enclosed in a brass frame with a suspension ring. The bulb of the tube is protected by a grid. A Réaumur scale from −16° to 80° with divisions into tenths of degrees is engraved on the plate. The half-degree lines are longer and end in dots, while the temperatures are marked every degree. The freezing point of water is marked "GLACE" and the boiling point "Eau Bouillante". The top of the plate bears the inscription "Mossy // quai Pelletier a // la Croix d'Or à Paris. 1783"; the bottom reads "divisé par Richer". In fact, the scale was engraved on the glass with a dividing engine made by Richer.

Middleton (1966), p. 119; Daumas (1989), p. 213

19917-0000- (Fig. 16)
Mercury thermometer
1786
MAM accession date: 1952
Carlo Artaria (fl. late 18th c.–early 19th c.)
22 × 1.4; 30 g
Glass, cork, paper, sealing wax

Mercury thermometer enclosed in a second larger tube with a cork at one end and a hanging thread at the other. In the tube there is a double paper scale from −23° to 53° with the temperatures marked every 5°. The scale also carries the handwritten inscription "Artaria a Mannheim 1786".

The Genevan Charles (or Carlo) Artaria was active in Mannheim as a meteorological instrument maker for the local academy of sciences.[7] Artaria collaborated with Rumford while the latter was based in Munich. This thermometer may therefore have been originally made for Rumford, not for Lavoisier.

Thompson (1805), pp. 448–449; Brown (1967), p. 138

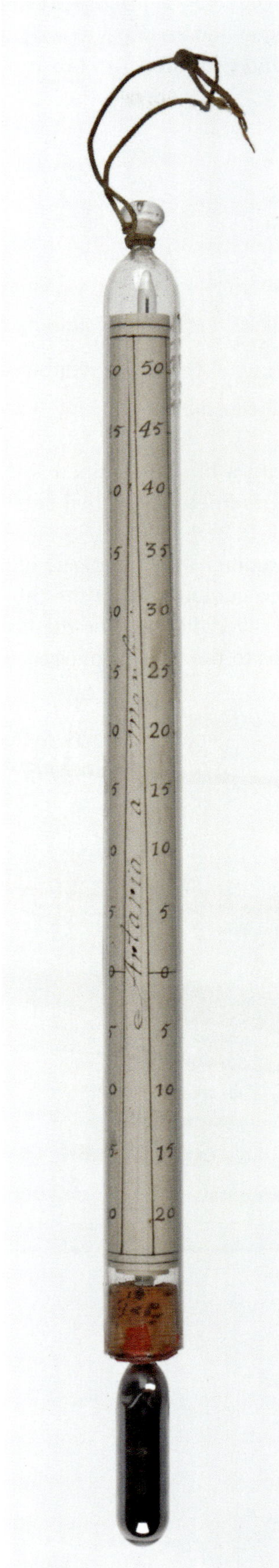

FIGURE 16
(inv. no. 19917-0000-)
© MAM/PHOTO FRANCK BOTTÉ

7 See also the barometer 19952-0000.

19908-0001-, 19908-0002-, 19908-0003-, 19908-0004-, 19908-0005-, 19908-0006-, 19908-0007- (Fig. 17)
Set of four thermometers and two rulers in a box
1786
MAM accession date: 1952
Nicolas Fortin (1750–1831)
19908-0001-: 39 × 1.2 × 1; 20 g
19908-0002-: 38.5 × 1.2 × 1; 30 g
19908-0003-: 32 × 2.5 × 1; 40 g
19908-0004-: 33.5 × 2.5 × 1; 50 g
19908-0005-: 40 × 2 × 1.7; 60 g
19908-0006-: 39 × 2 × 1.7; 60 g
19908-0007-: 38.5 × 10.5 × 7; 750 g
Glass, brass, mercury, wood, iron, fabric

Wooden box (19908-0007-) with a hinged lid lined with blue silk, containing four thermometers and two small rulers.

19908-0001: Mercury thermometer with spherical bulb. There are circular unnumbered graduations on the tube and a sliding nonius made of silvered brass. The nonius, which is fixed with two clamps to the tube, is marked 0, 5, 10, 15, 20, and the interval between 0 and 5 is divided into 25 parts. The back of the nonius (not silvered) has the inscription "Fortin Place Sorbonne A Paris 1786". There is a suspension silk loop at the top of the tube.

19908-0002-: Mercury thermometer with spherical bulb. Identical to 19908-0001-.

19908-0003-: Mercury thermometer with cylindrical bulb and bent tube.

19908-0004-: Mercury thermometer with cylindrical bulb and bent tube. Very similar to 19908-0003-.

19908-0005-: Glass scale from −23° to +85° with divisions every degree. One end has a hole to accommodate the mercury bulb of a thermometer. The opposite end has a brass hook with a small hanging thread for fixing the thermometer. Signed "Fortin a Paris".

19908-0006-: Glass scale from −35° to +120° with divisions every degree. One end has a hole to accommodate the mercury bulb of a thermometer. The opposite end has a brass hook with a small hanging thread for fixing the thermometer. Signed "Fortin a Paris". Very similar to 19908-0005-.

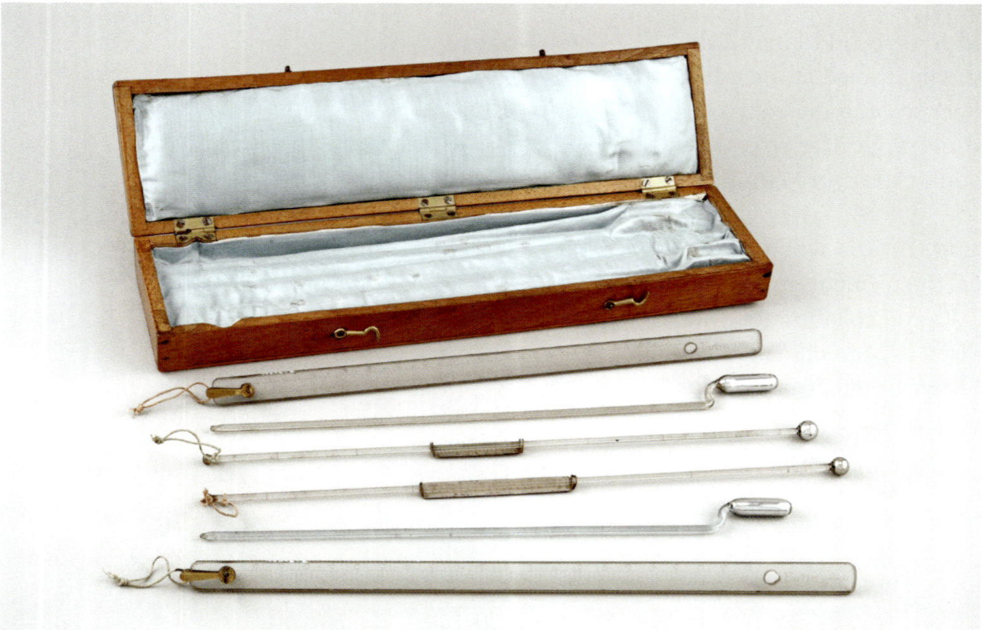

FIGURE 17
(INV. NOS. 19908-0001-, -0007-)
© MAM/PHOTO FRANCK BOTTÉ

19929-0000- (Fig. 18)
Mercury thermometer (broken)
1789
MAM accession date: 1952
Mossy (active ca. 1775–1821)
94.3 × 2.4 × 1.7; 130 g
Glass, paper

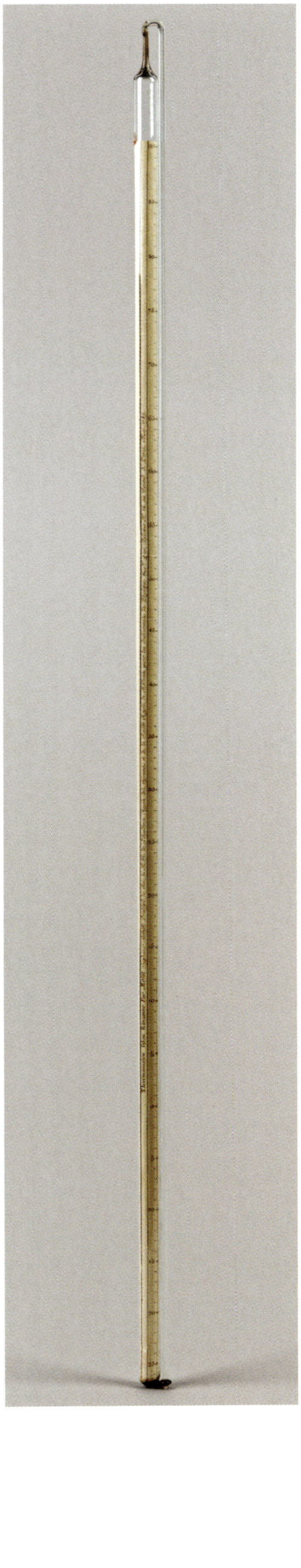

FIGURE 18
(inv. no. 19929-0000-)
© MAM/PHOTO FRANCK BOTTÉ

Thermometer composed of a capillary tube joined to a second parallel tube of larger diameter. The latter carries a paper scale from −26° to 90° with divisions every fifth of a degree and the temperatures marked every 5°. The bottom of the capillary is broken and the bulb is missing. The paper scale has the inscription "Thermometre selon Réaumur Par MOSSI Ingenieur Bréveté Privilégié du Roi; de MM. de l'Académie Royale des Sciences et de la Société Royale de Médecine pour les Instrumens de Physique &c.a Quai Pelletier N° 16 au Second à Paris Mai 1789".

19922-0000- (Fig. 19)
Bent mercury thermometer
ca. 1790
MAM accession date: 1952
30 × 1.5 × 10; 45 g
Glass, wood, mercury

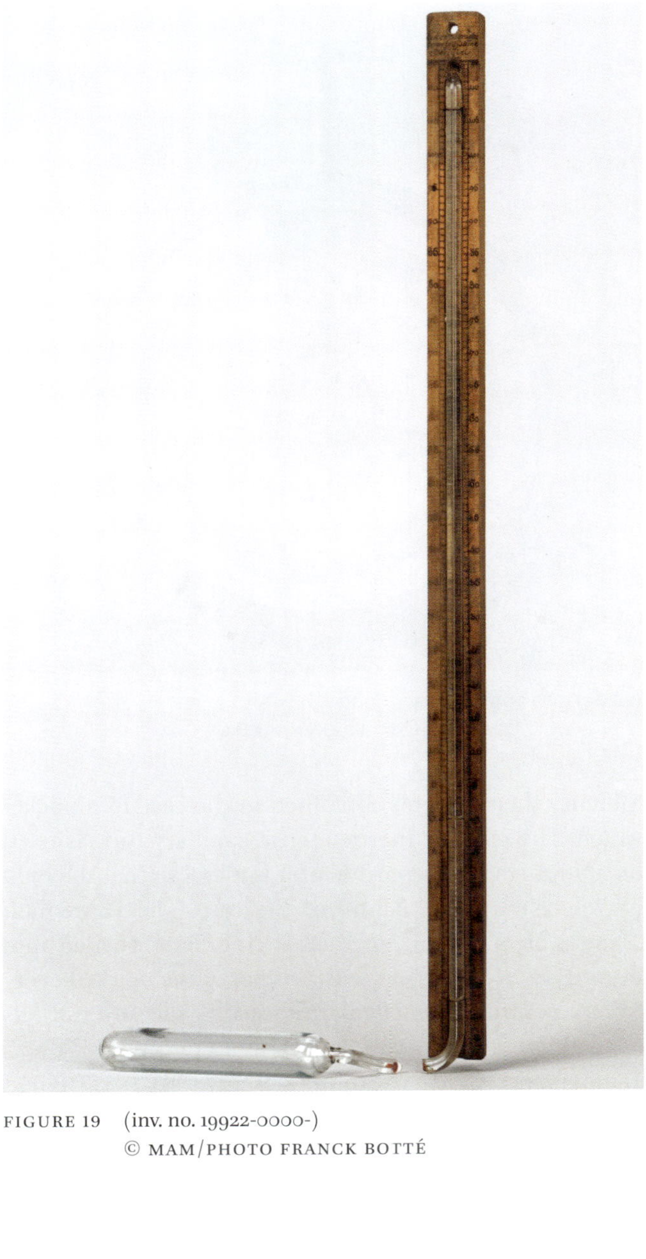

FIGURE 19 (inv. no. 19922-0000-)
© MAM/PHOTO FRANCK BOTTÉ

The tube of this mercury thermometer is fixed to a strip of wood marked with a barely legible scale. The elongated cylindrical bulb was originally perpendicular to the tube but the tube is now broken and the bulb is preserved with the instrument. The double scale is divided into degrees from 5° to 110°, with temperatures marked every 5°. The top of the scale carries the inscription "Thermomètre // selon le système // Décimal".

20577-0000- (Fig. 20)
Pocket thermometer
ca. 1790
MAM accession date: 1952
John Bleuler (ca. 1757–1829)
7.7 × 5.5 × 2.5; 60g
Steel, gilded brass, glass, enamel, mercury

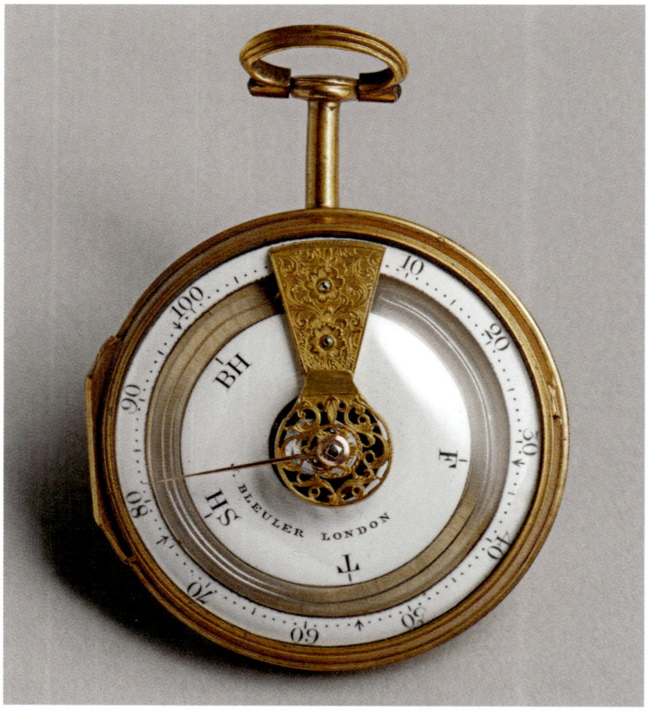

FIGURE 20 (inv. no. 20577-0000-)
© MAM/PHOTO FRANCK BOTTÉ

Mercury thermometer contained in the case of a pocket watch. The circular thermometric capillary tube is fixed on a brass ring but a portion of it is radial so that the bulb is concealed at the centre of the case. The Fahrenheit scale is on a second enamelled ring. It is divided into degrees from 8° to 100°, with temperatures marked every 10°. A central enamelled disk carries the inscriptions near different temperatures: "F" (32°, Freezing), "T" (55°, Temperate), "SH" (75° Summer Heat), "BH" (94° Blood Heat) and "J. BLEULER LONDON". An indicator pivots at the centre of the case under a decorated disk and can be rotated with a key (missing). It serves to mark the position reached by the mercury column.

Truchot (1879), p. 302

19945-0000- (Fig. 21)
Thermometer (broken)
Late 18th c.
MAM accession date: 1952
Pierre François Dumotiez (1743–1817), Louis Joseph Dumotiez (1757–1815)
38 × 2.5 × 1.5; 20 g
Glass, paper

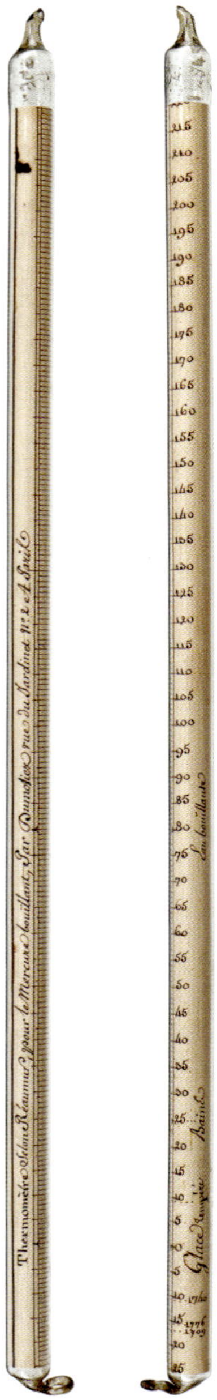

FIGURES 21A, 21B
(inv. no. 19945-0000-)
© MAM/PHOTO FRANCK BOTTÉ

Broken thermometer with a paper Réaumur scale divided into degrees from −25° to 220°, with temperatures marked every 5° The paper carries the inscriptions "1709" (−17°), "1740" (−15.5°), "Glace" (0°), "Temperé" (9.5°), "Bains" (25°), "Eau bouillante" (80°) and "Thermomètre Selon Réaumur, pour le Mercure boüillant, Par Dumotiez rue du Jardinet N° 2 à Paris".

19930-0000- (Fig. 22)
Minimum thermometer (alcohol)
Late 18th c.–early 19th c.
MAM accession date: 1952
30 × 7.7 × 1.5; 305 g
Glass, slate, spirit

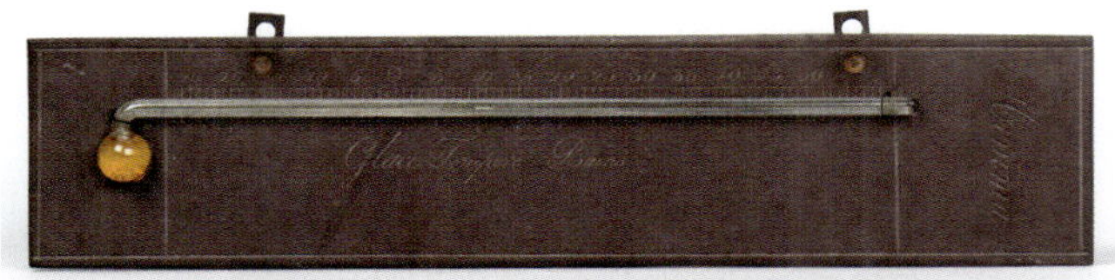

FIGURE 22 (inv. no. 19930-0000-)
© MAM/PHOTO FRANCK BOTTÉ

Minimum alcohol thermometer mounted horizontally on a slate plate. The tube, which is bent above the spherical bulb, contains a minuscule glass thread immersed at the end of the alcohol column. This is used as sliding index. The scale, engraved on the slate, is divided into degrees from −27° to 64°, with temperatures marked every 5°. There are also the inscriptions "Glace", "Tempere", "Bains" under the scale and "Minim." on the right side of the plate.
When the temperature decreases, the alcohol contracts and the glass index moves with it. With an increase of temperature, the alcohol expands but leaves the index at the point of maximum contraction. The index therefore indicates the minimum temperature. Before a new measurement, the index could be reset at the end of the alcohol column simply by inclining the thermometer.

This type of minimum thermometer (similar to the maximum one) was designed in 1794 by the Scottish physician, chemist and botanist Daniel Rutherford (1749–1819). It became immensely popular and is still widely used. Given the date of its introduction, it is uncertain whether the instrument belonged to Lavoisier or entered the collection after his death. The French inscriptions suggest that it did not belong to Rumford.

Nicholson, (1809) vol. 6, "Thermometer" entry; Middleton (1966), pp. 152–154

19909-0000-, 19909-0001-, 19909-0002-, 19909-0003-, 19909-0004-, 19909-0005- (Fig. 23)
Box with four mercury thermometers
Late 18th c.
MAM accession date: 1952
Box
19909-0005: 61 × 18.5 × 5; 2.28 kg
Thermometers
19009-0001-: 53 × 2.7 × 2.5; 70 g
19009-0002-: 50 × 2.7 × 2.5; 70 g
19009-0003-: 56 × 2.7 × 2.5; 110 g
19009-0004-: 56 × 2.7 × 2; 105 g
Glass, mercury, cork, fibre

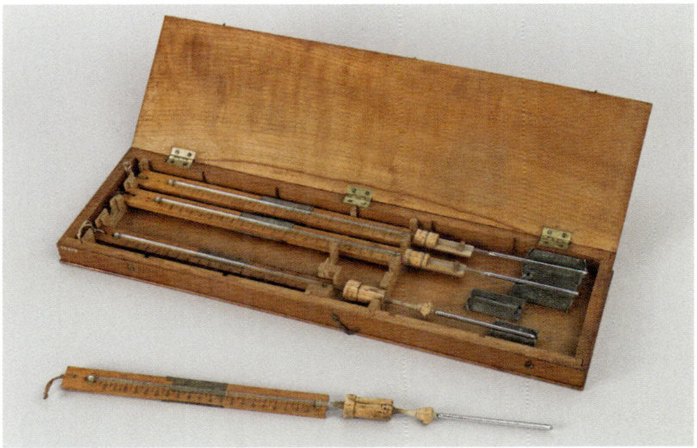

FIGURE 23 (inv. nos. 19909-0001-, -0005-)
© MAM/PHOTO FRANCK BOTTÉ

Set of four mercury thermometers with an elongated bulb preserved in a wooden box (two of the three hooks of the box are missing). The thermometers are all mounted on a wooden tablet with a handwritten scale and inserted in a cork. Each thermometer has a small loop of thread at the top.

The scales are different:

19009-0001-: from 20° to 225° with divisions every 5 degrees
19009-0002-: from 10° to 230° with divisions every 5 degrees
19009-0003-: from 30° to 225° with divisions every 5 degrees
19009-0004-: from 30° to 215° with divisions every 5 degrees

Along each scale is a crudely made brass sliding nonius with handwritten divisions for readings reading to 1°. Corks are inserted in the thermometric tubes.

19915-0001-, 19915-0002-, 19915-0003-, 19915-0004-, 19915-0005-, 19915-0006-, 19915-0007- (Fig. 24)
Set of 7 mercury thermometers
Late 18th c.
MAM accession date: 1952
19915-0001-: 22.5 × 1; 10 g
19915-0002-: 22.5 × 1.1; 10 g
19915-0003-: 22.5 × 0.5: 10 g
19915-0004-: 24 × 0.5; 10 g
19915-0005-: 24.5 × 0.5; 10 g
19915-0006-: 23 × 0.6; 20 g
19915-0007-: 23.5 × 0.6; 15 g
Glass, mercury

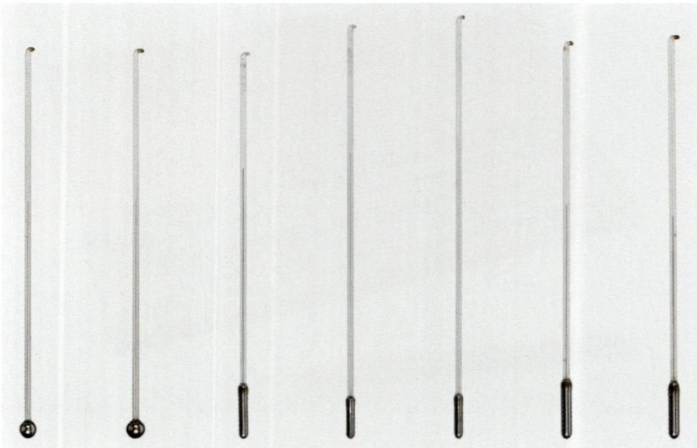

FIGURE 24 (inv. nos. 19915-0001-, -0007-)
© MAM/PHOTO FRANCK BOTTÉ

Set of seven very similar mercury thermometers. Two have a spherical bulb, and the five others a cylindrical one. On the thermometric tubes there are a few barely visible marks, but no scale.

19916-0000- (Fig. 25)
Mercury thermometer
Late 18th c.
MAM accession date: 1952
32.5 × 13 × 1.5; 50 g
Glass, mercury, wood, fibre

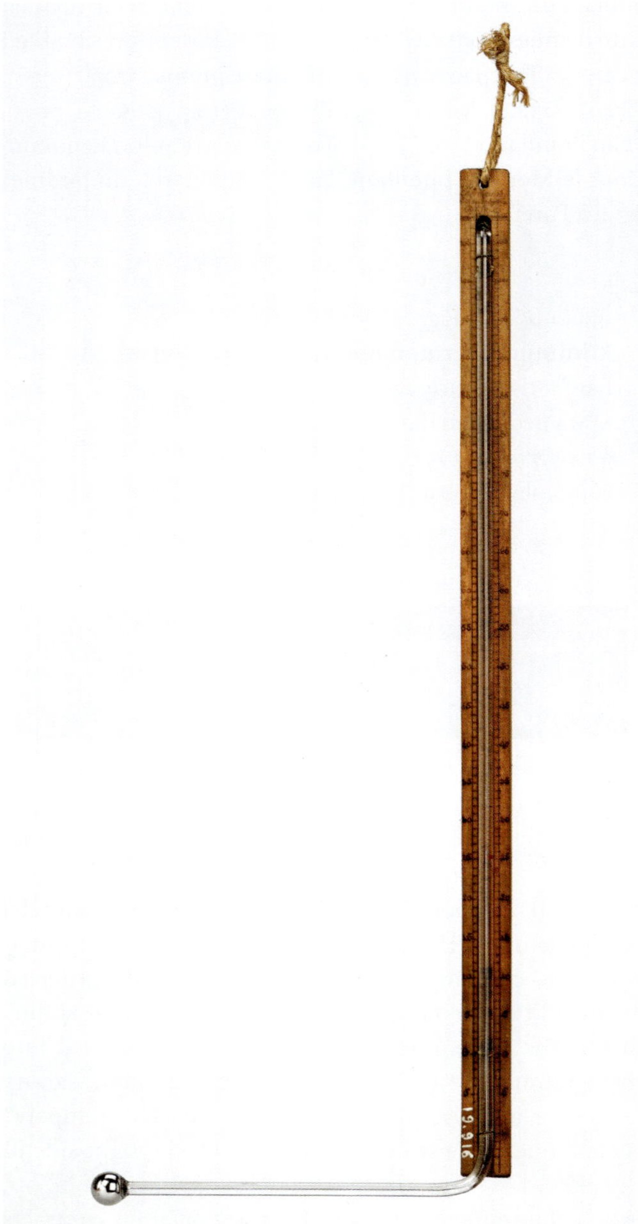

FIGURE 25 (inv. no. 19916-0000-)
© MAM/PHOTO FRANCK BOTTÉ

Mercury thermometer with a bent (L-shaped) tube and a spherical bulb. A wooden scale, which is fixed to the thermometric tube, is divided into degrees from −15° to 90° (?),[8] with markings every 5°. There is a thread for hanging the instrument.

8 The upper part of the scale is barely legible.

19928-0000- (Fig. 26)
Alcohol thermometer
Second half of 18th c.
MAM accession date: 1952
60 × 4.5 × 1.5; 190 g
Brass, glass, alcohol

Thermometer with a spiral-shaped bulb filled with slightly coloured alcohol and mounted on a long brass plate. A curved brass strip protects the spiral, but the plate has no scale or graduation.

19927-0000- (Fig. 27)
Alcohol thermometer
Second half of 18th c.
MAM accession date: 1952
50.5 × 4.2 × 1; 90 g
Wood, brass glass, alcohol

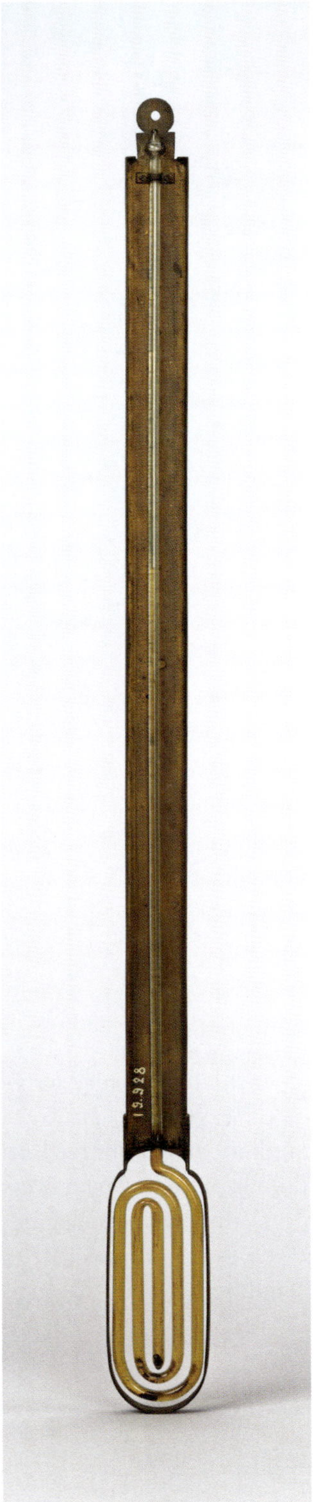

FIGURE 26
(INV. NO. 19928-0000-)
© MAM/PHOTO FRANCK BOTTÉ

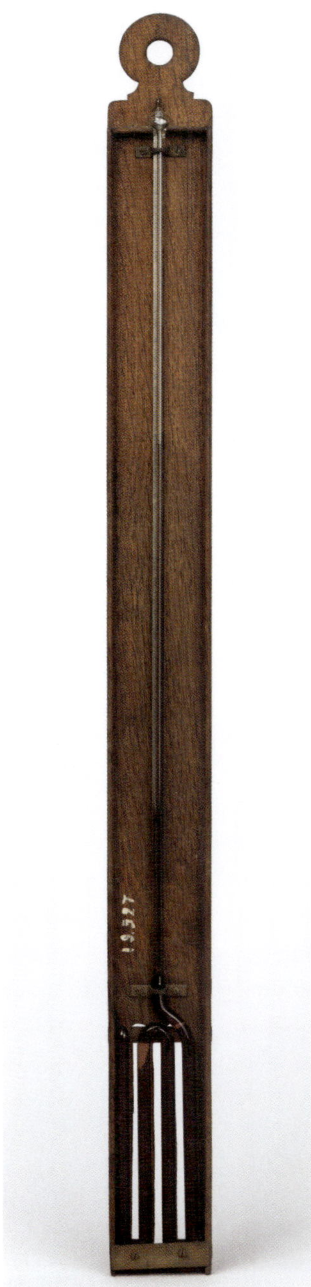

FIGURE 27
(inv. no. 19927-0000-)
© MAM/PHOTO FRANCK BOTTÉ

Thermometer with a long bulb bent several times and filled with dark red coloured spirit. It is mounted on a long wooden board in a frame with no scale or markings.

19921-0000- (Fig. 28)
Mercury thermometer
Second half of 18th c.
MAM accession date: 1952
66 × 4 × 2; 240 g
Brass, glass, mercury

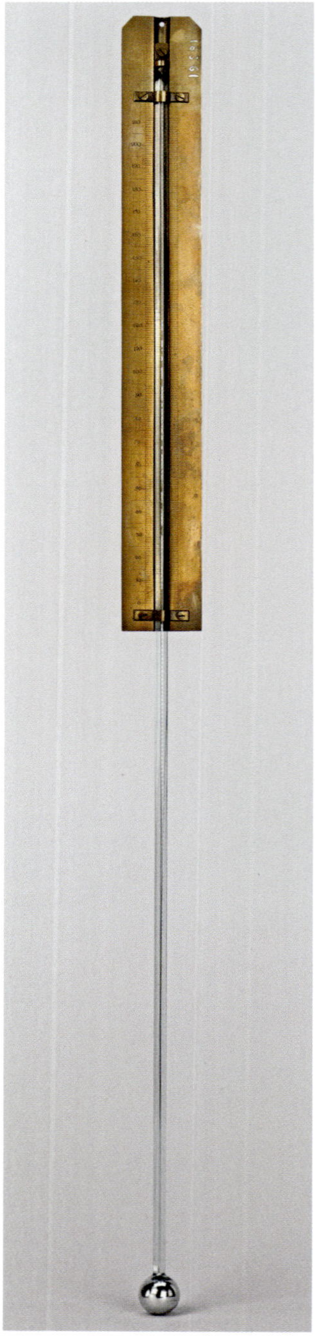

FIGURE 28
(inv. no. 19921-0000-)

Mercury thermometer mounted on a long brass plate bearing a Fahrenheit scale divided into degrees from 0° to 212°, with temperatures marked every 10°. Remarkably, the thermometric tube is more than twice as long as the scale.

19933-0000- (Fig. 29)
Mercury thermometer
Second half of 18th c.
MAM accession date: 1952
56 × 4 × 4; 345 g
Glass, brass, wood mercury

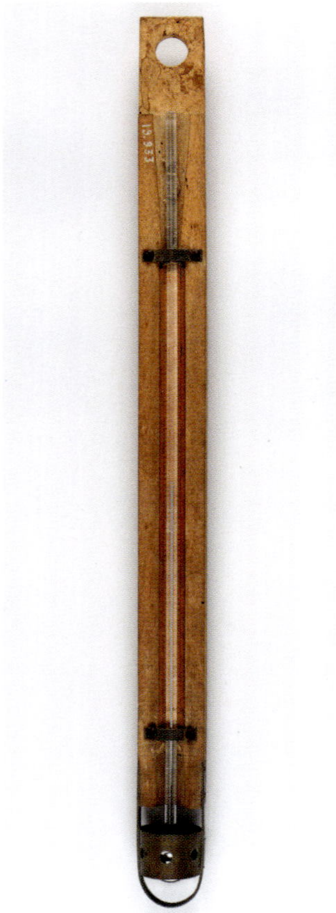

FIGURE 29
(inv. no. 19933-0000-)

Mercury thermometer with a capillary tube mounted on a wooden board. The spherical bulb is protected by a brass guard. The barely legible scale is divided into degrees (probably Fahrenheit) from 30° to 100° and also carries the markings "F", "T", "SH", "BH" and "FH".

19918-0000- (Fig. 30)
Mercury thermometer
Second half of 18th c.
MAM accession date: 1952
47 × 2 × 0.5; 110 g
Glass, mercury, brass

Mercury thermometer mounted on a brass plate with no scale (the only marking on it is a line). The capillary tube is broken and the bulb is separated from it.

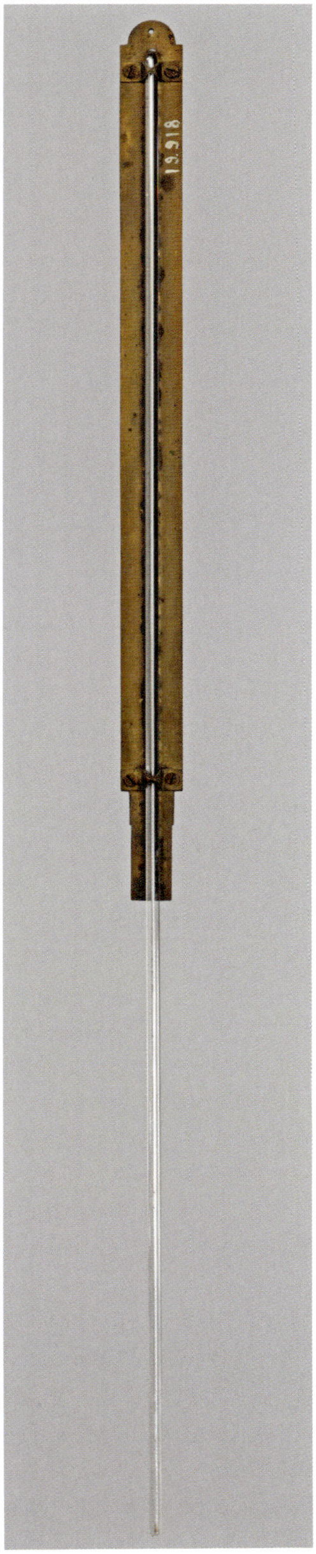

FIGURE 30
(inv. no. 19918-0000-)

19919-0000- (Fig. 31)
Mercury thermometer
Second half of 18th c.
MAM accession date: 1952
37.5 × 0.7 × 0.7; 20 g
Glass, mercury

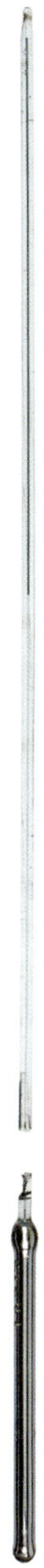

FIGURE 31
(inv. no. 19919-0000-)

Broken thermometer with an elongated bulb bearing the inscription "42 1/3" engraved with a diamond.

19920-0000- (Fig. 32)
Mercury (?) thermometer
Second half of 18th c.
MAM accession date: 1952
31.7 × 1 × 1.1; 30 g
Glass, ivory

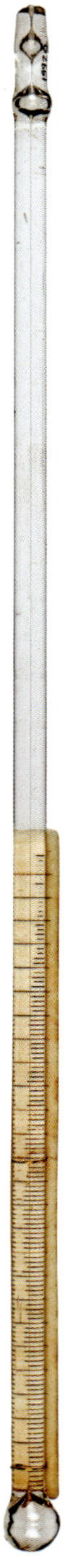

FIGURE 32
(inv. no. 19920-0000-)
© MAM/PHOTO FRANCK BOTTÉ

Empty thermometric tube with a small spherical bulb. There are two other small bulbs at the top of the capillary, which is broken.[9] A fragment of a divided ivory scale is attached to the tube with a wire.

19913-0000- (Fig. 33)
Tube for mercury thermometer
Second half of 18th c.
MAM accession date: 1952
50.5 × 2; 130 g
Glass

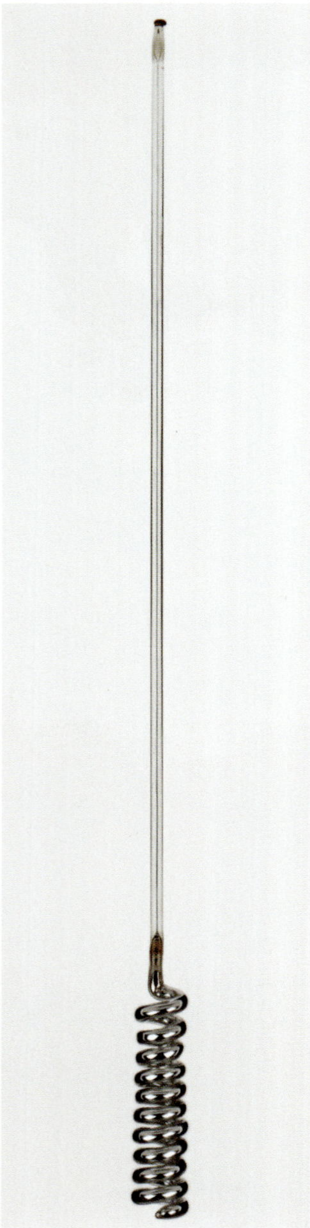

FIGURE 33
(inv. no. 19913-0000-)
© MAM/PHOTO FRANCK BOTTÉ

9 This thermometer was probably filled or had to be filled with mercury.

Mercury thermometer tube with a helical bulb and no scale.

19924-0000- (Fig. 34)
Tube for mercury thermometer
Second half of 18th c.
MAM accession date: 1952
54.5 × 1; 50 g
Glass, mercury

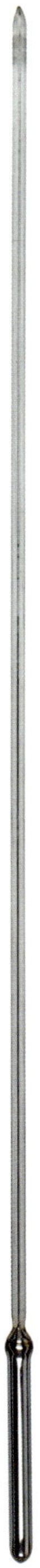

FIGURE 34
(inv. no. 19924-0000-)

Non-graduated tube with elongated bulb of a mercury thermometer.

19944-0001-, 19944-0002- (Fig. 35)
Two thermometric tubes (broken)
Late 18th c.
MAM accession date: 1952
19944-0001-: 28.2 × 0.8 × 0.7; 10 g
19944-0002-: 29.5 × 1.3 × 0.7; 10 g
Glass, mercury

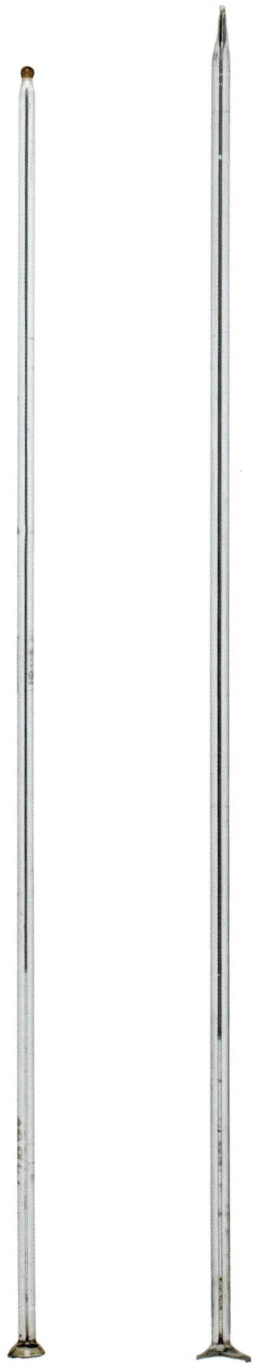

FIGURE 35
(inv. nos. 19944-0001-, 19944-0002-)

Two thermometric tubes with bulbs broken and missing. A few traces of mercury remain in the tubes. The tube -0002- bears the mark "No 31" (made with a diamond) near the top.

19946-0000- (Fig. 36)
Tube with thermometric scale (broken)
Late 18th c.
MAM accession date: 1952
Fragment 1: 35.7 × 1.1
Fragment 2: 27.8 × 1.1
Total mass: 55 g
Glass, paper

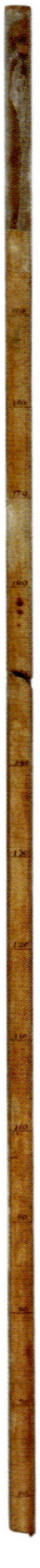

FIGURE 36
(inv. no. 19946-0000-)

Thermometric tube broken into two pieces. There are two paper strip scales glued to the front and back of the longest tubes. One is marked with lines and the other is numbered.

The intervals between 60, 70 and 80 are marked without sub-divisions; those from 80 to 220 are divided into ten parts. This tube may have been used as graduated slider for the thermometer 07547-0005-001.

Barometers

19949-0000- (Figs. 37a, 37b)
Double tube barometer
1779
MAM accession date: 1952
Pierre Bernard Mégnié (Mégnié le Jeune) (1751–1807)
174 × 43 × 39.5; 10.5 kg
Wood, silvered brass, brass, glass, mercury, ivory

This barometer is very similar to no. 08761-0000- and is one of the eight made by Mégnié for Lavoisier.[10] Two barometric tubes are fixed to brass chassis inserted in a wooden frame with a suspension ring. A silvered brass scale ranging from 21 to 29.2 Paris inches is divided into one-fourths of a line. A nonius moved by a rack-and-pinion system allows a reading precision of 1/25th of a line. The tube positions can be adjusted up and down slightly with a screw and threaded ring at the top of the brass frame. By immersing the bottoms of the tubes at a greater or lesser depth in the mercury, one can thus modify the mercury level in the cistern and adjust the zero. The cistern is made of ivory and enclosed in a cylindrical wooden case with two small glass windows. A mercury thermometer is fixed on the front of the instrument and its centigrade scale from −15° to +65° is divided every degree.

A brass plate at the top of the instrument has the inscription "Barometre // de // Megnié pour Mr de Lavoisier // de l'Academie R.le // des Sciences &c. // N° 2 // 1779". "n 2" is engraved at the top of the scale.

The barometer is hanging from an arched support which is fixed to a small four-legged table.

On the particular features and use of this type of barometer, see no. 08761-0000-.

Lavoisier (1779?); Cotte (1788), vol. 1, pp. 500–501; Truchot (1879); Middleton (1964), pp. 215 and 250; Bolle (1983), p. 86

10 The inventories of Fréschines record a barometer made by Mégnié numbered "14°" but we know that Lavoisier sent a few of his barometers made by Mégnié to other scientists; Cauchie (1900), p. 64.

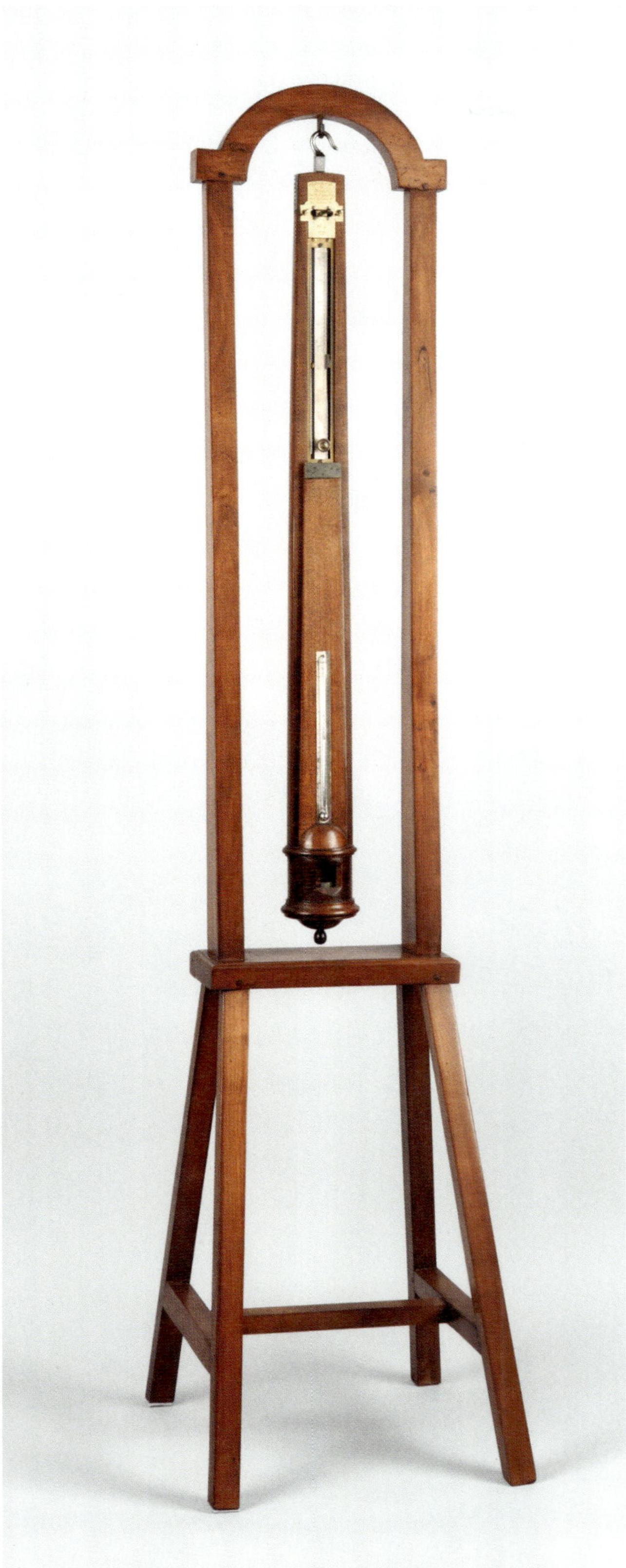

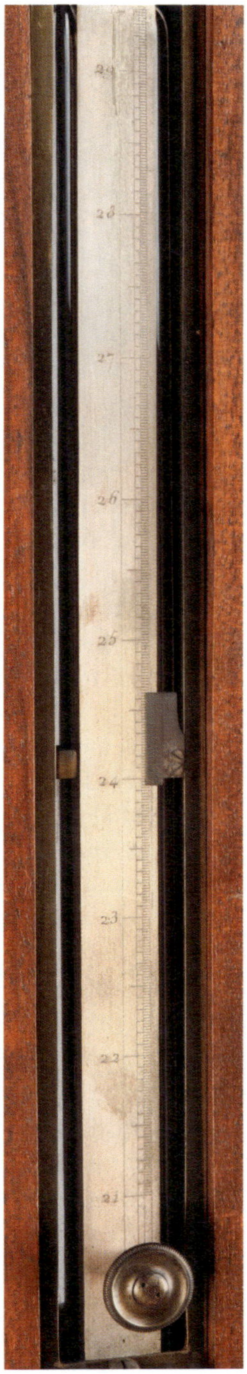

FIGURES 37A, 37B (inv. no. 19949-0000-)
© MAM/PHOTO FRANCK BOTTÉ

07658-0000- (Fig. 38)
Double tube barometer
1779
MAM accession date: 1867
Pierre Bernard Mégnié (Mégnié le Jeune) (1751–1807)
99 × 17 × 16.5; 3.12 kg
Wood, brass, iron, ivory, leather, mercury, glass

Two barometric tubes are inserted with brass holders in a wooden frame (with a suspension hook) mounted on a box, whose front side can be removed. At the top of the frame is a silvered brass scale from 25 to 20 Paris inches with divisions every half-line. A sliding nonius, moved by a rack-and-pinion mechanism, allows a reading precision of 1/50th of a line. The bottom ends of the tubes are in a leather-lined wooden cistern enclosed in the box. A flat C-shaped ivory plate with a vertical needle floated on the mercury. A reference horizontal line engraved on a silvered label indicated the position that had to be reached by the top of the needle in order to obtain a correct reading of the mercury columns. A screw on the back of the frame served to move an ivory cylinder in the cistern in order to regulate the mercury level. A mercury thermometer mounted on the frame has a scale from –15° to 85° with divisions every degree.

On the front of the box a trapezoidal silvered label bears the engraved inscription "BAROMETRE ayant appartenu // A // LAVOISIER // et donné à M.r le B.on Séguiér // Par // MONSIEUR DE CHAZELLES." The inscription shows that the instrument was donated by the de Chazelles family to Armand-Pierre Séguier (1803–1876).[11]

Mégnié made eight barometers for Lavoisier (see item 08761-0000-) incorporating improvements by the English instrument maker Jesse Ramsden (1835–1800).[12] The two independent tubes offered two independent readings. This allowed a double check of the atmospheric pressure and served as insurance against a defective barometric vacuum. Describing the best way to construct precision barometers, Lavoisier stated:

> [...] to ensure the accuracy of those who will build them, the barometer is a double one: there are two tubes on the same board. If the two tubes agree, it will be highly likely that the barometer is well made. If, on the contrary, the two tubes do not agree, that

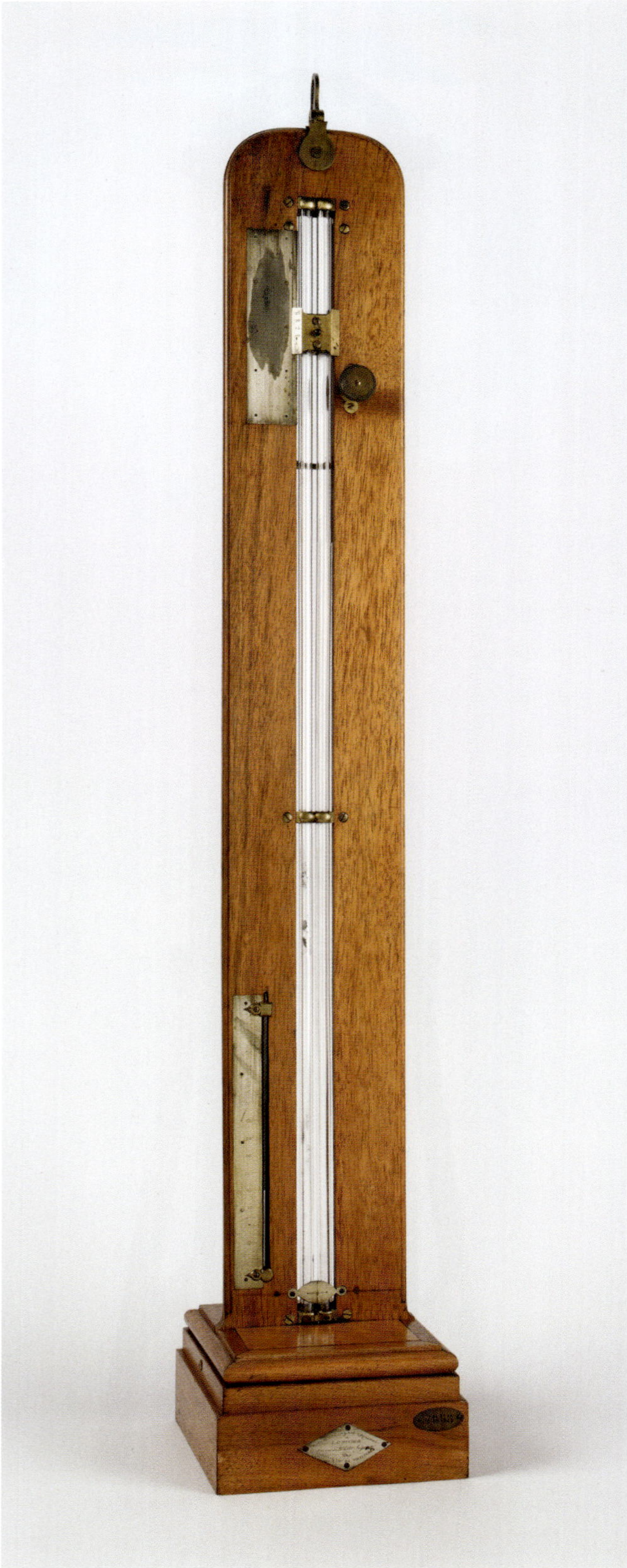

FIGURE 38 (inv. no. 07658-0000-)
© MAM/PHOTO FRANCK BOTTÉ

11 Baron Séguier was a lawyer, inventor and member of the Académie des Sciences.

12 Ramsden greatly improved the reading of the barometer with a better scale, nonius and index.

means the operation was not carried out correctly, and it will be necessary to boil the mercury again.[13]

Lavoisier (1779?); Cotte (1788), vol. 1, pp. 500–501; Truchot (1879), pp. 304–310; Daumas (1955), pp. 126–128; Middleton (1964), pp. 197, 215 and 250

08761-0000- (Fig. 39)
Double tube barometer
1779
MAM accession date: 1876
Pierre Bernard Mégnié (Mégnié le Jeune) (1751–1807)
102.5 × 11 × 9; 2.14 kg
Wood, glass, brass, ivory, mercury

This instrument presents several similarities with the double tube barometer no. 07658-0000- and is one of the eight made by Mégnié for Lavoisier. Two barometric tubes are fixed in a brass chassis inserted in a wooden frame with a brass suspension ring. A silvered brass scale from 21 to 29 Paris inches is divided into fourths of a line (every inch is divided into 48 parts). A sliding nonius moved by a rack-and-pinion mechanism allows a reading precision of 1/100th of a line. The tubes can be moved up and down slightly with a screw and threaded ring, placed at the top of the brass frame. As a result, by immersing the bottom of the tubes to a greater or lesser depth in the mercury, one can modify the mercury level in the cistern and thus adjust the zero. The cistern, made of ivory, is enclosed in a cylindrical wooden case with two small glass windows. A removable wooden plate, which partly conceals the tubes, carries a mercury thermometer with a brass silvered scale from 8° to 112° with divisions every degree. A brass plate at the top of the instrument has the inscription "Barometre // de // Megnié Pour Mr de Lavoisier // de l'Academie R.le // des Sciences &c. // N° 6 // 1779".[14]

This precision barometer required considerable craftsmanship from Mégnié, who also devised a special tool (a kind of rule) to determine the scale's exact position. In addition, Lavoisier altered one of the two barometric tubes in order to obtain a flat-surface barometer (*baromètre à*

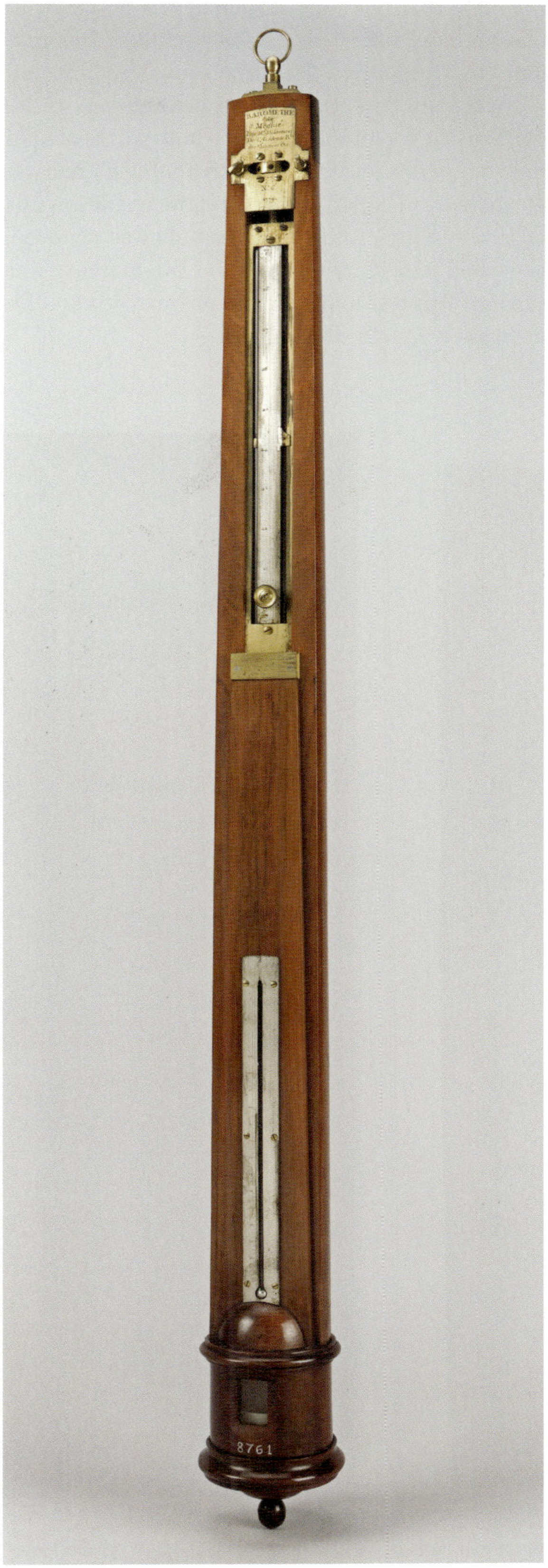

FIGURE 39 (inv. no. 08761-0000-)
© MAM/PHOTO FRANCK BOTTÉ

13 "… afin d'être assuré de l'exactitude de ceux qui les construiront, le baromètre est double: il y a deux tubes sur la même planche. Si ces deux tubes sont d'accord, il en résultera une très-grande probabilité que le baromètre soit bien fait: si au contraire les deux tubes ne s'accordent pas, il en résultera qu'on a mal opéré, et il faudra recommencer à faire bouillir le mercure". Lavoisier (1779?).

14 Truchot erroneously describes this barometer as number 2, whereas the inscription clearly indicates that it is number 6.

surface plane). This type of barometer was designed to have a flat mercury surface and thus eliminated the mercury meniscus. Obtaining a flat surface involved long and delicate operations. Essentially, it was necessary to solder a glass bulb on the top of the tube. The bulb was heated to a very high temperature and filled with boiling mercury. It was then cooled and boiled again – and the operation was repeated several times. At the end, the bulb was removed, the tube sealed with a blow lamp and the mercury was boiled again.[15] This technique did not prove to be very efficient and was eventually abandoned.

Lavoisier (1779?); Cotte (1788), vol. 1, pp. 500–501; Truchot (1879), pp. 304–310; Daumas (1955), pp. 126–128; Middleton (1964), pp. 197, 215 and 250

19954-0000- (Figs. 40a, 40b)
Syphon barometer
J.H.Tor...?
ca. 1750–1780
MAM accession date: 1952
95 × 11.5 × 4; 845 g
Wood, glass, brass, iron, paper

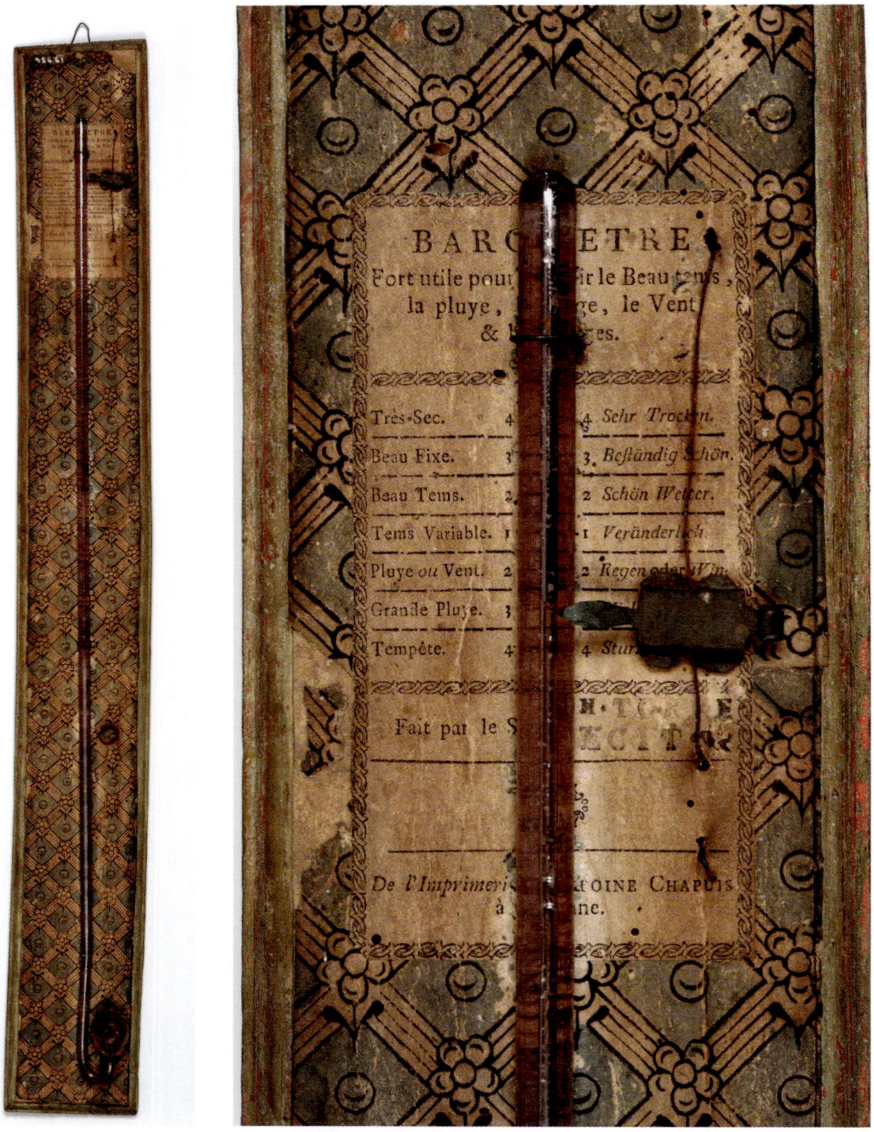

FIGURES 40A, 40B
(inv. no. 19954-0000-)
© MAM/PHOTO FRANCK BOTTÉ

15 The system was described by Lavoisier and in several other treatises. See for example: Monge, Cassini et al., *Dictionnaire de physique* (Paris, 1793) vol. 1, p. 107. In the multiple discussions on how to obtain a flat surface, it was suggested that the presence of mercury oxide could change the surface tension of the mercury. The problems related to the meniscus were investigated throughout the nineteenth century and several systems for improving the precision of mercury barometers and manometers were experimented.

Siphon barometric tube mounted on a wooden board covered with decorated paper. The tube is broken and the glass is badly deteriorated. On the top of the tube is an indicator sliding along an iron wire. A printed label carries the inscription "BAROMETRE // Fort Utile pour prévoir le Beau temps // la pluye la neige le Vent // & l' Orage [?]" as well as a bilingual list with the weather indications (with no graduated scale): "Très sec 4 4 Sehr Trocken // Beau Fixe 3 3 Beständig Schön // Beau Tems 2 2 Schön Wetter // Tems Variable 1 1 Veränderlich // Pluye ou Vent 2 2 Regen oder Wind // Grande Pluye 3 3 Viel Regen // Tempête 4 4 Sturm". The lower part of the label reads: "Fait par le Sieur J.H TOR […] // FECIT" and "De l'Imprimerie […] l'ANTOINE CHAPUIS à Lausanne".[16] The name of the instrument maker is not entirely visible.

The characteristics of this barometer show that it was a domestic instrument and not a true scientific apparatus.

Sigaud de La Fond (1775), vol. 2, p. 103 and plate IX, fig. 2

19952-0000- (Fig. 41)
Siphon barometer with thermometer (Mannheimer barometer)
1781
MAM accession date: 1952
95.5 × 15 × 5 (open); 865 g
Wood, glass, paper, brass, mercury

Portable siphon barometer and mercury thermometer inserted in a pair of hinged wooden boards forming a case. The mercury thermometer has a cylindrical bulb. A Réaumur scale from −20° to 80° is written on paper, with temperatures marked every 10°.

The siphon barometer, whose bulb is broken, has a brass scale from 22 to 29 Paris inches divided into lines. A sliding nonius allows a precision of 1/10th of a line. The top of the barometer bears the inscription "Carolus Theodor // Elector Palatinus // Musagetes 1781". A round paper label on the barometer board displays a uroborus, a masonic square and a Phrygian cap (all printed). The number "33" is inscribed on the outer part of the case.

Karl Theodor (1724–1799) was Prince Elector of the Palatinate and Bavaria. Encouraged by his chaplain Johan Jacob Hemmer, who was also a natural philosopher and linguist, he founded the *Societas Meteorologica Palatina* in 1780. The Society was the first to organise a network of 39 meteorological observatories and contributed to

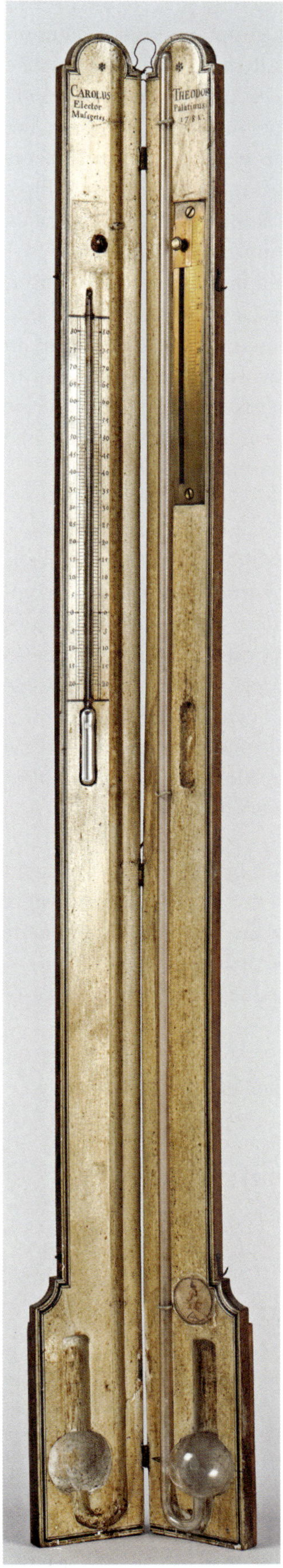

FIGURE 41
(inv. nc. 19952-0000-)
© MAM/PHOTO FRANCK BOTTÉ

16 Antoine Chapuis (1710?–1786) was a famous printer in Lausanne, where he opened his shop in 1737.

the development of modern meteorology. It distributed barometers, thermometers, hygrometers and declinometers to the observatories.[17] All these instruments, like this barometer in the Lavoisier collection, were made in Mannheim under the supervision of Hemmer, the Society's secretary. One of the best instrument makers working with Hemmer was Artaria, who was probably a member of the homonymous family of music and book printers and art dealers in Vienna.

Lavoisier had been interested in meteorological observation since 1770, when Hatton La Ganière began to send him regular meteorological reports. However, it was in 1784, according to Madame Lavoisier, that he thought to create a meteorological network reporting barometric observations taken from a series of barometers he had made by Mégnié. Madame Lavoisier writes as follows:

> In around 1784, Lavoisier believed it was very important to compile daily and accurate barometric observations in different parts of the globe. He commissioned the construction of twelve barometers, all in agreement, by the best workman. He placed one at the Paris Observatory, one in the Auvergne, he entrusted one to the consul in Baghdad, one was sent to Russia, another to Sweden. I do not know the outcome of this fine idea. Only the consul in Baghdad[18] send a few observations.[19]

One of these barometers made by Mégnié was brought by the abbé Claude Bertrand (1757–1792) to Dijon where Madame de Picardet, Guyton de Morveau's wife, made several observations (LC, vol. 5, pp. 156–157 and 163).
There is no documentary evidence of Lavoisier's use of the Mannheimer barometer.

Ephemerides (1783), pp. 57–90; Traumüller (1885); Bret (2008)

19948-0000- (Fig. 42)
Cistern barometer
Mossy (active ca. 1775–1821) and Jean-François Richer (1743–ca. 1820)
1789
MAM accession date: 1952
100.5 × 14 × 13; 4.16 kg
Brass, silvered brass, glass, mercury

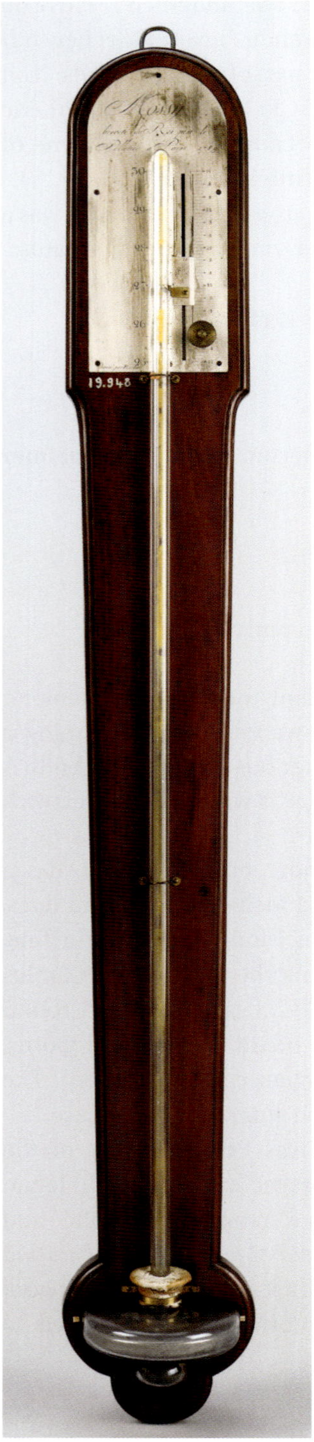

FIGURE 42
(inv. no. 19948-0000-)
© MAM/PHOTO FRANCK BOTTÉ

17 See also the declinometer in this collection, Inv. 20136-0000-.
18 The Abbé Beauchamp.
19 "Vers 1784 Lavoisier regarda comme très important d'établir dans différentes parties du globe des observations de baromètres journalières et exactes. Il fit construire par le meilleur ouvrier douze baromètres tous d'accord. Il en plaça un à l'observatoire de Paris, un en auvergne, il en confia un au soins du consul de bagdad, un fut envoyé en Russie, un en Suéde. J'ignore quel fut le succès de cette belle idee. Le consul de bagdad seul envoya quelques observations". Gillispie (1956), p. 59.

Cistern barometer mounted on a wooden board. The cistern is a large flat vessel with a bulb under it.[20] The scales are engraved on a silvered brass plate. One ranges from 25 to 30 Paris inches and is divided into lines. The second scale is also divided into lines, with the 4-line intervals marked "4", "8" and "12" respectively (repeated four times). A nonius sliding on the second scale, activated by a rack-and-pinion mechanism, allows a reading precision of 1/10th of a line. The silvered plate bears the inscriptions "Mossy // Breveté du Roi quai Le Pelletier // à Paris 1789" and "divisé par Richer".

Sigaud de La Fond (1775), vol. 2, p. 100 and plate IX, fig. 1; Chevalier, Fau (1854), vol. 1, p. 55 and plate 7, fig. 61; Daguin (1861), vol. 1, p. 281

19953-0000- (Fig. 43)
Siphon barometer
Early 19th c.
MAM accession date: 1952
Chabrol de Murol
94.5 × 11 × 4.5; mass 575 g
Glass, wood, mercury, cork, iron

Siphon barometer mounted on a wooden tablet with a scale printed on paper protected by a glass window. The top carries the inscription "BAROMETRE // Selon Toricelli // fait par Murol". The scale is divided into Paris inches from 26 inches and 4 lines to 29 inches and 8 lines. The scale also bears the following inscriptions: "Très sec" (29″), "Beau fixe" (28″ 8‴), "Variable" (28″), "Pluie" (27″ 8‴), "Grande pluie" (27″ 4‴), "Tempête" (27″), "Paris 1768" (26″ 8‴) and Paris "1821" (26″ 4‴). A metal wire acting as a pressure indicator slides on a thin vertical bar along the scale. The instrument, made in the nineteenth century, probably belonged to Madame Lavoisier.

Sigaud de La Fond (1775), vol. 2, p. 103 and plate IX, fig. 2; Chevalier, Fau (1854), vol. 1, p. 55 and plate 7, fig. 62; Daguin (1861), vol. 1, pp. 281–282

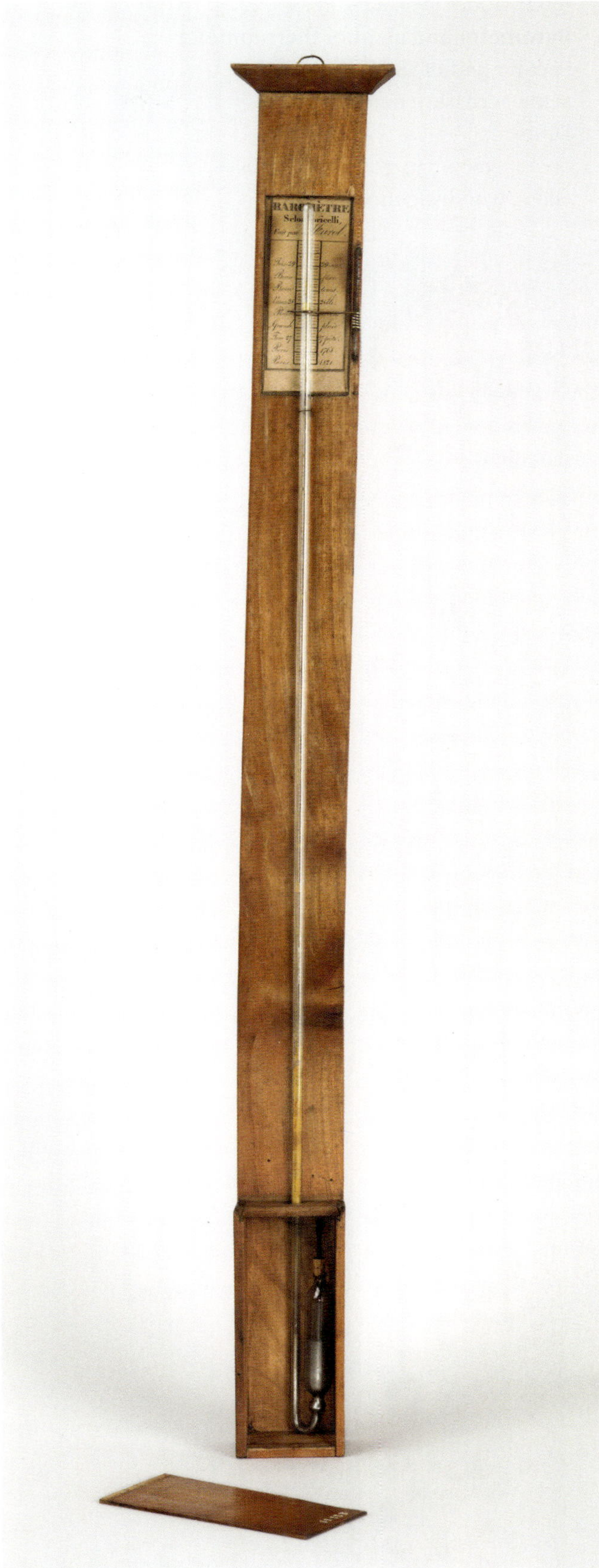

FIGURE 43 (inv. no. 19953-0000-)
© MAM/PHOTO FRANCK BOTTÉ

20 While the cistern has a large diameter compared with that of the tube, the variation in the height of the mercury column has very little influence on the mercury level in the cistern. In this thermometer, the tube is empty and only traces of mercury remain in the cistern.

19947-0000- (Fig. 44)
Barometer and alcohol thermometer
Second half of 18th c.
MAM accession date: 1952
Dinon
96 × 15.3 × 3; 730 g
Glass, wood, alcohol

Siphon barometer and Réaumur alcohol thermometer with spherical bulb mounted on a wooden board (in poor condition) decorated with some painted leaves and flowers. There is a barely visible inscription: "Barometre Thermometre de Reaumur par Dinon". The thermometer tube is broken at the top. The scale ranges from −18° to +41°. The writing is faded but the inscription "Glace" is still legible. The bulb of the barometric tube's siphon is broken.

Sigaud de La Fond (1775), vol. 2, p. 103 and plate IX, fig. 2

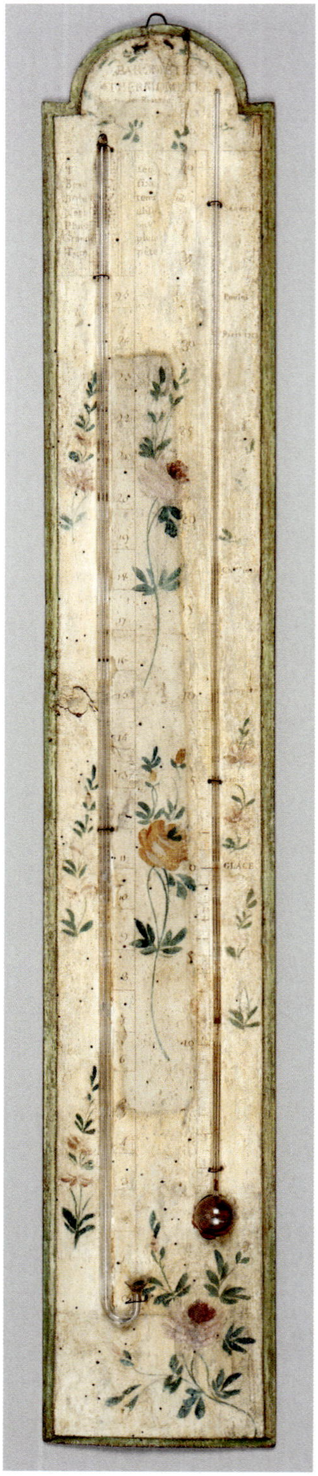

FIGURE 44
(inv. no. 19947-0000-)
© MAM/PHOTO FRANCK BOTTÉ

19950-0000- (Fig. 45)
Siphon barometer tube
Second half of 18th c.
MAM accession date: 1952
88.7 × 7 × 3; 115 g
Glass, cork, sealing wax

Barometric tube with a siphon closed by a cork and sealing wax. A glass rod with a ring at the top and a bulb at the bottom is inserted in the tube. A second, shorter siphon tube is connected to the lower part of the barometric tube. A line indicated the mercury level in the shorter tube needed for a correct reading of the scale. The level could be adjusted simply by pushing the glass rod into the siphon to a greater or lesser depth. In this way the level of the mercury in the open arm of the tube was maintained constant.

Truchot (1879), p. 310

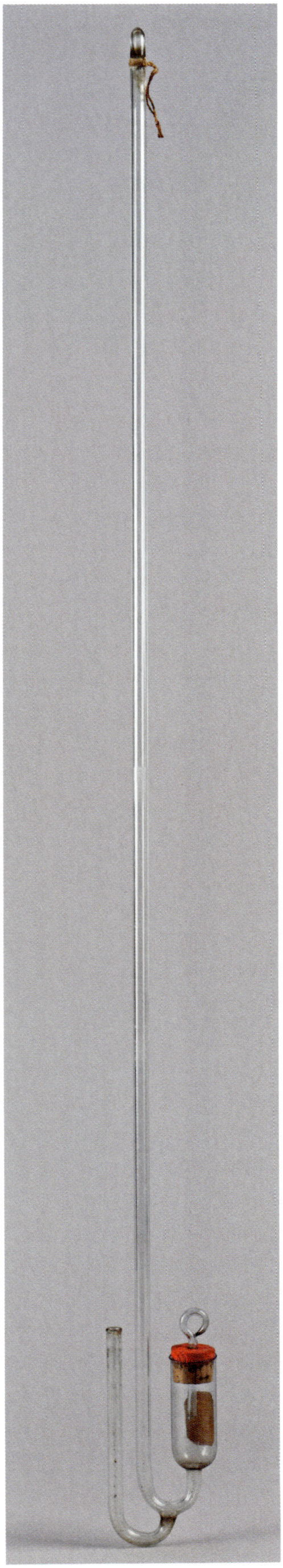

FIGURE 45
(inv. no. 19950-0000-)
© MAM/PHOTO FRANCK BOTTÉ

19955-0000- (Fig. 46)
Wooden tablet for barometer
Second half of 18th c.
MAM accession date: 1952
106 × 8.5 × 6.2; 620 g
Wood, brass

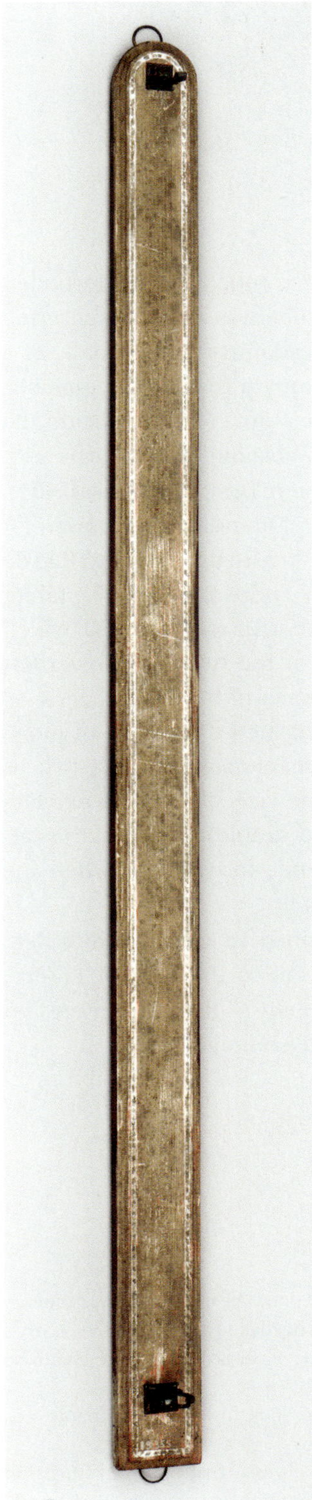

FIGURE 46
(inv. no. 19955-0000-)
© MAM/PHOTO FRANCK BOTTÉ

Long wooden tablet with traces of painted decoration and two brass supports. A barometric tube (missing) was originally mounted on the tablet.

Miscellaneous

19932-0000- (Fig. 47)
Hygrometer
ca. 1780
MAM accession date: 1952
Guillaume Charles Goubert (1747–1824)
43 × 6 × 2.5; 180 g
Glass, wood, copper

Hygrometer composed of a glass tube fixed on a wooden tablet. The lower end of the tube was originally inserted and sealed at the end of the calamus (hollow shaft) of a goose feather (now missing: only a fragment remains)[21] filled with mercury. The scale from 0° to 67° bears the inscriptions "Glace // humidité absolue" (0), "G. de humidité" (15), "Pluie" (28), "Tems moïen" (40), "Secheresse" (60), "Beau tems" (60). At the top: "Nouvel // higromètre // Comparable // Par Goubert // Artiste de la Société R.le de Medecine"; at the base: "Rüe St André des Arts". The tablet carries a fragment of a copper strip with a screw, which probably held the lower end of the tube. The lower left part of the tablet also shows traces of burning.

The instrument roughly indicated the degree of moisture of the air. Like many organic substances (such as bones, hair and bladders), the size of a feather's calamus varies with atmospheric humidity. When the latter decreases, the calamus tends to contract; when the humidity increases, it expands.[22]

The instrument was described in 1779 by Noël Retz (1758–1810?),[23] a naval physician to the king of France, who modified De Luc's hygrometer.[24] Retz was interested in meteorology and its influence on medicine.

Retz (1779); Goubert (1785), pp. 78–80

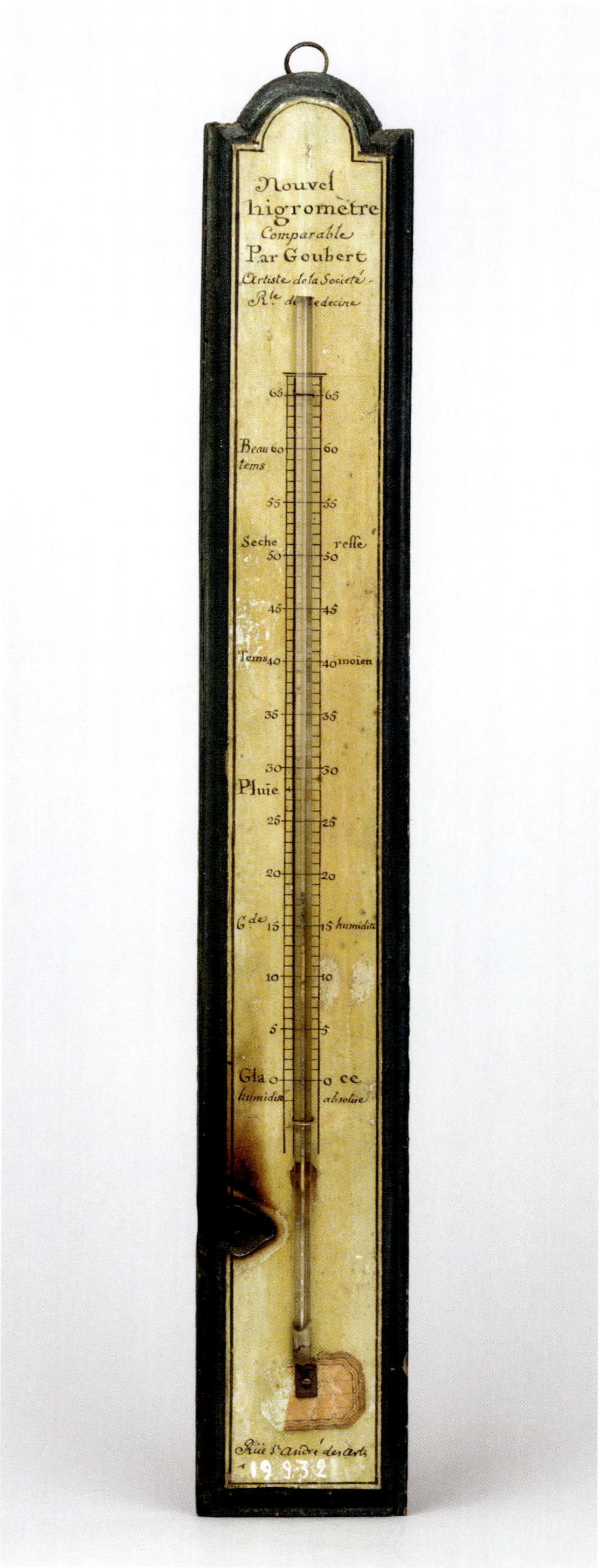

FIGURE 47 (inv. no. 19932-0000-)

21 The calamus had to be thinned out to be used in the hygrometer.
22 To measure humidity, one also needed to allow for the mercury's thermal expansion. The reading of the scale therefore had to be corrected.
23 Although generally known as Noël, his real given name was Hubert.
24 In De Luc's hygrometer, the small vessel that changed dimension according with atmospheric humidity was made of ivory.

20106-0000- (Fig. 48)
De Saussure hair hygrometer
1788
MAM accession date: 1952
Jean François Richer (1743–1820 ca.)
49 × 17.7 × 17.6; 2.46 kg
Brass, silvered brass, glass, wood, hair

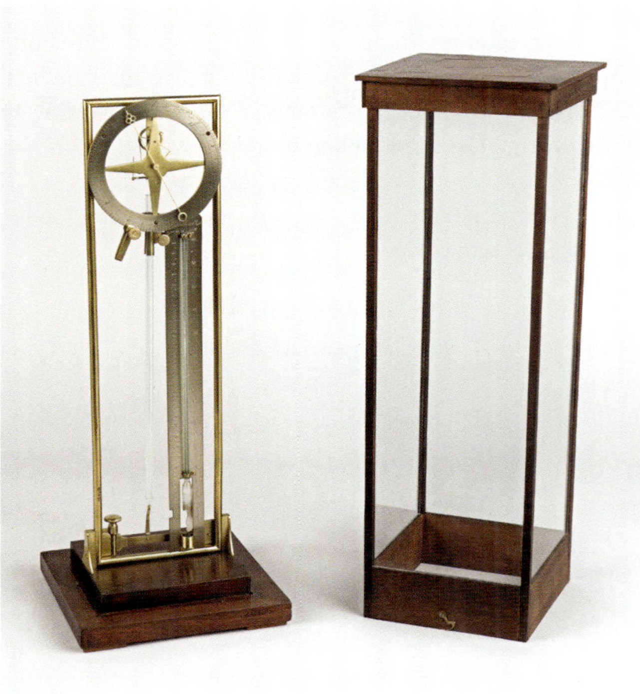

FIGURE 48 (inv. no. 20106-0000-)

A slightly modified version of the instrument described in 1783 by the Swiss physicist, meteorologist, geologist and explorer Horace-Bénédict de Saussure. De Saussure made extensive studies on hygrometry and designed two types of hair hygrometers. The "Grand hygromètre" is similar to the one described here, while the "Hygromètre portable", of simpler construction, was extremely popular between the late eighteenth early twentieth centuries.

The instrument is composed of a rectangular brass frame holding a ring scale and a thermometer whose bulb is missing. A hair protected by a vertical glass tube is fastened at the bottom to a small plier held by a lever. The lever's inclination can be varied with a vertical screw. Thanks to this system it is possible to set the position of the indicating hand. The upper end of the hair is fixed to a pulley connected to a pinion. The pinion is engaged in a toothed sector whose axle carries the indicator hand and a second larger pulley. A counterweight, which slides along two vertical pins, is connected to the pulley with a thread, and maintains the tension of the hair. The silvered scale is divided in 100 units with indications every 10 units. It also bears the inscriptions: "Sec" (near 0), "Humide" (near 100) and "Richer breveté du Roi N° 10". A thermometer is fixed on a brass plate carrying two scales divided into degrees; on the right, a Réaumur scale from −20° to 80°; on the left, a Fahrenheit scale from −10° to 212°. Temperatures are inscribed every 10°. The plate also bears the partly hidden inscriptions "Richer breveté du Roi à Paris 1788", "Fahrenheit" and "Réaumur". The instrument is mounted on a rectangular wooden base and protected by a glass case.

The hair's length varies with atmospheric humidity. These changes, multiplied by the rack-and-sector mechanism, are indicated by the hand along the scale.

De Saussure (1783); Middleton (1969), pp. 100–106; Archinard (1977)

20030-0000- (Fig. 49)
Storm glass
1780–1790
MAM accession date: 1952
24.5 × 2; 95 g
Glass, water, alcohol, camphor

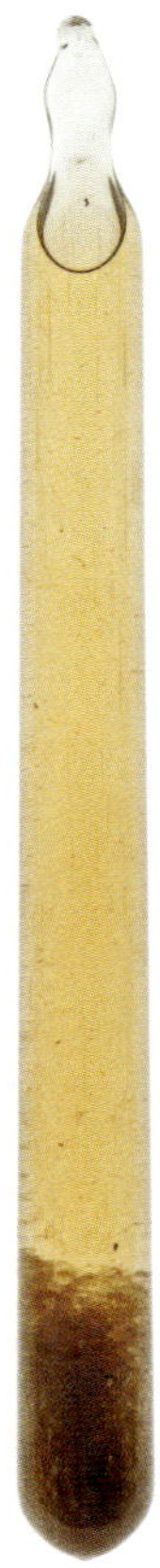

FIGURE 49
(inv. no. 20030-0000-)

The storm glass is a sealed glass tube partly filled with a solution of potassium nitrate, ammonium chloride, and alum in water mixed with a solution of camphor dissolved in alcohol.[25] The storm tube is supposed to indicate changes in the weather. As feathery crystal- or star-like crystals occasionally appear in the liquid, it was generally believed that these phenomena were reliable indicators for forecasting weather. The changes in the liquid were attributed to changes in temperature, light, wind or atmospheric electricity. However, the device – still sold today – is not a reliable scientific instrument.

Storm glasses became very popular in the second half of the nineteenth century thanks to the publicity given to them in Britain by Robert Fitzroy (1805–1865), head of the Meteorological Department of the Board of Trade, and by the famous instrument firm Negretti & Zambra. The origin of the storm glass is obscure and various inventors have been proposed. The device was presented in France in 1780 with the name of "prognostic" by Pierre Legaux (1748–1827), a lawyer in Metz. Legaux was also well known for a series of meteorological observations and for making important improvements to meteorological instruments. As Lavoisier knew Legaux's work, it is not surprising to find a storm glass in his collection.

Lavoisier (1779?); Cotte (1788), vol. 1, pp. 114–115; Collins (2004), pp. 9–23; McConnell (2006)

25 In the literature, one finds several different recipes for the mixture.

Chemistry

The chemical section of Lavoisier's instruments is divided into the following sections: furniture, heating apparatus and common laboratory tools, models, chemical apparatuses, chemical glassware, chemicals, minerals and various substances. Within each section we have arranged the instruments in chronological order, which was of particular importance for the chemical apparatuses. We have included in this section several instruments normally belonging to experimental physics – such as Papin's digester and hydrometers – because Lavoisier made a continuous effort to adapt them to chemical experiments and investigations.

Furniture

20574-0000- (Fig. 1a, 1b)
Large laboratory table
ca. 1775
MAM accession date: 1952
259.5 × 128 × 75; 85 kg
Wood, iron

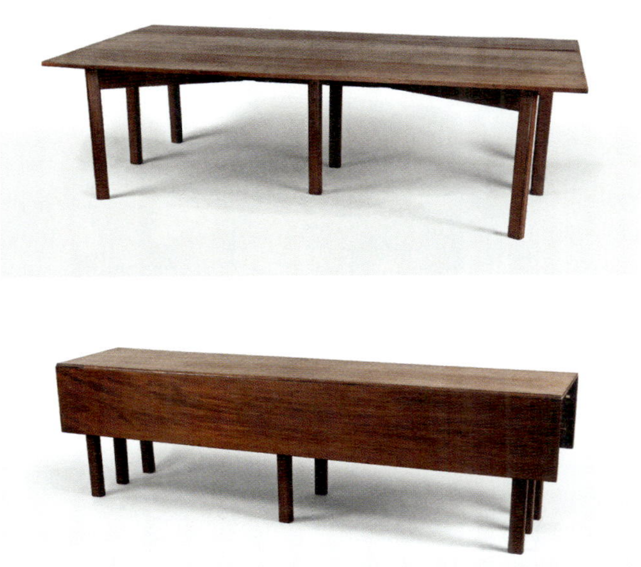

FIGURE 1 (inv. no. 20574-0000-)
© MAM/PHOTO FRANCK BOTTÉ

Lavoisier used what is known as the "Grande Table" for his experiments at the Arsenal laboratory. The table has two hinged longitudinal flaps which can be folded down, reducing the work surface. There are eight legs: one at the center of each side, and one at each corner to support the flaps. Chemical tables became increasingly important during the eighteenth century.

Morris (2015), pp. 52–57

20057-0000- (Fig. 2)
Glassblower's chair
Second half of 18th c.
MAM accession date: 1952
70 × 41 × 80 cm, 7.85 kg
Wood

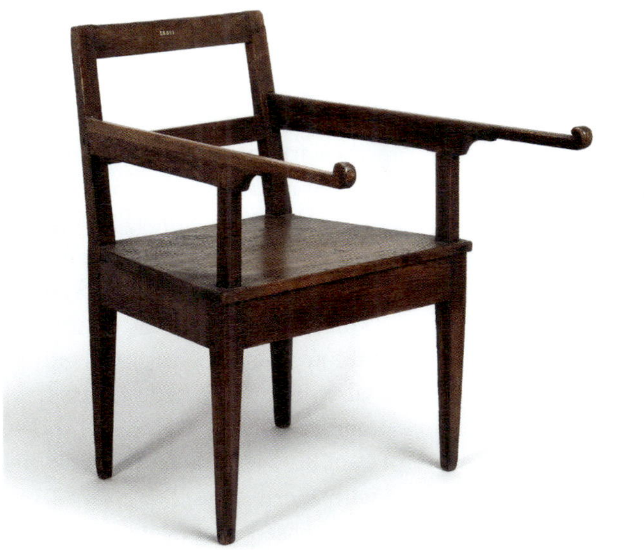

FIGURE 2 (inv. no. 20057-0000-)
© MAM/PHOTO FRANCK BOTTÉ

Wooden chair with seatback and long horizontal armrests. This type of chair, introduced towards the end of the sixteenth century, was used by glassblowers. The blowpipe, holding an incandescent glass mass at one end, was placed transversally on the armrest. The glassworker used one hand to roll the pipe back and forth on the armrest; with the other hand, he shaped the plastic glass mass using appropriate tools. If necessary, at the same time, a crouched glass-blower could blow air into the open end of the moving pipe. Lavoisier's chair is interesting because of its rather uncommon back, as shown in the plates devoted

 | DOI:10.1163/9789004511217_019

to glassworks (*la verrerie*) in Diderot and d'Alembert's *Encyclopédie* (vol. 27, Paris, 1772).

Although glass played a crucial role in eighteenth-century chemical laboratories, not much has been written on how chemists managed to make a vast array of glassware such as thick glass vessels, jars, bells, balloons, alembics, tubes, disks, thermometers, barometers and hydrometers. Many of these pieces were supplied by instrument makers or, more often, by glassmakers, but the growing specialization of chemical glassware made in-house production more common.

Lavoisier's interest in glass dated back to one of his earliest memoirs (1768–1770) on the supposed transmutation of water into earth and continued during the early 1770s with his experiments using large and small burning lenses for the combustion of diamond and metals. He also personally supervised the construction of sophisticated instruments made of different kinds of glass. During his scientific travels, he took the opportunity to inspect different glassworks.

Lavoisier had around 7,000 pieces of glassware in his laboratory. It is therefore not surprising to see that he set up a workshop for his production at the Arsenal. It is worth noting that during the early 1780s Lavoisier used the enameller lamp and perfected its use for the combustion of platinum and precious stones. Jean Nollet, Lavoisier's teacher, gave detailed instructions on how to use the enameller lamp in the fourth volume of his *Leçons de physique expérimentale* (lesson 14, plate VIII). This was conducive to the introduction of the blowpipe in eighteenth-century chemistry. Lavoisier was also very interested in glassmaking and in the chemical nature of glass. For this purpose, he followed the work of Paul Bosc d'Antic and, later in his career, he collaborated with Pierre Loysel, author of the first treatise on glassmaking based on Lavoisier's chemical theory. In 1788, Lavoisier promoted a prize for the best method to produce flint glass: "Prix extraordinaire proposé par ordre du Roi par l'Académie des sciences pour l'année 1788 (précis de ce qui a été fait pour perfectionner le flint-glass et le verre en général)" (LO, vol. 6, pp. 20–30). The making of glassware for chemical laboratories in the eighteenth century has been accurately described by Barrelet:

> Laboratory glassware was also worked with the enameller lamp. At faience shops in large towns, the craftsman would find the pipes and rods made of either "crystal" or "ordinary" glass, both supplied by the makers of small glassware. Crystal was reserved for the manufacture of larger-sized vessels. It was sold by the pound. The enameller could not soften large glass pieces or blow them with a torch. They would therefore be ordered from the glassworks themselves, generally through the faience merchant. But the intensive construction of glass apparatus is due to Lavoisier and the impulse he gave to chemistry in around 1790.[1]

Danger (1829); Bontemps (1868), 563; Barrelet (1953), pp. 116–120; Glocker (1992), pp. 55–57; Beretta (2012a); Beretta (2014a)

Heating Apparatus and Common Laboratory Tools

20055-0001- (Fig. 3)
Laboratory burner
Late 18th c.
MAM accession date: 1952
23 × 23 × 15.5; 2.82 kg
Brass

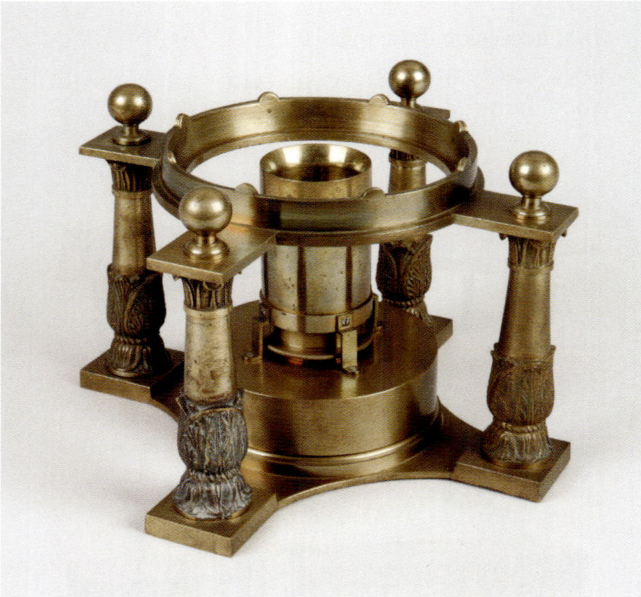

FIGURE 3 (inv. no. 20055-0001-)
© MAM/PHOTO FRANCK BOTTÉ

1 "La verrerie de laboratoire était également travaillée à la lampe d'émailleur; l'artisan trouvait chez le marchand faïencier des grandes villes les tubes et les baguettes soit en verre de 'cristal', soit en verre 'commun', tous deux fournis par les petites gobeleteries. Le cristal était réservé à la fabrication des vaisseaux d'une certain grandeur; il se vendait à la livre. L'émailleur ne pouvait amollir ni souffler à la lampe les grosses pièces de verre; on les commandait donc à la verrerie même, généralement par l'intermédiaire du marchand faïencier. Mais la construction intensive d'appareils en verre est due à Lavoisier et au développement qu'il donna à la chimie vers 1790". Barrelet (1953), pp. 116–120.

Oil burner with an X-shaped base with four columns decorated with acanthus leaves and surmounted by spheres.[2] In the centre of the base there is a cylindrical box with a central opening containing the wick and a concentric cylindrical funnel. A plug with a ring and a screw are joined to the burner but may not have actually belonged to it.

20055-0002- (Fig. 4)
Laboratory burner
Late 18th c.
MAM accession date: 1952
23.7 × 23.7 × 23.7; 2.91 kg
Tin, zinc, wood

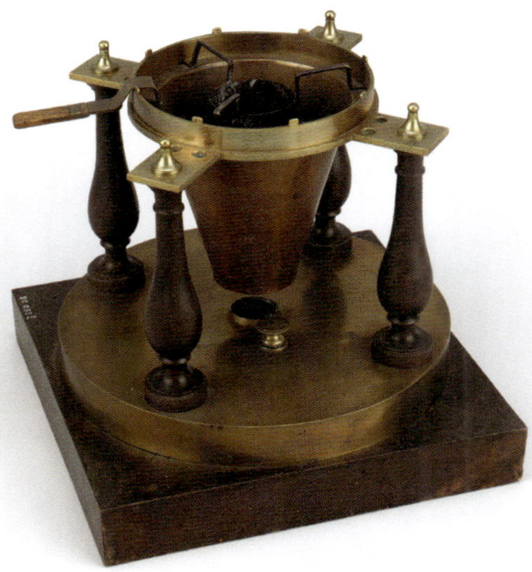

FIGURE 4 (inv. no. 20055-0002-)
© MAM/PHOTO FRANCK BOTTÉ

Laboratory oil (or alcohol) burner with a square wooden base and cylindrical brass vessel. On the vessel, four wooden turned columns support an annular frame with four arms. A conical vessel with a brass and wood handle is inserted in it. Three arms hold a cylindrical coaxial tube in the cone. The wick holder appears to be missing.

20055-0003-(Fig. 5)
Laboratory burner
Late 18th c.
MAM accession date: 1952
26.5 × 23 × 22.5; 2.72 kg
Brass, wood, fabric

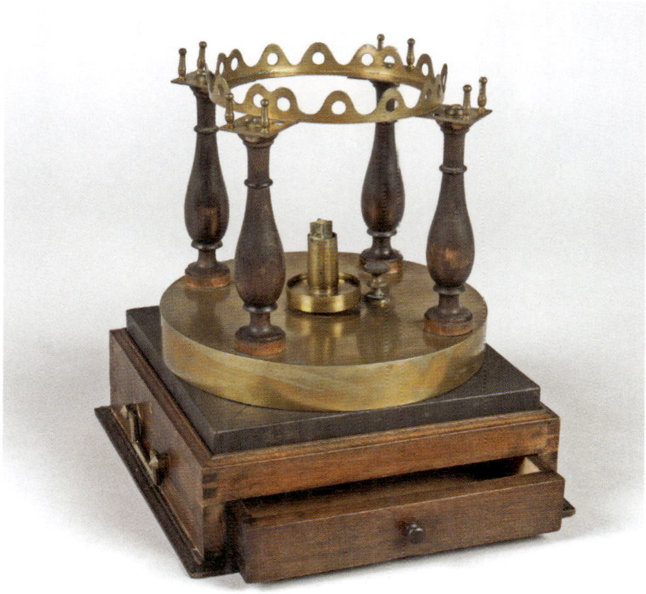

FIGURE 5 (inv. no. 20055-0003-)
© MAM/PHOTO FRANCK BOTTÉ

Laboratory oil (or alcohol) burner mounted on a wooden square box with two handles and a drawer. A flat cylindrical brass box has a central vertical tube with a ring that contains the wick in a metallic frame. A rotating knob next to the tube moves a lever for adjusting the wick position. Four turned wooden columns mounted on the box hold an annular support with decorated rim.

20055-0004- (Fig. 6)
Laboratory burner
Late 18th c.
MAM accession date: 1952
22.5 × 22.2 × 21.2; 1.63 kg
Wood, cork, porcelain, brass

Laboratory oil (or alcohol) burner with a square wooden base and four turned wooden columns. The base carries a cylindrical porcelain box with a tube for the wick. The tub has two small handles and is closed by a brass cap. The annular ring originally supported by the columns is missing.

2 This and the following burners could probably burn also spirit.

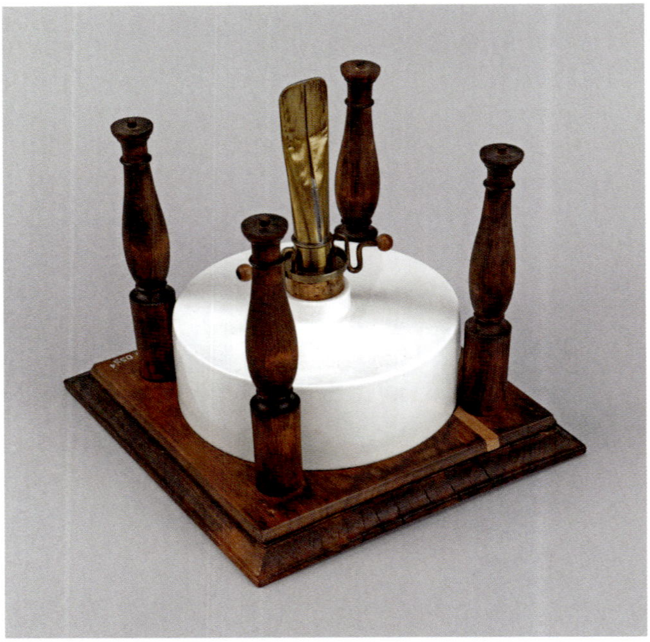

FIGURE 6 (inv. no. 20055-0004-)
© MAM/PHOTO FRANCK BOTTÉ

20055-0005- (Fig. 7)
Laboratory burner
Late 18th c.
MAM accession date: 1952
22.5 × 23.5 × 18; 2.14 kg
Wood, brass, tin, zinc? Cork, iron

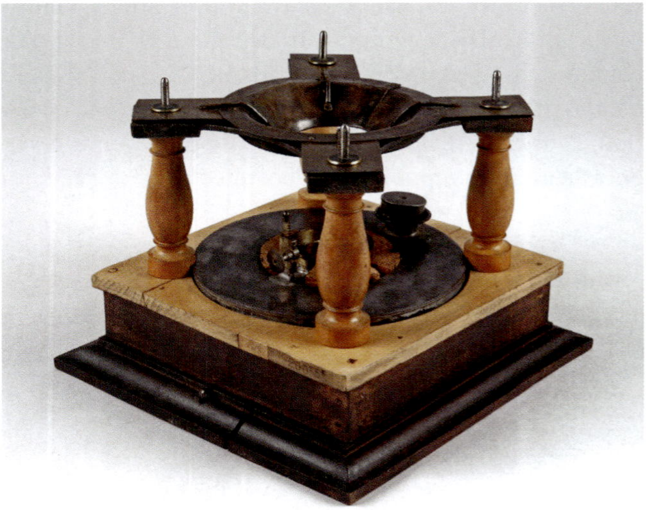

FIGURE 7 (inv. no. 20055-0005-)
© MAM/PHOTO FRANCK BOTTÉ

Laboratory oil (or alcohol) burner consisting of a annular shaped zinc vessel inserted in a wooden box. In the centre of the vessel there is a tube with the cylindrical wick. The tube is surrounded by pieces of cork. A rack-and-pinion mechanism with a rotating knob regulates the wick height. The burner is filled through an opening closed with a cork and a metal cap. A sliding rod connected to a plate acts as a valve, interrupting the flow of fuel to the wick. Four turned wood columns are inserted in the base and support a slightly conical iron ring with four arms. A similar brass ring is fixed to the iron ring.

20055-0006- (Fig. 8)
Laboratory burner
Late 18th c.
MAM accession date: 1952
20.2 × 15, 555 g
Tin, brass cork

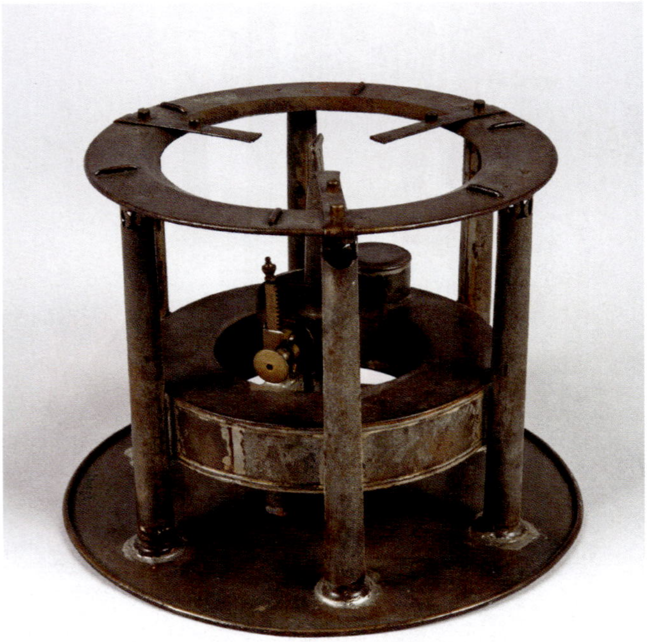

FIGURE 8 (inv. no. 20055-0006-)
© MAM/PHOTO FRANCK BOTTÉ

Laboratory alcohol lamp with a round tin plate and six columns supporting a ring. Between the columns there is an annular vessel connected to a coaxial tube through a cylindrical wick.

The wick height is adjusted with a rack-and-pinion mechanism. A metal cap with a cork and a vertical arm protects the wick. A second cap with a cork closes the hole for filling the lamp. On the top ring there are three

movable arms that serves to support a small vessel when it needs to be heated.
This type of lamp is similar to the so called Knight's assay lamp, which was invented by the ironmonger William Knight active in London at the end of the 18th century.

Francis (1846), no page (see the entry "Knight's assay lamp")

20055-0007- (Fig. 9)
Laboratory burner
Late 18th c.
MAM accession date: 1952
31.5 × 23.2 × 15.2; 690 g
Tin, brass, wood, cork.

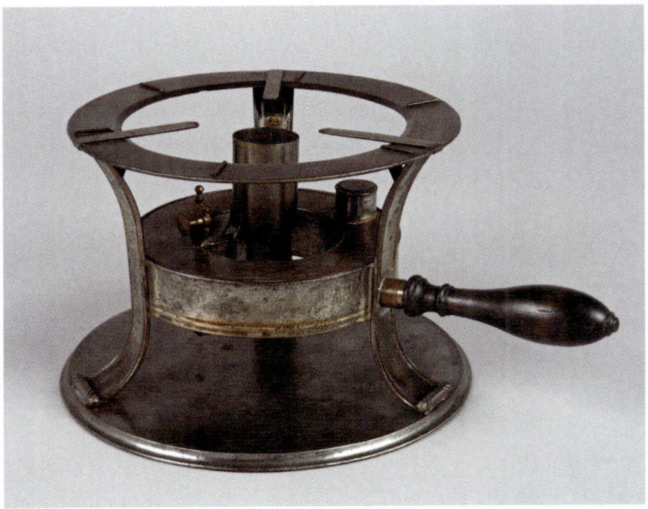

FIGURE 9 (inv. no. 20055-0007-)
© MAM/PHOTO FRANCK BOTTÉ

Laboratory oil (or alcohol) burner with a round brass base and three curved metal arms. These support a ring with three radial movable arms. An annular vessel with a wooden handle and the wick holder is inserted between the arms and is almost identical to the one in no. 20055-0006-. A funnel can be inserted on the wick.

Francis (1846), no page (see the entry "Knight's assay lamp")

20107-0000- (Fig. 10)
Portable burner
Second half of 18th c.
MAM accession date: 1952
34 × 21 × 19.5; 645 g
Brass, wood

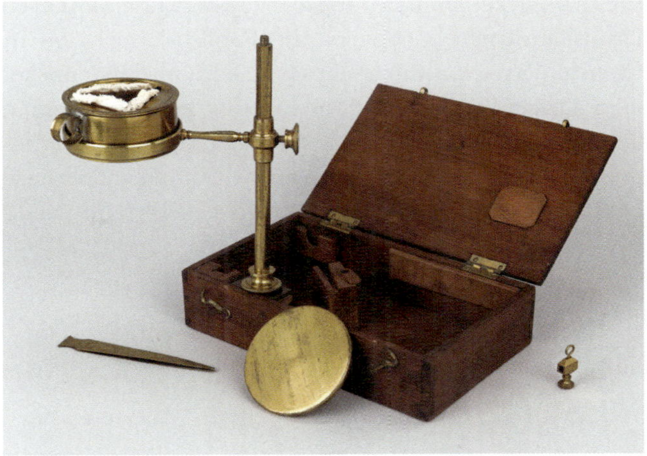

FIGURE 10 (inv. no. 20107-0000-)
© MAM/PHOTO FRANCK BOTTÉ

Small disassembled laboratory oil burner contained in a flat wooden box when not in use. The open box also serves as a stand: a brass bar is screwed vertically into it. A horizontal arm holding a ring is inserted in the bar and fastened with a screw. The lamp, placed in the ring, is a flat cylindrical box with a central triangular opening and three wicks (along the side of the opening). The box also contains a circular cap for the lamp and pliers.[3]

This burner was probably part of a portable laboratory to be used for mineralogical studies. The earliest such equipment was invented by the Swedish chemist Axel Fredrik Cronstedt, who, as early as 1758, prepared a set comprising a blowpipe, a burner, chemicals and reagents. His work was translated by Gustav von Engeström into English in 1770 with the title *An Essay towards a System of Mineralogy* with the addition of a *Description and use of a mineralogical pocket laboratory; and especially of the use of the blowpipe in mineralogy*. This handy equipment became quite fashionable and attracted the attention of chemists throughout Europe. In March 1771, Jean-Hyacinthe de Magellan (1722–1790), who traded English instruments in Europe, sent two sets of pocket chemical laboratories to the secretary of Jean-Philibert Trudaine de Montigny, Intendant des Finances and director of the Bureau du Commerce. At this time Trudaine de Montigny had one of the best-equipped chemical laboratories in France where Lavoisier conducted several of his crucial experiments in pneumatic chemistry until 1775. In addition to being very close to Trudaine, Lavoisier was also a correspondent of Magellan and may well have been the recipient of one of

3 The box also contains a small slider with a screw and hook, but they are probably not part of this set.

the two pocket laboratories sent by Magellan. Moreover, Lavoisier owned a copy of the English translation of Cronstedt's work, although in a later edition. Other pocket laboratories were made in the 1770s by Torbern Bergman and Louis-Bernard Guyton de Morveau. Lavoisier's burner must be the surviving item of a pocket laboratory. Because of his frequent experiments in the field, he certainly could have used it to his benefit.

Smeaton (1966); Beretta (1995); Beretta (2005); Home et al. (2017), vol. 1, pp. 160–162

19895-0000- (Fig. 11)
Régnier gunpowder tester
Late 18th c.
MAM accession date: 1952
Edme Régnier (1751–1825)
19.5 × 19.3 × 3.6; 500 g
Iron, brass fiber

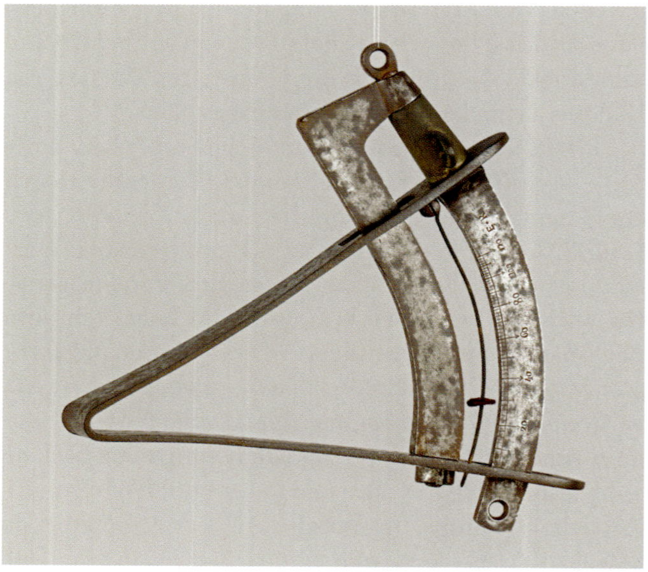

FIGURE 11 (inv. no. 19895-0000-)
© MAM/PHOTO FRANCK BOTTÉ

This instrument is used to measure the explosive force of a sample of gunpowder. Structurally, it is very similar to a V-spring scale. A sector with a scale from 0 to 120 (with unit division) is fixed to one end of the spring. Near the scale are the punched letters "E.R.", indicating the initials of the name of the inventor and maker. A small conical brass cannon with a small bowl on its side is also fixed to the same arm of the spring. A second sector (concentric to the first), fixed to the other arm, holds a cup-like vessel acting as a breech for the cannon. Between the sectors is a curved metal wire with a small sliding fabric disk serving as an indicator.

The instrument was suspended with a ribbon attached to it, and the arms of the spring had to be pressed together to separate the breech from the cannon. A gram of gunpowder to be tested was inserted into the breech, which was then gently closed and a small amount of powder was placed in the bowl. When the bowl was lit, the powder in the cannon exploded and the violent expansion of the gas pushed the breech and its arm off the spring. The spring movement also displaced the fabric disk, whose position after the explosion could be read on the scale,[4] indicating the gunpowder's force.

This powder tester (*éprouvette*) was devised at the end of the 18th century by the French inventor and gun- and instrument-maker Edme Régnier, who also was inspector of hand-firearms production under the Committee of Public Safety. Éleuthère-Irénée du Pont de Nemours (1771–1801), a student and assistant of Lavoisier before moving to the United States, acquired a Régnier tester in 1801 and ordered more both from Régnier and other makers.

Régnier (1798); Hachette (1828), pp. 34–35, plate 2, fig. 4; Timmerhans (1839), pp. 228–229, plate 3, fig. 4; Müller (1873), p. 214; Howard (2016)

20040-0000- (Fig. 12)
Crucible
1764–1768
MAM accession date: 1952
Brancas-Lauraguais manufacture
7.5 × 4.1 × 4; 60 g
Refractory earth

Small crucible made of refractory earth used in high-temperature chemical manipulations. The outer side is stamped with the two superimposed initials "B" and "L". The crucible appears to have been made at the short-lived porcelain manufacture founded by Louis-Léon-Félicité de Brancas, third Duc de Lauraguais (1733–1824) who, with the help of Jean Darcet, developed a new method of making porcelain from kaolin. Lauraguais was also a passionate amateur chemist and his laboratory was known to be one of the best-equipped in Paris. Brancas' discovery of kaolin was contested by Lavoisier's teacher, Jean-Étienne

4 Régnier's original tester had a scale with 60 divisions, each unit corresponding to a force of one kilogram.

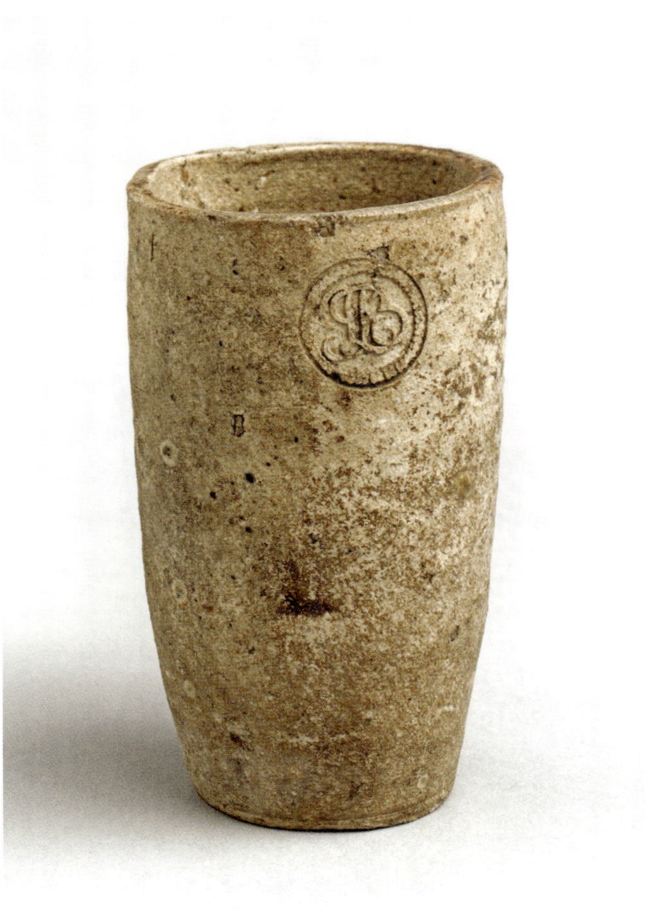

FIGURE 12 (inv. no. 20040-0000-)

Guettard, who claimed precedence for it at the Académie Royale des Sciences in 1765. Lavoisier's crucible appears to be the only surviving item from Brancas' manufacture.

Lavoisier rarely specified which kind of crucibles he used in his experiments. In his *Détail des expériences exécutées au moyen du grand verre ardent* (1772: LO, vol. 3, p. 314) he mentions the crucibles from Saint-Gobain as well as German ones. In a note on instructions for building a chemical laboratory written between 1773 and 1775, Lavoisier also mentioned Paris crucibles but, like most of his contemporaries, he seems to have preferred the German models (LO, vol. 5, pp. 335–339), of which Antoine Baumé was one of the main suppliers in Paris. Members of the furnace-makers' guilds (*fournalistes*) imported crucibles from Hesse, which were highly appreciated for their resistance to heat. It was only towards the end of the century that these were partly replaced by the clay products made by Josiah Wedgwood in England. Wedgwood became one of Lavoisier' suppliers in 1791 (LC, vol. 6, p. 314 and p. 323).

In his *Traité élémentaire de chimie* (1789), Lavoisier wrote:

> Chemists of all periods have therefore assigned great importance to procuring crucibles of highly refractory materials, that is, capable of resisting a most intense fire. The best are those made with very pure clay or porcelain earth. One must avoid using for this purpose clays mixed with silica or limestone soil, because they melt too easily. All the clays extracted in the near surroundings of Paris are in that category. As a result, the crucibles fabricated in this city melt at a rather low temperature, and are fit for only a small number of chemical operations. The clays from Hesse are fairly good, but the ones that should be preferred are those from Limoges, which seem to be absolutely infusible. In France there are a large number of clays suitable for making crucibles. One example is the clay used for the crucibles at the Saint-Gobain flat-glass works.[5]

However, none of those illustrated in plate VII (figs. 7, 8, 9, 10) resembles the one surviving at the Musée des Arts et Métiers.

Lavoisier (1775a); Lavoisier (1789), pp. 535–536 and LO, vol. 1, pp. 371–372; Chaffers (1891), p. 527; Chavagnac (1906), pp. 399–405; Martinón-Torres *et al.* (2006)

5 "Les chimistes de tous les âges ont en conséquence attaché une grande importance à se procurer des creusets de matières très-réfractaires, c'est-à-dire qui eussent la propriété de résister à un très-grand degré de feu. Les meilleurs sont ceux qui sont faits avec de l'argile très-pure ou de la terre à porcelaine. On doit éviter d'employer pour cet usage les argiles mélangées de silice ou de terre calcaire, parce qu'elles sont trop fusibles. Toutes celles qu'on tire aux environs de Paris sont dans ce cas; aussi les creusets qu'on fabrique dans cette ville fondent-ils à une chaleur assez médiocre, et ne peuvent-ils servir que dans un très-petit nombre d'opérations chimiques. Ceux qui viennent de Hesse sont assez bons, mais on doit préférer ceux de terre de Limoges, qui paraissent être absolument infusibles. Il existe en France un grand nombre d'argiles propres à faire des creusets; telle est celle, par exemple, dont on se sert pour les creusets de la glacerie de Saint-Gobain". Lavoisier (1789), pp. 535–536.

20159-0000- (Fig. 13)
Fire-making set
Second half of 18th c.
MAM accession date: 1952
12 × 7.3 × 2.3; 150 g
Tin, steel, flint stone, wood, sulphur, tinder

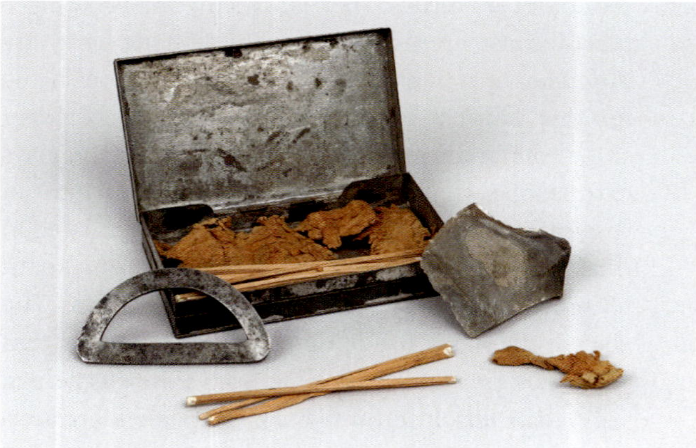

FIGURE 13 (inv. no. 20159-0000-)
© MAM/PHOTO FRANCK BOTTÉ

The set is contained in a tin box and includes a semi-circular fire steel, a flint stone, a tinder and 16 sulphur-tipped matches. The striker (made of steel or hard tempered iron) was used to repeatedly strike the flint stone (a hard sedimentary form of quartz – a variety of chert – but agate or jasper could also be used). The operation produced tiny white-hot particles of steel that were made to fall on a piece of tinder. The tinder consists of a spongy material derived from Fomes Fomentarius (tinder fungus), a fungus that grows on the bark of various trees, but many other easily flammable substances were used as well. The sparks ignited the tinder, from which fire was obtained with the sulphur-tipped wooden matches. This fire-making method was common for centuries until the first half of the nineteenth century.

The set was displayed at the Palais de la Découverte in 1943 at the exhibition marking the bicentenary of Lavoisier's birth.

Miller (1926); Lavoisier (1943); Watson (1930); O'Dea (1964)

20098-0000- (Fig. 14)
Papin digester
1775–1776
MAM accession date: 1952
François-Philippe Charpentier (1734–1819)
42.5 × 27.5 × 18; 13.43 kg
Brass, steel

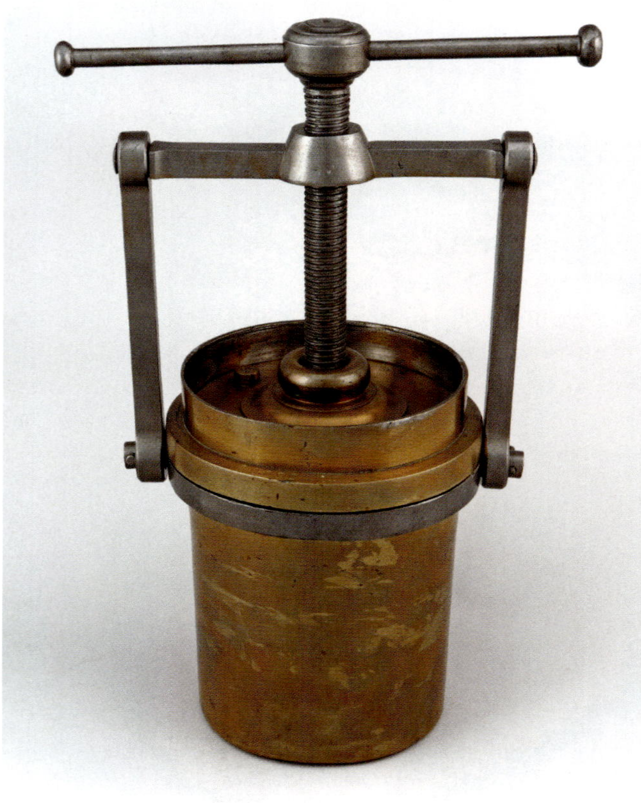

FIGURE 14 (inv. no. 20098-0000-)
© MAM/PHOTO FRANCK BOTTÉ

The digester is a simple kind of autoclave, the forerunner of the pressure cooker. It was described in 1679 by the French natural philosopher, mathematician and inventor Denis Papin.

It consists of a sturdy cylindrical vessel with a lid, both made of brass (tinned inside). The vessel holds a steel ring with two vertical arms and a crosspiece. A screw with a handlebar inserted in the crosspiece presses the lid firmly on the container.[6] A pressure safety valve (missing) was probably inserted in the hole in the lid.

The digester was heated by a burner or in an oven, then bones and water were placed in it. After cooling down, the digester was opened. The bones inside were now soft and friable, while the fat attached to them dissolved into

6 Cardboard gaskets were inserted between the vessel and the lid.

the water, forming a gelatine. The steam pressure kept the water liquid at temperatures far higher than 100° C. The digester was also used to dissolve resins in alcohol and prepare varnishes. In February 1776, Lavoisier purchased a digester from François-Philippe Charpentier, *Mécanicien du Roi*, for the substantial sum of 780 *livres*. He conducted several experiments with it, combining the use of a balance and a barometer. Charpentier's device was in many ways different and more elaborate than the one surviving in the Musée. In their 1794 inventory, Charles, Fortin and Lenoir referred to two Papin digesters, describing one of them as "Papin's very fine digester. It is the largest known" ("La très belle marmite de Papin. C'est la plus considérable que l'on connoisse"). This must have been Charpentier's. Lavoisier used a digester again in 1783 for experiments on the decomposition of water.

Papin (1681); LC, vol. 3, pp. 544–549 and pp. 559–560; Lavoisier (1781b); Sigaud de La Fond (1784), vol. 1, pp. 314–316 and plate XXIII, fig. 67; Ms Charles, Fortin, Lenoir (1794); Storni (2021)

20058-0000- (Fig. 15)
Protective mask
Second half of 18th c.
MAM accession date: 1952
15 × 14.5 × 11; 150 g
Iron, glass, fabric

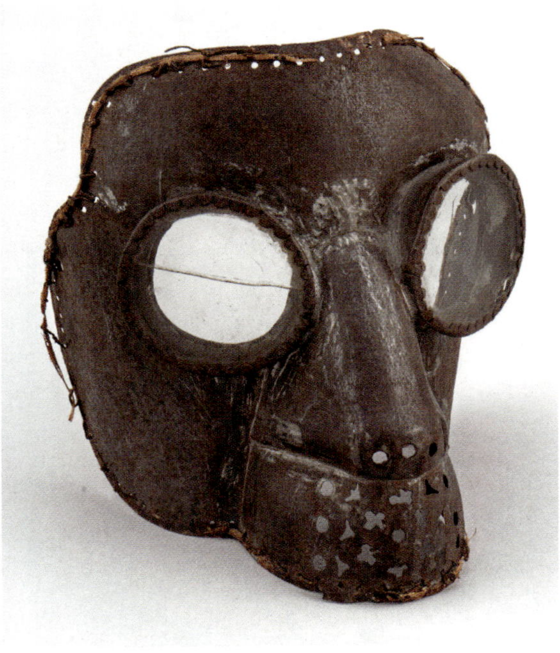

FIGURE 15 (inv. no. 20058-0000-)
© MAM/PHOTO FRANCK BOTTÉ

Mask made of a thin sheet of embossed iron. There are perforations for the nose and mouth, while the eyes are protected by two round glass disks (one of them is cracked) inserted in protruding rings. The interior of the mask was originally coated with leather or a similar material (missing).

On the basis of Madame Lavoisier's illustrations, Graham Lusk claimed that Armand Séguin used the mask in the respiration experiments of 1790–1791. While noting the perforations, Lusk argued that "the tube into which he breathed was introduced through either leather or rubber, which fell down from the nose piece of the mask and was fitted around the neck and around the head of the subject." However, Madame Lavoisier's drawings clearly show a full mask covering the entire face but with a different shape. Furthermore, Séguin, in his 1814 *mémoire*, describes the mask used in the respiration experiment as a "tête de cuivre" (copper head), which does not match this surviving model. Madame Lavoisier was an extremely accurate draftswoman and it seems impossible that she portrayed the same device now exhibited at the Musée.

It seems more likely that this mask was used to protect against fire and violent heat. It was displayed at the Musée Centennal in 1900 and at the Lavoisier exhibition in 1943.

Séguin (1814), p. 261; *Musée Centennal* (1900); Lusk (1925); Lavoisier (1943); Prinz (1992) pp. 68–70; Prinz (2005), p. 43

20229-0000- (Fig. 16)
Protective mask
Second half of 18th c.

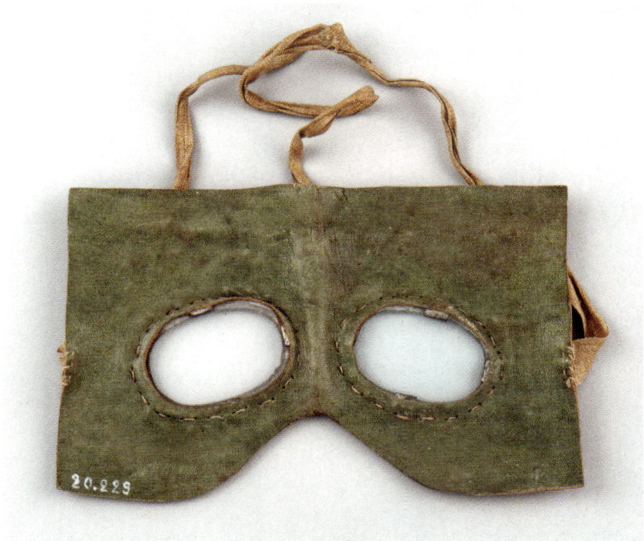

FIGURE 16 (inv. no. 20229-0000-)
© MAM/PHOTO FRANCK BOTTÉ

MAM accession date: 1952
22.5 × 20 × 14.3; 60 g
Leather, iron, glass, fabric

Protective mask made of soft green leather covering the eyes, nose and forehead. The eyes are protected by two round glass disks, one of them opaque.

Musée Centennal (1900); Lavoisier (1943)

20169-0000- (Fig. 17)
Mortar and pestle
Second half of 18th c.
MAM accession date: 1952
18 × 32 × 31; pestle height 17 diameter 6.5; 16.5 kg
Porphyry, porcelain

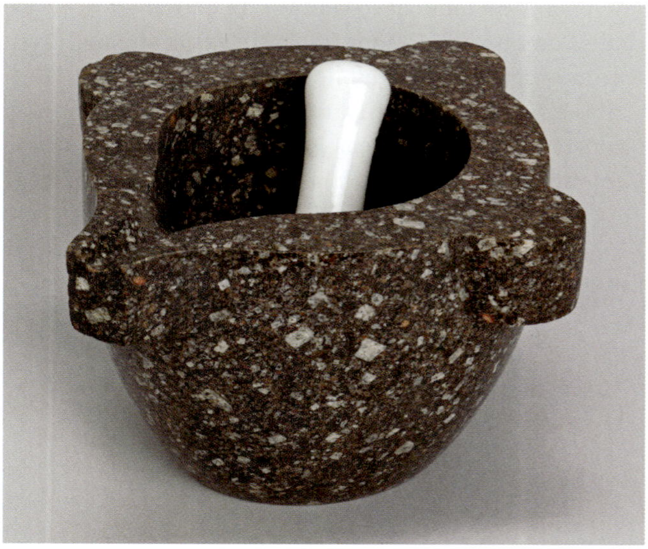

FIGURE 17 (inv. no. 20169-0000-)
© MAM/PHOTO FRANCK BOTTÉ

Heavy mortar made of porphyry with a porcelain pestle. The mortar has three protruding lobes for lifting and a beak with a groove for pouring the pulverized substance.

Lavoisier (1789), vol. 2, pp. 404–405, plate 1, fig. 4

20170-0000- (Fig. 18)
Mortar
Second half of 18th c.
MAM accession date: 1952
22 × 22 × 12.7; 4.71 kg
Marble

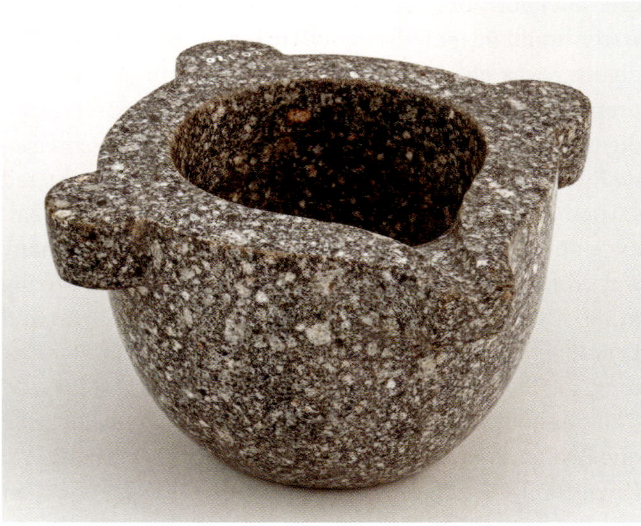

FIGURE 18 (inv. no. 20170-0000-)
© MAM/PHOTO FRANCK BOTTÉ

Marble mortar with three lugs and a pouring lip described by Lavoisier in his *Traité élémentaire de chimie*.

Lavoisier (1789), vol. 2, pp. 404–406, plate 1, fig. 2

20037-0001- (Fig. 19)
Pear-shaped glass
Second half of 18th c.
MAM accession date: 1952
9 × 3.8; 165 g
Glass

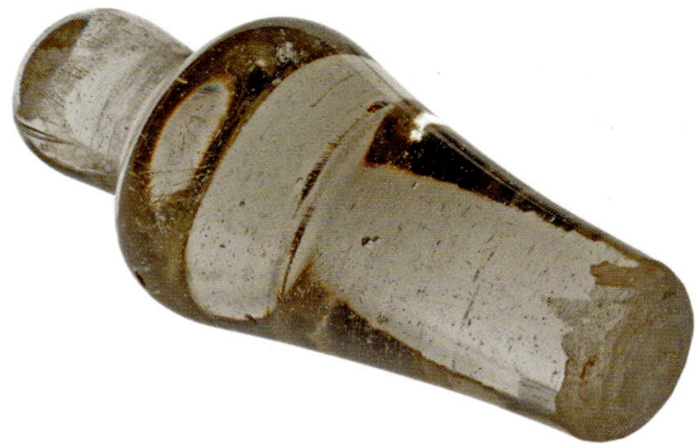

FIGURE 19 (inv. no. 20037-0001-)
© MAM/PHOTO FRANCK BOTTÉ

Pear-shaped glass probably used as a pestle in a mortar. In his *Traité élémentaire de chimie* (1789), Lavoisier discussed in considerable detail the different kinds of pestles to be used in a chemical laboratory. He remarked:

> The pestles used to grind bodies are also made of different materials. They are made of forged iron or copper, as in the first figure; of wood, as in figures 2 and 3; lastly, of glass, porcelain or agate, depending on the nature of the objects one wants to grind. A laboratory needs to have an assortment of these instruments of different sizes. Porcelain mortars and, even more so, glass mortars cannot be used for grinding per se, and would soon be in pieces, if one were to pound their contents carelessly with redoubled blows. It is by turning the pestle in the mortar, by crushing the molecules between the pestle and the mortar sides with skill and dexterity, that one can succeed in dividing the material.[7]

Lavoisier (1789), vol. 2, pp. 404–405 and plate I, figs. 1–5

20180-0000- (Fig. 20)
Basin
Second half of 18th c.
MAM accession date: 1952
37.5 × 35.5 × 15.5; 16.31 kg
Stone

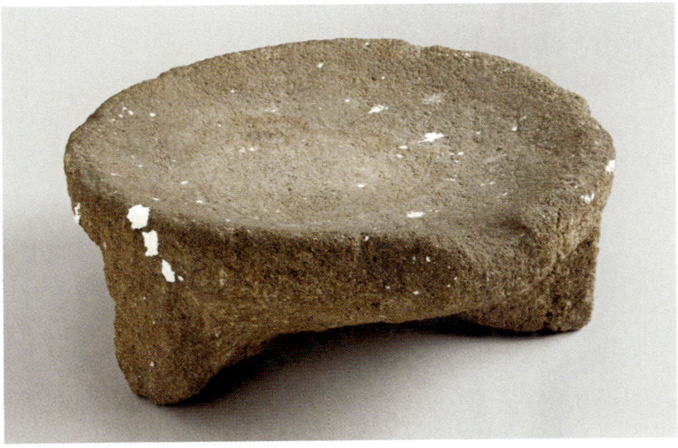

FIGURE 20 (inv. no. 20180-0000-)
© MAM/PHOTO FRANCK BOTTÉ

Three-footed round stone basin with a lateral beak. Probably used, with a stone or pestle, to crush and pulverize solid substances.

20062-0000- (Fig. 21)
Spoon
Second half of 18th c.
MAM accession date: 1952
14.2 × 3.2 × 1.5; 20 g
Silver

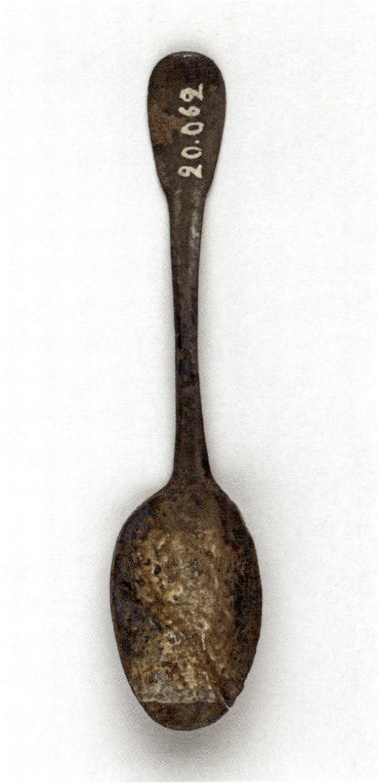

FIGURE 21
(inv. no. 20062-0000-)
© MAM/PHOTO FRANCK BOTTÉ

Worn, oxidized spoon. It appears to have been heated and used as a spatula. The stamped caduceus suggests that it was originally made for a physician. The spoon was displayed at the 1943 Lavoisier exhibition.

Lavoisier (1943)

7 "Les pilons dont on se sert pour triturer les corps sont aussi de différentes matières. Ils sont de fer ou de cuivre forgé, comme dans la figure première; de bois, comme dans les figures 2 et 3: enfin, de verre, de porcelaine ou d'agate, suivant la nature des objets qu'on veut triturer. Il est nécessaire d'avoir dans un laboratoire un assortiment de ces instruments de différentes grandeurs. Les mortiers de porcelaine, et surtout ceux de verre, ne peuvent pas être employés à la trituration proprement dite, et ils seraient bientôt en pièces, si on frappait dedans sans précaution, à coups redoublés. C'est en tournant le pilon dans le mortier, en froissant avec adresse et dextérité les molécules entre le pilon et les parois du mortier qu'on parvient à opérer la division". Lavoisier (1789), vol. 2, pp. 404–405.

20069-0000- (Fig. 22)
Scalpel
Late 18th c.
MAM accession date: 1952
16.7 × 2.3 × 1.7; 20 g
Steel, wood

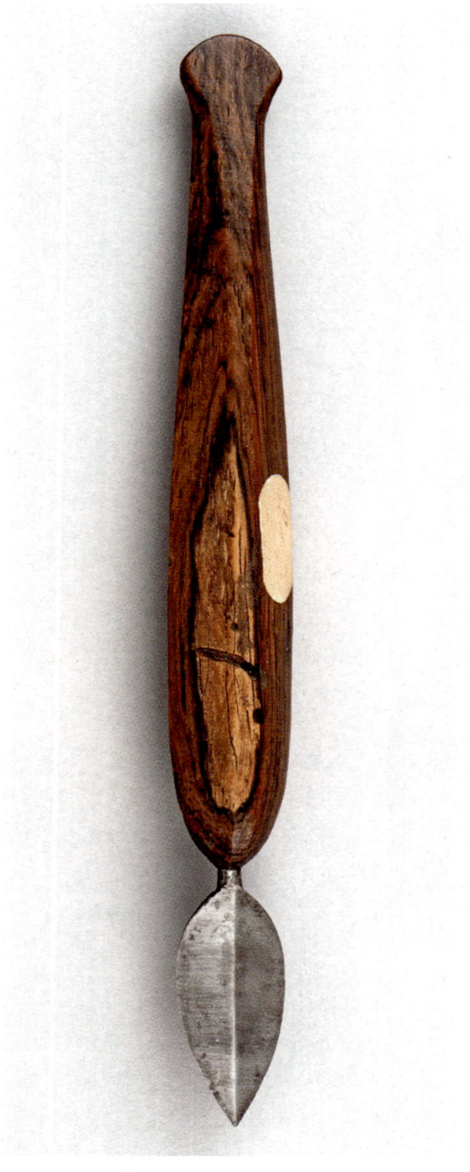

FIGURE 22 (inv. no. 20069-0000-)
© MAM/PHOTO FRANCK BOTTÉ

Laboratory knife with a wooden handle and a double-edged blade. There is a punch representing a crown.

20060-0000- (Fig. 23)
Laboratory spoon with ejecting mechanism
Second half of 18th c.
MAM accession date: 1952
25 × 2, 7 × 2.7; 95 g
Steel, ivory or bone

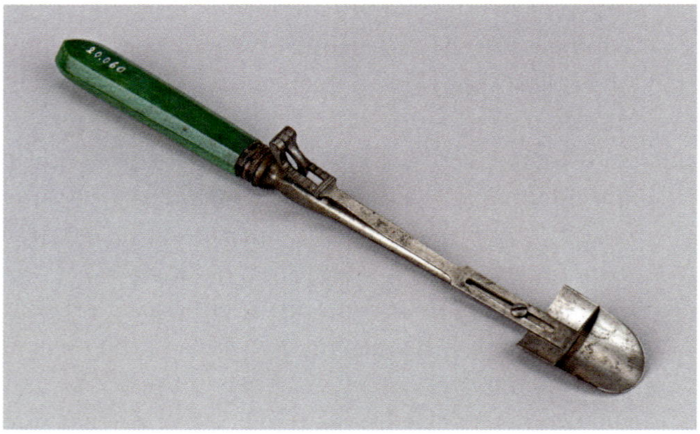

FIGURE 23 (inv. no. 20060-0000-)
© MAM/PHOTO FRANCK BOTTÉ

Steel spoon in the shape of a small shovel with a handle made of green-coloured ivory or bone.

A small lever serves to move a bar with a semi-circular sector perpendicular to the spoon and to eject the substance collected in it.

Lavoisier (1943)

20167-0000-, 20168-0000- (Fig. 24)
Kettle-shaped vessels
Second half of 18th c.
MAM accession date: 1952
24 × 14
Tin, brass, wood

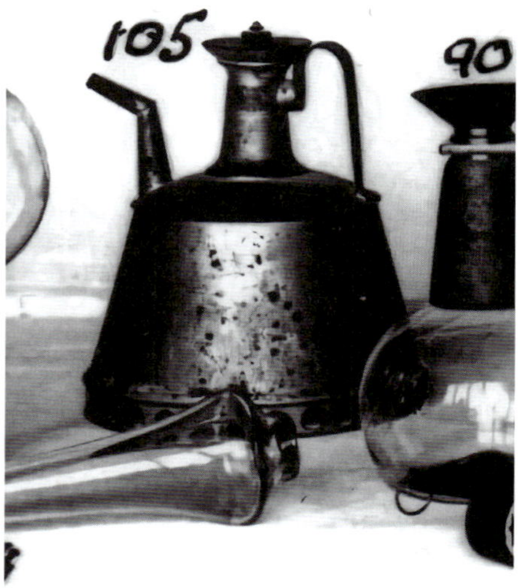

FIGURE 24 (inv. nos. 20167-0000-, 20168-0000-) PHOTO DOUGLAS MCKIE. COURTESY OF THE HAGLEY MUSEUM AND LIBRARY

Two almost identical kettle-shaped vessels, each with a perforated rim at the base. There is a stopper fitted with a small brass knob passing through a piece of wood (perhaps to insulate it from heat?).

These vessels were inventoried at the Musée des Arts et Métiers in 1952 as nos. 20167-0000- and 20168-0000- but their present location is unknown.

20165-0000-, 20166-0000- (Fig. 25)
Two stoneware bottles
Late 18th c.
MAM accession date: 1952
20165-0000: 39.5 × 14.5; 2.82 kg
20166-0000: 32 × 12; 1.68 kg
Stoneware, tin

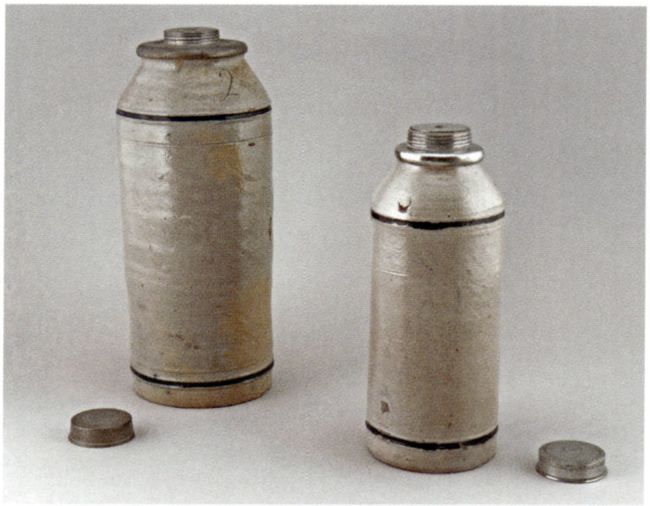

FIGURE 25 (inv. nos. 20165-0000-, 20166-0000-)
© MAM/PHOTO FRANCK BOTTÉ

Two cylindrical glazed stoneware bottles. Their necks have a threaded collar and a metal cap. The opening is also closed by a cork inserted between two metal disks that expand under pressure. The top disk has a threaded hole with a central pin. The bottles are marked "2" (no. 20165-0000-) and "3" (no. 20166-0000-).
These bottles were probably used to store mercury, a substance too heavy for glass bottles.

20146-0000- (Fig. 26)
Tin bottle
Late 18th c.
MAM accession date: 1952
17.5 × 10; 160 g
Tin

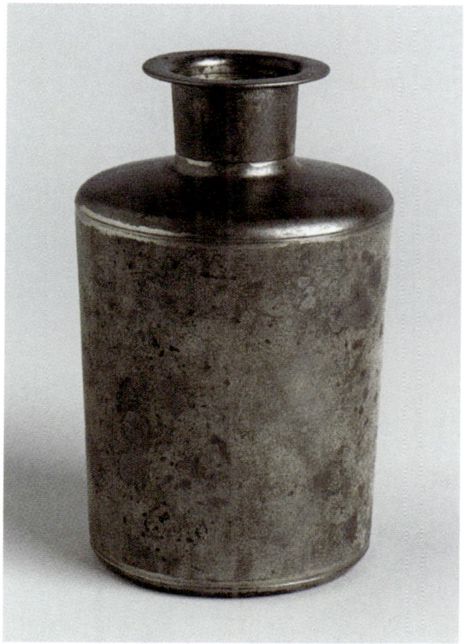

FIGURE 26 (inv. no. 20146-0000-)
© MAM/PHOTO FRANCK BOTTÉ

Cylindrical laboratory tin bottle with a short neck, probably used to store mercury.

20201-0000- (Fig. 27)
Metal vessel
Second half of 18th c.
MAM accession date: 1952
8.2 × 5.5 × 3.3; 65 g
Tinned brass

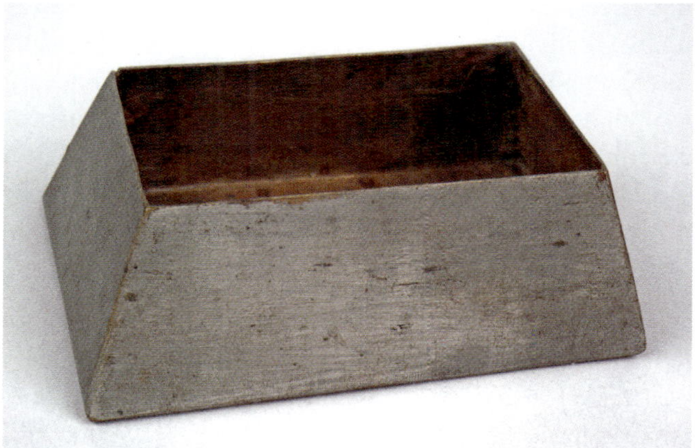

FIGURE 27 (inv. no. 20201-0000-)
© MAM/PHOTO FRANCK BOTTÉ

Small trapezoid-shaped metal vessel.

20200-0001-, 20200-0002-, 20200-0003-, 20200-0004- (Fig. 28)
Rod with hexagonal piece
Second half of 18th c.
MAM accession date: 1952
20200-0001-: 8.3 × 2.6; 10 g
20200-0002-: 6 × 13 × 11; 155 g
20200-0003-: 29.5 × 3.3 × 2.5; 85 g
20200-0004-: 1.5 × 35.2 × 3.5; 330 g
Brass

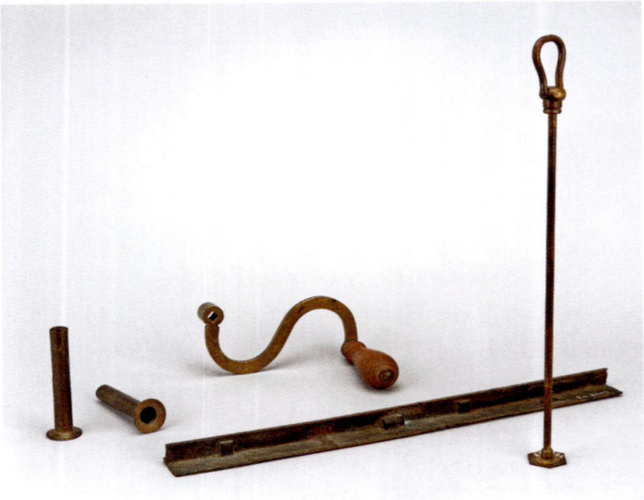

FIGURE 28 (inv. nos. 20200-0001-, -0004-)
© MAM/PHOTO FRANCK BOTTÉ

20200-0001-: Two identical brass tubes made of thin brass with a flange at one end.

20200-0002-: S-shaped brass crank with a wooden handle. It certainly belonged to an instrument that we were unable to identify.

20200-003-: Brass rod with a handle ring at one end and a small horizontal hexagonal plate at the other. A part of an unidentified apparatus, it was probably inserted in a glass bell for pneumatic experiments.

20200-0004-: Piece of brass profile, truncated at one end. Possibly part of an unidentified apparatus.

20204-0000- (Fig. 29)
Tap with hook
Second half of 18th c.
MAM accession date: 1952
14 × 7; 395 g
Brass

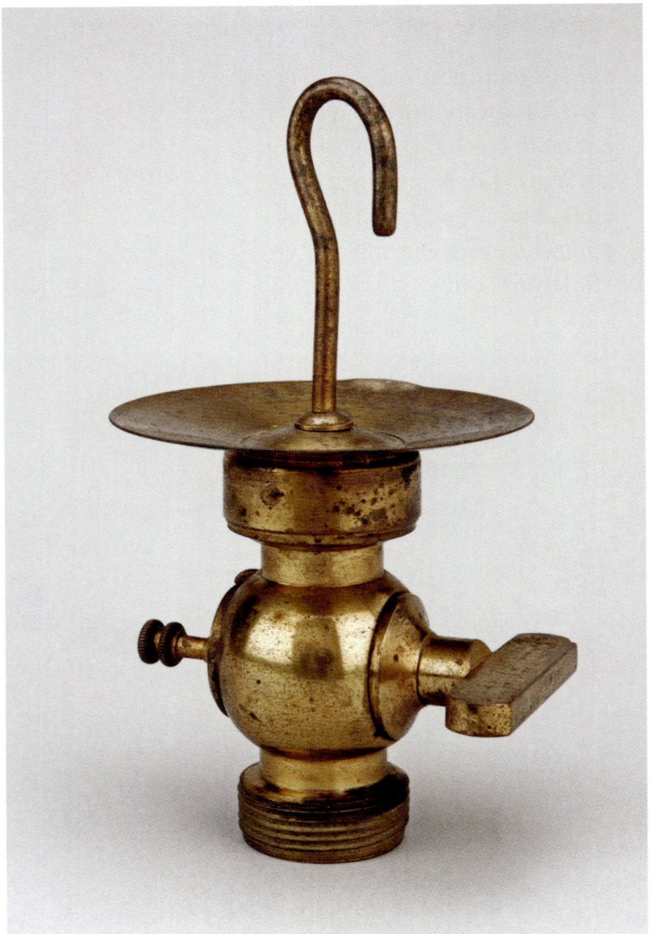

FIGURE 29 (inv. no. 20204-0000-)
© MAM/PHOTO FRANCK BOTTÉ

Threaded brass tap with a hook and a small cap. A small screw closes the opening. It was probably the upper closing part of a glass bottle for weighing gases.

20155-0000- (Fig. 30)
Pipe with window
Second half of 18th c.
MAM accession date: 1952
28.5 × 4 × 3.8; 110 g
Iron, glass

Pipe made of two sections of different diameter connected to a cone. The larger section has a hinged glass window. The use of this object is not clear. It may have been a lamp funnel.

20044-0000- (Fig. 31)
Laboratory stand
Late 18th c.
MAM accession date: 1952
61 × 31.5 × 15; 2 kg
Iron, brass

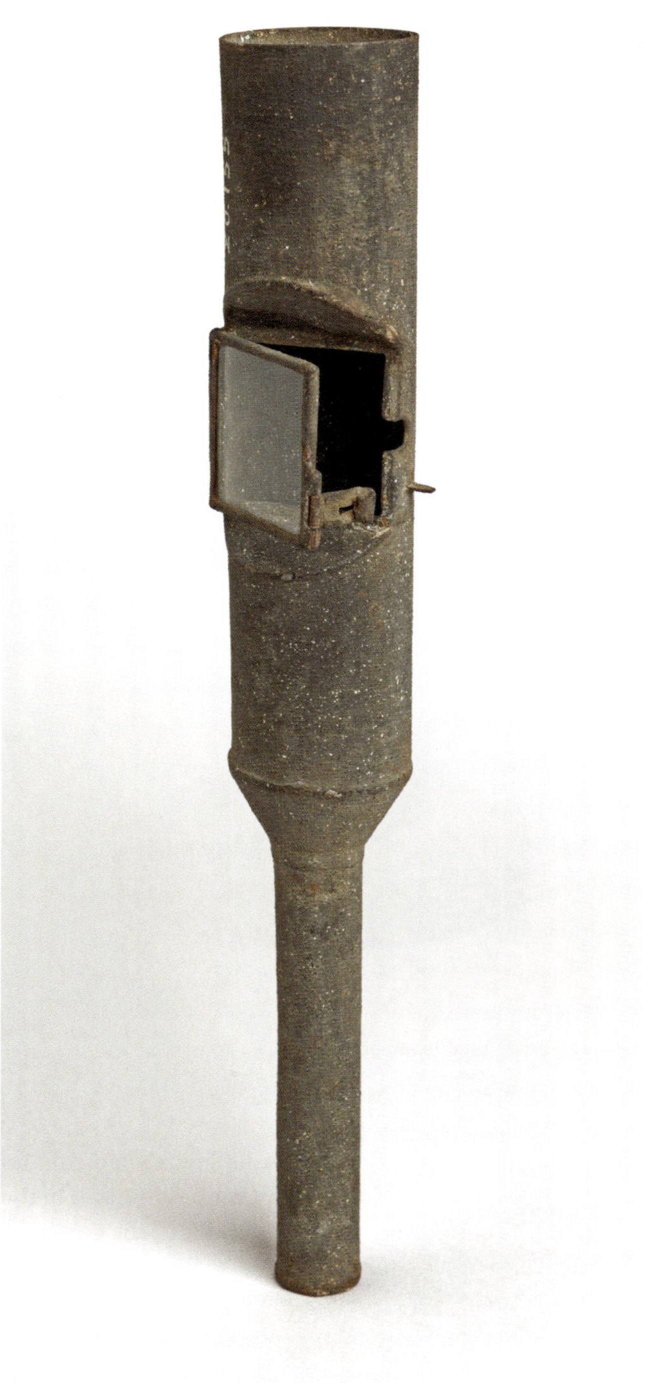

FIGURE 30 (inv. no. 20155-0000-)
© MAM/PHOTO FRANCK BOTTÉ

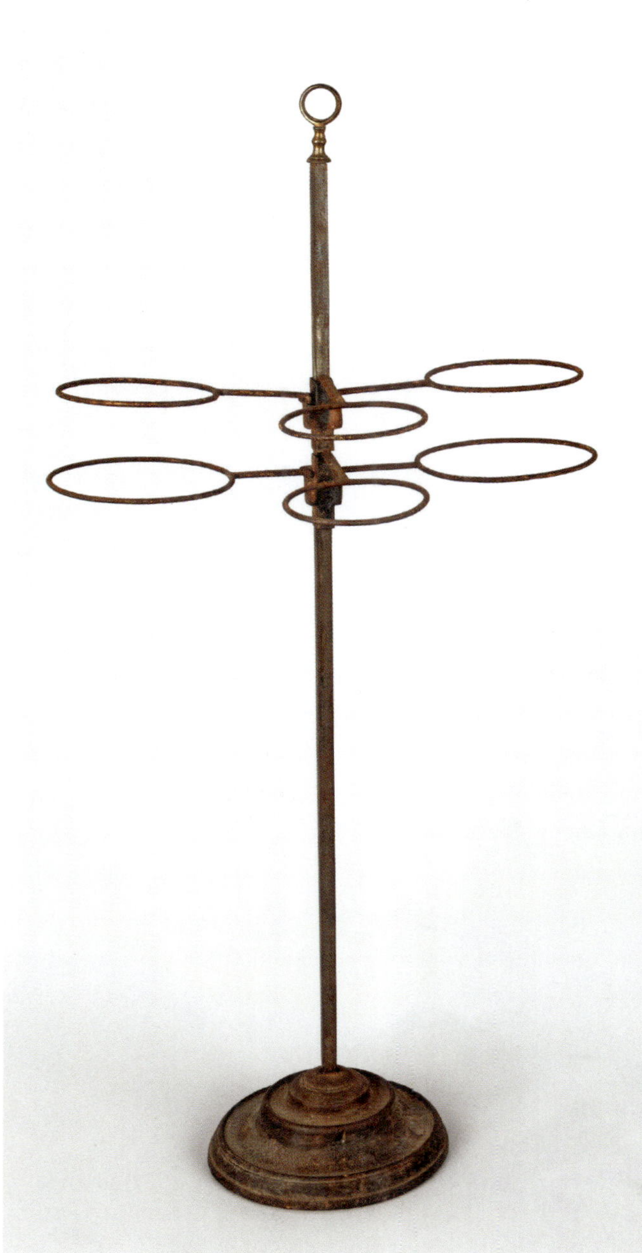

FIGURE 31 (inv. no. 20044-0000-)
© MAM/PHOTO FRANCK BOTTÉ

Round base supporting a vertical bar carrying two sliders. There are six iron rings of different diameters with a side-arm. The ends of two rings can be inserted in each slider.[8] A ring at the top of the bar serves to carry the instrument, which was used to hold retorts and other chemical glassware.

20045-0000- (Fig. 32)
Laboratory stand
Late 18th c.
MAM accession date: 1952
71 × 13.5; 4.27 kg
Brass, lead

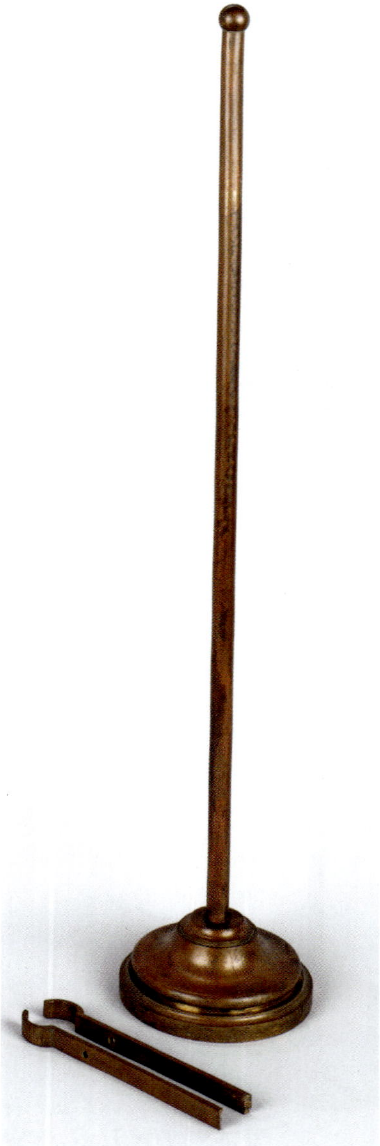

FIGURE 32 (inv. no. 20045-0000-) © MAM/PHOTO FRANCK BOTTÉ

Round iron base, loaded with a lead mass, supporting a vertical bar. It is similar to the stand no. 20044-0000-, but the sliders and rings are missing.

20046-0002- (Fig. 33)
Laboratory stand
Late 18th c.
MAM accession date: 1952
32, × 18.5; 555 g
Wood

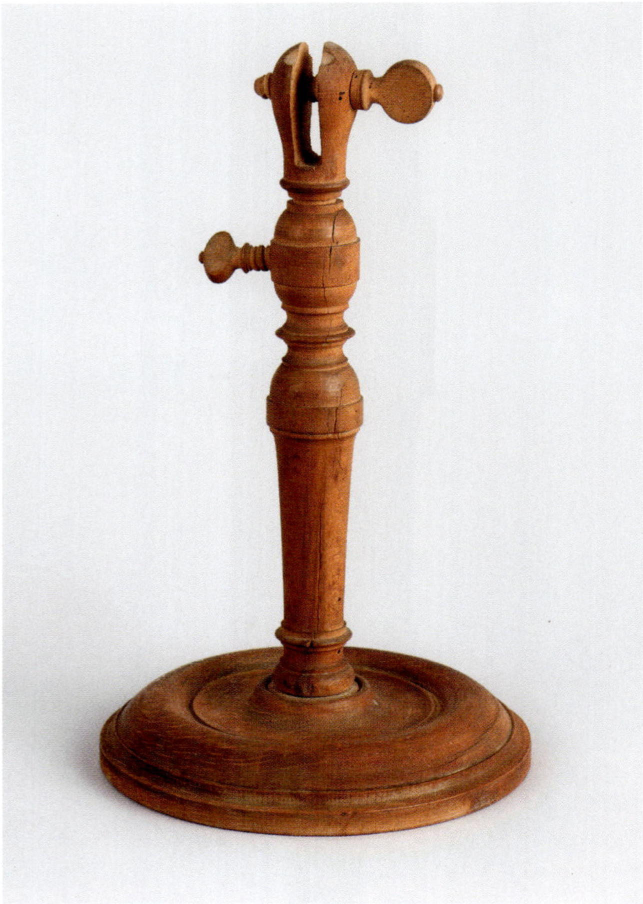

FIGURE 33 (inv. no. 20046-0002-) © MAM/PHOTO FRANCK BOTTÉ

Round wooden base supporting a hollow turned column with a screw. A vertical bar with a clamp is inserted in the column. The stand was used to hold glass tubes or other chemical apparatus.

8 One of the sliders is probably missing.

20046-0001- (Fig. 34)
Laboratory stand
Late 18th c.
MAM accession date: 1952
25.5 × 18.2; 540 g
Wood

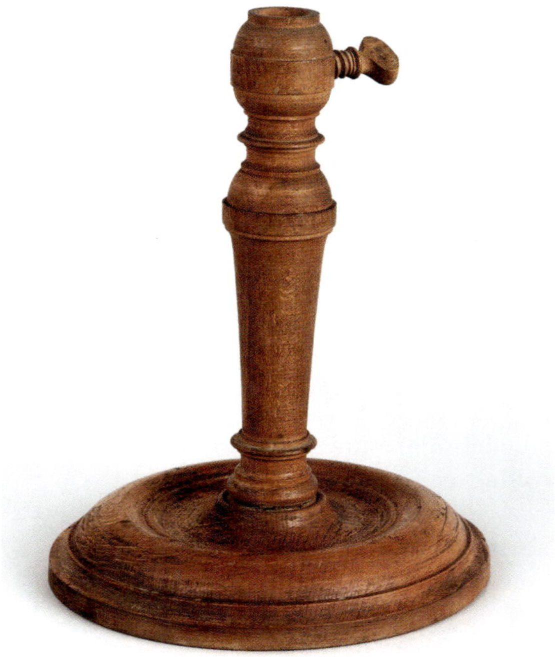

FIGURE 34 (inv. no. 20046-0001-)
© MAM/PHOTO FRANCK BOTTÉ

Very similar to the stand no. 20046-0002-. The bar with the clamp is missing.

20047-0000- (Fig. 35)
Laboratory stand
Late 18th c.
MAM accession date: 1952
36.5 × 32; 1.1 kg
Wood

Small round plate with a vertical bar inserted in a wooden tripod. The plate can be fixed at a desired height by a clamping screw in the tripod. The stand could hold various chemical apparatuses.

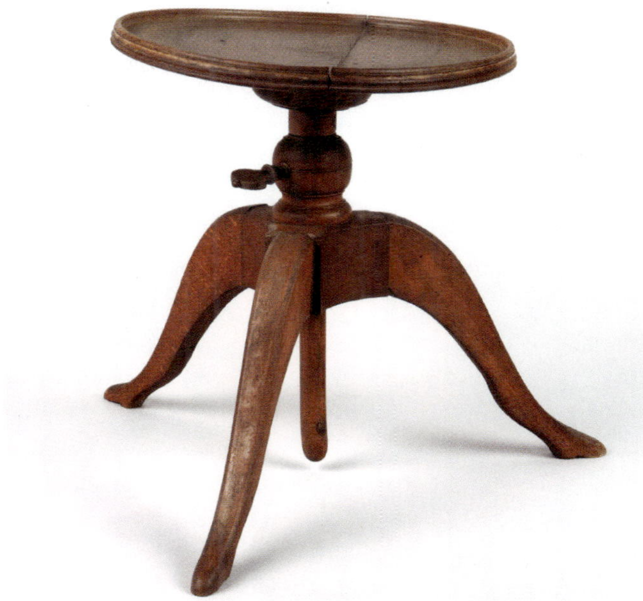

FIGURE 35 (inv. no. 20047-0000-)
© MAM/PHOTO FRANCK BOTTÉ

20038-0000-, 20039-0000-, C-2017-0064-, C-2017-0072-, C-2017-0073-, C-2017-0074-, C-2017-0075-, C-2017-0076- (Figs. 36a, 36b)
Straw rings
Second half of 18th c.
MAM accession date: 1952
20038-0000-: 20 × 8 × 5; 40 g
20039-0000-: 15.5 × 3.5; 55 g (Fig. 36a)
C-2017-0064-: 19 × 14 × 4; 70 g
C-2017-0072-: 17 × 3; 55 g
C-2017-0073-: 14.5 × 4; 70 g
C-2017-0074-: 4 × 14; 80 g
C-2017-0075-: 11 × 3.5; 40 g
C-2017-0076-: 20 × 4.3: 110 g
Straw, silk, fabric

20038-0000-: fragments of a straw ring originally in the Lavoisier collection and donated to the Museum in 1952. 20039-0000-: woven straw ring, originally in the Lavoisier collection and donated to the Museum in 1952.

The following items were found together with various instruments belonging to Lavoisier:

C-2017-0064-: ring covered with blue silk
C-2017-0072-: woven straw ring (together with no. 07547-0004-002-)
C-2017-0073-: woven straw ring (together with no. 07549-0000-)

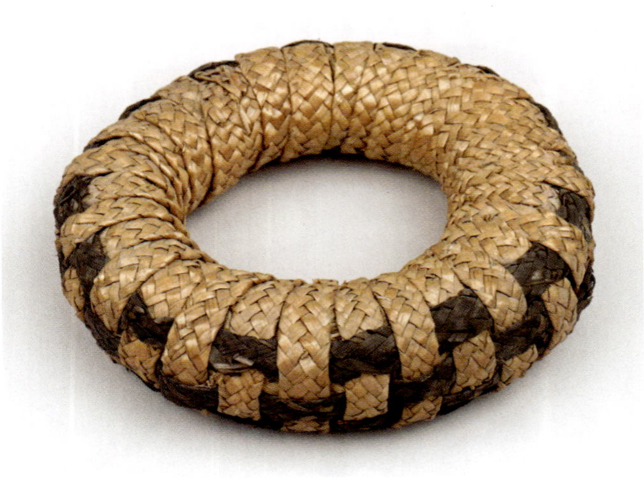

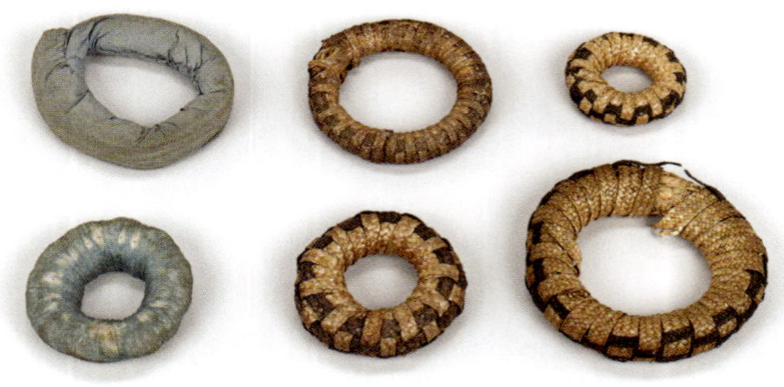

FIGURE 36A, 36B (inv. nos. 20039-0000-, C-2017-0064-, C-2017-0072-, C-2017-0073-, C-2017-0074-, C-2017-0075-, C-2017-0076-)
© MAM/PHOTO FRANCK BOTTÉ

C-2017-0074-: ring covered with blue silk (together with no. 19987-0000-)
C-2017-0075-: woven straw ring (together with no. 19988-0000-)
C-2017-0076-: woven straw ring (together with no. 19996-0000-)

These straw rings were used to support glass balloons and other round-bottom vessels.

20222-0001- (Fig. 37)
Straw rings
Second half of 18th c.
MAM accession date: 1952
20222-0001-: 28.5 × 6; 275 g
20222-0002-: 25 × 5; 230 g
20222-0003-: 2,5 × 14,5; 25 g.
Straw, fabric

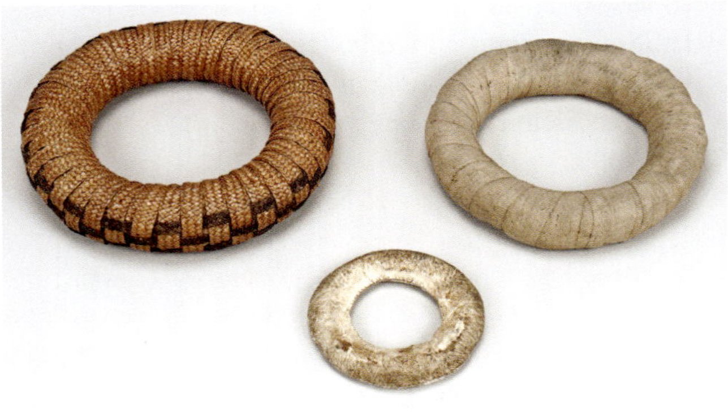

FIGURE 37 (inv. nos. 20222-0001-, -0002-, -0003-)
© MAM/PHOTO FRANCK BOTTÉ

Plaited straw rings used for supporting glass flasks and balloons. 20222-0002- and 20222-0003- are covered with fabrics.

Models

20171-0000-, 20176-0000-, 20177-0000- (Fig. 38)
Three models illustrating an apparatus by Lavoisier for distilling sea-water
ca. 1775
MAM accession date: 1952
20171-0000-: 40.5 × 28 × 23.5; 790 g
20176-0000-: 17.5 × 16 × 15.5; 585 g
20177-0000-: 9.3 × 8 × 6.5; 55 g
Wood, tin, brass, lead, cardboard, paper

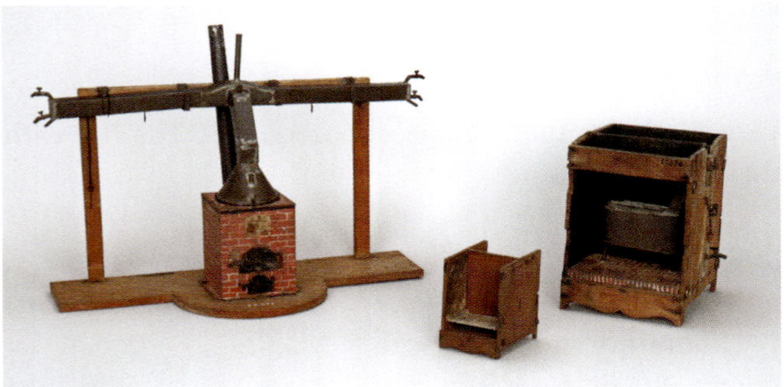

FIGURE 38 (inv. nos. 20171-0000-, 20176-0000-, 20177-0000-)
© MAM/PHOTO FRANCK BOTTÉ

Three models representing elements of a distilling apparatus to be used on ships to obtain fresh water from sea-water.

20171-0000-
This is the largest model. On a wooden board is an oven (painted to simulate brickwork) with an oblique metal

chimney and two doors: one to introduce the fuel, the other to extract the ashes. A boiler with a spherical dome is inserted in the oven and a quadrangular pipe connects it to the condenser. The latter comprises a quadrangular double-walled pipe supported by two vertical wooden pillars and a horizontal beam. The water vapour to be condensed entered into the central section of the pipe, while the water used to refrigerate the condenser circulated in the cavity between the two walls. There are two stopcocks and a nozzle at the end of the pipe. The stopcocks could be used to regulate the circulation of the refrigerating water, and the nozzle to collect the distilled water. The condenser was placed perpendicular to the ship's main axle. This made it possible to use either the left or the right portion of the condensing pipe during navigation, depending on the ship's tilt. This model bears the inscription "Lavoisier // p: BC // 1899".

20176-0000-
This model represents a room divided by a bulkhead into two kitchens, one for the crew, the other for the captain. The crew's kitchen contains the boiler used for cooking and to produce steam, conveyed to the condenser (not present in this model) by an inclined quadrangular pipe. A tube closed by a cork attached to a small chain was used to fill the boiler with water, and a stopcock served to empty it.

20177-0000-
This model shows a rectangular space divided by a wall. It is not certain what this represents. As one of the two spaces is lined with paper, it may have been the reservoir for the water used as refrigerant in the condenser.

In the early 1770s, Lavoisier designed and improved various apparatuses for producing fresh water on board a ship. After studying the industrial distilling devices used to produce wine spirit, he modified them to meet the specific constraints imposed by navigation. In particular, the apparatus had to be as compact as possible, it had to function whatever the ship's tilt, and its fuel consumption had to be low. Various versions of different sizes were tested.

Lavoisier (1775); Truchot (1879), pp. 310–311

20173-0000- (Fig. 39)
Model of pottery kiln
Second half of 18th c.
MAM accession date: 1952
26 × 16.5 × 10.5; 2.19 kg
Plaster

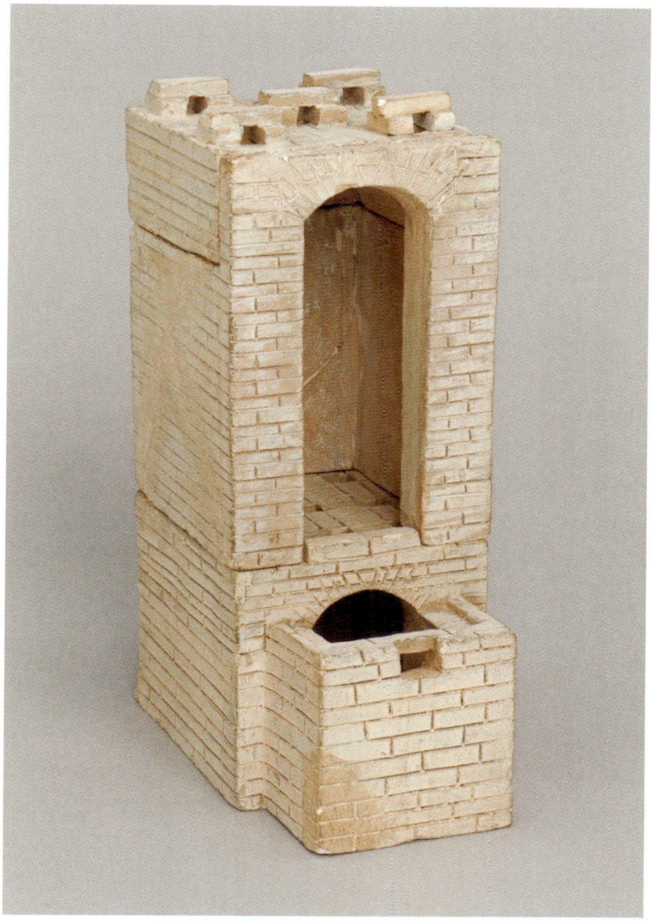

FIGURE 39 (inv. no. 20173-0000-)
© MAM/PHOTO FRANCK BOTTÉ

Plaster model of a rectangular pottery kiln. The model's surface represents the brickwork structure and consists of three separate parts that can be assembled. The parts, numbered 1, 2 and 3, consist of the kiln roof, the central section, where the pottery to be fired is placed, and the lower part, where the fuel is burned.

This model probably belonged to Rumford, who made extensive research to improve the construction and the efficiency of fireplaces, ovens and furnaces.

Thompson (1797)

Chemical Apparatus

07508-0001-, 07508-0002-, 07508-0003-, 07508-0004-, 07508-0005-, 07508-0006- (Fig. 40)
Constant-volume hydrometers
1767–1768
MAM accession date: 1864–1866
07508-0001-: 24 × 7; 690 g

07508-0002-: 25 × 7.5; 580 g
07508-0003-: 22.5 × 8.1 × 8.1; 1.41 kg
07508-0004-: 46 × 14; 4 kg
07508-0005-: 33 × 12; 2.21 kg
07508-0006-: 34.5 × 11.5; 2.19 kg
Brass, lead

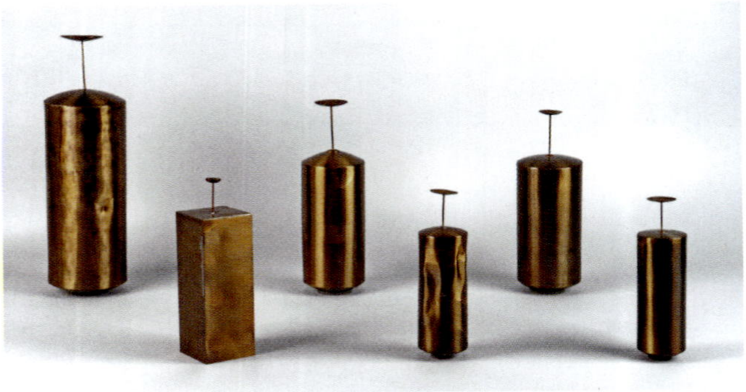

FIGURE 40 (inv. nos. 07508-0001-, -0006-) © MAM/PHOTO FRANCK BOTTÉ

Set of 6 constant-volume hydrometers.[9] These instruments were conceived by Lavoisier during his extensive research on the density of liquids. Five are composed of a brass cylinder whose concave bottom is made of lead and acts as ballast. Each cylinder holds a vertical brass rod supporting a small round pan.[10] The hydrometer 07508-0003- consists only of a parallelepiped floater partly filled with lead beads.

To determine the density (d) of a liquid, the constant-volume hydrometer (whose weight is W) is placed in water, and an additional weight (w) sufficient to sink a marking on the rod down to the water-line is placed on the pan. The instrument is then placed in the liquid whose density is to be determined and another weight (x) is placed on the pan so that the instrument sinks to the same marking. Lastly, the density is calculated with the formula: $d = (W + x)/(W + w)$.

One of these cylindrical hydrometers is depicted in the lower right corner of the famous double portrait of the Lavoisiers by Jacques-Louis David in 1788.

Lavoisier began to use the hydrometer systematically during his travels with Jean-Étienne Guettard between 1763 and 1767. He used the instrument to carry out various research studies on the different proportions of salt contained in mineral waters. Dissatisfied with traditional hydrometers, on 23 March 1768, when elected adjoint of the Académie royale des sciences, he read a memoir presenting a new type of device whose technical characteristics allowed the introduction of quantitative criteria into the chemical analysis of saline substances and of compounds in general. The role Lavoisier assigned to this instrument was, in fact, extremely ambitious, leading him to declare: "it is mainly in the art of combinations that knowing the specific weight of fluids can shed the most light. This part of chemistry is far less advanced than one thinks; we barely have its first elements."[11] Lavoisier's objective was therefore to adapt the hydrometer to the requirements of chemistry and, by means of its application, to explain the mechanism governing chemical combinations. At the end of his memoir, Lavoisier raised a series of questions that, as he saw it, a systematic application of the new instrument should have been able to answer:

> We combine an acid with an alkali every day, but how does the union of these two entities occur? Do the constituent molecules of the acid embed themselves in the alkali's pores, as M. de Lémery believed, or are the acid and alkali composed of different facets, one of which can mesh with the other or unite with it through mere contact, like the Magdeburg hemispheres? How do the acid and alkali hold up separately to water? Does the newly formed salt occupy only the pores of the water? And what is the source of the air that escapes with such ebullience in the moment when the combination occurs, and that, taking advantage of its natural elasticity, instantly occupies a space enormously greater than that of the two fluids from which it has issued? Did the air originally exist in the two mixtures? Was it in some way fixed, as Mr. Hales believed and as most physicists still do, or is it, so to speak, a fictitious air produced by the combination, as Mr. Eller believed? Chemistry, consulted on these different matters, will answer us with vacuous names of ratios, analogues, frictions […] that offer no ideas and whose only effect is to accustom the mind to content itself with mere words.[12]

9 In the constant-volume hydrometer, the volume of the instrument immersed in the liquid remains constant. For this purpose, its weight is adjusted by addition or subtraction. In the constant-weight hydrometer, the immersed volume changes while the instrument's weight remains the same.

10 The stems of hydrometers 07508-0005- and 07508-0006- have a mark.

11 "c'est principalement dans l'art des combinaisons que la connaissance de la pesanteur spécifique des fluides peut porter le plus de lumière. Cette partie de la chimie est beaucoup moins avancée qu'on pense; à peine en avons-nous les premiers éléments." Lavoisier (1768).

12 "Nous combinons tous les jours un acide avec un alcali; mais de quelle manière se fait l'union de ces deux êtres? Les molécules

Thus, it was not in 1772 – as Guerlac believed – but at least four years earlier that Lavoisier reached full awareness of the chemical role of air in chemical reactions. On the strength of his confidence in his hydrometer, Lavoisier launched a heavy attack on the traditional methods used by chemists to explain the origin of reactions. As an alternative, he proposed that the new device should be used, following a rigorously quantitative method, to resolve all the ambiguities and doubts chemists had had in the past. Contrary to what has been recently argued, Lavoisier never abandoned this method of investigation. In fact, in his *Registres de laboratoire*, he set out to determine the specific gravity of fluids during extremely important pneumatic investigations in 1772, 1773, 1774 (when he attempted to determine the specific gravity of the gas liberated by vegetables), 1777, 1784 (when he determined the specific gravity of hydrogen), 1785, 1786 and 1787. Regrettably, his experiments after 1788 are not documented in the surviving laboratory notebooks, but it is certainly not by chance that the first chapter of the third section of his *Traité élémentaire de chimie* opens with the examination of hydrometers – an eloquent sign that, as late as 1789, Lavoisier still regarded this instrument as crucially important. Regardless of the accuracy or simplicity of these instruments in the eighteenth century, it was primarily with their help that Lavoisier, as early as 1768, clearly raised the issue of the role of gases in chemical reactions and that he clearly outlined what, in a few years, would become his most important investigative program of chemical research.

constituantes de l'acide se logent-elles dans les pores de celles de l'alcali, comme le pensait M. de Lémery, ou bien l'acide et l'alcali sont-ils composés de différentes facettes dont l'une peut s'engrener avec l'autre ou s'unir par le simple contact, à la façon des hémisphères de Magdebourg? Comment l'acide et l'alcali tiennent-ils séparément à l'eau? Le nouveau sel qui s'est formé occupe-t-il seulement les pores de l'eau? Enfin d'où vient cet air qui s'échappe avec tant de vivacité dans le moment de la combinaison, et qui, jouissant de son élasticité naturelle, occupe sur-le-champ un espace énormément plus grand que celui des deux fluides dont il est sorti? Cet air existait-il primitivement dans les deux mixtes? Y était-il en quelque façon fixé, comme le pensait M. Hales et comme le pensent encore la plupart des physiciens, ou bien est-ce un air, pour ainsi dire factice et qui soit le produit de la combinaison, comme le pensait M. Eller? La chimie, consultée sur ces différents objets, nous répondra par des vains noms de rapports, d'analogues, de frottements [...] qui ne présentent aucune idée et qui n'ont d'autre effet que d'accoutumer l'esprit à se payer des mots". Lavoisier (1768).

Lavoisier (1768); Truchot (1879), pp. 314–315; Berthelot (1890), pp. 228–230, 245–248, 257, 261, 272–274, 288, 300, 304; Beretta (1994), pp. 35–45; Palmer (1998), pp. 124–210; Bensaude-Vincent (2000); Beretta (2001), pp. 31–34

20210-0000- (Fig. 41)
Constant-volume hydrometer
ca. 1768
MAM accession date: 1952
18 × 5.8; 280 g
Silver, lead

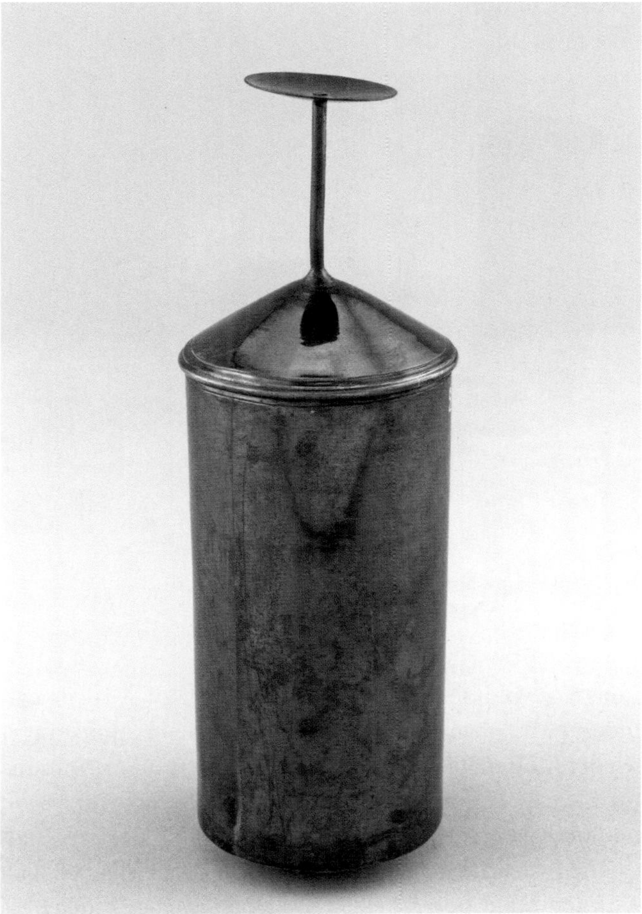

FIGURE 41 (inv. no. 20210-0000-)
© MAM/PHOTO FRANCK BOTTÉ

Small constant-volume hydrometer composed of a cylindrical silver vessel with a convex bottom and a conical top supporting a vertical rod with a small round plate. The hydrometer is weighted down with lead. It is punched with three silver hallmarks: a "c", a lily and a crown.

The instrument is similar in design and operation to hydrometers 7508-0001-6.

Lavoisier (1768); Truchot (1879), pp. 314–315; Berthelot (1890), pp. 228–230, 245–248, 257, 261, 272–274, 288, 300, 304; Beretta (1994), pp. 35–45; Palmer (1998), pp. 124–210; Bensaude-Vincent (2000); Beretta (2001), pp. 31–34

19959-0001-, 19959-0002-, 19959-0003- (Fig. 42)
Constant-volume hydrometer with container
ca. 1768
MAM accession date: 1952
Hydrometer with his leather bag: 20 × 10; 860 g
Container: 39.5 × 15 × 13; 4.21 kg
Cylindrical vessel: 22 × 11; 500 g
Silver, pewter, fabric

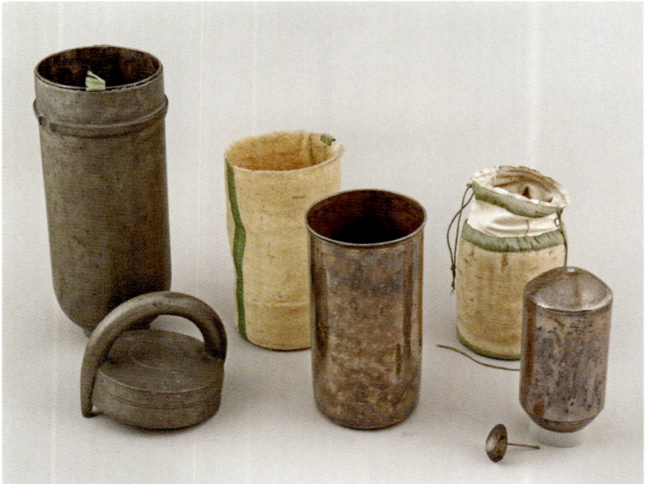

FIGURE 42 (inv. nos. 19959-0001-, -0003-)
© MAM/PHOTO FRANCK BOTTÉ

Constant-volume hydrometer composed of a silver cylinder with convex ends and a vertical stem supporting a small round plate. The bottom of the instrument contains tin ballast. This hydrometer works in exactly the same manner as nos. 07508-0001-0006-.

The instrument is preserved in a fabric and leather bag with a stuffed bottom, inserted in a silver cylinder protected by another fabric bag. The hydrometer and cylinder are placed in a second similar cylinder inserted into a pewter container (No. 19959-0002-) closed by a lid with a handle.[13] The bottom of the container carries a barely legible punched inscription "A … jervoisn[?] rue Honoré // n° 1896 A Paris". On the bottom of the internal silver cylinder there are three identical silver hallmarks.

13 The two silver cylinders with their fabric bags were numbered 19959-0003-.

The function of the two silver cylinders is unclear: they may have been used as containers for the liquid whose specific weight had to be determined. As their diameter is slightly larger than the hydrometer's, it was possible to use the latter with just a small quantity of liquid.

Lavoisier (1768); Truchot (1879), pp. 314–315; Berthelot (1890), pp. 228–230, 245–248, 257, 261, 272–274, 288, 300, 304; Beretta (1994), pp. 35–45; Palmer (1998), pp. 124–210; Bensaude-Vincent (2000); Beretta (2001), pp. 31–34

19960-0000- (Fig. 43)
Container for a hydrometer
ca. 1768
MAM accession date: 1952
27.5 × 12.5 × 12; 1.61 kg
Pewter

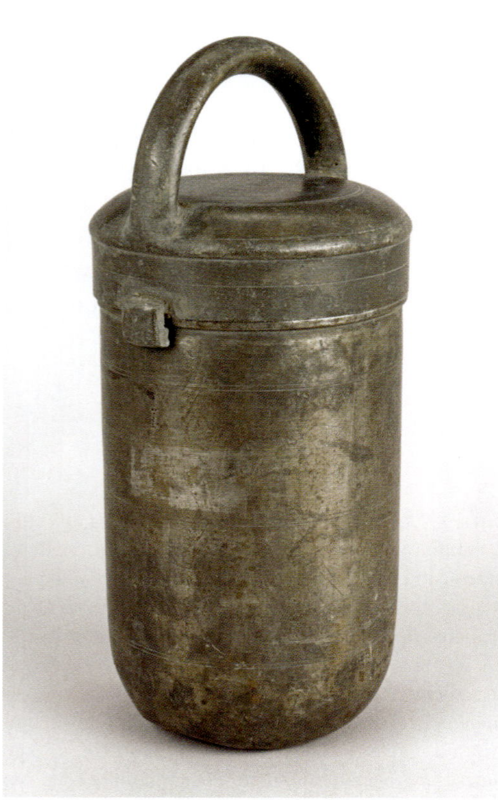

FIGURE 43 (inv. no. 19960-0000-)
© MAM/PHOTO FRANCK BOTTÉ

Cylindrical container for a hydrometer closed by a lid with a curved handle. The hydrometer originally stored in this container has not been identified. The item is marked with a punched "I" and has two hallmarks at the bottom.

19957-0000- (Fig. 44)
Constant-volume hydrometer
ca. 1768
MAM accession date: 1952
14 × 8.5; 140 g
Silver

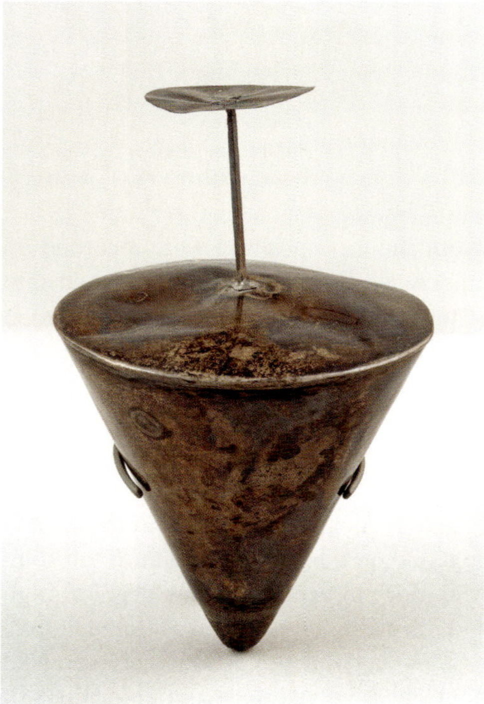

FIGURE 44 (inv. no. 19957-0000-)
© MAM/PHOTO FRANCK BOTTÉ

Constant-volume hydrometer composed of a silver cone with a vertical stem supporting a small round plate. The bottom of the instrument contains a tin ballast. This hydrometer works in exactly the same manner as nos. 07508-0001-0006-. Lavoisier selected the unusual conical shape because he was not fully satisfied with cylindrical hydrometers. These latter tended to capsize if the liquid whose specific weight had to be determined made it necessary to place too much weight on the plate.[14] The cone form greatly reduced this inconvenience. Conical hydrometers could measure a wider range of specific weights and, at the same time, required less liquid for the purpose.

Lavoisier (1768); Truchot (1879), pp. 314–315; Berthelot (1890), pp. 228–230, 245–248, 257, 261, 272–274, 288, 300, 304; Beretta (1994), pp. 35–45; Palmer (1998), pp. 124–210; Bensaude-Vincent (2000); Beretta (2001), pp. 31–34

19956-0001-, 19956-0002-, 19956-0003-, 19956-0004-, 19956-0005-, 19956-0006- (Fig. 45a, 45b)
Six constant-weight hydrometers
1770s
MAM accession date: 1952
Only 19956-00006- is signed: Mossy (active ca. 1775–1821)
19956-0001-: 20 × 3; 15 g
19956-0002-: 22 × 4; 20 g
19956-0003-: 22 × 4; 25 g
19956-0004: 21.5 × 3, 15 g
19956-0005-: 20 × 3.3; 25 g
19956-0006-: 15.2 × 1.3; 5 g
Glass, paper, mercury

These constant-weight hydrometers[15] served to measure the density of liquids on the basis of Archimedes' principle. The lower the fluid's density, the deeper a hydrometer of a given weight sinks into it. These hydrometers basically consist of a spherical floater with a stem on top containing a paper scale, while a short tube connects the bottom of the floater with a ballast composed of a bulb partly filled with mercury.

The graduations on the paper scales (from bottom to top) are the following:

19956-0001-: from 1030 to 800
19956-0002-: from 1030 to 780
19956-0003-: from 1010 to 770
19956-0004-: from 1020 to 800
19956-0005-: from 1000 to 800
19956-0006-: from 45 to 0.

The first five hydrometers have a centesimal scale of the specific weight, and Lavoisier carefully described how to trace their divisions.[16] The number 1000 on the scale marked the position of the instrument floating in distilled water at 10° Réaumur. He then repeatedly added some drops of mercury of known weights to the ballast and, afeter every addition, marked the hydrometer's floating position in the water with a silk thread. The distances

14 The instability was due to the fact that adding weight to the top plate raised the instrument's centre of gravity.

15 This model is also known as Boyle's hydrometer, after Robert Boyle, who described it in 1675. However, the principle behind the construction of the instrument was known since Antiquity.

16 Lavoisier indicated how to prepare the specific-weight scale as well as the scale indicating the volumes of liquid displaced by the hydrometers. From the latter, one could obtain the specific weight thanks to a conversion table.

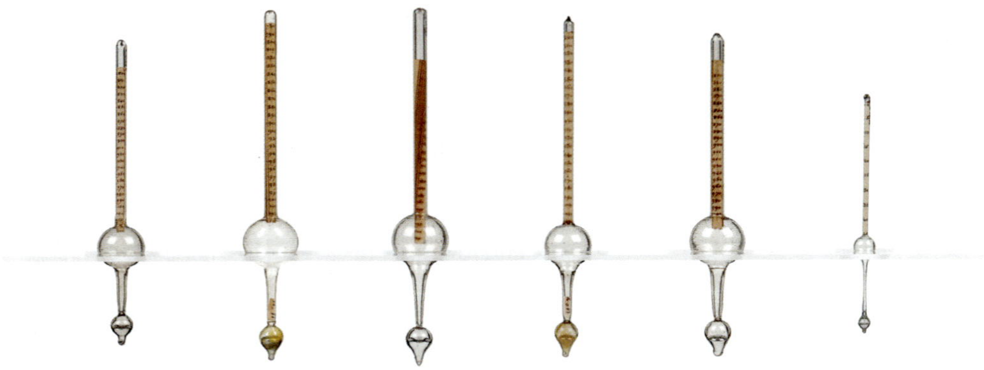

FIGURE 45A
(inv. nos. 19956-0001-, -0006-)
© MAM/PHOTO FRANCK BOTTÉ

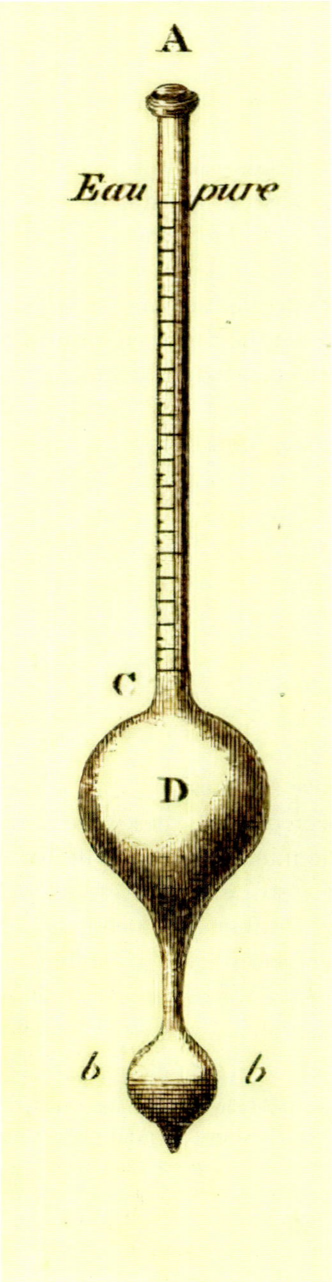

FIGURE 45B
Hydrometer published in LO, vol. 5, Pl. II, fig. 10

of the thread from the 1000 position were recorded on a calibrated paper scale that was then introduced into the stem of the instrument. Numbers smaller than 1000 indicate specific weights of liquids lighter than water; numbers greater than 1000 indicate specific weights of heavier liquids.

The sixth hydrometer (no. 19956-0006-) is similar to the others but has a Baumé scale ranging from 45 to 0 (from bottom to top). The French apothecary Antoine Baumé invented a scale based on the specific weight of salt solutions (sodium chloride) in water. Baumé's degrees originally represented the percent by mass of sodium chloride in water at 15° C. For example, 10° Baumé corresponded to the specific weight of a solution containing 10% of sodium chloride by mass of water.[17] The present hydrometer had to be used with a liquid with a specific weight heavier than water.[18] The scale bears the handwritten signature: "Pour les sels lessives et eaux fortes par Mossy Quay Pelletier AParis".

These hydrometers were probably used by Lavoisier to support his 1777 proposal to standardize weights and measures at the Régie des poudres (Lavoisier, 1777). In a work devoted to improving saltpetre production he proposed, among other things, a new method of graduating hydrometers. After praising the results obtained with this method, he remarked:

> The care taken to construct them has resulted in a perfect success, and they do not differ from each other by an eighth of a degree. We then turned to

17 Two simple approximate formulas make it possible to calculate the specific weight (SW) from the Baumé degrees (B°). SW = 145/(145 – B°) for a liquid denser than water; SW = 140 /(B° + 130) for a liquid less dense than water. (The coefficient varied slightly for different countries and different periods.)

18 For a liquid lighter than water, the 0 of the scale is positioned at the top of the instrument's stem.

M. Moussy [Mossy], maker of physics instruments for the Académie des sciences, to obtain a greater number; but he was requested not to release any of these instruments prior to its having been numbered and compared with the Arsenal standards – a test that will result in the issuance of a certificate to accompany each hydrometer. In this way, the public will be able to rely on the accuracy of the hydrometers supplied to it by M. Moussy: he resides at the Rue and Place Royale. Each of these instruments will be housed in a small tin-plate box, which will be enclosed in a slightly larger one serving as a case for it. When a test is to be conducted, the outer case will be filled up to a finger from the edge, the hydrometer will be dipped into it, and the degree will be observed. If, for example, the water surface reads twelve, we shall conclude that the liquid contains twelve per cent, as much of saltpetre as of salt and mother-water. And if we know from prior experiments that the waste from the workshop where we are working is one-quarter or one-third, we shall conclude that the liquid just tested contains eight or nine per cent of saltpetre.

For washing in laboratories, vessels of different shapes, earthenware pots, glass jars, etc. are used. To decant the liquid without stirring the deposit that has formed, a siphon is sometimes used. This instrument consists in a glass tube ABC (Fig. 45b), curved in B, whose branch BC must be a few inches longer than branch AB. To avoid having to hold it in one's hand, which can be tiring in some experiments, the siphon is run through a hole made in the middle of a small board DE. End A of the siphon must plunge into the liquid of jar FG, to the depth at which one wants to empty the jar. According to the hydrostatic principles on which the siphon effect is based, the liquid cannot flow into the jar unless the air inside the jar has been expelled. This is done by means of a small glass tube HI, hermetically sealed to branch BC. When one wants the liquid to flow from jar FG into jar LM using the siphon, one starts by plugging end C of the siphon branch BC with one's finger. One then sucks with one's mouth until all the air has been removed from the tube and replaced by the liquid. One then removes one's finger, and the liquid flows and continues to move from jar FG into jar LM.[19]

Lavoisier (1768); Baumé (1773), Lavoisier (1777), pp. 41–45 and plate 2, fig. 10; Beretta (1994)

19958-0000- (Fig. 46)
Constant-weight hydrometer
ca. 1770
10.5 × 4.4; 40 g
Silver, wood

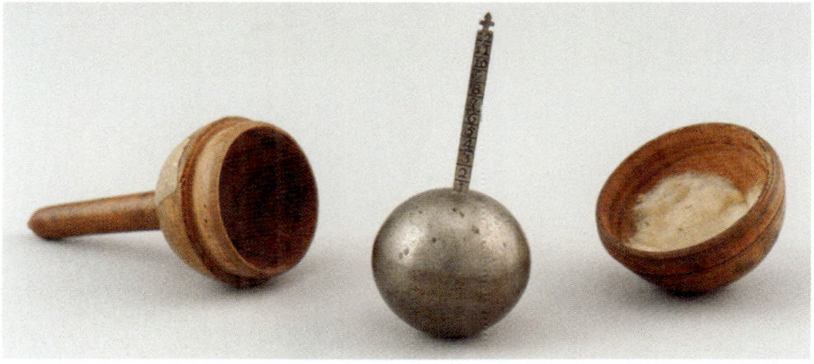

FIGURE 46 (inv. no. 19958-0000-)
© MAM/PHOTO FRANCK BOTTÉ

Spherical constant-weight hydrometer. It is composed of a silver sphere (with an interior ballast at its bottom) and a vertical flat stem surmounted by a lily. The scale on the rod ranges from 1 to 12. The instrument is contained in a turned spherical wooden box.

A constant-weight hydrometer is immersed in a liquid, whose specific weight (or concentration, if it consists of a solution) is indicated by the point on the scale at which the surface of the liquid touches the graduated stem.

19 "Le succès a répondu parfaitement aux soins qu'on avait pris pour les construire, et ils ne diffèrent pas entre eux d'un huitième de degré. On s'est ensuite adressé au sieur Moussy, constructeur d'instruments de physique de l'Académie des sciences, pour en avoir un plus grand nombre; mais on lui a imposé la condition de ne laisser sortir de ses mains aucun de ces instruments sans qu'il eût été préalablement numéroté et comparé avec les étalons de l'arsenal, de laquelle épreuve il sera délivré un certificat qui accompagnera chaque pèse-liqueur. Par ce moyen, le public pourra compter sur l'exactitude des pèse-liqueurs qui lui seront délivrés par le sieur Moussy: il demeure rue et place Royale. Chacun de ces instruments sera contenu dans une petite boite de fer-blanc, laquelle sera renfermée dans une un peu plus grande qui lui servira d'étui; lorsqu'on voudra faire une épreuve, on emplira celle-ci jusqu'à un doigt du bord; on y plongera le pèse-liqueur, et on observera le degré. Si, par exemple, la surface de l'eau répond à douze, on en conclura que la liqueur contient douze pour cent, tant de salpêtre que de sel et d'eau mère; et si l'on sait d'ailleurs par des expériences préalables que le déchet de l'atelier où l'on travaille est d'un quart ou d'un tiers, on en conclura que la liqueur qu'on vient d'essayer contient huit ou neuf pour cent de salpêtre." LO, vol. 5, pp. 428–429.

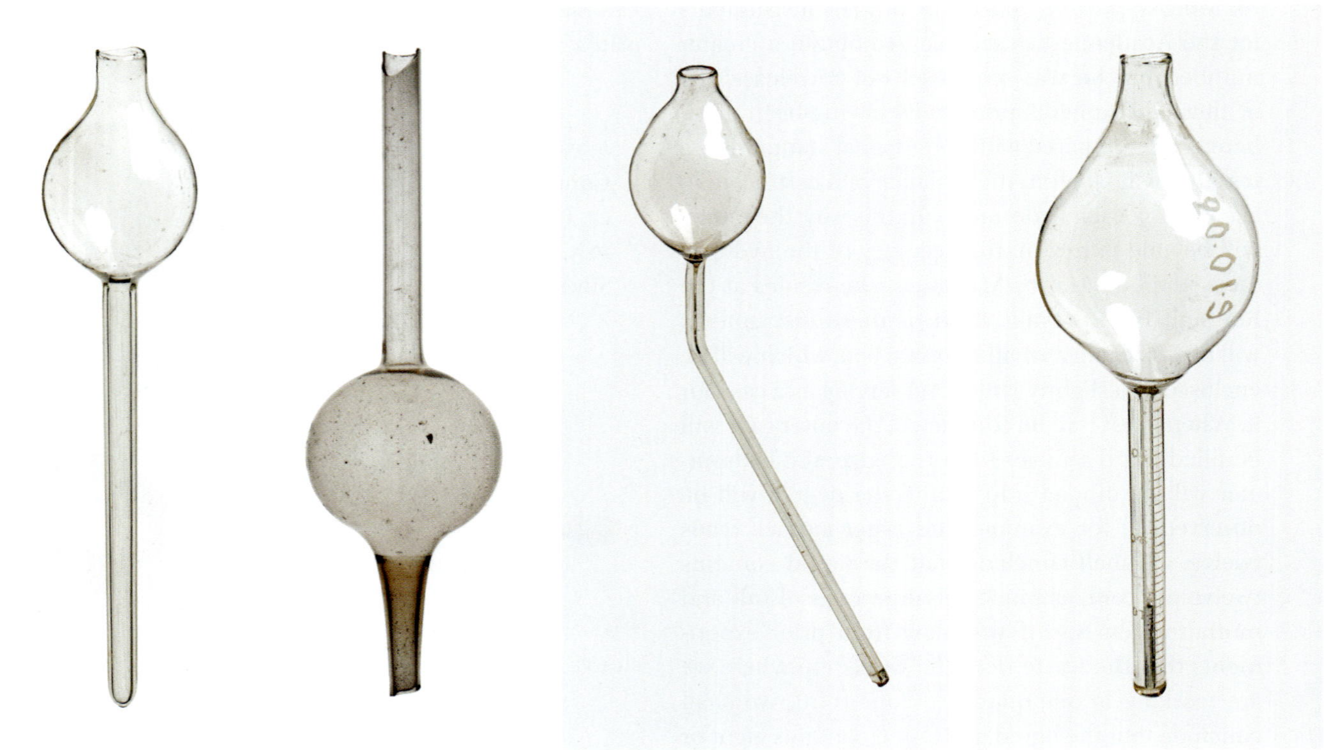

FIGURES 47A, 47B, 47C, 47D (inv. nos. 20015-000-, 20016-0000-, 20018-0000-, 20019-0000-)
© MAM/PHOTO FRANCK BOTTÉ

20015-000-, 20016-0000-, 20018-0000-, 20019-0000- (Figs. 47a, 47b, 47c, 47d)
Four glass bulbs with capillary tubes
Second half of 18th c.
MAM accession date: 1952
20015-0000-: 20 × 3.5; 20 g
20016-0000-: 12.5 × 3.5; 5 g
20018-0000-: 18.5 × 4.5 × 3.5; 10 g
20019-0000-: 12 × 3.8; 10 g
Glass

These items are similar. Their function is not clear but some appear to be parts of glass hydrometers.

20015-0000-: Glass bulb with an opening and a straight closed tube.

20016-0000-: Tube with glass bulb. The ends of the tube are broken. The tube has a handwritten label that reads "350".

20018-0000-: Glass bulb with an opening and a bent capillary tube. The bulb has a handwritten label that reads "357".

20019-0000-: Glass bulb with an opening and a capillary tube with traces of mercury. The tube carries an engraved scale from 0 to 35 (from top to bottom) with the indications "10", "20" and "30". Probably an unfinished hydrometer.

20152-0000- (Fig. 48)
Container with cap
Second half of 18th c.
MAM accession date: 1952
26 × 15; 570 g
Iron, cork, animal skin

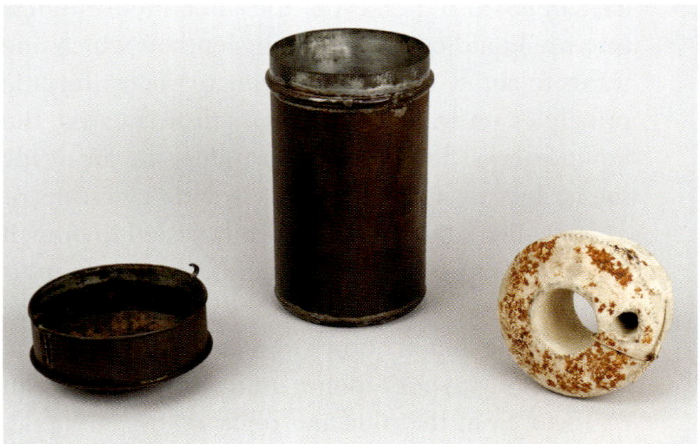

FIGURE 48 (inv. no. 20152-0000-)
© MAM/PHOTO FRANCK BOTTÉ

Cylindrical iron vessel with cap. Inside the cap there is a ring-shaped holder made of cork and coated with animal skin. The container was probably used for storing and carrying a fragile instrument, perhaps a hydrometer.

20024-0000- (Figs. 49a, 49b)
Laboratory siphon
Late 18th c.
MAM accession date: 1952
48 × 26 × 1; 60 g
Glass, cork, thread

Glass siphon composed of a V-shaped tube (broken into two pieces) to which a suction tube is soldered, running parallel to it. Some cork and a thread are inserted between the two tubes. The siphon is used to transfer a liquid from one container to another. To activate the siphon, one simply sucked on the suction tube while the opposite end of the V-shaped tube was immersed in a liquid. The suction tube also served as an essential safety device when poisonous or corrosive liquids had to be siphoned, for it prevented the risk of inhaling them. Lavoisier describes the use of the siphon in the *Traité élémentaire de chimie* as follows:

> For washing in laboratories, vessels of different shapes, earthenware pots, glass jars, etc. are used. To decant the liquid without stirring the deposit that has formed, a siphon is sometimes used. This instrument consists in a glass tube ABC (See Fig. 49b), curved in B, whose branch BC must be a few inches longer than branch AB. To avoid having to hold it in one's hand, which can be tiring in some experiments, the siphon is run through a hole made in the middle of a small board DE. End A of the siphon must plunge into the liquid of jar FG, to the depth at which one wants to empty the jar. According to the hydrostatic principles on which the siphon effect is based, the liquid cannot flow into the jar unless the air inside the jar has been expelled. This is done by means of a small glass tube HI, hermetically sealed to branch BC. When one wants the liquid to flow from jar FG into jar LM using the siphon, one starts by plugging end C of the siphon branch BC with one's finger. One then sucks with one's mouth until all the air has been removed from the tube and replaced by the liquid. One then removes one's finger, and the liquid flows and continues to move from jar FG into jar LM.[20]

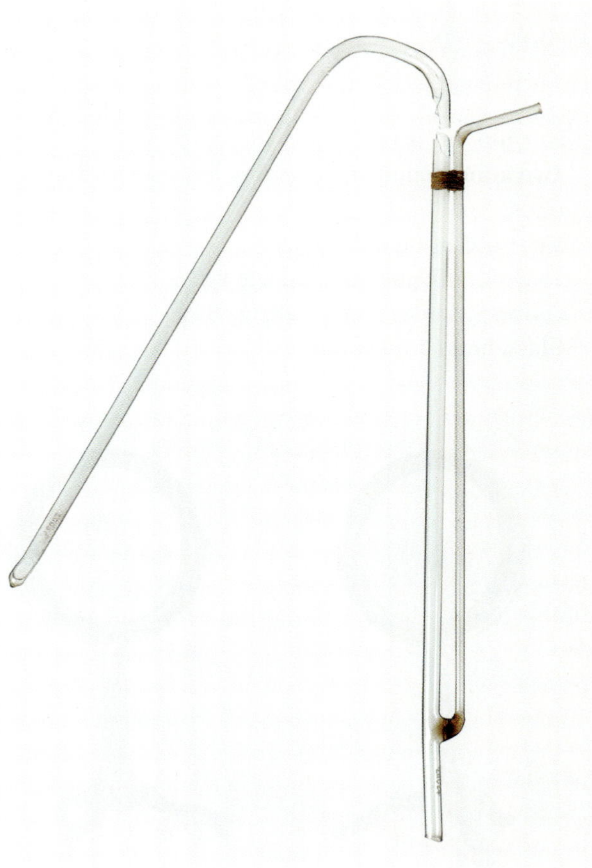

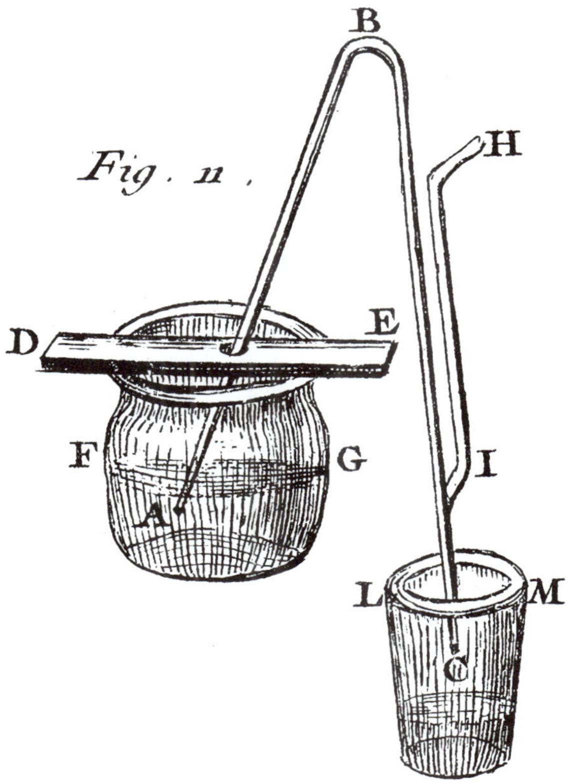

FIGURES 49A, 49B (inv. no. 20024-0000-)

20 "On se sert pour le lavage, dans les laboratoires, de vaisseaux de différentes formes, de terrines de grès, de bocaux de verre, etc.

Lavoisier (1789), vol. 2, pp. 411–412, plate II, Fig. 11; Chevallier (1819), pp. 68–71

20079-0001-, 20079-0002- (Fig. 50)
Two almost identical burning lenses with stand
1770s
MAM accession date: 1952
20079-0001-: 98 × 52 × 44; 14.6 kg
20079-0002-: 97 × 51.5 × 48; 12.38 kg
Glass, brass, iron wood

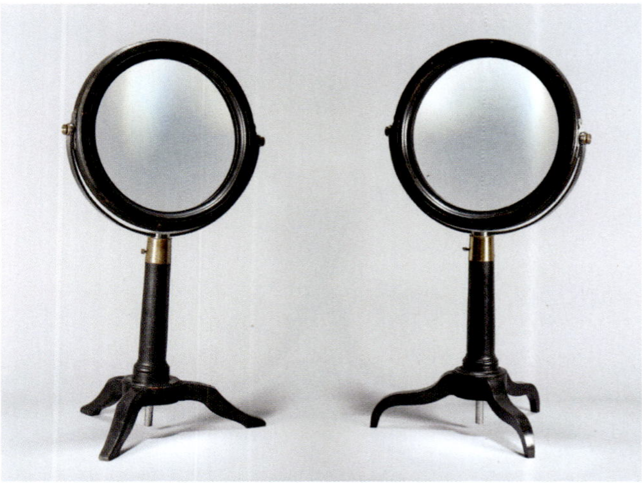

FIGURE 50 (inv. nos. 20079-0001-, -0002)
© MAM/PHOTO FRANCK BOTTÉ

Circular biconvex lenses mounted in a wooden frame and hinged on a semi-circular iron arc. The arc has a vertical shaft inserted into a vertical wooden column mounted on a tripod. The shaft can be clamped with a brass key in the brass collar at the top of the column.

This adjustable converging lens was used as a burning lens to focus sunlight on a specimen and studying the effect of heat on it.

Burning lenses had been used in chemistry since the seventeenth century. In his *Traicté de la Chymie* (Paris, 1669, vol. 2, p. 996) the French chemist Nicaise Le Febvre used a small burning lens to calcinate antimony. In the same period, Samuel Cottereau Duclos and Nicolas Lémery also noted that the burning lens was an extremely effective means for dissolving metals and could even reveal their inner structure. This widespread interest prompted the construction of several large burning mirrors in Paris in the second half of the seventeenth century. One with a diameter of 1.16 m was made by the engineer François Villette (1621–1698) in 1680. In 1701, on the recommendation of Wilhelm Homberg Philippe II, the Duc d'Orléans purchased a large burning mirror made by Ehrenfried Walther von Tschirnhaus (1651–1708) that could reach a temperature of 2000°. In the first decade of the eighteenth century, Wilhelm Homberg and his pupil Étienne-François Geoffroy used it systematically to examine the effect of extreme heat on metals. The research program with large burning lens was interrupted until 1772 when Louis-Claude Cadet de Gassicourt and Jacques-Mathurin Brisson sought the Académie's permission to use Tschirnhaus' mirror for a series of experiments on the combustion of diamonds and other precious stones. Cadet and Brisson were soon joined by Pierre-Joseph Macquer and Lavoisier. Some 200 experiments were performed at the Jardin de l'Infante at the Louvre between August 1772 and August 1773. For this series, another of Tschirnhaus' mirrors was borrowed from the Comte de la Tour d'Auvergne. In August 1772, Lavoisier drafted *Réflexions sur les expériences qu'on peut tenter à l'aide du miroir ardent* (LO, vol. 3, pp. 261–266), in which he suggested to his associates that they should perform experiments on heating metals and calces in closed vessels. This suggestion was not ignored but Lavoisier was unable to perform experiments on his own until late October–early November, when he had exclusive access to De La Tour d'Auvergne's lens (LO, vol. 1, pp. 599–600). It was only in Summer and Autumn 1773 that he had the apparatus ready to conduct his famous experiments on the reduction of minium. In later years, Lavoisier frequently used smaller burning lenses – such as those still in the MAM collection – to repeat the experiments on minium as well as on other metallic substances. He was certainly encouraged to make more regular use of burning lenses after Joseph Priestley noted in 1774 that "the most accurate method of procuring air from several substances,

Quelquefois, pour décanter la liqueur sans troubler le dépôt qui s'est formé, on emploie le siphon. Cet instrument consiste en un tube de verre ABC (pl. II, fig. 11), recourbé en B, et dont la branche BC doit être plus longue de quelques pouces que celle AB. Pour n'être point obligé de le tenir à la main, ce qui pourrait être fatigant dans quelques expériences, on le passe dans un trou pratiqué au milieu d'une petite planche DE. L'extrémité A du siphon doit être plongée dans la liqueur du bocal FG, à la profondeur jusqu'à laquelle on se propose de vider le vase. D'après les principes hydrostatiques sur lesquels est fondé l'effet du siphon, la liqueur ne peut y couler qu'autant qu'on a chassé l'air contenu dans son intérieur: c'est ce qui se pratique au moyen d'un petit tube de verre HI, soudé hermétiquement à la branche BC. Lors donc qu'on veut procurer, par le moyen du siphon, l'écoulement de la liqueur du vase FG dans celui LM, on commence par boucher avec le bout du doigt l'extrémité C de la branche BC du siphon; puis on suce avec la bouche, jusqu'à ce qu'on ait retiré tout l'air du tube et qu'il ait été remplacé par la liqueur: alors, on ôte le doigt, la liqueur coule et continue à passer du vase FG dans celui LM". LO, vol. 1, p. 299.

by means of heat, is to put them, if they will bear it, into phials full of quicksilver, with their mouths immersed in the same, and then throw the focus of a burning mirror upon them."

A new large lens, constructed thanks to a donation from Trudaine de Montigny, was used for several experiments by Lavoisier and his associates in October 1774.

The difficulty in controlling the quantity of heat produced with this kind of device prompted Lavoisier to search for a new and more efficient apparatus. On 26 March 1782, he finished designing a new kind of enameller's lamp that fuelled a blowpipe with measurable quantities of oxygen, producing higher temperatures than the burning lenses (Ms. Lavoisier (1772–1788), vol. 6, fol. 138).

Lavoisier (1774), 256–265 and fig. 8; Lavoisier (1774a); Priestley (1774), vol. 1, p. 13; Lavoisier (1782b); Daumas (1955), pp. 121–122; Guerlac (1961), pp. 157–161; Smeaton (1987); Lehman (2013); Lehman (2016)

20080-0000- (Fig. 51)
Burning mirror with stand
1770s
MAM accession date: 1952
176 × 81 × 60; 22 kg
Iron, wood, amalgamated glass

Large concave amalgamated glass mirror in a circular wooden frame. The frame is held by a semi-circular iron arc mounted on an ebonized wooden column with a tripod. Considering its dimensions, it was probably used as a burning mirror to focus solar rays on a substance in order to melt or calcinate it.

See historical note to 20079-0002-.

Lavoisier (1774), pp. 256–265 and fig. 8; Lavoisier (1774a); Priestley (1774), vol. 1, p. 13; Lavoisier (1782b); Daumas (1955), pp. 121–122; Guerlac (1961), pp. 157–161; Smeaton (1987); Lehman (2013)

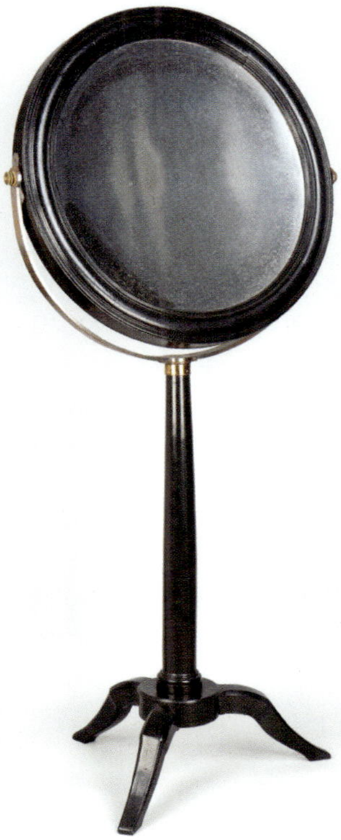

FIGURE 51 (inv. no. 20080-0000-)
© MAM/PHOTO FRANCK BOTTÉ

20081-0000- (Fig. 52a, 52b)
Concave mirror with stand
1770s
MAM accession date: 1952
82.5 × 48 × 28.5; 5.44 kg
Painted wood, amalgamated glass, iron

Concave amalgamated glass mirror in a circular wooden frame held by a semi-circular arc mounted on a turned wooden column resting on a circular base. The back of the frame is decorated with the image of a gilded anthropomorphic sun.

Lavoisier (1774), pp. 256–265 and fig. 8; Lavoisier (1774a); Priestley (1774), vol. 1, p. 13; Lavoisier (1782b); Daumas (1955), pp. 121–122; Guerlac (1961), pp. 157–161; Smeaton (1987); Lehman (2013)

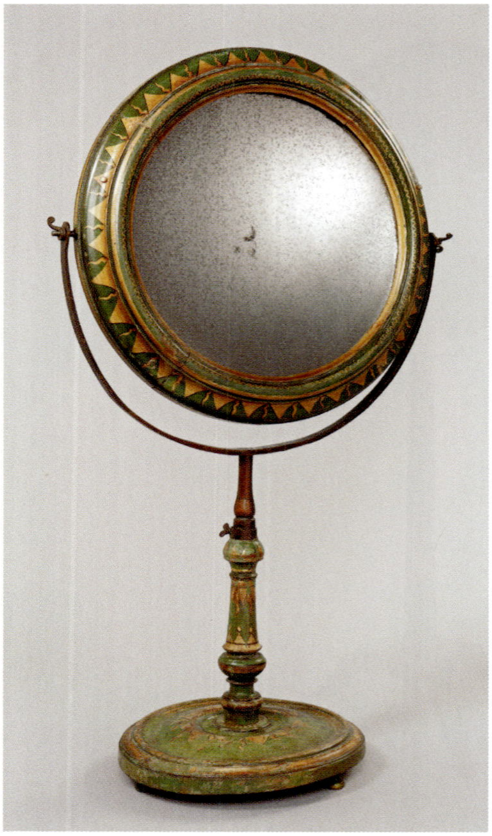

20082-0000- (Fig. 53)
Converging lens
1770s
MAM accession date: 1952
11 × 1; 140 g
Glass

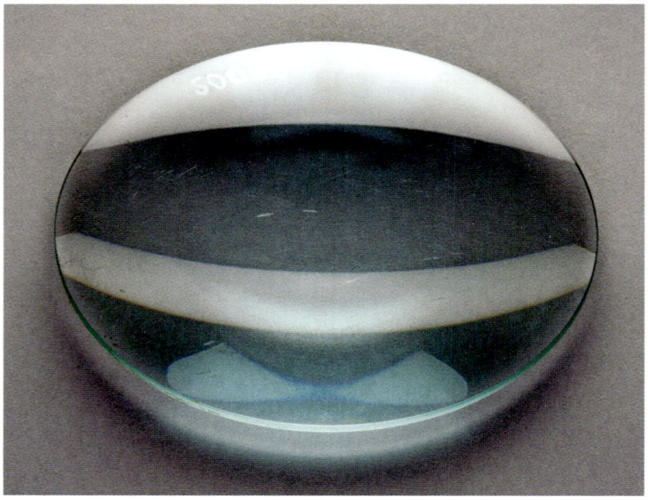

FIGURE 53 (inv. no. 20082-0000-)
© MAM/PHOTO FRANCK BOTTÉ

Unmounted biconvex lens. A paper label on the lens records the inventory number.
See historical note to 20079-0002-.

Lavoisier (1774), pp. 256–265 and fig. 8; Lavoisier (1774a); Priestley (1774), vol. 1, p. 13; Lavoisier (1782b); Daumas (1955), pp. 121–122; Guerlac (1961), pp. 157–161; Smeaton (1987); Lehman (2013)

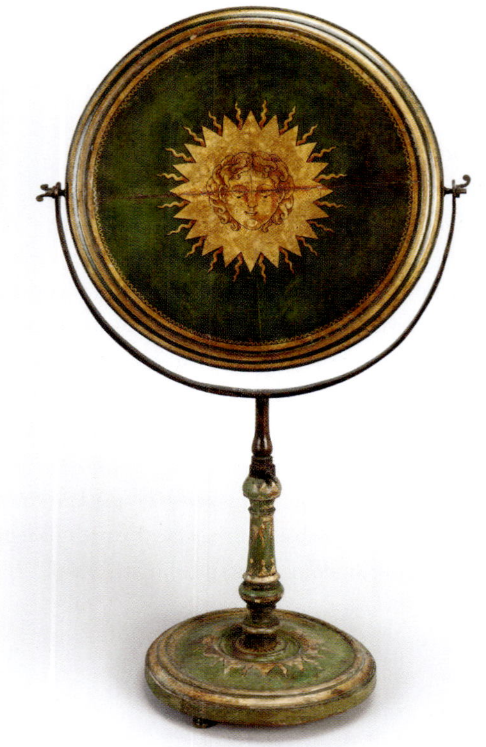

FIGURE 52A, 52B (inv. no. 20081-0000-)
© MAM/PHOTO FRANCK BOTTÉ

20083-0000- (Fig. 54)
Converging lens
1770s
MAM accession date: 1952
11 × 0.8; 115 g
Glass

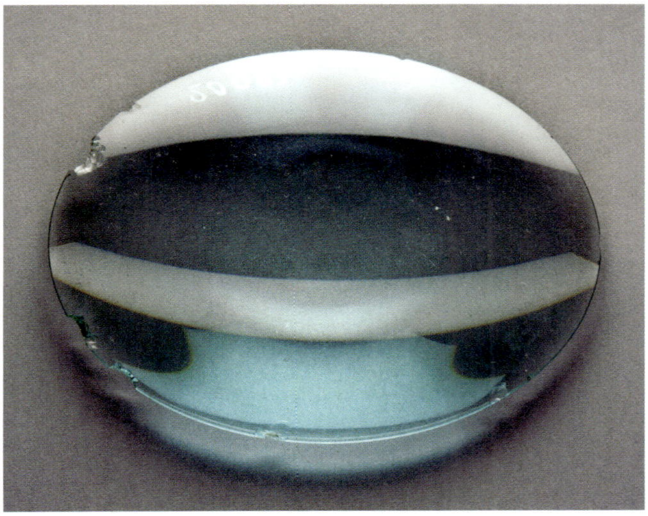

FIGURE 54 (inv. no. 20083-0000-)
© MAM/PHOTO FRANCK BOTTÉ

Unmounted biconvex lens with a chipped rim. There are two paper labels on the lens: one displays the inventory number, the other reads "… collection 175".
See historical note to 20079-0002-.

Lavoisier (1774), pp. 256–265 and fig. 8; Lavoisier (1774a); Priestley (1774), vol. 1, p. 13; Lavoisier (1782b); Daumas (1955), pp. 121–122; Guerlac (1961), pp. 157–161; Smeaton (1987); Lehman (2013)

08229-0001-, 08229-0002- (Fig. 55)
Converging lens with frame
Second half of 18th c.
MAM accession date: 1864–1866
08299-0002-: 44 × 33.5 × 10; 2.71 kg
08299-0001-: 25 × 1.5; 1.09 kg
Brass, glass

Biconvex lens mounted in a brass frame that pivots in a bracket with two butterfly screws. A tube attached to the frame enabled the instrument to be inserted at the top of a wooden tripod, now missing.[21] The lens is a modern replacement because the original one, preserved under

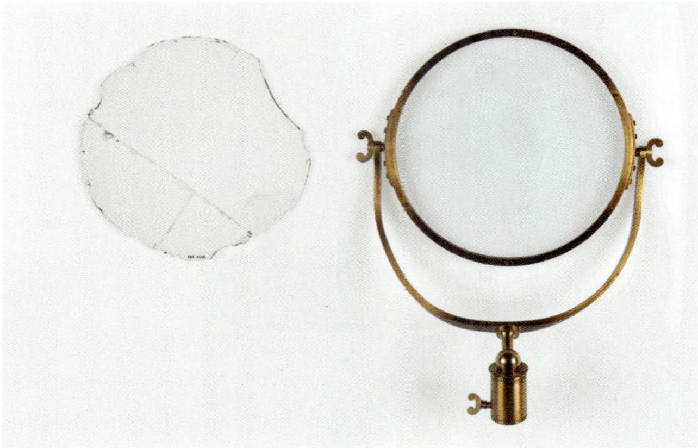

FIGURE 55 (inv. nos. 08229-0001-, -0002)
© MAM/PHOTO FRANCK BOTTÉ

inventory no. 08299-0001-, was broken into three pieces during a fire in the Museum in March 1973. It was probably used by Lavoisier as a burning lens while he was carrying out this type of experiment in the early 1770s.

Lavoisier (1774), pp. 256–265 and fig. 8; Lavoisier (1774a); Priestley (1774), vol. 1, p. 13; Lavoisier (1782b); Daumas (1955), pp. 121–122; Guerlac (1961), pp. 157–161; Smeaton (1987); Lehman (2013)

20003-0001-, 20003-0002- (Fig. 56a, 56b)
Mercury gasometer
1775–1782
MAM accession date: 1952
57 × 15 × 13.5; 3.38 kg
Brass, iron, glass

Glass cup supporting a brass collar with a one-way iron tap screwed on it. A second brass collar sealed to a cylindrical glass vessel is screwed on the upper part of the tap. The cylinder is closed at the top by a brass ring with a cracked glass disk.[22] On it there is a brass plate with a bent iron tube and a brass tap. A second short iron tube also penetrates vertically in the glass disk through the plate.

Examining the apparatus, we have noticed a clumsy restoration. The original glass cylinder was open at the top and the brass ring with the disk was placed on it. At an undetermined date, the cylinder was probably replaced by an inverted bottle whose bottom was not removed. As a result, the tube inserted in the brass plate on the glass disk cannot communicate with the inside of the bottle. The original glass cylinder may have been broken during the nineteenth century because in 1900 only the brass

21 Probably destroyed during the fire of 1973.

22 This element is inventoried under no. 20003-0002-.

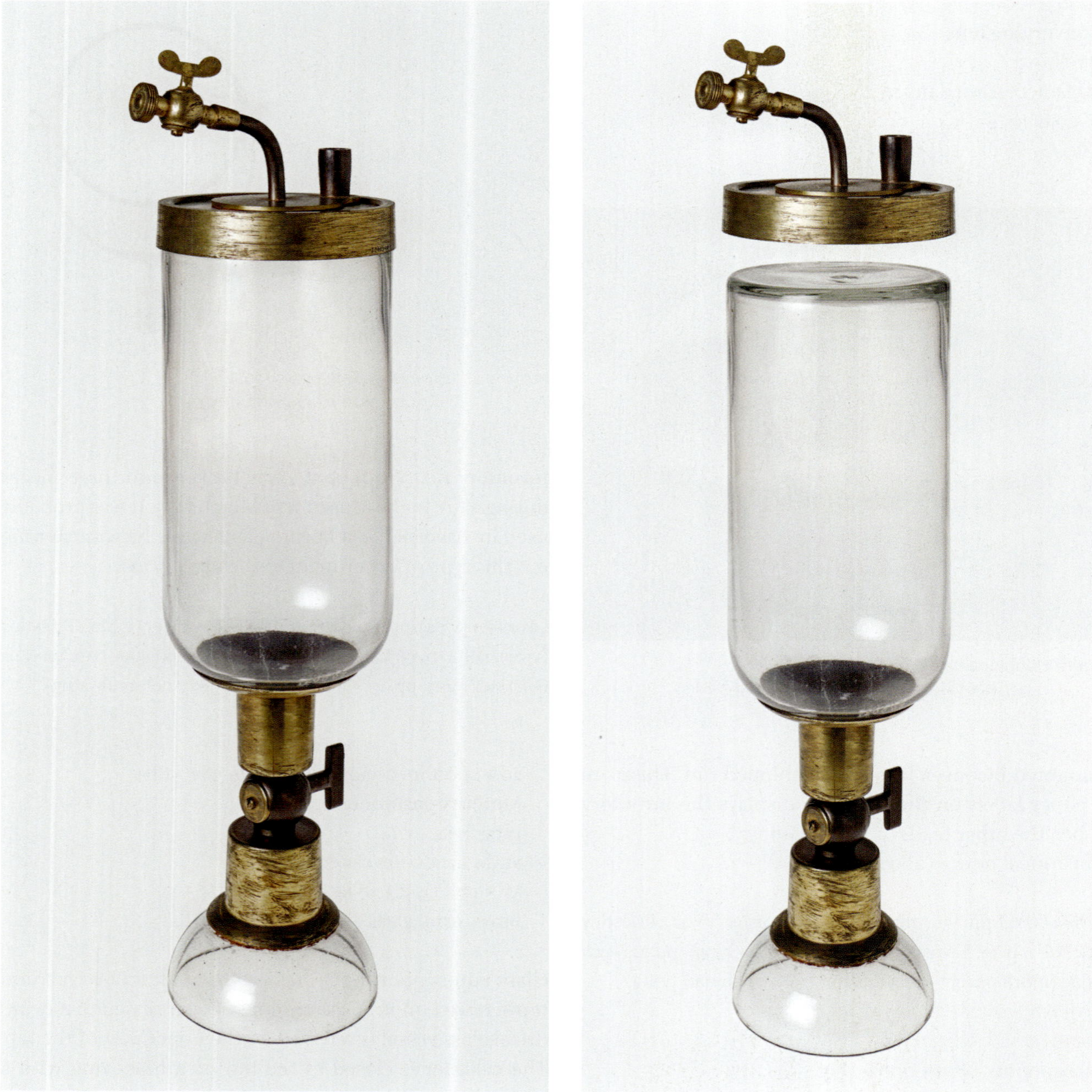

FIGURE 56A, 56B (inv. nos. 20003-0001-, -0002-)

collar was exhibited at the *Musée Centennal*. The apparatus was displayed in the 1943 exhibition and, from the photos available, it seems that the original had already been replaced by the present, rather absurd restoration. The short iron tube was originally connected to the bottom of the vessel with a glass pipe. The working apparatus was described in detail by Truchot but not by Lavoisier.

This special type of gasometer served to collect gases by using mercury. The gasometer was inserted in a pneumatic trough. The gas arriving in the trough from the bottom was collected by the cup acting as a funnel. The bottle was filled with water. When the iron tap was opened, the water fell into the trough and the gas filled the bottle. When the water was emptied out of the bottle, mercury would be gradually poured in from the top, enabling the oxygen to be released via the bent tube and the open tap.

This mercury gasometer is certainly one of the most iconic of Lavoisier's instruments, for it appears in the famous double portrait of the Lavoisiers by Jacques-Louis David in 1788. The representation of Lavoisier and the

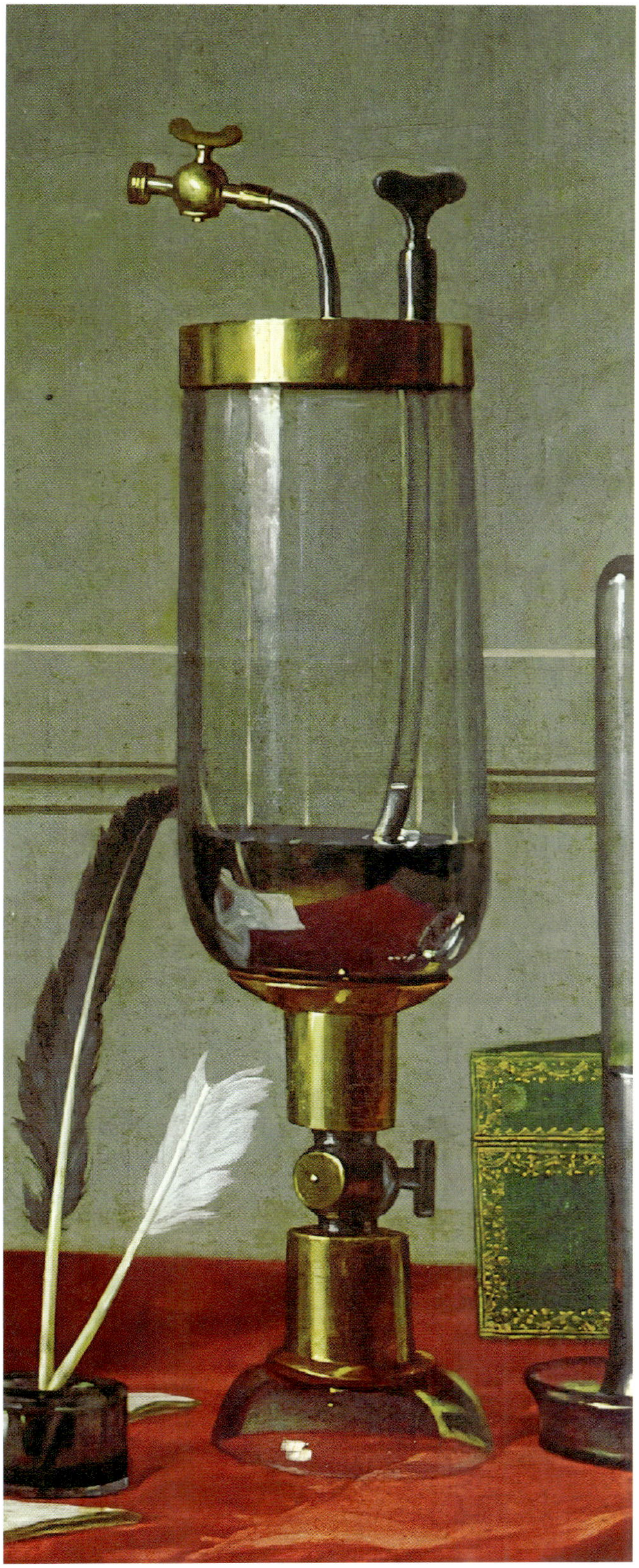

FIGURE 56C Detail of Jacques Louis David portrait of the Lavoisiers (1788)
COURTESY OF THE METROPOLITAN MUSEUM – NEW YORK

chemical instruments surrounding him while he is writing the last pages of the *Traité élémentaire de chimie* is not a purely random choice. It signifies more than the artist's intent to add decorative scientific symbols to a scene extolling the harmony of a prominent couple of the Parisian bourgeoisie. First, the instruments, starting from the hydrometer placed at the bottom right-hand side of the painting, seem to be arranged in a chronological order that represents the main stages of the chemical revolution. Second, the instrument placed on the work table almost beside Lavoisier is a special type of gasometer (Fig. 56c) that served to collect gases using mercury – a method that he had adopted in 1774 and thanks to which he was able to identify, in that very same year, the chemical properties of oxygen.

Truchot (1879); *Musée Centennal* (1900), p. 10; Lavoisier (1943); Beretta (2001), pp. 25–42

20160-0001-, 20160-0002-, 20161-0000-, 20162-0000-, 20163-0001-, 20163-0002- and 20129-0003- (Figs. 57a, 57b, 57c, 57d, 57e, 57f)
Six leather bags
Late 18th c.
MAM accession date: 1952
20160-0001-: 20.5 × 15 × 8; 215 g
20160-0002-: 22 × 14 × 4.5; 250 g
20161-0000-: 20.5 × 11 × 9; 240 g
20162-0000-: 18.5 × 11 × 2; 80 g
20163-0001-: 16.5 × 10.5 × 2.5; 100 g
20163-0002-: 17.5 × 9 × 5; 100 g
Leather, brass, thread

These very similar tooled leather bags were used for storing small quantities of gas.[23]

No. 20160-0001- has a (separated) brass stopcock which had been inventoried under no. 20129-0003-. No. 20161-0000- also has a stopcock fastened to the neck of the bag with a waxed thread.

23 Nos. 20160-0002- and 20162-0000- bear a printed label reading "Exposition Lavoisier // Palais de la découverte".

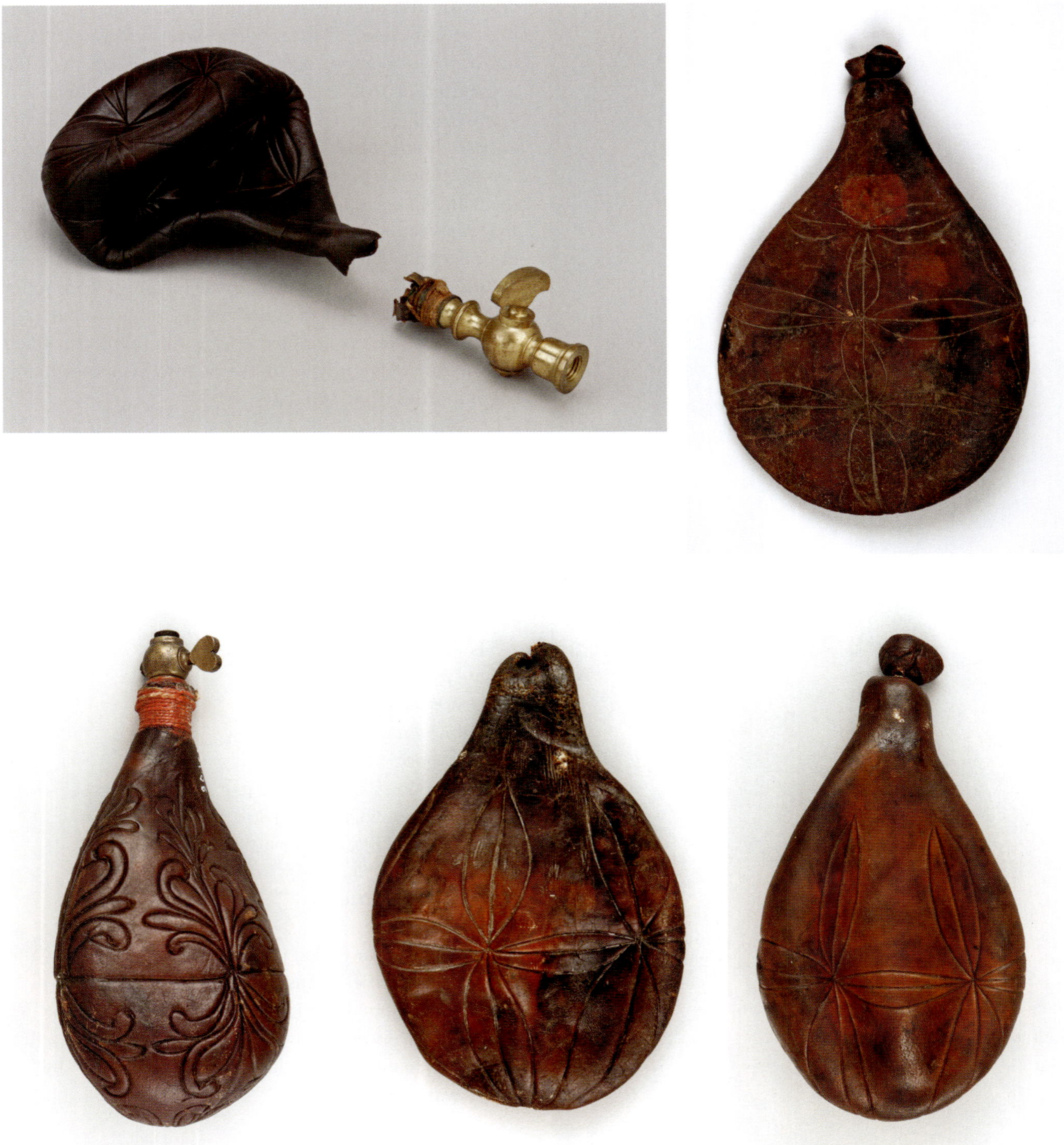

FIGURES 57A, 57B, 57C, 57D, 57E, 57F (inv. nos. 20160-0001-, 20160-0002-, 20161-0000-, 20162-0000-, 20163-0001-, 20163-0002- and 20129-0003-)

20102-0000- (Fig. 58)
Carbonic acid generator
Late 18th c.
MAM accession date: 1952
59 × 42 × 24; 2.8 kg
Tin, glass, brass

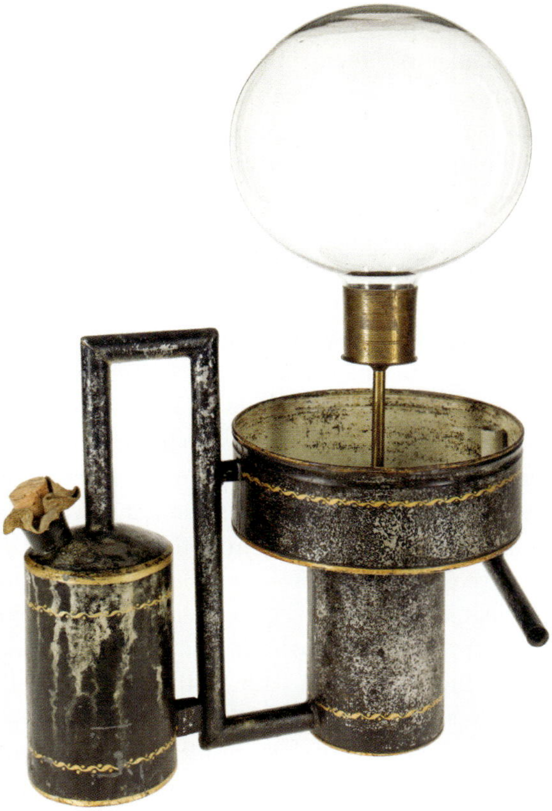

FIGURE 58 (inv. no. 20102-0000-)
© MAM/PHOTO FRANCK BOTTÉ

Gas generator. A cylindrical tin vessel with a short neck closed by a cork is connected with a double bent tube to a pneumatic trough. The latter is composed of a cylinder surmounted by a basin. An outflow pipe serves to evacuate excess water from the basin. A glass balloon with a brass collar and a pipe is screwed onto the opening of the tube channelling the gas. The tin parts of the apparatus are painted in blue with gilt decorations.

The substances were mixed in the cylindrical vessel and their reaction produced a gas (for example, an acid and calcium carbonate were mixed in order to produce carbon dioxide). The gas flowed through the tube into the pneumatic trough, where it could be collected. In its present state, the apparatus cannot work because the glass vessel cannot collect the gas. The gas was probably collected through a glass jar or bell.

20025-0000- (Fig. 59)
Blowpipe
ca. 1782
MAM accession date: 1952
38.5 × 23 × 7; 30 g
Glass

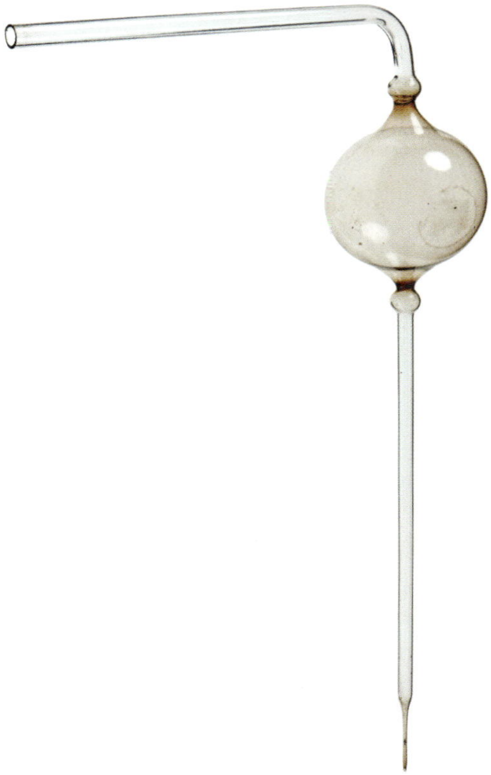

FIGURE 59 (inv. no. 20025-0000-)
© MAM/PHOTO FRANCK BOTTÉ

Blowpipe composed of an L-shaped tube with a glass bulb. One end of the tube is pointed, acting acts as a nozzle. By blowing in the pipe, one can direct the flame of a lamp towards an object to be soldered or a substance to be melted. For chemical analysis, glass blowpipes were generally replaced by brass ones, because glass was fragile and the pipe nozzle tended to melt in the flame. During the first half of the eighteenth century, Swedish chemists were the first to adapt the enameller's lamp to analytical chemistry and to connect it to a blowpipe, which became an extremely important instrument during the second half of the century. Despite its simplicity, it required considerable skill to be used effectively. Thanks to the blowpipe, Swedish chemists discovered several important elements such as nickel (1751), manganese (1774), molybdenum (1781) and tungsten (1783). French chemists also used the instrument but without the same results. In a letter of January 1769 to Torbern Bergman, Macquer lamented the difficulties he encountered:

> I am well acquainted with the use of the blowpipe that you have told me about. I have used it several times. It is a very convenient instrument for certain experiments or small-scale tests that allow for immediate contact with the inflammable principle, but I confess to you that I lack your talent for doing so without interruption, and even that it tires me considerably even if I catch my breath, which is why I seldom use it; this upsets me, for it is indeed a fine little device to carry in one's travels; it is very convenient to be able to carry one's melting oven, bellows, and supply of coal in one's pocket.[24]

Lavoisier used the blowpipe in a completely different way. In early February 1782, when Lavoisier had gained considerably more knowledge about oxygen, he began to work on a device for exploiting its combustive properties: hydrostatic bellows. (See chapter 2, fig. 13) This machine – of which only few fragments survive – consisted of a pneumatic trough in which the oxygen was collected; the trough was connected through copper pipes (1) to a large iron case in which the gas could be weighed with precision and (2) to an enameller's table. The table was fitted with a lamp to inflame a given quantity of oxygen, which was then directed, thanks to the blowpipe, onto the object or material to be fused. Lavoisier used the enameller's lamp and the blowpipe to observe – for the first time – the combustive properties of oxygen. On 5 June 1782, Lavoisier presented this device at the Académie royale des sciences, where he was able to fuse a specimen of platinum with remarkable speed. While the experiment did not lead to a scientific publication, it was reported in the Parisian press.

The success of this experiment showed that it was possible to overcome all the technical difficulties reported by Macquer in the use of the blowpipe. Lavoisier resorted to an ancient technology that he probably learned from Jean Nollet either while attending his lectures or by reading his *Leçons de physique expérimentale*. In this work (vol. 4, pp. 499 ff.), Nollet described the public demonstration made by Jean Raux, the King's enameller, before an officer of the court in 1739 and the apparatus he used to subject to the action of controlled fire and heat the small objects he wished to handle. As we can see from the plate published by Nollet (See chapter 2, fig. 16), the apparatus is very similar to that of Lavoisier and places the use of the blowpipe in an entirely new material context. Satisfied with the results he obtained with this new method of fusing substances, Lavoisier asked the assistance of the engineer Jean-Baptiste Meusnier de la Place to improve the device. This marked the first step towards the construction of the gasometer. With the improved instrument, Lavoisier subjected many precious stones – from the collections of his Académie colleagues and from jewellers – to the action of violent heat, i.e., fire fuelled by oxygen (see inv. no. 20185-0000- pp. 377–381). Although the results were not as spectacular as those obtained with platinum, Lavoisier was able to assess the nature of precious stones with new evidence. The publication of these results was delayed to 1785 when the *Mémoires* of the year 1782 were finally printed, but the experiments became widely known because they had been public and had been performed at the Arsenal in the presence of scientists and laymen. It was certainly because of this publicity and the interest they generated that, in July 1782, the astronomer Bochart de Saron suggested to Lavoisier that he should fill the blowpipe with a mixture of oxygen and hydrogen and determine whether the fusion of platinum could be further accelerated.

As Lavoisier himself later acknowledged (1789, vol. 2, 553), these devices are what led him, in 1783, to construct the first version of the gasometer and to perform experiments on the synthesis of water.

He described the glass blowpipe in the *Traité élémentaire de chimie* (1789). Glass blowpipes had the dual advantage that they could be produced cheaply and in any shape. As Berzelius pointed out, however: "Glass blowpipes are certainly less costly, and less liable to get dirty than those made of metal; but their brittleness and

24 "Je connois bien l'usage du chalumeau dont vous me parlés; je m'en suis servi quelques fois, c'est un instrument fort commode pour certaines experiences, ou essais en petit aux quels le contact immediat du principe inflammable n'est pas contraire, mais je vous avoue que je n'ai pas le même talent que vous pour le faire aller sans interruption, et même qu'il me fatigue beaucoup même en reprenant mon haleine, et que pour cette raison, je ne m'en sers que rarement; j'en suis faché, car c'est en effet un joli petit appareil a porter en voyage, il est bien commode de pouvoir porter dans sa poche, son fourneau a fondre, son souflet et sa provision de charbon". Bergman (1965), p. 235.

the fusibility of their beaks are so serious inconveniences, that they should never be used but in cases of necessity."

Lavoisier (1782); Lavoisier (1789), vol. 2, pp. 552–558, plate XII, fig. 15; Chevallier (1819), p. 71; Griffin (1827), pp. 267–268; Bergman (1965), p. 235; Jensen (1986); Weiss (2008); Abney Salomon (2019)

20013-0000- (Fig. 60)
Blowpipe
ca. 1782
MAM accession date: 1952
23 × 4 × 2; 15 g
Glass

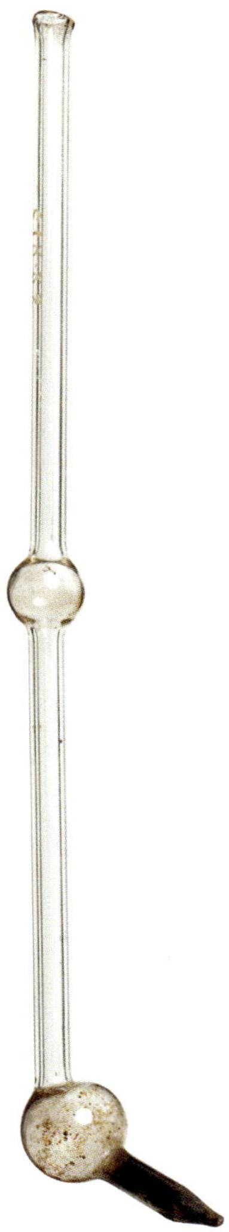

FIGURE 60 (inv. no. 20013-0000-)
© MAM/PHOTO FRANCK BOTTÉ

Glass blowpipe with two bulbs: a smaller one in the middle and a larger at one end. The latter has a short lateral nozzle. This blowpipe was probably used to blow and direct the flame of a lamp towards a substance to be melted or analysed. It differs from 20025-0000- and is described in the plates of the *Encyclopédie méthodique – Chymie.*

Lavoisier (1782); Lavoisier (1789), vol. 2, 552–558, plate XII, fig. 15; *Encyclopédie méthodique – Chymie* (1786–1815), vol. 6, plate 5, fig. 48; Chevallier (1819), p. 71; Griffin (1827), 267–268; Bergman (1965), 235; Jensen (1986); Weiss (2008); Abney Salomon (2019)

20123-0000- (Fig. 61)
Volta's pistol
ca. 1782
MAM accession date: 1952
10.7 × 6.5 × 17; 60 g
Tin, brass, glass

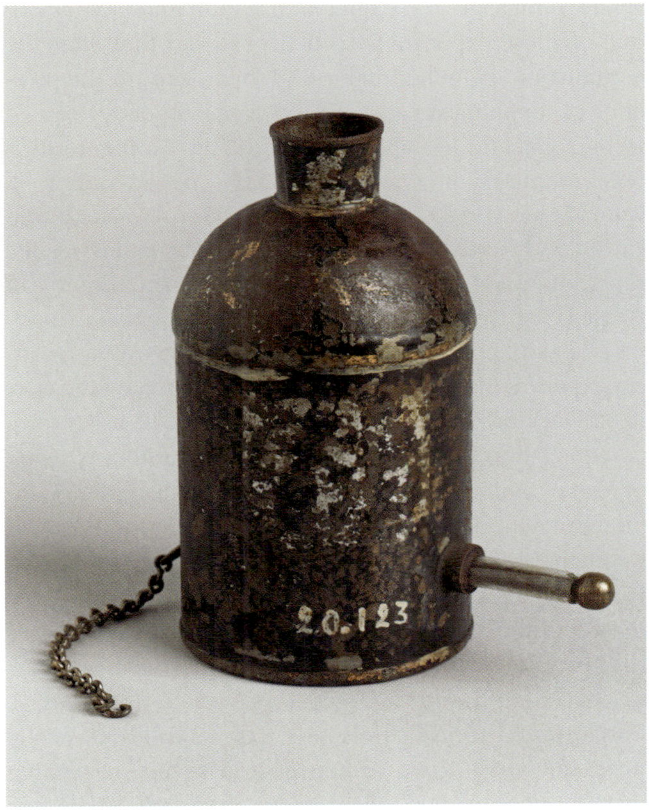

FIGURE 61 (inv. no. 20123-0000-)
© MAM/PHOTO FRANCK BOTTÉ

This instrument showed how a flammable gas (hydrogen and oxygen, or air and methane) could be detonated by an electric spark. Invented by the Italian physicist Alessandro Volta around 1777 following his discovery of "flammable air" (methane), he described several variants of it in his writings.

The "pistol" consists of a small tin bottle whose neck was closed by a cork (missing). An insulating glass tube containing a brass electrode is inserted into the lower part of the bottle. A short chain is attached to a ring fixed to the vessel; the ring was probably used to ground the apparatus. Only a few traces of the original paint remain on the tin.

The bottle was filled with a detonating gas and closed with the cork. A spark could be produced in the bottle by touching the electrode with a charged Leyden jar, an electrophorus or an electric machine. The spark detonated the gas with a loud bang and the explosion blew out the cork.

This instrument was used for the first time by Volta in spring 1777, when he used electrical discharges in a closed vessel to measure the degree of inflammability of airs. In one of these experiments, Volta observed that after the combustion of "inflammable air" (hydrogen) in the presence of "dephlogisticated air" (oxygen), the latter ceased to be a gas and left a dew in the vessel. As this residue was completely unexpected, Volta did not identify it as water. Interestingly, it was not until 1784, after the publication of the crucial memoirs by Cavendish and Lavoisier, that Volta understood the meaning of his experiment. In August 1784, the chemist Jean Darcet sent the Italian natural philosopher, on behalf of Lavoisier, a report on the latest experiments by Lavoisier and Jean-Baptiste Meusnier de la Place on the decomposition of water. In his accompanying letter, Darcet showed not only that he was aware of Volta's eudiometer but also that he realized its possible implications for Lavoisier's work on the synthesis of water. It is clear from Darcet's testimony that while in Paris in 1782, Volta had shown his eudiometer and made the experiments which allowed him to synthesise water. It is very unlikely that Volta showed these experiments to Darcet and not to Lavoisier, but even if this were so, Darcet would almost surely have told his colleague at the Académie about them. Moreover, it is interesting to note that the unpublished inventory of Lavoisier's laboratory, compiled in 1794, includes an "eudiomètre de Volta", one "pistolet de Volta en cuivre" and one "pistolet de Volta en fer blanc". Naturally, in addition to these instruments, we also find the electrical condenser used for the April 1782 experiments on the liberation of electricity through combustion, vaporisation and effervescence. Considering that Lavoisier and Laplace had been engaged in a series of experiments on the vaporisation of water since Volta's arrival in Paris, Lavoisier was likely inspired to investigate the nature of water from another perspective when he saw Volta's electrical demonstration on what he called the decomposition of inflammable air. Although there no conclusive evidence of this scenario, the numerous coincidences cited above combined with Lavoisier's intense program of experiments after 1782 suggest that we should look at the controversy on the nature of water within a wider perspective and a richer context.

However, in a letter to Van Marum of 1798, Volta claimed priority for the discovery as follows:

> I am very fond of chemistry, and particularly pneumatic chemistry, which I was the first to cultivate in Italy. You will probably know my discoveries and my experiments on inflammable air. Was I not the first to discover by a series of experiments with my pistol and my inflammable air lamp, and those with an eudiometer or apparatus to burn this air in closed vessels, mixed in different proportions of common air, or vital air, likewise my invention – who had discovered, I say, and found before 1781 that all inflammable air disappears and takes with it the destruction of a volume of vital air which is approximately half of its own volume? [...] This is the point at which I had arrived with my experiments several years before Lavoisier. [...] He therefore only verified or completed my discovery.

Since Volta did not recognize the composite nature of water until the late 1780s, none of his claims is historically grounded.

Volta (1949–1955), vol. 2, pp. 235–236; Forbes (1969–76), vol. 6, p. 360; Beretta (2000); Beretta (2001a); Bellodi et al. (2002), pp. 52–55

20004-0000- (Fig. 62)
Volta's electric eudiometer (incomplete)
ca. 1790
MAM accession date: 1952
27 × 7; 960 g
Brass, glass, sealing wax

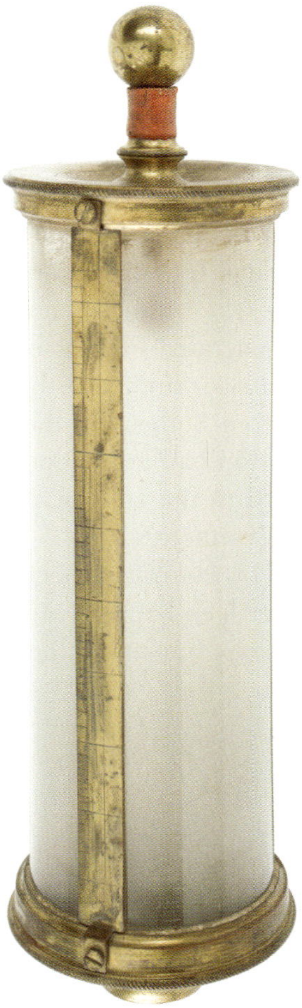

FIGURE 62 (inv. no. 20004-0000-)
© MAM/PHOTO FRANCK BOTTÉ

Vertical cylindrical glass vessel whose bottom is closed by a brass cap with a threaded hole. A tap with a small funnel – which was the base of the instrument and is now missing – was originally screwed onto the cap. The top of the cylinder is closed by a second threaded brass cap. In its centre there is an electrode with a small sphere and a wire (curved at the bottom and ending near the disk). This electrode is insulated from the cap by a glass tube covered with sealing wax. A vertical brass strip, fixed to the caps with two screws, carries a scale with nine divisions, each with five subdivisions.

This electrical eudiometer is similar to the one described by Alessandro Volta in 1790, made for testing the goodness of the air (*salubrità dell'aria*), i.e., its oxygen content. Volta developed these instruments after he discovered the "flammable air of the marshes" (*aria infiammabile delle paludi*: methane) and invented his electric pistol and electric lighter. An electric spark ignited a flammable gas such as methane or hydrogen in the eudiometer. The instrument was filled with water and its funnel was placed on a pneumatic trough also filled with water. Given quantities of regular air and "flammable air" were then introduced in the apparatus. Lastly, the mixture of gases was ignited by a spark[25] arching between the electrode and the upper cup of the eudiometer. If the air to be analysed contained some oxygen, the reaction[26] reduced the volume of the gases and caused the level of the water in the eudiometer to rise.

The reaction's intensity and, even more, the difference in water level before and after the ignition made it possible to measure the oxygen content in the air.

Volta (1790); Beretta (2000); Bellodi et al. (2002), pp. 73–83

19983-0000- (Fig. 63)
Cylindrical vessel
Last third of the 18th c.
MAM accession date: 1952
17 × 12; 750 g
Glass, brass

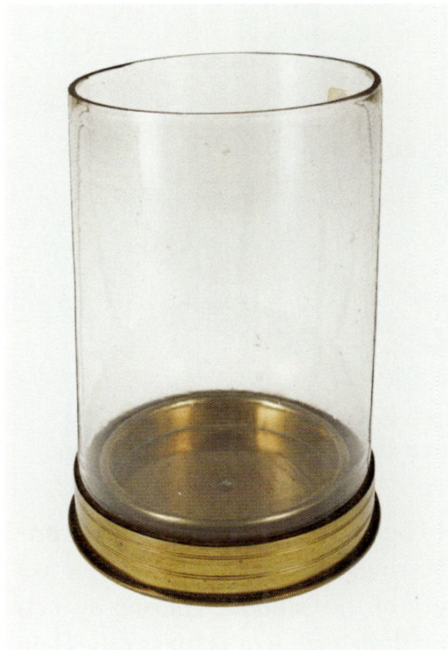

FIGURE 63
(inv. no. 19983-0000-)
© MAM/PHOTO
FRANCK BOTTÉ

25 The spark could be generated with a Leyden jar, an electrophorus or an electrostatic machine.

26 By using methane as flammable air the reaction was: $CH_4 + 2O_2 \rightarrow CO_2 + 2H_2O$.

Cylindrical glass vessel closed by a double-rimmed brass disk. In the centre of the disk there is a short-threaded tube, which was certainly screwed onto an unidentified instrument.[27]

The following apparatus, consisting of several pieces (see next four items up to p. 349), is a perfected version of Lavoisier's and Meusnier de la Place's gasometer, the device first used in 1783–1785 to perform the famous experiments on the synthesis of water. The original instrument set comprised two gasometers, a reaction balloon, a pneumatic pump (missing), and an electrical machine (missing).

In other experiments, the gasometers were connected to a pair of pneumatic troughs and their bell-jars for manipulating gases.

07548-0000- (Fig. 64)
Balloon for the synthesis of water
1785–1788
MAM accession date: 1864–1866
75.5 × 65 × 51; 4.65 kg
Glass, brass, iron, silk, paper

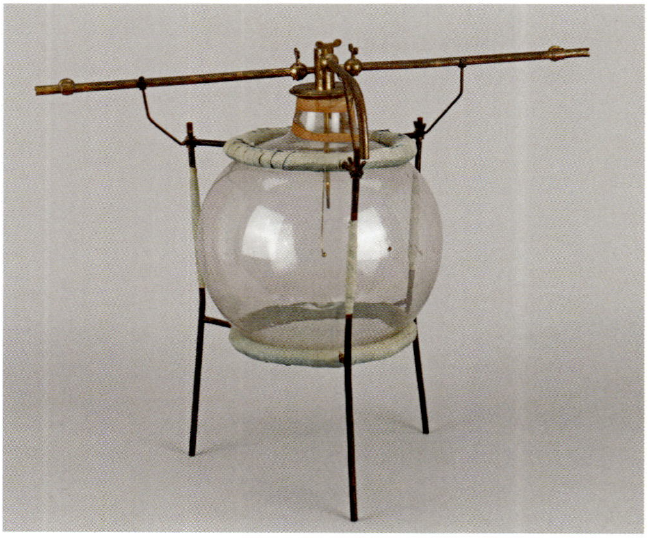

FIGURE 64 (inv. no. 07548-0000-)
© MAM/PHOTO FRANCK BOTTÉ

Glass balloon mounted between two stuffed rings covered with a silk ribbon. The rings are supported by three vertical iron legs. The neck of the balloon is closed by a circular brass cap holding three orthogonal brass tubes with stopcocks. Three iron arms fixed to the stand support the tubes. One of them, which was connected to the gasometer containing hydrogen, terminates in the centre of the balloon with a very fine nozzle. A second tube also joining in the nozzle was connected with the oxygen gasometer, while the third was connected with the pneumatic pump used to evacuate the air from the balloon. A wire electrode with a brass sphere at the top and a smaller sphere near the nozzle was inserted into the cap through a short insulating glass pipe. When the upper end of the electrode was connected with an electrostatic machine or a Leyden jar, it was possible to produce a spark between the lower end of the electrode and the nozzle of the hydrogen tube. The spark ignited the gas and started the combustion.

After the balloon was evacuated with a pneumatic pump, the oxygen from a gasometer was introduced into it. Next, the hydrogen from a second gasometer was introduced into the balloon through the tube inserted into it and was ignited by the spark. The hydrogen flame burned on the nozzle. The water generated by the combustion began to condense on the interior surface of the balloon, eventually falling to the bottom. Sometimes the combustion stopped and the operation had to be repeated. The volumes of the gases burned in the balloon were accurately measured by the two gasometers.

This apparatus was used by Lavoisier and the French mathematician and engineer Jean-Baptiste Meusnier de la Place (1754–1793) for their experiments on the synthesis of water.

For bibliography see next item.

07547-0001-001-, 07547-0002-001- (Figs. 65a, 65b, 65c)
Two gasometers
1787
MAM accession date: 1864–1866
Pierre Bernard Megnié (Megnié le Jeune) (1758–1807)
07547-0001-001-: 236 × 136 × 68; mass not measured
07547-0002-001-: 236 × 157 × 77; mass not measured
Wood, brass, steel, iron, glass, lead, mercury

Two gasometers used to store gases. The gasometer served to store and deliver a gas for various experiments. In particular, Lavoisier used it for the large-scale experiments on the synthesis of water. The quantity of gas stored in the gasometer was measured by reading the scale on the sector on the beam. This required accurately gauging the scale, a procedure performed by filling a bottle of a given volume several times with the air contained in the bell and by noting the value on the scale every time.[28] The

27 This vessel was perhaps the upper part of a Volta's lighter similar to the one preserved in the MAM under inventory no. 01627-0000-.

28 This operation required the use of a pneumatic trough (see nos. 07547-0001-002- and 07547-0002-002-).

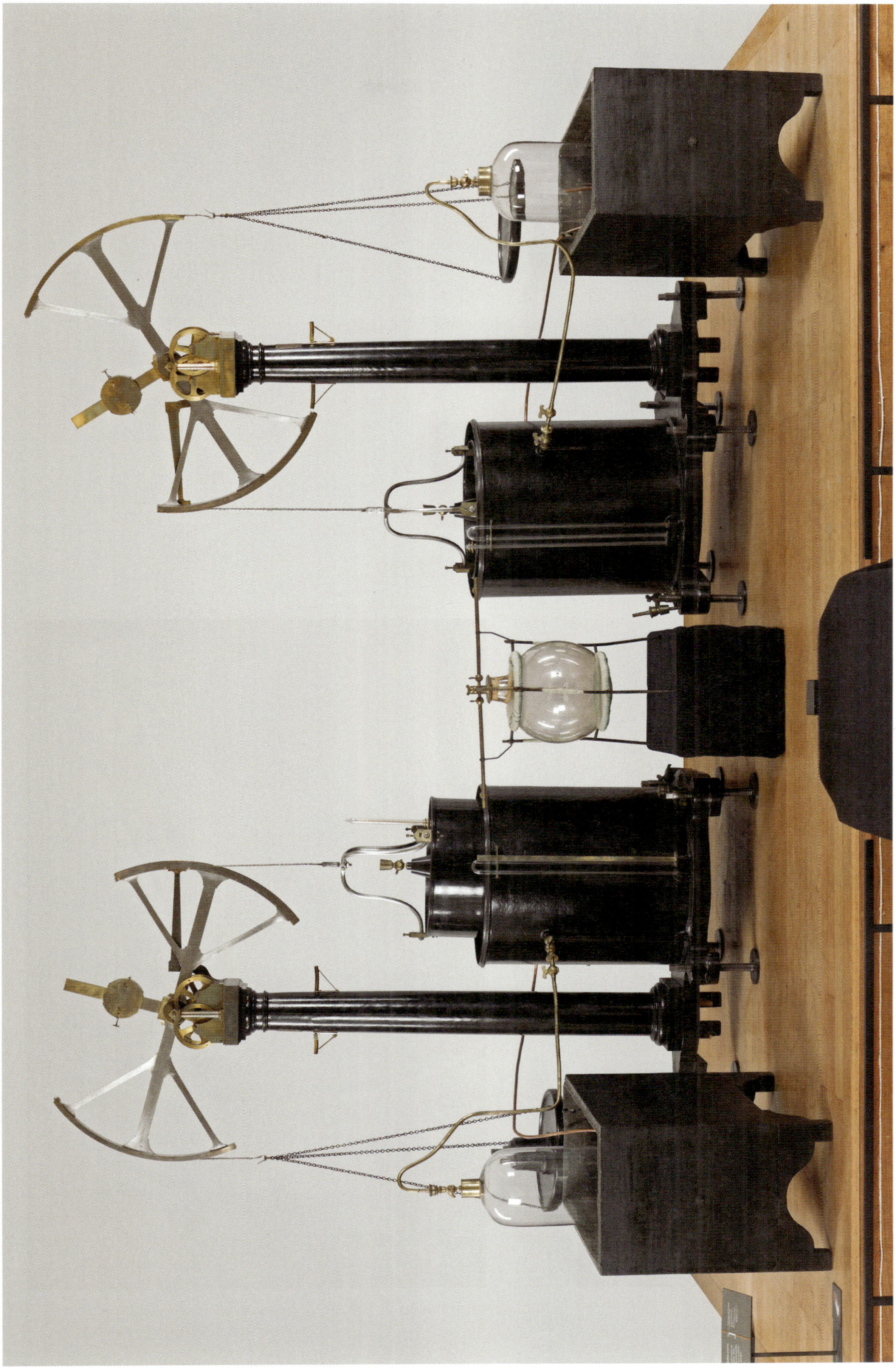

FIGURE 65A (inv. no. 07547-0001-001-)
© MAM/PHOTO FRANCK BOTTÉ

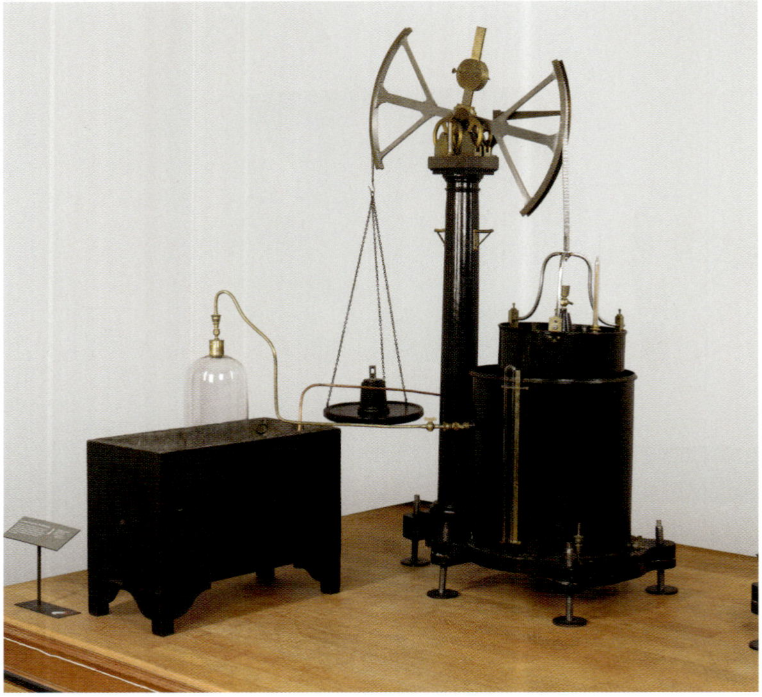

FIGURE 65B (inv. no. 07547-0001-001-)

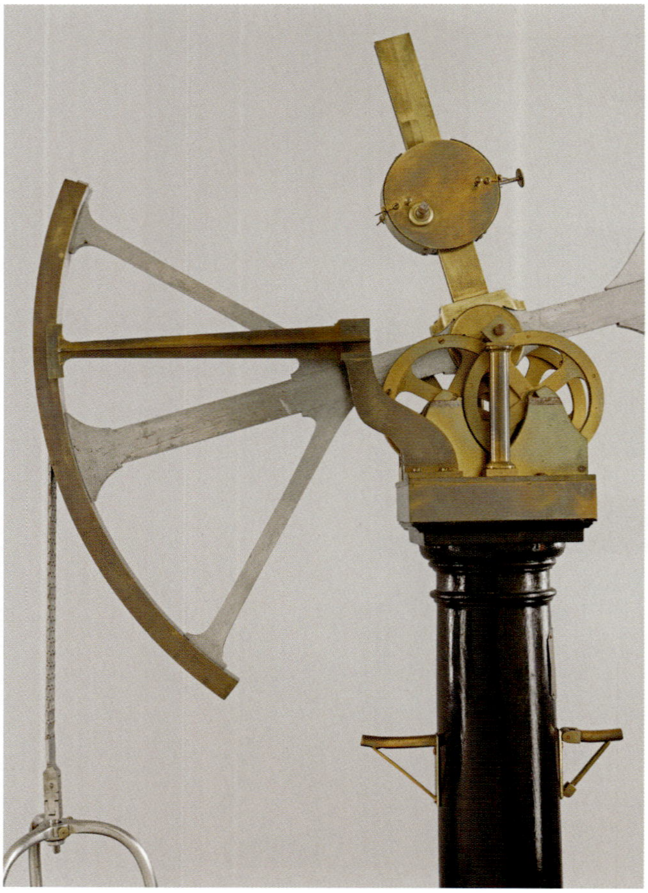

FIGURE 65C (inv. no. 07547-0001-001-)

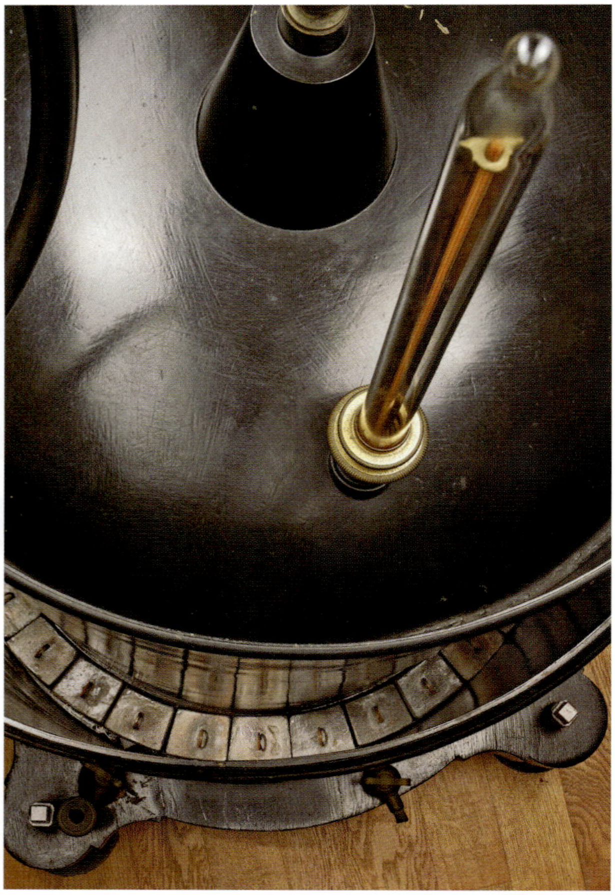

FIGURE 65D (inv. no. 07547-0001-001-)

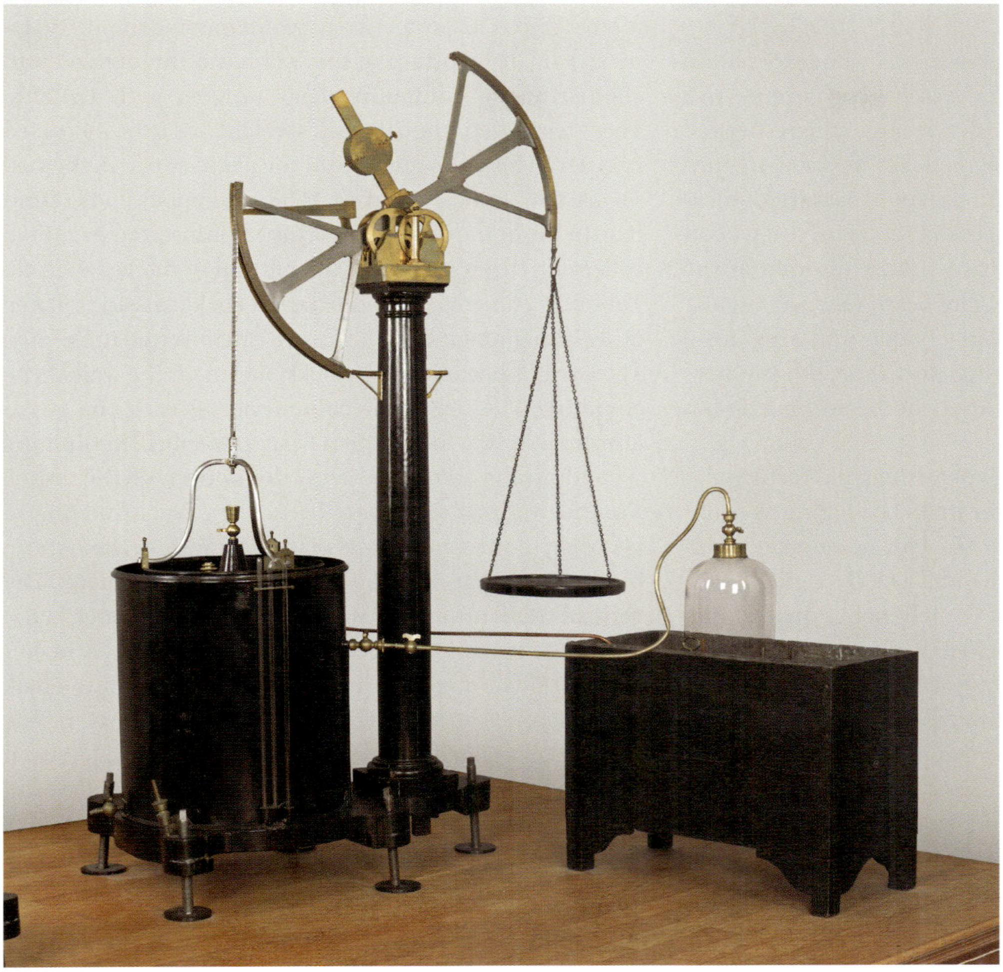

FIGURE 65E
(inv. no. 07547-0002-001-)

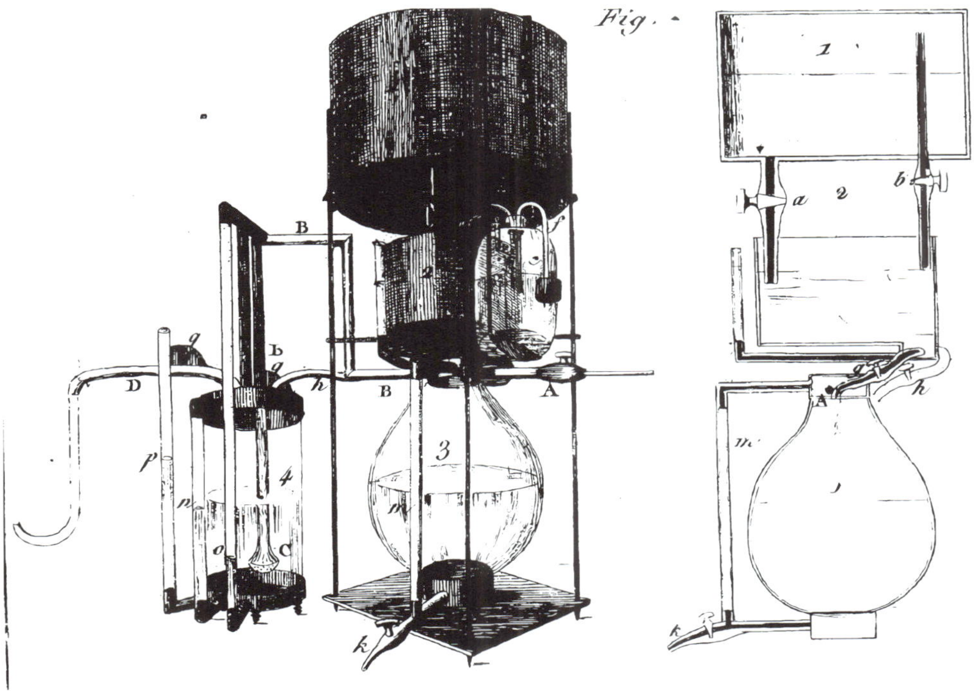

FIGURE 65F
Séguin's gasometer
FROM SÉGUIN (1798)

two gasometers are practically identical and the following description is valid for both.

The elements of the apparatus – a wooden column with a beam and the gasometer vessels – are fastened to a black-painted wooden base with five large square-headed levelling screws. A large T-shaped key is used to turn them. At the top of the column are four brass frictionless wheels.[29] The axle of the steel beam rests on the wheels and its end is held between two screws inserted at the top of two small brass pillars. The beam has two circular sectors at its ends. Two pairs of oblique iron arms fixed to the beam ensure their rigidity. Two short hinged brass arms fixed to the column delimit the beam's maximum rotation.

In the middle of the beam is an orthogonal rectangular brass bar with a sliding circular brass box acting as counterweight. Its position on the bar can be adjusted with a rack-and-pinion mechanism and set by tightening a screw.[30] The bar has a scale in French inches divided into lines. The tenth of line can be determined with a nonius fixed on the counterweight. By moving the counterweight along the bar one could exactly compensate the bell's loss of weight due to the Archimedean buoyant force when it was immersed in the water. Without this offset, the pressure of the gas in the bell would have decreased when the bell was immersed.

One section of the beam supports a flat steel chain[31] with a hook. Three smaller chains hanging from it support a round wooden plate on which weights could be placed. Another type of flat steel chain,[32] holding the top of a tripod, is attached to the other section. Its lower ends are inserted into three clamps fixed on the rim of the gasometer's movable copper bell. On one side of this section is a brass circular scale divided into half degrees. A nonius attached to the wooden column allows a reading precision of 3′ (arcminutes). The gasometer's bell is topped by a small conical dome with a brass funnel with a stopcock and, alongside it, a short brass collar with a vertical glass tube. The tube protects a mercury thermometer with a metal scale divided into degrees from 0° to 90°. The thermometer and its protective tube are missing in no. 07547-0002-001-. The bottom of the bell has a rim divided into compartments containing lead weights with handles. These were used when the bell's weight had to be increased in order to raise the pressure of the gas inside it. In the vessel are two vertical tubes (A, B) whose upper ends terminate in the dome of the bell.[33] Tube A is inserted in a small collecting cup is welded at the bottom of the vessel itself. Tube B is connected to an external cock (missing). Three other horizontal tubes (C, D and E) are inserted in the cup. Two are connected to two cocks placed on the wall of the container. The third tube communicates with one of the arms of a siphon formed by a ∩-shaped tube. The siphon's second arm is connected with the bottom of the vessel. Another vertical tube, parallel to the others (missing in no. 07547-0001-001-), was also connected with the bottom of the vessel. The system forms a sort of manometer. The vertical tubes always indicate the level of the water in the vessel, while the level in one of the siphon arms is higher or lower depending on the pressure of the air in the bell. The difference between the atmospheric pressure and the pressure in the bell corresponds to the difference between the two levels, which can be read on a brass scale of 12 inches divided into lines (missing in no. 07547-0002-001-). A tap serves to evacuate the water in the vessel.

The brass base of the frictionless wheels carries the inscription "Mégnié Le Jeune // AParis -1787-". A brass plate fixed to the column bears a second inscription: "Du Cabinet // de Lt A.^{ne} Lavoisier // Donné par // M.r & M.me Léon de Chazelles".

Mégnié began to construct the gasometers on 10 September 1785 and delivered them to Lavoisier on 11 July 1787 for a total price of 7554.6 *livres*.[34] This was the most expensive piece of apparatus owned by Lavoisier in his laboratory. Aside from the labour cost for assembling the machine (3216 *livres*), the most expensive item was the large quantity of copper used for its construction (937.16 *livres*).

Mégnié's improved gasometers came after those used by Lavoisier and Meusnier de la Place during the public experiments on decomposition and synthesis of water performed between 27 February and 1 March 1785. These original gasometers are now lost.[35] Maurice Daumas (1955) has shown that their construction began in late 1782 and

29 This type of frictionless system was introduced by the famous English clockmaker Henry Sully, who worked many years in France and installed a watch manufacture in Versailles. Since the end of the eighteenth century, the four-wheel frictionless bearing was commonly used in the Atwood machine for studying the law of motion of falling bodies.

30 Four hooks fasten the two disks that close the box.

31 Known as "Vaucanson's chain", after the French inventor Jacques de Vaucanson (1709–1782), it is characterised by a distinctive type of link.

32 Lavoisier pointed out that this chain was not "susceptible de s'allonger ni de se raccourcir lorsqu'elle est plus ou moins chargée" (Lavoisier (1789), vol. 2, p. 349).

33 When the bells are in their lowest position.

34 LC, vol. 5, pp. 52–53.

35 Daumas also pointed out that the old gasometers were no longer in Lavoisier's instrument collection in 1794 when the Comité d'instruction publique made the inventory of his laboratory in Boulevard de la Madeleine.

involved not only Mégnié and Meusnier de la Place but also Naudier and Fortin. The first stage in their construction was handled exclusively by Lavoisier in 1782, when he needed an apparatus that would allow him to use pure oxygen for the fusion of precious stones. Their production and the organization of the large-scale water experiments was a long process, beginning in 1783 and ending with the spectacular experiments on the decomposition and synthesis of water performed by Lavoisier and his associates at the Arsenal in early 1785. Although the experiments conducted with the old gasometers were extremely successful, Lavoisier must have been dissatisfied with the apparatus, for just a few months later he ordered Mégnié to make two new ones. Meusnier presented them before the Académie royale des sciences on 8 March 1787[36] and Lavoisier eventually described them in detail in the third part of his *Traité élémentaire de chimie*. It is uncertain whether Lavoisier actually used this instrument. At the end of his description Lavoisier remarked:

> The instrument I have just described, which I have called a gasometer, was built by M. Meignié the younger, engineer, maker of physics instruments, and holder of a royal patent. He brought uncommon care, exactitude and intelligence to the task. It is a valuable instrument because of its many possible applications, and because there are experiments nearly impossible to conduct without it. What makes it more expensive is that a single one is not enough: two are required in many cases such as the formation of water, that of nitrous acid, etc. It is an inevitable effect of the degree of perfection that chemistry is beginning to approach, that this requires costly and complicated instruments and apparatus. No doubt one must seek to simplify them, but this should not be at the expense of their ease of use and, above all, of their accuracy.[37]

In October 1793, the Italian instrument maker Francesco Benvenuti (1758–1831), also known as François Bienvenu, wrote a letter to Lavoisier with a view to replicating the experiments on the analysis and synthesis of water. Benvenuti asked eleven technical questions concerning the construction of the gasometer. Surprisingly, Lavoisier answered all the questions in a long letter, even admitting some faults in the machines. Interestingly, he also informed Benvenuti that Fortin had constructed two improved versions of the gasometer which were sent to Spain.

In 1798, Armand Séguin presented a new gasometer at the Institut. In a report published in the *Bulletin de la Société Philomathique de Paris*, we read:

> Citizen Séguin has conceived a gasometer, or instrument for measuring gases, which he intends to substitute for Lavoisier's gasometer; its purpose is to eliminate the need for the corrections that were required by barometric variations during experiments. By means of Cit. Seguin's barometer, the gases are kept in a state of constant density through an artificial and gradual compression that replaces the atmosphere's variable compression. The compression is achieved by freely introducing a quantity of water in the vessels intended to contain these gases.[38] (Fig. 65f)

Séguin's apparatus, which is considerably more complicated, did not met with success. Soon after Lavoisier's death in 1794, Séguin became a successful entrepreneur in the tanning industry and abandoned chemical research. In all likelihood, therefore, the apparatus he presented to the Institut in 1798 – probably in the hope of being elected member – had been conceived during the early 1790s. Despite his efforts, Séguin was never inducted.

36 "Mr Meusnier a presenté deux machines pour les expériences des airs. Commissaires MM Coulomb, Sage, Monge" (Procès-verbaux de l'Académie des sciences, vol. 108, 1788, p. 62v).

37 "L'instrument que je viens de décrire, et que j'ai nommé gazomètre, a été construit par M. Meignié le jeune, ingénieur, constructeur d'instruments de physique, breveté du roi. Il y a apporté un soin, une exactitude et une intelligence rares. C'est un instrument précieux par le grand nombre des applications qu'on en peut faire, et parce qu'il est des expériences à peu près impossibles sans lui. Ce qui le renchérit, c'est qu'un seul ne suffit pas; il le faut double dans un grand nombre de cas, comme dans la formation de l'eau, dans celle de l'acide nitreux, etc. C'est un effet inévitable de l'état de perfection dont la chimie commence à s'approcher, que d'exiger des instruments et des appareils dispendieux et compliqués: il faut s'attacher sans doute à les simplifier, mais il ne faut pas que ce soit aux dépens de leur commodité et surtout de leur exactitude". Lavoisier (1789), vol. 2, pp. 359–360.

38 "Le citoyen Séguin a imaginé un gazomètre, ou instrument propre à mesurer les gaz, qu'il propose de substituer au gazomètre de Lavoisier, et dont le but est de dispenser des corrections qu'exigeoient pendant le cours des expériences les variations barométriques. Au moyen du gazomètre du cit. Séguin on maintient les gaz dans un état de densité constant, par une compression artificielle et graduée substituée à la compression variable de l'atmosphère. La compression s'opère au moyen d'une quantité d'eau qu'on introduit à volonté dans les réservoirs destinés à contenir ces gaz". Séguin (1798).

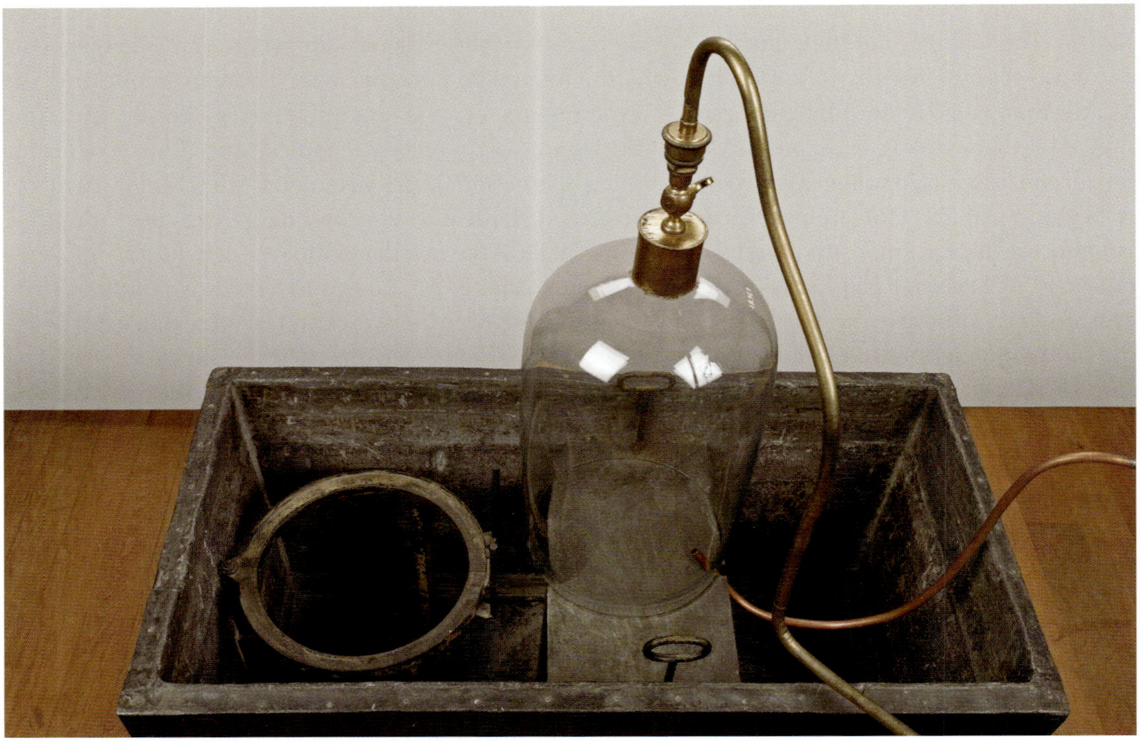

FIGURE 66A (inv. no. 07547-0001-002-)
© MAM/PHOTO FRANCK BOTTÉ

Lavoisier (1781a); Lavoisier, Meusnier de la Place (1781b); Lavoisier (1782); Lavoisier, Meusnier de la Place (1786); Lavoisier (1789), vol. 2, pp. 346–360 and plate VIII; Séguin (1798); Daumas, Duveen (1950); Duveen, Klickstein (1954); Daumas (1955), pp. 142–150; LC, vol. 4, pp. 299–303; LC, vol. 5, pp. 52–53; LC, vol. 7, pp. 396–398, pp. 405–408; Levere (1992); Levere (2000); Giormani (2001); Bret (2004a); Levere (2005); Bassani (2009), vol. 2, pp. 233–236

07547-0001-002-, 07547-0002-002- (Figs. 66a, 66b)
Two pneumatic troughs
ca. 1780
MAM accession date: 1864–1866
07547-0002-002-: 87 × 66 × 43; 57.5 kg
07547-0001-002-: 85.5 × 49.5 × 43; 57.5 kg
Wood, lead, iron

Two almost identical troughs. Both consist of a parallelepiped wooden box coated inside with lead. Each contains two supports: a rectangular plate with three feet and a pair of handles,[39] and a ring also with three feet. One foot is inserted in a holder with which to adjust the ring's vertical position. The ring is made of two pieces and can be opened.

The pneumatic trough was filled with water. On its supports, it was possible to place different vessels such as a balloon filled with water and turned upside down or a glass bell.[40] The gas to be collected arrived in the vessels through a pipe[41] connected to the gasometer. The gas displaced the water[42] and filled the vessel, where it could be collected.

For bibliography see previous item.

39 The handles are to be repaired.

40 Such as no. 07547-0002-004-, currently displayed with the troughs.

41 See no. 07547-0001-003-.

42 When the gas was highly soluble in water, it was common to use a smaller pneumatic trough (often made of marble) filled with mercury.

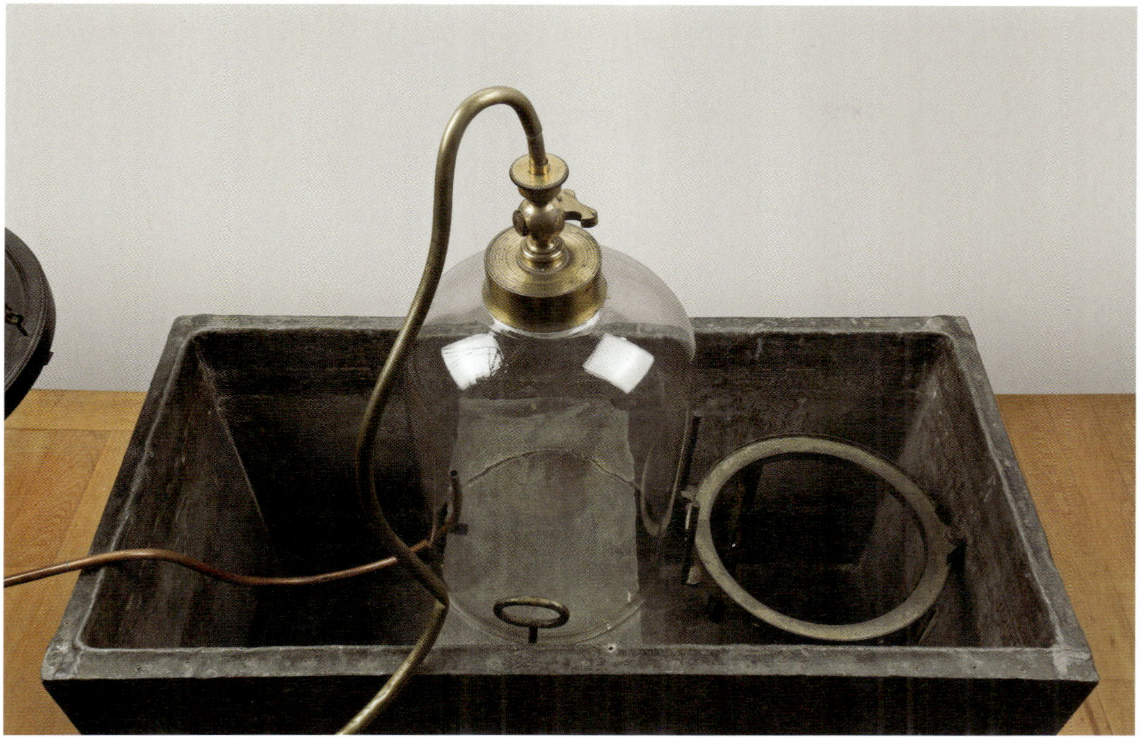

FIGURE 66B (inv. no. 07547-0002-002-)
© MAM/PHOTO FRANCK BOTTÉ

C-2017-0067- (Fig. 67)
Glass tube
Second half of 18th c.
26 × 2.3, 55 g
Glass

Glass tube probably protecting a thermometer of the large gasometer no. 07547-0001-001- or no. 07547-0002-001-.

MNHN OA 289 Muséum National d'Histoire Naturelle (Paris) (Fig. 68)

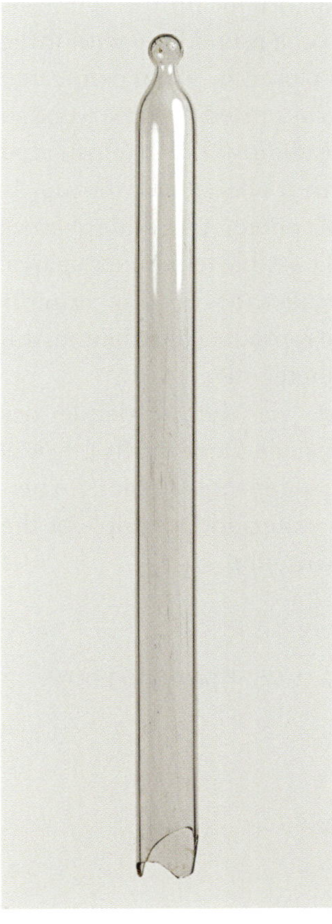

FIGURE 67
(inv. no. C-2017-0067-)
© MAM/PHOTO FRANCK BOTTÉ

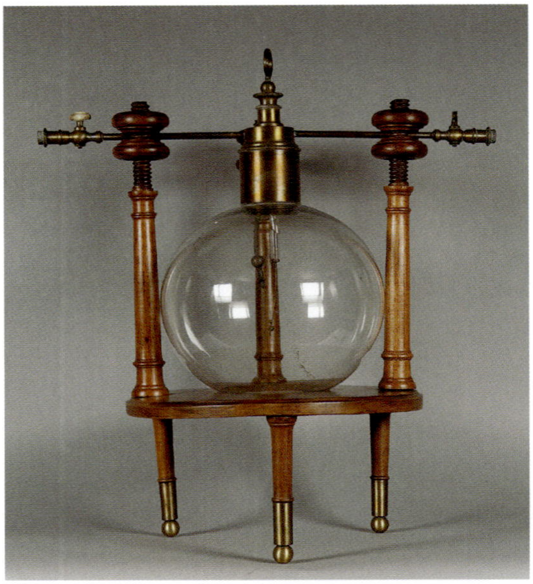
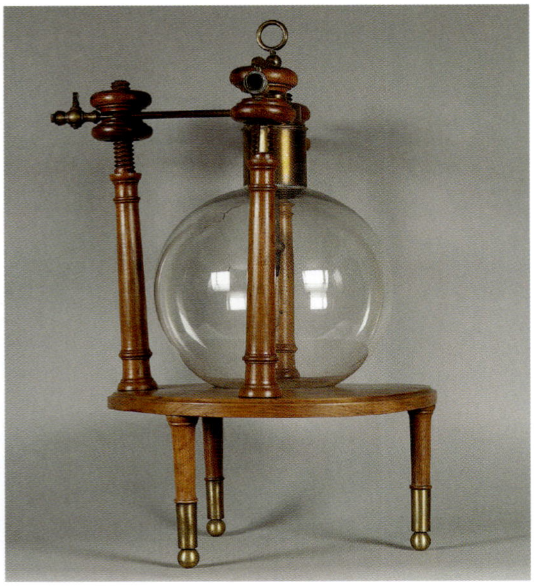

FIGURE 68 (inv. no. MNHN OA 289)
COURTESY OF MUSÉUM NATIONAL D'HISTOIRE NATURELLE (PARIS)

Balloon for the synthesis of water
ca. 1789
69 × 62
Glass, wood, brass

Glass balloon used for the experiment on the synthesis of water. A glass balloon rests on a round base with three feet. The top of the balloon is closed by a cap connected to three brass pipes, which are supported by three wooden columns with threaded rings. A sliding brass bar insulated in a glass pipe penetrates into the balloon from the top. At the bottom of the bar is a small sphere. A second sphere is fixed to a second bar, making it possible to produce sparks. This balloon is similar to no. 07548-0000- but was certainly made later. Such balloons were reproduced as demonstration apparatus well into the nineteenth century.

The balloon was donated, via the Académie des sciences, by Marie Anne Lavoisier's niece in 1854. It was eventually used by Gay-Lussac to repeat Lavoisier's experiments on the composition of water, and a sample of the water is still preserved in the Muséum.

07547-0005-001-07547-0005-002-, 07547-0005-003-
(Fig. 69)
Two thermometric tubes
Second half of 18th c.
MAM accession date: 1864–1866
07547-0005-001-: 27.5 × 2; 50 g
07547-0005-002-: 26.5 × 0.5; 10 g
07547-0005-003-: 26.5 × 0.5; 10 g
Glass

07547-0005-001-: closed glass tube probably protecting a thermometer in a Megnié gasometer.

07547-0005-002-, 07547-0005-003-: two capillary tubes of broken thermometers (the bulbs are missing).

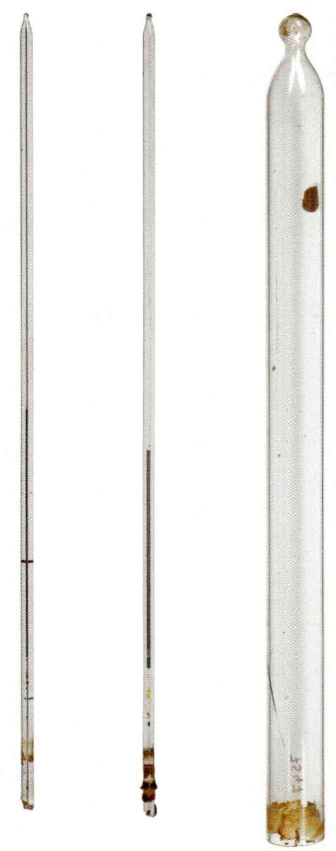

FIGURE 69 (inv. nos. 07547-0005-001-, -0003-)
© MAM/PHOTO FRANCK BOTTÉ

07547-0002-005- (Fig. 70)
Fragments of tubes
Second half of 18th c.
MAM accession date: 1864–1866
27 × 2.5
Glass, metal

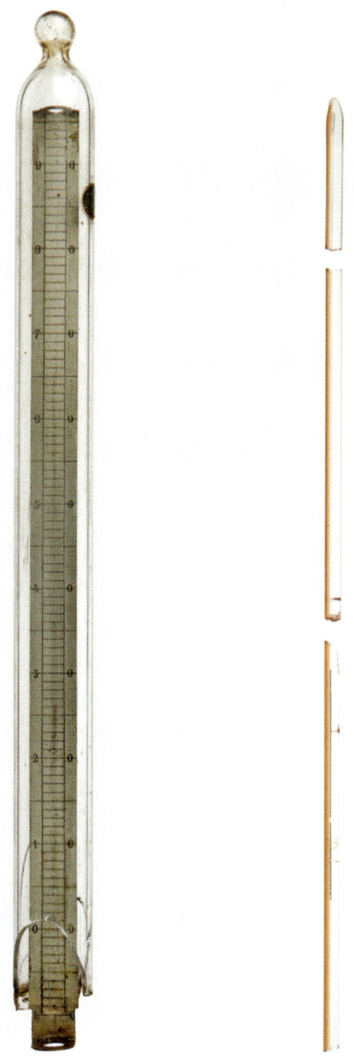

FIGURE 70 (inv. no. 07547-0002-005-)

These fragments probably belonged to the gasometer.

07550-0000-, 19998-0000- (Fig. 71)
Apparatus for vinous fermentation
1787–88
MAM accession date: 1864–1866
Nicolas Fortin (1750–1831)
147 × 103 × 65; 95 kg
Glass, brass, wood, straw

This apparatus was used by Lavoisier for studying "vinous fermentation". Known today as ethanol or alcoholic fermentation, it is an organic process through which sugars are transformed into ethanol and carbon dioxide by the action of yeast and specific bacteria.[43]

The apparatus is mounted on a long table carrying a brass label with the engraved inscription: "Du Cabinet de L.^{n} A.^{ne} Lavoisier. Donné par M. & Mme Leon de Chazelles".

A large glass balloon resting on a straw ring has a long neck connected in series with two horizontal glass tubes and four cylindrical gas washing bottles. The tubes are supported by a wooden tablet with two legs. The elements of the apparatus are connected by brass pipes, which are fastened to the glass tubes and to the cylindrical bottles with threaded flanges and brass collars. There is a tap between the balloon and the first pipe, and a second tap on the outlet pipe of the second tube. Each inlet and outlet pipe of the washing bottles has a tap. All the pipes connections consist of threaded collars.

The balloon was partly filled with the fermenting sugary mixture. The carbon dioxide developed by the process would drain water droplets, which were absorbed by the hydroscopic salt filling the two drying tubes. Eventually, the gas was completely absorbed by the limewater (alkali) contained in the four washing bottles. The last bottle was connected to a receiver in a pneumatic trough that collected only the air originally filling the apparatus.

The large glass balloon was probably added later to the apparatus and has a different inventory number: 19998-0000-.[44] The apparatus also comes with a curved brass connecting-piece with a tap and threaded rings – one large, one small – inventoried as no. 20129-0002-. A large table on which the experimental set-up can be placed was added later and it is not part of the original collection. (See chapter 5 Fig. 11)

43 The modern chemical formula of the process is $C_6H_{12}O_6$ (glucose) → $2C_2H_5OH$ (ethanol) + $2CO_2$ (carbon dioxide).

44 During the 1975 restoration, the neck of this balloon was shortened by 35 mm in order to fit with the rest of the apparatus!

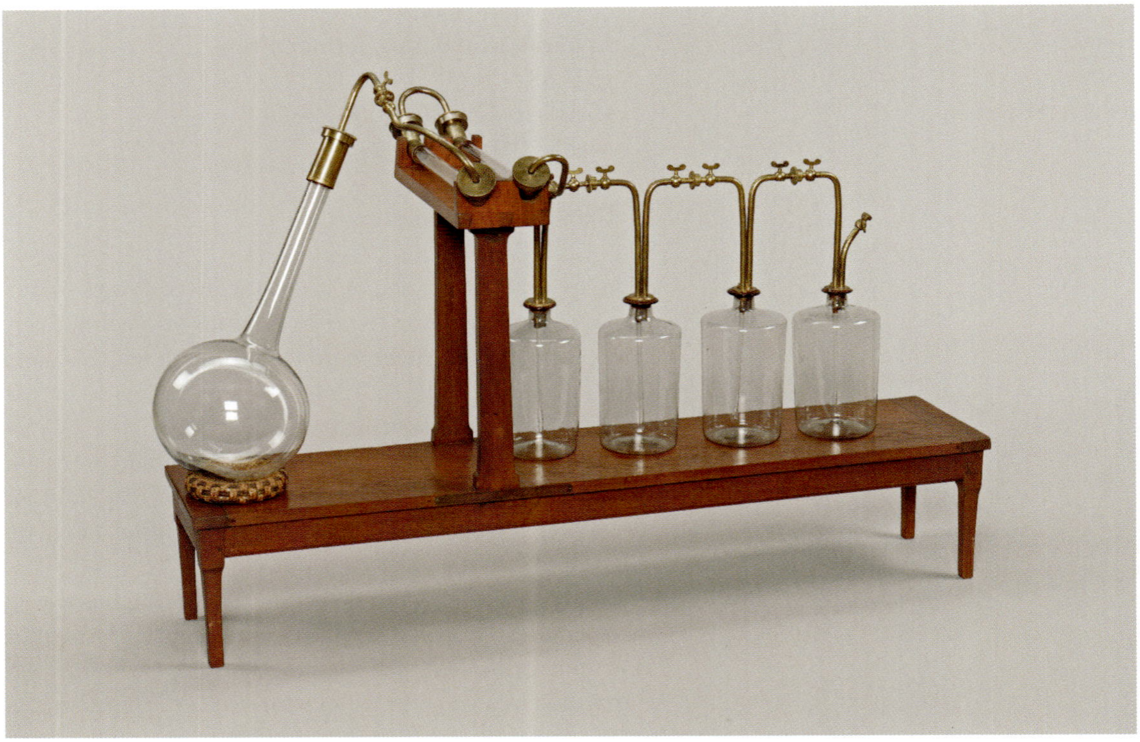

FIGURE 71 (inv. nos. 07550-0000-, 19998-0000-)
© MAM/PHOTO FRANCK BOTTÉ

Lavoisier began to work on alcoholic fermentation in January 1787 and he attached great importance to his experiments on the subject. Indeed, it was on this occasion that Lavoisier first stated what has become known as the law of conservation of mass during a chemical reaction. To reach this stage, he asked Fortin to build the present apparatus, which he eventually perfected (see next item). In the *Traité élémentaire de chimie* he declared:

> As a final remark on vinous fermentation, I should point out that it can offer a means to analyze sugar and, in general, plant substances liable to ferment. Indeed, as I already noted at the start of this article, I can regard the materials left to ferment and the result obtained after fermentation as an algebraic equation; and, by successively assuming that each element of the equation is unknown, I can extract a value from it and thus rectify experiment by calculation, and calculation by experiment. I have often taken advantage of this method to correct the initial results of my experiments, and to guide me in the precautions to be taken when I repeat them.[45]

He also applauded the efficiency of Fortin's apparatus: "One can see that, by means of this apparatus, one can determine with great precision the weight of the materials that have been left to ferment, and [the weight] of all the liquid or aeriform products that have been released from them".[46]

While Lavoisier had high expectations for his experiments, they proved to be a failure.

Lavoisier (1787), Lavoisier (1789), vol. 1, pp. 139–152; vol. 2, 461–464 and plate X; Daumas (1950), pp. 60–61; Holmes (1985), pp. 316–352

45 "Je terminerai ce que j'ai à dire sur la fermentation vineuse, en faisant observer qu'elle peut fournir un moyen d'analyse du sucre, et, en général, des substances végétales susceptibles de fermenter. En effet, comme je l'ai déjà indiqué au commencement de cet article, je puis considérer les matières mises à fermenter et le résultat obtenu après la fermentation comme une équation algébrique; et, en supposant successivement chacun des éléments de cette équation inconnus, j'en puis tirer une valeur et rectifier ainsi l'expérience par le calcul, et le calcul par l'expérience. J'ai souvent profité de cette méthode pour corriger les premiers résultats de mes expériences, et pour me guider dans les précautions à prendre pour les recommencer". Lavoisier (1789), vol. 1, p. 151.

46 "On voit qu'au moyen de cet appareil on peut connaître avec une grande précision le poids des matériaux mis à fermenter, et celui de tous les produits liquides ou aériformes qui s'en sont dégagés."

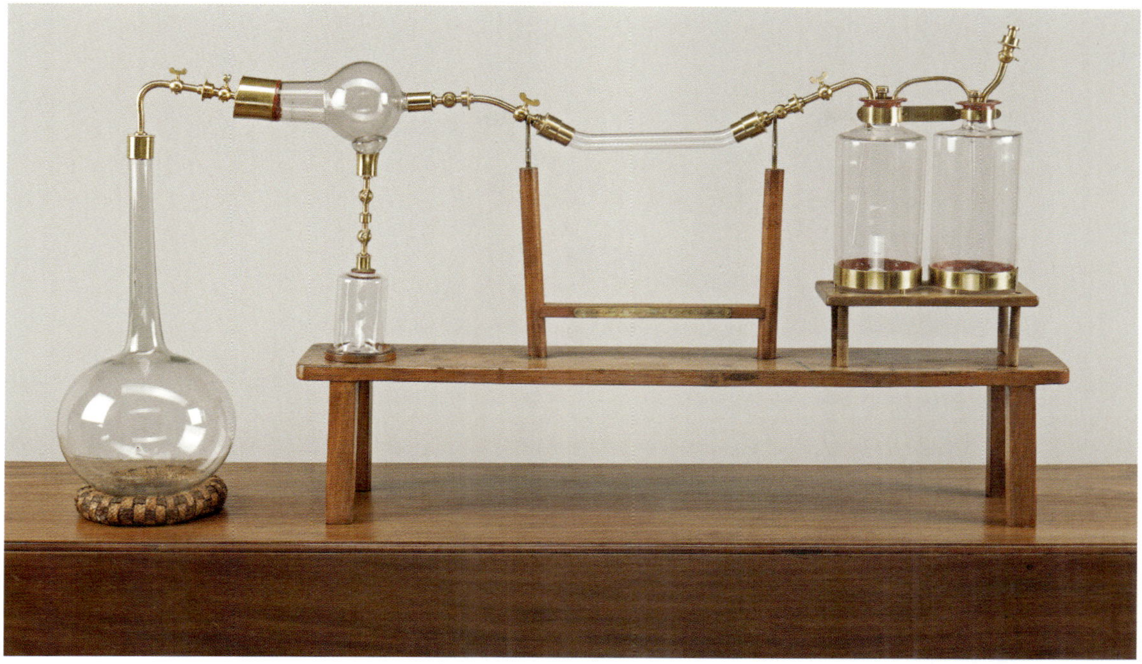

FIGURE 72 (inv. nos. 07551-0001-, 7551-0002-001-, 7551-0002-002-)
© MAM/PHOTO FRANCK BOTTÉ

07551-0001-, 7551-0002-001-, 7551-0002-002-[47] (Fig. 72)
Apparatus for vinous fermentation
1787–1788
MAM accession date: 1864–1866
Nicolas Fortin (1750–1831)
Apparatus:160 × 79.5 × 30; mass not measured
Table: 224 × 102 × 68; mass not measured[48]
Glass, brass, wood, straw, sealing wax

This final version of Lavoisier's apparatus for studying wine fermentation (ethanol or alcoholic fermentation) features improvements on the apparatus no. 07550-0000-.

A long-necked balloon, resting on a stuffed straw ring, is closed by a collar with a brass pipe and a tap. The pipe is connected to a second tap with a collar sealed to the largest neck of a three-necked pear-shaped balloon. A cylindrical bottle placed underneath this second balloon is connected to it with two taps, while the third neck of the balloon communicates with a long glass tube by means of a brass pipe and tap. The other end of the tube is connected with another pipe and a tap to the inlet pipe and tap of the first of two washing bottles. The outlet pipe of this first bottle is joined to the inlet pipe of the second bottle, whose its outlet pipe has another tap. The two bottles rest on a small wooden table, their bottoms are inserted in a brass collar, and a brass strip is fastened to their necks. The washing bottles are modern replacements because the original ones are broken. All the elements mentioned and a part of the large balloon rest on a longer wooden table on which two wooden columns with iron forks support the glass tube. A wooden crosspiece between the columns carries a brass label with the engraved inscription: "Du Cabinet de L.^{n} A.^{ne} Lavoisier. Donné par M. & Mme Leon de Chazelles". The table supporting the entire apparatus is not part of the original collection.

The apparatus worked like the one described under no. 07550-0000-. Lavoisier added to the apparatus the pear-shaped balloon and the bottle underneath in order to collect the abundant foam produced by the fermentation and prevent it from entering into the drying tube, which contained a hygroscopic substance. After drying, the carbon dioxide was totally absorbed by the limewater (alkali) that filled the washing bottles. When the second bottle was connected with a receiver in a pneumatic trough, the receiver could collect only the air that originally filled the apparatus. This is the apparatus illustrated in plate X of Lavoisier's *Traité élémentaire de chimie*.

See historical note on item 07550-000-, 19998-0000-.

Lavoisier (1787), Lavoisier (1789), vol. 1, pp. 139–152; vol. 2, pp. 461–464 and plate X; Daumas (1950), pp. 60–61; Holmes (1985), pp. 316–352

47 The last two do not correspond to the original glass recipients.
48 The table is not coming from Lavoisier's collection.

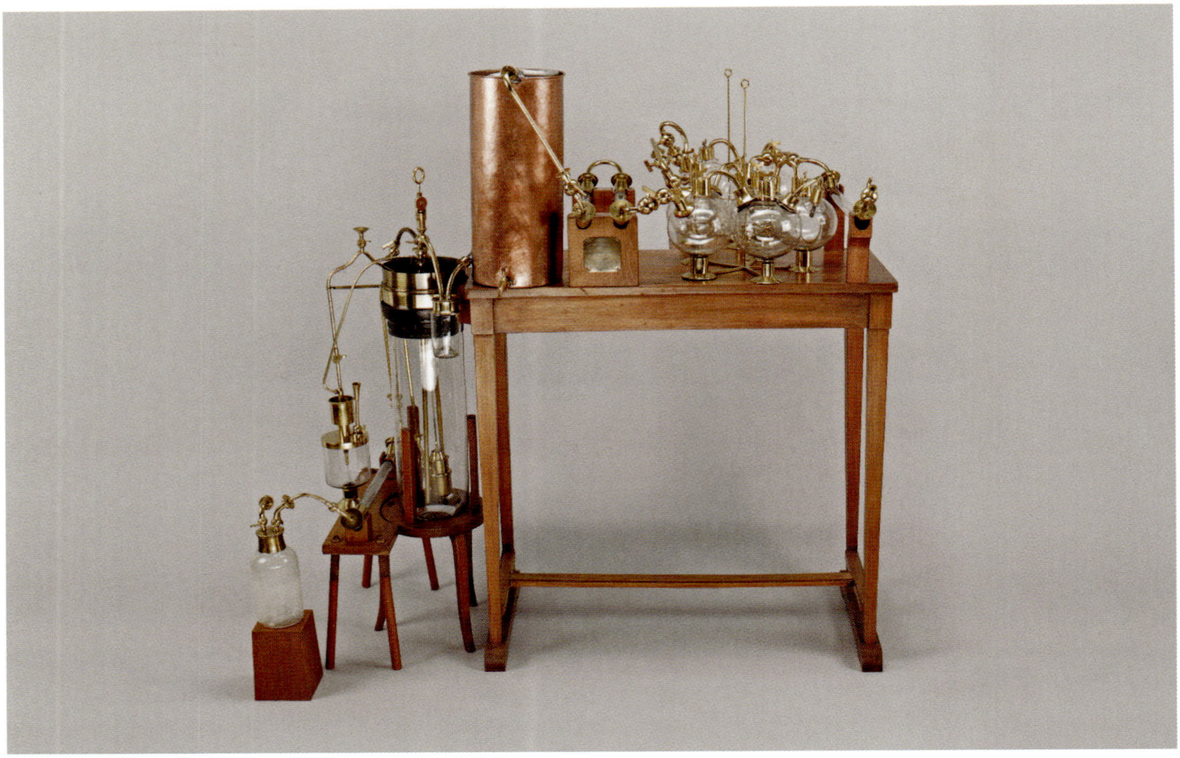

FIGURE 73A (inv. nos. 07549-0001-, -0006-)
© MAM/PHOTO FRANCK BOTTÉ

07549-0001-, 07549-0002-, 07549-0003-, 07549-0004--, 07549-0005- (Fig. 73a–d)
Apparatus for analysing the gases produced by the combustion of oils
1788
MAM accession date: 1864–1866
Nicolas Fortin (1750–1831)
155 × 147 × 75; mass not measured
Brass, glass, wood, iron, copper

This apparatus – one of the most complex and sophisticated conceived by Lavoisier – was used to study the gases produced by the combustion of oils.

It is composed of several elements.

The lamp burning the oil to be tested is placed at the bottom of a large cylindrical glass vessel (A) with an iron collar sealed at the top. Between the upper part of the collar and the vessel is a circular trough originally filled with mercury. The brass lid closing the vessel has an iron ferrule inserted into the trough. The mercury provided a hermetic seal for the vessel, and the lid could be easily removed and replaced.[49] In the centre of the lid is a tube with a stuffing box and a sliding rod ending with a ring. The rod served to regulate the wick height so as to ensure optimal com bustion of the oil, and could be moved up and down with a rack-and-pinion mechanism.

The brass lamp has a cylindrical wick. Two tubes pass through the lid and are connected with the lamp: one tube, with a stopcock, brings the oil to the wick; the other delivers the air needed for combustion.[50] The first tube is connected with a siphon to the bottle used as an oil reservoir. This bottle is placed on a small round table and mounted on a brass foot. It is closed by a lid with a stopcock and an elongated funnel, which was used to fill the bottle with oil. The siphon tube has two stopcocks. The one at the top of the tube has a funnel that served to fill the siphon with the oil in order to start the process. Another tube connects the oil-reservoir tube to the collar with the stuffbox of the movable rod.[51]

49 Thanks to this system, the vessel could be opened (and closed) without using any kind of putty.

50 This is in fact a small Argand lamp. The air passes through the centre of the wick and also around the outside of the cylindrical wick. The Genevan chemist and physicist François-Pierre-Ami Argand (1750–1803) patented his lamp in 1780. His system greatly improved the performance of oil lamps.

51 This tube is illustrated in table XI of Lavoisier's *Traité élémentaire de chimie*, although its use is not explained. It may have served to return excess oil to the oil reservoir. (Fig. 73d)

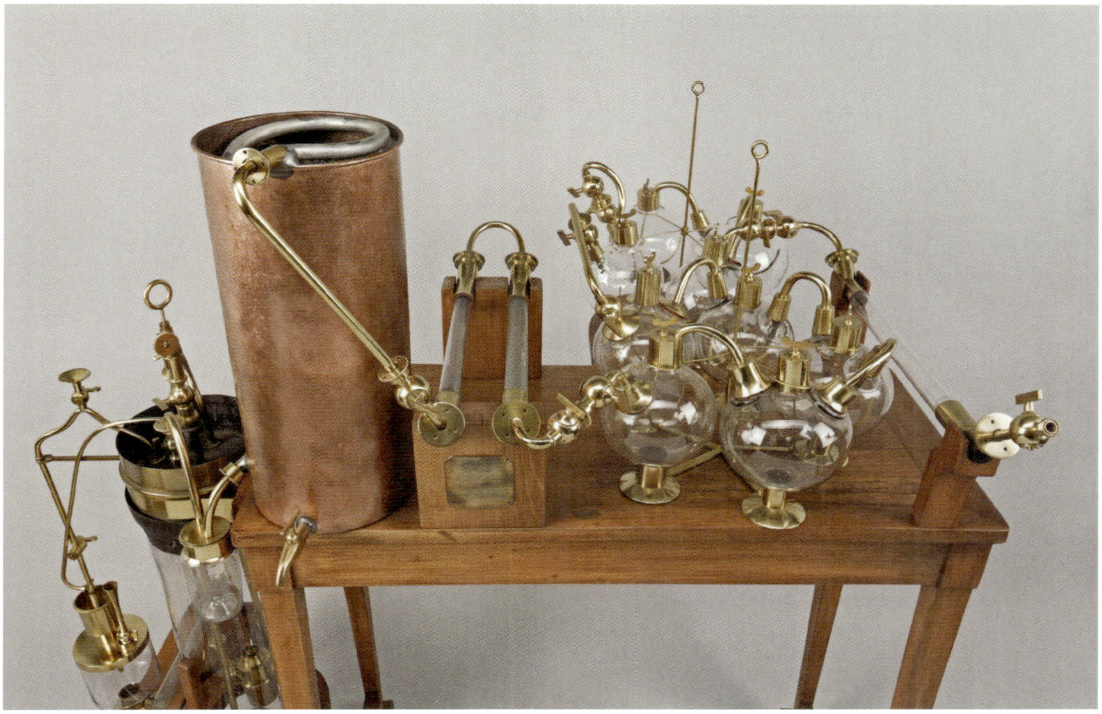

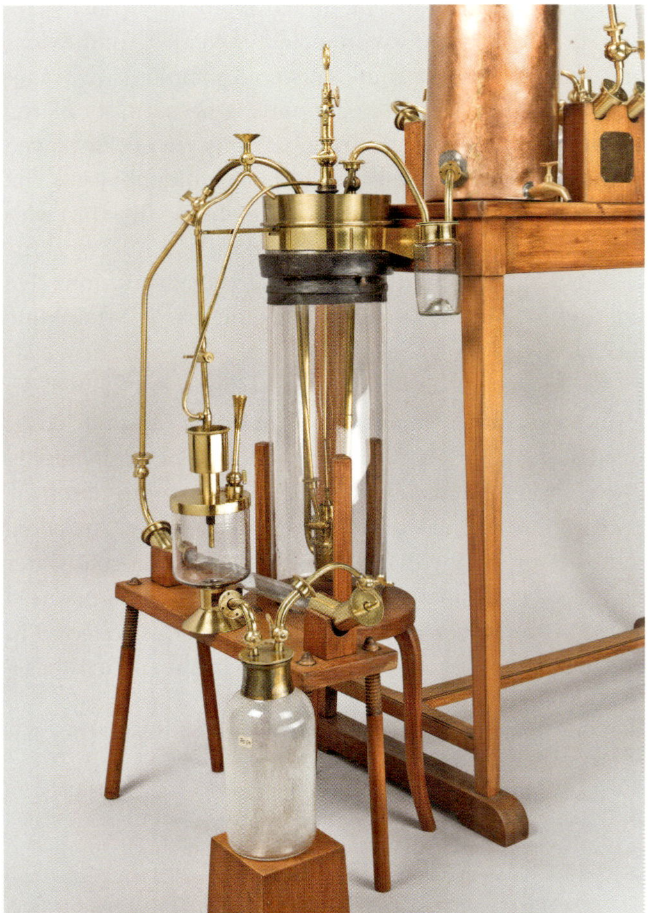

FIGURES 73B, 73C
(inv. nos. 07549-0001-, -0006-)

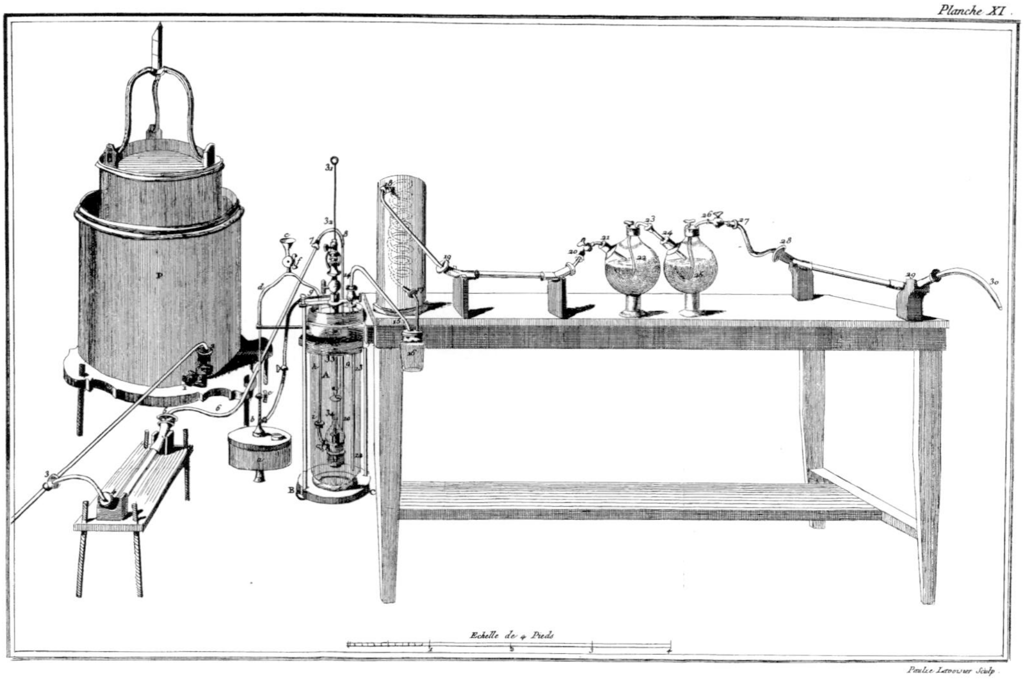

FIGURE 73D (inv. n. 07549-0000-). Lavoisier (1789), Plate XI

The tube delivering the air to the lamp has a stopcock and is connected with one end of a gas-drying tube (B), resting on a small table. This is a glass tube closed by brass collars. The other end of the drying tube is connected with a bottle (C) fitted with a brass lid carrying two short pipes with stopcocks.[52]

The outlet tube delivering the combustion gases runs to the cap of a small bottle (D). A second tube in the cap is connected to a copper serpentine contained in a cylindrical copper vessel with a tap at its bottom. The top end of the serpentine is connected to a pair of drying tubes (E and F) identical to (B). Both the inlet and outlet pipes have stopcocks.

The outlet pipe of the second tube is joined to the stopcock of the first of a series of five washing balloons. These rest on brass feet and are held together by a brass frame with a central vertical rod terminating with a ring. Each balloon has three openings.[53] The central one is closed by a cap with a small butterfly handle that rotates a brass arc for stirring the liquid in the bottle. The other two openings are also closed by caps with connecting pipes. There are stopcocks before the first bottle and after the last one.

A second group of four similar balloons[54] (no. 07549-0002-) was originally connected in series to the first and was placed on a rounded wooden tablet (no. 07549-0003-) fixed to the table supporting these parts of the apparatus.[55] (At present these balloons are not displayed because of lack of space in the museum cupboard.) The last balloon was connected to a final drying tube (G) with stopcocks (also identical to B, E and F).[56]

A brass plate on the table has the inscription: "Du Cabinet //de Lt A^{ne} Lavoisier // Donné par // Mr & Mme de Chazelles".

The apparatus was operated as follows: the flow of air from a gasometer[57] was washed in bottle C and dried by the deliquescent salt[58] contained in tube B. The air entered into vessel A and sustained the combustion of the lamp oil. The mixture of gases produced by the combustion flowed into the serpentine, which was cooled by the cold water that filled its copper vessel. The water vapour condensed in the serpentine and the resulting water was collected in bottle D. The gases were further dried by the salt in tubes E

52 This additional (gas-washing?) bottle is neither mentioned nor illustrated in Lavoisier's *Traité élémentaire de chimie*.

53 Only the first balloon has two necks. The connecting pipes are joined to the cap with the stirring device.

54 These balloons have two openings; one of each connecting pipe penetrates into the cap that holds the stirrer.

55 For simplicity, only two washing balloons are illustrated in table XI of the *Traité élémentaire de chimie*, but Lavoisier stated that he used at least nine of them. (Fig. 73d)

56 Together with this apparatus are also a glass balloon (no. 07549-0004--), a broken glass balloon (no. 07549-0000-) and the fragment of a balloon (no. 07549-0005-).

57 Lavoisier used two gasometers. When one was empty, it was possible to switch to the other without interrupting the experiment.

58 A salt that tends to absorb moisture from the air.

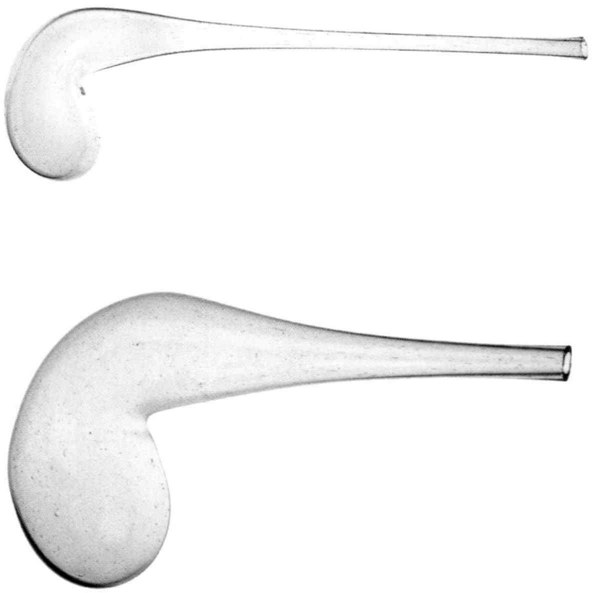

FIGURES 74A, 74B (inv. nos. 20011-0001-, 20011-0002-) © MAM/PHOTO FRANCK BOTTÉ

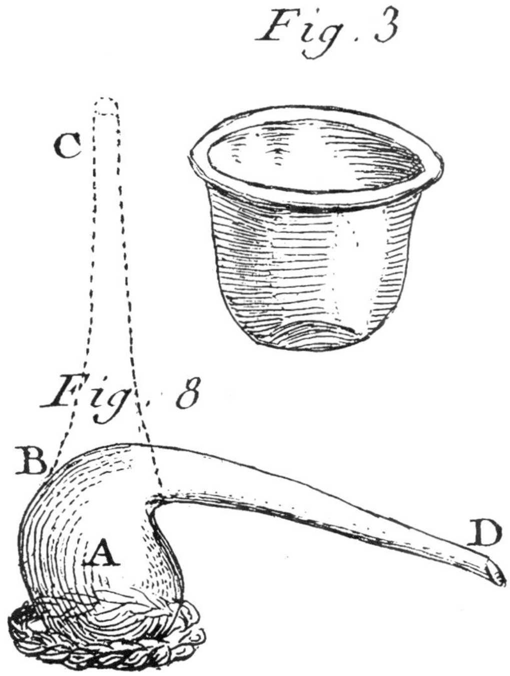

FIGURE 74C (inv. n. 20011). Lavoisier (1789), Plate III

and F and passed into the balloons. These were partly filled with a concentrated potash solution (potassium hydroxide) that absorbed all the carbon dioxide contained in the gas mixture. The last balloon was filled with limewater (solution of calcium hydroxide). The limewater remained clear and transparent if the gases were completely deprived of carbon dioxide. Next, the gases (essentially oxygen and nitrogen), which may have retained some moisture after passing through the solutions, were again dried by the salt in tube F. At the end, the remaining gases were stored in a gasometer (not part of this apparatus), and a sample was tested with potassium sulphide to determine the proportion of oxygen and nitrogen.

The apparatus was quite expensive (nearly 1,000 *livres*) and not easy to use. In a detailed description accompanying the invoice, dated August 1788 (LC, vol. 5, pp. 209–210), Fortin explained the function of its parts. The apparatus was delivered to Lavoisier in early autumn 1788, when he was about to complete the *Traité élémentaire de chimie*. While Lavoisier provided a comprehensive description of the machine, he probably did not have any time to make experiments with it before the book was printed in early spring 1789. It is likely that he used it only in late summer 1789 when the entire machine was finally ready. In a letter to Armand Séguin of 20 September, Lavoisier asked for help in mounting the apparatus.[59]

59 "Les appareils pour la combustion sont prêts et si vous le juges [*sic*] à propos nous prendrons un jour pour les monter." LC, vol. 6, p. 71.

Lavoisier (1789), vol. 2, pp. 493–503, plates XI and XII; Daumas (1955), pp. 152–153

Chemical Glassware

20011-0001-, 20011-0002- (Fig. 74)
Retorts
Second half of 18th c.
MAM accession date: 1952
20011-0001-: 34 × 10.5 × 6; 70 g
20011-0002-: 19.5 × 11 × 6; 54 g
Glass

Two glass retorts of different dimensions but of similar shape: a pear-shaped vessel with a downward-pointing neck. Retorts were very common laboratory glassware generally used for distillation. Lavoisier describes the use of this kind of retorts in his *Traité élémentaire de chimie* (1789):

> The simplest distilling apparatus of all is a bottle A (pl. III, fig. 8) (Fig. 74c), whose neck BC is bent in BD at the glassworks itself. This bottle or flask is called a retort. It is placed in a reverberatory furnace [...], or in a sand bed under a terracotta covering [...]. To collect and condense products, a vessel is fitted to the retort [...] and they are sealed together; sometimes, especially in pharmaceutical operations, a glass or earthenware cucurbit is used [...], topped

with its cap, or a glass alembic fitted with a cap, the two forming a single apparatus.[60]

Lavoisier (1789), vol. 2, pp. 443–444 and plate III, fig. 8

C-2017-0069- (Fig. 75)
Cucurbit
Second half of 18th c.
MAM accession date: 1864–1866
41 × 20; 1.32 kg
Glass

Large pear-shaped bottle made of thick green glass. Together with a *chapiteau* (distillation head) sealed on its neck, it formed an alembic. The cucurbit was used for evaporating and distilling.

Baumé (1773b), vol. 1, pp. cij–ciij, civ, plate VI; Lavoisier (1789), plate III, fig. 14; Principe, De Witt (2002), p. 17

36645-0000- (Fig. 76)
Alembic
Second half of 18th c.
MAM accession date: not documented
42 × 33.5 × 18; 915 g
Glass

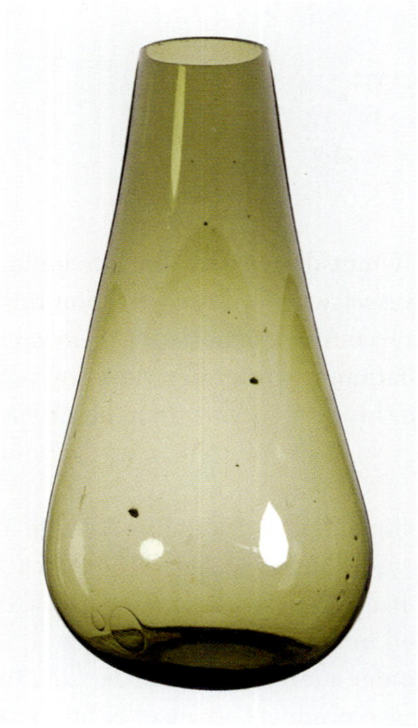

FIGURE 75 (inv. no. C-2017-0069-) © MAM/PHOTO FRANCK BOTTÉ

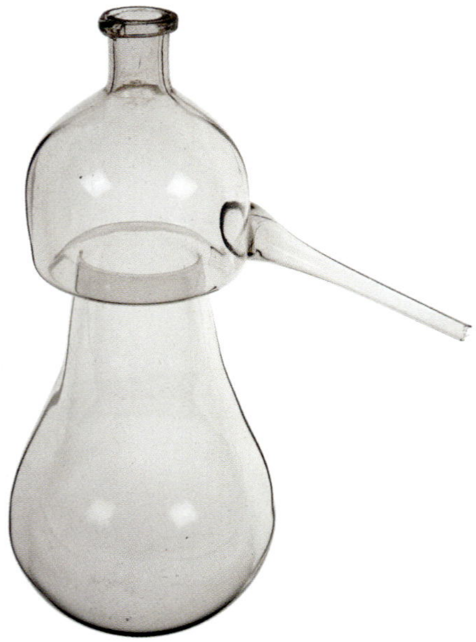

FIGURE 76 (inv. nos. 36645-0000-, -0006-) © MAM/PHOTO FRANCK BOTTÉ

Pear-shaped cucurbit made of thick transparent glass with its fitted *chapiteau* (distillation head) sealed on its neck, forming an alembic. The alembic was used for evaporating and distilling. Although this item is not directly traceable to the Lavoisier collection, there is circumstantial evidence in favor of it: the inventory number refers to an unidentified item; part of Lavoisier's glassware acquired by the MAM in 1864–1866 was not properly inventoried and was not identified;[61] and Lavoisier describes an almost identical cucurbit in his *Traité*.

Lavoisier (1789), pp. 443–444, plate III, fig. 12

60 "Le plus simple de tous les appareils distillatoires est une bouteille A (pl. III, fig. 8), dont on courbe, dans la verrerie même, le col BC en BD. Cette bouteille ou fiole porte alors le nom de cornue; on la place dans un fourneau de réverbère [...], ou au bain de sable sous une couverture de terre cuite [...]. Pour recueillir et pour condenser les produits, on adapte à la cornue un récipient [...] qu'on lute avec elle; quelquefois, surtout dans les opérations de pharmacie, on se sert d'une cucurbite de verre ou de grès [...], surmontée de son chapiteau, ou bien d'un alambic de verre auquel tient un chapiteau d'une seule pièce".

61 See Appendix 5, p. 161.

C-2017-0071- (Fig. 77)
Head of a cucurbit
Second half of 18th c.
MAM accession date: 1864–1866
43 × 40 × 24; 2.4 kg
Glass

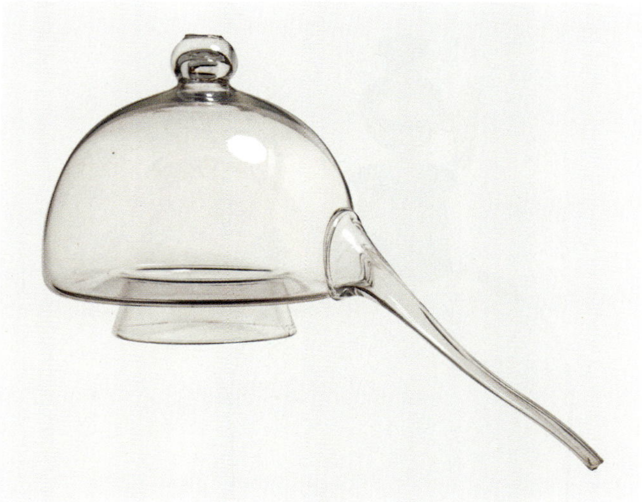

FIGURE 77 (inv. no. C-2017-0071-)
© MAM/PHOTO FRANCK BOTTÉ

Head (*chapiteau*) of a cucurbit. Semi-spherical glass piece with an internally salient rim and a long inclined and downward-sloping conical pipe. It is part of a simple distilling apparatus. The head closed the top of a cucurbit, which was the pot containing the liquid to be distilled. The vapours produced by the boiling liquid were collected and condensed in the head and the distilled liquid flowed through the tube into a receiver.

Lavoisier (1789), vol. 2, plate III, fig. 12

07547-0002-004- (Fig. 78)
Bell-jar
Mid-1780s
MAM accession date: 1864–1866
47.5 × 22; 3.27 kg
Glass, brass, iron, silk, paper

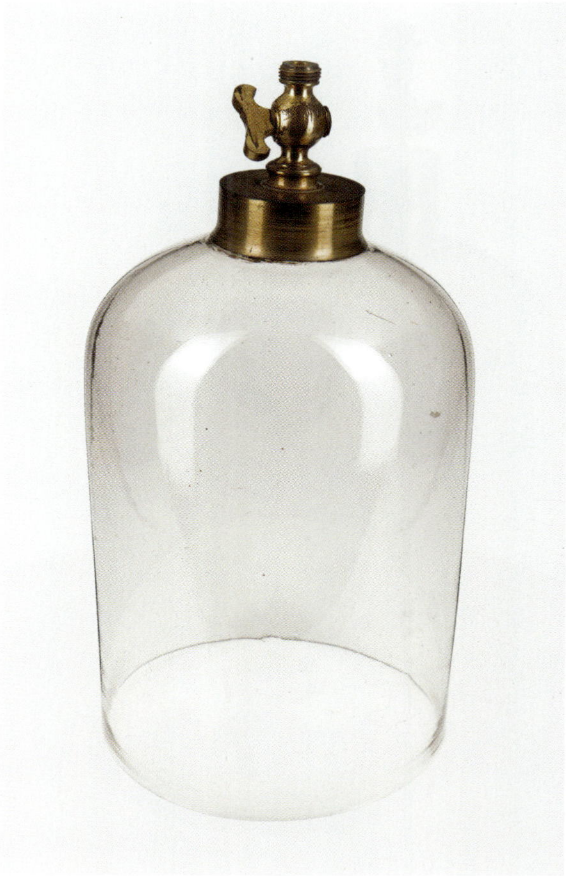

FIGURE 78 (inv. no. 07547-0002-004-)
© MAM/PHOTO FRANCK BOTTÉ

Cylindrical graduated bell-jar with a brass collar at the top. A stopcock is screwed onto the collar. This bell-jar, which is similar to no. 19963-0000-, was used to collect a gas in a pneumatic trough.[62] The gas could be used directly or transferred into a bladder (by sinking the balloon in the water of the trough) or a glass balloon previously evacuated with a pump.

62 This bell jar is currently displayed together with the trough no. 07547-0002-002-.

19963-0000- (Fig. 79)
Bell-jar
Late 18th c.
MAM accession date: 1952
54 × 23.5; 3.63 kg
Glass, brass

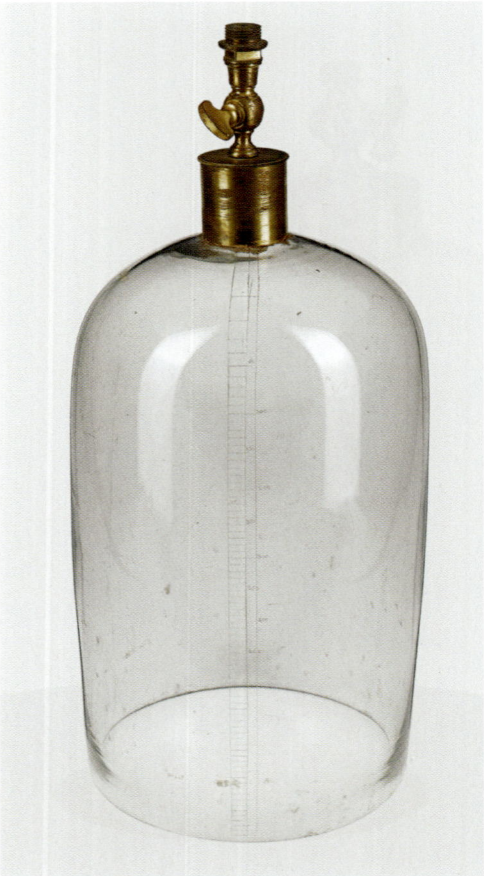

FIGURE 79 (inv. no. 19963-0000-)
© MAM/PHOTO FRANCK BOTTÉ

Cylindrical graduated bell-jar with a brass collar at the top. A stopcock is screwed onto the collar. The graduation from 0 to 700 (with divisions every 10 units) is engraved on the glass. The figures indicating the volume are marked every 50 divisions from the top down, but there are no figures from 450 to the base of the bell-jar.[63]

This bell-jar was used in a pneumatic trough for collecting gas and transferring it into a balloon (or bladder), which was screwed onto the stopcock.

Thénard (1835), pp. 7 and 22

19966-0000- (Fig. 80)
Bell-jar
Late 18th c.
MAM accession date: 1952
43.5 × 17; 1.65 kg
Glass, brass

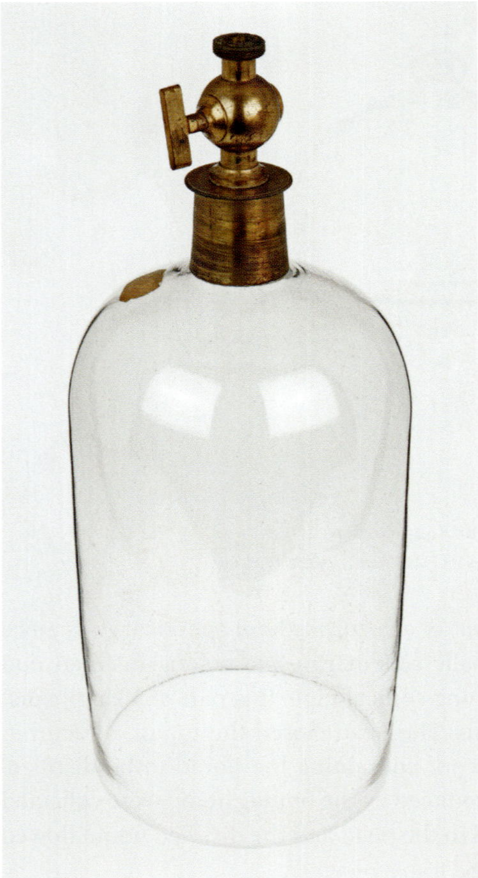

FIGURE 80 (inv. no. 19966-0000-)
© MAM/PHOTO FRANCK BOTTÉ

Cylindrical bell-jar with a brass collar at the top. A stopcock is screwed onto the collar. The instrument is similar to no. 19963-0000- and was used in the same way.

Thénard (1835), pp. 7 and 22

63 The jar was added to the gasometer 7547-0001-001- for decorative purposes.

19974-0000- (Fig. 81)
Bell-jar
Late 18th c.
MAM accession date: 1952
32 × 14; 1.16 kg
Glass, brass

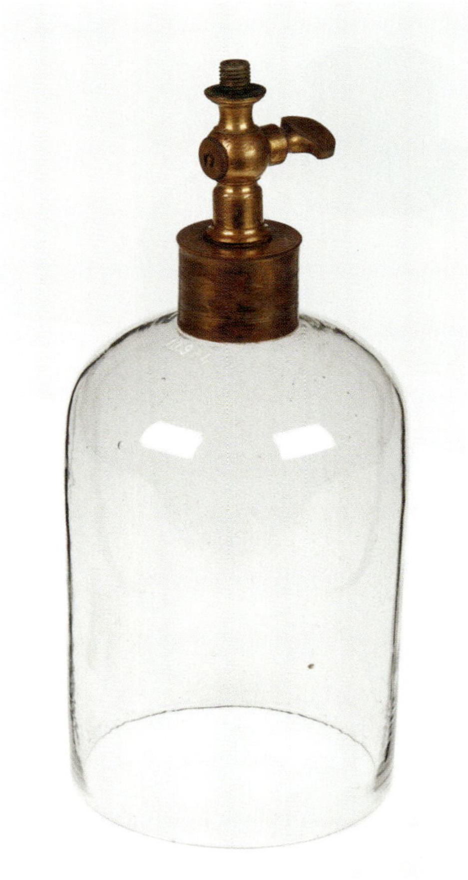

FIGURE 81 (inv. no. 19974-0000-)
© MAM/PHOTO FRANCK BOTTÉ

Cylindrical bell-jar with a brass collar at the top. A stopcock is screwed onto the collar. The instrument is similar to no. 19963-0000- and was used in the same way.

Thénard (1835), pp. 7 and 22

19973-0000- (Fig. 82)
Bell-jar
Late 18th c.
MAM accession date: 1952
30 × 13; 750 g
Glass, brass

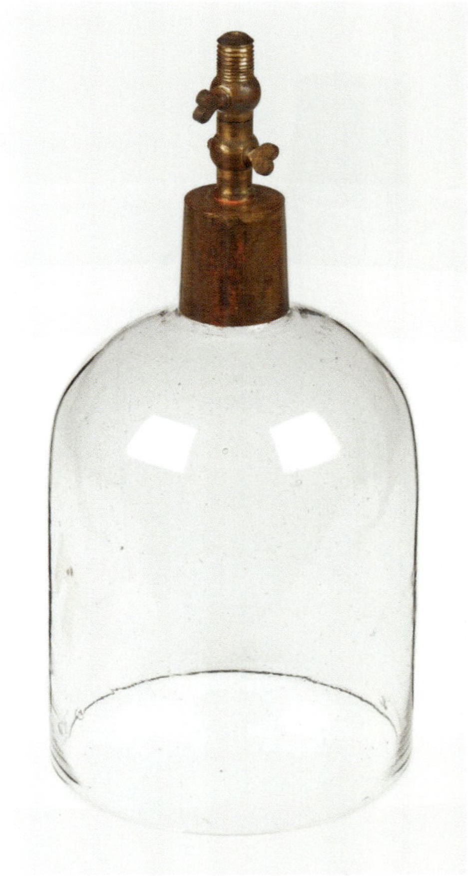

FIGURE 82 (inv. no. 19973-0000-)
© MAM/PHOTO FRANCK BOTTÉ

Cylindrical bell-jar with a brass collar at the top. Two stopcocks are screwed onto the collar, one above the other. The instrument is similar to no. 19963-0000- and was used in the same way.

Thénard (1835), pp. 7 and 22

19967-0000- (Fig. 83)
Bell-jar
Late 18th c.
MAM accession date: 1952
29.5 × 18.5; 1.24 kg
Glass, brass

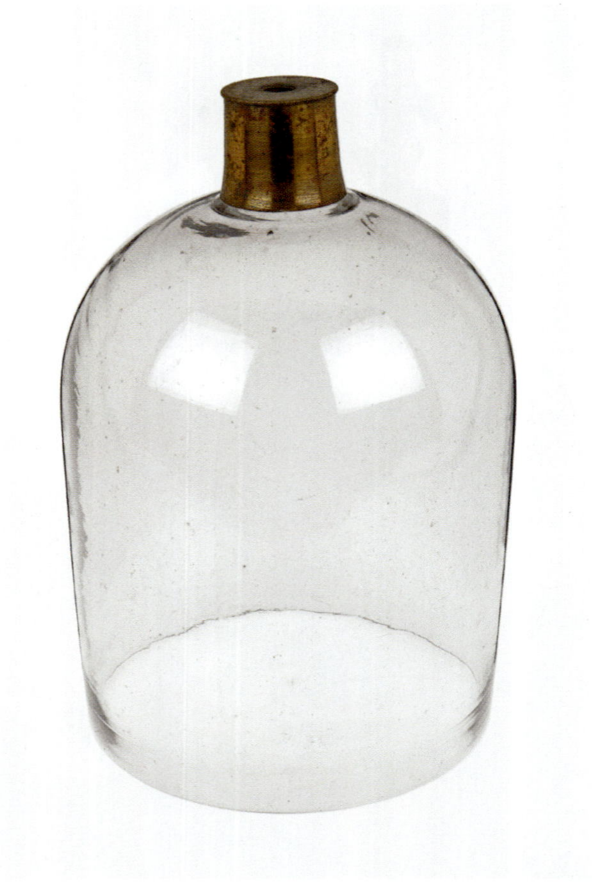

FIGURE 83 (inv. no. 19967-0000-)
© MAM/PHOTO FRANCK BOTTÉ

Cylindrical bell-jar with a brass collar at the top. A stopcock (missing) was originally screwed onto the collar. The instrument is similar to no. 19963-0000- and was used in the same way.

Thénard (1835), pp. 7 and 22

19971-0000- (Fig. 84)
Bell-jar
Late 18th c.
MAM accession date: 1952
34.5 × 17; 1.5 kg
Glass, brass

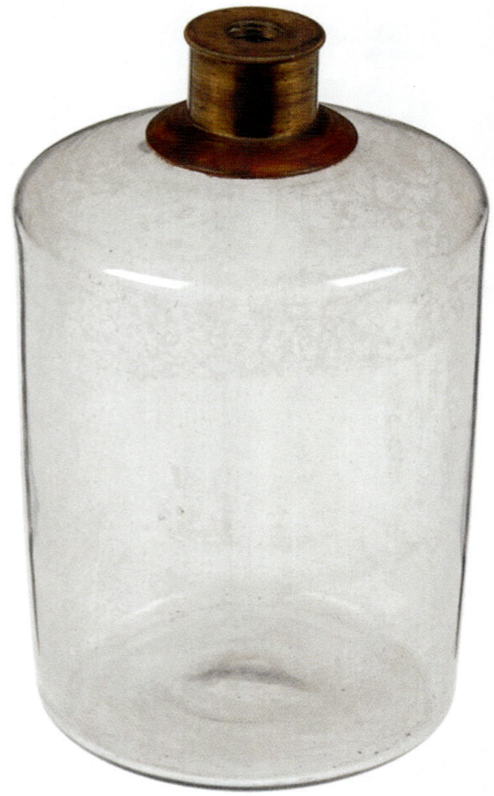

FIGURE 84 (inv. no. 19971-0000-)
© MAM/PHOTO FRANCK BOTTÉ

Cylindrical bell-jar with a brass collar at the top. A stopcock (missing) was originally screwed onto the collar. The instrument is similar to no. 19963-0000 and was used in the same way.

Thénard (1835), pp. 7 and 22

19964-0000- (Fig. 85)
Bell-jar
1780s
MAM accession date: 1952
34 × 14; 1.61 kg
Glass, brass

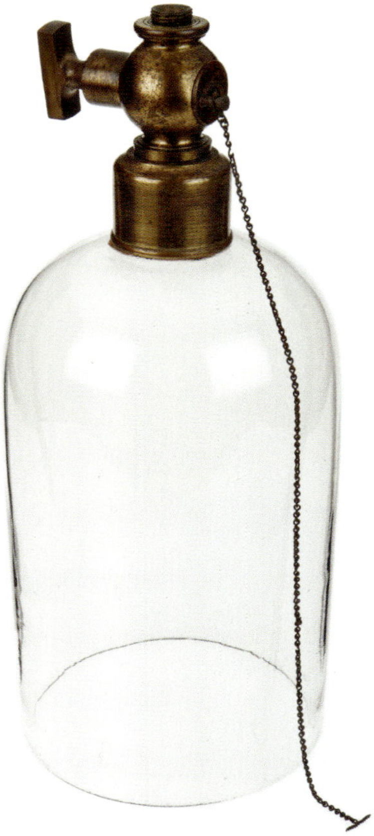

FIGURE 85 (inv. no. 19964-0000-)
© MAM/PHOTO FRANCK BOTTÉ

Cylindrical glass bell with a brass collar at the top. A stopcock is screwed onto the collar. A thin metal chain is attached to a peg inserted in the movable part of the cock. This bell-jar served to collect gases in a pneumatic trough and was used in the same way as no. 19963-0000-. An evacuated balloon could be screwed onto the stopcock and the gas filling the bell could be transferred into it.

Lavoisier (1789), vol. 2, plate V, figs. 10–12

19968-0000-, 19970-0000- (Fig. 86)
Bell-jars
Late 18th c.
MAM accession date: 1952
19968-0000-: 34 × 20.7; 1.38 kg
19970-0000-:28 × 21; 1.05 kg
Glass brass

19968-0000-: pear-shaped bell-jar with a brass collar at the top. A stopcock is screwed onto the collar. It was used with a pneumatic trough for collecting gas.

19970-0000-: Pear-shaped bell-jar with a brass collar at the top. A stopcock (missing) was originally screwed onto the collar. This bell jar is very similar to no. 19968-0000-.

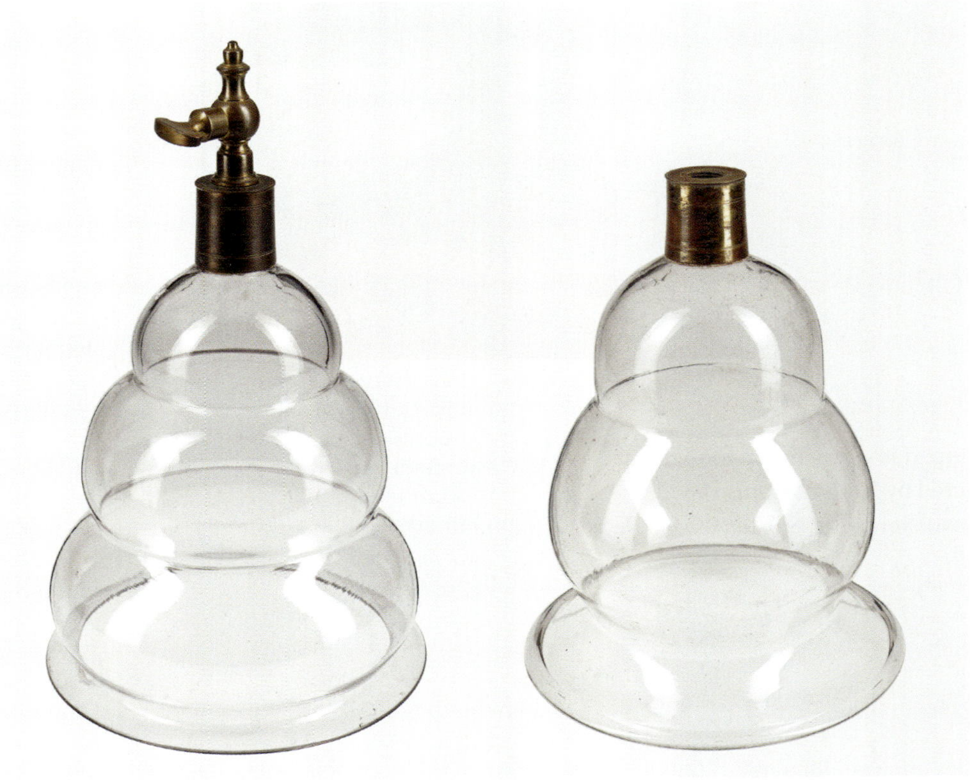

FIGURE 86
(inv. nos. 19968-0000-, 19970-0000-)
© MAM/PHOTO FRANCK BOTTÉ

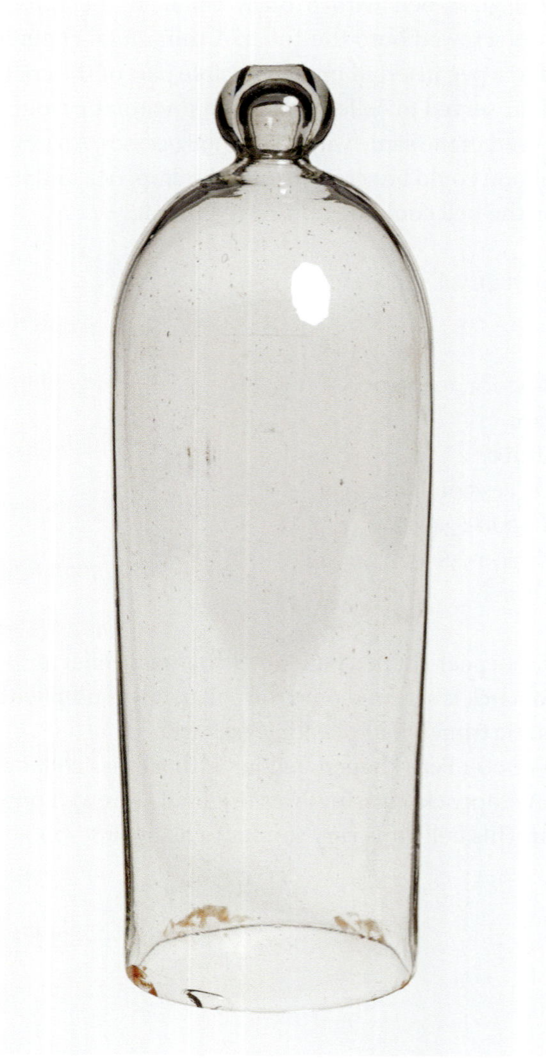

FIGURE 87A (inv. no. C-2017-0070-)
© MAM/PHOTO FRANCK BOTTÉ

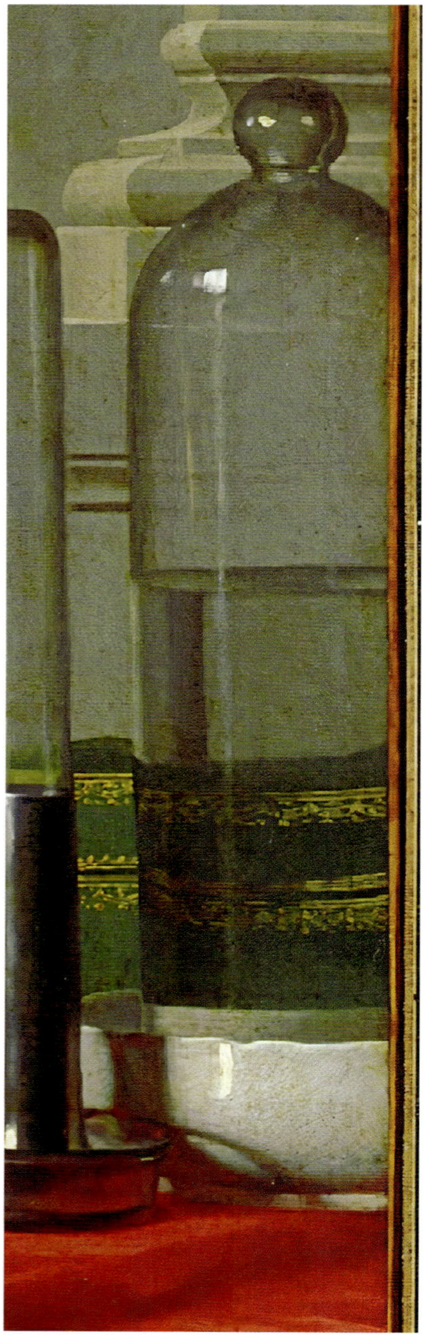

FIGURE 87B Detail of Jacques Louis David portrait of the Lavoisiers (1788)
COURTESY OF THE METROPOLITAN MUSEUM – NEW YORK

C-2017-0070- (Fig. 87a, 87b)
Bell jar
Second half of 18th c.
MAM accession date: 1864
28.7 × 9.2; 537 g
Glass

Glass bell jar with an knob on top. It is very similar, though smaller in size, to the one depicted by Jacques-Louis David in the double portrait of the Lavoisiers. (Fig. 87b)

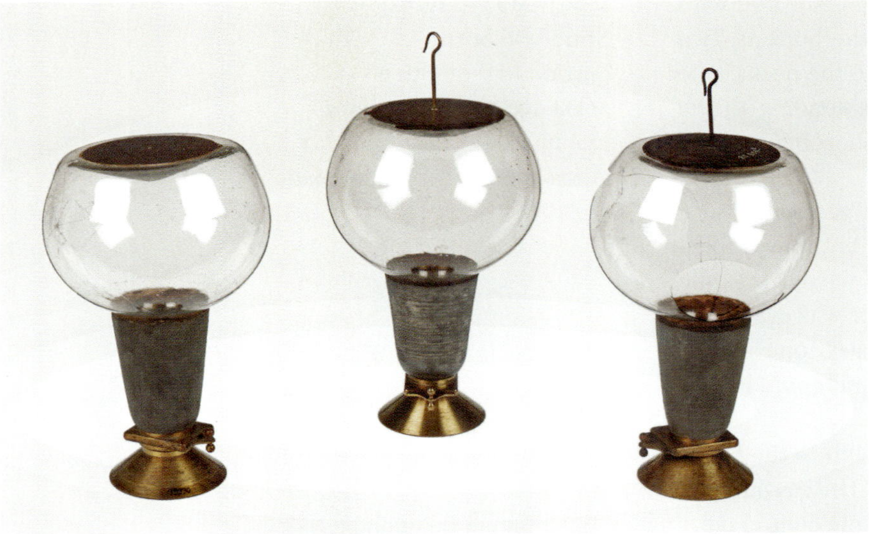

FIGURE 88 (inv. nos. 19990-0000-, 19991-0000-, 19992-0000-, 20206-0002-)
© MAM/PHOTO FRANCK BOTTÉ

19990-0000-, 19991-0000-, 19992-0000-, 20206-0002- (Fig. 88)
Three vessels for collecting gases
ca. 1770
MAM accession date: 1952
19990-0000-: 24 × 14; 1.73 kg
19991-0000-: 29 × 14: 1.88 kg
19992-0000-: 22.5 × 14; 1.54 k g
Glass, brass, lead, resin

Three nearly identical vessels. A brass funnel is joined to a flat square box with a sliding diaphragm. This is connected to the neck of a glass balloon. The neck is surrounded by a heavy lead collar. The concavity of the top of balloon 19990-0000- is filled with resin. A brass rod terminating in a hook is inserted in the resin at the top of balloon 19901-0000-. The balloon of 19992-0000- is broken and at its top there is a brass disk with a hook sealed to the glass with resin.[64]

There is no reference in Lavoisier's works to this kind of vessel, probably used with a pneumatic trough for collecting gases. The sliding diaphragm acted as a valve and the lead collar ensured the vessel's stability. The hook served to suspend the balloon to the balance arm in order to weigh the gas collected in it.

Truchot (1879), pp. 311–312; Leduc (1879), pp. 362–365

64 This brass piece was separated from the balloon and was wrongly inventoried as a single item as no. 20206-0002-.

20007-0000- (Fig. 89)
Glass vessel with three taps
Late 18th c.
MAM accession date: 1952
67.5 × 25 × 44; 6.42 kg
Glass, brass

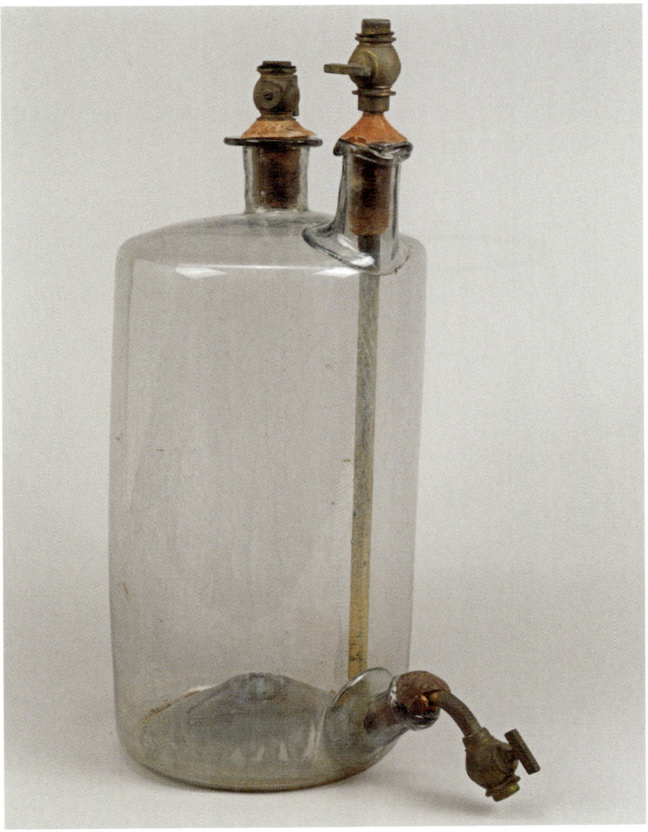

FIGURE 89 (inv. no. 20007-0000-)
© MAM/PHOTO FRANCK BOTTÉ

Cylindrical glass vessel similar to a Woulfe's bottle with two necks at the top and a third near the bottom. Two brass taps are inserted and cemented into the necks. One tap is connected to a long brass pipe extending to the bottom of the vessel. A third tap with a short bent pipe is inserted in the bottom neck.

The vessel may have been used as a gasometer or a gas-washing bottle. In the first case, the bottle was filled with water and the gas to be collected flowed into the vessel through the bottom tap. The water was evacuated through the vertical pipe. To use the gas, one simply refilled the vessel with water, so that the gas could be collected through the central tap.

To wash a gas, one simply introduced it into the bottle (filled with water) through the long pipe. The gas bubbled in the water and could be collected from the central tap. Lavoisier described a similar bottle without the bottom tap in his *Traité élémentaire de chimie*.

Lavoisier (1789), vol. 1, plate IV, fig. 14

20000-0000- (Fig. 90)
Large glass vessel
Second half of 18th c.
MAM accession date: 1952
38 × 20.5; 1.72 kg
Glass

Large pear-shaped vessels with two opposite openings of different diameters. The glass is cracked.

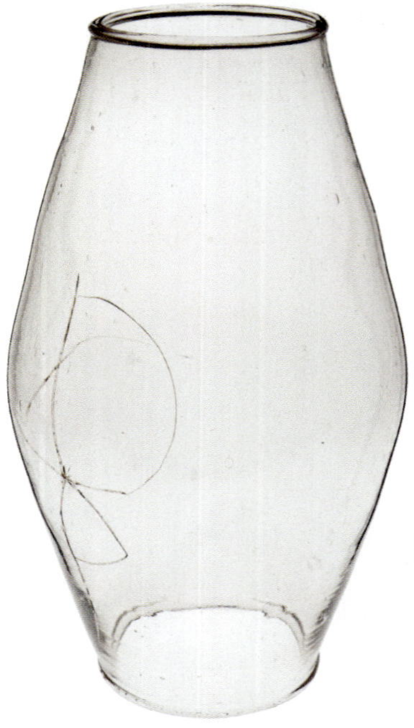

FIGURE 90 (inv. no. 20000-0000-) © MAM/PHOTO FRANCK BOTTÉ

19993-0000- (Fig. 91)
Spherical bottle
Second half of 18th c.
MAM accession date: 1952
21 × 16; 860 g
Glass, mercury, mercury oxide

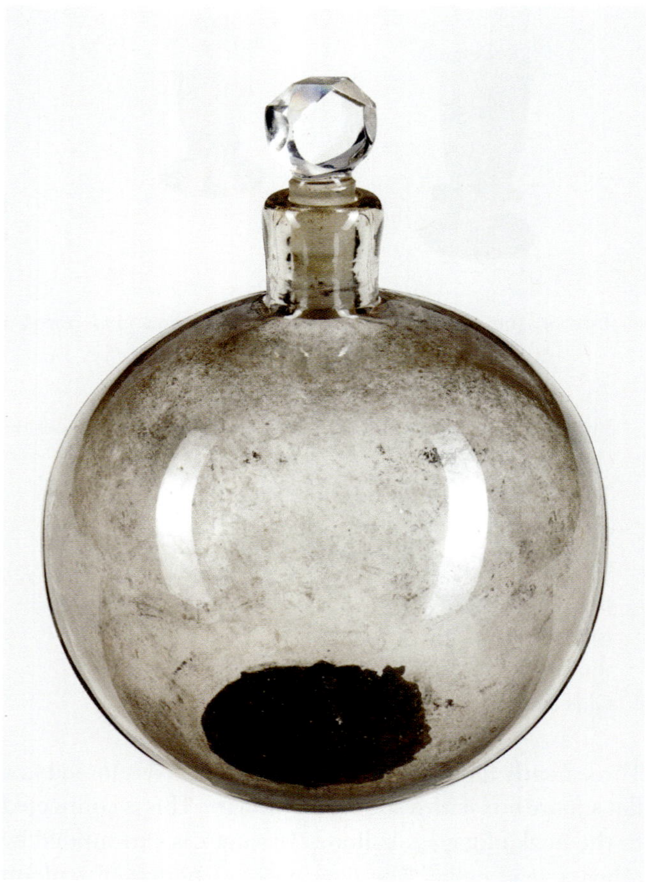

FIGURE 91 (inv. no. 19993-0000-) © MAM/PHOTO FRANCK BOTTÉ

Spherical glass vessel whose neck is closed by a ground-glass stopper. In reality, the stopper, which has a transverse hole, is the moving part of a glass tap and therefore does not belong to the bottle. The bottle contains a small amount of mercury (very soiled) and a solid grey substance (probably mercury oxide).

19977-0000- (Fig. 92)
Bottle with brass base
Late 18th c.
MAM accession date: 1952
25 × 13; 925 g
Glass, brass

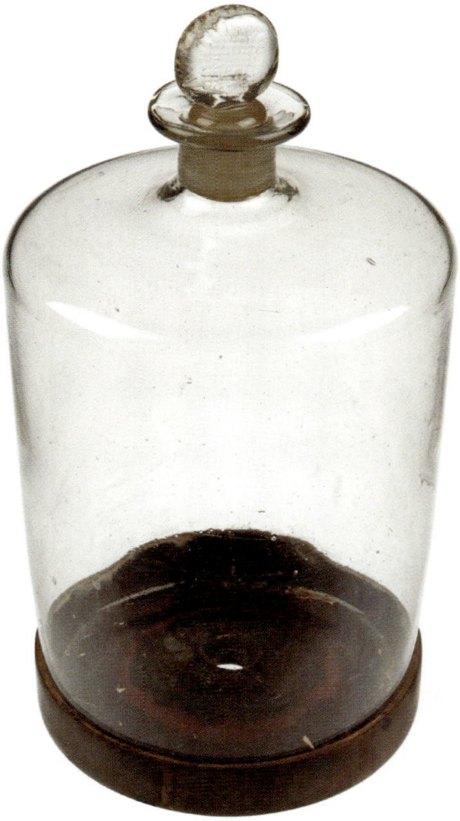

FIGURE 92 (inv. no. 19977-0000-)
© MAM/PHOTO FRANCK BOTTÉ

Cylindrical bottle fixed to a brass disk with salient ring. The bottom of the bottle has a perforation matching a threaded hole in the centre of the plate. The bottle neck is closed by a ground-glass stopper. The object was certainly part of an unidentified chemical or pneumatic apparatus.

McKie 81 (Fig. 93)
Bottle with brass base
Late 18th c.
Catalogued by Douglas McKie in 1952 but not included in the donation to the MAM
17 × 12
Glass, brass, wax

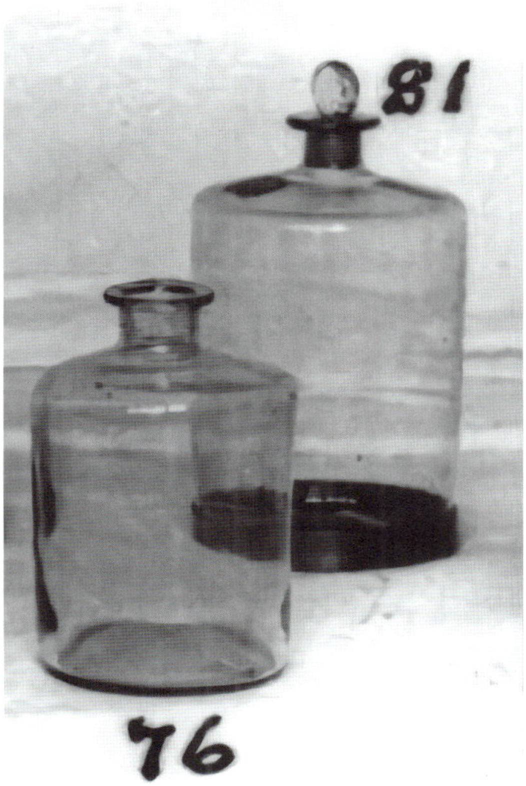

FIGURE 93 (inv. no. McKie 81)
PHOTO DOUGLAS MCKIE. COURTESY OF THE HAGLEY MUSEUM AND LIBRARY

Cylindrical glass bottle with ground-glass stopper. Lower end bound with brass ring. Brass base-piece with central hole and fixed with wax.

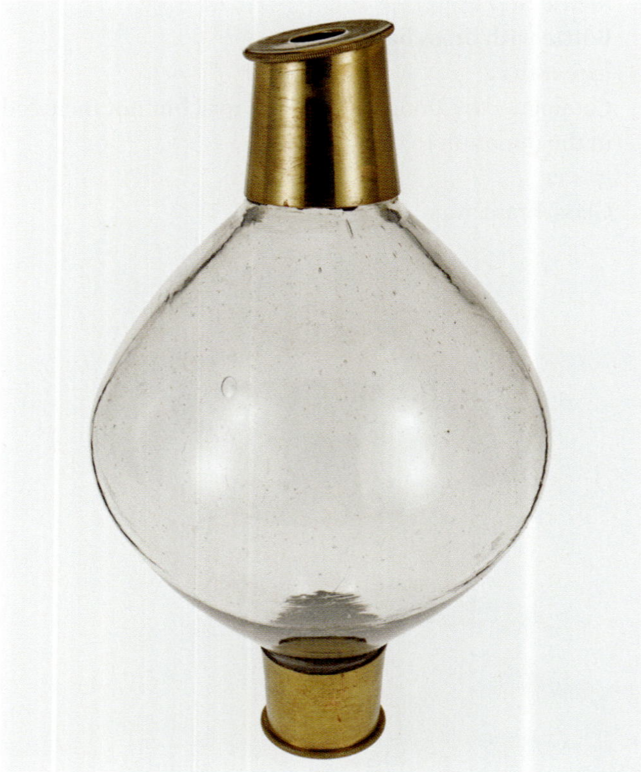

FIGURE 94 (inv. no. 19988-0000-)
© MAM/PHOTO FRANCK BOTTÉ

19988-0000- (Fig. 94)
Glass balloon
Second half of 18th c.
MAM accession date: 1952
21.5 × 12.5; 600 g
Brass, glass, traces of resin

Thick glass balloon with two necks, each enclosed in a brass cap with a threaded hole. The top of one of the caps is slightly inclined with respect to the vessel's symmetry axes. The balloon rests on a straw ring. It was certainly part of a more complex instrument.

C-2017-0068- (Fig. 95)
Glass balloon with three necks
Second half of 18th c.
MAM accession date: 1864–1866
31 × 23 × 5; 2 kg
Glass

Large green glass balloon with two diametrical necks and a third, smaller one perpendicular to them. The balloon rests on a straw ring.

Lavoisier (1789), vol. 2, plate III, fig. 22

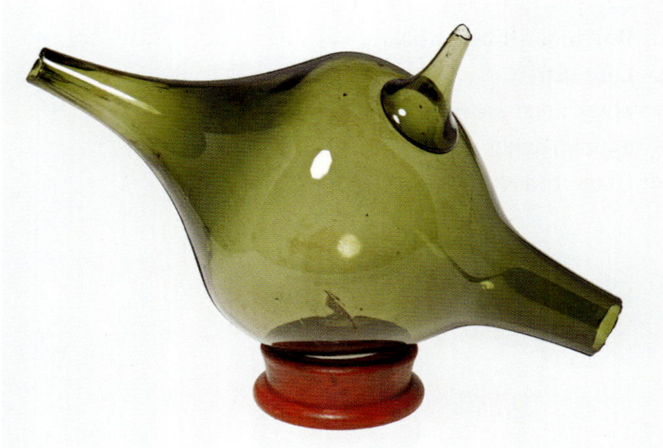

FIGURE 95 (inv. no. C-2017-0068-)
© MAM/PHOTO FRANCK BOTTÉ

19987-0000- (Fig. 96)
Glass balloon with three necks
Late 18th c.
MAM accession date: 1952
29 × 20 × 19.5; 1.16 kg
Glass

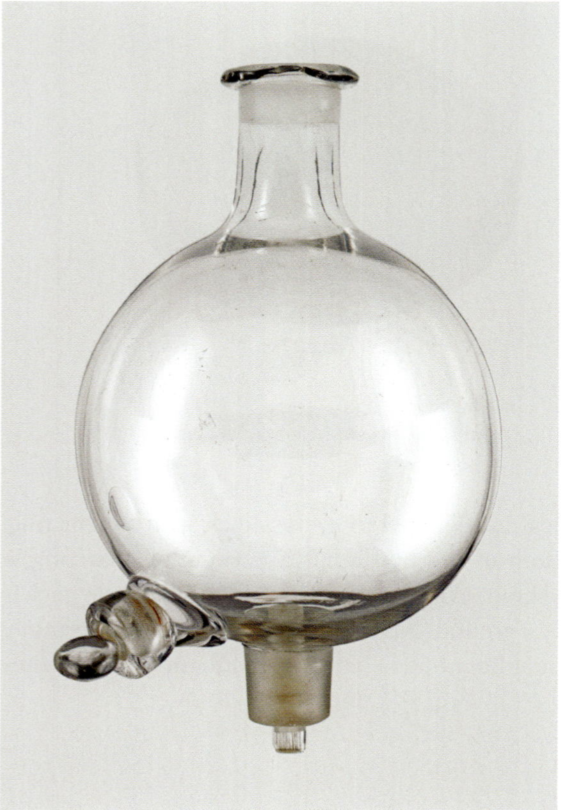

FIGURE 96 (inv. no. 19987-0000-)
© MAM/PHOTO FRANCK BOTTÉ

Glass balloon with three necks. One is longer than the other two, which are each closed by a ground-glass stopper. One of the stoppers has a small hole. The balloon was part of an unidentified apparatus. It stands on a straw ring covered in silk ribbon, together with inventory no. C-2017-0074-.

19989-0000- (Fig. 97)
Glass balloon
Late 18th c.
MAM accession date: 1952
29 × 16; 655 g
Brass, glass

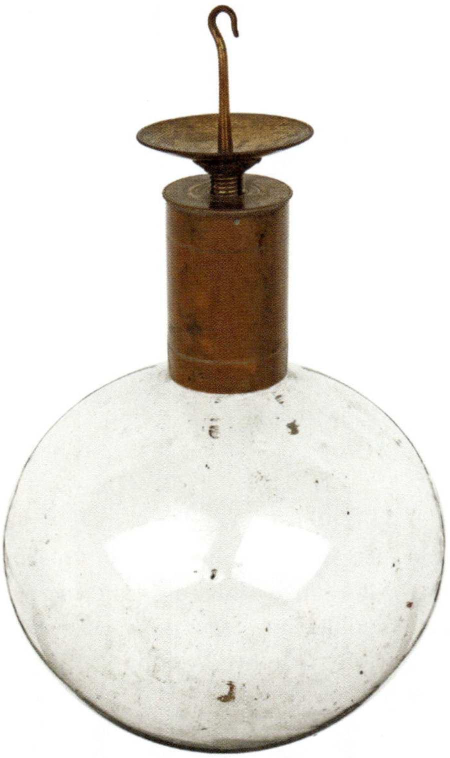

FIGURE 97 (inv. no. 19989-0000-)
© MAM/PHOTO FRANCK BOTTÉ

Glass balloon with a neck surrounded by a brass collar. A tap with a hook and a small conical funnel is screwed onto the neck. The tap does not seem to fit properly in the hole. The balloon was probably used for weighing gases and was therefore suspended from the arm of a precision balance.

19995-0000- (Fig. 98)
Glass balloon with cup
Second half of 18th c.
MAM accession date: 1952
22.5 × 11.5; 570 g
Glass

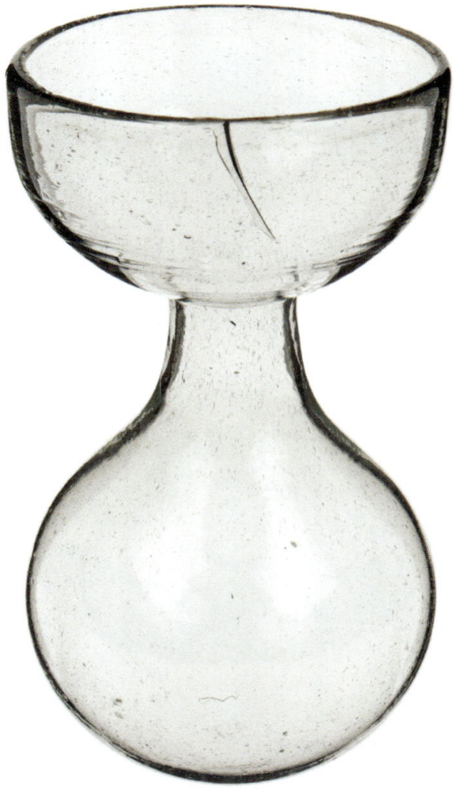

FIGURE 98 (inv. no. 19995-0000-)
© MAM/PHOTO FRANCK BOTTÉ

Thick glass balloon with a cup-shaped neck that may have been used as a funnel.

19986-0001-, 19986-0002- (Fig. 99)
Broken glass balloon
Second half of 18th c.
MAM accession date: 1952
42 × 35; 2.35 kg
Glass, brass, straw, silk

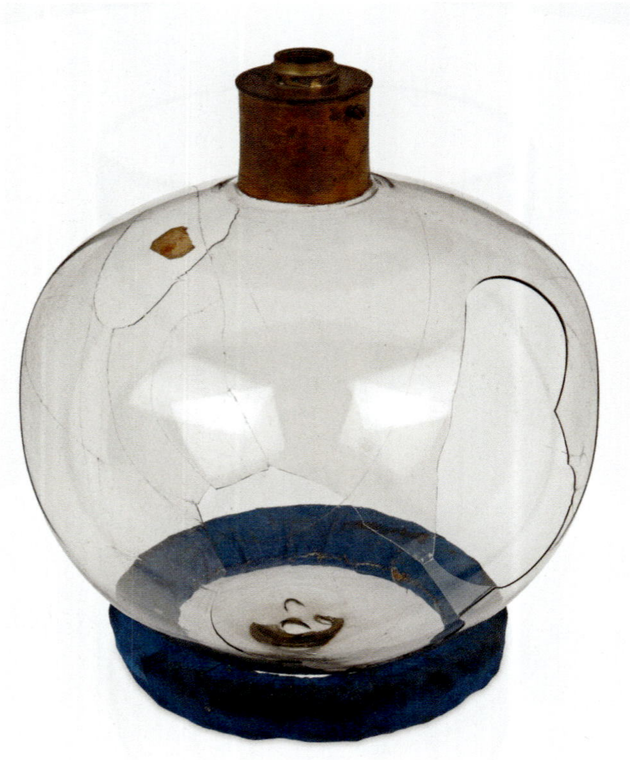

FIGURE 99 (inv. no. 19986-0001-, -0002)
© MAM/PHOTO FRANCK BOTTÉ

Large broken glass balloon whose neck is surrounded by a brass collar. The vessel was shattered after its accession by the museum. The largest fragment rests on a straw ring (19986-0002-) covered with blue silk. Originally, it was probably used for weighing gases.

19978-0000- (Fig. 100)
Leech jar
Late 18th c.
MAM accession date: 1952

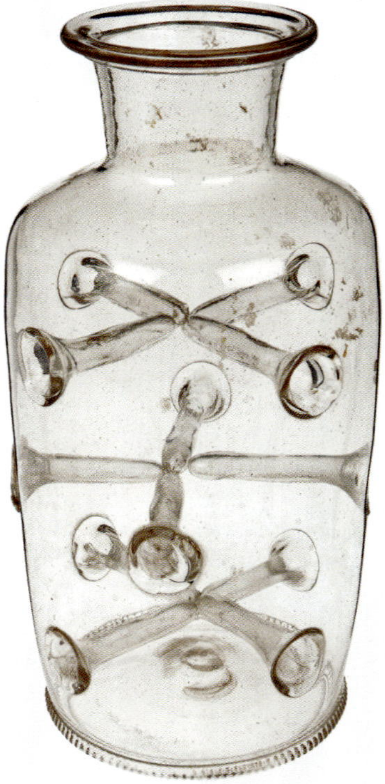

FIGURE 100 (inv. no. 19978-0000-)
© MAM/PHOTO FRANCK BOTTÉ

This jar was used for housing leeches. A species of blood-sucking worms, leeches have been used since ancient times to treat a wide range of diseases and medical conditions through bloodletting. It was during the nineteenth century that they were employed by physicians on a massive scale.

The jar has twelve inward-pointing hollow glass tubes and the bottom is decorated with a rim. The leeches could attach themselves to the tubes, while the opening at the top ensured a supply of fresh air.

Similar bottles, of later manufacture, may be found at the Science Museum in London (A651203) and in the collection of the Collegio Carlo Alberto in Moncalieri.

20001-0000- (Fig. 101)
Chilling carafe
Second half of 18th c.
MAM accession date: 1952
29 × 19; 1.03 kg
Glass

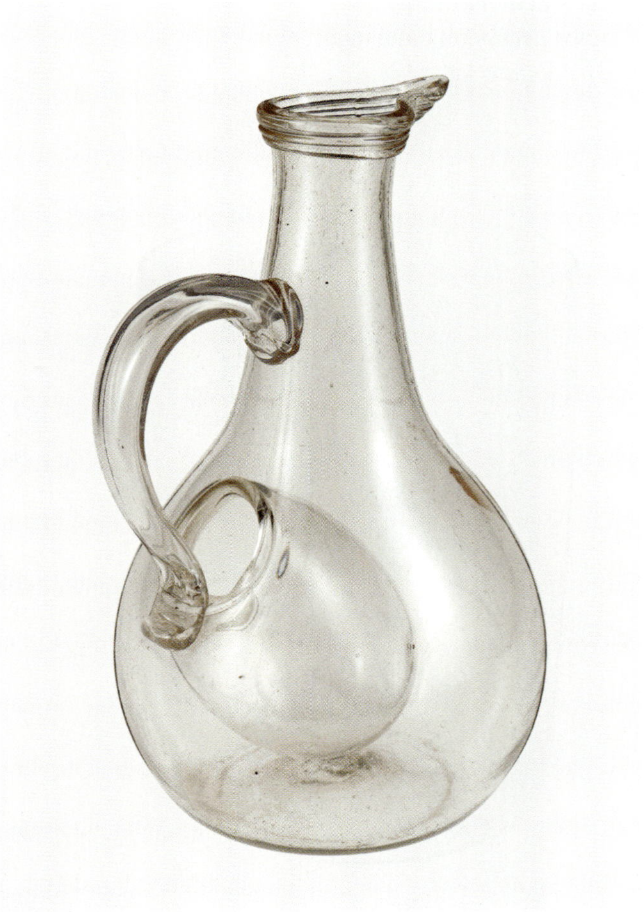

FIGURE 101 (inv. no. 20001-0000-)
© MAM/PHOTO FRANCK BOTTÉ

Glass carafe with a long neck and a handle. A lateral hole communicates with a rounded vessel inside the carafe. The liquid in the carafe is chilled by introducing crushed ice into the vessel.

19997-0000- (Fig. 102)
Thermally insulated glass flask
Late 18th c.
MAM accession date: 1952
36 × 30.5 × 24.5; 2.66 kg
Glass, cork, wood

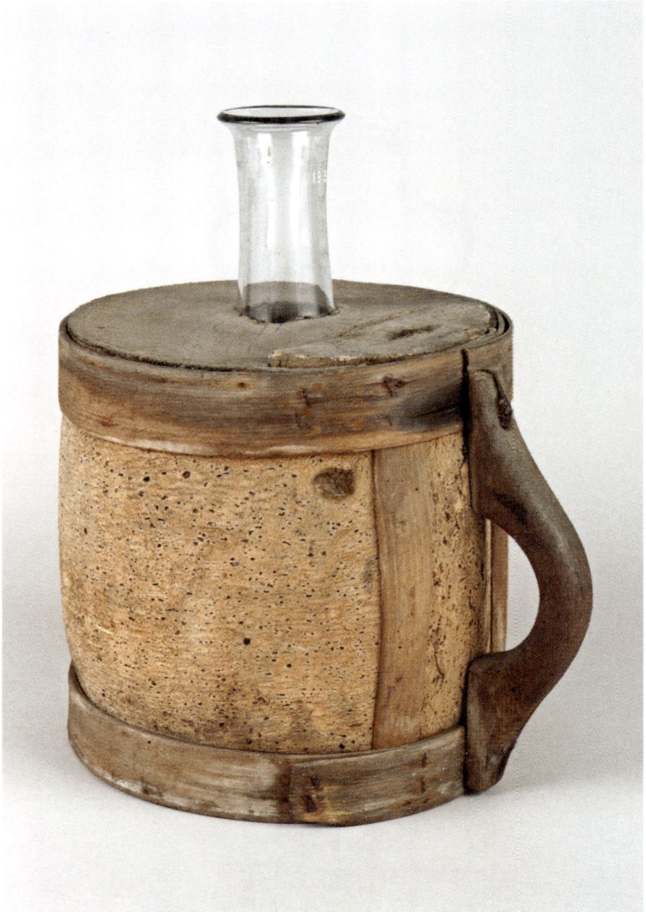

FIGURE 102 (inv. no. 19997-0000-)
© MAM/PHOTO FRANCK BOTTÉ

Glass bottle enclosed in cork insulation, surrounded by two wooden rings connected by a wooden handle. A part of the cork at the top can be removed to inspect the bottle. The cork insulation helped to keep the temperature of the liquid in the bottle constant.

20027-0000- (Fig. 103)
Long phial
Second half of 18th c.
MAM accession date: 1952
37.5 × 7.5; 325 g
Glass

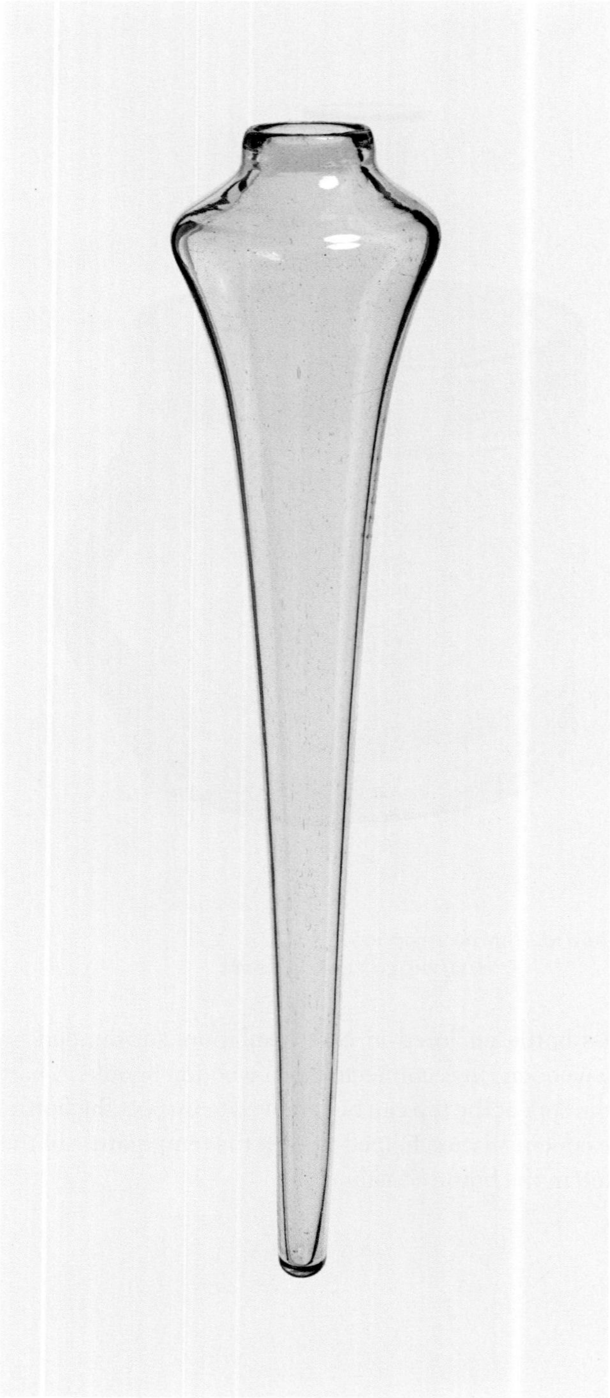

FIGURE 103 (inv. no. 20027-0000-)
© MAM/PHOTO FRANCK BOTTÉ

Long, narrow glass vessel in the form of a conical phial with an opening at the top. Its function is not clear. It may have been used as a decanter.

19984-0000- (Fig. 104)
Beaker
Second half of 18th c.
MAM accession date: 1952
27 × 20; 1.05 kg

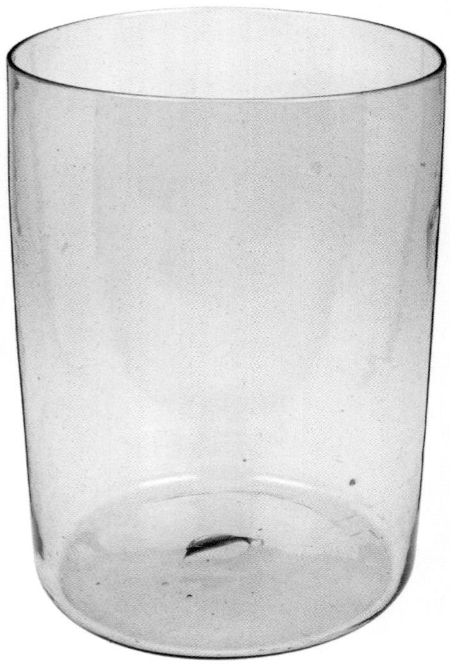

FIGURE 104 (inv. no. 19984-0000-)
© MAM/PHOTO FRANCK BOTTÉ

Large beaker commonly used in chemical laboratories.

19980-0000-, 19981-0000- (Figs. 105a, 105b)
Two beakers
Second half of 18th c.
MAM accession date: 1952
15 × 12.5; 370 g (19980-0000-) and 335 g (19981-0000-)
Glass

Two almost identical cylindrical glass vases.

19985-0000- (Fig. 106)
Beaker
Late 18th c.
MAM accession date: 1952
26 × 12; 560 g
Glass

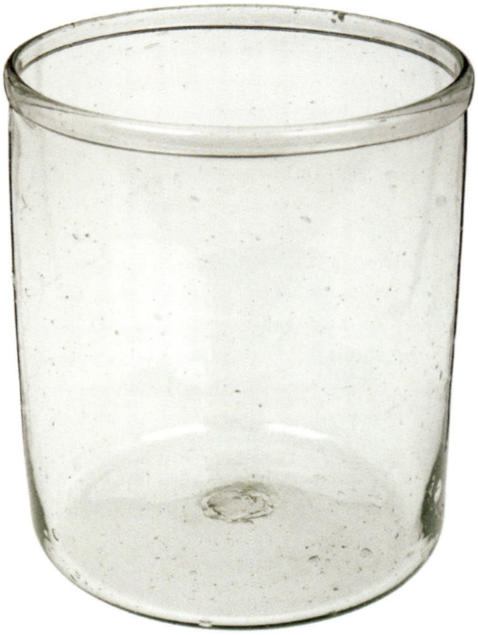

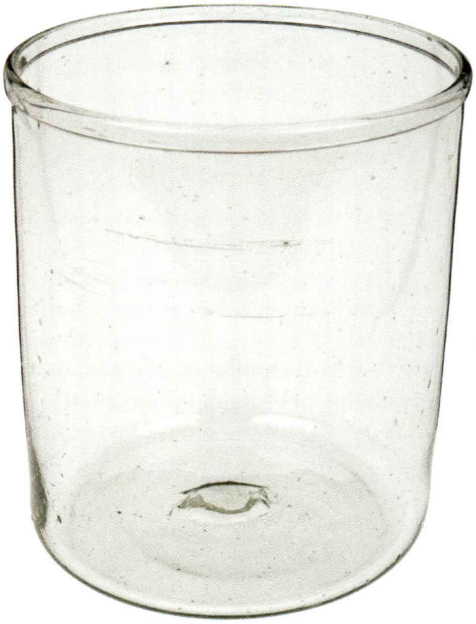

FIGURES 105A, 105B (inv. nos. 19980-0000-, 19981-0000-)
© MAM/PHOTO FRANCK BOTTÉ

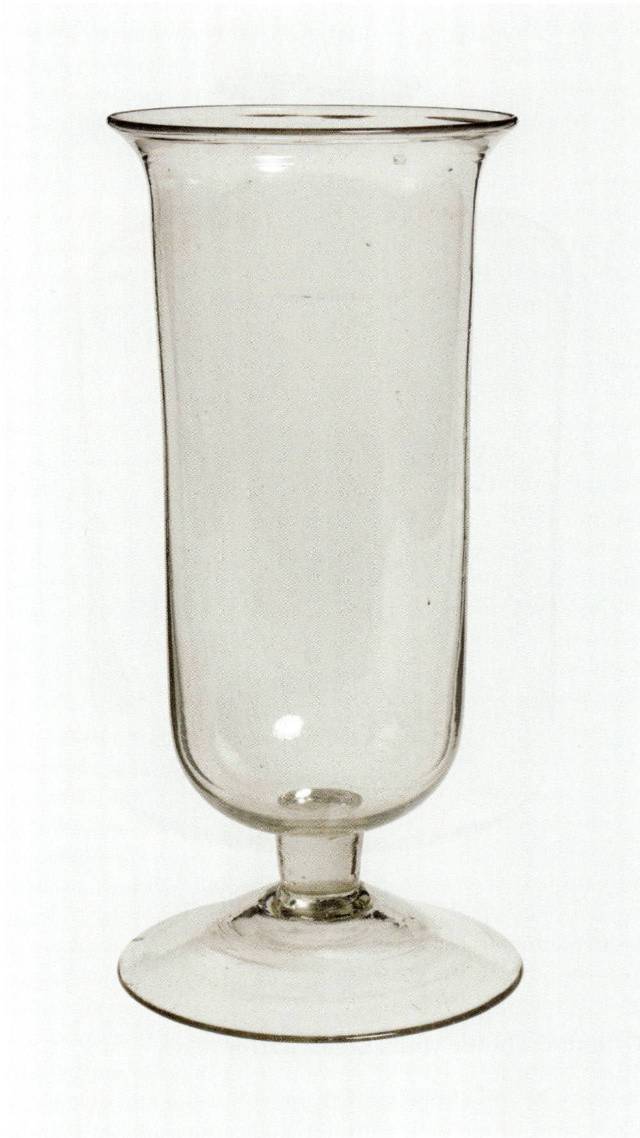

FIGURE 106 (inv. no. 19985-0000-)
© MAM/PHOTO FRANCK BOTTÉ

Tulip-shaped glass vessel (beaker) mounted on a short stem attached to a round base.

19979-0000- (Fig. 107)
Glass bottle
Second half of 18th c.
MAM accession date: 1952
17 × 10; 450 g
Glass

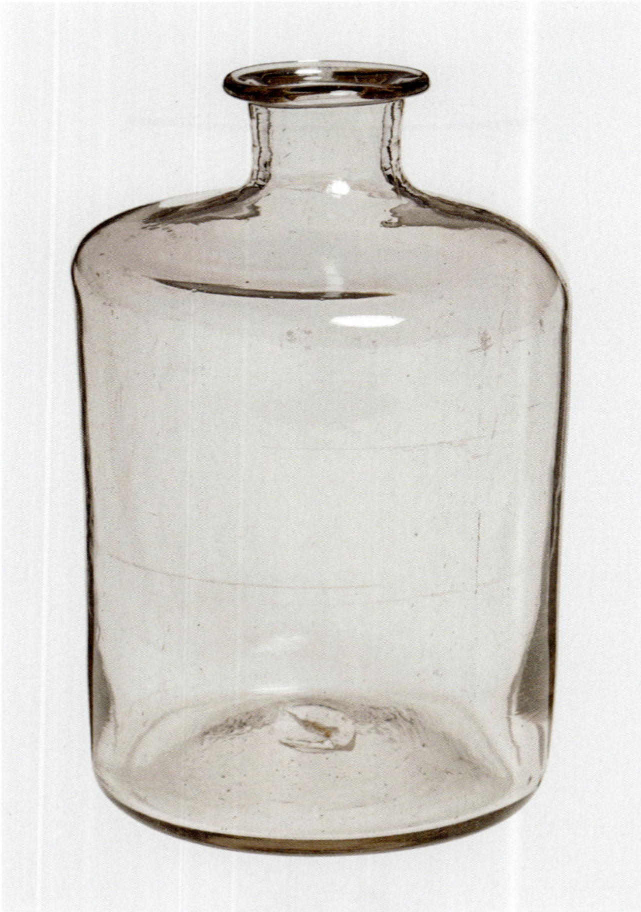

FIGURE 107 (inv. no. 19979-0000-)
© MAM/PHOTO FRANCK BOTTÉ

Cylindrical bottle with a short, narrow neck.

19982-0000- (Fig. 108)
Glass bottle
Late 18th c.
MAM accession date: 1952
18 × 10.5; 540 g
Glass, brass

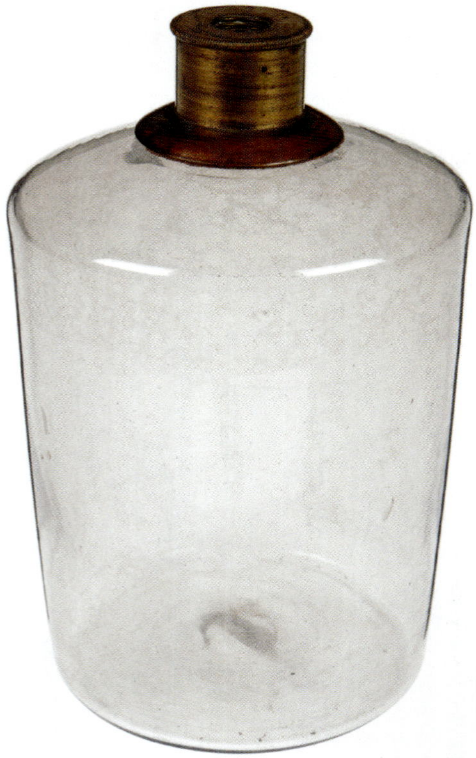

FIGURE 108 (inv. no. 19982-0000-)
© MAM/PHOTO FRANCK BOTTÉ

Cylindrical glass bottle. Its neck has a brass collar with a threaded hole. Probably part of an unidentified apparatus.

20041-0000- (Fig. 109)
Bowl
Second half of 18th c.
MAM accession date: 1952
4.5 × 10.5; 70 g
Wood

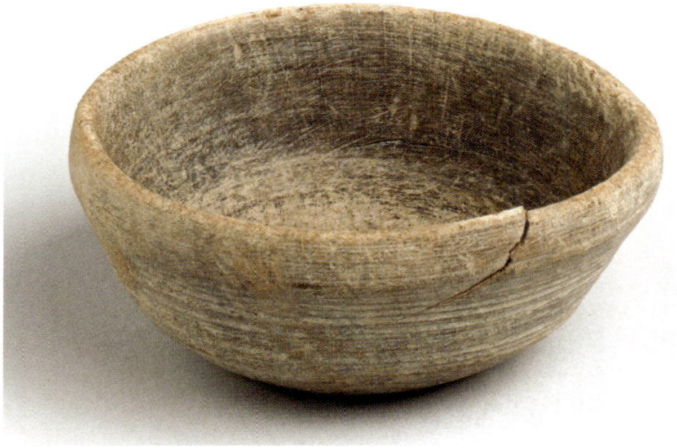

FIGURE 109 (inv. no. 20041-0000-)
© MAM/PHOTO FRANCK BOTTÉ

Small wooden bowl with a crack.

20023-0001-, 20023-0002-, 20023-0003-, 20023-0004-, 20023-0005- (Fig. 110)
Set of 5 test tubes
Second half of 18th c.
MAM accession date: 1952
20023-0001-: 11 × 1.2; 10 g
20023-0002-: 11.2 × 1; 10 g
20023-0003-: 10.7 × 1.4; 15 g
20023-0004-: 12.5 × 1.4; 20 g
20023-0005-: 8.5 × 1.9; 10 g
Glass

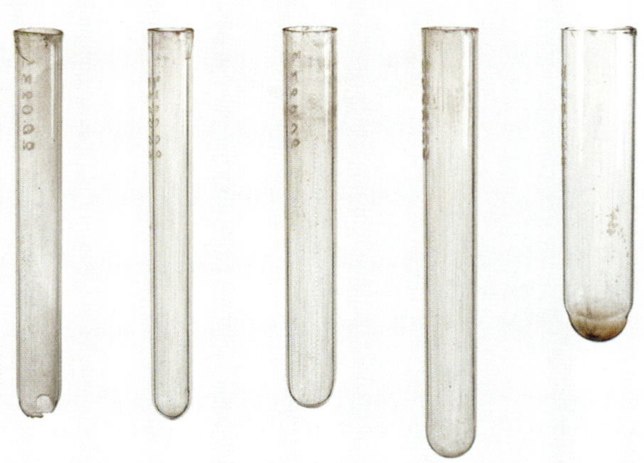

FIGURE 110 (inv. nos. 20023-0001-, -0005-)
© MAM/PHOTO FRANCK BOTTÉ

Set of five test tubes of thick glass. The bottom of tube 2023-0001- is broken.

20031-0001-, 20031-0002- (Fig. 111)
Set of 2 glass tubes in a wooden cylinder
Second half of 18th c.
MAM accession date: 1952
20031-0001-: 14 × 2.5; 40 g
20031-0002-: 14.5 × 2.5; 30 g
Glass, wood, paper

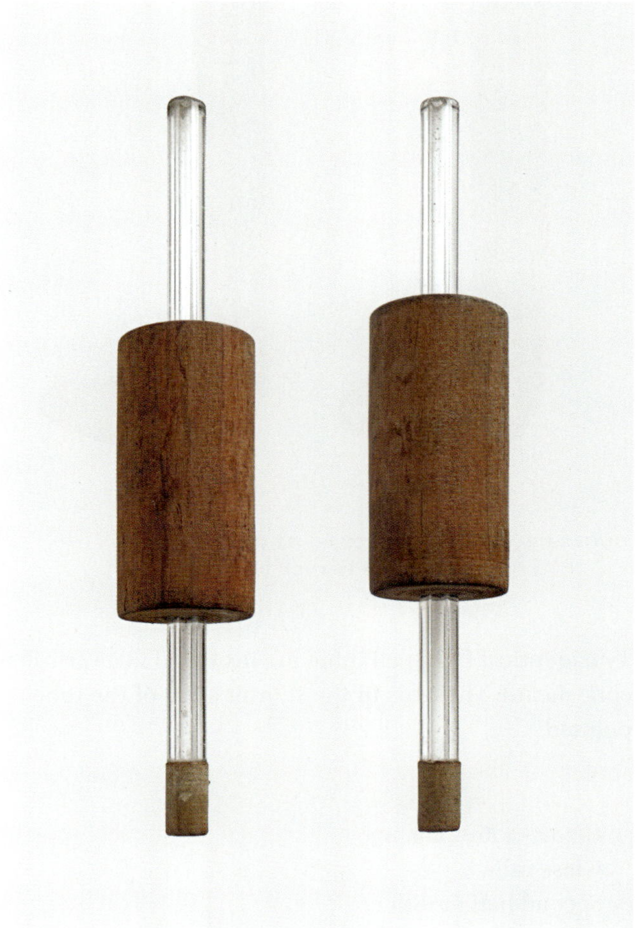

FIGURE 111 (inv. nos. 20031-0001-, -0002-)
© MAM/PHOTO FRANCK BOTTÉ

Two very similar glass tubes, each inserted into a shorter wooden cylinder. One end of each tube is wrapped in a paper strip.

20032-0001-, 20032-0002- (Fig. 112)
Set of 2 insulated glass tubes
Second half of 18th c.
MAM accession date: 1952
20032-0001-: 14.7 × 6 × 2.5; 20 g
20032-0002-: 15 × 6.5 × 2.5; 15 g
Glass, cork

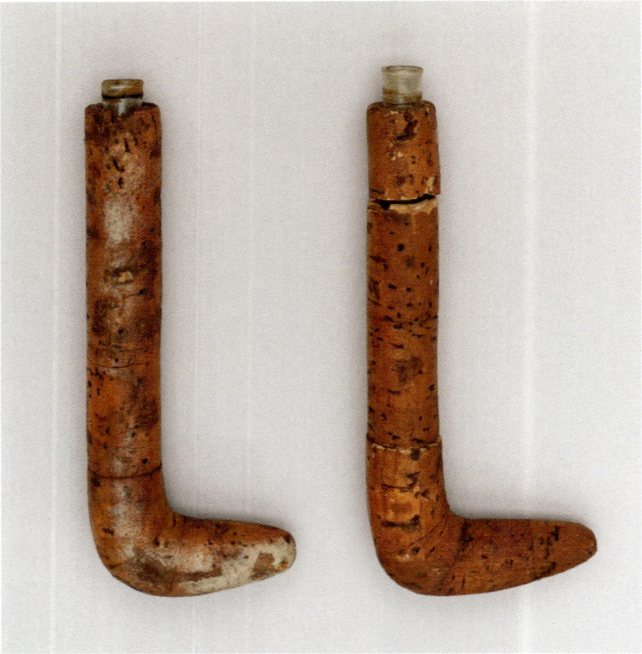

FIGURE 112 (inv. nos. 20032-0001-, -0002-)
© MAM/PHOTO FRANCK BOTTÉ

Two identical L-shaped tubes are insulated in a varnished cork sheath. The ends of the shorter arms of the tube are pointed.

C-2017-0061- (Fig. 113)
Glass tube
Second half of 18th c.
MAM accession date: 1952
16 × 2.5; 25 g
Glass

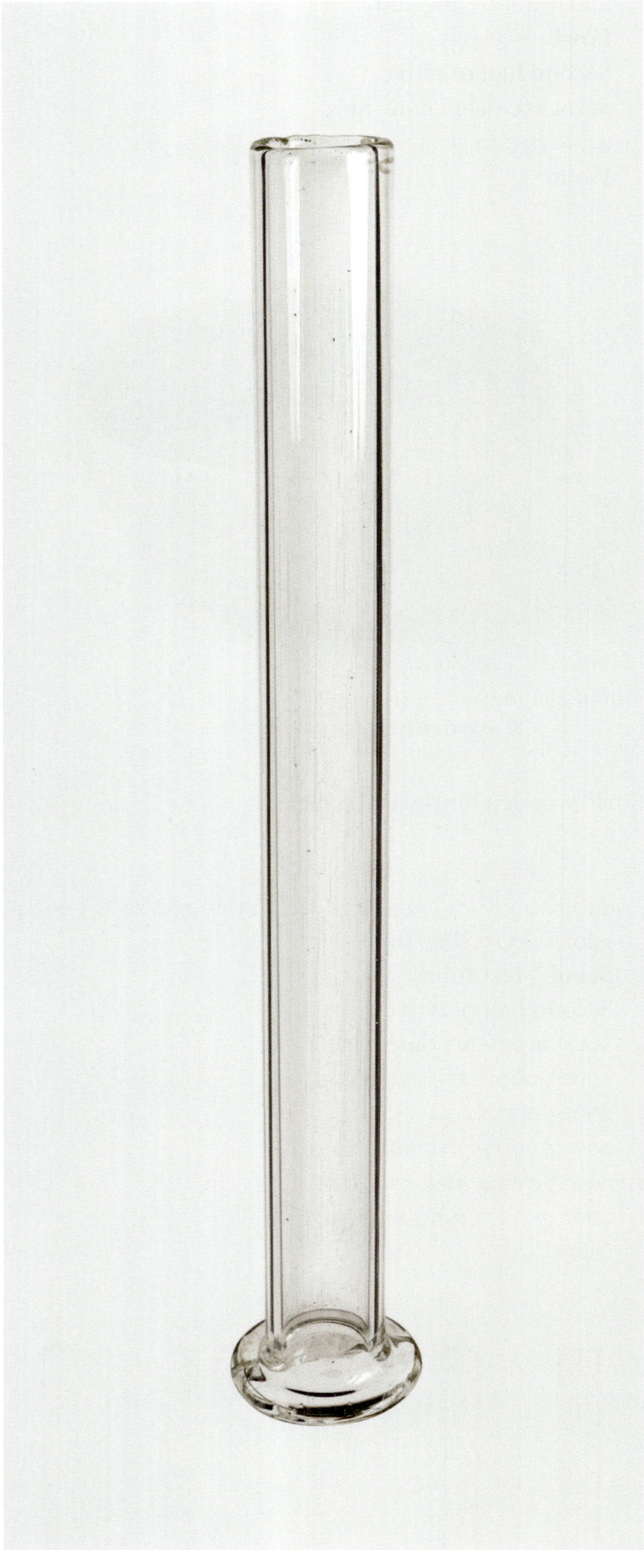

FIGURE 113 (inv. no. C-2017-0061-)
© MAM/PHOTO FRANCK BOTTÉ

Glass tube with a handwritten label: "Lavoisier ss n°".

C-2017-0062- (Fig. 114)
Glass tubes
Second half of 18th c.
MAM accession date: 1952
67 × 1.8 × 1.1; 75 g (dimensions for each tube)
Glass

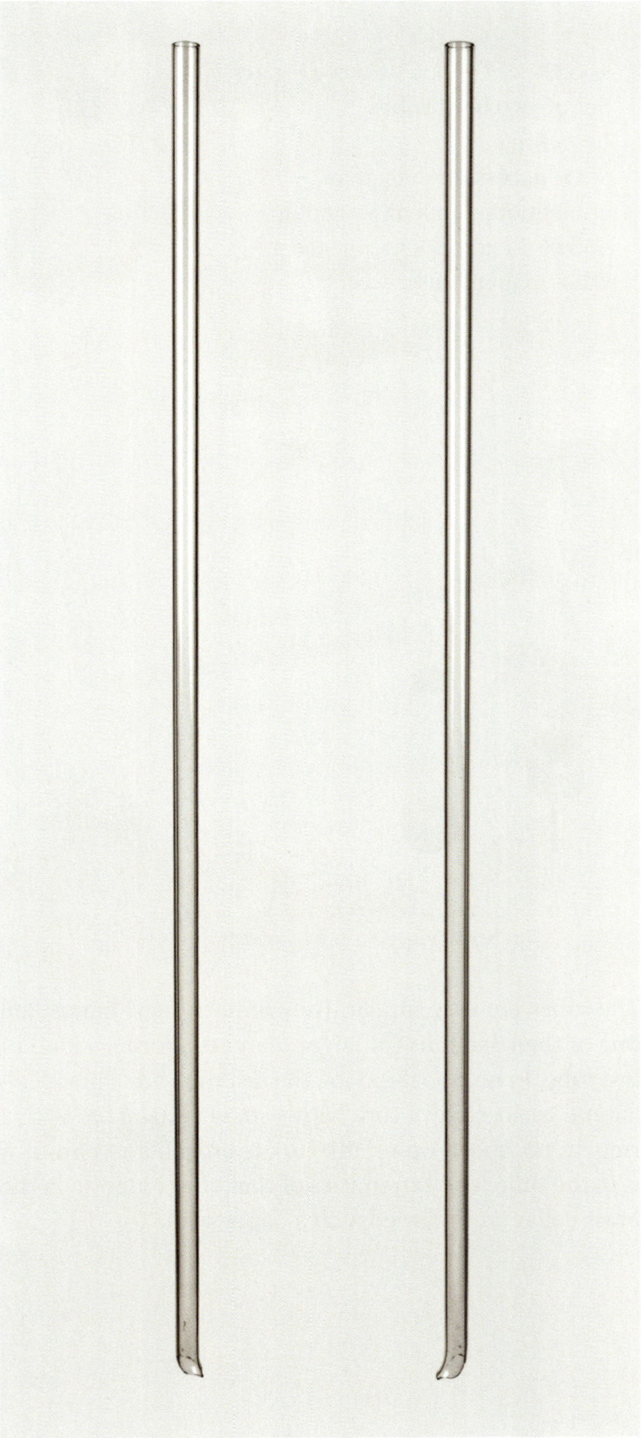

FIGURE 114 (inv. no. C-2017-0062-)
© MAM/PHOTO FRANCK BOTTÉ

Two identical glass tube with a handwritten label: "Lavoisier ss n°".

McKie 380
Boiling tube
Second half of 18th c. (?)
Catalogued by Douglas McKie in 1952 but not included in the donation to the MAM
12 × 2.3
Glass

Boiling tube, poorly blown.

20037-0002-, 20037-0003-, 20037-0004- (Fig. 115)
Three glass stopcocks
Last quarter of 18th c.
MAM accession date: 1952
20037-0002-: 10 × 7.7 × 4; 195 g
20037-0003-: 11 × 6.5 × 4.5; 240 g
20037-0004-: 9 × 6 × 3.7: 135 g

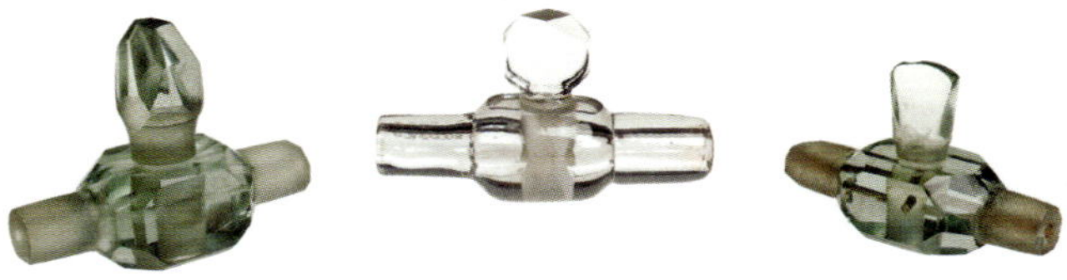

FIGURE 115 (inv. nos. 20037-0002-, -0004-)
© MAM/PHOTO FRANCK BOTTÉ

Three stopcocks made entirely of bevelled glass.

20006-0001-, 20006-0002- (Fig. 116)
Test tube
Late 18th c.
MAM accession date: 1952
67 × 7.5; 2.2 kg
Glass, brass

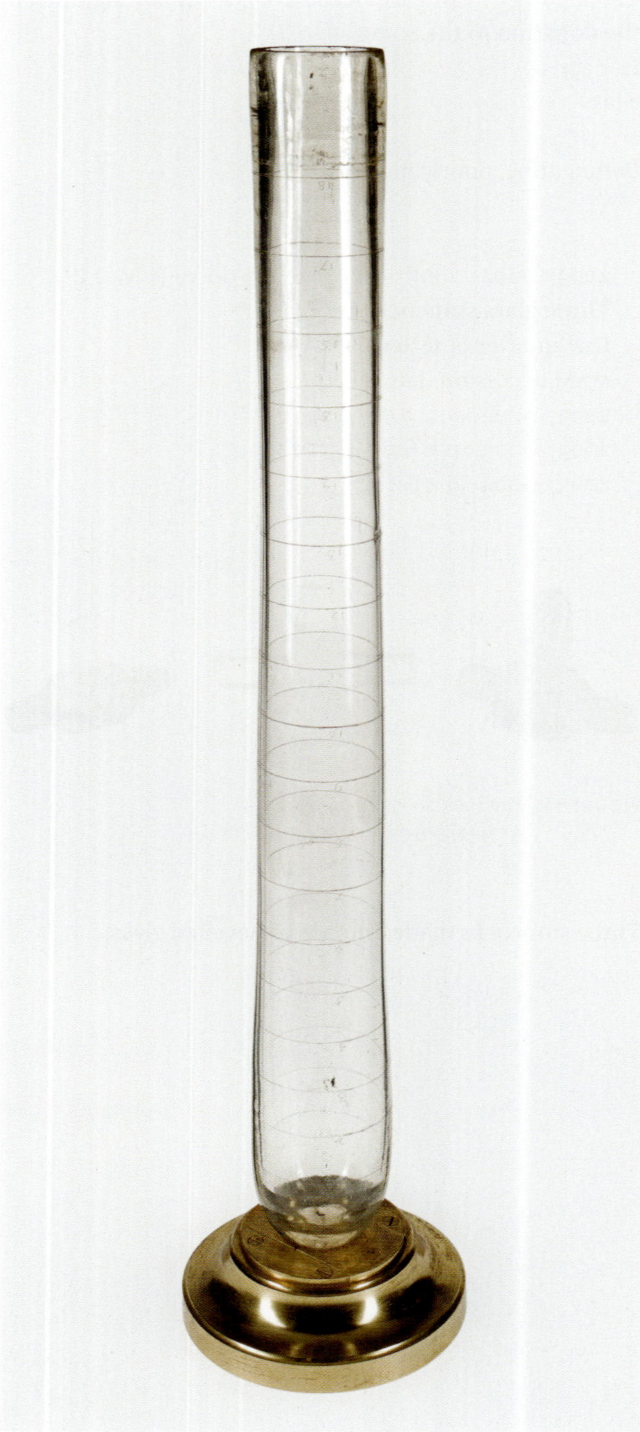

FIGURE 116 (inv. nos. 20006-0001-, -0002-)
© MAM/PHOTO FRANCK BOTTÉ

Large test tube with a scale with 18 engraved divisions[65] ranging from the top to the open bottom. The tube rests on a round brass base.[66] Its central hole is formed by two brass plates with semi-circular cuts. The distance between the plates can be slightly adjusted with a pair of screws. The tube was probably used in a pneumatic trough for collecting gases.

20028-0001-, 20028-0002- (Fig. 117)
Set of two bent tubes
Late 18th c.
MAM accession date: 1952
20028-0001-: 40 × 32 × 5; 275 g
20028-0002-: 50 × 32 × 5; 300 g
Glass, paper, brass, wax

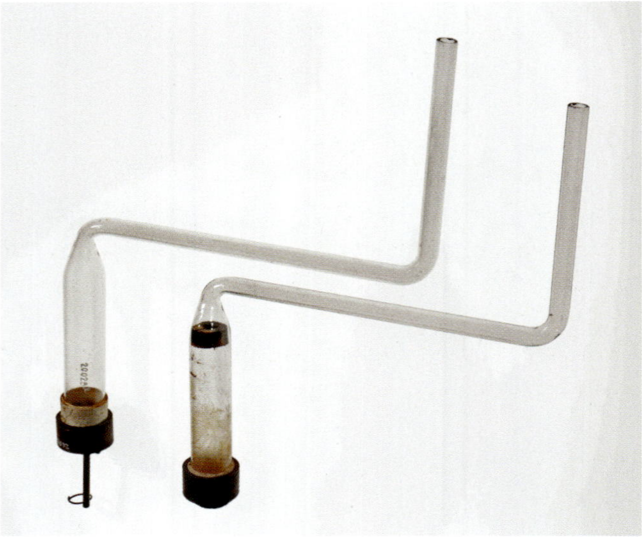

FIGURE 117 (inv. nos. 20028-0001-, -0002-)
© MAM/PHOTO FRANCK BOTTÉ

The tubes are very similar. They have two 90° bends and one of their sections (of larger diameter) forms a kind of test tube. In no. 20028-0000-, the test tube has a brass collar and is closed by a cork with a small vertical bar with a ring. In no. 20028-0001-, the cork is broken and part of it is in the tube. The extremities of the tubes entering in the brass collar are wrapped with a paper strip.

65 The divisions are circles engraved on the tube.
66 The brass base was separated from the test tube and catalogued under the number 20006-0002-.

20164-0000- (Fig. 118)
Case with pipettes
Late 18th c.
MAM accession date: 1952
14 × 3; 50 g
Wood, glass, resinous substance

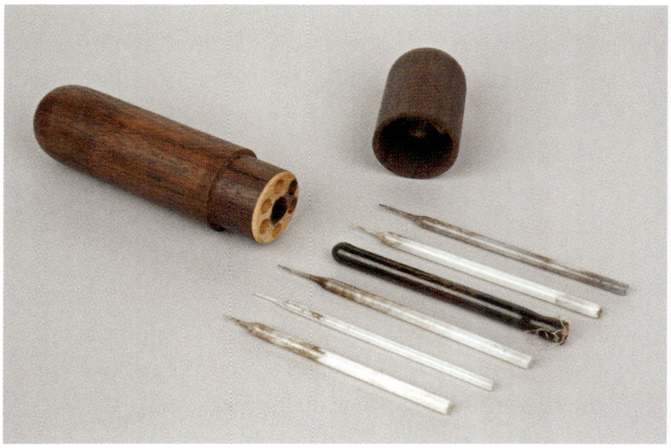

FIGURE 118 (inv. no. 20164-0000-)
© MAM/PHOTO FRANCK BOTTÉ

Four glass pipettes (for milk?) and a small test tube are inserted in the cavities of a solid cylindrical wooden case with rounded ends and cap. The test tube is sealed with wax and contains a brown, seemingly resinous substance. The box has a printed label with the words "Exposition Lavoisier // Palais de la découverte".

Lavoisier (1943)

20037-0005- (Fig. 119)
Glass cap
Second half of 18th c.
MAM accession date: 1952
6.6 × 4.5; 75 g

Conical blown-glass cap for an unidentified vessel.

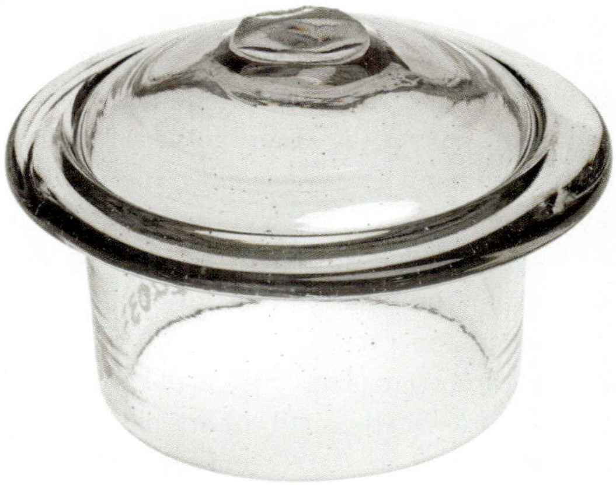

FIGURE 119 (inv. no. 20037-0005-)
© MAM/PHOTO FRANCK BOTTÉ

Chemicals, Minerals and Various Substances

20185-0000- (Fig. 120)
Box with samples of various substances
Mostly related to experiments conducted in 1782–1783
MAM accession date: 1952
20.7 × 9.5 × 6, mass 770 g
Wood, paper, minerals, pigments

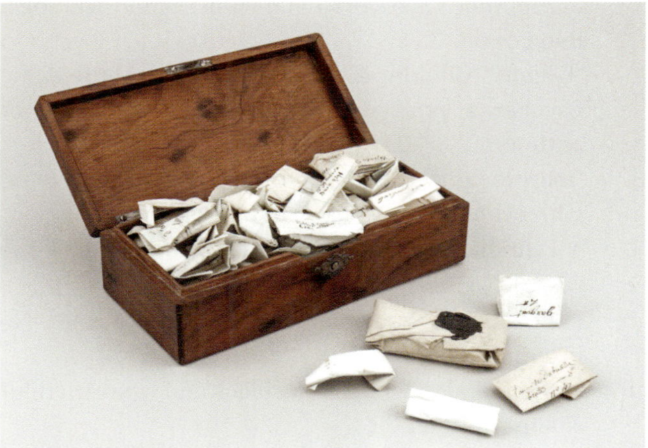

FIGURE 120 (inv. no. 20185-0000-)
© MAM/PHOTO FRANCK BOTTÉ

Wooden box with a hinged lid and a lock containing 66 different specimens of metals, alloys, minerals and precious stones, many of which were subjected to violent heat. Most are wrapped in pieces of paper bearing handwritten inscriptions, many of which are in Marie Anne Lavoisier's hand. Many of these stones were used by Lavoisier in

June–December 1782 and again in January 1783 when he subjected them to the action of a "lampe d'air dephlogistiquée", i.e., an oxygen-fuelled enameller's lamp. This instrument, combined with a blowpipe, was perfected by Meusnier de la Place with the introduction of a "soufflé hydrostatique", paving the way for the construction of the first gasometer The oxygen blowpipe yielded more accurate results for Lavoisier than the burning lens, with which he and other academicians had heated and combusted the same minerals (see historical description of item no. 20025-0000- on p. 335).

The dating and original description of Lavoisier's experiments are recorded in the sixth volume of his Registres de laboratoire (fols. 151r–192r). We were able to identify a few of the precious stones in Lavoisier's memoirs.

1. "plomb" "Cilindre de plomb": label in Marie Anne Lavoisier's hand with small cylindrical specimen of lead.
2. "gargont 1#": label in Marie Anne Lavoisier's hand with three stones inside.
3. "cristal": label in Marie Anne Lavoisier's hand with a specimen of a cylindrical rock crystal.
4. "zinc" "Cylindre de zinc": label in Marie Anne Lavoisier's hand with a cylindrical specimen of zinc.
5. "cuivre jaune" "cilindre de cuivre jaune": label in Marie Anne Lavoisier's hand with a cylindrical specimen of brass.
6. "etain fin" "cilindre d'étain pur fin, très fin": label in Marie Anne Lavoisier's hand with a cylindrical specimen of tin.
7. "Régule" "Cilindre de Régule": label in Marie Anne Lavoisier's hand with a cylindrical specimen of antimony.
8. "marbre noir": label in Marie Anne Lavoisier's hand with cylindrical specimen of black marble.
9. 3 cylindrical specimens, two of copper and one rusted. No inscriptions.
10. 1 cylindrical specimen of a black stone. No inscriptions.
11. 2 cylindrical specimens of a black and white marble? No inscriptions.
12. Fused specimen of unidentified mineral wrapped in blue paper. No inscriptions.
13. "fer": with a cylindrical specimen of rusted iron.
14. "cast iron": cylindrical specimen of rusted iron. Unidentified eighteenth-century inscription.
15. Cylindrical specimen of copper wrapped in paper. Later inscription: "swim in sulfac".
16. "marbre rouge": label in Léon de Chazelles' hand with a cylindrical specimen of red marble.
17. "topaze d'inde 1 #": label with Marie Anne Lavoisier's handwritten inscription. Experiment conducted on 6 January 1783. Inside the paper there is also the following manuscript note in Lavoisier's hand: "topaz dinde moins dur meme que le cristal de roche comme lemeraude".
18. "eau d'outre mer" "M de Lavoisier": label in Léon de Chazelles' hand.
19. "n° 43 fragmen de six petit rubis dont quelques uns sont spinet //2.890": label in Marie Anne Lavoisier's hand with 1 specimen. Experiment conducted on 6 January 1783.
20. "n° 39 Rubis du Brasil 3,291": label in Marie Anne Lavoisier's hand. Experiment conducted on 6 January 1783.
21. "n° 47 topaze de Brasil en canon 4.960": label in Marie Anne Lavoisier's hand. Lavoisier describes his experiment of 6 January 1783 on these two specimens of ruby and topaz as follows:

> I exposed a cylindrical Brazilian topaz weighing 4 grains 96/100 to heat for 3′35″. First, it split in two along the blades, then it swelled and started to bubble in places; then the material became increasingly softer, folded in on itself, and at that point its fusibility appeared to diminish; after cooling, the result was a fine-grained white substance resembling fine porcelain, but more vitrous than in the experiment on Saxon topaz: large cavities or bubbles could be seen inside it; its weight had been reduced to only 3 grains 89/100; the stone had thus lost 1 grain 7/100, i.e., more than one-fifth of its weight. I subjected a Brazilian ruby to the same trial, and, as the two stones are one and the same thing, the result was understandably similar. The Brazilian ruby that I used for the experiment weighed 3 grains 294/1000; it swelled like the topaz; the material then contracted on itself, taking on a roughly spherical shape, producing a very hard pellet with a beautiful white color, resembling fine-grain porcelain. At the end of the experiment, it weighed only 2 grains 750/1000; the loss was thus 544/1000 grains, i.e., almost exactly one-sixth.[67]

67 "J'ai exposé, pendant 3′ 35″, à l'action de la chaleur, une topaze du Brésil en canon, pesant 4 grains 96/100: elle s'est d'abord fendue en deux morceaux dans le sens des lames, puis elle a boursouflé et a commencé à bouillonner dans des endroits; ensuite la matière s'est ramollie de plus en plus, elle s'est rapprochée sur elle-même, et alors sa fusibilité a paru diminuer; ayant laissé refroidir, le résultat était une substance blanche d'un grain fin, ressemblant à une porcelaine fine, mais plus vitreuse que dans

22. "quartz porfirisé": label by Marie Anne Lavoisier and one specimen of quartz. The experiment was made on 6 January 1783:

> Exposed to fire at 6h10′5″, it began to take on a pasty consistency at 11′5″. When touched with an iron instrument, at 12′10″, it yielded without, however, having fully melted. We removed it shortly afterwards. After cooling, the piece produced a semi-vitrous, semi-transparent, colourless mass, shiny on the surface, and containing many small bubbles inside. The quartz had been porphyrised under sandstone grindstones, and had been given to me by M. Macquer.[68]

23. "Taupaze de Saxe 10 N° 45": label in Marie Anne Lavoisier's hand with five specimens of fused stone.

Lavoisier reported:

> I exposed a Saxony topaz weighing 2 grains 610/1000 to fire, for 2′25″: it began to split and swell, and bubbled up mainly towards the angles; the molecules then came together, and seemed less fusible than at the outset; the cooled result was nearly spherical; when broken, the interior resembled a fine porcelain biscuit for its whiteness and fine grain; it now weighed only 2 grains 168/1000 and had therefore lost 442/1000 grains, i.e., around one-sixth of its weight. Having repeated the same experiment on another Saxony topaz and having exposed it to the action of fire for 5′10″, I similarly observed that the matter became ever less fusible as it lost colour, leaving a very white, fine-grained substance resembling a small piece of quartz or very fine-grained porcelain.[69]

24. "Quartz porfirisé mélange avec spaths calacaire de St Marie aux Mines." Label in Marie Anne Lavoisier's hand with three specimens of fused mineral. Experiment conducted on 6 January 1783.
25. "groupes de lames de spaths phosophorique": label with Lavoisier's hand and one specimen of spath.
26. "saphir d'eau 6.910 n° 84": label in Marie Anne Lavoisier's hand with six specimens of fused stone. Experiment conducted on 6 January 1783.
27. "n° 42 hiacinthe brute 3 fragmens 8.758": label in Marie Anne Lavoisier's hand with five specimens. Experiment conducted on 6 January 1783.
28. "noyeau de spath phosophorique": label in Lavoisier's hand with one specimen.
29. "saphires dorient bleux 4 # n° 40": label in Marie Anne Lavoisier's hand with one fused specimen.
30. "14 hyacinthe du Puy par M Sage 4.508": label in Marie Anne Lavoisier's hand with ten specimens of fused stone. Experiment conducted on 6 January 1783:

> I exposed fourteen hyacinths from Le Puy, given to me by M. Sage, to the action of fire; together they weighed 4 grains 508/1000: in less than a minute they had been completely discoloured and had become white as porcelain; they now weighed only 4 grains 440/1000, so they had lost 68/1000. When these same stones were exposed longer to fire in another experiment, they agglutinated without, however, adhering very solidly together; their surface had become shiny: one can therefore see that hyacinth is somewhat similar to ruby for its fixity when exposed

l'expérience faite sur la topaze de Saxe: on apercevait de grandes cavités ou bulles dans son intérieur; son poids n'était plus que de 3 grains 89/100; ainsi cette pierre avait perdu 1 grain 7/100, c'est-à-dire plus d'un cinquième de son poids. J'ai fait subir la même épreuve à un rubis du Brésil, et, comme ces deux pierres ne sont qu'une seule et même chose, on conçoit que le résultat a été semblable. Le rubis du Brésil sur lequel j'ai opéré pesait 3 grains 294/1000; il s'est boursouflé comme la topaze; la matière s'est ensuite rapprochée sur elle-même, a pris la figure à peu près sphérique, et il est resté un globule d'un beau blanc, très-dur, et ayant l'apparence d'une porcelaine d'un grain fin: son poids, à la fin de l'expérience, n'était plus que de 2 grains 750/1000; ainsi la perte avait été de 544/1000 de grain, c'est-à-dire assez exactement d'un sixième". LO, vol. 2, pp. 447–448.

68 "Exposé au feu à 6h 10′ 5″, il a commencé à prendre une consistance pâteuse à 11′ 5″. L'ayant touché avec un instrument de fer, à 12′ 10″, il a cédé, sans être cependant parfaitement fondu. On l'a retiré peu de temps après. Refroidi, le morceau fournit une masse demi-vitreuse, demi-transparente, sans couleur, luisante à la surface, et contenant beaucoup de petites bulles dans son intérieur. Le quartz avait été porphyrisé sous des meules de grès, et m'avait été donné par M. Macquer". (LO, vol. 2, pp. 455–456).

69 "J'ai exposé au feu, pendant 2′ 25″, une topaze de Saxe du poids de 2 grains 610/1000: elle a commencé par se fendiller et se boursoufler, et elle a bouillonné principalement vers ses angles; les molécules se sont ensuite rapprochées, et ont paru moins fusibles que dans le premier instant; le résultat refroidi était presque sphérique; l'ayant cassé, l'intérieur ressemblait, pour la blancheur et pour la finesse du grain, à un beau biscuit de porcelaine; il ne pesait plus que 2 grains 168/1000 et avait, par conséquent, perdu 442/1000 de grain, c'est-à-dire environ un sixième de son poids. Ayant répété la même expérience sur une autre topaze de Saxe et l'ayant laissée exposée à l'action du feu pendant 5′10″, j'ai observé, de même, que la matière devenait de moins en moins fusible à mesure qu'elle se décolorait, et il est resté une substance très-blanche d'un grain fin, qui avait l'apparence d'un petit morceau de quartz ou de porcelaine d'un grain très-fin". LO, vol. 2, p. 447.

to fire – with the basic difference, however, that the colour remains very fast in ruby but is very easily destroyed in hyacinth.[70]

31. "taupaze du bresille brute 3 # n° 47": inscription in Marie Anne Lavoisier's hand with specimens of fused stones. Experiment conducted on 6 January 1783.
32. "N° 46 topaz dinde 3.982": label in Marie Anne Lavoisier's hand with one specimen of fused stone. Experiment conducted on 6 January 1783.
33. "n° 45 topaz de saxe 2.610": label in Marie Anne Lavoisier's hand with three specimens of fused stones. Experiment conducted on 6 January 1783.
34. "gargont": label in Marie Anne Lavoisier's hand with twelve specimens of stone.
35. "sable magnetique de bar // M de milly": label in Marie Anne Lavoisier's hand with one specimen of stone.
36. "prisme daupalle": label in Marie Anne Lavoisier's hand with one specimen.
37. "rubis de M Sage six fragmens 3,020": label in Marie Anne Lavoisier's hand with one fused specimen. Experiment conducted on 6 January 1783:

    From a rather large number of small pale rose rubies given to me by M. Sage, I chose six, weighing 3 grains 2/100 together; they were exposed to the violence of fire for 4′30″, after which they agglutinated and nearly formed a single mass; they now weighed 3 grains 5/100, i.e., their weight had increased by 3/100. One can therefore see that the increases or decreases in the ruby's weight due to the action of the most violent fire to which it has yet been exposed are almost imperceptible, and it is not impossible that these increases may have been caused by molecules of the charcoal ash combining with it during the operation.[71]

38. "Marbre blanc": Label in Marie Anne Lavoisier's hand with a cylindrical specimen.
39. "Ruby dorient 5 # n° 41": label in Marie Anne Lavoisier's hand with two fused specimens. Experiment conducted on 6 January 1783.
40. "schorll noir": label in Marie Anne Lavoisier's hand with seven specimens.
41. "melange de terre d'alun et de terre pesante": in Marie Anne Lavoisier's hand with one specimen.
42. "parties egales d'alun et de quartz porfirisé": In Marie Anne Lavoisier's hand with one specimen. Experiment conducted on 6 January 1783.
43. "tourmaline": label in Marie Anne Lavoisier's hand on the inner part with one specimen.
44. "alun [..]": in Lavoisier's hand with three specimens.
45. "parties egales de terre d'alun et de quartz porfirisé dans la proportion de 2 grains chacun": label in Marie Anne Lavoisier's hand with five specimens.
46. "magnesie et quartz profirisé": label in Marie Anne Lavoisier's hand with five specimens.
47. "n° 41 2 rubies dorient ensemble […] 5.240": label in Marie Anne Lavoisier's hand without specimen.
48. "quartz profirisé et terre pesante": label in Marie Anne Lavoisier's hand with twelve specimens.
49. "argille de Betheux": label in Marie Anne Lavoisier's hand with one specimen.
50. "terre d'auln 12 // terre calcaire 12 // sablon 12": in Marie Anne Lavoisier's hand with one specimen.
51. "4 petits rubies": in Marie Anne Lavoisier hand with one specimen.
52. "lave alteree par m de Faujas": Label in Marie Anne Lavoisier's hand with three specimens. Experiment conducted between December 1782 and January 1783.
53. "Magnesie et quarts porfirisé": label in Marie Anne Lavoisier's hand with six specimens. Experiment conducted on 6 January 1783.
54. "terre d'alun 24 grains // spath calcaire de St marie aux mines // exposé": label in Marie Anne Lavoisier's hand.

70 "J'ai exposé à l'action du feu quatorze hyacinthes du Puy, qui m'avaient été données par M. Sage; elles pesaient ensemble 4 grains 508/1000: en moins d'une minute elles ont été complètement décolorées et sont devenues d'un blanc de porcelaine; elles ne pesaient plus que 4 grains 440/1000 ainsi elles avaient éprouvé une perte de 68/1000. Ayant poussé plus longtemps ces mêmes pierres au feu dans une autre expérience, elles se sont agglutinées, sans cependant contracter une adhérence fort solide; elles étaient devenues luisantes à la surface: on voit donc que l'hyacinthe a quelque rapport avec le rubis par sa fixité au feu, mais avec cette différence essentielle cependant, que la couleur est très-fixe dans le rubis, et très-facile à détruire dans l'hyacinthe". LO, vol. 2, p. 446.

71 "Dans un assez grand nombre de petits rubis d'un rose pâle, qui m'avaient été donnés par M. Sage, j'en ai choisi six, qui pesaient ensemble 3 grains 2/100; ils ont été exposés pendant 4′ 30″ à la violence du feu, après quoi ils se sont trouvés agglutinés et presque réunis; ils pesaient alors 3 grains 5/100, c'est-à-dire qu'ils avaient reçu une augmentation de poids de 3/100. On voit donc que les augmentations ou les diminutions de poids qu'éprouve le rubis par l'action du feu le plus violent auquel il ait encore été exposé sont presque insensibles, et il ne serait pas impossible que ces augmentations fussent dues à des molécules de la cendre du charbon, qui se combinent avec lui pendant l'opération". LO, vol. 2, p. 445.

55. "magnesie du sel d'epsom": label in Marie Anne Lavoisier's hand with one specimen, was donated by Mr Godefroi.
56. "vitriol du cuivre": label in Marie Anne Lavoisier's hand with one specimen.
57. "chrysolite de M de Faujas": Label in Marie Anne Lavoisier hand with two specimens. Experiment conducted between December 1782 and January 1783.
58. "aimanthe": label in Marie Anne Lavoisier's hand with seven specimens.
59. "n° 50 emeraude en deux morceaux 2.170": label in Marie Anne Lavoisier's hand with four fused specimens.
60. "sablon .12 // terre d'alun .12 // terre calcaire 6 X": label in Marie Anne Lavoisier's hand with ten specimens.
61. "colcotar": label, wrapped in larger paper, in Marie Anne Lavoisier's hand with one specimen.
62. "Schroll vert melé d'un peu d'argille": label in Marie Anne Lavoisier's hand with three specimens.
63. "sablon blanc profirisé": label in Marie Anne Lavoisier's hand with three specimens.
64. "jacinthe du Puy de m Sage": label in Marie Anne Lavoisier's hand with six specimens.
65. "terre d'alun 12 // quartz 12 // terre calcaire 3": label in Marie Anne Lavoisier's hand with thirteen specimens. Cylindrical specimens of copper, iron, brass, ivory and a black stone are without labels but wrapped in paper.
66. "Couleurs d'Angleterre.' // Mr L'etourneau." Small envelope containing pigments from England, sealed with a black wax seal. With Marie Anne Lavoisier's inscription and Lavoisier's reference to the supplier (?).

Ms. Lavoisier (1772–1788), vol. 6

Lavoisier (1782); Lavoisier (1782a); Lavoisier (1782b); Meusnier (1782)

20197-0000- (Fig. 121)
Specimen of beryl
Second half of 18th c. (collected)
MAM accession date: 1952
5 × 3 × 3.5; 105 g

Sample of greenish crystal of beryl (beryllium aluminium cyclosilicate with the chemical formula $Be_3Al_2(SiO_3)_6$). On it is a paper label in Lavoisier's hand with the inscription: "Béryl". This sample probably belonged to the same set described in no. 20185-0000-.

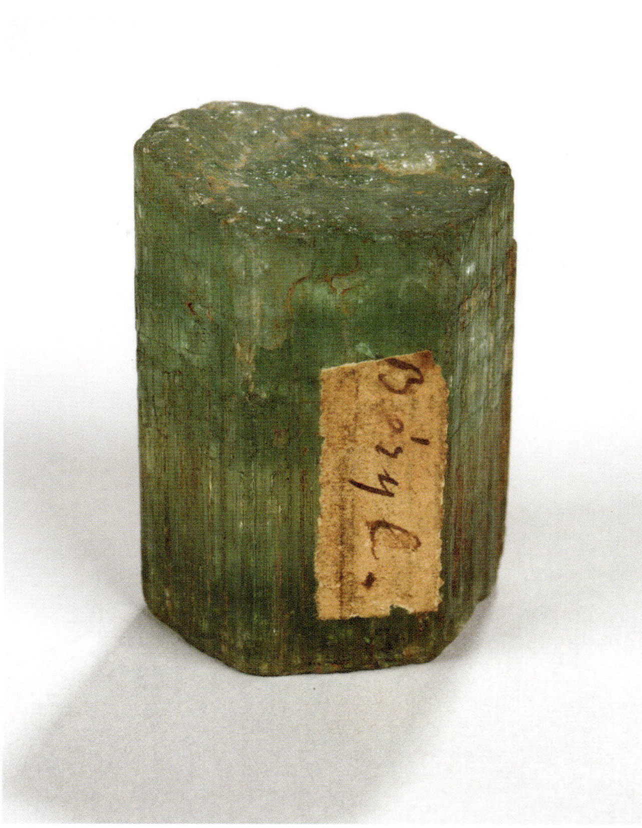

FIGURE 121 (inv. no. 20197-0000-)
© MAM/PHOTO FRANCK BOTTÉ

20192-0000- (Fig. 122)
Samples of iron and iron oxide
Second half of 18th (?)
MAM accession date: 1952
11 × 6.5 × 4; 70 g
Wool, paper, iron, iron oxide

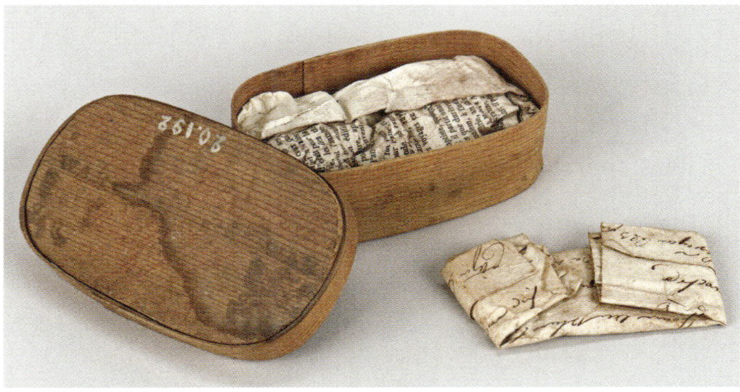

FIGURE 122 (inv. no. 20197-0000-)
© MAM/PHOTO FRANCK BOTTÉ

Oval wooden box containing two small packets. One of them consists of iron filings wrapped in a piece of a mid-nineteenth-century newspaper. The other consists of a dark red powder, probably iron oxide, wrapped in a manuscript letter (not addressed to Lavoisier). The wooden box is stamped with "WAR" or "WAK" and the figure of a lion.

19961-0000- (Fig. 123)
Phial of mercury
Second half of 18th c.
MAM accession date: 1952
8.2 × 3; 290 g.
Wood, brass, mercury, fabric

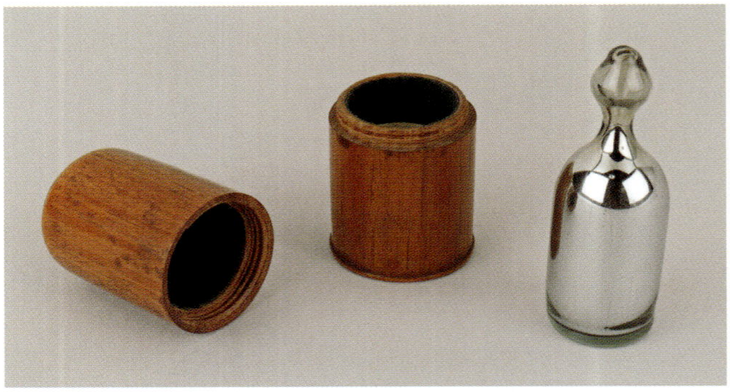

FIGURE 123 (inv. no. 19961-0000-)
© MAM/PHOTO FRANCK BOTTÉ

A sealed glass phial with a bulb on its neck containing mercury. The phial is housed in a wooden box with an internal baize lining.

19962-0000- (Fig. 124)
Bottle of mercury (missing)
Second half of 18th c.
MAM accession date: 1952
27 × 8
Glass, mercury

Brown eighteenth-century glass bottle containing mercury. Probably the only surviving bottle of Lavoisier's mercury. The bottle was lost when it was sent to be restored.

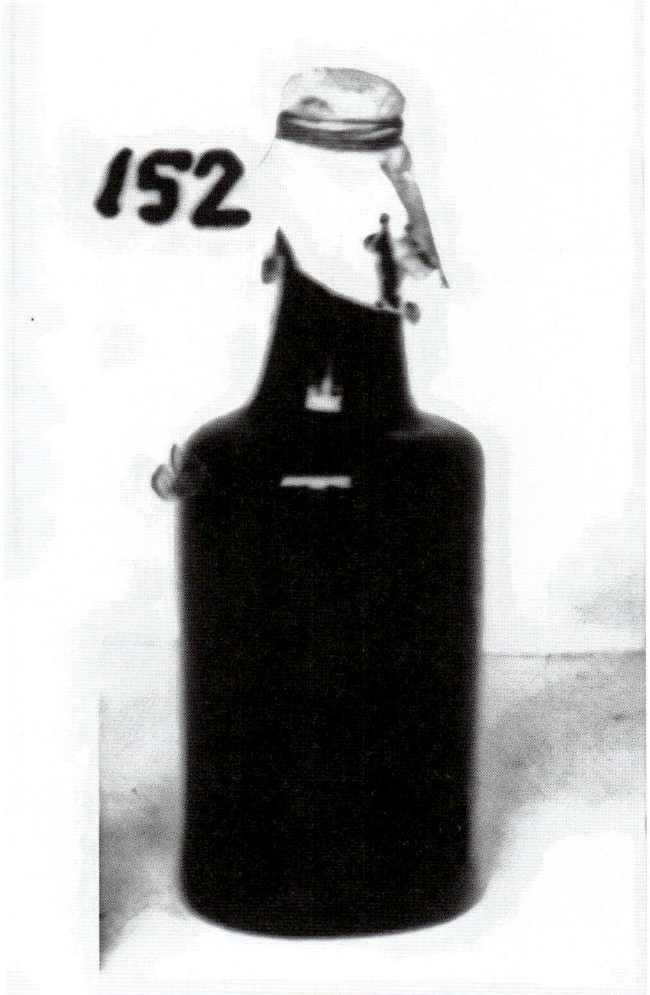

FIGURE 124 (inv. no. 19962-0000-)
PHOTO DOUGLAS MCKIE. COURTESY OF THE HAGLEY MUSEUM AND LIBRARY

Kroch Library, Cornell University, Ithaca (Fig. 125)
Lavoisier 4712 Box 29
Platinum salt-cellar with platinum spoon
1789–1790
Marc Etienne Janety (ca. 1750–ca. 1823)
Catalogued by Douglas McKie in 1952 but not included in the donation to the MAM
Box: diam 5.2 × 3.5
Diam. 2.4 to 4.05; length of slant side 2.2; overall length of spoon 5.3
Platinum

Platinum salt-cellar with platinum spoon in a thin greenish pillbox-shaped wooden case. One end of the case is marked "Lavoisier // PBC // 105" in ink. This was probably one of the two pieces presented by Lavoisier before the Académie des sciences in 1790 when he claimed that

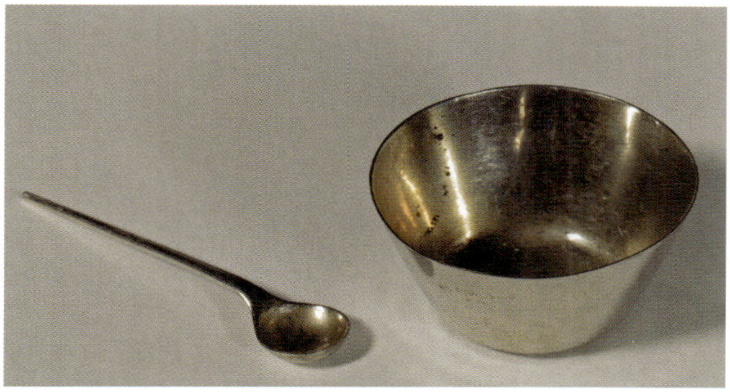

FIGURE 125 (Lavoisier 4712 Box 29)
COURTESY OF CORNELL UNIVERSITY LIBRARY – DIVISION OF RARE AND MANUSCRIPTS COLLECTIONS.

Janety had found a method to melt and work platinum *en grand*.

Lavoisier (1790); Lavoisier (1943); Smeaton (1997)

20187-0002- (Fig. 126)
Parcel with sample of saltpetre
Second half of 18th c. (collected)
MAM accession date: 1952
14.5 × 10 × 4; 75 g
Saltpetre, paper, thread

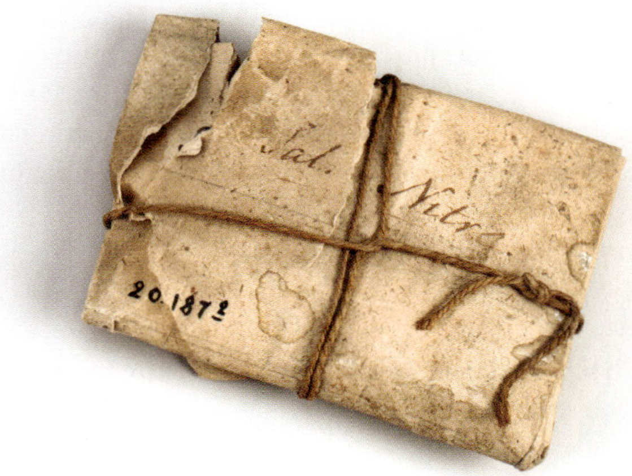

FIGURE 126 (inv. no. 20187-0002-)
© MAM/PHOTO FRANCK BOTTÉ

Small paper parcel containing a sample of saltpetre. Composed mainly of potassium nitrate, saltpetre was essential – with charcoal and sulphur – to the manufacture of black powder (the original form of gunpowder). The parcel bears the handwritten inscription "sal nitre", not a term Lavoisier would have used.

20188-0000- (Fig. 127)
Charcoal
Second half of 18th c. (collected)
MAM accession date: 1952
11 × 2.5 × 2.3; 20 g
Charcoal

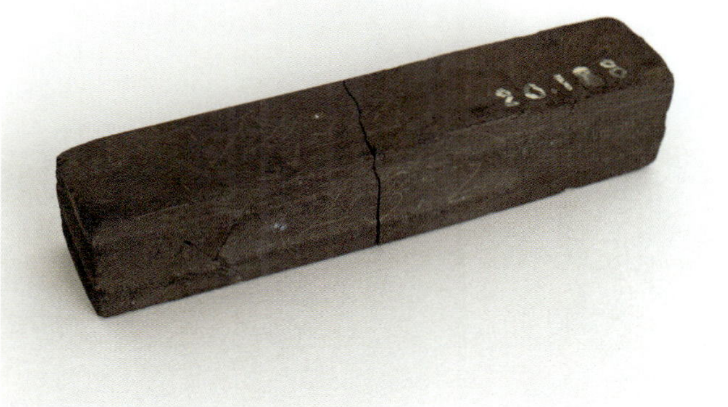

FIGURE 127 (inv. no. 20188-0000-)
© MAM/PHOTO FRANCK BOTTÉ

Parallelepiped block of fire charcoal, broken in two pieces. On it there are the faint inscriptions: "V 43.2", "sapin mâle", "n = 69.4 mâle".

20190-0001-, 20190-0002-, 20190-0003-, 20190-0004-, 20190-0005-, 20190-0006-, 20190-0007-, 20190-0008-, 20190-0009-, 20190-0010-, 20190-0011-, 20190-0012-, 20190-0013- (Fig. 128)
Set of mineral specimens
1760 (collected), acquired by Lavoisier in 1786 (?)
MAM accession date: 1952

Set of thirteen marble specimens and one porphyry specimen that were part of a much larger collection of minerals and fossils now preserved at the Muséum Lecoq in Clermont-Ferrand. The Muséum Lecoq also lists among its holdings the first part of a "Catalogue d'histoire naturelle" devoted to mineralogy. This folio document, in Lavoisier's hand, is unbound and probably formed part of a larger inventory of his natural-history cabinet – of which only the minerals, fossils and exceedingly few vegetable and animal specimens remain. Lavoisier's catalogue follows the standard classification proposed by the Swedish

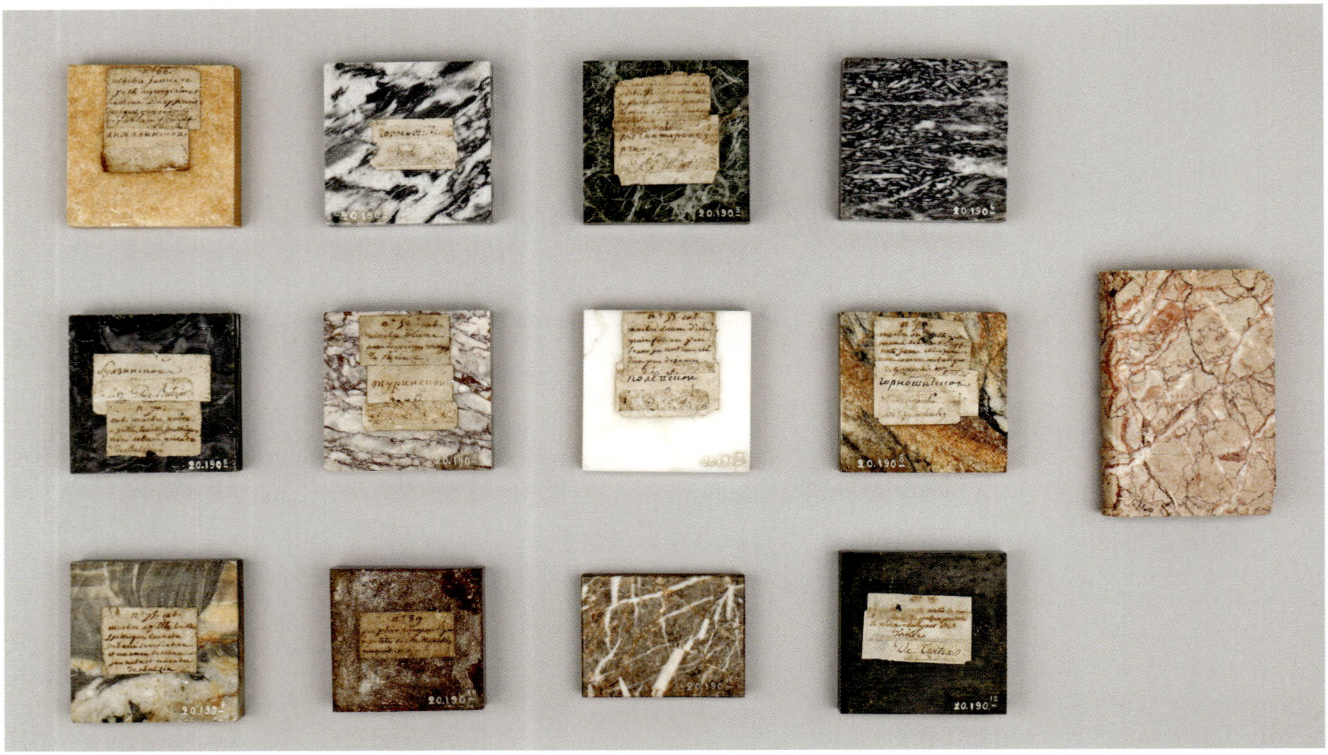

FIGURE 128 (inv. nos. 20190-0001-, -0013-)

chemist and mineralogist Johan Gottshalk Wallerius in his *Mineralogia* (Stockholm: Salvius, 1747), an extremely successful textbook that Lavoisier owned in the French translation (*Minéralogie*, 1753, 2 vols.). Wallerius' system, following Linnaeus, classified minerals in orders, genres and species – and so are the specimens preserved at the Musée des arts et métiers. Lavoisier collected most of the minerals of his collection, consisting of more than 3,000 items, during his mineralogical travels with his mentor Jean-Etienne Guettard in 1763–1769. When Guettard died in 1786, Lavoisier acquired a few specimens, books and manuscripts from his collection. Apart from this relatively late acquisition, post-1770 references to mineralogical surveying and collecting are rare and discontinuous.

A note by Guettard on some of the marble specimens listed here, most of which are from Russia, suggests that they belonged to Guettard and that they were therefore acquired by Lavoisier in 1786.

Like other minerals preserved in the Clermont-Ferrand collection, some of the present specimens have three labels. One, in Russian, is the oldest and seems to indicate the mineral's place of origin; one label, placed directly under the original, bears an inscription in Guettard's hand with a number and the name of the place; the third label refers to Lavoisier's description in his manuscript catalogue.

20190-0001-
Parallelepiped specimen of yellow-orange marble
11.2 × 11.2 × 2.2; 725 g

The specimen carries three labels, two in French and one in Russian. One has a partly faded inscription in Lavoisier's hand: "n°66. marbre jaune ou Spath a gros grains et Ecailleux. L'on appercoit dans quelques endroits des parties qui forment des angles." The second label in French, probably in Guettard's hand, reads "D'Akjal".

20190-0002-
Parallelepiped specimen of black and white marble
11.2 × 11.2 × 2.2; 710 g

The specimen carries an illegible label written in Russian.

20190-0003-
Parallelepiped specimen of black and green marble
11.2 × 11.2 × 2.2; 720 g

The specimen carries three labels, one in Russian and two in French. The first, in Lavoisier's hand, is partly cut and barely legible. The original text was: "n°69 tres beau

marbre verd veiné de partie blanche calcaire nuance de partie verdatre et noiratre de pierre ollaire ponctuée de mica. D'Ecaterimbourg. La partie argilleuse est dominante."

20190-0004
Parallelepiped specimen of black and white marble
11.1 × 11 × 2.2; 715 g

The original label, now missing, read: "marbre noir avec petites ecailles [...] veines et tachete de blanc de gozurchi".

20190-0005-
Parallelepiped specimen of black marble with white veins
11 × 11 × 2.2; 705 g

The specimen carries three labels, one in Russian, two in French. The one in Lavoisier's hand reads: "n° 70 cat. marbre noire a petites Ecailles Spathiques veiné de blan noiratre de Bulzin." The second label in French reads: "17 De Bulrin".

20190-0006-
Parallelepiped specimen of white marble with reddish veins
11 × 11 × 2.2; 725 g

The specimen carries three labels: one in Lavoisier's hand, the second in Russian, the third (probably Guettard's) no longer readable. Lavoisier's label reads: "n° 52. cat. marbre blanc avec des veines rouges de Turin [Poli]".

20190-0007-
Parallelepiped specimen of white marble
11.2 × 11.2 × 2.2; 710 g

The specimen carries three labels, one in Russian, two in French. Of these, one is in Lavoisier's hand and reads: n° 55. cat. marbre blanc d'un grain fin un peu transparent nuancé d'un peu de jaune. Pal. De". The second French label, probably Guettard's, reads: "4 De Pollisf".

20190-0008-
Parallelepiped specimen of yellow marble with black veins
11.2 × 11.2 × 2.2; 710 g

The specimen carries three labels, one in Russian, two in French. Of these, one is in Lavoisier's hand and is barely legible: "n°76. marbre ou Spath grains nuancé de blanc de differents jaune et de veine noire comme metallique de Gonochinskey [Gonochinskiy?]". The other French label reads: "1er De Cornochin ou Gonochinsky".

20190-0009-
Parallelepiped specimen of brown marble with green veins
11.2 × 11.2 × 2.2; 695 g

The specimen carries a label in Lavoisier's hand: "n°75. cat. marbre a petites Ecailles spathiques brunatre rubané de verdatre et nuancé de blanc jaunatre et noiratre de Bulzin".

20190-0010-
Parallelepiped specimen of porphyry
10 × 9.8 × 2.5; 705 g

The specimen carries a label in Lavoisier's hand: "n°89 porphir rouge antique a petites taches blanches rougeatres".

20190-0011-
Parallelepiped specimen of grey and beige marble
10.9 × 8.4 × 1.8, 430 g

20190-0012-
Parallelepiped specimen of green and black marble
11 × 11 × 2.2; 695 g

The specimen carries barely legible labels, one with the words "De Troitza", possibly in Guettard's hand.

20190-0013-
Specimen of book-shaped marble
MAM accession date: 1952
16.8 × 11 × 2.7; 1.34 kg

Specimen of reddish marble carved in the form of a book. It was probably used as a paperweight and bears the inscription: "TOM // I".

Pelucchi (2003); Beretta (2005); Pelucchi (2009); Pelucchi (2016)

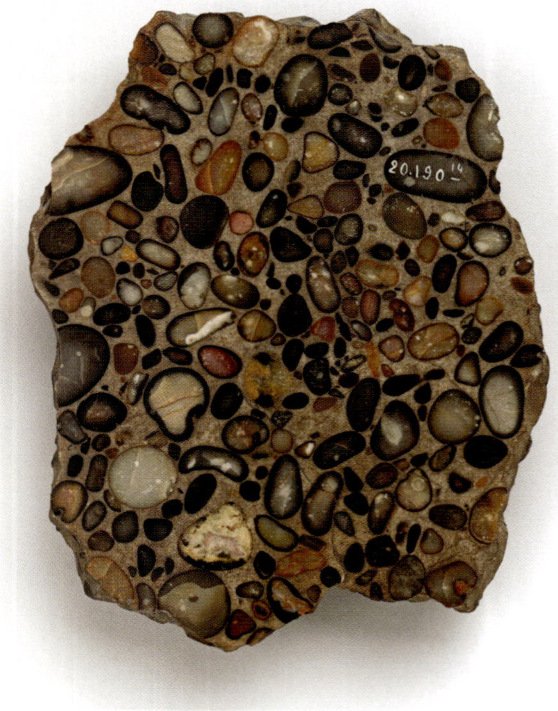

FIGURE 129 (inv. no. 20189-0000-)
© MAM/PHOTO FRANCK BOTTÉ

20189-0000- (Fig. 129)
Specimen of conglomerate rock
ca. 1760s (collected)
MAM accession date: 1952
24.5 × 20 × 4.5; 3.35 kg

Specimen of conglomerate rock of irregular shape with one polished face.

20182-0000- (Fig. 130)
Fragments of minerals and earthenware
Second half of 18th c.
MAM accession date: 1952
Box: 21 × 14 × 5.5; 1.2 kg
Wood, mineral substances

Modern wooden box[72] with lid containing 28 samples of minerals and earthenware.

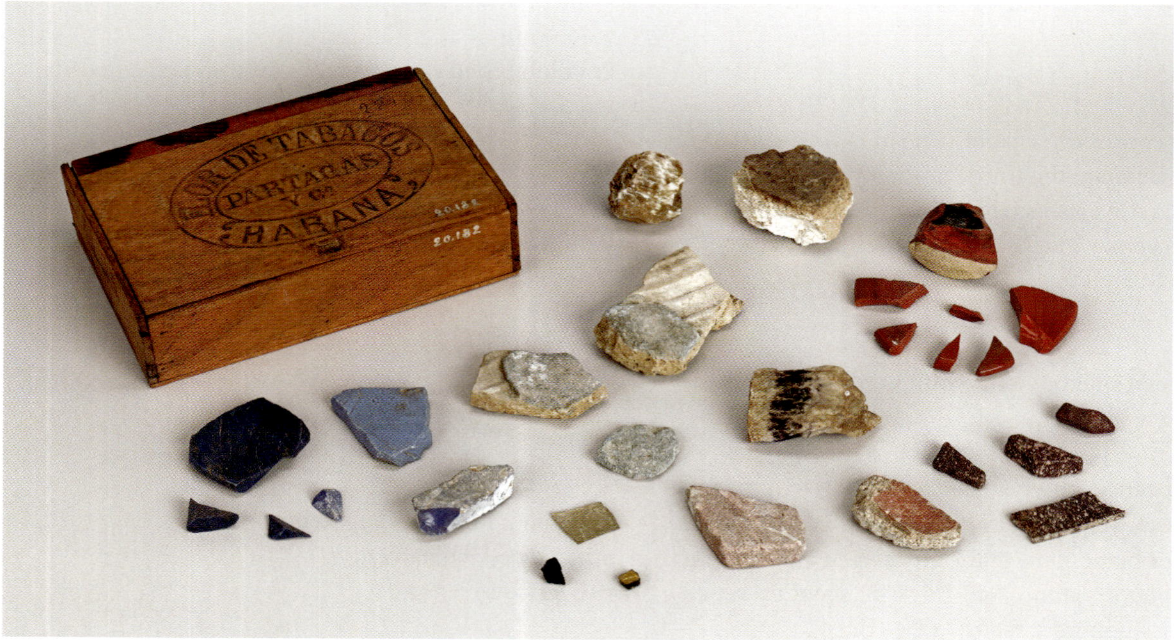

FIGURE 130 (inv. no. 20182-0000-)
© MAM/PHOTO FRANCK BOTTÉ

72 The inscription "FLEUR DE TABACOS // PARTAGAS N 6 // HABANA" is printed on the box.

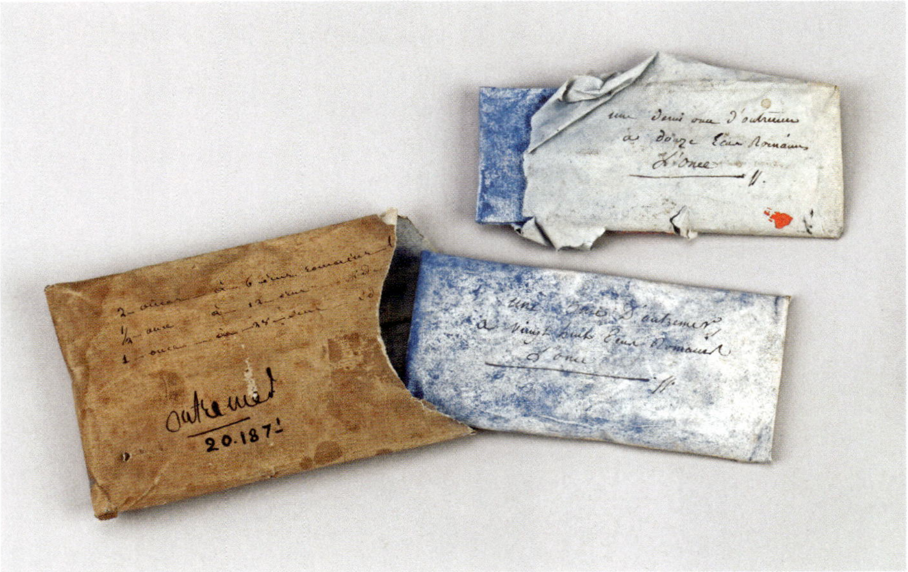

FIGURE 131 (inv. no. 20187-0001-)
© MAM/PHOTO FRANCK BOTTÉ

20187-0001- (Fig. 131)
Parcel with sample of ultramarine
Second half of 18th c.
MAM accession date: 1952
15 × 9 × 2.5; 60 g
Ultramarine, paper, sealing wax

Flat paper parcel containing two envelopes with samples of the blue pigment called ultramarine. Ultramarine, a very expensive substance, was obtained by grinding lapis lazuli[73] to powder. Synthetic ultramarine was introduced in 1826.

There are the following handwritten inscriptions by Marie Anne Lavoisier: on the parcel, "2 onces a 6 écus romain l'once" // "1/2 once a 12 écus id" // "1 once a 28 écus id"; on the first envelope, "une demie once d'outremer // à 12 écus romains l'once"; on the second envelope, "une once d'outremer // à ving huit écus romains // l'once".

The parcel is sealed with the de Chazelles family seal.

20184-0001-, 20184-0002-, 20184-0003-, 20184-0004- (Fig. 132)
Four rock specimens
MAM accession date: 1952
20184-0001-: 9 × 4 × 2.5; 150 g
20184-0002-: 4.5 × 4 × 3; 80 g
20184-0003-: 3.5 × 2.5 × 2.3; 30 g
20184-0004-: 5.5 × 4 × 2.5; 75 g.
Mineral

Four pebbles: one of quartz (20184-0001-), two of yellowish marble (20184-0002- and -0004-), and one of an unidentified grey stone (20184-0003-).

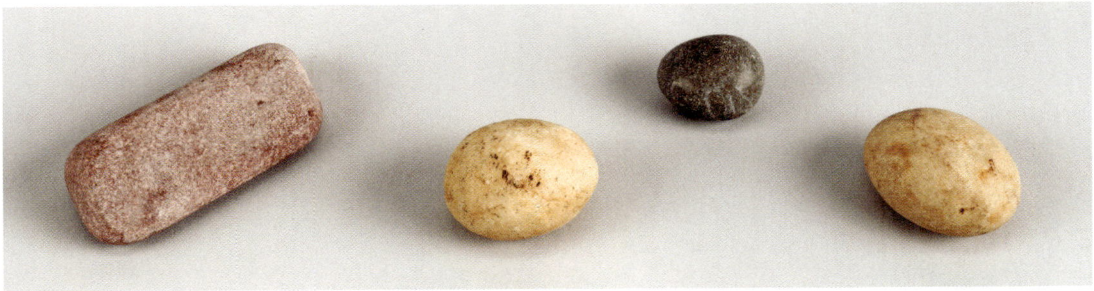

FIGURE 132 (inv. nos. 20184-0001-, -0004-) © MAM/PHOTO FRANCK BOTTÉ

73 Lapis lazuli is composed of the cubic blue mineral called lazurite and pyrites. Lazurite is a complex sodium silicate containing sulphur.

20194-0000- (Fig. 133)
Quartz crystals
1760s (collected)
MAM accession date: 1952
19 × 15.5 × 11; 1.24 kg.

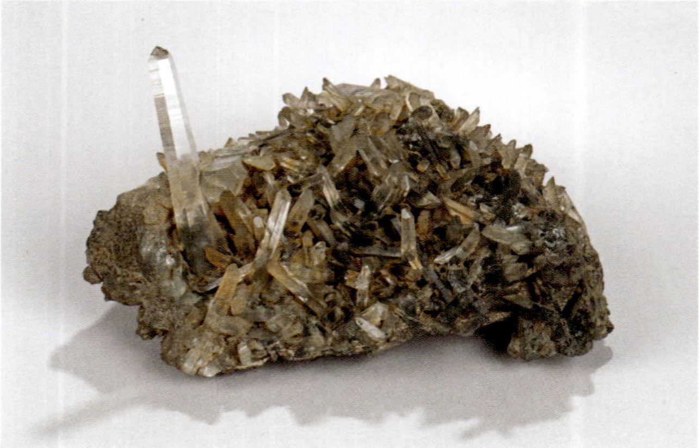

FIGURE 133 (inv. no. 20194-0000-)
© MAM/PHOTO FRANCK BOTTÉ

A block of quartz covered with numerous fine quartz crystals.

20195-0000- (Fig. 134)
Stone axe
Neolithic?
MAM accession date: 1952
25 × 9.5 × 5; 1.33 kg

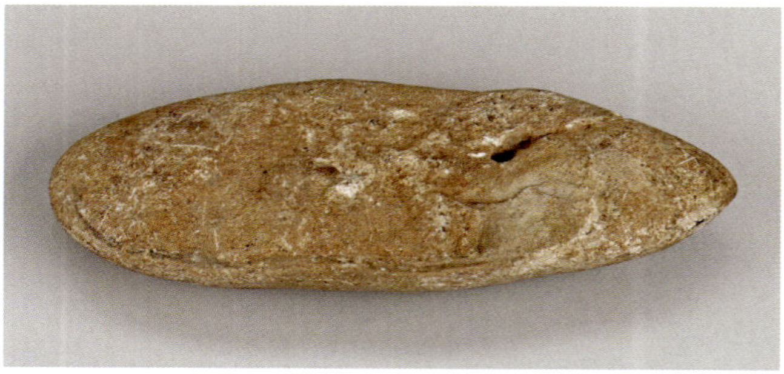

FIGURE 134 (inv. no. 20195-0000-)
© MAM/PHOTO FRANCK BOTTÉ

Oval-shaped stone supposedly used as an axe in the Neolithic age.

20191-0000- (Fig. 135)
Box with white powder
Second half of 18th c.
MAM accession date: 1952
4.2 × 2.5; 10 g
Cardboard, white powder

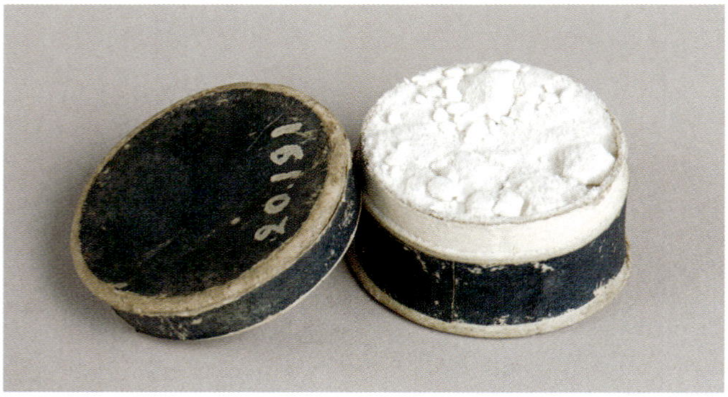

FIGURE 135 (inv. no. 20191-0000-)
© MAM/PHOTO FRANCK BOTTÉ

Cylindrical green cardboard box containing an unidentified white powder.

20186-0000- (Fig. 136)
Parcel with samples of cinnamon
1787
MAM accession date: 1952
24.5 × 8 × 4; 75 g
Cinnamon, paper, hemp (twine), sealing wax

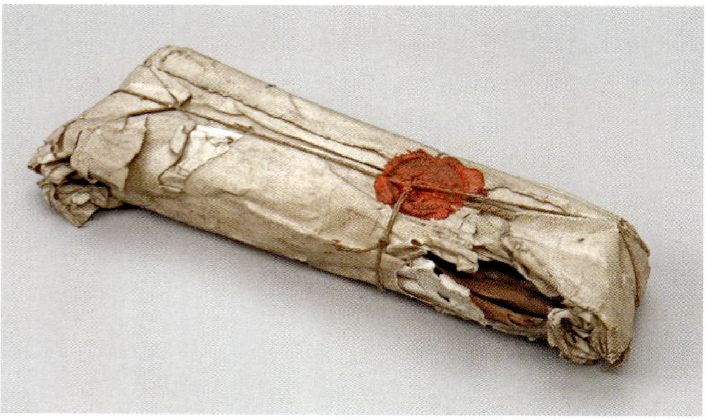

FIGURE 136 (inv. no. 20186-0000-)
© MAM/PHOTO FRANCK BOTTÉ

Several cinnamon sticks are wrapped in a paper parcel closed with twine and a royal seal of red wax. The parcel bears the inscription: “Canelle de la Guiane // française, cueillie en 1787”. The seal displays the coat of arms of the

king of France (with the crown and three lilies) and one can read "... DE LA GUIANE".
Although Lavoisier's works make no reference to cinnamon, he was asked in June 1788 to analyse and distil a sample of clove from Guyane.

LO, vol. 6, pp. 74–81

20051-0000- (Fig. 137)
Strips of cork
Second half of 18th c.
MAM accession date: 1952
38.5 × 11 × 7.2; 495 g
Cork, fibre, vellum

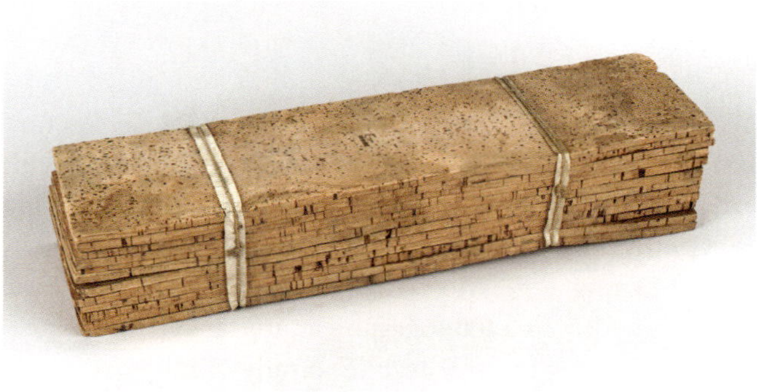

FIGURE 137 (inv. no. 20051-0000-)
© MAM/PHOTO FRANCK BOTTÉ

Bundle of twelve rectangular strips of cork. These are tied together with two strings with pieces of vellum under them. The strips are marked with an "F". Cork was commonly used in laboratories as thermal insulating material.

Miscellaneous

This section includes items that are part of the collection donated in 1952 but do not have a direct connection with Lavoisier's scientific research – except for the engraving illustrating the experiment on diamonds with Trudaine's large burning lens. Most of the artefacts are furniture, utensils, spectacles, iconographic material and one instrument belonging to the Chazelles family.

20576-0000- (Figs. 1a, 1b)
Roll top desk
1774
MAM accession date: 1956
Jean Caumont (1736–1800)
Closed: 192.5 × 116.5 × 94.5; 173 kg
With the additional tables opened: 270 × 116.5 × 122
Mahogany and others woods, gilded bronze, iron, leather

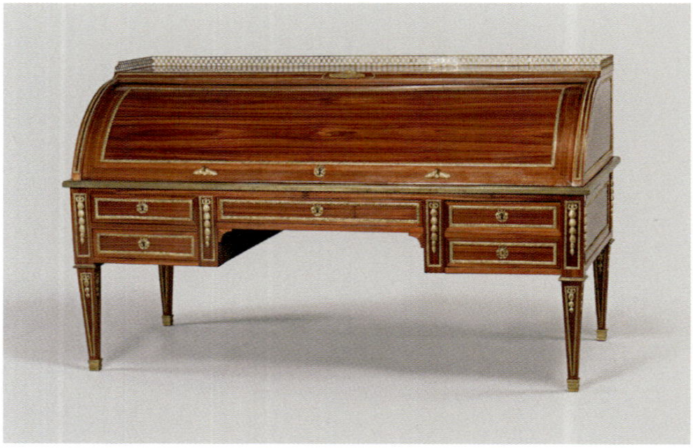

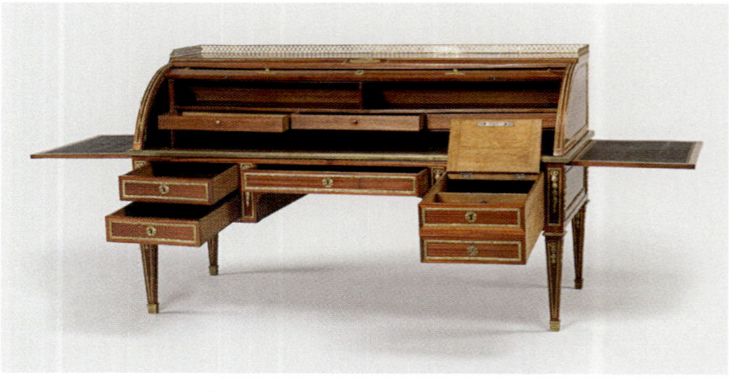

FIGURES 1A, 1B (inv. no. 20576-0000-)
© MAM/PHOTO FRANCK BOTTÉ

Lavoisier's roll top desk (*bureau à cylindre*). The large four-leg writing desk is veneered in mahogany and is enhanced by a series of ormolu mountings.[1] The gallery top of the desk can be closed by the cylindrical top and has two sections for storing documents and three drawers. The tooled black leather writing surface can be pulled out. Two similar additional sliding tables are inserted into the left and right sides of the desk. The lower part of the desk has two drawers (with locks) on the left and a longer one at the centre. The single large drawer (with lock) on the right appears as a double drawer from the outside. In fact, it conceals a safe closed by a wooden lid with an iron lock. The desk comes with three keys, one of which does not fit. Under the right side between the legs is the barely legible inscription "CAUMONT". Jean Caumont (1736–1800) was a famous French cabinet maker who built several items of furniture in Louis XVI and Transition style; they were elegant and extremely well produced in a remarkable variety of materials.

Ledoux-Lebard (1989) p. 116; Monnier (2012) p. 131

20575-00001, 20575-00002 (Fig. 2)
Two console tables
Last quarter of 18th c.
MAM accession date: 1956
Georges Jacob (1739–1814)
20575-0001-: 162 × 108 × 73; 62 kg
20575-0002-: 101 × 108 × 72.5; 62 kg
Mahogany, brass, iron

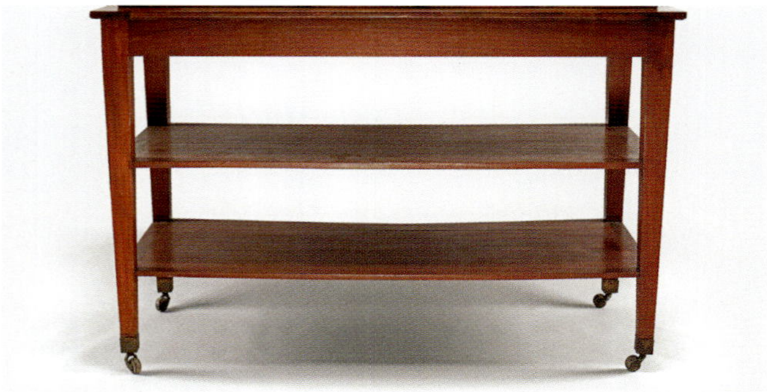

FIGURE 2 (inv. nos. 20575-0001-, -0002-)
© MAM/PHOTO FRANCK BOTTÉ

1 Two vertical iron bars, added later, serve as additional legs to improve the desk's stability.

 | DOI:10.1163/9789004511217_020

Two almost identical console tables. Both have two shelves fixed between the legs under the working plane. The tables are mounted on castors inserted in the legs.

Both are stamped "G-JACOBS". Georges Jacob (1739–1814), the most renowned cabinet maker of his time, received the title of *maître* (master) in 1765. An innovator, he was one of the first French cabinet makers to use mahogany. He produced many carved, gilded and painted pieces of furniture for the French royal châteaux.

Lefuel (1923)

20061-0000- (Fig. 3)
Candle holder
Second half of 18th c.
MAM accession date: 1952
10.5 × 4.5 × 4.5; 45 g
Silver

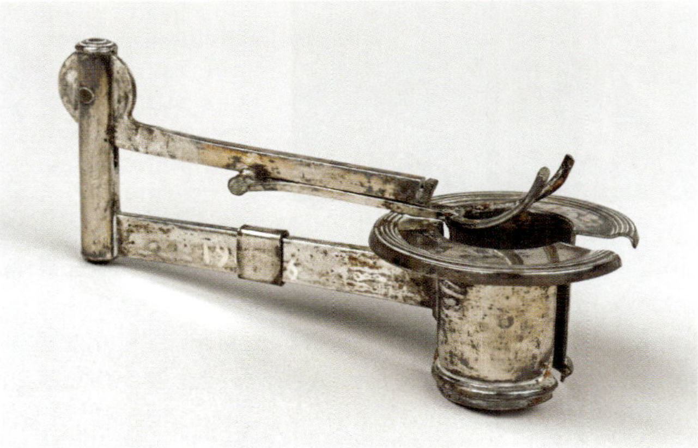

FIGURE 3 (inv. no. 20061-0000-)
© MAM/PHOTO FRANCK BOTTÉ

Candle holder composed of two arms, each supporting a semi-cylindrical housing at one end. The opposite ends of the arms are fixed to a vertical bar. A sliding square-shaped band serves to fasten the arms together and hold the candle. A movable lever with a fork-shaped end is inserted in the bar. The fork was probably used to remove the remains of a burnt candle.

20211-0000- (Fig. 4)
Candle snuffer
Late 18th c.
MAM accession date: 1952
P.W. Rotton
18 × 12 × 4; 120 g
Steel

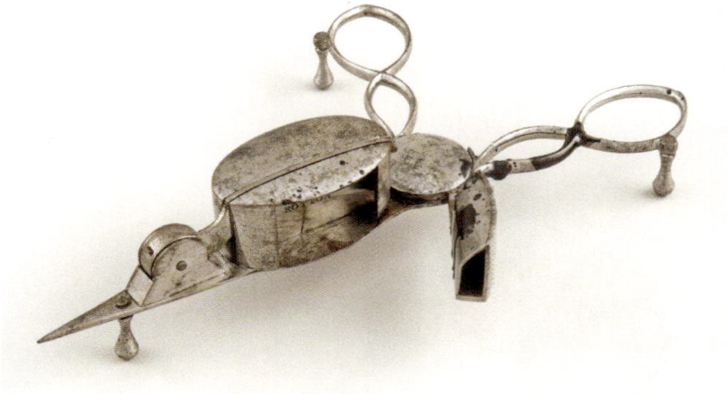

FIGURE 4 (inv. no. 20211-0000-)
© MAM/PHOTO FRANCK BOTTÉ

Scissors-like device with two flat blades and a receptacle. This tool was used to maintain efficient burning by trimming the candle wick without extinguishing the flame. A small receptacle catches the trimmed bit of wick. The candle snuffer became obsolete after the invention of self-snuffing wicks, which curled out of the flame when charred. The signature "P W Rotton" and a feather are engraved on the blade. P.W. Rotton was an English cutler active between the late eighteenth and early nineteenth centuries. Domestic objects signed by Rotton, such as corkscrews and lancets, are preserved in public and private collections.

Flook (2008)

20054-0000- (McKie 213)
Spirit lamp
18th c.
MAM accession date: 1952
11 × 8 × 7
Brass

Small brass ornamental spirit lamp. This item was stolen, together with inv. no. 20096, in February 1984. The only description left was that made by Douglas McKie in 1952.

20059-0000- (Fig. 5)
Set of four small spoons
Second half of 18th c.
MAM accession date: 1952
12 × 2 × 3.5; 40 g; (a single spoon: 10.3 × 2.7 × 1.4; mass 10 g)
Silver, cardboard, leather

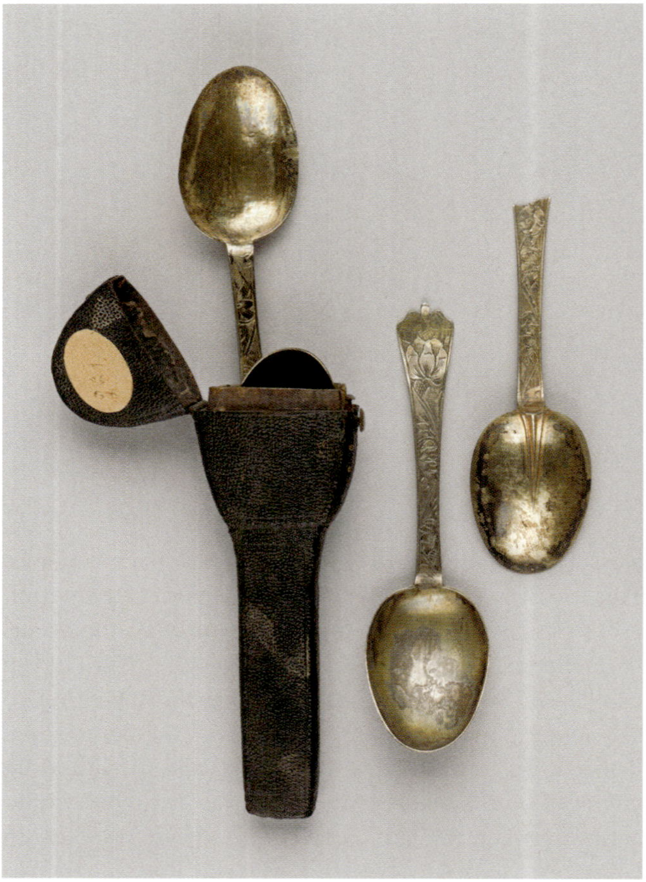

FIGURE 5 (inv. no. 20059-0000-)
© MAM/PHOTO FRANCK BOTTÉ

Set of four silver spoons with a decorated handle. One handle is broken with a missing piece. The set is housed in a cardboard and leather case carrying a small label with the inscription "267".

Kroch Library, Cornell University (Ithaca) (Fig. 6)
Lavoisier 4712 Box 29
Travel knife/fork
Second half of 18th c.
14.2 × 1.2
Box: 15.2 × 2.4
Gold, wood, leather, iron

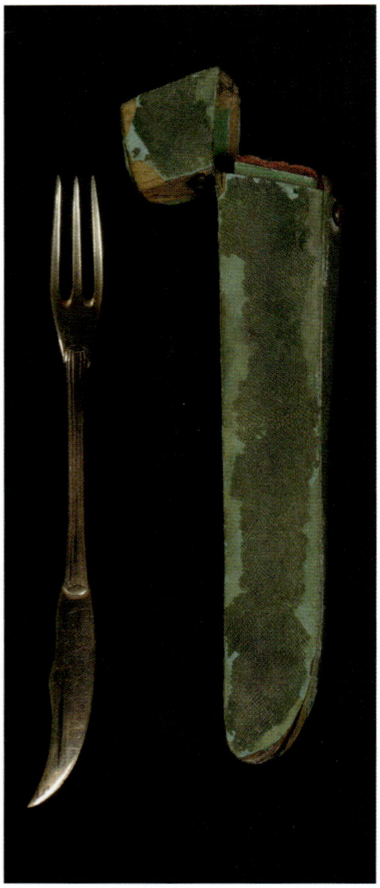

FIGURE 6
(Lavoisier 4712 Box 29)
COURTESY OF CORNELL UNIVERSITY LIBRARY – DIVISION OF RARE AND MANUSCRIPTS COLLECTIONS

Gold travel knife/fork with its original box opened at upper end. The box is coated with green leather. The interior is lined with red linen. There is an unidentifiable punch on the back of the fork. A similar set and an almost identical box is preserved at the Musée Carnavalet – Paris (inv. no. Ancien Fonds – OM4153).

The object, possibly the property of Marie Anne Lavoisier, was exhibited in 1943.

Lavoisier (1943)

McKie 420
Long-cased clock
18th c. (?)
Catalogued by Douglas McKie in 1952 but not included in the donation to the MAM

Long-cased clock, *pendule à gaine*, or grandfather clock

McKie 279 (Fig. 7)
Diptych dial
1654?
Ivory
Catalogued by Douglas McKie in 1952 but not included in the donation to the MAM

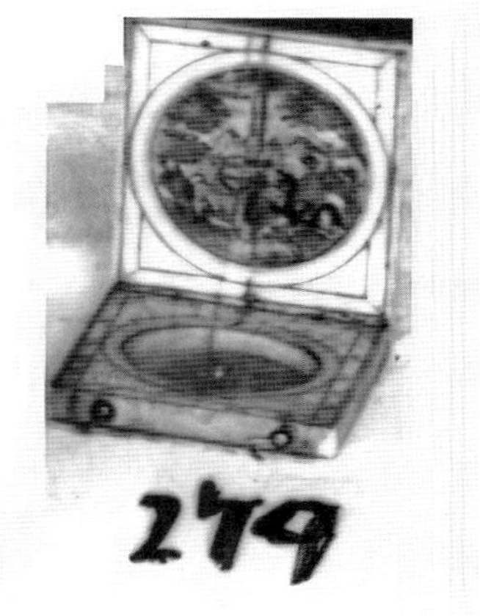

FIGURE 7
(inv. no. McKie 279)
PHOTO DOUGLAS MCKIE. COURTESY OF THE HAGLEY MUSEUM AND LIBRARY

Diptych dial with a coloured hunting scene on the inside of the upper hinged lid.

The marking "1654" is not necessarily a date.

This item is now part of the collection of the late Pierre S. Dupont.

Lavoisier (1943)

McKie 275 (Fig. 8)
String gnomon sundial
ca. 1750
Catalogued by Douglas McKie in 1952 but not included in the donation to the MAM
Pierre Le Maire
8.8 × 5.7
Silver, leather, fabric, agate

Sundial with compass mounted in a silver case by "Pre le Maire Paris", with a plummet and four levelling screws. The plummet runs on a limb that can be raised vertically on a spring and retracted to lie on the face of the compass. The case is rectangular. The outer case is covered with leather and lined with green velvet. The Roman numerals I to XII are arranged along the edges of the rectangular silver case and the latitudes of various cities are similarly marked on the underside of the base. Below the first series of Roman numerals is a series of Arabic numerals, with a second series of Roman numerals below them. The upright is graduated from 40 to 55 on the inner surface. The needle is mounted on agate; also on the dial is a fixed needle indicating magnetic north, i.e., giving declination. The latitudes shown on the underside are: "Londres 51.31′ Lille 50.40′ // Paris 48.51′ Brest 48.28′ Strasburg [sic] 48.30′ Maseille [sic] 43.10′ Monpellier [sic] 48.37′ [should be 43] Toulouse 43.37′ la Rochelle 46.10′ Lyon 45.45′ // Monse 51.31′ Anvers 51.35′".

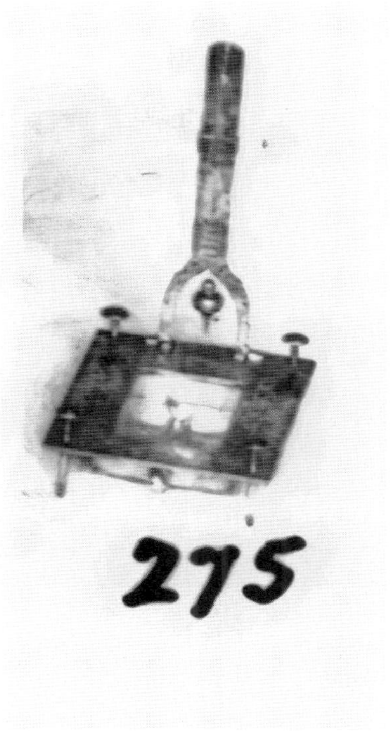

FIGURE 8
(inv. no. McKie 275)
PHOTO DOUGLAS MCKIE. COURTESY OF THE HAGLEY MUSEUM AND LIBRARY

This item is now part of the collection of Lammot du Pont Copeland.

Lavoisier (1943)

McKie 280 (20144-0000-) (Fig. 9)
Cubic sundial
Second half of the 18th c.
David Beringer (1756–1821)
10.7 × 8.8 (base); 5.5 (stand); 7 × 6 (cube)
Wood

Compass mounted in base, carrying upright with movable cube in wood on hinged wooden joint. Marked "D. Beringer". Dials and designs on face of cube.

Catalogued by Douglas McKie in 1952 but eventually excluded from the donation to the MAM. Its present location is unknown.

Lavoisier (1943)

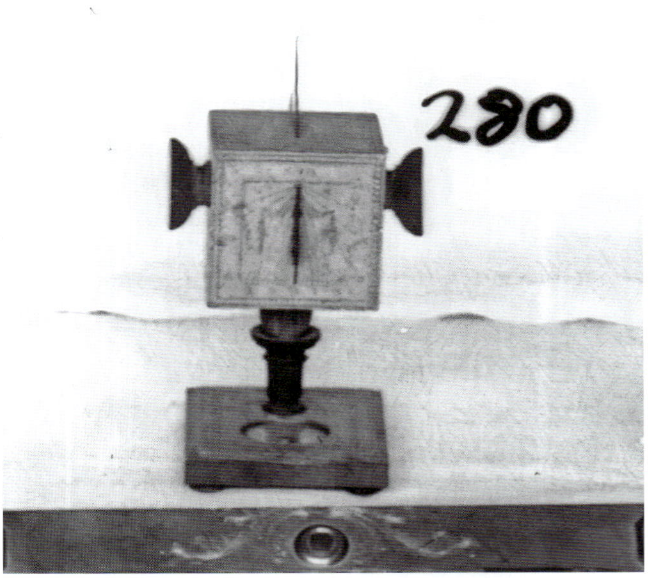

FIGURE 9 (inv. no. McKie 280, 20144-0000-)
PHOTO DOUGLAS MCKIE. COURTESY OF THE HAGLEY MUSEUM AND LIBRARY

20143-0000- (Fig. 10)
Sandglass
Second half of 18th c.
MAM accession date: 1952
13.2 × 7.1; 150 g
Glass, thread, wood

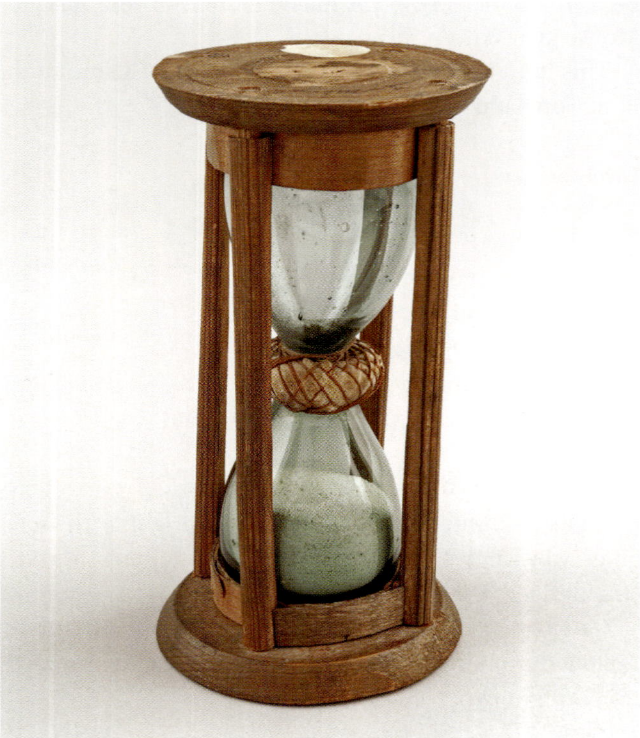

FIGURE 10 (inv. no. 20143-0000-)
© MAM/PHOTO FRANCK BOTTÉ

Sandglass composed of two pear-shaped glass bottles, partly filled with sand and held between two wooden disks and four small vertical columns. The junction of the two necks of the bottles is protected by a padded ring wrapped in thread. One disk has the inscription "3 minutes", indicating the time taken by the sand to pass from the top bottle to the bottom one.

Lavoisier (1943)

Kroch Library, Cornell University (Ithaca) (Fig. 11)
Lavoisier 4712 Box 29e
Oracles Merveilleux (magnetic game)
Second half of 18th c.
Larger box: 12.5 × 10 × 4.
Smaller box: 8 × 8 × 2.
Wood, cardboard, paper, magnet

The "magnetic oracle" was a mathematical recreation based on the property of magnetism. There are several descriptions of this type of game, which became extremely popular during the second half of the eighteenth century. The game combined the use of a circular box divided into 8–10 magnetized compartments with 60 cards showing the oracle's responses, mostly concerning romantic themes. Lavoisier's specimen is not complete and all the cards are missing. There are two wooden boxes. The smaller one contains a magnetic needle and is externally decorated with the inscription "Oracles merveilleux". The larger one, divided in two, is decorated on the upper side with a circle bearing eight different inscriptions: "les Dents // le Sabot // le Coeur // le Lacet // le Lit // le Cocq // la Carte // l'Amour". At the centre of the circle is a movable paper angel protected by transparent glass. The lower part of the larger box has a small drawer with a handle.
The magnetic game was inventoried among Lavoisier's instruments in 1794 (see Appendix 3).

Guyot (1774), vol. 3, pp. 7–26 and Plate 1

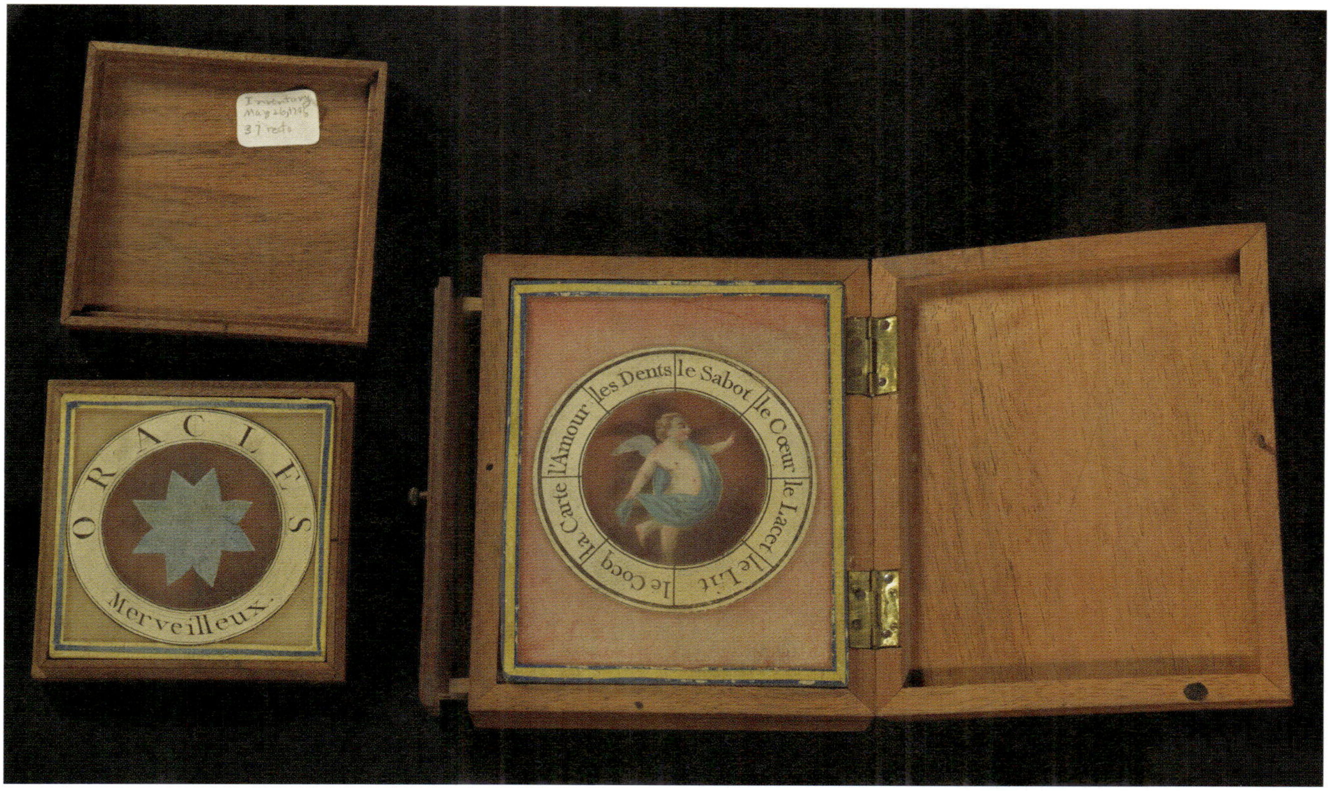

FIGURE 11 (Lavoisier 4712 Box 29e)
COURTESY OF CORNELL UNIVERSITY LIBRARY – DIVISION OF RARE AND MANUSCRIPTS COLLECTIONS

20147-0001-, 20147-0002- (Fig. 12)
Foot warmers
Second half of 18th c.
MAM accession date: 1952
32 × 12 × 7.5; 680 g (20147-0001-) and 630 g (20147-0002-)
Tin, wood

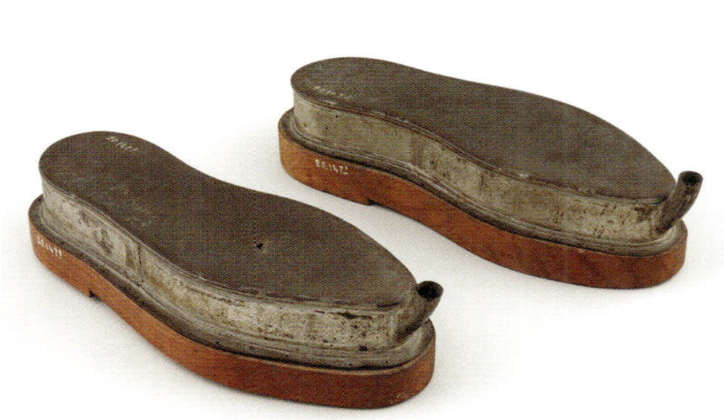

FIGURE 12 (inv. Nos. 20147-0001-, -0002-)
© MAM/PHOTO FRANCK BOTTÉ

Each foot warmer is composed of a wooden sole to which a foot-shaped flat tin vessel made of tin is fixed. A short tube acts as a funnel to fill the vessel with hot water.

20048-0000- (Fig. 13)
Knife with serrated blade
Second half of 18th c.
MAM accession date: 1952
42.5 × 2.7; 150 g
Steel, wood, brass

Knife composed of a steel shaft ending in a short, curved and serrated steel blade inserted in a wooden handle. The steel carries the inscription "POISSON A. CRÉPY" with a star.

The tool was probably used for cutting glass tubes.

20233-0000- (Fig. 14)
Upholstery hammer
Second half of 18th c.
MAM accession date: 1952
29 × 14.5 × 2; 290 g
Iron alloy, wood

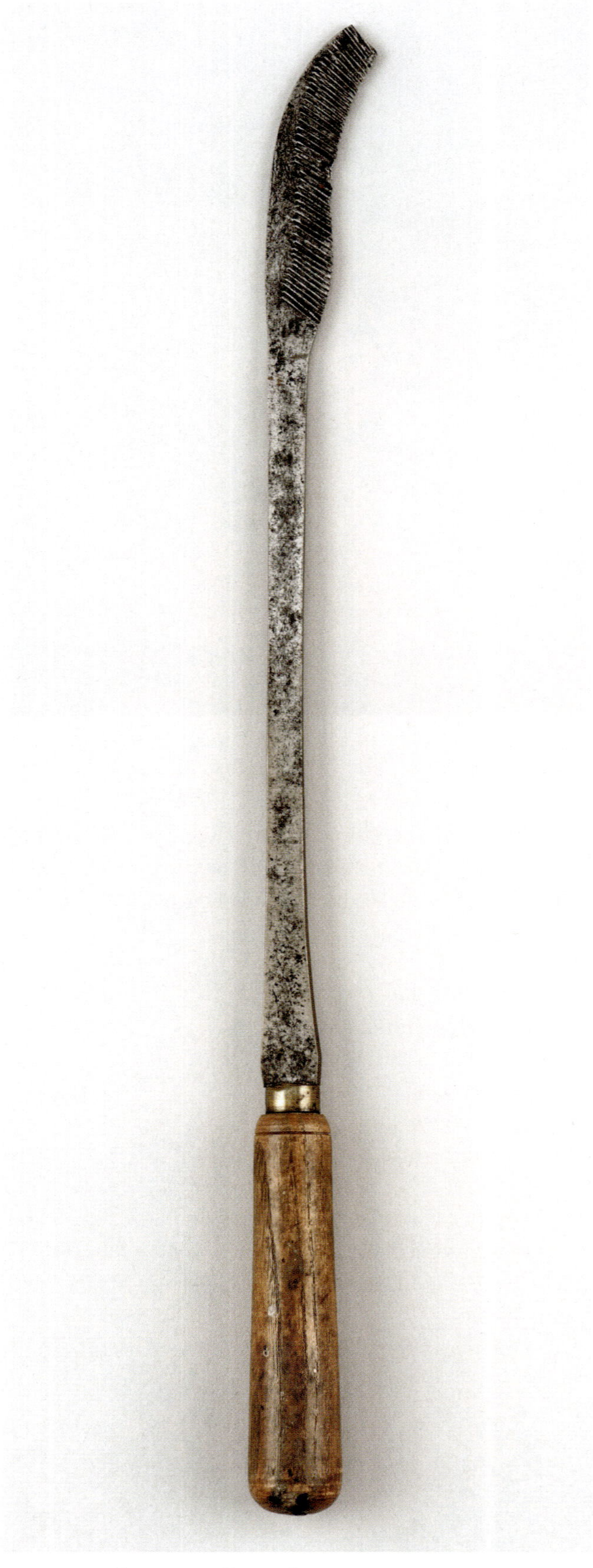

FIGURE 13 (inv. no. 20048-0000-)
© MAM/PHOTO FRANCK BOTTÉ

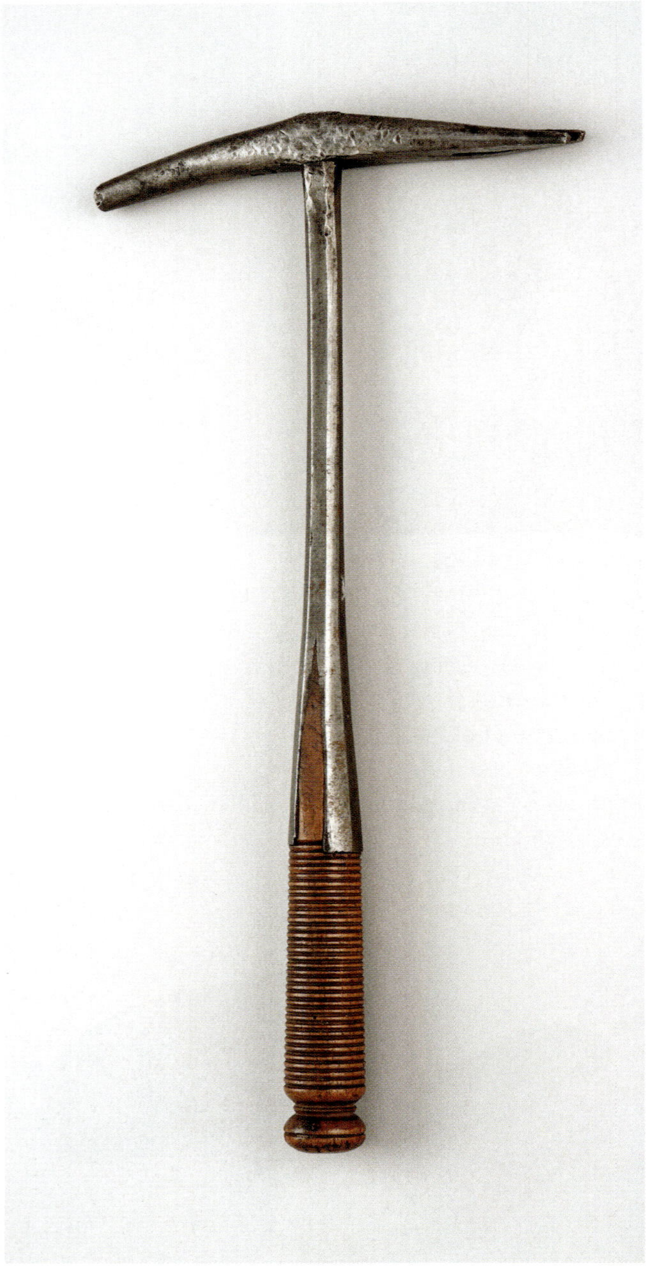

FIGURE 14 (inv. no. 20233-0000-)
© MAM/PHOTO FRANCK BOTTÉ

This type of hammer (also called a tack hammer) was used to fasten upholstery fabric to furniture frames with tacks or nails. It bears the inscription "3".

20234-0000 (Fig. 15)
Scraper
Late 18th c.
MAM accession date: 1952
Edme Régnier
21 × 11.5 × 6; 70 g
Steel, brass, wood

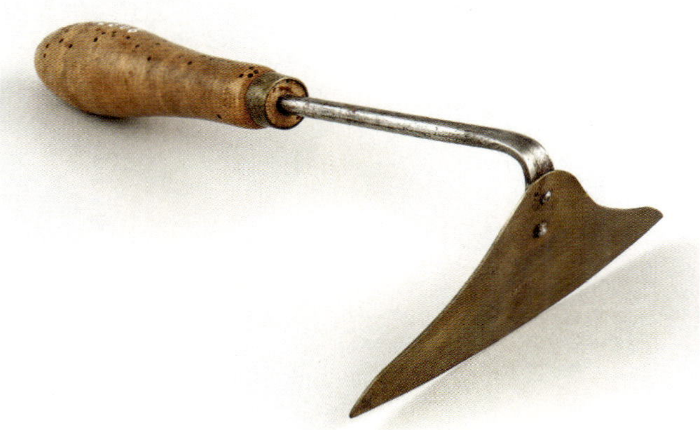

FIGURE 15 (inv. no. 20234-0000-)

Small scraper composed of a steel shaft ending in a perpendicular triangular blade inserted in a wooden handle. The tool, which bears the inscription "REGNIER", was probably made by the engineer and clock maker Edme Régnier (1751–1825), who also invented a popular spring dynamometer.

20230-0000- (Fig. 16)
Pen with wooden case
Late 18th c. (?)
MAM accession date: 1952
14.8 × 1.8; 20 g
Wood, steel

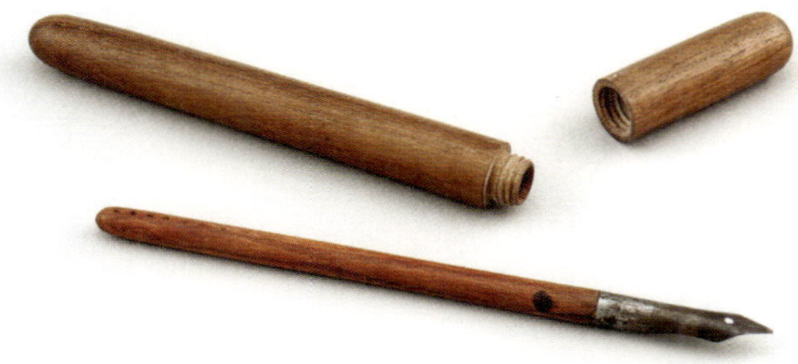

FIGURE 16 (inv. no. 20230-0000-)

Wooden pen with a steel nib inserted in a cylindrical wooden case. The shaft is inscribed with "12" at one end and decorated with a series of dots at the other end.

McKie 122
Ring sizer and box
Second half of 18th c.
Catalogued by Douglas McKie in 1952 but not included in the donation to the MAM
Box: 14.1 × 23.8 × 7.2
Ivory (bone?), brass, wood

Wooden box containing a ring sizer. The printed label inside the lid reads: "Richer // Ingénieur en Instrumens de Mathématique, // ci devant breveté du Roi, // Par L'Académie des Sciences, // Présentement Père et Fils, // Boulevard Saint-Antoine, N° 61, à Paris".

The box contains an ivory-handled brass ring sizer marked with numbers and fractions indicating different diameters. The ring sizer is held by movable pegs.

20072-0000- (Fig. 17)
Pen and inkwell
Second half of 18th c.

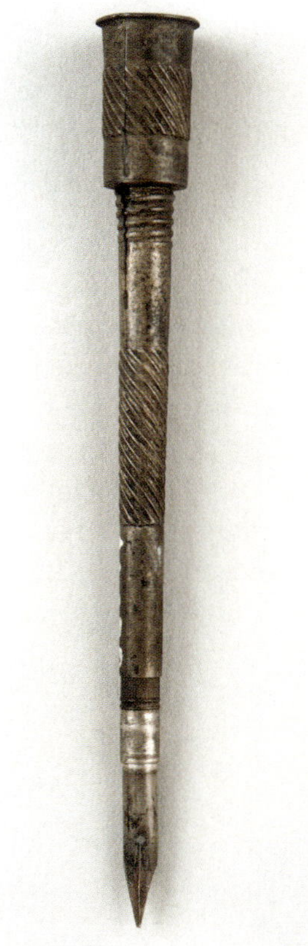

FIGURE 17
(inv. no. 20072-0000-)

MAM accession date: 1952
11.7 × 1.2; 10 g
Alloy, silver

Writing pen with a small inkwell. When not in use, the two elements can be screwed together.

20071-0000- (Fig. 18)
Quill pen cutter
Late 18th c.
MAM accession date: 1952
John Wilkes
1.5 × 6.5 × 2.8; 20 g (with the box)
Brass, ivory, cardboard, leather

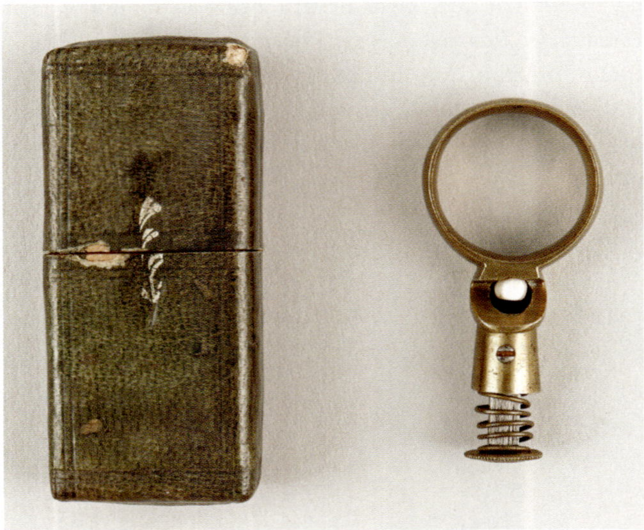

FIGURE 18 (inv. no. 20071-0000-)
© MAM/PHOTO FRANCK BOTTÉ

Quill pen cutter composed of a brass ring holding a short cylinder with an ivory peg and a stud with a spring. The cutter carries the inscription "Wilkes". It is housed in a cardboard case coated with decorated green leather.

The instrument served to cut a quill tip with precision so that it could be used as a nib. A few other specimens of this type are known and were probably patented by John Wilkes (late eighteenth-early nineteenth c.), an English professional pen cutter and author of *The art of making pens scientifically*.

Wilkes (1799)

20067-0000- (Fig. 19)
Glass filter
Second half of 18th c.
MAM accession date: 1952
5 × 3.5, mass 10 g
Glass, tortoiseshell

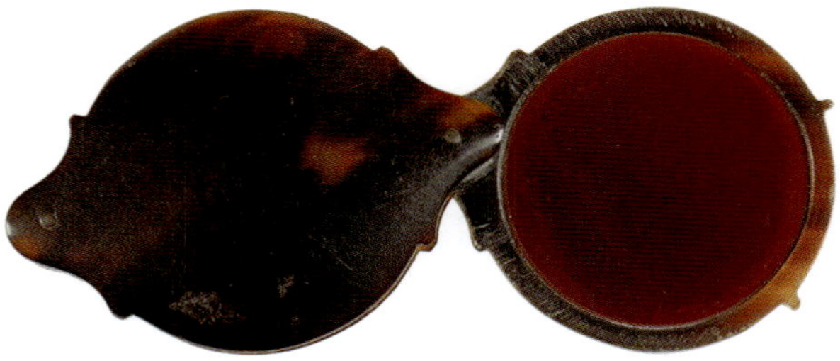

FIGURE 19 (inv. no. 20067-0000-)
© MAM/PHOTO FRANCK BOTTÉ

Brown glass filter that swivels in its tortoiseshell cover.

Lavoisier (1943)

20084-0000- (Fig. 20)
Magnifying lens with handle
Second half of 18th c.
MAM accession date: 1952
19.7 × 10.5 × 1.5; 130 g
Brass, wood, glass

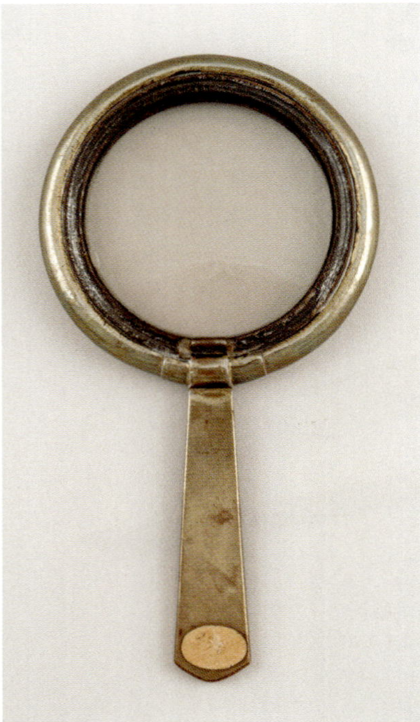

FIGURE 20
(inv. no. 20084-0000-)
© MAM/PHOTO
FRANCK BOTTÉ

Magnifying lens inserted in a pair of wooden rings held in a brass frame, with a brass handle. Focal length: ca. 28 cm.

20087-0000- (Fig. 21)
Compound lens magnifier
Late 18th c.
MAM accession date: 1952
Nöel Simon Carochez (1740 *c.*–1813)
7.5 × 4.5; 120 g
Silver, glass, brass, leather

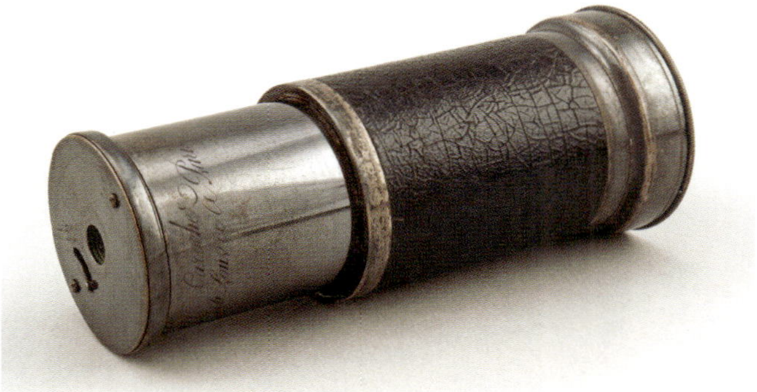

FIGURE 21 (inv. no. 20087-0000-)
© MAM/PHOTO FRANCK BOTTÉ

Two-tube magnifier covered in leather with silver mountings. Both the eyepiece and the objective are simple converging lenses. The tube bears the inscription "Carochez // Au Louvre Paris".

20065-0000- (Fig. 22)
Double lens magnifier
Second half of 18th c.
MAM accession date: 1952
16.8 × 4 × 1.4; 20 g
Tortoise, glass, silver, silk

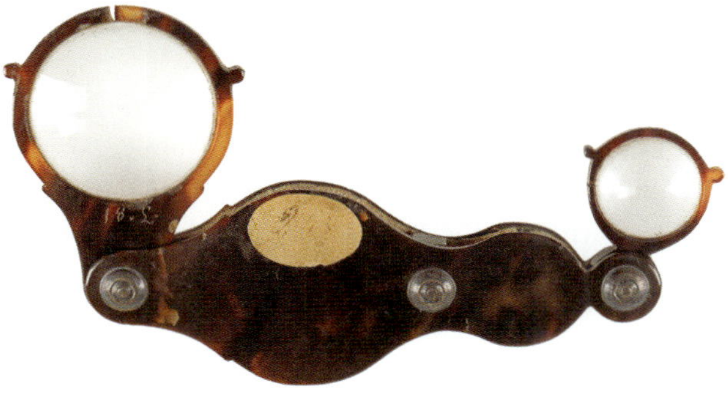

FIGURE 22 (inv. no. 20065-0000-)
© MAM/PHOTO FRANCK BOTTÉ

Two horn-rimmed biconvex lenses of different diameter can swivel out from their horn cover, where they are protected when not in use. The large lens is scratched and part of its horn rim is missing. Under this lens is the inscription "18.1".

Lavoisier (1943)

20066-0000- (Fig. 23)
Lorgnette (hand spectacles)
Late 18th c.
MAM accession date: 1952
20.7 × 2.6 × 0.6
Tortoise, glass, silver

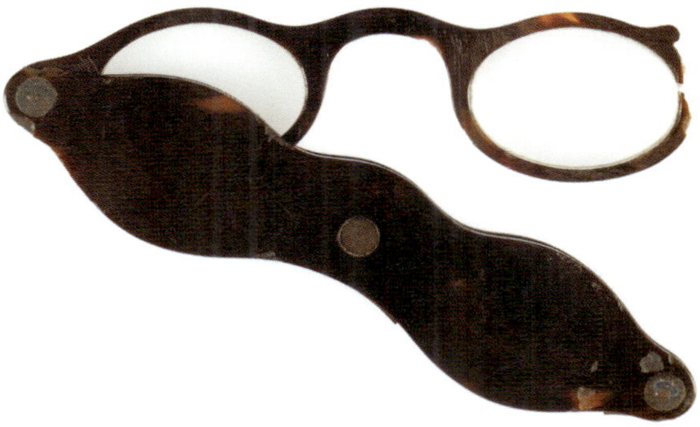

FIGURE 23 (inv. no. 20066-0000-)
© MAM/PHOTO FRANCK BOTTÉ

Lorgnette mounted in tortoiseshell chipped in a few places.

20068-0001-, 20068-0002- (Fig. 24)
Spectacles
Late 18th c.
MAM accession date: 1952
0001-: 9.3 × 4 × 0.3; 10 g
0002-: 8.9 × 4 × 0.3; 10 g
Glass, non-ferrous metal, fabric

Two almost identical spectacles (pince-nez type) with two convex lenses mounted in a metal frame. They were used to correct short-sightedness.

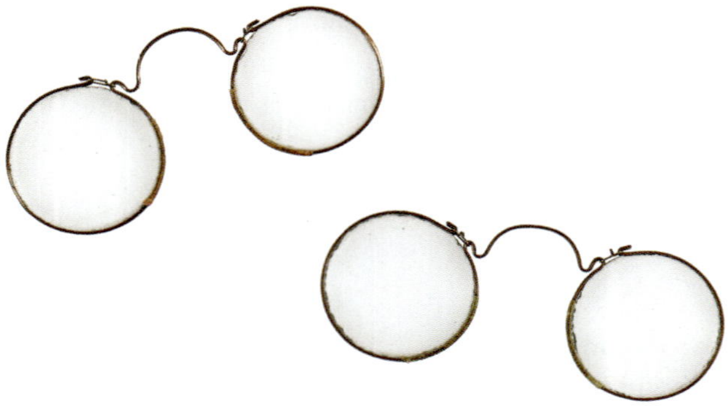

FIGURE 24 (inv. nos. 20068-0001-, -0002-)
© MAM/PHOTO FRANCK BOTTÉ

20193-0000- (Fig. 25)
Chinese snuff bottle
Second half of 18th c.
MAM accession date: 1952
12.5 × 4.5; 210 g
Tin?, paper

Small tin snuff bottle lined with yellow paper and painted with a Chinese pattern. It has two nested caps. The outer cap is decorated with a polyhedral blue glass. The bottle still contains powdered tobacco.

Snuff is a smokeless tobacco made from pulverised tobacco leaves. It is inhaled or "snuffed" into the nasal cavity, delivering a swift hit of nicotine and a lasting flavoured scent. Tobacco was introduced by the Portuguese to the court at Beijing during the second half of the sixteenth century. The use of snuff and snuff bottles spread through the upper class, and by the end of the seventeenth century it had become a part of social ritual. Snuff bottles soon became an object of beauty and a way to represent status and wealth.

As a Fermier Général, Lavoisier was also responsible for the administration of the tobacco trade and of the Ferme de Tabac.

Doré (1994)

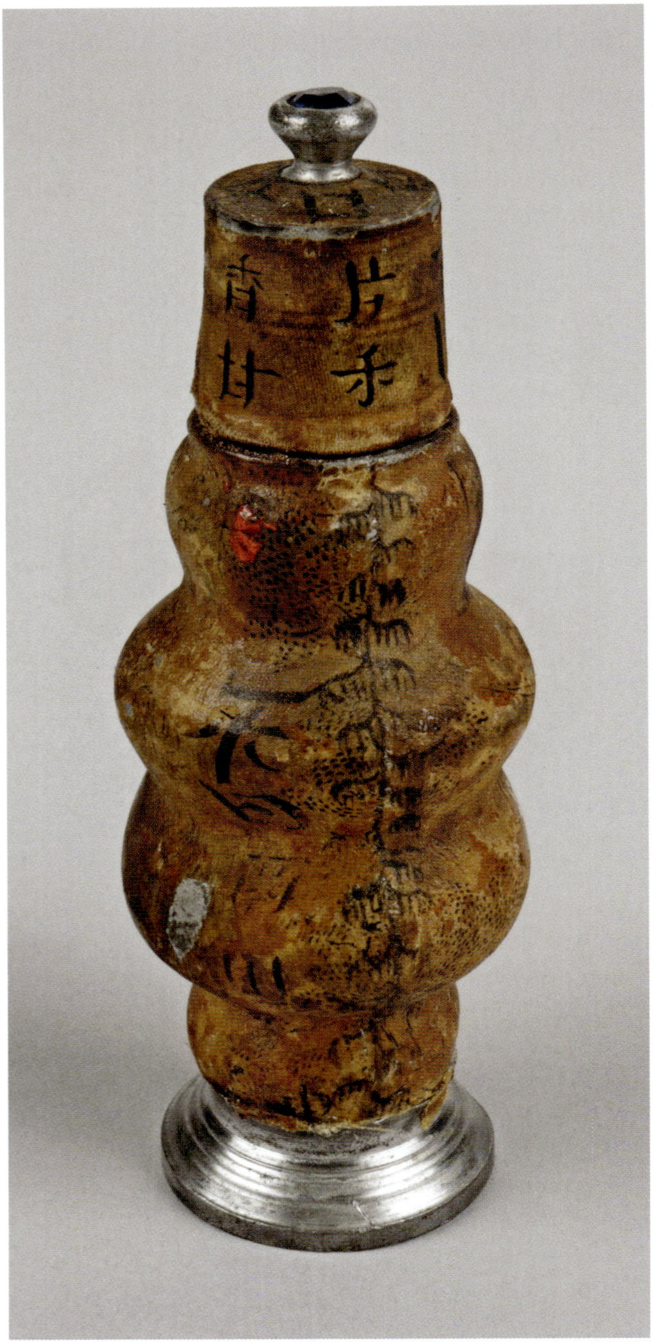

FIGURE 25 (inv. no. 20193-0000-)
© MAM/PHOTO FRANCK BOTTÉ

McKie 121
Marble tobacco jar
18th c.
Catalogued by Douglas McKie in 1952 but not included in the donation to the MAM
18 × 25
Marble, brass

Marble vessel. The lid is hinged with decorated gilt-wire brass hinges. The jar has side handles. The inside of the jar is hollowed out to form two cylindrical containers with lids, fitted with tags, to hold tobacco. There is a lock, but no key.

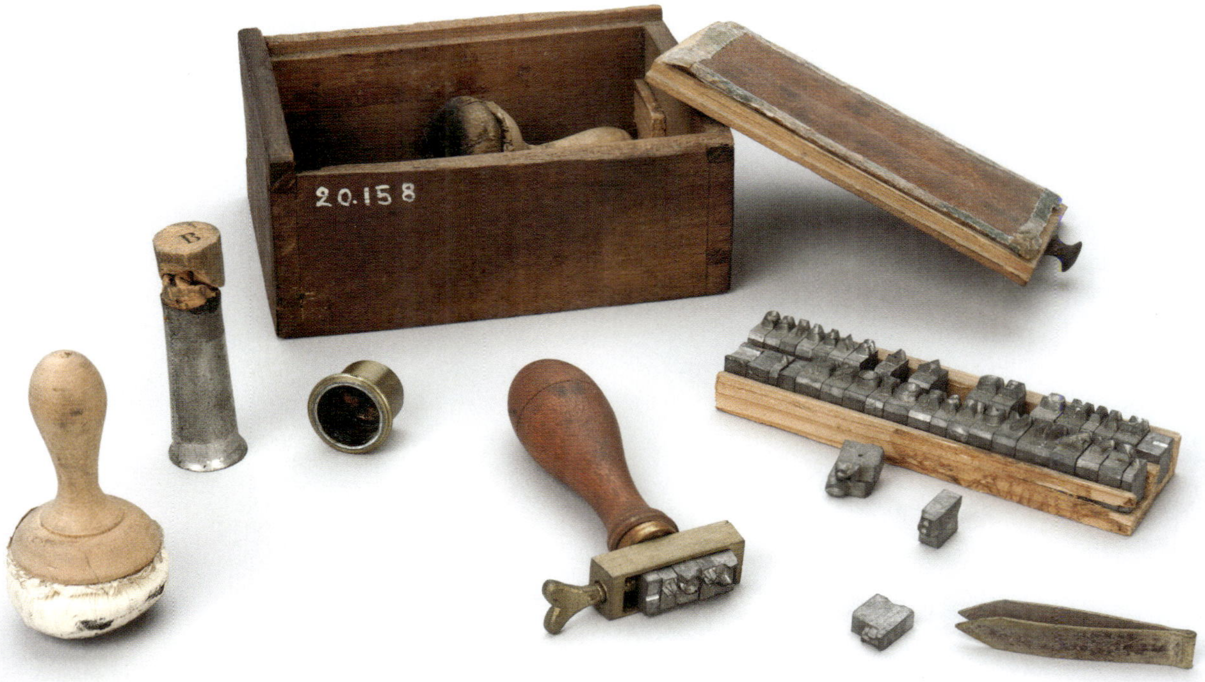

FIGURE 26 (inv. no. 20158-0000-)
© MAM/PHOTO FRANCK BOTTÉ

20158-0000- (Fig. 26)
Hand printing set
Late 18th c.
MAM accession date: 1952
Box: 11.5 × 8.2 × 5.5; 349 g
Lead alloy, tin, brass, cork, wood, leather

The set, housed in a wooden box, served to print a short line of alphanumeric fonts.
It includes:

a) A series of 45 type fonts (letters, numbers, spaces, punctuation marks) in a wooden frame.
b) A cylindrical tin ink container closed by a cork bearing the letter "B".
c) Two ink pads with wooden handles.
d) A hand printing tool with a wooden handle and a brass press with a screw. The fonts can be aligned and fixed in the press.
e) A leather plate to be used for stamping.
f) A brass collar, the use of which is unknown.

20056-0000- (Fig. 27)
Portable grinding wheel
1770–1780
MAM accession date: 1952
23.7 × 30.5 × 28.5; 7.35 kg
Wood, iron, grindstone, brass

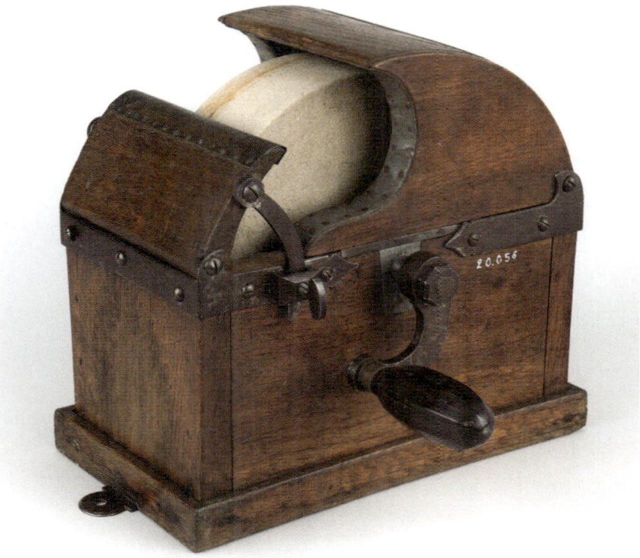

FIGURE 27 (inv. no. 20056-0000-)
© MAM/PHOTO FRANCK BOTTÉ

A grinding wheel made of natural stone (sandstone?) is enclosed in a wooden box reinforced with iron strips. The wheel can be rotated thanks to a crank with a wooden handle. In front of the wheel is a wooden board, partly covered with metal foil, whose tilt can be adjusted with a pair of metal arcs and screws. On opposite sides of the base are two protruding irons with holes for fastening the machine to a table.

The grinding wheel is used for sharpening small tools such as knives and lancets. Their angle on the wheel can be adjusted by varying the board's tilt.

20063-0000- (Fig. 28)
Reamer
Second half of 18th c.
MAM accession date: 1952
15 × 1.5 × 1; 25 g
Ivory, steel, cardboard, leather

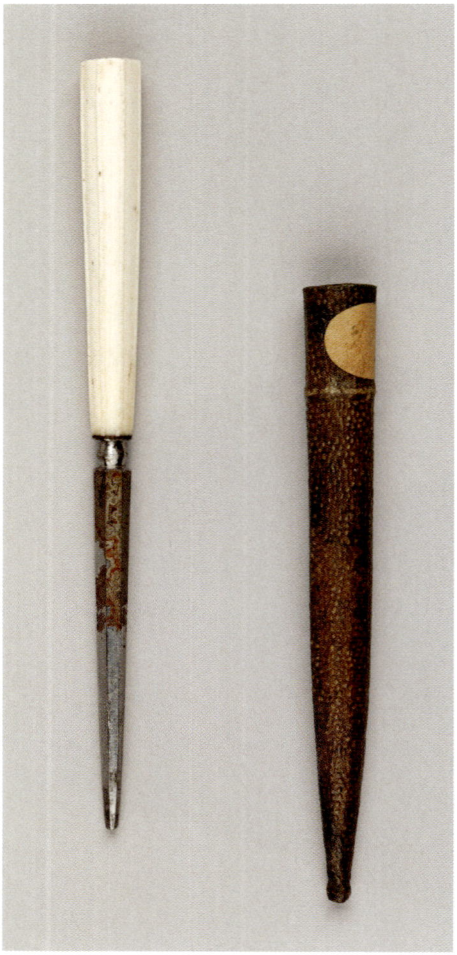

FIGURE 28 (inv. no. 20063-0000-)
© MAM/PHOTO FRANCK BOTTÉ

Tapered steel hand reamer with an ivory handle. It is housed in a leather pouch.

The tool was used for cleaning burrs from a drilled hole, or to enlarge a hole.

20042-0000- (Fig. 29)
Calibre
Second half of 18th c.
MAM accession date: 1952
27.5 × 51 × 2; 50 g
Wood, steel

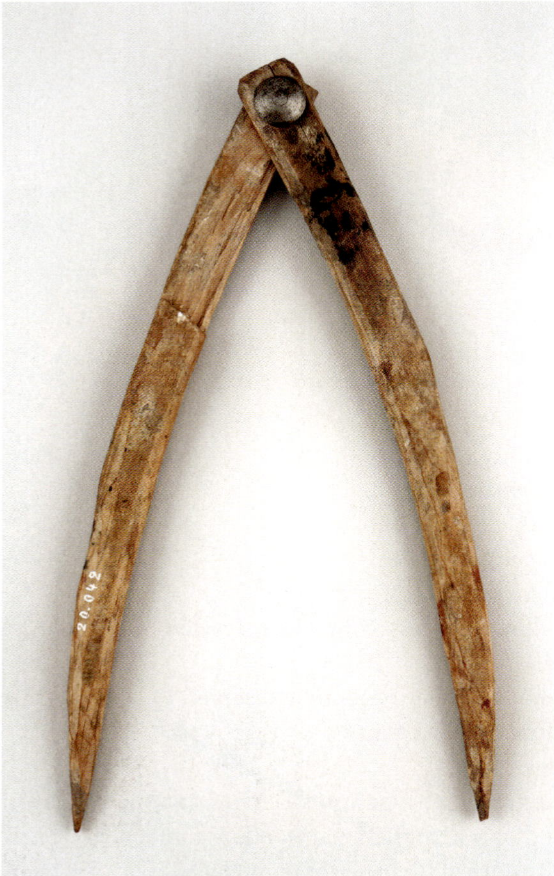

FIGURE 29 (inv. no. 20042-0000-)
© MAM/PHOTO FRANCK BOTTÉ

Calibre composed of two wooden arcs hinged together with a steel screw. This rudimentary device was used to take and transfer linear measurements.

20183-0000- (Fig. 30)
Set of six wax seals
Second half of 18th c.
MAM accession date: 1952
12.5 × 1.5; 105 g (a single seal)
Sealing wax

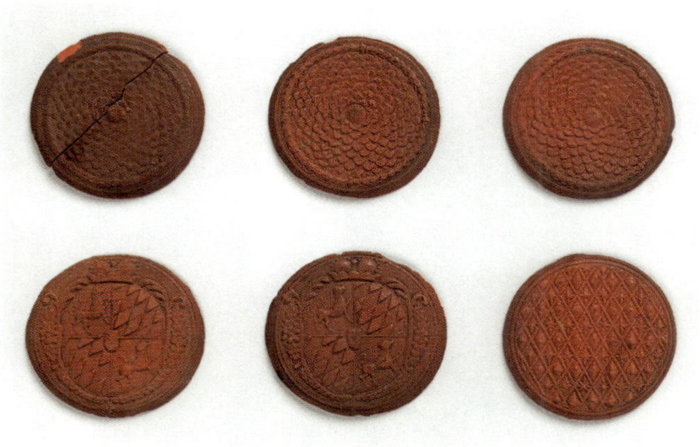

FIGURE 30 (inv. no. 20183-0000-)
© MAM/PHOTO FRANCK BOTTÉ

Six large red wax seals. Three are decorated with a concentric leaf pattern (one of these seals is broken in two), two reproduce a coat of arms and the sixth is decorated with a rhomboidal pattern.

20078-0000- (Fig. 31)
Coffee-making set with Rumford's percolator
ca. 1810–1812
MAM accession date: 1952
48 × 30 × 27; 4.26 kg
Tin, iron, leather, wood, brass

The set is placed in a tin box with a hinged lid. The box contains rings and fittings for placing the various items of the set. The box is enclosed in a leather case with straps and buckles and a handle.

The set includes the following items:

a) A tin coffee percolator. It is composed of a coffee pot with a lid (with a small wooden knob), a curved handle and a spout. There is a small flat hinged cover at the top of the handle. The filter (placed in the pot) consists of a cylindrical funnel with a perforated disk at the bottom. A cylindrical container with two folding handles is inserted under the filter, which remains suspended in the pot. A perforated disk with a handle serves as the rammer to press the coffee powder in the filter.

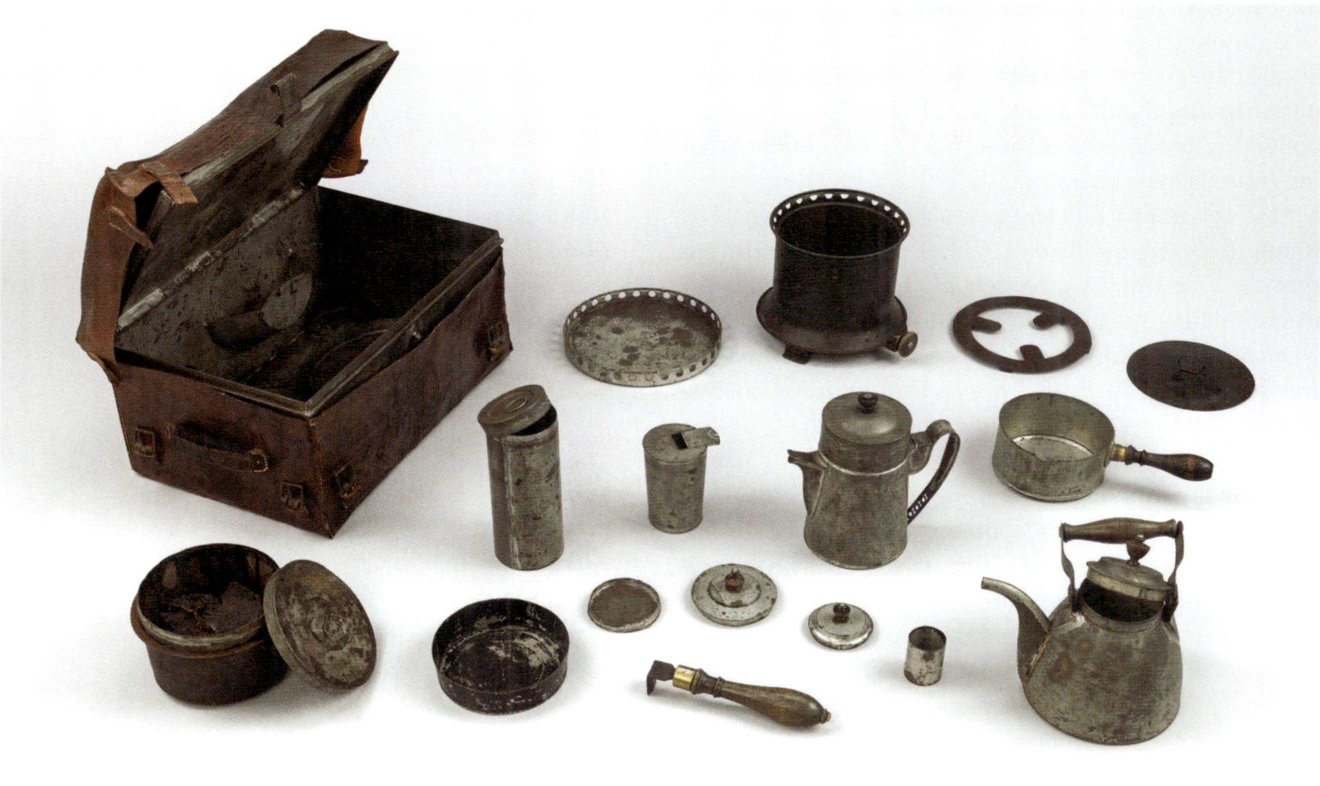

FIGURE 31 (inv. no. 20078-0000-)
© MAM/PHOTO FRANCK BOTTÉ

b) A tin kettle with a hinged wooden handle and a lid with a wooden knob.
c) A cylindrical blackened tin stove on a base a with perforated collar and a wooden handle. The latter can be removed when the item is placed in the box. The stove probably contained a spirit lamp, now missing. A flat ring with three protruding arms was put on the stove to support the kettle. A wooden knob is inserted in a square piece on the bottom of the stove. Its function could not be determined.
d) A cylindrical box closed by a lid with a handle ring inserted in the stove. It contains a brown substance that was the fuel to heat the coffee.
e) A tin pan with a wooden handle. The latter can be unscrewed when the item is placed in the box. The pan can be placed on a flat cylindrical box.
f) A cylindrical tin box with a sliding lid at the top.
g) A cylindrical tin box with a lid with a handle ring.
h) A small cylindrical tin box.
i) A circular lid with a handle ring (probably for the stove).
j) Two tin circular lids with wooden knobs.

The percolator worked as follows: the ground coffee was pressed with the rammer in the filter mounted on its container. Both were inserted in the coffee pot. The hot water was poured on the coffee and percolated in the container. To keep the drink hot, hot water was also poured in the pot through the opening near its handle.

Until the late eighteenth century, coffee was prepared as and infusion or decoction. The first percolators were not introduced until around 1800. The boiling (or almost boiling) water circulated through a small chamber holding ground coffee beans.

One of the first efficient percolators was designed by Rumford during the first decade of the nineteenth century. As a convinced promoter of the use of coffee for its stimulating quality, Rumford invented his percolator when he was working for the prince-elector of Bavaria and was conducting successful research to improve existing cooking, heating and lighting methods.

Apparently, it is no coincidence that Rumford's percolator remained among Marie Anne Lavoisier's belongings. In March 1787, Lavoisier's laboratory assistant, Jean-Henri Hassenfratz, presented Madame Lavoisier with a new model of *cafetière* that made it possible to boil coffee like an herbal tea more efficiently than with the standard machine. Hassenfratz also included two drawings.

LC, vol. 5, pp. 25–26; Thompson (1812)

20076-0000- (Fig. 32)
Travel sewing kit
End 18th c.
MAM accession date: 1952
9.5 × 2; 20 g
Ebony, ivory, thread

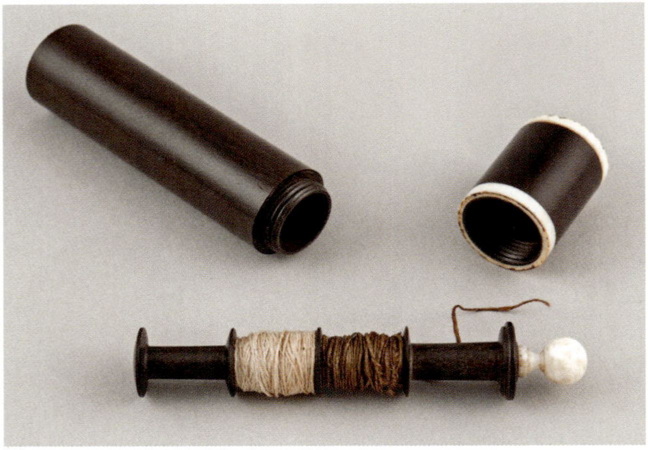

FIGURE 32 (inv. no. 20076-0000-)
© MAM/PHOTO FRANCK BOTTÉ

Travel sewing kit housed in a cylindrical box with a cap bound in ivory. A multiple spool in four sections still holds two bobbins of thread (white and brown). An ivory knob closes the hollow space in the centre of the spool, where the needles (missing) were kept.

Lavoisier (1943)

McKie 290
Globe
Catalogued by Douglas McKie in 1952 but not included in the donation to the MAM

Terrestrial globe on a stand.

McKie 296
Bronze bar
1763
Catalogued by Douglas McKie in 1952 but not included in the donation to the MAM
9.5 × 1.7
Bronze

Short bronze bar. The base carries the following inscription: "Portion de la matiere employee a la fonte de la statue pedestre du Roy coulée à l'Arcenal [*sic*] pour la ville de Reims le 29. Janv. 1763. 7 Part. de cuivre rouge et 6 part. de jeune". Top surface rough. The pedestrian statue of Louis XV was commissioned by the city of Reims from Jean-Baptiste Pigalle in 1757. The French artist was paid a fabulous sum for the artefact. The statue was cast at the Arsenal in January 1763 using a new process, in the presence of a delegation from Reims. It is unclear how this piece ended up in Lavoisier's collection.

19884-0000- (Fig. 33)
Engraving
1775
MAM accession date: 1952
Charpentier François Philippe
56 × 38.5 × 1.5; 1.74 kg (with frame)
35 × 24.5 (engraving alone)
Paper, glass, wood

Engraving glued on cardboard in a glazed gilt wooden frame. There are two inscriptions on the back of the cardboard: "Madame de Chazelles" and "Palais de la découverte". Under the engraving is the following inscription:

> Dessein en Perspective d'une Grande Loupe, formée par 2 Glaces de 52 po. de diam. Chacune, coulées à la Manufacture Royale de S.t Gobin, courbées et travaillées sur une portion de Sphere de 16 pieds de diam. Par M.r de Berniere, Controlleur des Ponts et Chaussées, et ensuite opposées l'une à l'autre par la concavité. L'espace lenticulaire qu'elles laissent entr'elles a été rempli d'esprit de vin, il y a quatre pieds de diam. et plus de 6 pouc. d'épaisseur au centre. Cette Loupe a été construite d'après le desir de l'Académie Roiale des Sciences aux frais et par les soins de Monsieur De Trudaine, Honoraire de cette Académie, sous les yeux de Messieurs de Montigny, Macquer, Brisson, Cadet et Lavoisier,

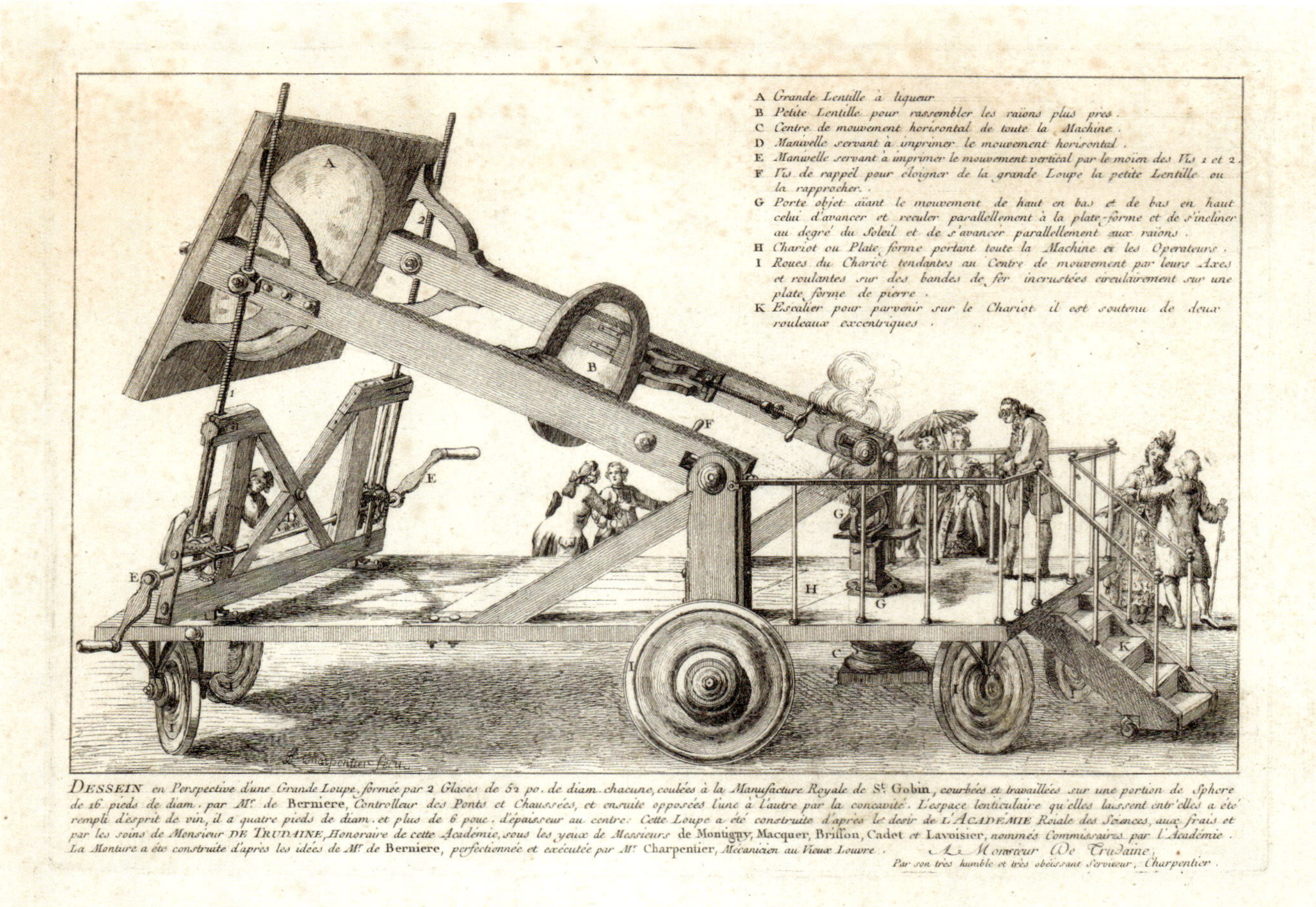

FIGURE 33 (inv. no. 19884-0000-)
© MAM/PHOTO FRANCK BOTTÉ

nommés Commissaires par l'Académie. La Monture a éte construite d'après les idées de M.r de Berniere, perfectionnée et exécutée par M.r Charpentier, Mécanicien au Vieux Louvre.

A Monsieur De Trudaine,

Par son très humble et très obéissant Serviteur, Charpentier.

(Perspective Drawing of a Large [Burning] Glass, composed of 2 Lenses of 52 inches in diameter. Each, cast at the Royal Manufactory of Saint-Gobain, was bent and shaped on a portion of a Sphere of 16 feet in diameter by M. de Bernière, Contrôleur des Ponts et Chaussées, and they were then assembled together on their concave sides. The lenticular space between them was filled with wine spirit. The diameter is four feet and the thickness at the centre more than six inches. This [Burning] Glass was built at the request of the Royal Academy of Sciences at the expense of and under the supervision of Monsieur de Trudaine, Honorary Member of the Academy, in the presence of Messrs. de Montigny, Macquer, Brisson, Cadet and Lavoisier, appointed Commissioners by the Academy. The Frame, built in conformity with the plans of M. de Bernière, was improved and produced by M. Charpentier, Mechanic at the Vieux Louvre.

To Monsieur de Trudaine,

By his very humble and very obedient Servant, Charpentier.)

The engraving, signed by Charpentier, illustrates the 1774 large-scale public experiment of the combustion of diamond made with Trudaine's large burning glass. The setting of the picture was probably the Jardin de l'Infante at the Louvre. Trudaine's lens was composed of "two spherical caps of 1.5 metres in diameter, the interstice of which was filled with 140 Paris pints (133 litres) of wine spirit" (Lehman, 2013). The burning lens was constructed at the expense of Jean-Charles-Philibert Trudaine de Montigny (1733–1777) and under the supervision of a committee of the Académie royale des sciences consisting of Brisson, Cadet, Lavoisier, Macquer and Trudaine. The lens was made at Saint Gobain on a machine designed by Claude Bernière with improvements by Charpentier. Owing to bad weather, the lens could not effectively be used until 1775. On the whole, the experiments made with it were not particularly innovative but they did not go unnoticed, and the Parisian press reported extensively on an event that attracted public attention.

Lavoisier (1774a); Lavoisier (1943); Duveen, Klickstein (1954a), p. 29; Lehman (2013)

McKie 21
Engraving
1784
Catalogued by Douglas McKie in 1952 but not included in the donation to the MAM

Three copper engraving blocks illustrating shells, potatoes and beetles respectively. One has the engraver's name and a date: "Robert 1784". Together with many other blocks, these were probably prepared to illustrate a work by Guettard.

19882-0000- (Figs. 34a, 34b)
Sanguine drawing of a nude man with his back turned
1786
MAM accession date: 1952
Marie Anne Lavoisier (1758–1836)
86 × 61 × 1.5: 4.12 kg
Paper, glass wood, cardboard

Sanguine drawing showing a study of a nude man turning his back. The drawing is mounted on a piece of cardboard inserted in a glazed wooden frame. It carries a handwritten annotation by the painter Jacques-Louis David: "jusqu'a present / je suis on ne peut plus / content David / le 2 mars 1786". Marie Anne Lavoisier was a skilled painter and a pupil of David. She made two portraits of Benjamin Franklin in 1787. In the early 1790s, she painted four portraits of her husband, his assistants and herself in the laboratory at the Arsenal while performing the experiments on respiration and perspiration. In 1788, she illustrated Lavoisier's *Traité élémentaire de chimie* (1789) and signed its 13 engraved plates. There is evidence that Marie Anne Lavoisier continued to be interested in the figurative arts and perhaps in painting after 1794. In September 1805, she received from a Bachelier an invoice for a drawing course she attended sometime before. During the same period, she was engaged in preparing new drawings of chemical apparatuses for the plates to illustrate the planned edition of Lavoisier's *Mémoires de chimie et de physique*. The work was never completed.

Since many of the paintings and drawings inventoried during the confiscation of Lavoisier's properties were sold

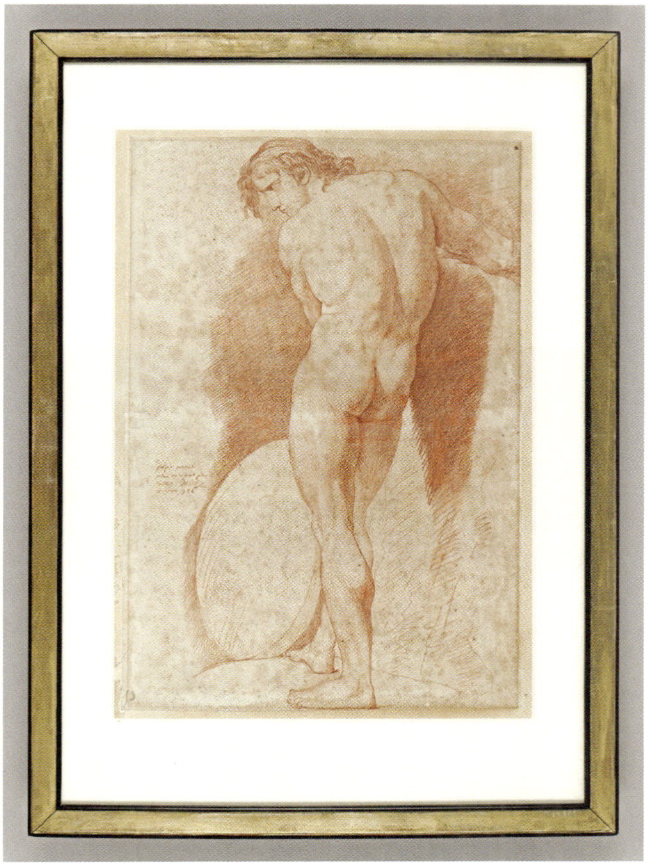

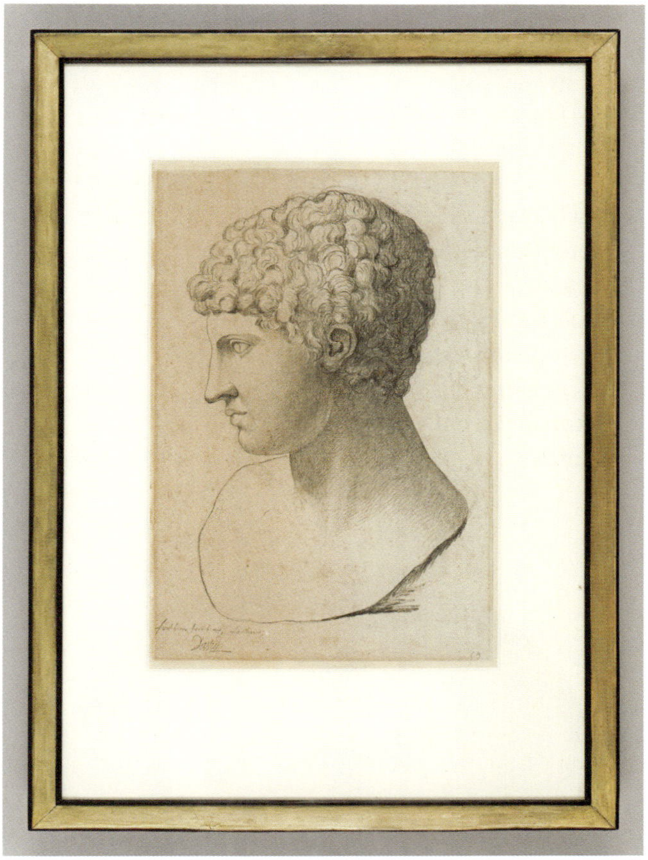

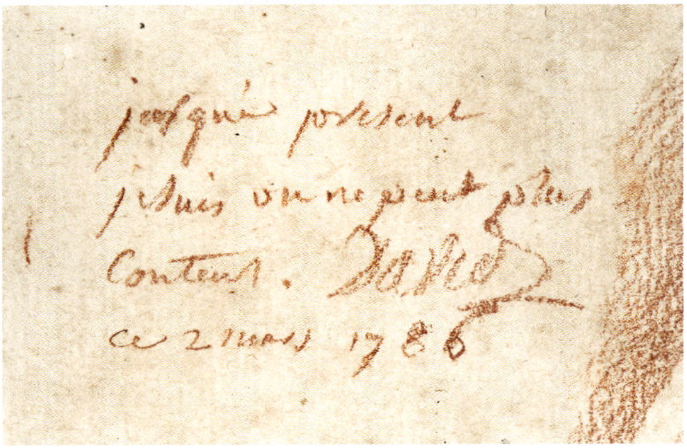

FIGURES 34A, 34B (inv. no. 19882-0000-)
© MAM/PHOTO FRANCK BOTTÉ

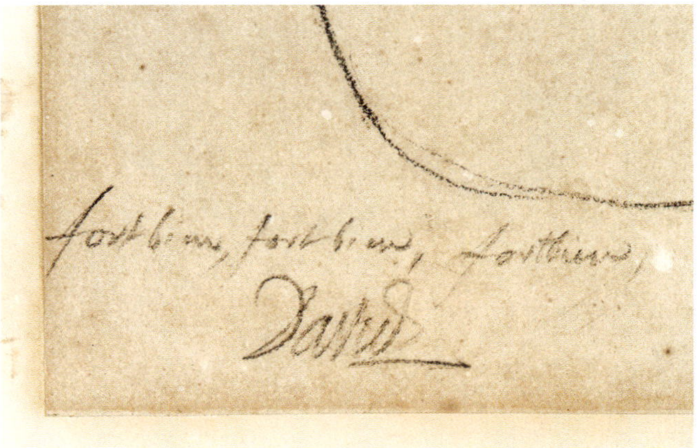

FIGURES 35A, 35B (inv. no. 19883-0000-)
© MAM/PHOTO FRANCK BOTTÉ

or dispersed, a number of unrecorded iconographic items portraying the French chemist were likely produced during his lifetime by professional artists and by his wife.

Beretta (2001); Beretta (2012)

19883-0000- (Figs. 35a, 35b)
Pencil drawing of a male head
1786
MAM accession date: 1952
Marie Anne Lavoisier (1758–1836)
Glass, wood, cardboard, paper
85.7 × 61 × 3.7: 3.81 kg

Pencil drawing of a male head by Marie Anne Lavoisier, in a glazed wooden frame. An autograph annotation by the painter Jacques-Louis David reads "fort bien, fort bien, fort bien.". The signature is hidden by the passe-partout. See historical note to no. 19882-0000-.

Beretta (2001); Beretta (2012)

20077-0000- (Fig. 36)
Medallion with a portrait of Lavoisier
1843–1853
MAM accession date: 1952
Pierre-Jean David d'Angers (1788–1856)
Eck et Durand (1843–1863)
Diameter 16.5; thickness 1.8; 225 g
Bronze

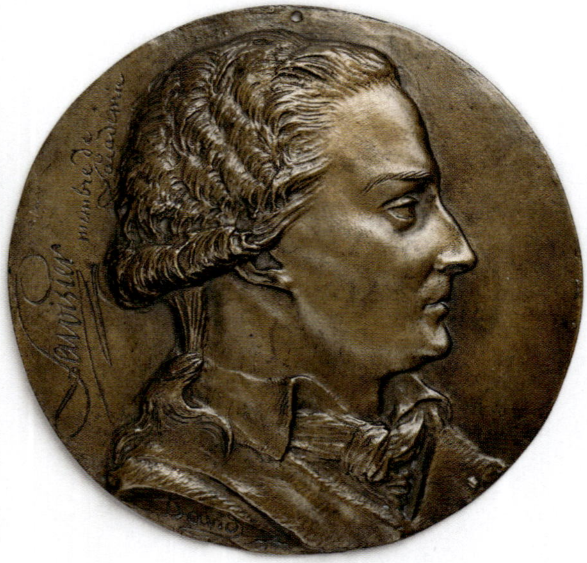

FIGURE 36 (inv. no. 20077-0000-)
© MAM/PHOTO FRANCK BOTTÉ

Bronze medallion representing Lavoisier's profile. The front displays the moulded inscriptions "Lavoisier" (reproducing the scientist's signature), "membre de l'académie" and the artist's name "David". The inscription on the back reads "Eck and Durand". A paper label on the back carries the number "261".

The medallion, made by the prolific sculptor and medallist David d'Angers (1788–1856), was probably inspired by the portrait by the French painter Marie-Renée-Geneviève Brossard de Beaulieu (1755-after 1835). The moulder Jean-Georges Eck (1795–1863) and the chiseller Pierre Durand (1794–1880) were partners in Paris between 1843 and 1853.

Beretta (2001), p. 94

20108-0000- (Figs. 37a, 37b)
Chevalier's camera lucida
After 1834
MAM accession date: 1952
Charles Chevalier (1804–1859)
Camera: 49 × 8 × 4; 160 g
Case 25.3 × 5.5 × 3.3; 250 g
Brass, glass, wood, fabric

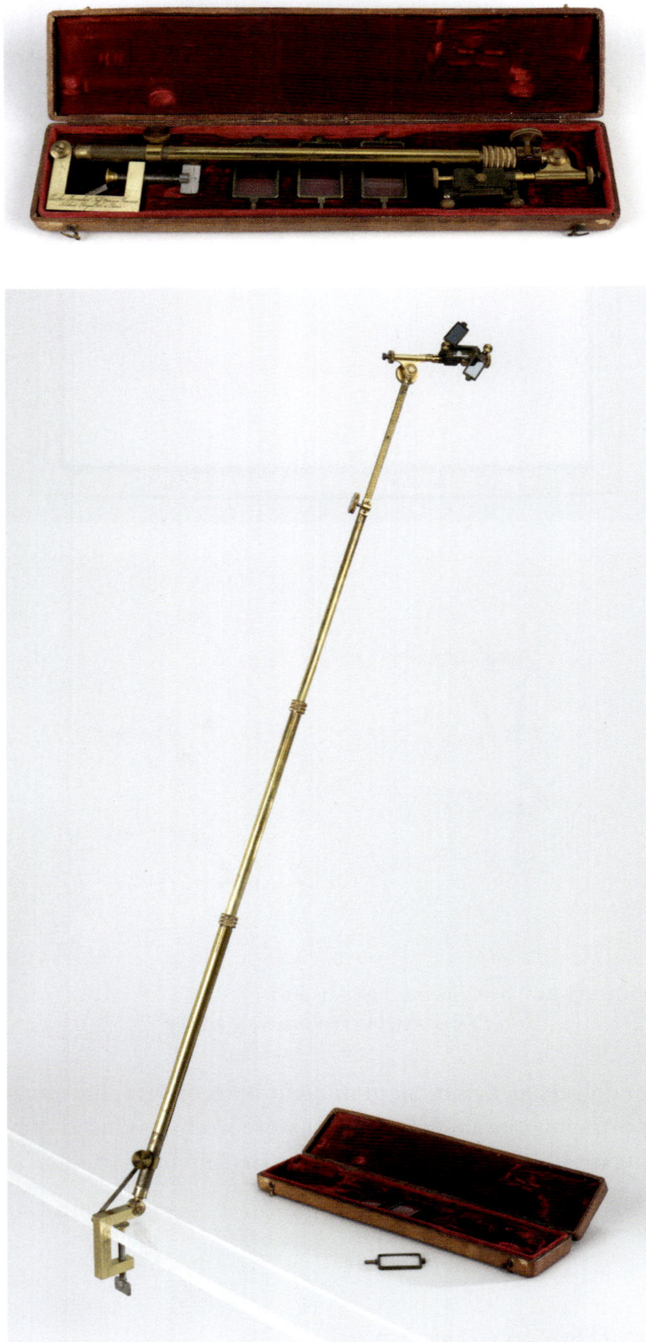

FIGURES 37A, 37B (inv. no. 20108-0000-)
© MAM/PHOTO FRANCK BOTTÉ

Camera lucida with a telescopic arm mounted on a quadrangular clamp with a blocking screw.

The arm's tilt can be adjusted by means of a jaw (with a screw) sliding on the arm. On top of the camera is an adjustable cross-arm whose position can be set by another screw. The cross-arm holds a small metal box containing the optical system, which consists of a prism and a plane parallel glass. The camera comes with three square lenses and two filters, which could be placed in front of the optical system. The device is contained in a case internally lined with red velvet.

The clamp bears the inscription: "Charles Chevalier Ing r Opticien Breveté / Palais Royal 163 à Paris".

The camera lucida had to be fixed on a table covered with a sheet of paper and properly oriented. The user looking into the window of the box could see the drawing paper with the image of the subject (such as a person or landscape) superimposed in front of it. The image could thus be easily copied. The lens served to adjust the subject's size to the drawing paper. The filters were used to dim excess luminosity of the subject or of the paper.

This instrument was very popular throughout the nineteenth century and was widely used by artists, architects, portraitists, watercolourists and others.

In 1834, the French instrument maker Jacques Louis Vincent Chevalier (1770–1841) published a description of an improved version of the camera lucida presented in 1818 by the Italian optician and astronomer Giovan Battista Amici (1786–1863). Jacques Louis Vincent's son, Charles Chevalier (1804–1859), opened his own business in 1831.

Chevalier (1834)

Fragments

This section includes parts of apparatuses and broken pieces that we were unable to identify.

20157-0000- (Fig. 1)
Case (empty)
Second half of 18th c.
MAM accession date: 1952
17.5 × 3 × 2.7; 30 g
Leather, cardboard

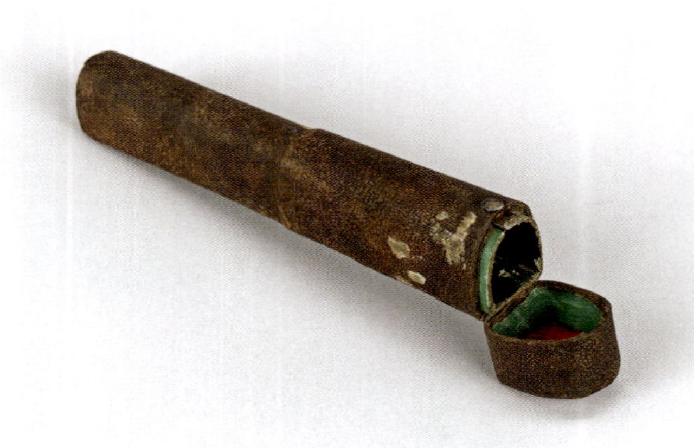

FIGURE 1 (inv. no. 20157-0000-)
© MAM/PHOTO FRANCK BOTTÉ

Empty elongated leather case covered by cardboard, with a hinged cover and a spring catch. It probably contained a bubble level.

20196-0001- (Fig. 2)
Piece of resin
Second half of 18th c
MAM accession date: 1952
7 × 7 × 4.5; 100 g
Piece of resin (shellac, copal?). It is unclear why it was catalogued with a series of geometrical models.

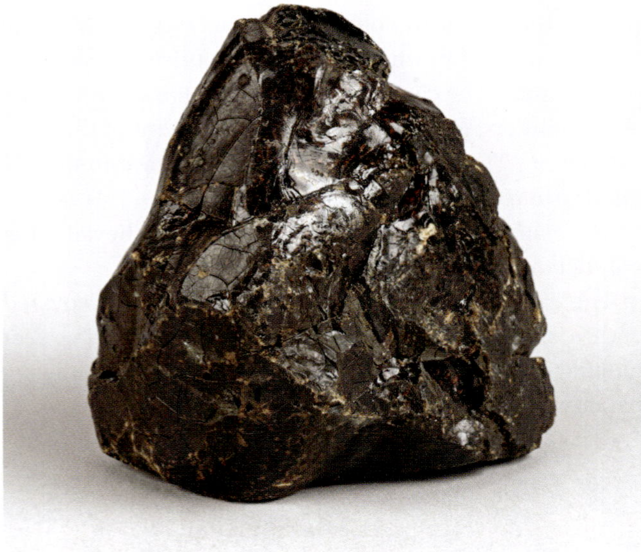

FIGURE 2 (inv. no. 20196-0001-)
© MAM/PHOTO FRANCK BOTTÉ

C-2017-0066- (Fig. 3)
Fragment of resin
Second half of 18th c.
MAM accession date: ?
5.2 × 1.5 × 1.5; 5 g
Resin

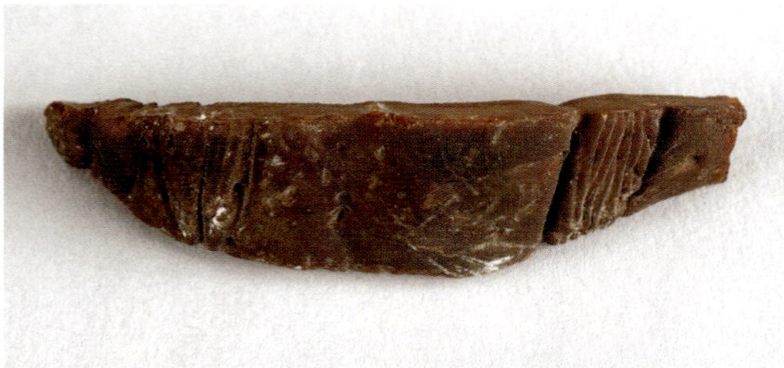

FIGURE 3 (inv. no. C-2017-0066-)
© MAM/PHOTO FRANCK BOTTÉ

Oblong fragment of semi-plastic material (natural resin?).

 | DOI:10.1163/9789004511217_021

20198-0001-, 20198-0002- (Fig. 4)
Two threaded keys
Second half of 18th c.
MAM accession date: 1952
6 × 4.5 × 1.8; 45 g
Tin, brass

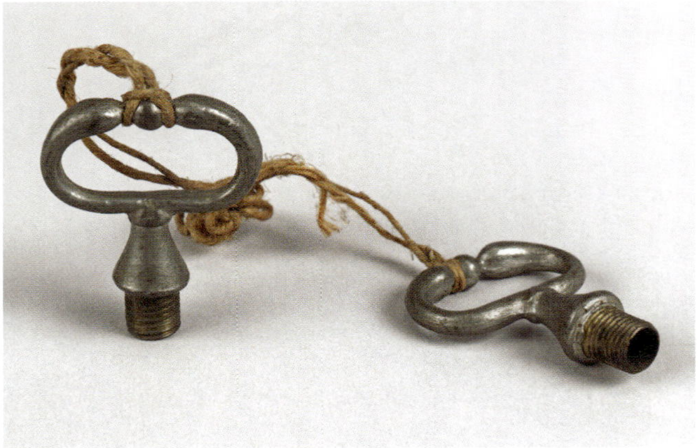

FIGURE 4 (inv. nos. 20198-0001-, -0002-)
© MAM/PHOTO FRANCK BOTTÉ

Two identical keys with an oval tin bow and a hollow threaded brass stem. They were probably part of an unidentified instrument.

20221-0001-, 20221-0002-, 20221-0003- (Fig. 5)
Three butterfly screws
Second half of 18th c.
MAM accession date: 1952
20221-0001- and 20221-0002-: 8 × 3.7 × 1.2; 35 g
20221-0003-: 20 × 4.2 × 1.6; 95 g
Brass

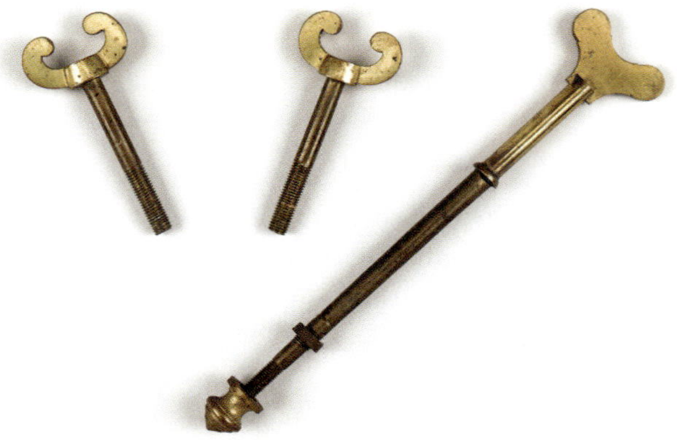

FIGURE 5 (inv. nos. 20221-0001-, -0003-)
© MAM/PHOTO FRANCK BOTTÉ

Two of these butterfly screws are identical, the third is longer. A small ring and a cap are screwed on the threaded end of the third. These screws belonged to unidentified instruments.

20223-0000- (Fig. 6)
Bar with two screws[1]
Second half of 18th c.
MAM accession date: 1952
31.5 × 5 × 1.3; 200 g
Copper

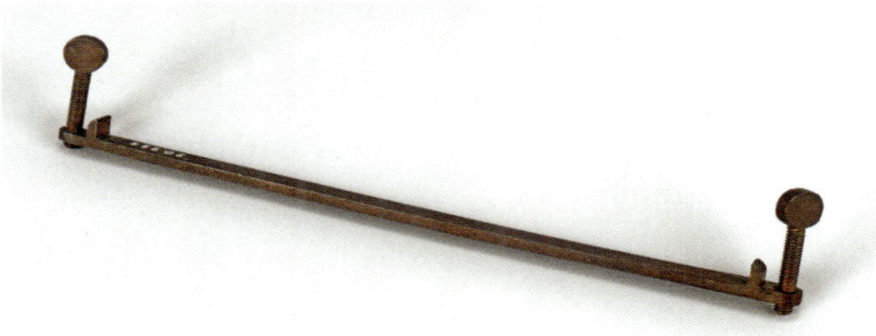

FIGURE 6 (inv. no. 20223-0000-)
© MAM/PHOTO FRANCK BOTTÉ

Quadrangular copper bar with two levelling screws and two uprights at its ends. Possibly used for testing spirit levels.

C-2017-0065- (Fig. 7)
Two brass bars
Second half of 18th c.
MAM accession date: 1952
17 × 1.5 × 1.2; 75 g
16.5 × 3 × 1.5; 80 g
Brass

1 The items arrived in the Museum with the Lavoisier collection but is not mentioned in any list of the de Chazelles collection.

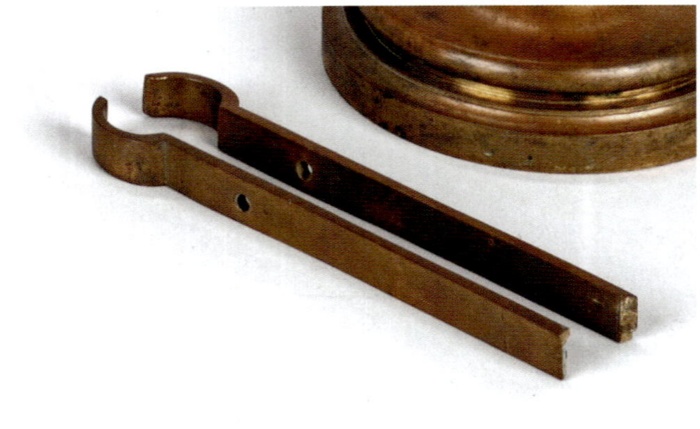

FIGURE 7 (inv. no. C-2017-0065-)

Two flat bars (one slightly bent), each with a hole and a curved end. They were probably used as holders in an unidentified instrument.

McKie 123
Brass part of instrument in a box
1808
Catalogued by Douglas McKie in 1952 but not included in the donation to the MAM
Jean-François Richer (1743–ca. 1820)
Brass, wood (?)

Box containing a brass support for an unidentified instrument invented by Rumford. The support has a scale marked 0–200 and a slider. The box is inscribed "Inv. par Mr le Comte de Rumford // Ex par Richer // à Paris 1808".

20172-0000- (Fig. 8)
Copper sphere
Second half of 18th c.
MAM accession date: 1952
32 × 10; 460 g
Copper, brass, glass, sealing wax

Hollow copper sphere with a straight brass pipe. The end of the pipe contains a broken glass tube. A paper label on the sphere bears the handwritten number "96".

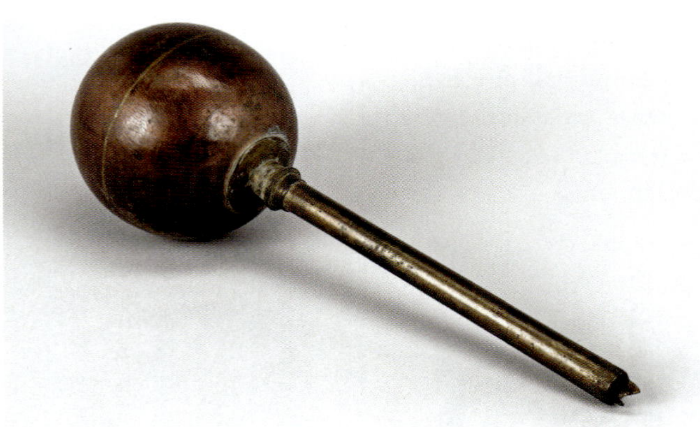

FIGURE 8 (inv. no. 20172-0000-)

This item (perhaps a small boiler) is a part of an unidentified instrument.

20218-0000- (Fig. 9)
Brass sphere
Second half of 18th c.
MAM accession date: 1952
28 × 10; 640 g
Brass

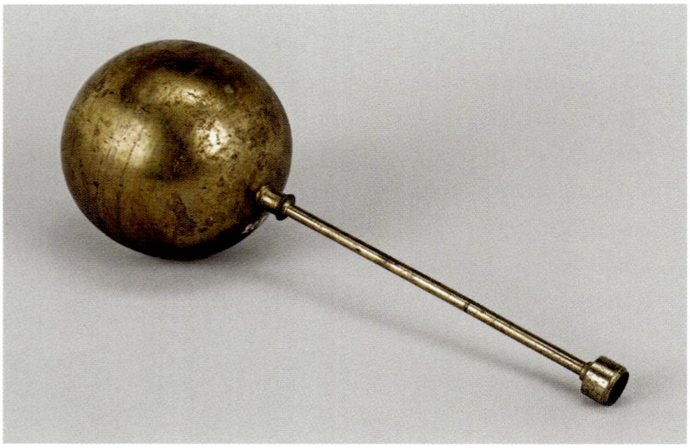

FIGURE 9 (inv. no. 20218-0000-)

Hollow brass sphere fixed to a thin straight tube, whose open end has an internally threaded collar. It is probably part of an unidentified apparatus (for hydrostatic experiments?).

20199-0000- (Fig. 10)
Cylinder with plunger
Second half of 18th c.
MAM accession date: 1952
18.5 × 5; 110 g
Iron, brass cork

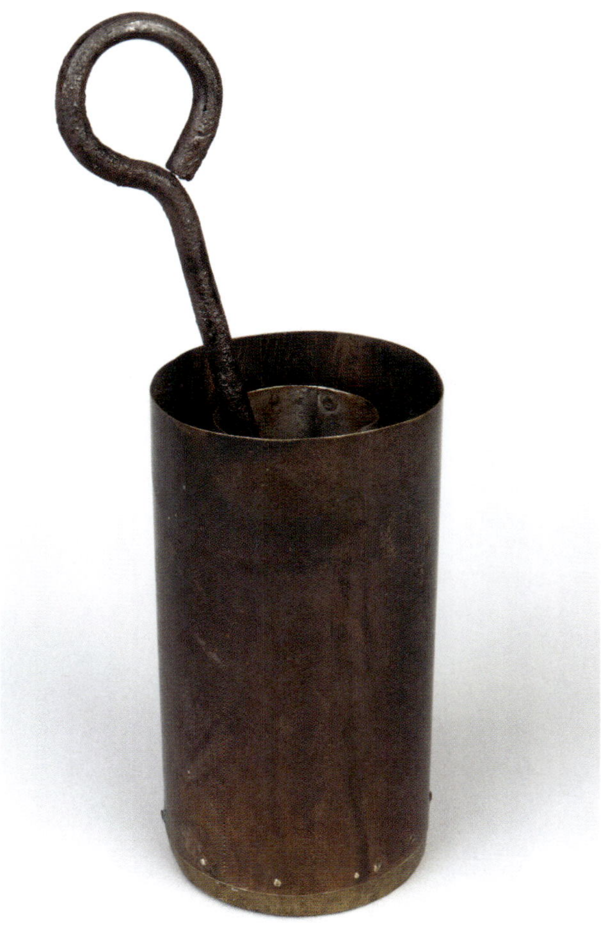

FIGURE 10 (inv. no. 20199-0000-)
© MAM/PHOTO FRANCK BOTTÉ

Iron plunger with a cork gasket and a rod ending in a loop is inserted in a thin brass cylinder. The latter is enclosed and riveted in a second iron cylinder.

20129-0001- (Fig. 11)
Round piece (base?)
Second half of 18th c.
MAM accession date: 1952
12.5 × 5; 1.28 kg
Brass

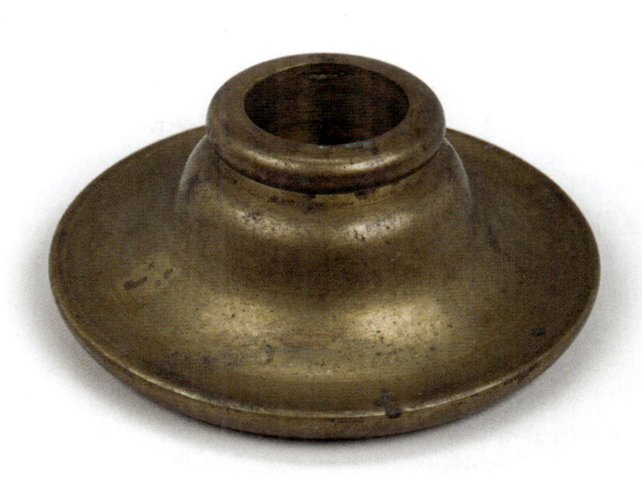

FIGURE 11 (inv. no. 20129-0001-)
© MAM/PHOTO FRANCK BOTTÉ

Round brass piece with a partially threaded hole in its centre. It was probably part of an unidentified instrument (a base?).

20225-0000- (Fig. 12)
Brass cap
Second half of 18th c.
MAM accession date: 1952
2.7 × 1.6; 25 g
Brass

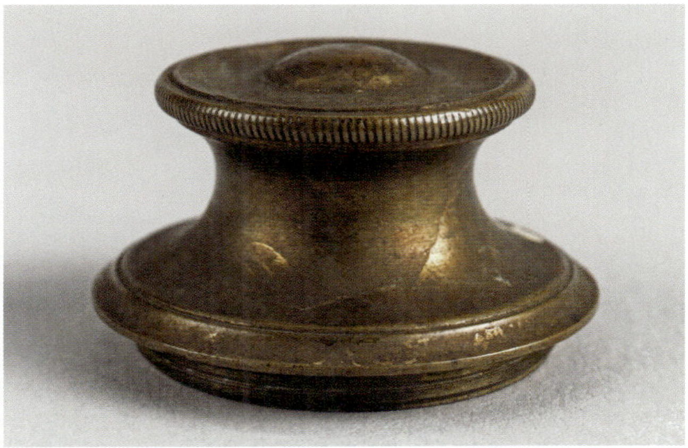

FIGURE 12 (inv. no. 20225-0000-)
© MAM/PHOTO FRANCK BOTTÉ

Threaded brass cap belonging to an unidentified instrument.

20070-0000-
Ebony rod
Second half of 18th c.
12.5 × 0.5; 10 g
Ebony

Ebony rod with threaded ends, one of which is protected by a cap. The item's function is not clear. After its cataloguing the item was lost and could not be photographed.

McKie 415
Rosewood strip
Second half of 18th c. (?)
Catalogued by Douglas McKie in 1952 but not included in the donation to the MAM
h. (?); base 4.8 × 5.4
Wood, brass

Long rosewood strip with brass-bound ends, divided in one side along its whole length by a line scratched on the wood, with transverse scratches at intervals of 16.15 cm, some of these transverse marks being made right across and others only half-way. There are twelve of these intervals, the last ending on the brass end-piece with 2.1 cm to spare.

20035-0000- (Fig. 13)
Box with glass tubes
Second half of 18th c.
MAM accession date: 1952
Box: 55.5 × 6 × 4; 715 g
Wood, glass leather, brass, thread

Wooden box containing 26 fragments of glass tubes, capillaries, and other glass bits and pieces. The box also contains a small leather cushion and is subdivided by two small shaped wooden boards. Originally used for a fragile instrument, now missing.

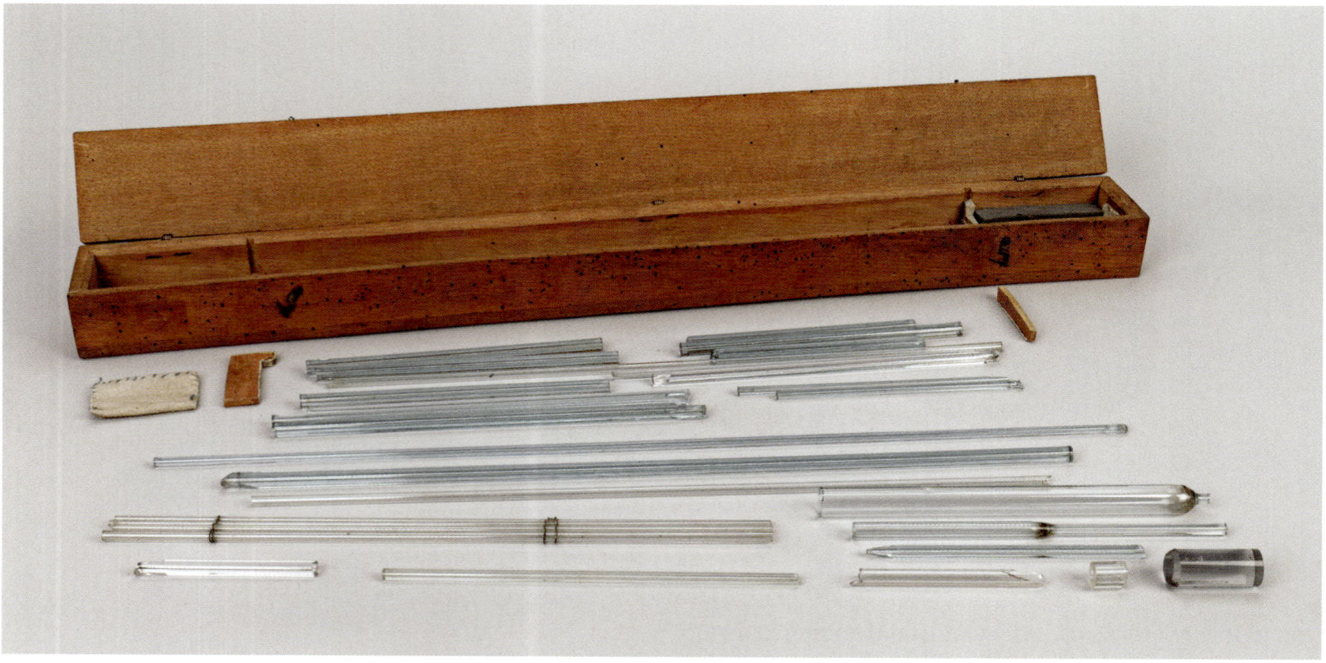

FIGURE 13 (inv. no. 20035-0000-)
© MAM/PHOTO FRANCK BOTTÉ

20208-0001-, 20208-0002- (Fig. 14)
Square glass plates
Second half of 18th c.
MAM accession date: 1952
20208-0001-: 16.7 × 16.7 × 0.9; 610 g
20208-0002-: 16.8 × 16.8 × 0.9; 630 g
Glass

FIGURE 14 (inv. nos. 20208-0001-, -0002-)
© MAM/PHOTO FRANCK BOTTÉ

Two almost identical square plates made of frosted glass. One (20208-0002-) has a handwritten numerical table.

20209-0001-, 20209-0002-, 20209-0003-, 20209-0004-, 20209-0005-, 20209-0006-, 20209-0007- (Fig. 15)
Round glass plates
Second half of 18th c.
MAM accession date: 1952
20209-0001-: 6 × 0.4; 30 g
20209-0002-: 6.7 × 0.7; 65 g
20209-0003-: 6.5 × 0.7; 60 g
20209-0004-: 6.5 × 0.8; 80 g
20209-0005-: 7 × 0.9; 70 g
20209-0006-: 7.2 × 0.5; 60 g
20209-0007-: 7.5 × 0.9; 95 g
Glass

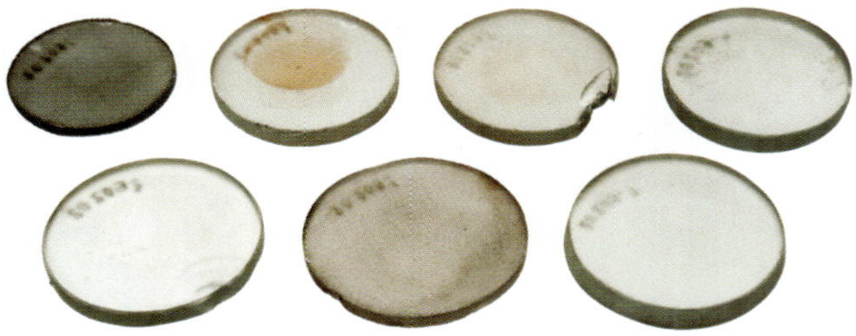

FIGURE 15 (inv. nos. 20209-0001-, -0007-)
© MAM/PHOTO FRANCK BOTTÉ

Seven round plates made of frosted glass. Nos. 20209-0003- and 20209-0005- are chipped, while nos. 20209-0001- and 20209-0006- are made of a darker smoked glass.

20207-0001-, 20207-0002-, 20207-0003-, 20207-0004-, 20207-0005- (Fig. 16)
Round glass plate
Second half of 18th c.
MAM accession date: 1952
20207-0001-: 10 × 0.4; 85 g
20207-0002-: 10.7 × 0.2; 40 g
20207-0003-: 12 × 0.4; 130 g
20207-0004-: 14 × 0.6; 235 g
20207-0005-: 15 × 0.4; 195 g
Glass

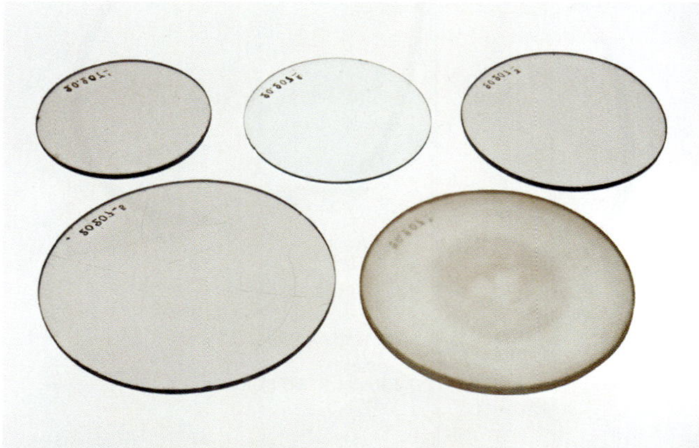

FIGURE 16 (inv. nos. 20207-0001-, -0005-)
© MAM/PHOTO FRANCK BOTTÉ

Four round glass plates. Nos. 20207-0001- and 20207-0003- are slightly grey, no. 20207-0004- is made of frosted glass, while the surface of no. 20207-0005- shows traces of geometrical engravings.

20113-0000- (Fig. 17)
Holder
Second half of 18th c.
MAM accession date: 1952
15.3 × 12.7 × 11.5; 30 g
Iron

Three curved arms are connected to a vertical stem with a ring at the top. Probably used to hold a piece of an unidentified apparatus.

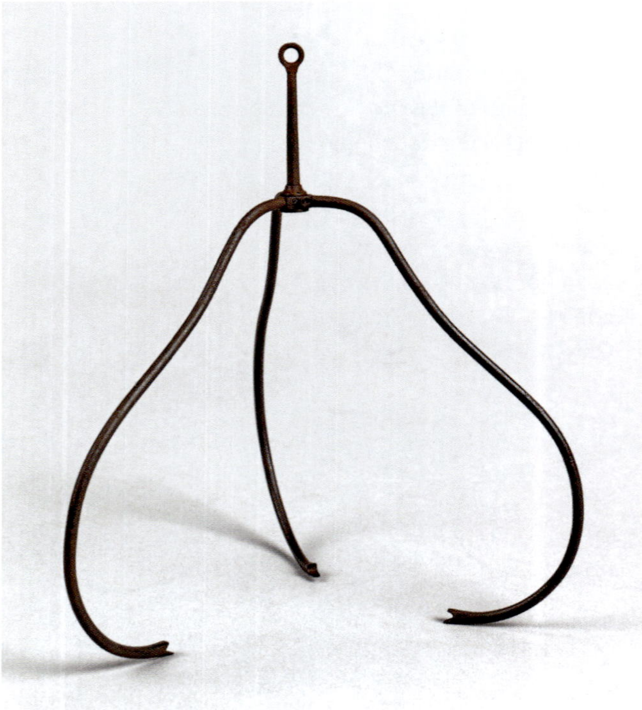

FIGURE 17 (inv. no. 20113-0000-)
© MAM/PHOTO FRANCK BOTTÉ

20036-0000- (Fig. 18)
Oval glass
Second half of 18th c.
MAM accession date: 1952
5 × 3.7 × 0.2; 5 g
Glass

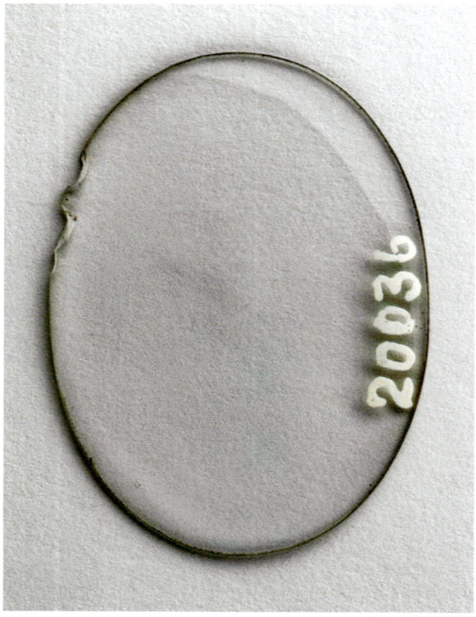

FIGURE 18 (inv. no. 20036-0000-)
© MAM/PHOTO FRANCK BOTTÉ

An oval-shaped concave glass probably used as a cover for a miniature.

19975-0000- (Fig. 19)
Large glass tube
Mid-18th c.
MAM accession date: 1952
40.5 × 15.5; 1.34 kg
Glass, brass

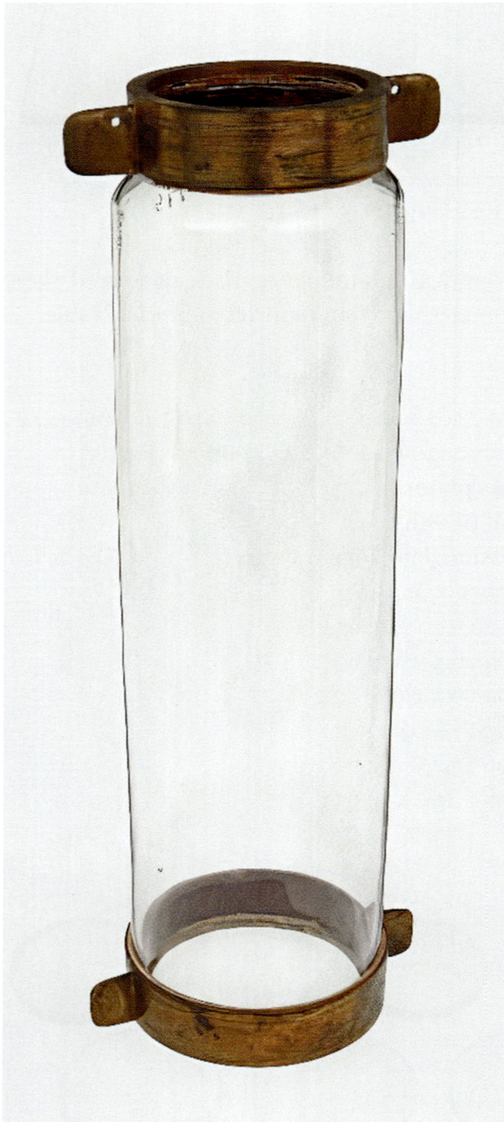

FIGURE 19 (inv. no. 19975-0000-)
© MAM/PHOTO FRANCK BOTTÉ

Cylindrical glass tube with brass rings with protruding ears at its ends. Part of an unidentified apparatus.

20034-0004- (Fig. 20)
Capillary tube with side-arm
Second half of 18th c.
MAM accession date: 1952
11.5 × 3.5 × 1.8; 10 g
Glass

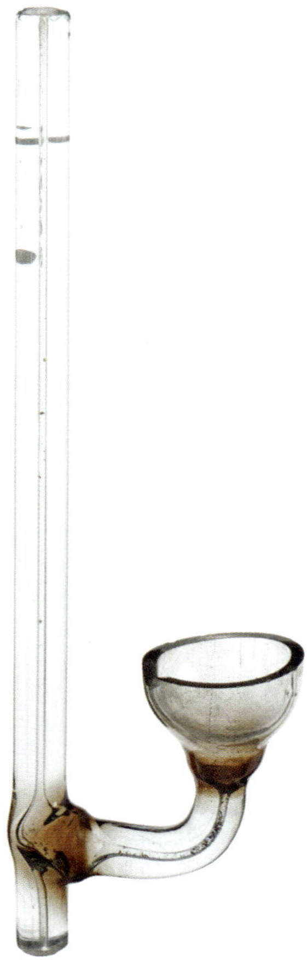

FIGURE 20 (inv. no. 20034-0004-)
© MAM/PHOTO FRANCK BOTTÉ

Capillary tube with a bent side-arm and a small funnel.

20020-0000- (Fig. 21)
Glass bulb
Second half of 18th c.
MAM accession date: 1952
10 × 6; 15 g
Glass

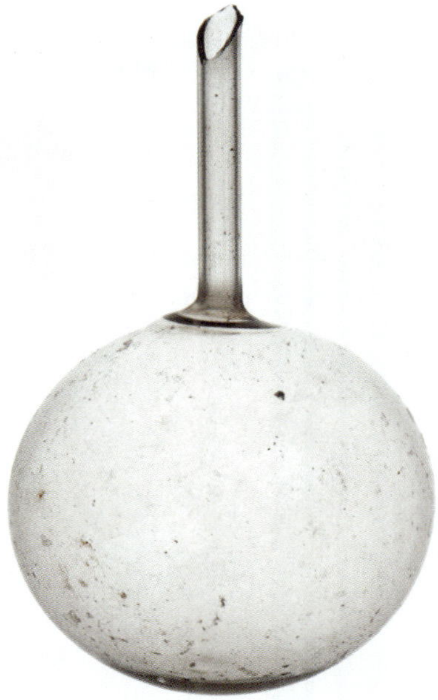

FIGURE 21 (inv. no. 20020-0000-)
© MAM/PHOTO FRANCK BOTTÉ

Glass bulb with a short broken tube.

20017-0000- (Fig. 22)
Glass bulb with capillary
Second half of 18th c.
MAM accession date: 1952
15 × 3.7; 30 g
Glass

Small glass bulb with a capillary tube.

McKie 387
Capillary tube
Second half of 18th c.
Catalogued by Douglas McKie in 1952 but not included in the donation to the MAM
12 × 1.7

Capillary tube of thick glass (pin-hole capillary) with flat bulb, and broken at both ends.

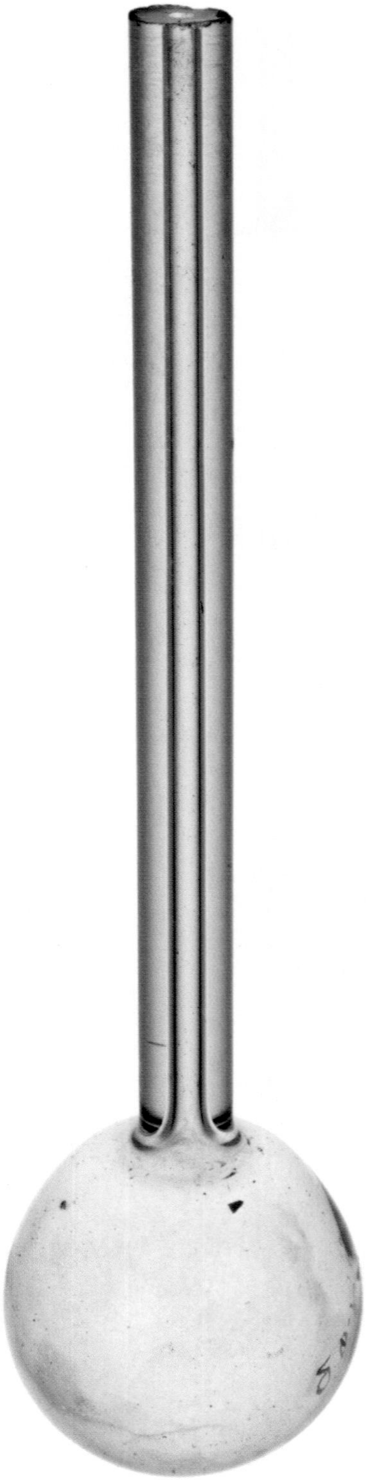

FIGURE 22 (inv. no. 20017-0000-)
© MAM/PHOTO FRANCK BOTTÉ

20031-0010- (Fig. 23)
Test tube
Second half of 18th c.
MAM accession date: 1952
21 × 3.5; 60 g
Glass

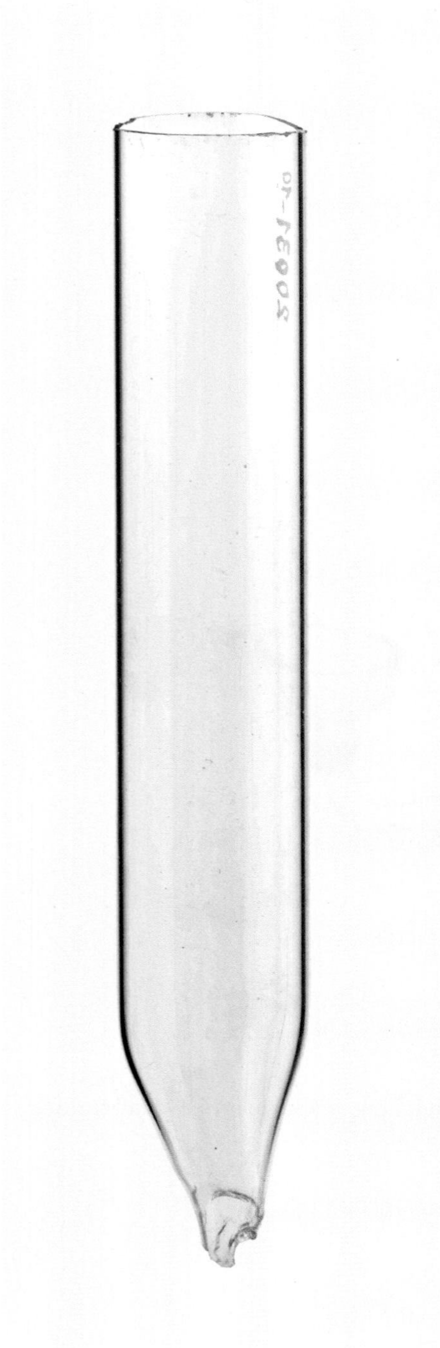

FIGURE 23 (inv. no. 20031-0010-)
© MAM/PHOTO FRANCK BOTTÉ

Glass test tube with an irregular conical bottom.

20034-0003- (Fig. 24)
Curved tube
Second half of 18th c.
MAM accession date: 1952
14 × 4.5 × 0.5; 5 g
Glass

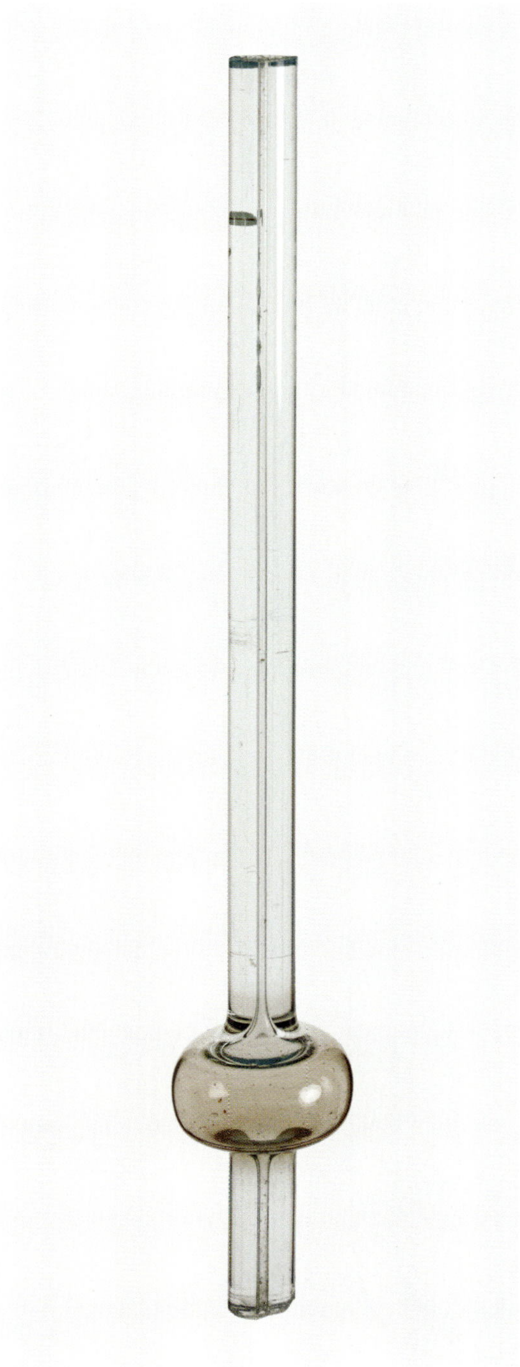

FIGURE 24 (inv. no. 20034-0003-)
© MAM/PHOTO FRANCK BOTTÉ

Hook-shaped curved glass tube.

20034-0002- (Fig. 25)
L-shaped tube
Second half of 18th c.
MAM accession date: 1952
11.7 × 9.3 × 0.7
Glass

FIGURE 25 (inv. no. 20034-0002-)
© MAM/PHOTO FRANCK BOTTÉ

L-shaped glass tube. The diameters of the perpendicular arms are different. One is a capillary tube.

20130-0000- (Fig. 26)
Tube fragment
Second half of 18th c.
MAM accession date: 1952
8.5 × 2; 60 g
Glass, brass

Short fragment of glass tube with a perforated brass cap. Probably a pillar of an unidentified instrument.

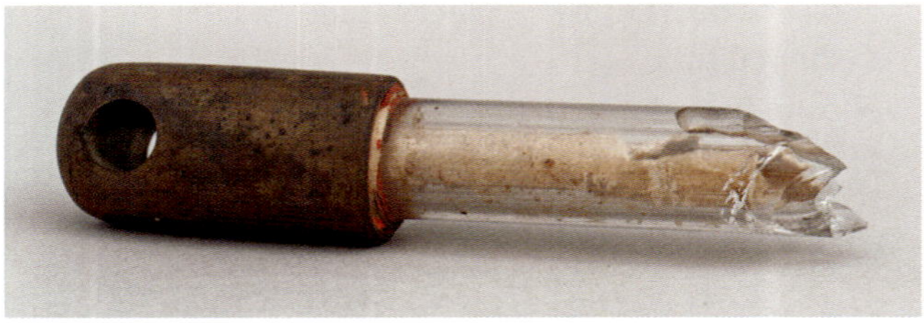

FIGURE 26 (inv. no. 20130-0000-)
© MAM/PHOTO FRANCK BOTTÉ

20014-0000- (Fig. 27)
Tube with capillary
Second half of 18th c.
MAM accession date: 1952
28.2 × 1.4; 15 g
Glass

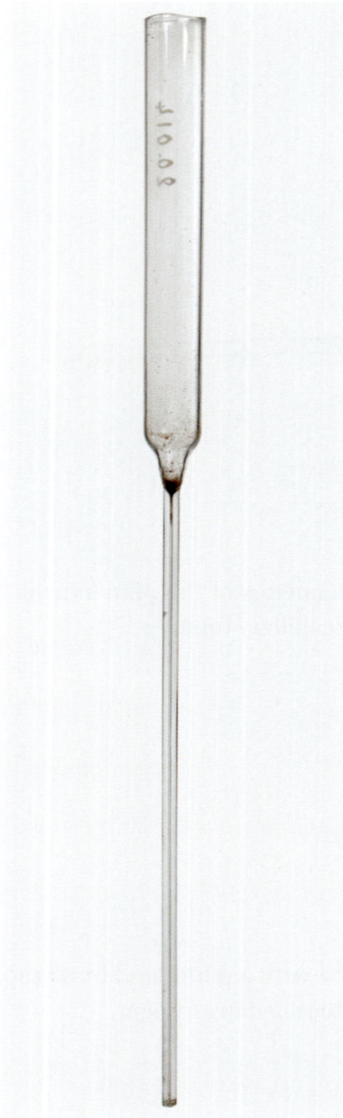

FIGURE 27
(inv. no. 20014-0000-)
© MAM/PHOTO FRANCK BOTTÉ

Glass tube ending with a capillary of smaller diameter.

20010-0001- (Fig. 28)
U-shaped tube
Second half of 18th c.
MAM accession date: 1952
33.5 × 6 × 2; 105 g
Glass

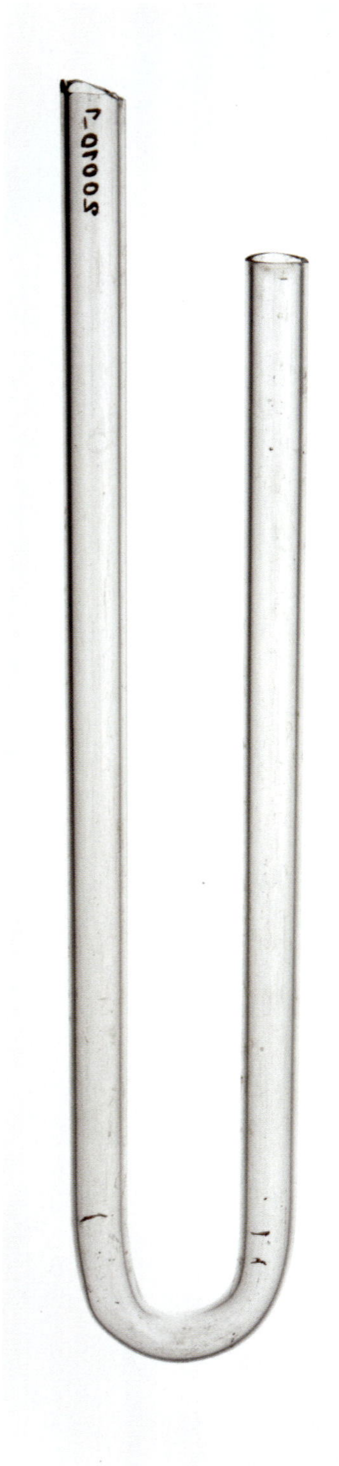

FIGURE 28
(inv. no. 20010-0001-)
© MAM/PHOTO FRANCK BOTTÉ

U-shaped tube whose arms have slightly different lengths.

20034-0001- (Fig. 29)
Capillary tube
Second half of 18th c.
MAM accession date: 1952
12.2 × 1.7; 10 g
Glass

FIGURE 29 (inv. no. 20034-0001-)
© MAM/PHOTO FRANCK BOTTÉ

Capillary tube with a bulb near one of its ends.

20031-0003-, 20031-0004- (Fig. 30)
Two tubes
Second half of 18th c.
MAM accession date: 1952
20031-0003-: 24 × 1.2; 55 g
20031-0004-: 33 × 2.3; 55 g

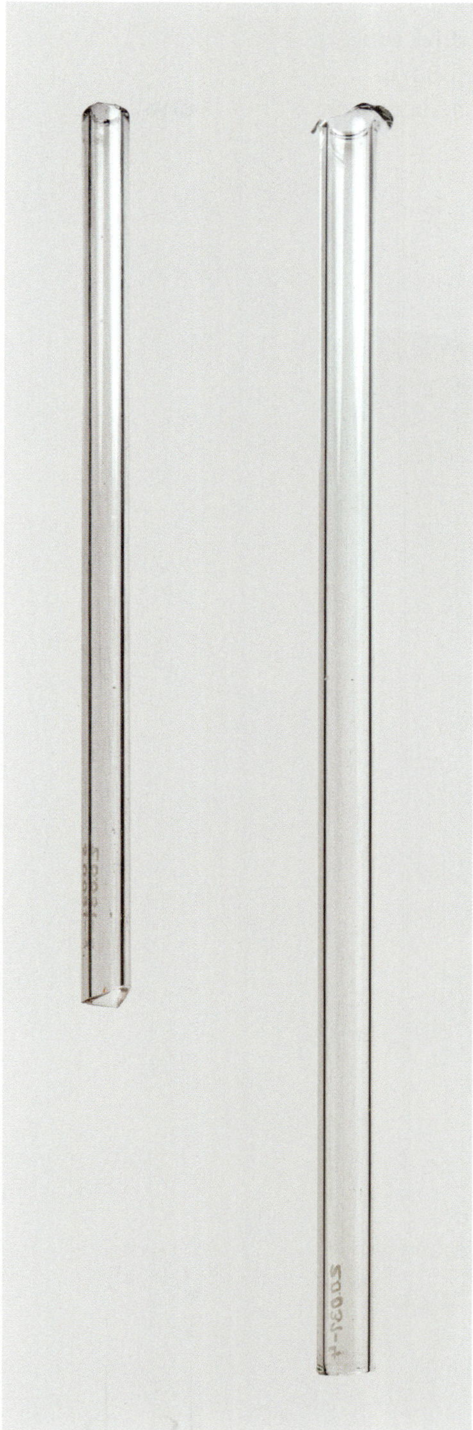

FIGURE 30 (inv. nos. 20031-0003-, -0004-)
© MAM/PHOTO FRANCK BOTTÉ

Two tubes of different diameters. No. 20031-0004- shows traces of a broken bulb at one end.

20031-0005- (Fig. 31)
Fragment of thick tube
Second half of 18th c.
MAM accession date: 1952
10.5 × 2, 25 g
Glass

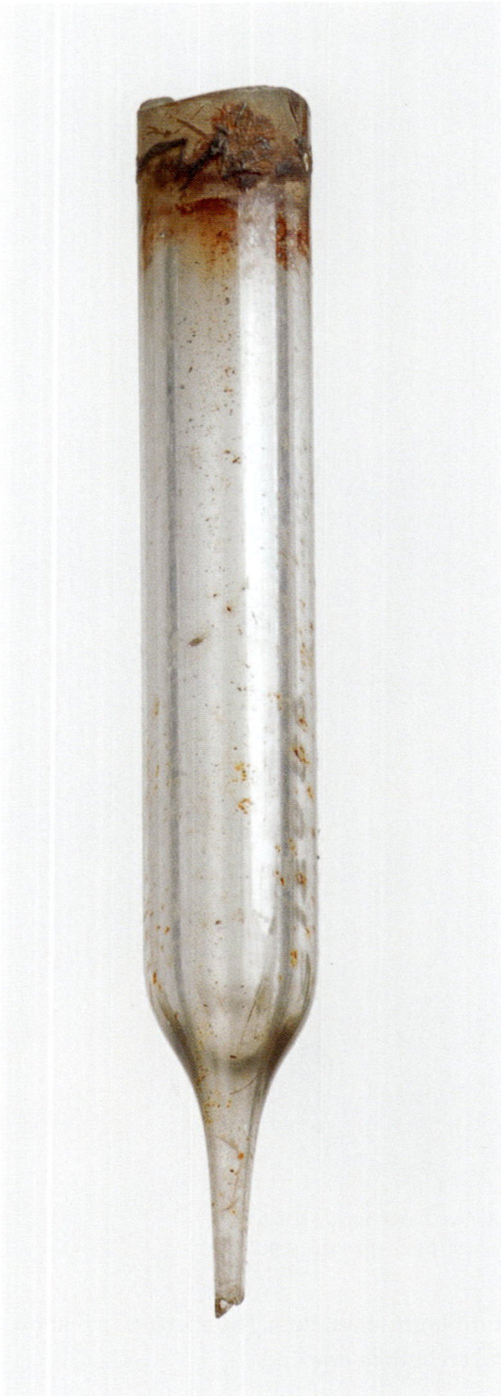

FIGURE 31 (inv. no. 20031-0005-)
© MAM/PHOTO FRANCK BOTTÉ

Short fragment of thick glass tube with a pointed end.

20010-0002- (Fig. 32)
U-shaped tube
Second half of 18th c.
MAM accession date: 1952
20 × 4 × 1.5; 50 g
Glass, brass

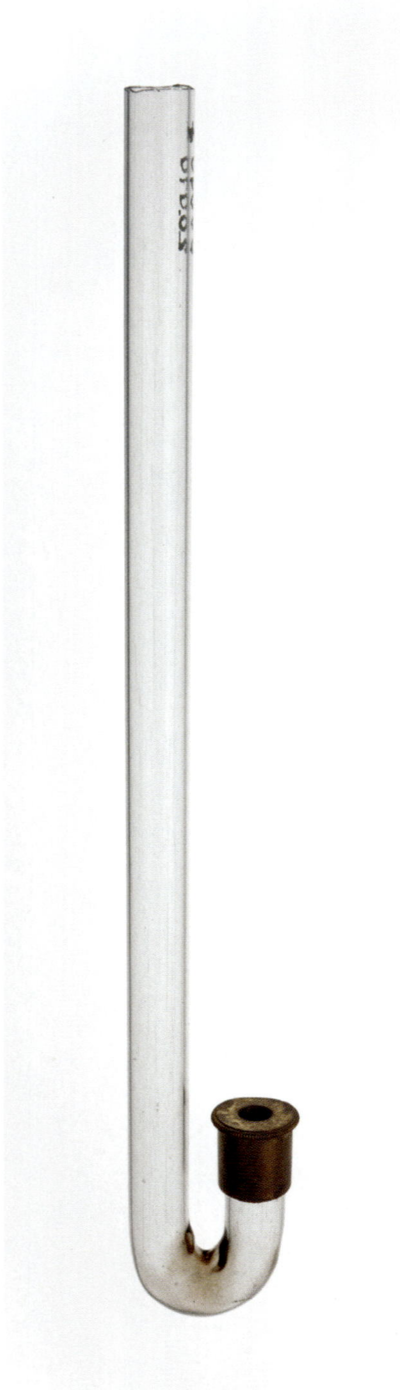

FIGURE 32 (inv. no. 20010-0002-)
© MAM/PHOTO FRANCK BOTTÉ

U-shaped tube whose short arm has a brass collar.

20031-0007- (Fig. 33)
Test tube with a phial
Second half of 18th c.
MAM accession date: 1952
18 × 0.7; 35 g
Glass, cork

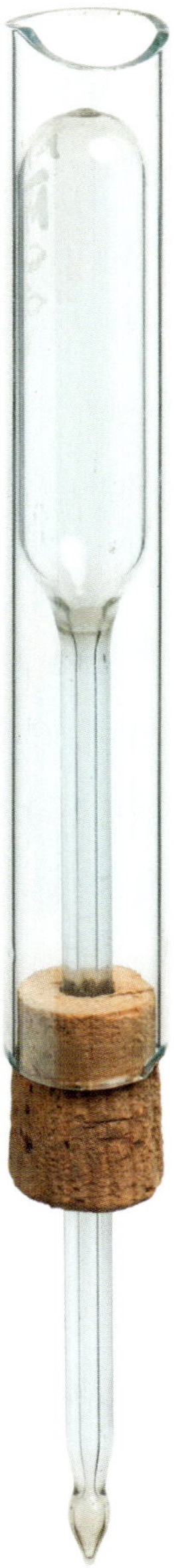

FIGURE 33
(inv. no. 20031-0007-)
© MAM/PHOTO FRANCK BOTTÉ

Glass phial with a capillary tube inserted in a test tube closed by a cork. The capillary passes through the cork.

20031-0009- (Fig. 34)
Test tube
Second half of 18th c.
MAM accession date: 1952
12 × 2.4; 30 g
Glass

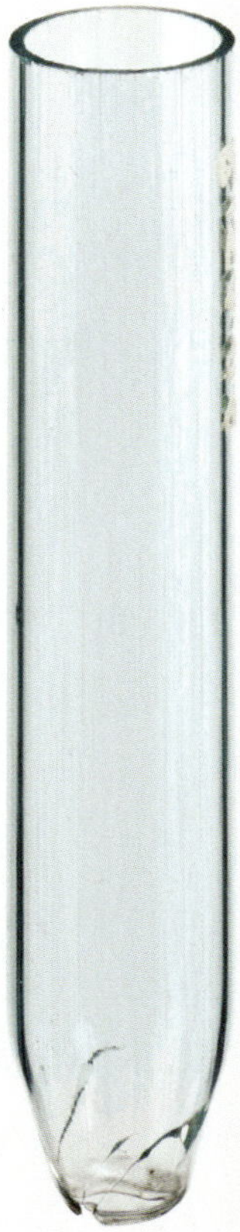

FIGURE 34
(inv. no. 20031-0009-)
© MAM/PHOTO FRANCK BOTTÉ

Test tube with broken bottom.

20034-0005- (Fig. 35)
Capillary tube
Second half of 18th c.
MAM accession date: 1952
23 × 1.5 × 0.4, 5 g
Glass

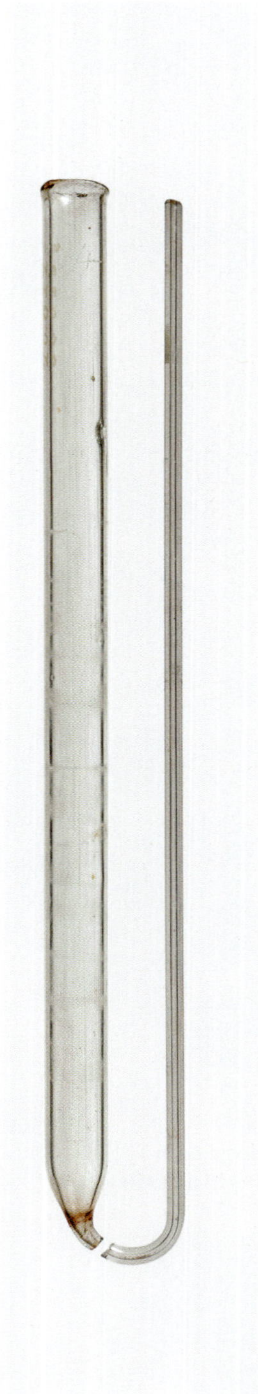

FIGURE 35
(inv. no. 20034-0005-)

Capillary tube with a bent end.

20031-0008- (Fig. 36)
Fragment of tube
Second half of 18th c.
MAM accession date: 1952
9.5 × 1; 5 g
Glass

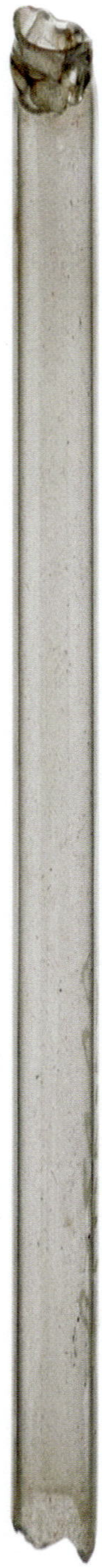

FIGURE 36
(inv. no. 20031-0008-)

Short fragment of glass tube.

20026-0000- (Fig. 37)
U-shaped tube
Second half of 18th c.
MAM accession date: 1952
21 × 7 × 5.5; 235 g
Brass, glass

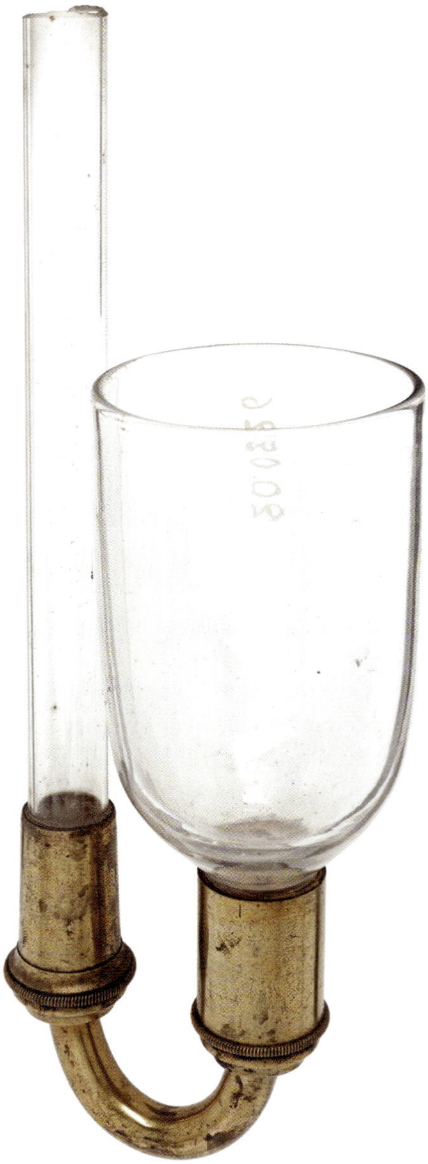

FIGURE 37 (inv. no. 20026-0000-)
© MAM/PHOTO FRANCK BOTTÉ

U-shaped brass tube with two collars screwed on its ends. One collar is joined to a glass pipe, the other to a cylindrical funnel. Probably part of an unidentified instrument.

20037-0006- (Fig. 38)
Glass stopper
Second half of 18th c.
MAM accession date: 1952
8 × 7.5; 260 g
Glass

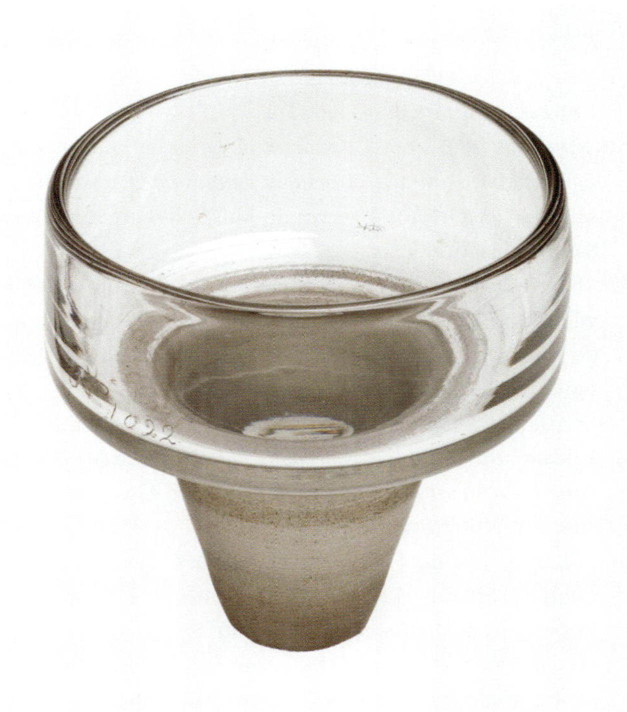

FIGURE 38 (inv. no. 20037-0006-)
© MAM/PHOTO FRANCK BOTTÉ

Cylindrical glass stopper with a heavy ground-glass conical base.

General Bibliography

Manuscript Sources

MS Afzelius (1785). Johan Afzelius, *Förteningar på de till Chemiska Professiones in Uppsala hörande smalingar, hvilka Kongl. Academien, dels före, dels under Framl. Hr: Profess. Och Riddaren Bergmans tjenstetid inkjöpt dels också sedermera af Dess Enkefru blifvit till Kongl. Academien uplåtne*, Uppsala University Library. MS D 1469 Fol.

MS *Appareils* (1864–1866). *Appareils de Lavoisier que possède le Conservatoire des arts et métiers donnés par l'Académie des sciences (1864–1866)*. Réserves du Musée des arts et métiers, Saint-Denis. DOC 2224.

MS Baron (1750–1760). Hyacinthe-Théodore Baron d'Hénouville, *Experiences, conjectures et réflexions sur l'augmentation de poids des chaux metalliques*. Bibliothèque du Muséum national d'histoire naturelle – Paris. MS 181.

MS Berthollet, Leblanc (1794). Claude Louis Berthollet, Nicolas Leblanc, *Inventaire des objects de chimie qui se sont trouvés dans la demeure du condamné Lavoisier*. Archives nationales – Paris. F17A/1337 Dossier 5.

MS Charles, Lenoir, Fortin (1794). Jacques Alexandre César Charles, Etienne Lenoir, Nicolas Fortin, *Inventaire des Instrumens de physique et de chymie du cabinet de Lavoisier cy devt. fermier générale et de l'Académie des sciences, Boulevard de La Madeleine*. Archives nationales – Paris. F 17/1219 Dossier 10.

MS Chazelles, Dupin (1836). *Correspondence between Léon de Chazelles and Charles Dupin*. Archives de l'Académie des sciences de Paris. Fonds Lavoisier, Dossier Biographique, carton 2.

MS Chazelles (1846). Léon de Chazelles, *Inventaire des papiers de Mr. De Lavoisier*. Archives de l'Académie des sciences de Paris. Fonds Lavoisier, Dossier Biographique, carton 2.

MS *Collections* (1880). *Conservatoire des arts et métiers. Collections*, vol. 1 (1880). Réserve du Musée des arts et métiers, Saint-Denis, main courante (MC2).

MS *Détail* (1756). *Détail de touts batimens de l'Arsenal de Paris ordonnée par Monseigneur le Comte d'Argenson ministre et secrétaire d'État de la guerre l'année 1756*. Bibliothèque de l'Arsenal – Paris. MS 6485.

MS Dupin (1836). Charles Dupin, *Letter to Léon de Chazelles* (27 June 1836). Archives de l'Académie des sciences de Paris. Fonds Lavoisier, Dossier Biographique, carton 2, Correspondance 1835–1853.

MS Fontana (1788). [Felice Fontana], *Invoice*. Archivio R. Museo di fisica e storia naturale (Firenze). Museo Galleo – Florence. Carteggio della Direzione, 1773–1789, aff. 104.

MS Gabriel (1804). Gabriel, *Estimation d'une maison size à Paris à l'angle des Rues des Bons Enfans et Neuve des Bons Enfans, en face du Jardin de la maison de Toulouse, suivant sa valeur et au cours des ventes publiques des maisons de Paris. Aujourd'hui vingt sept floréal au douze*. Archives de l'Académie des sciences – Paris.

MS Geoffroy (1795). Étienne Geoffroy Saint-Hilaire, *Letter to Madame Lavoisier*. Archives nationales – Paris. F17A/1076 Dossier 11.

MS Gondouin (1796). Pierre Charles Gondouin, *Inventaire du citoyen Lavoisier fait par le citoyen Gondouin notaire, 7 prairial An IV* [26 May 1796] ,92 fols. Cornell University, Kroch Library, Ithaca, NY. Lavoisier/Mss./8.11.

MS Gondouin (1796b). Pierre Charles Gondouin, *Inventaire après décès d'Antoine Laurent Lavoisier, membre de l'académie des sciences. 26 prairial an IV* [26 mai 1796], 83 fols. Archives nationales – Paris. MC/ET/XCIX/754.

MS Guerlac-Perrin (1939–1987). Guerlac-Perrin, *Lavoisier Archive, 1939–1987. Notes on manuscript sources for the study of French chemist Antoine Laurent Lavoisier, compiled by Henry Guerlac, Carl Perrin, and Rhoda Rappaport*. Division of Rare and Manuscript Collections, Cornell University Library, Ithaca, NY. 4710. [8 boxes and 11 notebooks].

MS Lavoisier (1764). Antoine-Laurent Lavoisier, *Journal d'experiences commencé le 2 juillet 1764. Expériences sur le gypse*. Archives de l'Académie des sciences – Paris. Dossier Lavoisier 1388-03.

MS Lavoisier (1765). Antoine-Laurent Lavoisier, *Journal d'experiences commencé le 15 janvier 1765*. Archives de l'Académie des sciences – Paris. Dossier Lavoisier 1388-04.

MS Lavoisier (1765a). Antoine-Laurent Lavoisier, *Journal d'experiences commencé le 1er mars 1765*. Archives de l'Académie des sciences – Paris. Dossier Lavoisier 1388-05.

MS Lavoisier (1765b). Antoine-Laurent Lavoisier, *Experiences sur les meches plattes (1765)*. Archives de l'Académie des sciences – Paris. Dossier Lavoisier 1371-1.

MS Lavoisier (1767). Antoine-Laurent Lavoisier, *Note des articles envoyés a Bourbonnes les Bains le sammedy 13 juin 1767*. Archives de l'Académie des sciences – Paris. Dossier Lavoisier 100-R-6.

MS Lavoisier (1767a). Antoine-Laurent Lavoisier, *Experiences faittes pour determiner la quantitè d'air necessaire a l'entretien de la flamme*. Archives de l'Académie des sciences – Paris. Dossier Lavoisier 1371-5.

MS Lavoisier (1767b). Antoine-Laurent Lavoisier, *Experiences sur differentes especes d'huille (1767)*. Archives de l'Académie des sciences – Paris. Dossier Lavoisier 1371-6.

MS Lavoisier (1767c). Antoine-Laurent Lavoisier, *Experiences sur la comparaison des Lumières*. Archives de l'Académie des sciences – Paris. Dossier Lavoisier 1371-7.

MS Lavoisier (1767d). Antoine-Laurent Lavoisier, *Experiences faittes sur les chandelles en janvier et fevrier 1767* [January–February 1767]. Archives de l'Académie des sciences – Paris. Dossier Lavoisier 1371-10.

MS Lavoisier (1767–1768). Antoine-Laurent Lavoisier, *Analyse des eaux (1767–1768)*. Archives de l'Académie des Sciences – Paris. Dossier Lavoisier 1376.

MS Lavoisier (1767–1788). Antoine-Laurent Lavoisier, *Carnets de voyage*. Archives de l'Académie des sciences – Paris. Fonds Lavoisier, Carnets de voyage 1–17. [17 vols].

MS Lavoisier (1772–1788). Antoine-Laurent Lavoisier, *Registres de laboratoire*. Archives de l'Académie des sciences – Paris. Fonds Lavoisier, Registres de laboratoire, 1–12. [12 vols].

MS Lavoisier (1775). Antoine-Laurent Lavoisier *et al.*, *Copie du Registre tenue par les commissaires nommés par l'Academie Royale des Sciences en conséquence des ordres de sa Majesté pour le jugement du prix et la recherche sur la fabrication du Salpêtre*. Archives de l'Académie des sciences – Paris. Fonds Lavoisier 837.

MS Lavoisier (1781–1782). Antoine-Laurent Lavoisier, *Agricultural experiments at Freschines*. Archives de l'Académie des sciences – Paris. Dossier Lavoisier 129-A-B-C-D-E-F-G.

MS Lavoisier (1786). Antoine-Laurent Lavoisier, *Refflexions pour servir a l'eloge de M Guettard*. Archives de l'Académie des sciences – Paris. Dossier Lavoisier 1967/13.

Lavoisier, Marie Anne, see *Paulze-Lavoisier*

MS McKie (1947–1952). Douglas McKie, *Correspondence with Pierre Samuel Dupont*. Hagley Museum & Library, Wilmington, DE. 19807.

MS McKie (1952). Douglas McKie, *Lavoisier scientific instruments and card index of Chazelles' Collection*. Hagley Museum & Library, Wilmington, DE. 19807. (Accession no. 1983.218).

MS Monnet (1785). Antoine Grimoald Monnet, *Tableau historique et philosophique de l'origine et des progrès de la chymie et de la minéralogie en France* Bibliothèque de l'École des Mines – Paris. MS 18–19. [2 vols].

MS *Musée Lavoisier* (1935). *Le Musée Lavoisier*. Archives de l'Académie des Sciences – Paris, Fonds Lavoisier, Dossier Biographique.

MS Paulze-Lavoisier (1835). *Marie Anne Paulze-Lavoisier, Note manuscrite d'un envoi des manuscrits par Mad.e de Rumford à M Arago*. Archives de l'Académie des sciences – Paris. Dossier Biographique Lavoisier, carton 2.

MS Paulze-Lavoisier (1836). Alexis Vavin, *Inventaire après décès de Paulze, Marie Anne-Pierrette, 3 mars 1836*. Archives nationales – Paris. MC/RE/XCIX/12.

MS *Plan du Petit Arsenal* (ca 1750). *Plan du Petit Arsenal*. Bibliothèque nationale de France. BNF Va 419 J Ft4 4 arron. 15 quartier.

MS *Plan de l'Arsenal de Paris* (ca 1780). *Plan de l'Arsenal de Paris*. Bibliothèque historique de la Ville de Paris. 4-PLA-1768 (RES).

MS Pommier (2007). Aimé Pommier, *Catalogue des mesures de longueur, 2007*. Réserve du Musée des arts et métiers – Paris.

MS *Procès-verbaux de l'Académie des sciences*. Archives de l'Académie des sciences – Paris.

MS Register (1864). *Register of copies of letters of the Académie des Sciences 1861–1874*. Archives de l'Académie des sciences – Paris.

MS Thompson (1814). Pierre Amable Viault, *Inventaire après décès de Benjamin comte de Rumford, 6 octobre 1814*. Archives nationales – Paris. MC/RE/XXXV/19.

Printed Sources

Abbri (1984). Ferdinando Abbri, *Le terre l'acqua le arie: La rivoluzione chimica del Settecento* (Bologna: Il Mulino, 1984).

Abney Salomon (2019). Charlotte A. Abney Salomon, "The Pocket Laboratory: The Blowpipe in Eighteenth-Century Swedish Chemistry", *Ambix*, 66 (2019): 1–22.

Adams (1785). George Adams, *An Essay on Electricity Explaining the Theory and Practice of That Useful Science* (London: Tycho Brahe's Head, 1785).

Adams (1797). George Adams, *Geometrical and Graphical Essays Containing a General Description of the Mathematical Instruments Used in Geometry, Civil and Military Surveying, Levelling, and Perspective: With Many New Practical Problems* (London: J. Dillon and Co, 1797).

Albury (1972). William R. Albury, *The Logic of Condillac and the Structure of the French Chemical and Biological Theory, 1785–1801* (PhD thesis, The Johns Hopkins University, Baltimore, MD, 1972).

Anastasi (1884). Auguste Anastasi, *Nicolas Leblanc, sa vie, ses travaux, et l'histoire de la soude artificielle* (Paris: Hachette, 1884).

Anderson (1978). Robert G.W. Anderson, *The Playfair Collection and the Teaching of Chemistry at the University of Edinburgh 1713–1858* (Edinburgh: The Royal Scottish Museum, 1978).

Anderson (2013). Robert G.W. Anderson, "Chemistry Laboratories, and How They might be Studied", *Studies in History and Philosophy of Science*, 44 (2013): 669–675.

Andrieu (1937). B. Andrieu, "Le microscope de Lavoisier", *Bulletin de la Société française de microscopie*, 6 (1937): 16–19.

Anonymous (1894). Anonymous, "Lavoisier's Scientific Apparatus", *The Daily Graphic* (9 June 1894).

Anonymous (1902). Anonymous, *1851–1901: Cinquantenaire scientifique de M. Berthelot* (Paris: Gauthiers-Villars, 1902).

Anonymous (1929). Anonymous, *Centenaire de Marcelin Berthelot, 1827–1927* (Paris: 1929).

Antonelli (2021). Francesca Antonelli, *Scrittura, sociabilità e strategie di persuasione: Marie Anne Paulze-Lavoisier, secrétaire (1758–1836)* (PhD thesis, University of Bologna – EHESS Paris, 2021).

Arbey (1952). L. Arbey, "Le premier thermomètre de Lavoisier de l'Observatoire de Paris", *L'astronomie*, 66 (1952): 80–81.

Archinard (1977). Margarida Archinard, "L'apport genevois à l'hygrométrie", *Gesnerus*, 34 (1977): 242–262.

Avis (1770). "Avis sur les leçons de chymie de feu M. Guillaume-François Rouelle, maître apothicaire, démonstrateur de chymie au Jardin du roi, associé des Académies royales des sciences de Paris & de Stockholm, & de l'Académie électorale d'Erford", *Journal de médecine, de chirurgie et de pharmacie*, 33 (October 1770): 381–382.

Avis (1775). "Avis. Baromètres & thermomètres: Cappy fils & Mossy, neveu du feu Sr Cappy, constructeurs des instruments de physique de l'Académie des sciences, suivant le brevet qui leur en a été expédié le trois mai dernier, continuent de faire & vendre des barometres, thermometres concernant la physique, en leur demeure à Paris, rue & place Royale", *Mercure de France* (August 1775) : 199–200.

Bachaumont, see *Mémoires (1777–1789)*.

Babelon (1970). Jean Pierre Babelon, "La Palais de l'Arsenal à Paris: Étude architecturale et essai de répertoire iconographique critique", *Bulletin monumental*, 128/4 (1970): 267–310.

Badash (1964). Lawrence Badash, "Joseph Priestley's Apparatus for Pneumatic Chemistry", *Journal of the History of Medicine* (April 1964): 139–155.

Baetjer (2019). Katharine Baetjer, *French Paintings in the Metropolitan Museum of Art from the Early Eighteenth Century through the Revolution* (New York, NY: The Metropolitan Museum, 2019).

Ballot (1923). Charles Ballot, *L'introduction du machinisme dans l'industrie française* (Lille-Paris, 1923).

Banks (2007). *Scientific Correspondence of Sir Joseph Banks, 1765–1820*, edited by Neil Chambers, (London: Pickering and Chatto, 2007), 6 vols.

Barletti (1794). Carlo Barletti, "Della percossa dell'acciarino nell'aria rarefatta", *Opuscoli scelti sulle scienze e sulle arti*, 17 (1794): 214–216.

Barrelet (1953). James Barrelet, *La verrerie en France de l'époque gallo-romaine à nos jours* (Paris: Larousse, 1953).

Bassani (2009). Angelo Bassani, *Per la storia della Facoltà di Scienze in Italia: la Chimica a Padova dalla caduta di Venezia alla II guerra mondiale (1797–1943). Con un contributo sugli antecedenti settecenteschi di Virgilio Giormani* (Padova: Cleup, 2009), 2 vols.

Basso Ricci *et al.* (1997). Maria Basso Ricci, Lili Caffarella, Antonio Meloni, Pasquale Tucci, *Due secoli di strumenti geomagnetici in Italia (1740–1971)* (Bologna: Editrice Compositori, 1997).

Batiffol (1929). Louis Batiffol, "Le mail de l'arsenal au XVIIIe siècle", *La cité, bulletin trim. de la Société historique et archéologique des IVe et IIIe arrondissements de Paris*, 56 (1929): 5–22.

Baumé (1773). Antoine Baumé, *Elements de pharmacie théorique et pratique* (Paris: Samson, 1773).

Baumé (1773b). Antonie Baumé, *Chymie expérimentale et raisonnée* (Paris: Didot le jeune, 1773), 3 vols.

Baumé (1775). Antoine Baumé, *Prix courants des préparations de chymie et de pharmacie qui se trouvent à Paris, chez Baumé, apothicaire, rue Coquillière* (Paris, 1775).

Baumé (1797). Antoine Baumé, *Éléments de pharmacie*, 6th ed. (Paris: Samson, 1797).

Beaudouin, Brenni, Turner (2018). Denis Beaudouin, Paolo Brenni, Anthony Turner, *Dictionary of precision Instrument-makers and related craftsmen. A Bio-bibliographical Dictionary of precision Instrument-makers and related craftsmen in France & Switzerland*, 1430–1960 (2018): https://omekas.obspm.fr/s/dictionarypim/page/prolegomena (accessed 6 August 2021).

Béchamp (1871). Antoine Béchamp, *La France et la Prusse: Lavoisier et M. Liebig* (Paris, 1871).

Bedel (1986). Charles Bedel, "Les cabinets de chimie", in René Taton, ed., *Enseignement et diffusion des sciences en France au XVIIIe siècle*, 2nd ed. (Paris: Hermann, 1986), pp. 647–652.

Belin (1992). Pierre Belin, "Un collaborateur d'Antoine-Laurent Lavoisier à l'Hôtel de l'Arsenal: Jean-Baptiste Meusnier (1754–1793)", in Michelle Goupil, ed., *Lavoisier et la révolution chimique* (Paris: Sabix, 1992), pp. 263–293.

Belhoste (2011). Bruno Belhoste, *Paris savant: Parcours et rencontres au temps des Lumières* (Paris: Colin, 2011).

Bellodi *et al.* (2002). Giuliano Bellodi, Fabio Bevilacqua, Gianni Bonera, Lidia Falomo, *Gli strumenti di Alessandro Volta: Il gabinetto di fisica dell'Università di Pavia* (Milan: Hoepli, 2002).

Bennett (1789). Abraham Bennett, *New Experiments on Electricity, Wherein the Causes of Thunder and Lightning as well as the Constant State of Positive or Negative Electricity in the Air or Clouds, are Explained* (Derby: J. Dewry, 1789).

Bennett, *et al.* (1982). Jim A. Bennett, Olivia Brown, *The Complete Surveyor* (Cambridge: Whipple Museum of the History of Science, 1982).

Bensaude-Vincent (1990). Bernadette Bensaude-Vincent, "A View of the Chemical Revolution through Contemporary Textbooks: Lavoisier, Fourcroy and Chaptal", *British Journal for the History of Science*, 23 (1990): 456–460.

Bensaude-Vincent (1993). Bernadette Bensaude-Vincent, *Lavoisier: Mémoires d'une révolution* (Paris: Flammarion, 1993).

Bensaude-Vincent (2000). Bernadette Bensaude-Vincent, "'The Chemist's Balance for Fluids': Hydrometers and their Multiple Identities, 1770–1810", in Frederic Lawrence Holmes, Trevor Levere, eds., *Instruments and Experimentation in the History of Chemistry* (Cambridge, MA and London: The MIT Press, 2000), pp. 153–183.

Bensaude-Vincent, Blondel (2008). Bernadette Bensaude-Vincent, Christine Blondel, eds., *Science and Spectacle in the European Enlightenment* (Aldershot: Ashgate, 2008).

Beretta (1994). Marco Beretta, *A New Course in Chemistry: Lavoisier's First Chemical Paper* (Florence: Leo. S. Olschki, 1994).

Beretta (1995). Marco Beretta, *Bibliotheca Lavoisieriana: The Catalogue of the Library of Antoine Laurent Lavoisier* (Florence: Leo S. Olschki, 1995).

Beretta (1996). Marco Beretta, "I philosophes e la chimica: All'origine del materialismo scientifico", in Marco Beretta, Felice. Mondella, Maria Teresa Monti, eds., *Per una storia critica della scienza* (Milan: Cisalpino, 1996), pp. 11–48.

Beretta (1999–2010). Marco Beretta, ed., *Panopticon Lavoisier* (1999–2010): http://moro.imss.fi.it/lavoisier/ (accessed 6 August 2021).

Beretta (2000). Marco Beretta, "Pneumatics vs. Aerial Medicine? Salubrity and Respirability of Air at the End of the 18th Century", *Nuova Voltiana: Studies on Volta and his Time*, 2 (2000): 49–72.

Beretta (2001). Marco Beretta, *Imaging a Career in Science: The Iconography of Antoine Laurent Lavoisier* (Canton, MA: Science History Publications, 2001).

Beretta (2001a). Marco Beretta, "Lavoisier and his Last Printed Work: The *Mémoires de physique et de chimie* (1805)", *Annals of Science*, 58 (2001): 327–356.

Beretta (2001b). Marco Beretta, "From Nollet to Volta: Lavoisier and Electricity", *Revue d'histoire des sciences*, 54/1 (2001): 29–52.

Beretta (2003). Marco Beretta, "Lavoisier's Collection of Instruments: A Checkered History", in Marco Beretta, Paolo Galluzzi, Carlo Triarico, eds., *Musa Musaei: Studies on Scientific Instruments and Collections in Honour of Mara Miniati* (Florence: Leo S. Olschki, 2003), pp. 313–334.

Beretta (2004). Marco Beretta, "Sur les traces du laboratoire de Lavoisier", *La revue: Musée des arts et métiers*, 41 (2004): 14–23.

Beretta (2005). Marco Beretta, "Collected, Analyzed, Displayed: Lavoisier and Minerals", in Marco Beretta, ed., *From Private to Public: Natural Collections and Museums* (Sagamore Beach, MA: Science History Publications, 2005), pp. 113–140.

Beretta (2009). Marco Beretta, "Big Chemistry: Lavoisier's Design and Organisation of his Laboratories", in Marta Lourenço, Ana Carneiro, eds., *Spaces and Collections in the History of Science* (Lisbon: Museum of Science of the University of Lisbon, 2009), pp. 65–80.

Beretta (2011). Marco Beretta, "Rinman, Diderot, and Lavoisier: New Evidence Regarding Guillaume François Rouelle's Private Laboratory and Chemistry Course", *Nuncius – Journal of the Material and Visual History of Science*, 26 (2011): 355–379.

Beretta (2011a). Marco Beretta, "The Changing Role of the Historiography of Chemistry in Continental Europe since 1800", *Ambix*, 58 (2011): 257–276.

Beretta (2012). Marco Beretta, "Imaging the Experiments on Respiration and Transpiration of Lavoisier and Séguin: Two Unknown Drawings by Madame Lavoisier", *Nuncius – Journal of the Material and Visual History of Science*, 27 (2012): 163–191.

Beretta (2012a). Marco Beretta, "The Rise and Fall of the Glassmaker Paul Bosc d'Antic (1753–1784)", *Annals of Science*, 69 (2012): 375–393.

Beretta (2014). Marco Beretta, "Between the Workshop and the Laboratory: Lavoisier's Network of Instrument Makers", *Osiris*, 29 (2014): 197–214.

Beretta (2014a). Marco Beretta, "Unveiling Glass' Mysteries: Lavoisier, Loysel and the First Chemical Treatise on Glass (1765–1799)", in Ursula Klein, Carsten Reinhardt, eds, *Objects of Chemical Inquiry* (Sagamore Beach, MA: Science History Publications, 2014), pp. 1–20.

Beretta (2014b). Marco Beretta "Transmutations and Frauds in Enlightened Paris: Lavoisier and Alchemy", in Marco Beretta, Maria Conforti, eds., *Fakes!? Hoaxes, Counterfeits and Deception in Early Modern Science* (Sagamore Beach, MA: Science History Publications, 2014), pp. 69–108.

Beretta (2019). Marco Beretta, "*Lavoisier and the French Chemical Heritage at the Musée Centennal of the Paris World Exhibition (1900)*", in Elena Canadelli, Marco Beretta, Laura Ronzon, eds., *Behind the Exhibit: Displaying Science and Technology at World's Fairs and Museums in the Twentieth Century* (Washington, DC: Smithsonian Institution Scholarly Press, 2019), pp. 2–14.

Beretta (2021). Marco Beretta, "Laboratory and Technology", in Matthew Eddy, Ursula Klein, eds., *Cultural History of Chemistry – The Eighteenth Century* (London: Bloomsbury, 2021), pp. 71–91.

Beretta (2022). Marco Beretta, "The Cost of Lavoisier's Laboratory" in Liliane Hilaire-Pérez, Catherine Lanoë eds., *Les sciences et les techniques, laboratoire de l'Histoire. Mélanges en l'honneur de Patrice Bret* (Paris : Presses des Mines, collection Histoire, sciences, techniques et sociétés, 2022), pp. 165–181.

Bergman (1781). Torbern Bergman, "Mémoire sur le chalumeau à souder, & son usage dans l'analyse, surtout des substances minérales", *Observations sur la physique*, 18 (1781): 207–221 and 467–480.

Bergman (1784). Torbern Bergman, *Manuel du minéralogiste, ou Sciagraphie du règne minéral, distribué d'après l'analyse chimique* (Paris: Cuchet, 1784).

Bergman (1965). [Tobern Bergman], *Torbern Bergman's Foreign Correspondence*, eds. Göte Carlid, Johan Nordström (Stockholm: Almqvist & Wiksell, 1965).

Berthelot (1890). Marcellin Berthelot, *La révolution chimique, Lavoisier: Ouvrage suivi de notices et extraits des registres inédits de laboratoire de Lavoisier* (Paris: Alcan, 1890).

Bertucci (2001). Paola Bertucci, "A Philosophical Business: Edward Nairne and the Patent Medical Electrical Machine", *History of Technology*, 23 (2001): 41–58.

Bertucci (2017). Paola Bertucci, *Artisanal Enlightenment: Science and Mechanical Arts in the Old Regime* (New Haven, CT: Yale University Press, 2017).

Bins de Saint-Victor (1809). Jacques-Maximilien Benjamin Bins de Saint-Victor, *Tableau historique et pittoresque de Paris depuis les Gaulois jusqu'à nos jours* (Paris: Nicole, 1809), 2 vols.

Bion (1716). Nicolas Bion, *Traité de la construction et des principaux usages des instrumens de mathématique* (Paris: Veuve Boudot, 1716).

Biot (1816). Jean Batiste Biot, *Traité de physique expérimentale et mathématique* (Paris: Deterville, 1816), 3 vols.

Biot (1824). Jean-Baptiste Biot, *Précis de physique expérimentale* (Paris: Deterville, 1824), vol. 2.

Birembaut (1958). Arthur Birembaut, "La contribution de Réaumur à la thermométrie", *Revue d'histoire des sciences et de leurs applications*, 11 (1958): 302–329.

Bodman (1949–1952). Gösta Bodman, "Sven Rinmans reseanteckningar 1746–1747", *Dædalus* (1949): 71–85; (1950): 53–63; (1952): 104–115.

Bolle (1983). Bert Bolle, *Barometers in Beeld* (Lochem: Tijdstroom, 1983).

Bontemps (1868). Georges Bontemps, *Guide du verrier: Traité historique et pratique de la fabrication des verres, cristaux, vitraux* (Paris: Librairie du dictionnaire des arts et manufactures, 1868).

Bothereau (2018). Benjamin Bothereau, *A la lanterne! Modes d'existence d'un objet banal entre imaginaire technique et politique: Invention, économie urbaine, publics et circulations du "réverbère", Paris, Barcelone, XVIIIe siècle*. Thèse de doctorat (Paris: EHESS, 2018), 2 vols.

Bottée, Riffault (1811). Jean-Joseph-Auguste Bottée, Jean Renée Denis Alexandre Riffault, *Traité de l'art de fabriquer la poudre à canon, contenant: L'extraction du salpêtre, son raffinage; la fabrication du salin ... Précédé d'un exposé historique sur l'établissement du service des poudres et salpêtres en France* (Paris: Leblanc, 1811).

Bouchot (1887). Henry Bouchot, *Histoire anecdotique des métiers avant 1789* (Paris, 1887).

Bourguet (2002). Marie-Noëlle Bourguet, "Landscape with Numbers: Natural History, Travel and Instruments in the late Eighteenth and early Nineteenth Centuries", in Marie-Noëlle Bourguet, Christian Licoppe and H. Otto Sibum eds., *Instruments, Travel and Science: Itineraries of precision from the Seventeenth to the Twentieth century* (London: Routledge, 2002), pp. 96–125.

Bouvet (1937). Maurice Bouvet, *Histoire de la pharmacie en France* (Paris: Éditions Occitania, 1937).

Brachner (1983). Alto Brachner, ed., *G.F. Brander 1713–1783: Wissenschaftliche Instrumente aus seiner Werkstatt* (Munich: Deutsches Museum, 1983).

Brander (1779). Georg F. Brander, *Beschreibung eines magnetischen Declinatorii und Inclinatorii: Nebst der Anweisung, wie man sich dieser Instrumente bedienen soll* (Augsburg: Klett und Frank, 1779).

Brenni (2002). Paolo Brenni, "Jean Antoine Nollet and Physics Instruments", in Lewis Pyenson, Jean-François Gauvin, ed., *The Art of Teaching Physics: The Eighteenth-Century Demonstration Apparatus of Jean Antoine Nollet* (Sillery, Québec: Les éditions du Septentrion, 2002), pp. 11–27.

Brenni (2003). Paolo Brenni, "Volta's Electric Lighter and its Improvements: The Birth, Life and Death of a Peculiar Scientific Apparatus which became the First Electric Household Appliance", in Marco Beretta, Paolo Galluzzi, Carlo Triarico, eds., *Musa Musaei Studies on Scientific Instruments and Collections in Honour of Mara Miniati* (Florence: Leo Olschki, 2003), pp. 371–394.

Brenni, see Beaudouin, Brenni, Turner (2018).

Bret (1992). Patrice Bret, "Une tentative d'exploitation militaire de la recherche en chimie: Berthollet et la poudre de muriate oxygéné de potasse (1787–1794)", Michelle Goupil ed., *Lavoisier et la Révolution chimique* (Paris: Sabix – Ecole Polytechnique, 1992), pp. 195–238.

Bret (1994). Patrice Bret, "Lavoisier, la recherche appliquée et l'enseignement de la chimie à la Régie des poudres", in *Seminario internazionale per il bicentenario della scomparsa di Antoine Laurent Lavoisier* (Rome: Accademia nazionale delle scienze detta dei XL, 1994), pp. 231–250.

Bret (1994a). Patrice Bret, "Lavoisier à la régie des poudres: Le savant, le financier, l'administrateur et le pédagogue", *La vie des sciences/Comptes rendus de l'Académie des sciences, sér. générale*, 11/4 (1994): 297–317.

Bret (1995). Patrice Bret, "La enseñanza durante la Revolución química en el Arsenal: el curso de Gengembre en la Escuela de pólvora (1785)", in Patricia Aceves Pastrana ed., *Las ciencias químicas y biológicas en la formación de un mundo nuevo* (México: Universidad autónoma metropolitana, 1995), pp. 49–62.

Bret (1996). Patrice Bret, "Lavoisier et l'apport de la chimie académique à l'industrie des poudres et salpêtres", *Archives internationales d'histoire des sciences*, 46 (1996): 57–74.

Bret (2000). Patrice Bret, "Des essais de la monnaie à la recherche et à la certification des métaux: Un laboratoire modèle au service de la guerre et de l'industrie (1775–1830)", *Annales historiques de la Révolution française*, 320 (2000): 137–148.

Bret (2004). Patrice Bret, "Lavoisier à la régie des poudres: Le savant, le financier, l'administrateur et le pédagogue", HAL (2004): 1–22: https://halshs.archives-ouvertes.fr/halshs-0000 2883 (accessed 6 August 2021).

Bret (2004a). Patrice Bret, "Un bateleur de la science: Le 'machiniste-physicien' François Bienvenu et la diffusion de Franklin et Lavoisier", *Annales historiques de la Révolution française*, 338 (2004): 95–127.

Bret (2005). Patrice Bret, "« Devenir des héros ». Le mémoire inédit de Lavoisier au concours d'éloquence de l'académie de Besançon en 1761", *Dix-Huitième Siècle*, 37 (2005): 329–346.

Bret (2008). Patrice Bret, "Les promenades littéraires de Madame Picardet: La traduction comme pratique sociale de la science au XVIIIe siècle", in Pascal Duris ed., *Traduire la science: hier et aujourd'hui* (Pessac: Maison des Sciences de l'Homme d'Aquitaine, 2008) <http://books.openedition.org/msha/8775>.

Bret (2012). Patrice Bret, "Lavoisier et le Bureau de consultation des arts et métiers", LC, vol. 7, pp. 551–562.

Bret (2019). Patrice Bret, "Instruments of Knowledge and Power in a Colonial Context: Scientific Instruments during the French Occupation of Egypt, 1798–1801", in Neil Brown, Silke Ackermann, Feza Günergün ed., *Scientific Instruments between East and West* (Leiden: Brill, 2019), pp. 206–228.

Bret, Lanoë (2006). Patrice Bret, Catherine Lanoë, "La formation d'un espace de travail entre sciences et arts et métiers: Le laboratoire du chimiste et le laboratoire du parfumeur au XVIIIe siècle", in Maurice Hamon, ed., *Le travail avant la révolution industrielle* (Paris: Éditions. du CTHS, 2006), pp. 139–154.

Brown (1967). Sanborn C. Brown, *Benjamin Thompson – Count Rumford* (Oxford: Pergamon Press, 1967).

Brugnatelli (1997). Luigi Valentino Brugnatelli, *Diario del viaggio in Svizzera e Francia con Alessandro Volta* (Milan: Cisalpino, 1997).

Buffon (1817–1818). Georges-Louis Leclerc de Buffon, *Œuvres complètes* (Paris: Rapet, 1817–1818), 12 vols.

Büsching (1768–1779). Anton Friedrich Büsching, *Nouveau traité de géographie* [*suivi de*] *Géographie universelle, traduit de l'allemand* (Strasbourg: Gosse et Pinet, 1768–1779), 12 vols.

Busser (1769–1773). Johan Benedict Busser, *Utkast till beskrifning om Upsala* (Uppsala: J. Edman, 1769–1773), 2 vols.

Cadet de Gassicourt (1805). Louis Claude Cadet de Gassicourt, *Éloge d'Antoine Baumé* (Brussels: Weissenbruch, 1805).

Cartier (1783). [Cartier], *Observations sur l'aréomètre ou Pèse-liqueur de comparaison fabriqué par le sieur Cartier, suivant les principes de M. Perou, inspecteur général de la jauge … et établi par les lettres patentes du 3 août 1771 pour règle dans la vérification des différentes eaux-de-vie, par autres lettres patentes du 13 février 1782, et par arrêt de la cour des Aides du 28 mars 1783* (Paris: Knapen et fils, 1783).

Catalogue (1889). *Catalogue général officiel: Exposition rétrospective du travail et des sciences anthropologiques … Section II: Arts libéraux / Exposition universelle internationale de 1889 à Paris* (Lille: Danel, 1889).

Catalogue (1952). *Catalogue of Printed Works by and Memorabilia of Antoine-Laurent Lavoisier 1743–1794: Tuesday, February 19 through Sunday, March 30, 1952* (New York, NY: The Grolier Club, 1952).

Cauchie (1900). A. Cauchie, "Le domaine de Freschines sous Lavoisier", *Mémoires de la Société des Sciences et Lettres de Loir-et-Cher* (1900): 47–71.

Chaffers (1891). William Chaffers, *Marks and Monograms on European and Oriental Pottery and Porcelain with Historical Notices of Each Manufactory* (London: Reeves and Turner, 1891).

Chaldecott (1953). J.A. Chaldecott, "The Zograscope or Optical Diagonal Machine", *Annals of Science*, 9 (1953): 315–322.

Chaldecott (1968). J.A. Chaldecott, "Scientific Activities in Paris in 1791: Evidence from the Diaries of Sir James Hall for 1791, and Other Contemporary Records", *Annals of Science*, 24 (1968): 21–52.

Chandler (1969). Bruce Chandler, "Three Nürnberg Compassmacher: Hans Troschel the Elder, Hans Troschel the Younger and David Beringe", *Metropolitan Museum Journal*, 2 (1969): 211–216.

Chapin (1971). Seymour L. Chapin, "The Persistent Ghost: The Artisan Laroche", *Technology and Culture*, 12 (1971): 69–74.

Chappey (2004). Jean-Luc Chappey, "Enjeux sociaux et politiques de la 'vulgarisation scientifique' en révolution (1780–1810)", *Annales historiques de la révolution française*, 338 (2004): 11–51.

Chaptal (1819). Jean-Antoine-Claude Chaptal, *De l'industrie* (Paris: Renouard, 1819), vol. 2.

Chavagnac (1906). Xavier-Roger-Marie de Chavagnac, *Histoire des manufactures françaises de porcelaine* (Paris: A. Picard et fils, 1906).

Chen (2005). Xian Chen, "Visual Photometry in the Early 19th Century: A 'Good' Science with 'Wrong' Measurements", in Jed Z. Buchwald, Allan Franklin, eds., *Wrong for Right Reasons* (Dordrecht: Springer, 2005), pp. 161–183.

Chevalier, Fau (1854). Charles Chevalier, Julien Fau, *Nouveau manuel complet du physicien-préparateur ou Description d'un cabinet de physique* (Paris: Librairie encyclopédique de Roret, 1854), 2 vols.

Chevalier (1834). Jacques Louis Vincent Chevalier, *Notice sur la chambre claire* (*camera lucida*) (Paris, 1834).

Chevallier (1819). Jean-Gabriel-Auguste Chevallier, *Essai sur l'art de l'ingénieur en instrumens de physique expérimentale en verre* (Paris: Mme Huzard, 1819).

Chayette-Cheval (2018). *Sciences, techniques, horlogerie* [auction catalogue] (Paris: Chavette-Cheval, 25 June 2018).

Chladni (1809). Ernst F.F. Chladni, *Traité d'acoustique* (Paris: Courcier, 1809).

Choullier (1883). Ernst Choullier, "Les Trudaines", *Revue de Champagne et de Brie*, 14 (1883): 263–272.

Clay, Court (1975). R.S. Clay, T.H. Court, *The History of the Microscope* (London: The Holland Press, 1975 [1st ed. 1932]).

Clifton (1995). Gloria Clifton, *Directory of British Scientific Instrument Makers 1550–1851* (London: Zwemmer, National Maritime Museum, 1995).

Cochu (1777). Joseph-Félicité Cochu, *Résumé pour les salpêtriers de Paris contre le sieur Micault de Courbeton, commissaire général et fermier des poudres et salpêtres* (Paris: Hérissant, 1777).

Collins (2004). Philips R. Collins, *Bizarre Barometers and Other Unusual Weather Forecasters* (Trowbridge: Baros Book, 2004).

Comparato (2010). Guillaume Comparato, *L'homme les fluides et la vie: Physique, médecine et universalisme au cœur des années Mesmer*, Université Pierre MENDES FRANCE (Grenoble II) Master 1 "Sciences humaines et sociales" Mention Histoire et Histoire de l'art Spécialité Histoire des échanges culturels internationaux Septembre 2010 Mémoire présenté sous la direction du Professeur Gilles Bertrand.

Comparato (2018). Guillaume Comparato, *Barthélemy Faujas de Saint-Fond, parcours d'un homme de sciences mondain au tournant des Lumières (1741–1819)* (PhD thesis, Université Grenoble Alpes, 2018), 2 vols.

Condorcet (1799). Jean-Antoine-Nicolas-Caritat de Condorcet, "Eloge de Guettard", in idem, *Éloges des académiciens de l'Académie royale des sciences*, vol. 4 (Paris: Vieweg, 1799), pp. 178–214.

Conservatoire (1882). *Conservatoire national des arts et métiers: Catalogue des collections*, 7th ed. (Paris: Dunot, 1882).

Conservatoire (1904). *Le Conservatoire national des arts et métiers: 1794–1904* (Paris: Imprimerie nationale, 1904).

Contant (1952). Jean Paul Contant, *Contribution à l'histoire de l'enseignement de la pharmacie: L'enseignement de la chimie au Jardin royal des plantes de Paris* (Cahors: Coueslant, 1952).

Coornaert (1968). Emile Coornaert, *Les corporations en France avant 1789* (Paris: Éditions ouvrières, 1968).

Cotte (1788). Louis Cotte, *Memoires sur la météorologie: Pour servir de suite et de supplément, au traité de météorologie publiée en 1772* (Paris: Imprimerie royale, 1788), 2 vols.

Coulomb (1777). Charles-Augustin Coulomb, "Recherches sur la meilleure manière de fabriquer les aiguilles aimantée", *Mémoires de l'Académie royale des sciences de Paris* (1777 pub. in 1780): 166–264.

Coulomb (1785). Charles-Augustin Coulomb, "Description d'une boussole dont l'aiguille est suspendue par un fil de soie", *Mémoires de l'Académie royale des sciences de Paris* (1785 pub. in 1788):560–568.

Crawforth (1987). Michael A. Crawforth, *Handbook of Weighing Instruments* (Oxford: International Society of Antique Scale Collectors, 1987).

Crell (1789). Lorenz Crell, ed., "Vom H[er]rn von E.** in Paris", *Chemische Annales*, 1 (1789): 519.

Cronstedt (1770). Axel Fredrik Cronstedt, *An Essay Towards a System of Mineralogy* (London: Dilly, 1770).

Crosland (1967). Maurice Crosland, *The Society of Arcueil. A View of French Science at the time of Napoleon I* (London: Heinemann, 1967).

Crosland (1969). Maurice Crosland, ed., *Science in France in the Revolutionary Era Described by Thomas Bugge* (Cambridge, MA: The MIT Press, 1969).

Crosland (2005). Maurice Crosland, "Early Laboratories c. 1600–c. 1800 and the Location of Experimental Science", *Annals of Science*, 62/2 (2005): 233–253.

Cuvier (1812). Georges Cuvier, *Recherches sur les ossements fossiles des quadrupèdes* (Paris: Deterville, 1812), 4 vols.

Cuvier (1819). Georges Cuvier, "Lavoisier", in *Biographie Universelle Ancienne et Moderne* (Paris: Michaud, 1819), vol. 23.

Daguin (1861). Adolphe Daguin, *Traité élémentaire de physique théorique et expérimentale* (Toulouse: E. Privat, 1861–1862), 4 vols.

Danger (1829). T.-P. Danger, *L'art du souffleur à lampe* (Paris: Bachelier, 1829).

Daumard, Furet (1961). Adeline Daumard, François Furet, *Structures et relations sociales à Paris au milieu du XVIIIe siècle* (Paris: Colin, 1961).

Daumas (1941). Maurice Daumas, *Lavoisier* (Paris: Gallimard, 1941).

Daumas (1949). Maurice Daumas, *Conservatoire nationale des arts et métiers: Catalogue du musée section JA, astronomie* (unpublished typescript, Paris, 1949).

Daumas (1950). Maurice Daumas, "Les appareils d'expérimentation de Lavoisier", *Chymia*, 5 (1950): 45–62.

Daumas (1952). Maurice Daumas, "Le corps des ingénieurs brevetés en instruments scientifiques (1787)", *Archives internationales d'histoire des sciences*, 5 (1952): 86–96.

Daumas (1955). Maurice Daumas, *Lavoisier théoricien et expérimentateur* (Paris: PUF, 1955).

Daumas (1959). Maurice Daumas, "The Lavoisier Collection at the Conservatoire National des Arts et Métiers", *Journal of Chemical Education*, 36/5 (1959): 231–232.

Daumas (1960). Maurice Daumas "Laboratoriumsgaräte aus dem Lavoiser-Saal des Conservatoire National des Arts et Métiers in Paris", *Die BASF*, 10/5 (1960): 181–187.

Daumas (1989). Maurice Daumas, *Scientific Instruments of the Seventeenth and Eighteenth Centuries and Their Makers* (London: Portland, 1989 [1st (French) ed. 1953]).

Daumas, Duveen (1950). Maurice Daumas, Denis I. Duveen, "Lavoisier's Relatively Unknown Large-Scale Decomposition Synthesis of Water, February 27 and 28, 1785", *Chymia*, 3 (1950): 113–129.

Davy (1955). René Davy, *Contribution à l'étude des origines de la droguerie pharmaceutique et de l'industrie du sel ammoniac en France: L'apothicaire Antoine Baumé (1728–1804)* (Cahors: Imprimerie A. Coueslant, 1955).

de Clercq (1997). Peter de Clercq, *The Leiden Cabinet of Physics: A Descriptive Catalogue* (Leiden: Museum Boerhaave, 1997).

De Gaulle (1839). Jules de Gaulle, *Nouvelle histoire de Paris et de ses environs* (Paris: Pourrat, 1839).

Dekker (1999), Elly Dekker, *Globes at Greenwich. A Catalogue of the Globes and Armillary Spheres in the national Maritime Museum, Greenwich* (Oxford, New York: Oxford University Press, National Maritime Museum, 1999).

De la Cité (1902). Jehan De La Cité, *L'arsenal et la Bastille* (Paris: Didot, 1902).

Delacroix, Porte (1995). Alain Delacroix, Catherine Porte, "Les balances et gazomètres de Lavoisier", in Christiane Demeulenaere-Douyère, ed., *Il y a 200 ans Lavoisier* (Paris: Tec Doc Lavoisier, 1995), pp. 289–298.

Delahante (1880). Adrien Delahante, *Une famille de finance au XVIIIe siècle. Mémoires, correspondance et papiers de famille réunis et mis en ordre par M. Adrien Delahante* (Paris: J. Hetzel et Cie., 1880), 2 vols.

Delamarche (1800). Charles François Delamarche, *Les usages de la sphère et des globes célestes et terrestres* (Paris: for the author, 1800).

Delambre (1806–1810). Jean Baptiste Delambre, *Base du système métrique décimal, ou Mesure de l'arc du méridien* (Paris, 1806–1810), 4 vols.

Delaunay (2017). Bernard Delaunay, *La pensée technique de l'Académie royale des sciences (1699–1750)*, thèse de doctorat (Paris: Université Paris 1 Panthéon-Sorbonne, 2017), 2 vols.

Dellebarre (1777). Louis-François Dellebarre, *Mémoire sur les differences de la construction & des effects du microscope de MLF Dellebarre, d'avec tous ceux qu'on faits précédement* (La Haye, 1777).

Delorme (1950). Suzanne Delorme, "Une famille de grands commis de l'état, amis des sciences, au XVIIIe siècle: Les Trudaine", *Revue d'histoire des sciences*, 3/2 (1950): 101–109.

De Luc (1784). Jean André De Luc, *Recherches sur les modifications de l'atmosphère* (Paris: Veuve Duchesne, 1784).

Delunel (1805). M. Delunel, *Éloge de Baumé, lu à la séance publique du 28 brumaire an 14* (Paris: Couturier, 1805).

Demachy (1790). Jean-François Demachy, "Examen impartial de la nouvelle doctrine des chimistes modernes ou pneumatiques", in *Tribut de la Société nationale des neuf sœurs* (Paris: Onfroy, 1790), pp. 116–123.

Demachy (1794). Jean-François Demachy, "Examen d'un traité élémentaire de chymie, présenté dans un ordre nouveau; par M. Lavoisier. 2 vol. in-8vo", *L'esprit des journaux* (Janvier 1794): 236–300.

Demeulenaere-Douyère (1994). Christiane Demeulenaere-Douyère, "A propos d'une entreprise intellectuelle: La publication des *Œuvres* et de la *Correspondance* de Lavoisier", *La vie des sciences, comptes rendus, série générale*, 11 (1994): 319–332.

Denton (2003). Elizabeth Denton, *Sir James Hall, un homme de science dans la Révolution Édition critique et commentaire de son journal de voyage en France (avril–août 1791)* (Thesis, École national des chartes, Paris, 2003).

d'Hollander (1999). Raymond d'Hollander, *L'Astrolabe, histoire, théorie, pratique* (Paris: Institut océanographique, 1999).

Diderot (1969–1972). Denis Diderot, "Note historique sur Michel Vanloo, peintre, et le chimiste Rouelle", in idem, *Œuvres complètes, éd. chronologique*, vol. 9 (Paris: Club Français du Livre, 1969–1972), pp. 600–601.

Diebknecht (1716). Johann Georg Diebknecht, Johann Michael Heusinger, *De noctiluca mercuriali sive de luce quam argentum vivum in tenebris fundit* (Giessen: Müller, 1716).

Doré (1994). Christophe Doré, ed., *La Ferme du tabac: Hommage à Antoine Laurent Lavoisier* (Paris: Musée-Galerie de la Seita, 1994).

Dorveaux (1918). Paul Dorveaux, "Antoine Baumé", *Bulletin de la Société d'histoire de la pharmacie*, 19 (1918): 345–352.

Dorveaux (1919). Paul Dorveaux, "Les grands pharmaciens: VII. Quinquet (1745–1803)", *Revue d'histoire de la pharmacie*, 21 (1919): 1–14.

Dorveaux (1935). Paul Dorveaux, "Apothicaires membres de l'Académie des sciences XI: Balthazar-Georges Sage", *Revue d'histoire de la pharmacie*, 23 (1935): 152–166.

Dorveaux (1935a). Paul Dorveaux, "Apothicaires membres de l'Académie des sciences: Balthazar-Georges Sage (suite et fin)", *Revue d'histoire de la pharmacie*, 23 (1935): 216–232.

Duchesne (1801). Henri Gabriel Duchesne, *Dictionnaire de l'industrie, ou Collection raisonnée des procédés utiles dans les sciences et dans les arts: Contenant nombre de secrets curieux et intéressants pour l'économie et les besoins de la vie; l'indication de différentes expériences à faire; la description de plusieurs jeux très-singuliers et très-amusants; les notices des découvertes et inventions nouvelles; les détails nécessaires pour se mettre à l'abri des fraudes et falsifications dans plusieurs objets de commerce et de fabrique: Ouvrage également propre aux artistes, aux négocians et aux gens du monde* (Paris: Poignée, 1801), 6 vols.

Dujarric de la Rivière (1954). René Dujarric de la Rivière, *E.-I. Du Pont de Nemours élève de Lavoisier* (Paris: Librairie des Champs-Élysées, 1954).

Dumas (1837). Jean Baptiste Dumas, *Leçons sur la philosophie chimique: Professées au Collège de France* (Paris, 1837).

Dumas (1862). Jean Baptiste Dumas, "Remarques à l'occasion d'une note de M. Becquerel concernant des manuscrits de

Lavoisier existant dans la bibliothèque d'Orléans, " *Comptes rendus de l'Académie des sciences* (1862): 713–714.

Dupré (2014). Sven Dupré, ed., *Laboratories of Art: Alchemy and Art Technology from Antiquity to the 18th Century* (Dordrecht, London and New York, NY: Springer, 2014).

Duveen (1951). Denis I. Duveen, "Antoine Laurent Lavoisier (1743–1794): A Note Regarding his Domicile During the French Revolution", *Isis*, 42/3 (1951): 233–234.

Duveen (1958). Denis I. Duveen, "Antoine-Laurent Lavoisier and the French Revolution – IV", *Journal of Chemical Education*, 35/9 (1958): 470–471.

Duveen (1965). Denis I. Duveen, *Supplement to a Bibliography of the Works of Antoine Laurent Lavoisier (1743–1794)* (London: Dawsons of Pall Mall, 1965).

Duveen, Klickstein (1954). Denis I. Duveen, Herbert S. Klickstein, "A Letter from Berthollet to Blagden Relating the Experiments for a Large-Scale Synthesis of Water Carried Out by Lavoisier and Meusnier in 1785", *Annals of Science*, 10 (1954): 58–62.

Duveen, Klickstein (1954a), Denis I. Duveen, Herbert S. Klickstein, *A Bibliography of the Works of Antoine Laurent Lavoisier (1743–1794)* (London: Dawsons & Sons, 1954).

Ehrmann (1787). Friedrich Ludwig Ehrmann, *Essai d'un art de fusion à l'aide de l'air du feu, ou air vital / par Mr. Ehrmann…; traduit de l'allemand par M. de Fontallard & revu par l'auteur; suivi des mémoires de Mr. Lavoisier … sur le même sujet* (Strasbourg: chez Jean George Treuttel, 1787).

Eklund (1975). Jon Eklund, *The Incompleat Chymist* (Washington, DC: Smithsonian Institution Press, 1975).

Ellis (1871). George Ellis, *Memoir of Sir Benjamin Thompson, Count Rumford: With Notices of His Daughter* (Boston, MA, 1871).

Emptoz (1995). Gérard Emptoz et al., *Catalogue des collections de mécanique Collections Alexandre Clair* (Le Puy en Velay: Musée Crozatier, 1995).

Encyclopédie (1751–1772). *Encyclopédie ou Dictionnaire raisonné des sciences, des arts et des métiers, par une société de gens de lettres* (Paris: Barrias, 1751–1772), 26 vols: http://encyclopedie.uchicago.edu/ (accessed 6 August 2021).

Encyclopédie (1778). *Encyclopédie ou Dictionnaire raisonné des sciences, des arts et des métiers, par une société de gens de lettres: Nouvelle édition* (Geneva: Pellet, 1778), 36 vols.

Encyclopédie méthodique – Chymie (1786–1815). *Encyclopédie méthodique: Chymie, pharmacie et métallurgie* (Paris: Panckoucke, 1786–1815), 6 vols.

Ephemerides (1783). *Ephemerides societati meteorologicae palatinae, historia et observationes anni 1781* (Mannheim: Officina Novae Societatis Typographicae, 1783).

État (1776). *État de médecine, chirurgie et pharmacie en France, en Europe pour l'année 1776* (Paris: Didot, 1776).

Fahrenheit (1724). Daniel Gabriel Fahrenheit, "Aræometri novi descriptio & usus a D.G. Fahrenheit", *Philosophical Transactions*, 384 (1724): 140–141.

Faure (2002). Juliette Faure, *L'Arsenal de Paris: Histoire et chroniques* (Paris: L'Harmattan, 2002).

Fell (2000). Ulrike Fell, *Disziplin, Profession und Nation: Die Ideologie der Chemie in Frankreich vom zweiten Kaiserreich bis in die Zwischenkriegszeit* (Leipzig: Leipziger Universitätsverlag, 2000).

Flook (2008). Ron Flook, *The London Knife Book: An A-Z Guide to London Cutlers, 1820–1945* (Slough: Antique Knives, Limited, 2008).

Forbes (1969–1976). R.J. Forbes, ed., *Martinus van Marum: Life and Work* (Haarlem: H.D. Tjeenk Willink & zoon, 1969–1976), 6 vols.

Fors (2003). Hjalmar Fors, *Mutual Favours: The Social and Scientific Practice of Eighteenth-Century Swedish Chemistry* (PhD thesis, Uppsala University, 2003).

Fourcroy (1796). Antoine-François Fourcroy, *Notice sur la vie et les travaux de Lavoisier, précédée d'un discours sur les funérailles et suivie par un ode sur l'immortalité de l'âme* (Paris: Feuille du cultivateur, an IV [1796]).

Fournier (2003). Marianne Fournier, *Early Microscopes: A Descriptive Catalogue* (Leiden: Museum Boerhaave, 2003).

Fox (2016). Robert Fox, "Science, Celebrity, Diplomacy: The Marcellin Berthelot Centenary, 1927", *Revue d'histoire des sciences*, 69/1 (2016): 77–115.

Francis (1846). George William Francis, *The Dictionary of the Arts, Sciences, and Manufactures* (London: W.Brittain, 1846).

Francoeur (1835). Louis Benjamin Francoeur, *Géodésie, ou Traité de la figure de la terre et de ses parties; comprenant la topographie, l'arpentage, le nivellement; la géomorphie terrestre et astronomique, la construction des cartes* (Paris: Bachelier, 1835).

Franklin (1862). Alfred Franklin, *Recherches historiques sur le Collège des quatre-nations* (Paris: Auguste Aubry, 1862).

Franklin (1867–1873). Alfred Franklin, *Les anciennes bibliothèques de Paris* (Paris: Imprimerie impériale, 1867–1873), 3 vols.

Franklin (1905). Alfred Franklin, *Dictionnaire historique des arts, métiers et professions* (Paris: Welter, 1905), 2 vols.

Franklin (1959–2018). Benjamin Franklin, *The papers of Benjamin Franklin* (New Haven: Yale University Press, 1959–2018), 43 vols.

Fraser (1780–1805). William Fraser, *The New Pedometer which when Wore in the Pocket not only Ascertains the Space of the Nearer Walks but Likewise which Way* (London, ca. 1780–1805). Printed advertisement (cf. History of Science Museum, Oxford, inventory no. 14387).

Ganot (1894). Adolphe Ganot, *Traité élémentaire de physique*, 21st ed. (Paris: Hachette, 1894).

Gardiner (1737). William Gardiner, *Practical Surveying Improved or Land Measuring According to the Present Most Correct Methods* (London, 1737).

Gehler (1831). Johann Samuel Traugott Gehler, *Physikalisches Wörterbuch, oder, Versuch einer Erklärung der vornehmsten Begriffe und Kunstwörter der Naturlehre: Mit kurzen Nachrichten von der Geschichte der Erfindungen und Beschreibungen der Werkzeuge begleitet in alphabetischer Ordnung* (Leipzig: E.B. Schwickert, 1831), 6 vols.

Gille (1963). Bertrand Gille, *Documents sur l'état de l'industrie et du commerce de Paris et du département de la Seine (1778–1810)* (Paris: Imprimerie municipale, 1963).

Gillispie (1956). Charles C. Gillispie, "Notice biographique de Lavoisier par Madame Lavoisier", *Revue d'histoire des sciences et de leurs applications*, 9/1 (1956): 52–61.

Gillispie (1980). Charles C. Gillispie, *Science and Polity in France at the End of the Old Regime* (Princeton, NJ: Princeton University Press, 1980).

Gillispie (1983). Charles C. Gillispie, *The Montgolfier Brothers and the Invention of Aviation 1783–1784: With a Word on the Importance of Ballooning for the Science of Heat and the Art of Building Railroads* (Princeton, NJ: Princeton University Press, 1983).

Giormani (2001). Virgilio Giormani "1793: Vincenzo Dandolo e l'insegnamento della nuova chimica al teatro La Fenice di Venezia", in Paolo Mirone, ed., *Atti del IX Convegno nazionale di storia e fondamenti della chimica* (Roma: Accademia dei XL, 2001), pp. 159–169.

Gires (2016). Francis Gires ed., *Encyclopédie des instruments de l'enseigement de la physique du XVIIIe au milieu du XXe siècle* (Niort: ASEISTE, 2016), 3 vols.

Glocker (1992). Winfried Glocker, *Glastechnik* (Munich: Deutsches Museum, 1992).

Goebel *et al.* (2003). Manfred Goebel *et al.*, *Der Pantograph in historischen Veröffentlichungen des 17.bis 19. Jahrhunderts* (Halle: Universitäts- und Landesbibliothek Sachsen-Anhalt, 2003).

Golinski (1992). Jan Golinski, *Science as Public Culture: Chemistry and Enlightenment in Britain 1760–1820* (Cambridge: Cambridge University Press, 1992).

Golinski (1994). Jan Golinski, "Precision Instruments and the Demonstrative Order of Proof in Lavoisier's Chemistry", *Osiris* (1994): 30–47.

Golinski (1995). Jan Golinski, "'The Nicety of Experiment': Precision of Measurement and Precision of Reasoning in Late Eighteenth-Century Chemistry", in M. Norton Wise, ed., *The Values of Precision* (Princeton, NJ: Princeton University Press, 1995), pp. 72–91.

Good, Multhauf (1987). Gregory Good, Robert P. Multhauf, *A Brief History of Geomagnetism and a Catalogue of the Collections of the National Museum of American History* (Washington, DC: Smithsonian Institution Press, 1987).

Goubert (1785). Jean Pierre Goubert, *Description et usage des baromètres, thermomètres et autres instrumens météorologiques* (Dijon: J.B. Capel, 1785).

Graham (1981). J.T. Graham, *Scales and Balances* (Aylesbury: Shire Oublication, 1981).

Griffin (1827). John Joseph Griffin, *A Practical Treatise on the Use of the Blowpipe, in Chemical and Mineral Analysis: Including a Systematic Arrangement of Simple Minerals, Adapted to Aid the Student in His Progress in Mineralogy, by Facilitating the Discovery of the Names of Species* (Glasgow: R. Griffin & Company, 1827).

Grimaux (1888). Edouard Grimaux, *Lavoisier 1743–1794, d'après sa correspondance, ses manuscrits, ses papiers de famille et d'autres documents inédits* (Paris: F. Alcan 1888).

Grimm (1812–1813). Friedrich Melchior Grimm, *Correspondance littéraire, philosophique et critique, adressée a un souverain d'Allemagne* (Paris: Longchamps, 1812–1813), 16 vols.

Grison (1996). Emmanuel Grison, *L'étonnant parcours du républicain J.H. Hassenfratz (1755–1827)* (Paris: Les presses de l'École des Mines, 1996).

Grouchy, Cottin (1906–1907). Emmanuel Henri Grouchy, Paul Cottin, ed., *Journal inédit du duc de Croÿ, 1718–1784* (Paris: Flammarion, 1906–1907), 4 vols.

Guerlac (1951). Henry Guerlac, "The Continental Reputation of Stephen Hales", *Archives internationales d'histoire des sciences*, 15 (1951): 393–404.

Guerlac (1956). Henry Guerlac, "A Note on Lavoisier's Scientific Education", *Isis*, 47 (1956): 211–216.

Guerlac (1961). Henry Guerlac, *Lavoisier – The Crucial Year: The Background and Origin of his First Experiments on Combustion in 1772* (Ithaca, NY: Cornell University Press, 1961).

Guerlac (1963). [Henry Guerlac], ed., *Antoine Laurent Lavoisier: An Exhibition* (Ithaca, NY: The Cornell University Library, 1963).

Guerlac (1975). Henry Guerlac, *Antoine-Laurent Lavoisier: Chemist and Revolutionary* (New York, NY: Scribner's Sons, 1975).

Guerlac (1979). Henry Guerlac, "The Lavoisier Papers – A Checkered History", *Archives internationales d'histoire des sciences*, 29 (1979): 95–100.

Guettard (1756). Jean Etienne Guettard, "Description minéralogique des environs de Paris", *Mémoires de l'Académie royale des sciences de Paris* (1756 pub. in 1762): 217–258.

Guettard (1768–1783). Jean Etienne Guettard, *Mémoires sur différents parties des sciences et des arts* (Paris: Prault, 1768–1783), 5 vols.

Guettard, Monnet (1780). Jean Etienne Guettard, Antoine Grimoald Monnet, *Atlas et descriptions minéralogiques de la France, entrepris par ordre du Roi, par MM. Guettard & Monnet. Publiés par M. Monnet, d'après ses nouveaux voyages. Première partie, comprenant le Beauvoisis, la Picardie, le Boulonnais, la Flandre française, le Soissonnais, la Lorraine Allemande, une partie de la Lorraine Française, le pays Messin, & une partie de la Champagne* (Paris: Didot l'ainé, 1780).

Guichard (1937). Marcel Guichard, *Essai historique sur les mesures en chimie*, vol. 2 (Paris: Hermann, 1937)

Guillerme (2007). André Guillerme, *La naissance de l'industrie à Paris: Entre sueurs et vapeurs 1780–1830* (Sayssel: Champ Vallon, 2007).

Guyot (1774). Edme-Gilles Guyot, *Nouvelles récréations physiques et mathématiques* (Paris: Gueffier, 1774), 4 vols.

Hackmann (1978). Willem D. Hackmann, *Electricity from Glass: The History of the Frictional Electrical Machine 1600–1850* (Alphen aan den Rijn: Sijthoff & Noordhoff, 1978).

Hackmann (1978). Willem D. Hackmann, "Eighteenth-Century Electrostatic Measuring Devices", *Annali dell'Istituto e Museo di storia della scienza di Firenze*, 3/2 (1978): 3–58.

Hachette (1828). Jean Nicolas Pierre Hachette, *Traité élémentaire des machines* (Paris: Corby, 1828).

Hales (1727). Stephen Hales, *Vegetable Staticks, or, An Account of Some Statical Experiments on the Sap in Vegetables: Being an Essay Towards a Natural History of Vegetation: Also, a Specimen of an Attempt to Analyse the Air, by a Great Variety of Chymio-Statical Experiments; which were Read at Several Meetings before the Royal Society* (London: Innys, 1727).

Hambly (1988). Maya Hambly, *Drawing Instruments 1580–1980* (London: Sotheby's Publications, 1988).

Hammond (1981). John H. Hammond, *The Camera Obscura – A Chronicle* (Bristol: Adam Hilger Ltd, 1981).

Hanin (1768). Hanin, "Un peson a ressort présenté par le sieur Hanin serrurier", *Mémoires de l'Académie royale des sciences* (1765 pub. in 1768): 135–136.

Hanin (1791). Hanin, "Twenty Guineas were Given, as a Bounty to Monsieur Hanin", *Transaction of the Society Instituted at London for the Encouragement of Arts, Manufactures and Commerce*, 11 (1791): 151–152.

Hassenfratz (1786). Jean Henri Hassenfratz, "Mémoire sur les chalumeaux à bouche, à soufflets et à l'air déphlogistiqué", *Obervations sur la physique*, 28 (1786): 345–352.

Hauksbee (1705). Francis Hauksbee, "Several Experiments on the Mercurial Phosphorus made before the Royal Society", *Philosophical Transactions*, 24 (1705): 2129–2135.

Heering (2005). Peter Heering, "Weighing Heat: The Replication of the Experiments with the Ice-Calorimeter of Lavoisier and Laplace", in Marco Beretta, ed., *Lavoisier in Perspective* (Munich: Deutsches Museum, 2005), pp. 27–41.

Hellmann (1932). Doris Clarisse Hellman, "John Bird (1709–1776): Mathematical Instrument-Maker in the Strand", *Isis*, 17 (1932): 127–153.

Henley (1772). William Henley, "An Account of a New Electrometer", *Philosophical Transactions*, 62 (1772): 359–364.

Hilaire-Pérez (2000). Liliane Hilaire-Pérez, *L'invention technique au siècle des Lumières* (Paris: Albin Michel, 2000).

Hillaret (1964). Jacques Hillaret, *Dictionnaire historique de Paris*, 2nd ed. (Paris: Les éditions de minuit, 1964), 2 vols.

Holland, Stöhr (2013). Reiner Holland, Gerhard Stöhr, *Thermometer, Skalen und deren Vater* (Riedlingen: Freunden alter Wetterinstrumente/G. Stöhr, 2013).

Holmes (1985). Frederic Lawrence Holmes, *Lavoisier and the Chemistry of Life: An Exploration of Scientific Creativity* (Madison, WI: The University of Wisconsin Press, 1985).

Holmes (1998). Frederic Lawrence Holmes, *Antoine Lavoisier – The Next Crucial Year* (Princeton, NJ: Princeton University Press, 1998).

Holmes (2000). Frederic Lawrence Holmes, "The Evolution of Lavoisier's Chemical Apparatus", in Frederic Lawrence Holmes, Trevor Levere, *Instruments and Experimentation in the History of Chemistry* (Boston, MA: The MIT Press, 2000), pp. 137–152.

Home *et al.* (2017). Roderick W. Home, Isabel M. Malaquias, Manuel F. Thomaz, eds., *For the Love of Science: The Correspondence of J.H. de Magellan (1722–1790)* (Bern: Peter Lang, 2017), 2 vols.

Howard (2016). Robert A. Howard, "Realities and Perception in the Evolution of Black Powder Making", in Brenda J. Buchanan, ed., *Gunpowder, Explosives and the State: A Technological History* (Oxford and New York, NY: Routledge, 2016), pp. 21–41.

Hübner (2010). Marita Hübner, *Jean André Deluc (1727–1817): Protestantische Kultur und moderne Naturforschung* (Göttingen: Vandenhoeck & Ruprecht, 2010).

Hufbauer (1982). Karl Hufbauer, *The Formation of the German Chemical Community, 1720–1795* (Berkeley, CA: University of California Press, 1982).

Huzard (1836). Jean Baptiste Huzard, *Nécrologie: Madame la comtesse de Rumfort [sic], veuve de l'illustre chimiste Lavoisier, est morte avant hier matin (jeudi 11 février), l'âge de 81 ans, en son hôtel, rue d'Anjou-Saint-Honoré, N° 39* (Extrait du Moniteur du Commerce du Samedi 13 février 1836, N° 993).

Jacomy (1994). Bruno Jacomy, "Le laboratoire de Lavoisier", *La revue: Musée des arts et métiers*, 6 (1994): 17–22.

Jacques (1985). Jean Jacques, "Le cours de chimie de G.-F. Rouelle recueilli par Diderot", *Revue d'histoire des sciences*, 38 (1985): 43–53.

Jenemann (1997). Hans R. Jenemann, *The Chemist's Balance: Die Waage des Chemikers* (Frankfurt a.M.: DACHEMA, 1997).

Jensen (1986). William B. Jensen, "The Development of Blowpipe Analysis", in J.T. Stock and M.V. Orna, eds., *The History and Preservation of Chemical Instrumentation* (Dordrecht: D. Reidel Pub. Co., 1986), pp. 123–149.

Jèze (1765). Jèze, *État ou Tableau de la ville de Paris* (Paris: Prault père, 1765).

Kersaint (1966). Georges Kersaint, *Antoine François de Fourcroy (1755–1809): Sa vie et son œuvre* (Paris: Muséum d'Histoire Naturelle, 1966).

Kindleberger (1984). Charles P. Kindleberger, *A Financial History of Western Europe* (London: Allen & Unwin, 1984).

Klein (2007). Ursula Klein, "Apothecary-Chemists in Eighteenth-Century Germany", in Lawrence M. Principe, ed., *New Narratives in Eighteenth-Century Chemistry* (Dordrecht: Springer, 2007), pp. 97–137.

Klein (2009). Ursula Klein, "Chemical and Pharmaceutical Laboratories Before the Professionalization of Chemistry", in Marta Lourenço, Ana Carneiro, eds., *Spaces and Collections in the History of Science* (Lisbon: Museum of Science of the University of Lisbon, 2009), pp. 3–12.

Klein (2012). Ursula Klein, ed., "Artisanal-Scientific Experts in Eighteenth-Century France and Germany", *Annals of Science*, 69/3 (2012).

Klein, Spary (2010). Ursula Klein, Emma C. Spary, eds., *Materials and Expertise in Early Modern Europe* (Chicago, IL: University of Chicago Press, 2010).

Knight (1744 and 1746). Gowin Knight, "Account of Some Magnetical Experiments", *Philosophical Transactions*, 43 (1744): 161–161; 44 (1746): 657–672.

Knoefel (1984). Peter K. Knoefel, *Felice Fontana: Life and Works* (Trento: Società di studi trentini di scienze storiche, 1984).

Lacour (2019). Pierre-Yves Lacour, *La République naturaliste. Collections d'histoire naturelle et Révolution française (1789–1804)* (Paris: Publications scientifiques du Muséum, 2019).

Lalande (1795). Jérôme Lalande, "Notice sur la vie et les ouvrages de Lavoisier", *Magasin encyclopédique*, 5 (1795): 174–188.

Landriani (1995), Marsilio Landriani, *Ricerche fisiche intorno alla salubrità dell'aria: Con un articolo di Alessandro Volta sull'eudiometria. A cura di Marco Beretta* (Florence: Giunti, 1995).

Langins (1987). Janis Langins, *La République avait besoin de savants* (Paris: Belin, 1987).

Langlois (1744). Claude Langlois, *Description et usage du pantographe, autrement appelé singe, changé & perfectionné par C. Langlois, ingénieur du Roi & de l'Académie royale des sciences pour les instrumens de mathématiques* (Paris, 1744).

Lanterne magique (1990). *Lanterne magique et fantasmagorie* (Paris: CNAM – Musée national des techniques, 1990).

Lavoisier, Correspondance, see *LC*.

Lavoisier, Œuvres, see *LO*.

Lavoisier (1765). Antoine-Laurent Lavoisier, "Mémoire sur les différents moyens qu'on peut employer pour éclairer une grande ville", LO, vol. 3, pp. 1–77.

Lavoisier (1767). Antoine-Laurent Lavoisier, "De la nature des eaux d'une partie de la Franche-Comté, de l'Alsace, de la Lorraine, de la Champagne, de la Brie et du Valois", LO, vol. 3, pp. 145–162.

Lavoisier (1768). Antoine-Laurent Lavoisier, "Recherches sur les moyens les plus sûrs, les plus exacts et les plus commodes de déterminer la pesanteur spécifique des fluides, soit pour la physique, soit pour le commerce", LO, vol. 3, pp. 427–450.

Lavoisier (1772). Antoine-Laurent Lavoisier, "Expériences sur la pesanteur de différentes eaux et principalement de celle de mer, depuis Cadix jusqu'aux cotes de Californie", in Jean-Baptiste Chappe d'Auteroche, *Voyage en Californie pour l'observation du passage de venus sur la disque du soleil* (Paris: Jombert, 1772), pp. 47–52. This contribution, unknown to Duveen's and Klickstein bibliography of Lavoisier's work, was drawn from an uncompleted manuscript draft and was published in LO, vol. 3, pp. 456–460.

Lavoisier (1773). Antoine-Laurent Lavoisier, "Premier mémoire sur la nature de l'eau, et sur les expériences par lesquelles on a prétendu prouver la possibilité de son changement en terre", *Mémoires de l'Académie royale des sciences* (1770 pub. in 1773): 73–82; reprinted in LO, vol. 2, pp. 1–11.

Lavoisier (1774). Antoine-Laurent Lavoisier, *Opuscules physiques et chymiques* (Paris: Durand, 1774), reprinted in LO, vol. 1, pp. 437–666.

Lavoisier (1774a). Antoine-Laurent Lavoisier, "Premier Essai du Grand Verre ardent de M. Trudaine Etabli au Jardin de l'Infante, au commencement di mois d'Octobre de l'année 1774. Par M.rs Trudaine de Montigny, Macquer, Cadet, Lavoisier & Brisson", *Mémoires de l'Académie Royale des Sciences de Paris* (1774 pub. in 1778): 62–72, reprinted in LO, vol. 3, pp. 274–283.

Lavoisier (1775). Antoine-Laurent Lavoisier, "Mémoire sur une nouvelle méthode distillatoire appliquée à la distillation des eaux-de-vie et à celle de l'eau de mer", LO, vol. 4, pp. 717–767.

Lavoisier (1775a). Antoine-Laurent Lavoisier, "État des vaisseaux et ustensiles nécessaires pour monter un laboratoire de chimie", LO, vol 5, pp. 335–339.

Lavoisier (1776). Antoine-Laurent Lavoisier, "Observations sur le froid de l'hiver de 1776; documents réunis par Lavoisier", LO, vol. 3, pp. 387–420.

Lavoisier (1777). Antoine-Laurent Lavoisier, *Instruction sur l'établissement des nitrières et sur la fabrication du salpêtre, publiée par ordre du roi: Par les régisseurs généraux des poudres et salpêtres* (Paris: Imprimerie Royale, 1777), reprinted in LO, vol. 5, pp. 391–460.

Lavoisier (1777a). Antoine-Laurent Lavoisier, "Mémoire sur la proportion d'acide et de base qui entre dans la composition des différents sels neutres", LO, vol. 5, pp. 256–266.

Lavoisier (1778). Antoine-Laurent Lavoisier, *Introduction et plan d'un deuxième volume des Opuscules physiques et chimiques*, LO, vol. 5, pp. 267–270.

Lavoisier (1779). Antoine-Laurent Lavoisier, "Rapport sur une machine pneumatique du Sieur Fortin", LO, vol. 4, p. 327.

Lavoisier (1779a). Antoine-Laurent Lavoisier, "Mémoire sur la combustion en général", *Mémoires de l'Académie Royale des Sciences de Paris* (1777 pub. in 1780):592–600, reprinted in LO, vol. 2, pp. 225–233.

Lavoisier (1779?). Antoine-Laurent Lavoisier, "Mémoire sur la construction des baromètres à surface plane", LO, vol. 3, pp. 753–758.

Lavoisier, Meusnier de la Place (1781a). Antoine-Laurent Lavoisier, Jean-Baptiste Meusnier de la Place, "Mémoire où l'on prouve par la décomposition de l'eau, que ce fluide n'est point une substance simple, & qu'il y a plusieurs moyens d'obtenir en grand l'air inflammable qui y entre comme principe constituant. Par M.rs Meusnier et Lavoisier", *Mémoires de l'Académie Royale des Sciences de Paris* (1781 pub. in 1784 : 269–283, reprinted in LO, vol. 2, pp. 360–374.

Lavoisier (1781b). Antoine-Laurent Lavoisier, "Mémoire dans lequel on a pour objet de prouver que l'eau n'est point une substance simple, un élément proprement dit, mais qu'elle est susceptible de décomposition et de recomposition" *Mémoires de l'Académie Royale des Sciences de Paris* (1781 pub. in 1784), reprinted in LO, vol. 2, pp. 334–359.

Lavoisier (1781c). Antoine-Laurent Lavoisier, "Mémoire sur la manière d'éclairer les salles de spectacle", *Mémoires de l'Académie Royale des Sciences de Paris* (1781 pub. in 1784): 409–420, reprinted in LO, vol. 3, pp. 91–100.

Lavoisier (1781d). Antoine-Laurent Lavoisier, "Mémoire sur l'électricité qu'absorbent les corps qui se réduisent en vapeurs", *Mémoires de l'Académie Royale des Sciences de Paris* (1781 pub. in 1784): 292–294, reprinted in LO, vol. 2, pp. 374–376.

Lavoisier (1782). Antoine-Laurent Lavoisier, "Mémoire sur un moyen d'augmenter considérablement l'action du feu et de la chaleur, dans les opérations chimiques", *Mémoires de l'Académie Royale des Sciences de Paris* (1782 pub. in 1785): 457–465, reprinted in LO, vol. 2, pp. 423–432.

Lavoisier (1782a) Antoine-Laurent Lavoisier, "Mémoire sur l'effet qui produit pour les pierres précieuses un degré de feu trés-violent", *Mémoires de l'Académie Royale des Sciences de Paris* (1782 pub. in 1785): 476–485, reprinted in LO, vol. 2, pp. 441–450.

Lavoisier (1782b). Antoine-Laurent Lavoisier, "De l'action du feu, animé par l'air vital, sur les substances minérales les plus réfractaires", *Mémoires de l'Académie Royale des Sciences de Paris* (1783 pub. in 1786); 563–614, reprinted in LO, vol. 2, 451–502.

Lavoisier, Laplace (1783). Antoine Laurent Lavoisier, Pierre Simon Laplace, *Mémoire sur la chaleur. Lû à l'Académie Royale des Sciences, le 28 Juin 1783. Par M.rs Lavoisier & De La Place, de la même Académie*. Separately printed memoir, published in August 1783. The same memoir was also published in the *Mémoires de l'Académie Royale des Sciences de Paris* (1780 pub. in 1784): 355–408 and reprinted in LO, vol. 2, pp. 283–333.

Lavoisier (1783a). Antoine-Laurent Lavoisier, "De l'action du calorique sur les corps liquides, depuis le terme de leur liquéfaction, jusqu'à celui de leur vaporisation", LO, 2, pp. 773–782.

Lavoisier (1785). Antoine-Laurent Lavoisier, "Thermomètre des caves de l'Observatoire: Précautions prises pour construire et graduer ce thermomètre", LO, vol. 3, pp. 421–425.

Lavoisier, Meusnier de la Place (1786). Antoine-Laurent Lavoisier, Jean-Baptiste Meusnier de la Place, "Développement des dernières expériences de la décomposition et recomposition de l'eau, faites par MM. Lavoisier et Meusnier, de l'Académie des Sciences", *Journal Polytype des sciences et des arts* (27 February 1786), 1: 21–44, reprinted in LO, vol. 5, pp. 320–334.

Lavoisier (1787). Antoine-Laurent Lavoisier, "Mémoire sur la fermentation spiritueuse", LO, vol. 3, pp. 777–790.

Lavoisier (1787a). Antoine-Laurent Lavoisier, "Résultats de quelques expériences d'agriculture, et réflexions sur leurs relations avec l'économie politique", LO, vol. 2, pp. 812–823.

Lavoisier (1789). Antoine-Laurent Lavoisier, *Traité élémentaire de chimie* (Paris: Cuchet, 1789), 2 vols.

Lavoisier (1790). Antoine-Laurent Lavoisier, "Observations sur le platine", *Annales de chimie*, 5 (1790): 137–141; reprinted in LO, vol. 5, pp. 351–353.

Lavoisier (1791). Antoine-Laurent Lavoisier, *De la richesse territoriale du royaume de France* (1791) (Paris: Éditions du Comité des travaux historiques et scientifiques, 1988).

Lavoisier (1792). Antoine-Laurent Lavoisier, "Mémoire sur la hauteur des montagnes des environs de Paris (1792)", LO, vol. 5, pp. 205–213. This essay was drafted in 1771 but read by Lavoisier before the Académie only in November 1792.

Lavoisier (1805). Antoine-Laurent Lavoisier, *Mémoires de physique et de chimie* (Paris: 1805 ca), reprinted with introduction by Marco Beretta and additional plates (Bristol: Thoemmes Press, 2004), 2 vols.

Lavoisier (1943). *Lavoisier: Exposition présentée à l'occasion du deuxième centenaire de Lavoisier* (Paris: Palais de la Découverte, 1943).

Lavoisier (1963). *Antoine Laurent Lavoisier: An exhibition* (Ithaca, NY: Cornell University Library, 1963).

Lavoisier (1994). *Il ya 200 ans Lavoisier: Exposition* (Paris: Institut de France – Académie des Sciences, 1994).

Lavoisier, Laplace (1982). Antoine-Laurent Lavoisier, Pierre Simon Laplace, *Memoir on Heat: Read to the Royal Academy of Sciences, 28 June 1783*, transl., introd. and notes by Henry Guerlac (New York, NY: Neale Watson Academic Publications, 1982).

LC. Antoine-Laurent Lavoisier, *Correspondance* (Paris: Albin Michel, Belin, Académie des sciences, Hermann, 1955–2012), 7 vols.

Ledoux-Lebard (1989). Denise Ledoux-Lebard, *Le mobilier français du XIXe siècle: 1795–1889: Dictionnaire des ébénistes et des menuisiers* (Paris: Édition de l'Amateur, 1989).

Leduc (1879). A. Leduc, "Les instruments de Lavoisier", *La Nature*, I sem. (1879): 362–365.

Lefèvre (1803). A. Lefèvre, *Nouveau traité géométrique de l'arpentage* (Paris: Duprat, 1803).

Lefuel (1923). Hector-Martin Lefuel, *Georges Jacob, ébéniste du XVIIIe siècle* (Paris: A. Morancé, 1923).

Lehman (2006). Christine Lehman, *Gabriel François Venel (1723–1775). Sa place dans la chimie du XVIIIe siècle*. PhD thesis – Université Paris X (Paris-Nanterre, 2006).

Lehman (2008). Christine Lehman, "Between Commerce and Philanthropy: Chemistry Courses in Eighteenth-Century Paris", in Bernadette Bensaude-Vincent, Christine Blondel, eds., *Science and Spectacle in the European Enlightenment* (Aldershot: Ashgate, 2008), pp. 103–116.

Lehman (2012). Christine Lehman, "Pierre-Joseph Macquer: An Eighteenth-Century Artisanal-Scientific Expert", *Annals of Science*, 69/3 (2012): 307–333.

Lehman (2013). Christine Lehman, "Alchemy Revisited by the Mid-Eighteenth Century Chemists in France: An Unpublished Manuscript by Pierre-Joseph Macquer", *Nuncius – Journal of the Material and Visual History of Science*, 28 (2013): 165–216.

Lehman (2016). Christine Lehman, "What is the 'True"Nature of Diamond'", *Nuncius – Journal of the Material and Visual History of Science*, 31 (2016): 361–407.

Lehman (2019). Christine Lehman, "Les lieux d'activité du chimiste Pierre-Joseph Macquer (1718–1784): Laboratoires et instruments", *Revue d'histoire des sciences*, 72/2 (2019): 221–254.

Lehman, Pépin (2009). Christine Lehman, François Pépin, eds., "La chimie et *l'Encyclopédie*", *Corpus: Revue de philosophie*, 56 (2009): 5–255.

Lemay (1934), Pierre Lemay, "Les habitations de Lavoisier", *Le courrier médical*, 84/35 (1934): 591–592.

Lenglen (1936). M. Lenglen, *Lavoisier agronome* (Douai: Lunven, 1936).

Le Roux (2011). Thomas Le Roux, *Le laboratoires des pollutions industrielles: Paris 1770–1830* (Paris: Albin Michel, 2011).

Levere, see *Turner*.

Levere (1992). Trevor H. Levere, "Balance and Gasometer in Lavoisier's Chemical Revolution", in Michelle Goupil, ed., *Lavoisier et la révolution chimique* (Paris: Sabix, 1992), pp. 313–332.

Levere (2000). Trevor H. Levere, "Measuring Gases and Measuring Goodness", in Frederic L. Holmes, Trevor H. Levere, eds., *Instruments and Experimentation in the History of Chemistry* (Cambridge, MA: The MIT Press, 2000), pp. 105–135.

Levere (2005). Trevor H. Levere, "Lavoisier's Gasometer and Others: Research, Control and Dissemination", in Marco Beretta, ed., *Lavoisier in Perspective* (Munich: Deutsches Museum, 2005), pp. 53–67.

Lewis (1763). William Lewis, *Commercium Philosophico-technicum, or, The Philosophical Commerce of Arts: Designed as an Attempt to Improve Arts, Trades, and Manufactures* (London: H. Baldwin, 1763).

Lilti (2005). Antoine Lilti, *Le monde des salons: Sociabilité et mondanité à Paris au XVIII*e *siècle* (Paris: Fayard, 2005).

Liste (1814). *Liste des personnes qui ont péri par jugement du tribunal révolutionnaire, depuis le 26 août 1792 jusqu'au 13 juin 1794 (25 prairial an II): Et dont les corps ont été inhumés dans le terrain de l'ancien cimetière de la Madelaine, situé rue d'Anjou, faubourg Saint-Honoré, appartenant à présent à M. Descloseaux* (Paris: J.-R. Lottin de Saint-Germain, 1814).

LO. Antoine-Laurent Lavoisier, *Œuvres* (Paris: Imprimerie impériale, Imprimerie nationale, 1862–1893), 6 vols.

Lourenço, Carneiro (2009). Marta Lourenço, Ana Carneiro, eds., *Spaces and Collections in the History of Science* (Lisbon: Museum of Science of the University of Lisbon, 2009).

Lusk (1921). Graham Lusk, "Some Influences of French Science on Medicine", *The Journal of the American Medical Association*, 76/1 (1 January 1921): 1–8.

Lusk (1925). Graham Lusk, "Mementoes of Lavoisier: Notes on a Trip to Chateau de la Canière", *The Journal of the American Medical Association*, 85/16 (17 October 1925): 1246–1247.

Macquer (1753). Pierre Joseph Macquer, *Elémens de chymie théorique* (Paris, 1753).

Macquer (1777). Pierre-Joseph Macquer, "Laboratory", in idem, *A Dictionary of Chemistry* (London: Cadell, 1777), vol. 2, [pages not numbered].

Macquer (1778). Pierre Joseph Macquer, *Dictionnaire de chimie*, 2nd ed. (Paris: Barrois, 1778), 4 vols.

Macquer (1766). Philippe Macquer, ed., *Dictionnaire portatif des arts et métiers: Contenant en abrégé l'histoire, la description et la police des arts et métiers, des fabriques et manufactures de France et des pays étrangers* (Paris: Lacombe, 1766), 2 vols.

Magellan, see *Home*.

Magellan (1777). Jean Hyacinthe de Magellan, "Lettre contenant la description d'un appareil de crystal pour faire, en peu de minutes & à peu de frais des eaux minérales semblables à celles de Pyrmont, Spa, Saltzer, &c.", *Observations sur la physique*, 9 (1777): 280–287.

Marcelin (2004). Frank Marcelin, *Dictionnaire des fabricants français d'instruments de mesure au XV*e *au XIX*e *siècle* (Aix-en-Provence: Galerie Frank Mercelin, 2004).

Marinoni (1751). Giovanni Jacopo Marinoni, *De re iconographica* (Vienna: Leopold Kaliwoda, 1751).

Marinoni (1775). Giovanni Jacopo Marinoni, *De re ichnometrica veteri, ac nova* (Vienna: Leopold Kaliwoda, 1775).

Marivetz, Goussier (1780–1787). Étienne-Claude Marivetz, Louis-Jacques Goussier, *Physique du monde* (Paris: Quillau, 1780–1787), 7 vols.

Martinón-Torres *et al.* (2006). Marcos Martinón-Torres, Thilo Rheren, Ian C. Freestone, "Mullite and the mystery of the Hessian wares", *Nature*, 444/23 (2006): 437–438.

McConnell, Collins (2006). Anita McConnell, Philips R. Collins, "Will the True Originator of the Storm Glass Please Own Up", *Ambix*, 53 (2006): 67–75.

McConnell (2007). Anita McConnell, *Jesse Ramsden (1735–1800): London's Leading Scientific Instrument Maker* (Aldershot: Ashgate, 2007).

McKie (1956). Douglas McKie, "Priestley's Laboratory and Library and Others of His Effects", *Notes and Records of the Royal Society*, 12 (1956): 114–136.

Mémoire (1764). *Mémoire pour le Sieur La Planche* (Paris, 1764).

Mémoires (1777–1789), *Mémoires secrets pour servir à l'histoire de la République des lettres en France depuis 1762 jusqu'à nos jours par feu Bachaumont* (London: Adamson, 1777–1789), 36 vols.

Mercier (1976). Pierre Mercier, "Armand François Séguin (1765–1835)", *Bulletin de la Section d'histoire des usines Renault*, 7/2 (1976): 218–233.

Meusnier de la Place (1782). Jean-Batiste-Marie Charles Meusnier de la Place, "Description d'un appareil propre à manœuvrer différentes espèces d'air, dans les expériences qui en exigent des volumes considérables, par un écoulement continu parfaitement uniforme et variable à volonté, et donnant, à chaque instant, la mesure des quantités d'air employées, avec toute la précision qu'on peut désirer, par M. Meusnier", in *Mémoires de l'Académie royale des sciences de Paris* (1782 pub. in 1785):466 and ff.; reprinted in LO, 2, 432–440.

Michaud (1811–1837). Louis-Gabriel Michaud, ed., *Biographie universelle ancienne et moderne: Histoire par ordre alphabétique de la vie publique et privée de tous les hommes* (Paris: Beck, 1811–1837), 85 vols.

Michel (1976). Henri Michel, *Traité de l'astrolabe* (*Reimpression with a New Introduction by Francis Maddison and Corrigenda*) (Paris: A. Brieux, 1976).

Middleton (1964). William Edgar Knowles Middleton, *The History of the Barometer* (Baltimore, MD: Johns Hopkins Press, 1964).

Middleton (1966). William Edgar Knowles Middleton, *A History of the Thermometer and its Use in Meteorology* (Baltimore, MD: John Hopkins Press, 1966).

Middleton (1969). William Edgar Knowles Middleton, *Invention of the Meteorological Instruments* (Baltimore, MD: Johns Hopkins Press, 1969).

Miller (1926). Christy Miller, *The Bryant and May Museum of Fire-Making Appliances: Catalogue of the Exhibits* (London: Bryant & May Ltd. Appliances, 1926).

Monnet (1797). Antoine-Grimoald Monnet, *Démonstration de la fausseté des principes des nouveaux chymistes, pour servir de supplément au Traité de la dissolution des métaux* (Paris: H.J. Jansen, 1797).

Monnier (2012). Raymonde Monnier, *Le faubourg Saint Antoine (1789–1815)* (Paris: Société des études Robespierristes, 2012).

Montjoye (1800). F.L.C. de Montjoye, *Eloge historique de Jaen-Baptiste-Gaspard Bochart de Saron* (Paris: Le Normant, ca 1800).

Morris (2015). Peter J. Morris, *The Matter Factory: A History of the Chemistry Laboratory* (London: Reaktion Books Ltd, 2015).

Morrison (2007). James E. Morrison, *The Astrolabe* (Rehoboth Beach, DE: Janus, 2007).

Morrison-Low, Schechner, Brenni (2017). Alison D. Morrison-Low, Sara J. Schechner, Paolo Brenni, eds., *How Scientific Instruments have Changed Hands* (Leiden: Brill, 2017).

Morvan Becker (2010). Frédéric Morvan Becker, *L'école gratuite de dessin de Rouen, ou La formation des techniciens au XVIIIe siècle* (PhD thesis, Université de Paris VIII – Saint-Denis, 2010).

Müller (1973). H.G. Müller, "A Brief History of Powder Testers", in Robert Held, ed., *Armors and Armor Annual* (Northfield, IL: Digest Books, 1973), vol. 1, pp. 206–215.

Musée Centennal (1900). *Musée Centennal de la classe 87: Arts chimiques et pharmacie* (Saint-Cloud: Belin, 1900).

Musée National des Techniques (1990). *Musée national des techniques, inventaire des poids collection des poids et mesures* (Paris: Cnam, 1990).

Nairne (1777). Edward Nairne, *An Account of Some Experiments Made with an Air-pump on Mr. Smeaton's Principle* (London: Nichols, 1777).

Newton Harvey (1957). E. Newton Harvey, *A History of Luminescence from the Earliest Time until 1900* (Philadelphia, PA: The American Philosophical Society, 1957).

Niceron (1638). Jean François Niceron, *La perspective curieuse ou magie artificielle des effets merveilleux* (Paris: Pierre Billaine, 1638).

Nicholson (1809). William Nicholson, *The British Encyclopedia, or, Dictionary of Arts and Sciences: Comprising an Accurate and Popular View of the Present Improved State of Human Knowledge* (London: C. Whittingham, 1809), 6 vols.

Nollet (1743–1764). Jean Nollet, *Léçons de physique expérimentale* (Paris: d'Acquin, Guérin, Delatour, 1743–1764), 6 vols.

Nollet (1764). Jean Nollet, *Léçons de physique expérimentale*, 6th ed. (Paris: Guerin, 1764), 6 vols.

Nollet (1770). Jean Nollet, *L'Art des expériences, ou Avis aux amateurs de la physique, sur le choix, la construction et l'usage des instruments; sur la préparation et l'emploi des drogues qui servent aux expériences* (Paris: P.-E.-G. Durand, neveu, 1770), 3 vols.

Nooth (1775). John Mervin Nooth, "The Description of an Apparatus for Impregnating Water with Fixed Air; and of the Manner of Conducting that Process", *Philosophical Transactions*, 65 (1775): 59–66.

Oberkirch (1854). Henriette-Louise de Waldner de Freundstein d'Oberkirch, *Mémoires sur la cour de Louis XVI et la société française: Avant 1789* (Brussels: Meline, Cans et Compagnie 1854), 2 vols.

O'Dea (1964). William T. O'Dea, *Making Fire: A Science Museum Illustrated Booklet* (London: H.M. Stationery Office, 1964).

Oesper (1936). Ralph E. Oesper, "Priestley, Lavoisier and Trudaine de Montigny", *Journal of Chemical Education*, 13 (1936): 403–412.

Owne (1922). George A. Owne, *A Treatise of Weighing Machines* (London: C. Griffin & Company, 1922).

Palmer (1998). Louise Yvonne Palmer, *The Early Scientific Work of Antoine Laurent Lavoisier: In the Field and in the Laboratory* (Ph.D. dissertation, Yale University, New Haven, 1998).

Panopticon Lavoisier, see *Beretta (1999–2010)*.

Papin (1681). Denis Papin, *A New Digester or Engine for Softening Bones Containing the Description of its Make and Use in these Particulars: Viz. Cookery, Voyages at Sea, Confectionary, Making of Drinks, Chymistry, and Dying* (London: J.M. for Henry Bonwicke, 1681).

Partington (1961–1970). James Riddick Partington, *A History of Chemistry* (London: Macmillan, 1961–1970), 4 vols.

Passemant (1764). Claude Passemant, *Description et usage des télescopes, microscopes, divers ouvrages et inventions* (Paris, 1764).

Payen (1969). Jacques Payen, *Capital et machine à vapeur au XVIII*[e] *siècle: Les frères Périer et l'introduction en France de la machine à vapeur de Watt* (Paris: Mouton, 1969).

Pelucchi (2003). Stéphane Pelucchi, "Histoires parallèles: Histoire de la collection de minéralogie d'Antoine Laurent Lavoisier", *Nuncius – Journal of the Material and Visual History of Science*, 18 (2003): 705–732.

Pelucchi (2009). Stéphane Pelucchi, *Le cabinet d'histoire naturelle de Lavoisier. Première partie: Les bocaux de minéralogie* (Clermont Ferrand: Muséum d'histoire naturelle Henri-Lecoq, 2009).

Pelucchi (2009a). Stéphane Pelucchi, "Le cabinet d'histoire naturelle d'Antoine-Laurent Lavoisier", *Sparsae*, Hors-série n°4 (2009): 85–102.

Pelucchi (2016). Stéphane Pelucchi, "La collection du cabinet d'histoire naturelle de Lavoisier: Sa place dans son œuvre scientifique", *Revue d'histoire des sciences*, 69 (2016): 153–169.

Pépin (2010). François Pépin, "Diderot: La chimie comme modèle d'une philosophie expérimentale", *Dix-huitième siècle*, 42 (2010): 445–472.

Perkins (2010). John Perkins, "Chemistry Courses, the Parisian Chemical World and the Chemical Revolution, 1770–1790", *Ambix*, 57 (2010): 27–47.

Perrin (1981). Carleton E. Perrin, "The Triumph of the Antiphlogistians", in Harry Woolf ed., *The Analytic Spirit. Essays in the History of Science in Honor of Henry Guerlac*, (Ithaca and London: Cornell University Press, 1981), pp. 40–63.

Perrin (1989). Carleton E. Perrin, "The Lavoisier-Bucquet Collaboration: A Conjecture", *Ambix*, 36 (1989): 5–13.

Pictet (1996–2004). Marc-Auguste Pictet, *Correspondance: Sciences et techniques*, ed. René Sigrist (Geneva: Slatkine, 1996–2004), 4 vols.

Piot (2019), Yann Piot, *Jean-Antoine Nollet, artisan expérimentateur. Un discours technique au XVIII*[e] *siècle* (Paris: Garnier, 2019).

Poirier (1996). Jean Pierre Poirier, *Lavoisier. Chemist, Biologist, Economist* (Philadelphia, PA: University of Pennsylvania Press, 1996).

Poncelet (1755). Polycarpe Poncelet, *Chimie du gout et de l'odorat, ou, Principes pour composer facilement, & à peu de frais, les liqueurs à boire, & les eaux de senteurs* (Paris: Imprimerie de P.G. Le Mercier, 1755).

Potofsky (2009). Allan Potofsky, *Constructing Paris in the Age of Revolution* (London: Palgrave Macmillan, 2009).

Pouillet (1856). M. Pouillet, *Éléments de physique expérimentale et de météorologie* (Paris: Hachette, 1856), 2 vols.

Powers (2014). John C. Powers, "Measuring Fire: Herman Boerhaave and the Introduction of Thermometry into Chemistry", *Osiris*, 29 (2014): 158–177.

Priestley (1767). Joseph Priestley, *History and Present State of Electricity, with Original Experiments* (London: J. Dodsley and J. Johnson, B. Davenport, 1767).

Priestley (1774–1777). Joseph Priestley, *Experiments and Observations on Different Kinds of Air* (London: Johnson, 1774–1777), 3 vols.

Priestley (1790). Joseph Priestley, *Experiments and Observations on Different Kinds of Air* (Birmingham: Pearson, 1790), 3 vols.

Priestley (1800). Joseph Priestley, *The Doctrine of Phlogiston Established and that of the Composition of Water Refuted*, 1st ed. (Northumberland: for the author, 1800).

Priestley (1803). Joseph Priestley, *The Doctrine of Phlogiston Established and that of the Composition of Water Refuted*, 2nd ed. (Northumberland: Bryne, 1803).

Principe, De Witt (2002). Lawrence M. Principe, Lloyd De Witt, *Transmutations: Alchemy in Art* (Philadelphia, PA: Chemical Heritage Foundation, 2002).

Prinz (1992). Johann Peter Prinz, *Die experimentelle Methode der ersten Gasstoffwechseluntersuchungen am ruhenden und quantifiziert belasteten Menschen (A.L. Lavoisier und A. Séguin 1790): Versuch einer kritischen Deutung* (Sankt Augustin: Academia Verlag, 1992).

Prinz (2005). Johann Peter Prinz, "Lavoisier's Experimental Method and his Research on Human Respiration", in Marco Beretta, ed., *Lavoisier in Perspective* (Munich: Deutsches Museum, 2005), pp. 43–52.

Pujoulx (1801). Jean-Baptiste Pujoulx, *Paris à la fin du XVIII*[e] *siècle* (Paris: Mathé, 1801).

Pullins, Mahon, Centeno (2021). David Pullins, Dorothy Mahon, Silvia A. Centeno, "The Lavoisiers by David: Technical Findings on Portraiture at the Brink of Revolution", *The Burlington Magazine*, 163 (september 2021): 780–791.

Pyenson, Gauvin (2002). Lewis Pyenson, Jean-François Gauvin, *The Art of Teaching Physics: The Eighteenth-Century Demonstration Apparatus of Jean Antoine Nollet* (Sillery, Québec: Les éditions du Septentrion, 2002).

Quenard (1796). Philippe Quenard, "Lavoisier", in *Portraits des personnages célèbres de la révolution [texte imprimé] / par François Bonneville; avec tableau historique et notices de P. Quenard* (Paris: for the author, 1796), vol. 2.

Rappaport (1960). Rhoda Rappaport, "G.-F. Rouelle: 18th-Century Chemist and Teacher", *Chymia*, 6 (1960): 68–101.

Rappaport (1961). Rhoda Rappaport, "Rouelle and Stahl – The Phlogistic Revolution in France", *Chymia*, 7 (1961): 73–102.

Rappaport (1973). Rhoda Rappaport "Lavoisier's Theory of the Earth", *British Journal for the History of Science*, 6 (1973): 247–260.

Rappaport (1994). Rhoda Rappaport, "Baron d'Holbach's campaign for German (and Swedish) Science", *Studies on Voltaire and the Eighteenth Century*, 323 (1994): 225–246.

Rapport (1819). *Rapport du jury central sur les produits de l'industrie française* (Paris: Imprimerie royale, 1819).

Rapport (1819a). "Rapport fait par M. Régnier, au nom du Comité des arts mécaniques, sur une nouvelle balance pour la pesée des grains et farines, exécutée par M. Chemin, balancier mécanicien, rue de la Ferronnerie, à Paris", *Bulletin de la société d'encouragement pour l'industrie nationale*, 181 (1819): 213–219.

Ray (2015). Meredith K. Ray, *Daughters of Alchemy: Women and Scientific Culture in Early Modern Italy* (Cambridge, MA: Harvard University Press, 2015).

Raymond-Latour (1836). Jean-Michel Raymond-Latour, *Souvenirs d'un oisif* (Lyon: Ayné fils, 1836).

Réaumur (1716–1727). René-Antoine Ferchault de Réaumur, "Réflexions sur l'utilité dont l'Académie des sciences pourroit être au royaume, si le royaume luy donnoit les secours dont elle a besoin", in E. Maindron, *L'Académie des sciences* (Paris: Alcan, 1888), pp. 104–105.

Réaumur (1731). René-Antoine Ferchault de Réaumur, "Règle pour construire des thermomètres dont les degrés sont comparables", *Mémoire Académie royale des sciences* (1730 pub. in 1731): 452–507.

Réaumur (1733). René-Antoine Ferchault de Réaumur, "Second mémoire sur la construction des thermomètres dont les degrés sont comparables", *Mémoire Académie royale des sciences* (1731 pub. in 1733): 250–296.

Régnier (1798). Edme Régnier, "Description et usage d'une éprouvette portative à peson", in *Mémoire explicatif d'un dynamomètre et d'autres machines* (Paris: Imprimerie de la République, Brumaire an 7 [1798]), pp. 29–31.

Restif de La Bretonne (1776). Nicolas-Edme Restif de La Bretonne, *Le Paysan perverti* (La Haye, 1776), 2 vols.

Retz (1779). Noël Retz, *Traité d'un nouvel hygromètre comparable* (Paris: Méquignon, 1779).

Robens *et al.* (2014). Eric Robens *et al.*, *Balances: Instruments, Manufacturers, History* (Heidelberg and New York, NY: Springer, 2014).

Roberts (1991). Lissa Roberts, "A Word and the World: The Significance of Naming the Calorimeter", *Isis*, 82 (1991): 198–222.

Robertson (1831). Étienne-Gaspard Robertson, *Mémoires récréatifs, scientifiques et anecdotiques du physicien-aéronaute E.G. Robertson* (Paris: for the author, 1831), 2 vols.

Rocca, Launay (2018). Patrick Rocca, Françoise Launay, "La dynastie Langlois – Lordelle – Canivet – Lennel, « fabricateurs » d'instruments de mathématiques à Paris au XVIIIe siècle", *Artefact*, 7 (2018): 151–186.

Roche (1981). Daniel Roche, *Le peuple de Paris* (Paris: Éditions Aubier-Montaigne, 1981).

Rouelle (1744). Guillaume-François Rouelle, "Mémoire sur les sels neutres, dans lequel on propose une division méthodique de ces sels, qui facilite les moyens pour parvenir à la théorie de leur crystallisation", *Mémoires de l'Académie royale des sciences* (1744 pub. 1748): 353–364.

Rouelle (1759). Guillaume-François Rouelle, *Cours d'expériences chymiques* (Paris: Grange, 1759): https://gallica.bnf.fr/ark:/12148/bpt6k618568.texteImage (accessed 6 August 2021).

Rumford, Count of, see *Thompson, Benjamin*.

Sage (1784). Balthazard-Georges Sage, *Description méthodique du cabinet de l'École royal des mines* (Paris: Imprimerie royale, 1784).

Sage (1813). Balthazard Georges Sage, *Opuscules de physique* (Paris: Didot, 1813).

Sage (1813a). Balthazard Georges Sage, *Exposé sommaire des principales découvertes faites dans l'espace de cinquante-quatre années* (Paris: Didot, 1813).

Sage (1818). Balthazard Georges Sage, *Notice biographique* (Paris: Didot, 1818).

Sage (1820). Balthazard Georges Sage, *Supplément à la notice biographique de B.G. Sage* (Paris: Didot, 1820).

Salleron (1864). Jules Salleron, *Notice sur les instruments de précision construits par J. Salleron* (Paris: 14 Rue Pavée au Marais, 1864).

Saule, Arminjon (2010). Béatrix Saule, Catherine Arminjon eds., *Science & curiosités à la cour de Versailles* (Paris: Edition de la Réunion des Musées nationaux, 2010).

Saussure (1779–1796). Horace Bénédict de Saussure, *Voyages dans les Alpes: Précédés d'un essai sur l'histoire naturelle des environs de Genève* (Neufchâtel-Geneva: Fauche, 1779–1796), 4 vols.

Saussure (1783). Horace Bénédict de Saussure, *Essais sur l'hygrométrie* (Neufchâtel: Samuel Fauche, père et fils, 1783).

Scheler (1960). Lucien Scheler, *Lavoisier et la révolution française* (Paris: Hermann, 1960), 2 vols.

Sée (1925). Henri Sée, *L'évolution commerciale et industrielle de la France sous l'ancien régime* (Paris: Giard, 1925).

Séguin (1798). [Armand Séguin], "Sur le nouveau gazomètre du C. Séguin", *Bulletin des sciences par la Société philomatique*, 10 (Janvier 1798): 75–76.

Séguin (1814). Armand Séguin, "Mémoire sur la salubrité et l'insalubrité de l'air atmosphérique, dans ses divers degrés de pureté", *Annales de chimie*, 89 (1814): 251–272.

Selig (2013). Robert Selig, "Eighteenth-Century Currencies", *The Brigade Dispatch*, 43 (2013): 16–32.

s'Gravesande (1742). Willem Joacob s'Gravesande, *Physices elementa mathematica, experimentis confirmata* (Leiden: Johannes Arnold Langerak, 1742), 2 vols.

Sigaud de La Fond (1775). Joseph-Aignan Sigaud de La Fond, *Description et usage d'un cabinet de physique expérimentale* (Paris: Gueffier, 1775), 2 vols.

Sigaud de La Fond (1784). Joseph-Aignan Sigaud de La Fond, *Description et usage d'un cabinet de physique expérimentale* (Paris: Gueffier, 1784), 2 vols.

Sigaud de La Fond (1785). Joseph-Aignan Sigaud de La Fond, *Précis historique et expérimental des phénomènes électriques* (Paris: Rue et Hôtel Serpente, 1785).

Simon (2005). Jonathan Simon, *Chemistry, Pharmacy and Revolution in France 1777–1809* (Aldershot: Ashgate, 2005).

Smeaton (1962). William A. Smeaton, *Fourcroy: Chemist and Revolutionary (1755–1809)* (Cambridge: Heffers & Sons, 1962).

Smeaton (1966). William A. Smeaton, "The Portable Chemical Laboratories of Guyton de Morveau, Cronstedt and Göttling", *Ambix*, 13 (1966): 84–91.

Smeaton (1975). William A. Smeaton, "The Early History of Laboratory Instruction in Chemistry at the École Polytechnique", *Annals of Science*, 10 (1975): 224–234.

Smeaton (1987). William A. Smeaton, "Some Large Burning Lenses and their Use by Eighteenth-Century French and British Chemists", *Annals of Science*, 44/3 (1987): 265–276.

Smeaton (1997). William A. Smeaton, "Bertrand Pelletier, master pharmacist: His Report on Janety's preparation of malleable platinum", *Platinum Metals Review*, 41 (1997): 86–88.

Smith (1979). Johan Graham Smith, *The Origins and Early Development of Heavy Chemical Industry in France* (Oxford: Oxford University Press, 1979).

Souvenirs de Lavoisier (1956). *Souvenirs de Lavoisier: Catalogue des livres, autographes pièces d'archive, portrait de Lavoisier peinture de l'école française, objets variés, objets de vitrine, siège et meubles estampillés des maitres ébénistes, important bureau à cylindre d'époque Louis XVI par J. Caumont, tapis, provenant du célèbre physicien Lavoisier et appartenant à Madame de V …* (Paris: Hotel Drouot, 7 March 1956).

Spary (2014). Emma C. Spary, *Feeding France: New Sciences of Food 1760–1815* (Cambridge: Cambridge University Press, 2014).

Sparrow (1958). Wilfrid James Sparrow, "Count Rumford's Journal", *Archives internationales d'histoire des sciences*, 42 (1958): 15–20.

Stock (1969). John T. Stock, *Development of the Chemical Balance* (London: Her Majesty's Stationery Office, 1969).

Stone (1971). Lawrence Stone, "Prosopography", *Dedalus*, 100/1 (1971): 46–79.

Storni (2021). Marco Storni, "Denis Papin's digester and its eighteenth-century European circulation", *The British Journal for the History of Science* Published online by Cambridge University Press (11 October 2021): 1–21.

Taylor (1966). Eva Germaine R. Taylor, *The mathematical practitioners of Hanoverian England 1714–1840* (Cambridge: for The Institute of Navigation at the University Press, 1966).

Terrall (2015). Mary Terrall, "Masculine Knowledge, the Public Good, and the Scientific Household of Réaumur", *Osiris*, 30 (2015): 182–201.

Thébaud-Sorger (2007). Marie Thébaud-Sorger, *L'Aérostation au temps des Lumières* (Rennes: Presses Universitaires de Rennes, 2007).

Thébaud-Sorger (2018). Marie Thébaud-Sorger, "Capturing the Invisible: Heat, Steam and Gases in France and Great Britain, 1750–1800", in Lissa Roberts, Simon Werrett eds., *Compound Histories. Materials, Governance and Production, 1760–1840* (Leiden: Brill, 2018), pp. 85–105.

Thénard (1835). Louis-Jacques Thénard, *Traité de chimie élémentaire, théorique et pratique* (Paris: Chrochard, 1835), atlas.

Thiéry (1784). Luc-Vincent Thiéry, *Almanach du voyageur a Paris: Contenant une description exacte & intéressante de tous les monumens, chefs-d'œuvres des artes, etablissemens utiles, & autres objets de curiosité que renferme cette capitale: Ouvrage utile aux citoyens & indispensable pour l'etranger* (Paris: Hardouin, 1784).

Thompson (1794). Benjamin Thompson, "Experiments of the Relative Intensities of the Light Emitted by Luminous Bodies", in *The Collected Works of Count Rumford,* ed. Sanborn C. Brown (Cambridge, MA: Harvard University Press, 1970), vol. 4, pp. 1–51.

Thompson (1798). Benjamin Thompson, "Nouvelles expériences et observations relatives à la propagation de la chaleur dans les liquides", in *The Collected Works of Count Rumford,* ed. Sanborn C. Brown (Cambridge, MA: Harvard University Press, 1968), vol. 1, pp. 147–217.

Thompson (1804). Benjamin Thompson, "Inquiry Concerning the Nature of Heat, and the Mode of its Communications", *Philosophical Transactions of the Royal Society*, 94 (1804): 77–182.

Thompson (1805). Benjamin Thompson, "Historical Review of the Various Experiments of the Author on the Subject of Heat", in *The Collected Works of Count Rumford,* ed. Sanborn C. Brown (Cambridge, MA: Harvard University Press, 1968), vol. 1, pp. 443–496.

Thompson (1812). Benjamin Thompson, "Of the Excellent Quality of Coffee, and the Art of Making it in the Highest Perfection", in *The Collected Works of Count Rumford,* ed. Sanborn C. Brown (Cambridge, MA: Harvard University Press, 1970), vol. 5, pp. 263–315.

Thompson (1813). Benjamin Thompson, *Recherches sur la chaleur développée dans la combustion et dans la condensation des vapours* (Paris: Éverat, 1813).

Timmerhans (1839). G. Timmerhans, *Essai d'un traité élémentaire d'artillerie* (Liège: Librairie A. Leroux, 1839).

Traumüller (1885). Friedrich Traumüller, *Die Mannheimer meteorologische Gesellschaft* (1780–1795) (Leipzig: Dürrsche Buchhandlung, 1885).

Tresca (1864). Henri Édouard Tresca, "Compte rendu de la soirée scientifique du 29 octobre 1864, au Conservatoire des arts et métiers", *Annales du conservatoire Impériale des arts et métiers* (1864): 205–296.

Tresca (1869) Henri Édouard Tresca, "Le Conservatoire impériale des arts et métiers en 1849 et en 1869", *Annales du Conservatoire impériale des arts et métiers* (1869): 321–333.

Tron (1996). Hélène Tron, "L'enseignement de la chimie a l'École polytechnique 1794–1805: Les années fastes", *Bulletin de la Sabix*, 15 (1996): 1–49.

Truchot (1879). Pierre Truchot, "Les instruments de Lavoisier: Relation d'une visite à La Canière, où se trouvent réunis les appareils ayant servi à Lavoisier", *Annales de chimie*, 5th series, 18 (1879): 289–319.

Turgot (1739). Michel-Étienne Turgot, *Plan de Paris, dessiné par Louis Bretez, gravé par Claude Lucas, et écrit par Aubin* (Paris, 1739).

Turner, Levere (1973). Gerard L'E. Turner, Trevor H. Levere, "Van Marum Scientific Instruments in Teyler's Museum", in Forbes (1969–1976), vol. 4.

Turner (1989). Gerard L'E. Turner, *The Great Age of Microscope* (Bristol and New York, NY: Adam Hilger, 1989).

Turner (1987). Anthony Turner, *Early Scientific Instruments Europe 1400–1800* (London: Sotheby's Publications, 1987).

Turner A. (1989). Anthony J. Turner, "Paper, Print and Mathematics: Philippe Danfrie and the Making of Mathematical Instruments in Late-16th-Century Paris", in Christine Blondel, Françoise Parot, Anthony J. Turner, Mari Williams, eds, *Studies in the History of Scientific Instruments* (London and Paris: Centre de Recherche en histoire des sciences et des techniques de la cité des sciences et de l'industrie, 1989), pp. 22–42.

Turner A.(1989a). Anthony J. Turner, *From Pleasure and Profit to Science and Security: Etienne Lenoir and the Transformation of Precision Instrument-Making in France 1760–1830* (Cambridge: The Whipple Museum of the History of Science, 1989).

Turner (2018). Anthony J. Turner, *Mathematical Instruments in the Collection of the Bibliothèque Nationale de France* (East Grinstead: BNF Éditions/Brepols, 2018).

Turner, see Beaudouin, Brenni, Turner (2018).

Van Damme (2005). Stéphane Van Damme, *Paris, capitale philosophique de la fronde à la révolution* (Paris: Odile Jacob, 2005).

Van Klooster (1946). H.S. Van Klooster, "The Lavoisier Bicentenary Exhibition in Paris", *Journal of chemical education*, 23/5 (1946): 210–212.

Van Leempoel (2005). Koenraad van Leempoel, *Astrolabes at Greenwich: A Catalogue of the Astrolabes in the National Maritime Museum, Greenwich* (Oxford: Oxford University Press, 2005).

Van Marum, see *Forbes (1969–1976)*.

Veau Delaunay (1809). Claude Veau Delaunay, *Manuel de l'électricité* (Paris: for the author, 1809).

Viel (1995a). Claude Viel, "Lavoisier avait-il un laboratoire autre que celui de l'Arsenal", *Revue d'histore de la pharmacie*, 42 (1995): 369–373.

Viel (1995b). Claude Viel, "Le salon et le laboratoire de Lavoisier à l'Arsenal, cénacle où s'élabora la nouvelle chimie", *Revue d'histore de la pharmacie*, 42 (1995): 255–266.

Volta (1784). Alessandro Volta, "Del mondo di rendere sensibile la più debole elettricità sia naturale, sia artificiale", *Opuscoli scelti sulle scienze e sulle arti*, 7 (1784): 145–162.

Volta (1790). Alessandro Volta, "Descrizione dell'eudiometro ad aria infiammabile", *Annali di chimica e storia naturale*, 1 (1790): 171–231.

Volta (1918–1929). Alessandro Volta, *Opere* (Milan: Hoepli, 1918–1929), 7 vols.

Volta (1949–1955). Alessandro Volta, *Epistolario* (Bologna: Zanichelli, 1949–1955), 5 vols.

Watson (1930). Warren N. Watson, *Early Fire-Making Methods and Devices: From the Stone Age until the Introduction of the Match* (Published by Self-published, 1930).

Webster and Webster (1998). Roderick Webster, Marjorie Webster, *Western Astrolabes* (Historic Scientific Instruments in the Adler Planetarium and Astronomy Museum, vol. 1) (Chicago, IL: Adler Planetarium and Astronomy Museum, 1998).

Weeks (1933). Mary Elvira Weeks, *The Discovery of the Elements* (Easton, PA: Journal of Chemical Education, 1933).

Weelen (1943). J.-E. Weelen, "Les loisirs d'un grand savant: Lavoisier à Fréchines", *L'illustration* (27 novembre 1943): 342.

Weiss (2008). Alfred Weiss, "Die Entwicklung der Lötrohr-Probierkunde", *Der Steirische Mineralog*, 22 (2008): 16–21.

Wernimont (2019). Jacqueline Wernimont, *Numbered Lives Life and Death in Quantum Media* (Cambridge Mass., London: The MIT Press, 2019).

Wilkes (1799). John Wilkes, *The Art of Making Pens Scientifically*, 2nd ed. (London: J. Vigevena, 1799?).

Wolf (1902). Charles Joseph Etienne Wolf, *Histoire de l'Observatoire de Paris de sa fondation à 1793* (Paris: Gauthier-Villars, 1902).

Wolfe (1999). John Wolfe, *Brandy, Balloons & Lamps: Ami Argand (1750–1803)* (Carbondale, IL: Southern Illinois University Press, 1999).

Wurtz (1869). Charles Adolphe Wurtz, *Histoire des doctrines chimiques depuis Lavoisier jusqu'à nos jours* (Paris: Hachette, 1869).

Young (1792). Arthur Young, *Travels During the Years 1787, 1788 and 1789* (Bury St. Edmunds: Rackham, 1792).

Index of Inventory Numbers

Musée des arts et métiers

Cornell University, Kroch Library

Harvard University, Collection of Historical Scientific Instruments

Muséum National d'Histoire Naturelle – Paris

Observatoire de Paris

Index of Names

Note: The name of Antoine Laurent Lavoisier has been omitted from the index.